NEUROMETHODS

T0155343

Series Editor
Wolfgang Walz
University of Saskatchewan,
Saskatoon, SK, Canada

For further volumes:
http://www.springer.com/series/7657

Computational Modeling of Drugs Against Alzheimer's Disease

Edited by

Kunal Roy

Department of Pharmaceutical Technology, Jadavpur University, Kolkata, India

Editor
Kunal Roy
Department of Pharmaceutical Technology
Jadavpur University
Kolkata, India

ISSN 0893-2336 ISSN 1940-6045 (electronic)
Neuromethods
ISBN 978-1-4939-8475-6 ISBN 978-1-4939-7404-7 (eBook)
DOI 10.1007/978-1-4939-7404-7

© Springer Science+Business Media LLC 2018
Softcover reprint of the hardcover 1st edition 2017
This work is subject to copyright. All rights are reserved by the Publisher, whether the whole or part of the material is concerned, specifically the rights of translation, reprinting, reuse of illustrations, recitation, broadcasting, reproduction on microfilms or in any other physical way, and transmission or information storage and retrieval, electronic adaptation, computer software, or by similar or dissimilar methodology now known or hereafter developed.
The use of general descriptive names, registered names, trademarks, service marks, etc. in this publication does not imply, even in the absence of a specific statement, that such names are exempt from the relevant protective laws and regulations and therefore free for general use.
The publisher, the authors and the editors are safe to assume that the advice and information in this book are believed to be true and accurate at the date of publication. Neither the publisher nor the authors or the editors give a warranty, express or implied, with respect to the material contained herein or for any errors or omissions that may have been made. The publisher remains neutral with regard to jurisdictional claims in published maps and institutional affiliations.

Printed on acid-free paper

This Humana Press imprint is published by Springer Nature
The registered company is Springer Science+Business Media, LLC
The registered company address is: 233 Spring Street, New York, NY 10013, U.S.A.

Preface to the Series

Experimental life sciences have two basic foundations: concepts and tools. The *Neuromethods* series focuses on the tools and techniques unique to the investigation of the nervous system and excitable cells. It will not, however, shortchange the concept side of things as care has been taken to integrate these tools within the context of the concepts and questions under investigation. In this way, the series is unique in that it not only collects protocols but also includes theoretical background information and critiques which led to the methods and their development. Thus it gives the reader a better understanding of the origin of the techniques and their potential future development. The *Neuromethods* publishing program strikes a balance between recent and exciting developments, like those concerning new animal models of disease, imaging, and in vivo methods, and more established techniques, including immunocytochemistry and electrophysiological technologies. New trainees in neurosciences still need a sound footing in these older methods in order to apply a critical approach to their results.

Under the guidance of its founders, Alan Boulton and Glen Baker, the *Neuromethods* series has been a success since its first volume published through Humana Press in 1985. The series continues to flourish through many changes over the years. It is now published under the umbrella of Springer Protocols. While methods involving brain research have changed a lot since the series started, the publishing environment and technology have changed even more radically. *Neuromethods* has the distinct layout and style of the Springer Protocols program, designed specifically for readability and ease of reference in a laboratory setting.

The careful application of methods is potentially the most important step in the process of scientific inquiry. In the past, new methodologies led the way in developing new disciplines in the biological and medical sciences. For example, physiology emerged out of anatomy in the nineteenth century by harnessing new methods based on the newly discovered phenomenon of electricity. Nowadays, the relationships between disciplines and methods are more complex. Methods are now widely shared between disciplines and research areas. New developments in electronic publishing make it possible for scientists that encounter new methods to quickly find sources of information electronically. The design of individual volumes and chapters in this series takes this new access technology into account. Springer Protocols makes it possible to download single protocols separately. In addition, Springer makes its print-on-demand technology available globally. A print copy can therefore be acquired quickly and for a competitive price anywhere in the world.

Saskatoon, SK, Canada ***Wolfgang Walz***

Preface

According to an estimate published in the World Alzheimer Report 2016 (https://www.alz.co.uk/research/world-report-2016), there were 46.8 million people worldwide living with dementia (loss of memory and other cognitive abilities) in 2015, and this number will reach 131.5 million in 2050. The total estimated worldwide cost of dementia is US$ 818 billion, and it will become a trillion-dollar disease by 2018. People living with dementia have poor access to appropriate healthcare, even in most high-income country settings, where only around 50% of people living with dementia receive a diagnosis. In low- and middle-income countries, less than 10% of cases are diagnosed.

Alzheimer's disease is the most common form of dementia. It is a lethal neurological disorder due to progressive degeneration of various parts of the brain [1]. It gradually destroys learning skills, thinking, memory, and finally the ability to carry out basic activities of daily living. *Alzheimer's disease* (AD) is *multifactorial, involving genetic and environmental factors,* and it apparently involves several different etiopathogenic mechanisms. It has been identified as a protein misfolding disease due to the accumulation of abnormally folded amyloid beta proteins in the brains of Alzheimer's patients. The root cause as well as treatment of AD is unknown, while several hypotheses exist such as the cholinergic hypothesis, amyloid hypothesis, tau hypothesis, etc. Currently, four FDA-approved drugs (rivastigmine, galantamine, donepezil, memantine) are available in the market, and these only provide symptomatic relief. Although research has revealed a number of biological targets, specific drug molecules against these targets showing a complete cure of the disease still remain unknown. Thus, it is of timely need to develop an effective treatment strategy against the disease.

Computational modeling techniques including quantitative structure–activity relationship (QSAR), pharmacophore mapping, homology modeling docking, virtual screening, and other cheminformatics approaches play a vital role in the finding and optimization of leads in any drug discovery program. Computational modeling helps to understand the important molecular features contributing to the binding interactions with the target proteins thus facilitating the design of new potential compounds and prediction of activity of designed compounds which have not yet been tested. These approaches can save time, money, and more importantly animal sacrifice in the complex, long, and costly drug discovery process.

This volume under the *Neuromethods* series describes different computational methods encompassing ligand-based approaches (QSAR, pharmacophore), structure-based approaches (homology modeling, docking, molecular dynamics simulation), and combined approaches (virtual screening) with applications in anti-Alzheimer drug design. Different background topics like molecular etiologies of Alzheimer's disease, targets for new drug development, and different cheminformatics modeling strategies have also been covered for completeness. Special topics like multi-target drug development, natural products, protein misfolding, and nanomaterials have also been included in connection with computational modeling of anti-Alzheimer drug development.

Chapter 1 authored by *Awanish Kumar and Ashwini Kumar* gives an introduction to Alzheimer's disease and presents an overview of the pathophysiology and disease etiologies. Different promising targets for anti-Alzheimer drug development are also discussed.

Chapter 2 authored by *Sergi Gómez-Ganau, Jesús Vicente de Julián Ortiz, and Rafael Gozalbes* gives a brief introduction to different computational drug design approaches, mainly QSAR, molecular docking, pharmacophore development, and molecular dynamics simulations. Some recent applications of these techniques for design of anti-Alzheimer agents are also briefly mentioned.

Chapter 3 authored by *Agostinho Lemos, Rita Melo, Irina S. Moreira, and M. Natália D. S. Cordeiro* focuses on the role of G-protein-coupled receptors (GPCRs) in the pathogenesis of AD. Different structure- and ligand-based in silico approaches and their applicability on the development of small molecules that target various GPCRs potentially involved in AD are discussed in this chapter.

Chapter 4 authored by *Livia Basile* discusses on virtual screening as a tool to select quickly and economically compounds endowed with optimum physicochemical, pharmacokinetic, and biological properties for the development of new potential drug candidates. The use of virtual screening for the design of specific ligands for targets related with AD is discussed with examples.

Chapter 5 authored by *Eugene V. Radchenko, Vladimir A. Palyulin, and Nikolay S. Zefirov* discusses on molecular field topology analysis (MFTA) as a QSAR method designed to model the activities mediated by small molecules binding to biotargets using the local physicochemical descriptors reflecting the major types of ligand–target interactions. The design of potential anti-Alzheimer and other neuroprotective compounds based on the MFTA structure–activity models is described.

Chapter 6 contributed by *Irini Doytchinova, Mariyana Atanasova, Georgi Stavrakov, Irena Philipova, and Dimitrina Zheleva-Dimitrova* describes a docking-based technique for the designing of galantamine derivatives with dual-site binding fragments – one blocking the catalytic site and another blocking the peripheral anionic site of acetylcholinesterase. The study illustrates the efficiency of the docking-based design of galantamine derivatives as acetylcholinesterase inhibitors.

Chapter 7 authored by *Odailson Santos Paz, Thamires Quadros Froes, Franco Henrique Leite, and Marcelo Santos Castilho* focuses on the importance of computational tools in the hit identification and lead optimization steps with reference to anti-Alzheimer drug development. The theory and application of different computational tools for development of BACE-1 inhibitors as anti-Alzheimer agents are described.

Chapter 8 contributed by *Yoshio Hamada and Kenji Usui* discusses a novel "electron-donor bioisostere" concept in drug discovery study and reports the design of potent BACE1 inhibitors using this concept. The authors suggest that a quantum chemical interaction, such as σ–π interaction or π–π stacking, plays a critical role in BACE1 inhibition mechanism.

Chapter 9 authored by *Maricarmen Hernández Rodríguez, Leticia Guadalupe Fragoso Morales, José Correa Basurto, and Martha Cecilia Rosales Hernández* focuses on inhibition of Aβ production and aggregation as an important therapeutic strategy to design compounds against AD. This chapter analyzes the applications of molecular docking and molecular dynamics simulations to estimate the binding energy and binding pose of compounds that could avoid Aβ42 aggregation.

Chapter 10 authored by *Praveen P. N. Rao and Deguo Du* focuses on computational tools to study and design amyloid aggregation inhibitors and modulators. The authors describe an in silico method that uses the Aβ hexapeptide-derived steric-zipper octamer assembly as an alternative and effective model to predict the binding interactions of planar small molecule libraries with complex Aβ structures including oligomers, protofibrils, and fibrils.

Chapter 11 authored by *Kailas Dashrath Sonawane and Maruti Jayaram Dhanavade* deals with the applications of computational methods to investigate structure–function relationship of enzyme Aβ complex and to design new lead molecules to control amyloid beta peptide levels in Alzheimer's disease. Different molecular modeling methods like energy minimization, molecular docking, molecular dynamics simulation, virtual screening, binding free energy, solvent-accessible surface calculations, etc. have been discussed in this connection.

Chapter 12 authored by *Prabu Manoharan and Nanda Ghoshal* focuses on gamma secretase (GS) as an attractive therapeutic strategy to slow down the pathological progression of AD. This chapter provides with a detailed discussion on a QSAR-guided fragment-based virtual screening method for GS inhibitor design and identification.

Chapter 13 authored by *Carlos Navarro-Retamal and Julio Caballero* focuses on tau hyperphosphorylation, which is related to the formation of amyloid plaques and neurofibrillary tangles in Alzheimer's disease patients. Molecular modeling studies on complexes between proline-directed protein kinases, which are responsible for tau hyperphosphorylation, and their inhibitors are described in this chapter.

Chapter 14 authored by *Mange Ram Yadav, Mahesh A. Barmade, Rupesh V. Chikhale, and Prashant R. Murumkar* focuses on roles of protein kinases in the hyperphosphorylation of tau protein in AD. The computational studies carried out on various protein kinases in search of potential anti-Alzheimer agents are critically discussed.

Chapter 15 contributed by *Ádám A. Kelemen, Stefan Mordalski, Andrzej J. Bojarski, and György M. Keserű* focuses on 5-hydroxytryptamine receptor 6 (5-HT_6R) as a target for drug development for alleviating cognitive, learning, and memory deficits related to Alzheimer's disease. The authors discuss the use of ligand- and structure-based methods for the design of new 5-HT_6R antagonists and highlight advantages and limitations of corresponding approaches and computational modeling tools in the field of 5-HT_6R drug design.

Chapter 16 contributed by *Dionysia Papagiannopoulou and Dimitra Hadjipavlou-Litina* focuses on radionuclide imaging techniques and molecular imaging agents for the in vivo detection of amyloid plaques in Alzheimer's disease. The authors review in this chapter computational modeling studies performed on PET and SPECT imaging agents in connection with Alzheimer's disease.

Chapter 17 authored by *Manika Awasthi, Swati Singh, Sameeksha Tiwari, Veda P. Pandey, and Upendra N. Dwivedi* discusses the recent advances in the application of cheminformatics methods to quantify the chemical diversity and structural complexity of natural products and analyze their arrangement in chemical space with respect to the treatment of AD. The authors also discuss the advancement in virtual screening to systematically identify bioactive compounds from natural products databases and the progress in target identification methods to discover molecular targets of compounds from natural origin with reference to AD.

Chapter 18 authored by *Luciana Scotti and Marcus T. Scotti* discusses in silico studies reported on natural products with reference to anti-AD drug development. The authors also discuss multi-target QSAR models which can be used to predict activity or classify compounds as actives or inactives against different targets, such as proteins (amyloid-A4 protein (ABPP), glycogen synthase kinase-3 alpha, glycogen synthase kinase-3 beta (GSK-3β), monoamine oxidase B (MAO-B), and presenilin-1 (PSN-1)).

Chapter 19 contributed by *Akhil Kumar and Ashok Sharma* focuses on dual- or multi-target inhibitors which halt multiple disease-causing pathways and improve the disease conditions. The authors describe various computational methods to screen and identify top hits and molecular dynamics to ensure the affinity in terms of binding free energy of the receptor–ligand complex to design multi-target-directed ligands for Alzheimer's disease.

Chapter 20 authored by *Gerald H. Lushington, Frances E. S. Parker, Thomas H. W. Lushington, and Nora M. Wallace* focuses on misfolding of central nervous system proteins into energetically favored physiologically dysfunctional forms leading to several neurological disorders. The authors discuss the fundamental physiological issues that cause neuropathies and suggest the ways in which molecular docking and molecular dynamics simulations can be brought to bear in formulating testable hypotheses that can form a basis for the systematic formulation of a new generation of medicines.

Chapter 21 contributed by *R. Navanietha Krishnaraj, Dipayan Samanta, and Rajesh K. Sani* discusses the potential of nanomaterials for the treatment of AD. This chapter addresses the advantages of computational analysis of ADME (absorption, distribution, metabolism, and excretion) characteristics, the druglikeness of nanomaterials, and the role of molecular docking techniques for assessing the therapeutic efficacy of nanomaterials.

I am sure that this collection of 21 chapters will be helpful to researchers working in the field of anti-Alzheimer drug research. I am especially thankful to the series editor Prof. Wolfgang Walz for his help during the development of this book and to the publisher for bringing out this volume.

Kolkata, India ***Kunal Roy***

Contents

Contributors

MARIYANA ATANASOVA • *Faculty of Pharmacy, Medical University of Sofia, Sofia, Bulgaria*
MANIKA AWASTHI • *Bioinformatics Infrastructure Facility, Center of Excellence in Bioinformatics, Department of Biochemistry, University of Lucknow, Lucknow, India*
MAHESH A. BARMADE • *Faculty of Pharmacy, Kalabhavan Campus, The Maharaja Sayajirao University of Baroda, Vadodara, Gujarat, India*
LIVIA BASILE • *Department of Drug Sciences, University of Catania, Catania, Italy*
JOSÉ CORREA BASURTO • *Laboratorio de Modelado Molecular y Diseño de Fármacos, Escuela Superior de Medicina, Instituto Politécnico Nacional, Ciudad de Mexico, Mexico*
ANDRZEJ J. BOJARSKI • *Department of Medicinal Chemistry, Institute of Pharmacology, Polish Academy of Sciences, Krakow, Poland*
JULIO CABALLERO • *Centro de Bioinformática y Simulación Molecular, Facultad de Ingeniería, Universidad de Talca, Talca, Chile*
MARCELO SANTOS CASTILHO • *Programa de Pós-graduação em Biotecnologia, Universidde Estadual de Feira de Santana, Feira de Santana, BA, Brazil; Faculdade de Farmácia da Universidade Federal da Bahia, Salvador, BA, Brazil*
RUPESH V. CHIKHALE • *Faculty of Pharmacy, Kalabhavan Campus, The Maharaja Sayajirao University of Baroda, Vadodara, Gujarat, India*
M. NATÁLIA D.S. CORDEIRO • *LAQV@REQUIMTE, Department of Chemistry and Biochemistry, Faculty of Sciences, University of Porto, Porto, Portugal; (MNDSC) LAQV@REQUIMTE, Department of Chemistry and Biochemistry, Faculty of Sciences, University of Porto, Porto, Portugal*
MARUTI JAYARAM DHANAVADE • *Department of Microbiology, Shivaji University, Kolhapur, Maharashtra, India*
IRINI DOYTCHINOVA • *Faculty of Pharmacy, Medical University of Sofia, Sofia, Bulgaria*
DEGUO DU • *Department of Chemistry and Biochemistry, Florida Atlantic University, Boca Raton, FL, USA*
UPENDRA N. DWIVEDI • *Bioinformatics Infrastructure Facility, Center of Excellence in Bioinformatics, Department of Biochemistry, University of Lucknow, Lucknow, India*
THAMIRES QUADROS FROES • *Programa de Pós-graduação em Biotecnologia, Universidade Estadual de Feira de Santana, Feira de Santana, BA, Brazil; Faculdade de Farmácia da Universidade Federal da Bahia, Salvador, BA, Brazil*
NANDA GHOSHAL • *Structural Biology and Bioinformatics Division, CSIR-Indian Institute of Chemical Biology, Kolkata, India*
SERGI GÓMEZ-GANAU • *ProtoQSAR SL, Centro Europeo de Empresas Innovadoras (CEEI), Parque Tecnólogico de Valencia, Paterna, Valencia, Spain*
RAFAEL GOZALBES • *ProtoQSAR SL, Centro Europeo de Empresas Innovadoras (CEEI), Parque Tecnólogico de Valencia, Paterna, Valencia, Spain*
DIMITRA HADJIPAVLOU-LITINA • *Department of Pharmaceutical Chemistry, School of Pharmacy, Faculty of Health Sciences, Aristotle University of Thessaloniki, Thessaloniki, Greece*
YOSHIO HAMADA • *Faculty of Frontiers of Innovative Research in Science and Technology, Konan University, Chuo-ku, Kobe, Japan*

MARTHA CECILIA ROSALES HERNÁNDEZ • *Laboratorio de Biofísica y Biocatálisis, Escuela Superior de Medicina, Instituto Politécnico Nacional, Ciudad de Mexico, Mexico*
ÁDÁM A. KELEMEN • *Medicinal Chemistry Research Group, Research Center for Natural Sciences, Hungarian Academy of Sciences, Budapest, Hungary*
GYÖRGY M. KESERŰ • *Medicinal Chemistry Research Group, Research Center for Natural Sciences, Hungarian Academy of Sciences, Budapest, Hungary*
R. NAVANIETHA KRISHNARAJ • *Department of Biotechnology, National Institute of Technology Durgapur, Durgapur, India; Department of Chemical and Biological Engineering, South Dakota School of Mines and Technology, Rapid City, SD, USA*
AKHIL KUMAR • *Biotechnology Division, CSIR-Central Institute of Medicinal and Aromatic Plants, CIMAP, Lucknow, India*
ASHWINI KUMAR • *Department of Biotechnology, National Institute of Technology Raipur, Raipur, Chhattisgarh, India*
AWANISH KUMAR • *Department of Biotechnology, National Institute of Technology Raipur, Raipur, Chhattisgarh, India*
FRANCO HENRIQUE LEITE • *Laboratório de Modelagem Molecular, Departamento de Saúde, Universidade Estadual de Feira de Santana, Feira de Santana, BA, Brazil*
AGOSTINHO LEMOS • *LAQV@REQUIMTE, Department of Chemistry and Biochemistry, Faculty of Sciences, University of Porto, Porto, Portugal; GIGA Cyclotron Research Centre In Vivo Imaging, University of Liège, Liège, Belgium*
GERALD H. LUSHINGTON • *LiS Consulting, Lawrence, KS, USA*
THOMAS H.W. LUSHINGTON • *LiS Consulting, Lawrence, KS, USA*
PRABU MANOHARAN • *Centre of Excellence in Bioinformatics, School of Biotechnology, Madurai Kamaraj University, Madurai, India*
RITA MELO • *CNC-Centre for Neuroscience and Cell Biology, Faculty of Medicine, University of Coimbra, Coimbra, Portugal; Centre for Nuclear Sciences and Technologies, Instituto Superior Técnico, University of Lisbon, Bobadela, LRS, Portugal*
LETICIA GUADALUPE FRAGOSO MORALES • *Laboratorio de Biofísica y Biocatálisis, Escuela Superior de Medicina, Instituto Politécnico Nacional, Ciudad de Mexico, Mexico*
STEFAN MORDALSKI • *Department of Medicinal Chemistry, Institute of Pharmacology, Polish Academy of Sciences, Krakow, Poland*
IRINA S. MOREIRA • *CNC-Centre for Neuroscience and Cell Biology, Faculty of Medicine, University of Coimbra, Coimbra, Portugal; Bijvoet Center for Biomolecular Research, Faculty of Science-Chemistry, Utrecht University, Utrecht, The Netherlands; (ISM) CNC – Center for Neuroscience and Cell Biology, Faculty of Medicine, University of Coimbra, Coimbra, Portugal*
PRASHANT R. MURUMKAR • *Faculty of Pharmacy, Kalabhavan Campus, The Maharaja Sayajirao University of Baroda, Vadodara, Gujarat, India*
CARLOS NAVARRO-RETAMAL • *Centro de Bioinformática y Simulación Molecular, Facultad de Ingeniería, Universidad de Talca, Talca, Chile*
JESÚS VICENTE DE JULIÁN-ORTIZ • *Departamento de Química-Física, Facultad de Farmacia, Universidad de Valencia, Burjassot, Valencia, Spain*
VLADIMIR A. PALYULIN • *Department of Chemistry, Lomonosov Moscow State University, Moscow, Russia; Institute of Physiologically Active Compounds RAS, Moscow Region, Russia; Institute of Organic Chemistry RAS, Moscow, Russia*
VEDA P. PANDEY • *Bioinformatics Infrastructure Facility, Center of Excellence in Bioinformatics, Department of Biochemistry, University of Lucknow, Lucknow, India*

DIONYSIA PAPAGIANNOPOULOU • *Department of Pharmaceutical Chemistry, School of Pharmacy, Faculty of Health Sciences, Aristotle University of Thessaloniki, Thessaloniki, Greece*

FRANCES E.S. PARKER • *LiS Consulting, Lawrence, KS, USA*

ODAILSON SANTOS PAZ • *Programa de Pós-graduação em Biotecnologia, Universidde Estadual de Feira de Santana, Salvador, BA, Brazil*

IRENA PHILIPOVA • *Institute of Organic Chemistry with Centre of Phytochemistry, Bulgarian Academy of Sciences, Sofia, Bulgaria*

EUGENE V. RADCHENKO • *Department of Chemistry, Lomonosov Moscow State University, Moscow, Russia; Institute of Physiologically Active Compounds RAS, Moscow Region, Russia; Institute of Organic Chemistry RAS, Moscow, Russia*

PRAVEEN P. N. RAO • *School of Pharmacy, Health Sciences Campus, University of Waterloo, Waterloo, ON, Canada*

MARICARMEN HERNÁNDEZ RODRÍGUEZ • *Laboratorio de Biofísica y Biocatálisis, Escuela Superior de Medicina, Instituto Politécnico Nacional, Ciudad de Mexico, Mexico*

DIPAYAN SAMANTA • *Department of Biotechnology, National Institute of Technology Durgapur, Durgapur, India*

RAJESH K. SANI • *Department of Chemical and Biological Engineering, South Dakota School of Mines and Technology, Rapid City, SD, USA*

LUCIANA SCOTTI • *Federal University of Paraíba, Health Center, João Pessoa, PB, Brazil*

MARCUS T. SCOTTI • *Federal University of Paraíba, Health Center, João Pessoa, PB, Brazil*

ASHOK SHARMA • *Biotechnology Division, CSIR-Central Institute of Medicinal and Aromatic Plants, CIMAP, Lucknow, India*

SWATI SINGH • *Bioinformatics Infrastructure Facility, Center of Excellence in Bioinformatics, Department of Biochemistry, University of Lucknow, Lucknow, India*

KAILAS DASHRATH SONAWANE • *Structural Bioinformatics Unit, Department of Biochemistry, Shivaji University, Kolhapur, Maharashtra, India; Department of Microbiology, Shivaji University, Kolhapur, Maharashtra, India*

GEORGI STAVRAKOV • *Faculty of Pharmacy, Medical University of Sofia, Sofia, Bulgaria*

SAMEEKSHA TIWARI • *Bioinformatics Infrastructure Facility, Center of Excellence in Bioinformatics, Department of Biochemistry, University of Lucknow, Lucknow, India*

KENJI USUI • *Faculty of Frontiers of Innovative Research in Science and Technology, Konan University, Chuo-ku, Kobe, Japan*

NORA M. WALLACE • *LiS Consulting, Lawrence, KS, USA*

MANGE RAM YADAV • *Faculty of Pharmacy, Kalabhavan Campus, The Maharaja Sayajirao University of Baroda, Vadodara, Gujarat, India*

NIKOLAY S. ZEFIROV • *Department of Chemistry, Lomonosov Moscow State University, Moscow, Russia; Institute of Physiologically Active Compounds RAS, Moscow Region, Russia; Institute of Organic Chemistry RAS, Moscow, Russia*

DIMITRINA ZHELEVA-DIMITROVA • *Faculty of Pharmacy, Medical University of Sofia, Sofia, Bulgaria*

Part I

An Introduction to the Disease

Chapter 1

Alzheimer's Disease Therapy: Present and Future Molecules

Awanish Kumar and Ashwini Kumar

Abstract

Alzheimer's disease (AD) is one of the most common neurodegenerative disorders and a cause of progressive dementia worldwide. It is generally attributed to multiple genetic factors and, thus, is genetically heterogeneous. The two basic pathological features of AD are extra-neuronal plaques of misfolded β-amyloid proteins and intraneuronal neurofibrillary tangles of hyperphosphorylated *tau* protein. On therapeutic front, presently there are only two targets for AD, namely, acetylcholinesterase inhibitors and NMDA receptor antagonists that improve the cognitive functions. But these drugs do not act in ameliorating the pathological causes behind AD. Therefore, active research is the need of the hour toward AD treatment. In designing novel drugs or modifying existing molecules, computational approaches have proved to be very useful in saving time and money. Virtual screening, modeling, and docking are being widely used for the last few years by researchers globally. These have indeed helped with a lot of promising compounds on the desk. With continuous efforts in bringing down the problem of AD, researchers have targeted some widely known neuronal targets such as muscarinic/nicotinic acetylcholine receptors, tau hyperphosphorylation, beta-secretase enzyme, and β-amyloid plaques, while many new targets such as sigma-1, α-secretase, histamine H3 receptor, and Lingo-1 have been identified and molecules targeting them are being developed. Apart from synthesizing chemical entities, several natural compounds have been extracted and tested for AD. Compounds such as flavonoids, curcumin, alkaloids, and terpenoids have shown promising activity against various targets. Thus, owing to the helpful hand of computational biology and natural treasures, several promising molecules are on the front, and many candidates are in different clinical trial phases that give positive hopes in near future.

Key words Muscarinic receptor, Nicotinic receptor, Tau protein, β-Secretase, Sigma-1 receptor, Anti-amyloid

1 Introduction

The story actually started in the year 1907 when Alois Alzheimer reported a woman with rapid and progressive memory deterioration and psychiatric disturbances. She was reported dead 4 years later. Alzheimer's disease (AD) is an irreversible and progressive neurodegenerative disease that slowly results in the state of dementia. The primary pathological findings are the extracellular β-amyloid plaques (Aβ) and intracellular neurofibrillary tangles (NFTs). The two basic types of AD reported originally are early-onset or

Kunal Roy (ed.), *Computational Modeling of Drugs Against Alzheimer's Disease*, Neuromethods, vol. 132, DOI 10.1007/978-1-4939-7404-7_1, © Springer Science+Business Media LLC 2018

familial AD (FAD) and late-onset sporadic AD (SAD). The former is said to be hereditary and it increases the risk of AD in first-degree relatives of the patients and starts early in life. It is generally thought to occur due to a mutation in three genes, namely, amyloid precursor protein (APP), presenilin 1 (PSEN 1), and presenilin 2 (PSEN 2). The sporadic form generally starts late and is said to be caused due to the interplay between environmental and genetic reasons. The gene apolipoprotein E ε4 (ε4 allele of ApoE gene) is said to be the major factor for SAD. The familial AD has a strong genetic relation with links on chromosomes 21, 19, 14, and 1 [1, 2]. The current population with dementia worldwide has been estimated to be approximately 47 million while it has been projected to rise to 75 million by the year 2030 [3]. According to Alzheimer's Association, the basic symptoms of AD are memory loss, difficulty in solving problems, difficulty doing work at home, confusion with remembering time and dates, the problem in speaking or writing, and loss of ability to retrace events and steps [4]. The pathogenesis of AD is primarily based on the extracellular accumulation of beta-amyloid (Aβ), which is a degradation product of amyloid precursor protein (APP) and intracellular deposition of hyperphosphorylated tau protein leading to NFTs. Some typical features of the brain of AD patient include neuronal and dendritic loss, dystrophic neuritis, granulovacuolar degeneration, and cerebrovascular amyloids. The pathological classification of AD is based on two major criteria, namely, CERAD and Braak criterion. This new classification describes AD into preclinical AD and dementia due to AD. Where CERAD emphasized the presence of Aβ as the primary feature of AD, Braak criterion emphasizes on the presence and progression of NFTs as the primary feature. The mostly favored Braak classification suggests the AD progression into three parts: stages I–II (degeneration of transentorhinal region), stages III–IV (limbic area), and stages V–VI (degeneration of isocortical area). The majority of patients at stages I–II are considered to be free of cognitive impairment, while those at stages V–VI have a severe cognitive impairment [1]. On the other hand, Aβ was found to inhibit the neural mitochondrial activity, inhibit the synaptic path, inhibit the proteasome pathway, and alter the intraneural calcium level. Mild cognitive impairment (MCI) is generally said to be present in the early case of pathological neurodegeneration, where the patient is said to have cognitive dysfunction but cannot be classified with dementia. In view of the report by Castellani et al., MCI comes somewhere between normal aging and AD. Thus, patients who demonstrate MCI may be seen as preclinical AD case, and those who have a defined dementia could be said to come under "dementia due to AD criteria" [1]. The two most widely accepted biochemical reasons for cognitive impairment in AD patients are a continuous loss of acetylcholine (ACh) due to hyperactive acetylcholinesterase enzyme (AChE) and hyperactive

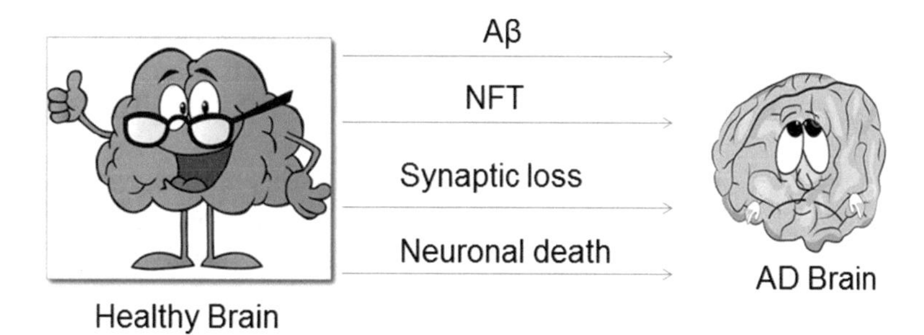

Fig. 1 A cartoon representation of different factors contributing to the changes in the brain of an AD patient

N-methyl-D-aspartate (NMDA) glutamate receptors. Since it is widely known that Ach is primarily responsible for cognition and behavioral aspect, loss of Ach results in cognitive impairment. Hyperactive glutamate NMDA receptor leads to increased influx of calcium ions (neuronal excitotoxicity) and increased free radical generation that becomes detrimental to the neurons [5] (Fig. 1).

The therapeutic approaches currently available for AD are acetylcholinesterase inhibitors and N-methyl-D-aspartate (NMDA) receptor antagonists. Donepezil, galantamine, rivastigmine, and tacrine are the currently marketed AChE inhibitors that improve the cholinergic activity of the brain, while memantine is the only available NMDA receptor antagonist in the clinical use. Apart from these small molecule drugs, cerebrolysin is a peptide drug prepared from purified proteins from the brain and is supposed to mimic the effects of endogenous nerve growth factor. The drug is already approved in many European nations. The drug has been proposed to promote neuronal differentiation and plasticity and improvement in cognitive functions too. The mechanisms suggested are interaction with the inhibitory pathways (GABA) and inhibition of Aβ formation. The recommended dose is 5–30 ml/day for approximately 20 days as intravenous administration while lower doses as intramuscular administration [6]. Among various experimental and trial candidates, there are β-secretase inhibitors, γ-secretase inhibitors, and α-secretase enhancers which reduce the production and burden of beta-amyloids in the brain. On the other hand, tramiprosate and epigallocatechin gallate (EGCG) decrease the aggregation of misfolded beta-amyloid proteins. Certain monoclonal antibodies such as bapineuzumab and solanezumab have been tested in clinical trials for clearance of beta-amyloid. Glycogen synthase kinase-3 (GSK-3) has been shown to accelerate the phosphorylation of *tau* protein. Tideglusib is a glycogen synthase kinase-3 (GSK-3) inhibitor which is under clinical trial for AD. Another such GSK-3 inhibitor is methylthioninium chloride which was found to improve cognitive function in AD patients in phase II trial [6]. Certain muscarinic and nicotinic ACh receptor agonists are in clinical trials which restore the cholinergic activity and have shown significant improvement in AD patients. One of the examples is EVP-6124 by Elan Pharmaceuticals which is in phase II trial [5].

Since drug discovery and development is a time-consuming and very costly process, bioinformatics and computational biology approaches have made the initial selection of potent drug molecules easy. These computational biology approaches have become very crucial right from lead identification to optimization and more. Docking, one of the most applied computational processes, involves the analyses of virtual binding interactions between a proposed ligand and the target receptor (protein). The process starts with preparation of ligand and receptor structures, making the interaction of ligand to the active site of the receptor possible, comparing the scoring functions and structural changes and other features. There are a lot of free and licensed tools for performing molecular docking. Some of the very popular docking tools are Schrodinger Suite, Discovery Studio, and FlexX among the commercial tools, while AutoDock and DOCK are the most trusted freeware available [7]. Apart from the docking functions, computational biology approaches have led the researchers to have an idea of structure-activity relationship (SAR) and pharmacokinetic properties (absorption, distribution, metabolism, excretion, and toxicity or ADMET) of the potential ligands. Other widely used approaches in computer-based drug design are pharmacophore modeling, de novo drug design, and fragment-based drug design that are indeed very cost-effective designing tools. Application of various computational tools, thus, helps save time that was spent in traditional combinatorial chemistry screening experiments. The two most used approaches in drug designing are structure-based drug designing (SBDD) and ligand-based drug designing (LBDD). SBDD involves the knowledge of the receptor structure upon which an array of ligands is tested. On the other hand, LBDD involves the use of a space which consists of many similar ligands known to bind to a particular receptor [8]. Computational approaches have given various potential drug candidates against the conventional and promising targets against AD. Molecules such as flavonoids, carbamates, pyridonepezil, coumarin derivatives, and many more have been found to be great lead compounds which can be propagated against Alzheimer's disease. Many molecules such as MK-8931 (against β-secretase, Merck), TAK-070 (against β-secretase, Takeda Pharmaceuticals), and LMTX (against tau hyperphosphorylation, TauRx Inc.) have been brought into different phases of clinical trials [5].

1.1 Pathophysiology of AD

As said in the previous section, the two original and major hallmarks of AD are the extracellular β-amyloid plaque and intraneuronal NFTs. The amyloid precursor protein or APP is cleaved by three different enzymes in two separate pathways. The initial two enzymes are α-secretase and β-secretase while γ-secretase comes later in the picture. When APP is cleaved by α-secretase followed by γ-secretase, soluble fragments of 40 amino acid length are generated. But when the same APP is cleaved by β-secretase followed by γ-secretase, the amyloidogenic Aβ fragments are generated which are 42 amino acids long. These Aβ fragments are hydrophobic and tend to aggregate creating extra-neuronal plaques. With an increase in the aggregation of these insoluble fragments, the toxicity increases which leads to synaptic dysfunction and gradual neuronal death. Strong evidence suggests a genetic predisposition of a person toward AD. Autosomal dominant mutation in presenilin genes is the strongest empirical reason for AD. The two presenilin genes are presenilin 1 (PS1) present on chromosome 14 and presenilin 2 (PS2) on chromosome 1. The presenilin protein forms the active site of the APP-degrading enzyme γ-secretase. The PS1 or PS2 mutation shifts the APP cleavage toward the amyloidogenic pathway. Another mutation is said to occur in the APP in which the sites for β-secretase and γ-secretase become prominent and undergo increased cleavage [9]. It was also reported that the amygdala region of the brain was worst hit with the pathological features of AD and have the highest deposition of Aβ and NFTs. Among other regions of the brain, hippocampus had high NFT features compared to the neocortex which had higher plaques and lesser NFTs. Several theories have been put up relating to the synaptic dysfunction in AD such as Aβ toxicity and oxidative stress. The large presence of activated microglial cells have also been reported in the senile plaques, and it holds the theory of AD being an inflammatory condition [1]. In a review by Niedowicz et al., it was reported that in normal circumstances, Aβ oligomer is degraded by two enzymes, namely, insulin-degrading enzyme (IDE) and neprilysin (NEP). Their activity and expression are seen to be significantly decreased in aging and diseased brain. Particularly, the activity of NEP has been reported to be significantly decreased in the cerebrospinal fluid of early AD patients [9]. The second hallmark feature of AD is the neurofibrillary tangles. The NFTs are the aggregates of hyperphosphorylated microtubule protein *tau*. It was shown in an experiment that wild-type mouse developed NFTs in the neurons when injected with phosphorylated tau protein. This hyperphosphorylation of tau protein is said to be a result of certain hyperactive kinase enzymes such as glycogen synthase kinase-3β (GSK-3β), an extracellular signal-related kinase 2 (ERK2) and cyclin-dependent kinase 5 (CDK5). Therefore, these enzymes are being considered as potential targets against AD [10]. NFTs were primarily found in the cerebral neocortex and Ammon's horn regions of the brain with their significant presence in hypothalamus too. Apart from the mutated tau protein, several biochemical and immunological analyses have confirmed the presence, in the helical NFTs, of other biochemical molecules such as tropomyosin, elastase, trypsin, proteoglycans, acute-phase proteins, and others [1].

Genetic predisposition has been found to be quite significant in AD patients. The gene for APP is found on chromosome 21 and is said to give a link between the incidences of AD in the patients with Down syndrome. According to a recent review published in Nature Reviews Neurology, the ε4 allele of apolipoprotein E (ApoE) is considered to be the strongest risk factor for both early- and late-onset AD. ApoE is a transporter protein of serum cholesterol that has been significantly found in both Aβ and NFTs. Researchers have reported that approximately 40.7% ε4 carriers developed Aβ plaques as compared to only 8% noncarrier population. APO ε4 is said to promote AD by accelerating the formation and accumulation of Aβ plaques. APO ε4 has also been said to increase the risk of cerebral amyloid angiopathy (CAA) and related cerebral bleeding [11]. The presence of ε4 allele increases the risk of development of AD by 7 times as compared to the presence of ε3 allele [10]. Another set of genes that predisposes very high risk of AD is the *presenilin 1* (PSN1) and *presenilin 2* (PSN2) on chromosome 14 and chromosome 1, respectively. Mutations in these two genes favor the synthesis and deposition of beta-amyloid around the neurons. Presenilin mutation in early-onset or familial AD displays high Aβ depositions around neurons and largely in the blood vessels. Apart from the above-stated genetic reasons, there are a few other gene mutations which show a predisposition toward AD-like *CLU* (encodes protein clusterin) and *PICALM* (phosphatidylinositol-binding clathrin assembly protein or *CALM*). Clusterin protein in wild form suppresses Aβ deposition, while a mutation in PICALM protein disturbs the synaptic function [1].

Based on these abovementioned causes of AD, there have been many theories postulated for the disease. The cholinergic theory, which was initially suggested, states that the AD brains show a significant reduction in the neurotransmitter acetylcholine. There is a direct relation shown between the levels of senile plaques and reduction of the enzyme choline acetyltransferase or CAT (an enzyme that catalyzes the formation of ACh) in the neocortex of the brain and the similar relation between NFT and CAT enzyme in the hippocampus region. The cortico-cortical hypothesis suggests a degeneration of cerebral cortex. Another research displayed a decrease in SP6, a synaptic marker, in the brain of AD patients. Another hypothesis suggests that the disease-causing Aβ can be released by one neuron through exocytosis and enter another one through endocytic process creating the chain of pathological features. The mitochondrial theory suggests that swelling and distortion in mitochondria represent an early sign in AD. There could be a deficiency of enzymes involved in carbohydrate metabolism such as pyruvate dehydrogenase and phosphofructokinase that themselves are present in mitochondria [12]. The mitochondrial defect would also result in strong oxidative stress that becomes detrimental to the neuronal cells. It was reported that oxidative stress starts much before the formation of beta-amyloid plaque that is suggestive of mitochondrial trigger of AD. It was reviewed that in AD, cytochrome oxidase enzyme, an important component of electron-transport chain (ETC), was significantly damaged and decreased. Another striking feature reported was the presence of mtDNA (mitochondrial DNA) and cytochrome oxidase in neuronal cytoplasm and significantly decreased mitochondria in those neurons. Apart from neurons, cytochrome oxidase was also found to be very low in platelets from AD patients [13]. On the path of oxidative stress, it was reported that the oxidized nucleotide bases in nuclear and mitochondrial DNA were significantly higher in AD patients as compared to age-matched control patients. Comparing nuclear and mitochondrial DNA, it was also reported that oxidization in mtDNA was almost ten times higher than that in nuclear DNA [14]. It was also reported that mitochondria in neurons of AD patients have increased membrane permeability, reduced membrane potential, and greater reactive oxygen species (ROS) forming ability. Evidences suggest that Aβ precursor protein (APP) and Aβ are present in the mitochondrial membrane interacting with the mitochondrial proteins and enzymes in neurons of AD patients [15].

2 Promising Drug Targets

2.1 Muscarinic and Nicotinic ACh Receptors

Muscarinic (mAChR) and nicotinic (nAChR) receptors are the two types of ACh receptors found on the neurons. mAChRs are reported to be responsible for cognitive functions such as learning process, motor control, and memory [5]. The muscarinic receptors have been classified into five subtypes, viz., M_1–M_5, and are found throughout the central nervous system and peripheral nervous system. M1, M3, and M5 are the $G\alpha q$ G-protein type and mobilize calcium ions and initiate inositol triphosphate (IP3) production. On the other hand, M2 and M4 consist of $G\alpha i$ components and inhibit cAMP production. Nicotinic AChRs are also an acetylcholine-binding ion channel. nAChRs are formed of different subunits where mammals have 16 genes that encode these subunits. Neuronal nAChRs are formed from α-subunits (α2–α10) and β-subunits (β2–β4) where the most abundant are α7, α4β2, and α3β4 subunits. Several analyses have revealed a loss of both muscarinic and nicotinic ACh receptors in the brain of AD patients. Since binding of ACh to these two receptors is responsible for positive cognitive effects, these receptors have been looked as potential candidates to develop agonists that can confer cognitive functions to the AD brains. Apart from the cognitive effect, nicotine, the natural ligand of nAChRs, has been shown to protect the neurons from beta-amyloid-induced toxicity. Thus, activation of nAChRs can have a neuroprotective effect too [16, 17]. It was also reported that the cholinergic hypofunction accelerates the formation of beta-amyloid. These plaques, in turn, disrupt further cholinergic signaling. The action of M1-specific agonists can help in two ways – firstly the ACh signaling will improve, and secondly, the APP processing will not produce the Aβ. This improvement of cognitive function is due to the fact that M1 subtype is the major cholinergic receptor found in the hippocampus and cerebral cortex region which is the major site responsible for cognitive effects. The promotion of non-amyloidogenic pathway is said to be due to the preference of α-secretase pathway [18]. About two decades back, Bodick et al. published randomized trial results of the compound xanomeline or {3(3-hexyloxy-1,2,5-thiadiazol-4-yl)-1,2,5,6-tetrahydro-1-methylpyridine)} that was a M1 receptor agonist. It was the first selective M1/M4 mAChR agonist to enter the clinical trials. Xanomeline improved various AD-related cognitive and behavioral changes such as delusion, hallucination, agitation, and vocal outbursts. But due to significant adverse events, it was discontinued and was not approved. The major adverse effects observed in the trial were nausea (in 51% patients), vomiting (42% patients), dyspepsia (24% patients), increased salivation (24% patients), and increased diaphoresis (76% patients). In approximately 13% of the patients, hepatic and biliary transaminases were found to be elevated [19, 20]. After the failure of xanomeline, the M1/M4 receptor agonist-based research came to a halt. After the xanomeline episode, Fisher et al. used AF267B, AF102B, and AF150S on AD pathophysiology. AF102B treatment reduced the Aβ load significantly in the AD patients. The molecule AF102B (cevimeline, Evoxac) is the first selective M1 agonist approved by FDA for Sjogren's syndrome and available in the USA and Japan [21]. Heptares Therapeutics has multiple molecules in pipeline against AD. A M1 mAChR agonist is presently being studied in phase Ib clinical trial. Another agonist for M4 mAChR is currently in the preclinical study, while a molecule that acts as a dual M1/M4 agonist is also in the preclinical development stage (http://www.heptares.com/pipeline/). Positive allosteric modulators (PAMs) are the molecules that bind to the allosteric site where they have no effect on their own but increase the action of endogenous ligands (agonists) to the receptors. The first PAM for AD was benzyl quinolone carboxylic acid (BQCA) that finds very high binding affinity at the allosteric site of mAChR and increases the action of ACh by approximately 129-folds. Apart from M1, two selective M4 receptor agonist molecules were developed, namely, VU10010 and LY2033298, where the former was a PAM, while the later a simple agonist. But due to unsuitable physiochemical properties, VU10010 was not continued to preclinical and clinical studies. Further modification of VU10010 gave two improved compounds VU0152100 and VU0152099 with improved pharmacokinetic properties and also exhibited much better brain penetration in animal models [22]. Among the nAChRs, α4β2 particularly enhances the ACh release and effect. Another nAChR subtype which is the center of attraction for AD treatment is the α7 nAChR. Galantamine is the FDA-approved anti-Alzheimer's drug that also significantly activates α4β2 nAChR. There are other agonists for both the subtypes in different phases of development, such as ABT-418, RJR-2403, GTS-21, A-582941, and AR-R1779 [23]. Other evidences show that the marketed AChEIs also act as allosteric modulators of nAChR and thus exhibit their neuroprotection effect both by increasing the concentration of ACh and reducing the burden of Aβ and activation of α7 nAChRs. AstraZeneca developed a molecule AZD3480 which was highly selective for α4β2 nAChRs, and Daiichi Sankyo developed nefiracetam that enhanced the activity of α4β2 nAChRs [17]. Elan Pharmaceuticals has developed a molecule named EVP-6124 (encenicline) which is an agonist for α7 nAChRs [5].

2.2 Tau Protein Hyperphosphorylation

As stated in the previous section, hyperphosphorylated tau protein is the major constituent of the intraneuronal NFTs. Tau phosphorylation has been attributed to two major kinases, namely, GSK-3 and CDK-5. Though tau phosphorylation is required for various normal physiological functions such as microtubule formation, neurite growth, and axonal transport, the hyperphosphorylation of tau is attributed to the pathological effects. It has been reported that the GSK-3β isoform is the primary culprit for the development of tauopathy in AD brain. Not only responsible for the formation of NFTs, the GSK-3β isoform also facilitates the amyloidogenic processing of APP. Thus, the first strategy against the accumulation of hyperphosphorylated tau protein was to inhibit these enzymes [24, 25]. LMTM [methylthioninium chloride (MTC)], developed by TauRx Therapeutics Ltd., is considered to be the best molecule today that inhibits the aggregation of hyperphosphorylated tau protein and is currently in phase III trial. Another molecule TPI-287 by Cortice Bioscience, currently in phase I trial, is indicated as a tubulin-binding or microtubule-stabilizing drug. It is a derivative of taxane molecule, and unlike other taxanes, it can effectively cross the blood-brain barrier (BBB). There are other molecules which are potential vaccine candidates for AD presently in different phases of clinical trial. RO 7105705 [phase I, AC Immune SA, Genentech, Hoffmann-La Roche], ACI-35 [phase I, AC Immune SA, Janssen], AADvac-1 [phase II, Axon Neuroscience SE], and C2N 8E12 [phase I, AbbVie, C2N Diagnostics, LLC] are such immunotherapy candidates targeting tau protein in one way or the other. ACI-35 is a liposome-based vaccine that consists of synthetic hyperphosphorylated tau protein that can elicit a strong internal immune response, while RO 7105705 is an antibody that inhibits the extracellular tau and stops tau relay across the neurons (http://www.alzforum.org/therapeutics) .

2.3 Beta-secretase Enzyme

As it is known and described in the earlier sections, the transmembrane protein APP undergoes two different cleavage and processing pathways among which one is amyloidogenic while other is not. The cleavage of APP by β-secretase (BACE1) followed by γ-secretase generates Aβ which forms insoluble plaques, while the Aβ fragments are not generated if the initial enzyme is α-secretase followed by γ-secretase action. Thus, the strategy to inhibit BACE1 would result in a significant decrease in the Aβ load and would result in improved cognitive functions [26]. Some of the most discussed BACE1 inhibitors are MK8931 (Merck & Co.) and AZD3293 or LY3314814 (AstraZeneca and Eli Lilly & Co.). Both the molecules are currently in phase III trial. Other molecules in different phases of clinical trial are CTS21166 (phase I, CoMentis), E2609 (phase II, Eisai/Biogen Idec), HPP854 (phase I, High Point), PF-05297909 (phase I, Pfizer), TAK070 (phase I, Takeda), and VTP-37948 (phase I, Vitae/Boehringer Ingelheim). A very promising candidate by Lilly, LY2886721, was withdrawn after phase II trials due to abnormal liver biochemistry [27]. Although the BACE1 inhibition in animals and knockout animal models showed significantly improved reduction in Aβ load and related pathologies, it also led to the symptoms and features such as axonal defects, increase in seizures, schizophrenia-like behavior, and loss of myelin sheath. A recent and important research revealed certain functions of BACE1 enzyme such as its critical requirement in maintaining dendritic feature in the adult brain, maintaining neuronal plasticity, and synaptic communication, and its physiological level is necessary for normal cognition and memory. Thus, a careful and time-based dosing of BACE1 inhibitors is the need of the hour against AD [28].

2.4 Other Novel Targets for Anti-Alzheimer's Drug Development

Among the new targets for AD, the sigma-1 receptor is being seen as one of the best target candidates. Sigma-1 is a highly conserved membrane protein found in the membrane of endoplasmic reticulum (ER). It is mostly situated at the junction of ER and mitochondria. Sigma-1 is found extensively in various cells of CNS. Being a chaperone protein, sigma-1 translocates between different organelles when the ligand binds to it. One method by which sigma-1 exerts neuroprotective effect is by regulating the intraneuronal calcium homeostasis. Another piece of evidence suggests that sigma-1 activation attenuates the glutamate excitotoxicity by inhibiting NMDA receptors in AD brain. Sigma-1 activation has also shown to attenuate the reactive oxygen and nitrogen species that can cause macromolecular damage. AD brains have shown decreased expression of sigma-1 receptors. Though the mechanisms of sigma-1-mediated neuroprotective effects are not completely understood, speculation and certain pieces of evidence include regulation of intraneuronal calcium homeostasis, modulation of Bcl-2 gene and related caspase levels, and reduction of oxidative stress [29]. Sigma-1 has also been stated to help significantly in neurite growth, thus helping in normal inter-neuronal communication. Combining the data from different research, it has been shown that activation of sigma-1 receptor exerts neurotransmitter release, inositol 1,4,5-triphosphate (IP3)-mediated Ca^{2+} signaling across the cells, cell differentiation and plasticity, and neuronal survival [30]. Anavex 2-73, a molecule developed by Anavex Life Sciences Corp., is currently in phase II trial. It is said to be a mixed ligand for both sigma-1 and muscarinic receptors. Some recent studies also suggest that Anavex 2-73 also inhibit the tau hyperphosphorylation.

The enhancement of α-secretase is another potential area of AD. It has been speculated that increasing the effect of this enzyme would enhance the non-amyloidogenic processing pathway of APP that results in the soluble fragments known as soluble APPα or sAPPα. Etazolate basically is a GABA modulator that also significantly enhances α-secretase enzyme shifting to the production of sAPPα. Another molecule, bryostatin-1, activates protein kinase C which upregulates the α-secretase pathway [6]. Another very promising approach in AD treatment is the anti-amyloid therapy that includes Aβ production minimization, reducing Aβ aggregation, and enhancing Aβ clearance. BACE1 inhibition and α-secretase enhancement come under the first approach. The presence of Aβ does not ensure neuronal toxicity rather it must aggregate to form plaques. Glycosaminoglycans (GAG) have been shown to strongly bind Aβ to form plaques. Tramiprosate is a GAG mimetic that is shown to inhibit Aβ aggregation. But this drug was terminated in phase III trial because it failed to demonstrate any cognitive improvement when compared to the placebo group. Scyllo-inositol (ELND005) is an inositol molecule that also inhibits the Aβ aggregation. The compound has shown considerable positive effects in phase II trials and is being carried forward to phase III trials. As a serendipity, it was found that immunization with Aβ in mouse led to vaccination-like effects and enhanced pathological Aβ clearance. The first immunotherapy molecule AN1792 was terminated in phase II trial due to the occurrence of meningoencephalitis as the primary adverse effect. Other anti-amyloid vaccines CAD106 (Novartis) and V950 (Merck) are under phase II trial. Passive immunization approach in which anti-Aβ antibody is infused in patients further protects against the T-cell-mediated inflammatory response generated by vaccination of Aβ itself. Bapineuzumab (AAB-001) is a fully humanized monoclonal antibody against the N-terminal of Aβ. But its intravenous (IV) infusion displayed vasogenic edema and microhemorrhages, and its further trials of IV administration are terminated, but the subcutaneous route is being investigated. Another monoclonal antibody Solanezumab has also failed in clinical trials. Active immunization is always preferred to the passive immunization since the former reduces the cost of booster dosing which is very common with passive immunization [31].

3 Approved Drugs and Targets in AD

3.1 Acetylcholinesterase Inhibitors

Cholinesterase inhibitors were the first class of drugs that were marketed for the treatment of AD. AD brain displays significant loss of cholinergic neurons and a decreased activity of acetylcholine-synthesizing enzyme choline acetyltransferase (CAT). Acetylcholinesterase (AChE) is responsible for the breakdown of the neurotransmitter acetylcholine which renders the cognitive function to the brain. Thus, acetylcholinesterase inhibitors (AChEIs) increase the concentration of ACh molecules and help maintain the cognitive functions. Apart from inhibiting AChE, the inhibitors have been shown to mediate the processing pathways of APP and decrease the deposition of beta-amyloid plaques. The second function of AChEIs is that AChE enzyme has a peripheral anionic site (PAS) that helps in the aggregation of Aβ. The common AChEIs marketed currently are donepezil (trade name Aricept), rivastigmine (trade name Exelon), galantamine (trade name Reminyl and Nivalin), and tacrine. Tacrine is now rarely used due to the significant occurrence of hepatotoxicity. Another molecule Huperzine A has been approved in China but not in the European countries and the USA [6, 31].

3.2 NMDA Receptor Antagonists

The primary neurotransmitter present in the hippocampus and cortex of the brain is glutamate. NMDA receptors are the most prominent receptors on which the glutamate binds. Glutamate is also said to be responsible for cognition and intellectual behavior. AD patients show increased oxidation of the enzyme glutamine synthetase (GS) that leads to increased production of glutamate. Over-activation of NMDA receptor by glutamate results in excitotoxicity that leads to neural death. This excitotoxicity is the result of increased calcium influx, increased nitric oxide production, mitochondrial dysfunction, and accelerated neuronal apoptosis. Memantine is the only approved and marketed drug that blocks the NMDA receptor and, thus, prevents the neuronal damage from glutamate excitotoxicity. Another mentioned role of memantine is the prevention of *tau* hyperphosphorylation, thus decreasing the formation of NFTs. Fixed dose combination of donepezil and memantine has also been approved and marketed and exhibited great cognitive improvement in AD patients [31, 32].

4 In silico Approach Decipher New Molecules

Computational approaches have provided drug discovery a very helpful hand. In the case of AD, several compounds have been generated which have an inhibitory action on targets such as AChE, BACE1, Aβ aggregation, mAChRs, and GSK-3β. Flavonoids and derivatives have been seen as very effective inhibitors of AChE and BACE1. Certain flavonoids have been shown to have better AChE inhibitory activity than the marketed drug rivastigmine and donepezil. A phenolic compound isolated from the plant *Larrea tridentata* has shown to have effective AChE inhibitory activity and also prevented aggregation of Aβ. Certain pyridopyrimidine derivatives have been shown to have AChE inhibitory activity approximately 2.5 times higher than the marketed drug galantamine. Pyridonepezil and 6-chloro-pyridonepezil, derivatives of the marketed drug donepezil, have much better and selective inhibition action against AChE. Piperazine derivatives and 6-chloro-pyridonepezil were found to have AChE inhibitory activity at both active site and peripheral anionic site [5]. Another potent molecular skeleton is coumarin whose derivative 4-hydroxycoumarin demonstrated significant AChE inhibition. Recently, 4-aminoquinoloine was found to have significant AChE inhibitory activity with an average IC50 of 0.60 μM. There are other compounds such as hydroxyethylene and hydroxyethylamine which show dual inhibition against AChE and BACE1. In terms of BACE1 inhibition, acylguanidine and its derivatives have been considered as the most preferred BACE1 inhibitors currently in experimental stages. On the natural product front, flavonoids and flavanols have been identified as a strong inhibitor of BACE1 activity [5]. One of the first reports of synthesis of acylguanidine structures come from a research published a decade ago. Cole et al. synthesized several acylguanidines and found five of them to be a great inhibitor of the BACE1 enzyme, with a lowest IC50 value of 110 nM. An aminoimidazole derivative has also shown great inhibition of BACE1 with an IC50 value of 7.4 μM [33]. Statin and norstatine-based moieties have been investigated as potential BACE1 peptidomimetic inhibitors. Some of these molecules displayed an IC50 value of 0.20 μM. Other peptidomimetic inhibitors of BACE1 have a parenteral group of hydroxyethylene and hydroxyethylamine [26]. Structure-based drug design or SBDD has recently led to the synthesis of iminopyrimidinone compounds which were a great inhibitor of BACE1 with an IC50 value of 11 μM [26]. 3-hydroxy-1H-quinazoline-2,4-dione derivatives, 1-benzyl-1,2,3,4-tetrahydro-b-carboline, and 3-substituted-1H-indoles are some of the compounds synthesized and found to be a great antagonist of NMDA receptor. NMDA receptor consists of different subunits, namely, NR1, NR2A, NR2B, NR2C, and NR2D. Glycine has been shown to bind to NR1 subunit and co-activate NMDA receptor. Antagonists are being searched that could block the binding of glycine to the NR1 subunit [5]. In one of the recent review articles, a synthetic pyrimidine derivative [6-(4-pyridyl) pyrimidin-4(3H)-one] was reported as a strong inhibitor of GSK-3β that could reduce the hyperphosphorylation and further aggregation of tau protein [31].

5 Current Status of Experimental Drugs

Experimental drugs, as can be commonly said, are the potential drug candidates which are being investigated for a particular disease condition but have not reached the clinical trial stages. Several research groups around the world are engaged in deciphering promising targets and compounds that can be tested in Alzheimer's disease. The molecules described in the previous section (Sect. 4), are the experimental candidates against the targets established in AD. In terms of targets or receptors, histamine H3 receptors are being investigated as a promising target for AD. The H3 receptors are histamine autoreceptors which have a vast presence in CNS, particularly in the areas responsible for cognition and memory like hippocampus, cortex, and hypothalamus. Studies have revealed that activation of H3 receptors inhibited the release of ACh, while antagonists for H3 receptor increased the release of ACh neurotransmitter. On the level of expression, it was found that the expression of the H3 receptor was found to be significantly high in AD brains. Esbenshade et al. have described various H3 receptor antagonists such as ABT-239 (2-ethylaminobenzofuran), BF2-649 (piperidinylpropoxyalkylphenyl), JNJ-5207852/JNJ-10181457 (Johnson & Johnson), GSK189254 (benzodiazepine derivative by GSK), and MK-0249 (phase II trial, Merck) [34]. Some other promising targets recently identified for AD treatment are the GABA transporter GAT3/4 in the reactive astrocytes of the dentate gyrus, *Lingo-1*, α2-adrenergic receptor, PERK or the stress-induced kinase in endoplasmic reticulum, and sirtuins [35–37]. *Lingo-1* is a transmembrane protein abundant in oligodendrocyte, and neurons are found to be a negative modulator of functions such as axonal integrity and oligodendrocyte differentiation. Lingo-1 is reported to favor the BACE1-induced cleavage of APP favoring the formation of Aβ. It also activated pathways that hampered the neuronal survival. An anti-Lingo-1 antibody, BIIB033 produced by Biogen, is currently in phase II trial for multiple sclerosis, but it can prove to be a useful candidate against AD [36]. In a recent report, it was shown that activation of α2-adrenergic receptors in brain favored the generation of Aβ. Gannon et al. have shown that treatment of AD mice model with α2 receptor antagonist idazoxan significantly reduced the Aβ burden and improved cognitive functions too. Thus, inhibition of α2 receptor could be an effective treatment option against AD [38]. Sirtuin protein SIRT1 has been found in vitro and in vivo to reduce the accumulation and deposition of Aβ [39] (Fig. 2).

Alzheimer's Disease	
Conventional targets	Novel targets
Acetylcholinesterase enzyme	GSK-3 enzyme
NMDA Receptor	BACE 1 enzyme
	α-secretase enzyme
	α2 adrenergic receptor
	H3 receptor
	Sigma-1 receptor
	Lingo-1
	PERK enzyme

Fig. 2 Different targets for anti-Alzheimer's drug development

6 Current Status of Drugs in Clinical Trial

A large number of molecules fail to enter market since they are withdrawn even at phase III stage if (a) severe adverse event(s) is/are reported. For example, a very promising drug, encenicline, was put on hold for AD by FDA in phase III trial after severe gastrointestinal adverse effects. Currently, there are several drugs against AD in different phases of clinical trials. Based on extensive literature survey, we have tried to compile various such drugs for AD and presented them in previous sections of this book. According to a detailed report published in 2016, there are a total of 93 drugs in different phases of clinical trial. Of them, 24 are in phase III trial, 45 in phase II trial, and 24 agents in phase I trial [40]. For a detailed view, the readers can visit the following websites:

(a) http://www.alzforum.org/therapeutics
(b) http://www.heptares.com/pipeline/
(c) http://www.alz.org/research/science/alzheimers_treatment_horizon.asp
(d) http://taurx.com/pipeline.html

Some of the molecules in different phases of clinical trial have been listed in Table 1.

Table 1
Drug molecules against Alzheimer's disease in clinical trials

S. No.	Name of the molecule	Mechanism	Clinical trial status	Company	References
1.	Aducanumab	Anti-amyloid monoclonal antibody	Phase III	Biogen	Cummings et al. [40]
2.	AZD 3293	BACE inhibitor	Phase III	AstraZeneca	Cummings et al. [40]
3.	CAD 106	Amyloid vaccine	Phase III	Novartis	Cummings et al. [40]
4.	CNP 120	BACE inhibitor	Phase III	Novartis	Cummings et al. [40]
5.	MK 8931	BACE inhibitor	Phase III	Merck & Co.	Cummings et al. [40]
6.	Solanezumab	Anti-amyloid antibody	Phase III	Eli Lilly	Cummings et al. [40]
7.	TRx0237	Anti-tau	Phase III	TauRx	Cummings et al. [40]
8.	ANAVEX 2-73	Sigma-1 receptor agonist	Phase II	Anavex Life Science	Cummings et al. [40]
9.	BAN 2401	Anti-amyloid antibody	Phase II	Eisai	Cummings et al. [40]
10.	CNP 520	BACE inhibitor	Phase II	Novartis	Cummings et al. [40]
11.	E2609	BACE inhibitor	Phase II	Eisai	Cummings et al. [40]
12.	JNJ-54861911	BACE inhibitor	Phase II	Janssen	Cummings et al. [40]
13.	TPI-287	Tau inhibitor	Phase I	Cortice Biosciences	http://www.alzforum.org/therapeutics, https://clinicaltrials.gov/ct2/show/NCT01966666
14.	RO 7105705	Vaccine	Phase I	AC Immune SA, Genentech, Hoffmann-La Roche	http://www.alzforum.org/therapeutics
15.	ACI-35	Vaccine	Phase I	AC Immune SA, Janssen	http://www.alzforum.org/therapeutics
16.	AADvac-1	Vaccine	Phase II	Axon Neuroscience	http://www.alzforum.org/therapeutics
17.	CTS21166	BACE inhibitor	Phase I	CoMentis	Vassar [27]
18.	E2609	BACE inhibitor	Phase II	Eisai/Biogen	Vassar [27]
19.	PF-05297909	BACE inhibitor	Phase I	Pfizer	Vassar [27]
20.	VTP-37948	BACE inhibitor	Phase I	Vitae/Boehringer Ingelheim	Vassar [27]

7 Alzheimer's Disease Treatment: The Natural Way

Extensive researches are being conducted across the globe to identify natural compounds for the treatment of AD. In an extensive report published in 2010, it was reviewed that curcumin, the natural compound in curcumin, has significant Aβ inhibiting property and could also destabilize the Aβ plaques to generate soluble components [41]. The property was also seen in curcumin derivative rosmarinic acid. It has been speculated that curcumin actually binds the free Aβ and prevents the aggregation into plaques. Curcumin also possesses significant antioxidant activity and can prevent the ROS generation in the brain. Resveratrol, a polyphenolic flavonoid found in grapes and red wine, has been shown to have various cellular protective activities since long. Resveratrol has been shown to reduce the production of Aβ in vitro, protect the Aβ-induced cytotoxicity, penetrate the BBB, reduce the malondialdehyde levels in animal models, and improve the cognitive function and memory in animal models of AD. Green tea, one of the most widely marketed sources of natural antioxidants, has been reported to increase the activity of α-secretase, thereby increasing the non-amyloidogenic pathway processing of APP. It has also been shown to inhibit BACE1 and accumulation of Aβ plaques. It also protects the neurons from damage caused due to Aβ fibrils. Green tea catechins have also significant protective role by inhibiting the hyperphosphorylation of tau protein [41]. In another extensive review published recently, sufficient evidences have been cited for anti-AD and neuroprotective effect of various natural compounds [42]. Alkaloids, terpenoids, and shikimate-based compounds have been shown to be AChE inhibitors. Shikimate-derived compounds, polyketides, terpenoids, and alkaloids also have significant BACE1 inhibitory activity. These classes of compounds also have shown to greatly inhibit the aggregation of Aβ and promoting their clearance. Bryostatin, a PKC activator isolated from *Bugula neritina*, has been shown to enhance the α-secretase activity, and it reduces the mortality of AD mice models. Although it activates PKC, it has been found to be non-carcinogenic and non-tumorigenic. Manzamine A, an alkaloid isolated from sponge *Haliclona*, has been shown to be a great inhibitor of GSK-3, and it effectively reduces the NFT formation. Palinurin, a sesterterpene isolated from a sponge, has significant GSK-3 inhibitory action, and it has been patented by Spanish biotechnology company NeuroPharm. Other marine compounds such as hymenialdisine (from a sponge) and indirubin (from mollusks) have been shown to be a specific inhibitor of the GSK-3 enzyme. Among indirubin derivatives, 6-bromo-substituted indirubin (6-bromoindirubin-3-oxime) has been a great specific GSK-3 inhibitor [42, 43]. In a recent report, rosmarinic acid was extracted from *Salvia sclareoides* and was found to prevent aggregation of Aβ [44]. A recent report suggests the AChE inhibitory activity of the phenolic extracts of the plant *Phyllanthus acidus* [45]. In a most recent experiment published, the aqueous extracts of *Securidaca longipedunculata* root and *Olax subscorpioidea* leaf were found to significantly inhibit AChE enzyme [46]. These plants were used as the traditional medicines or folk medicines against AD symptoms but did not have scientific evidences. Among the two, the former had higher phenol content and was better inhibitor of AChE. Their extracts also exhibited strong antioxidant activities in vitro [46].

In terms of safety, large array of compounds were reported to be safe in different in vivo studies; some were even found to be safe in humans too. For example, Huperzine A, as discussed in a previous section, is extracted from *Huperzia serrata* and has been approved for use in China. Since, many compounds discussed above, such as curcumin, are present in daily diet in a large part of the world, minimal adverse effects can be presumed. But many compounds showing promising for AD are just of plant or herbal origin and need extensive in vitro and in vivo research before being termed "safe." Though we still need a large amount of experimental results to claim the safety of natural compounds to be effective against AD, we can still hope them to be comparatively safer as compared to synthesized chemical drugs.

8 Conclusion

Alzheimer's disease (AD) is probably the biggest cause of dementia in society and a gradually fatal neurodegenerative disease. It is characterized by various pathophysiological traits such as extracellular plaque formation, intracellular NFTs, cognitive decline, and neural death. These features alter the normal physiology of the brain and are detrimental to the cognitive behavior and day-to-day activities. Aging has the widest contribution to AD, but strong genetic evidences are also present. Apart from that, the genetic predispositions are very often hereditary. Where AChE and NMDA represent the classical targets, promising receptors such as GSK-3 enzyme, mAChRs and nAChRs, H3 receptors, α2-adrenergic receptors, and *Lingo-1* have also been widely tested and presented as suitable anti-AD targets. Presently, only four drugs are marketed and approved for AD, namely, galantamine, rivastigmine, donepezil, and memantine. The former three are AChE inhibitors, while the last one is a NMDA antagonist. Almost 93 drugs are in different phases of clinical trials and can be expected to be marketed in near future. Apart from these, a number of molecules and derivatives have been tested in experimental setups. There are molecules which are derivatives of drugs which were withdrawn from different stages of development, while many natural compounds have also been isolated and tested in vitro and in vivo. Plant-extracted alkaloids, terpenes, and flavonoids are some of the most exploited natural molecules targeted at different AD pathologies. Apart from their protective and therapeutic roles in AD, many natural compounds are strong antioxidants. The authors hope that this chapter would provide the readers a basic understanding of the pathophysiology of Alzheimer's disease, various targets of AD, disease-modifying agents presently in the market, and the scope of research in therapeutic approaches in AD.

References

1. Castellani RJ, Rolston RK, Smith M (2010) Alzheimer disease. Dis Mon 56(9):484–546
2. Vilatela MEA, Lopez ML, Gomez PY (2012) Genetics of Alzheimer's disease. Arch Med Res 43:622–631
3. Cummings J, Aisen PS, DuBois B, Frölich L, Jack CR, Jones RW et al (2016) Drug development in Alzheimer's disease: the path to 2025. Alzheimers Res Ther 8:39
4. Alzheimer's Association (2013) Alzheimer's disease facts and figures. Alzheimer's Association, Chicago
5. Kumar A, Nisha CM, Silakari C, Sharma I, Anusha K, Gupta N et al (2016) Current and novel therapeutic molecules and targets in Alzheimer's disease. J Formos Med Assoc 115:3–10
6. Herrmann N, Chau SA, Kircanski I, Lanctôt KL (2011) Current and emerging drug treatment options for Alzheimers disease: a systematic review. Drugs 71:2031–2065
7. Kitchen D, Decornez H, Furr J, Bajorath J (2004) Docking and scoring in virtual screening for drug discovery: methods and applications. Nat Rev Drug Discov 3:935–949
8. Ferreira LG, Dos Santos RN, Oliva G, Andricopulo AD (2015) Molecular docking and structure-based drug design strategies. Molecules 20(7):13384–13421
9. Niedowicz DM, Nelson PT, Paul Murphy M (2011) Alzheimer's disease: pathological mechanisms and recent insights. Curr Neuropharmacol 9:674–684
10. Ballard C, Gauthier S, Corbett A, Brayne C, Aarsland D, Jones E (2011) Alzheimer's disease. Lancet 377:1019–1031
11. Liu C-C, Liu C-C, Kanekiyo T, Xu H, Bu G (2013) Apolipoprotein E and Alzheimer disease: risk, mechanisms and therapy. Nat Rev Neurol 9:106–118
12. Armstrong RA (2013) What causes alzheimer's disease? Folia Neuropathol 3:169–188
13. Swerdlow RH, Burns JM, Khan SM (2010) The Alzheimer's disease mitochondrial cascade hypothesis. J Alzheimers Dis 20:S265–S279
14. Moreira PI, Carvalho C, Zhu X, Smith MA, Perry G (2010) Mitochondrial dysfunction is a trigger of Alzheimer's disease pathophysiology. Biochim Biophys Acta Mol basis Dis 1802:2–10
15. Onyango IG, Dennis J, Khan SM (2016) Mitochondrial dysfunction in Alzheimer's disease and the rationale for bioenergetics based therapies. Aging Dis 7:201–214
16. Fisher A (2012) Cholinergic modulation of amyloid precursor protein processing with emphasis on M1 muscarinic receptor: perspectives and challenges in treatment of Alzheimer's disease. J Neurochem 120:22–33
17. Buckingham S, Jones A (2009) Nicotinic acetylcholine receptor signalling: roles in Alzheimer's disease and amyloid neuroprotection. Pharmacol Rev 61(1):39–61
18. Fisher A, Michaelson DM, Brandeis R, Haring R (2000) M1 muscarinic agonists as potential disease- modifying agents in Alzheimer's disease. Ann N Y Acad Sci 920:315–320
19. Bodick NC, Offen WW, Levey AI, Cutler NR, Gauthier SG, Satlin A et al (1997) Effects of xanomeline, a selective muscarinic receptor agonist, on cognitive function and behavioral symptoms in Alzheimer disease. Arch Neurol 54:465–473
20. Bymaster FP, Whitesitt CA, Shannon HE, Delapp N, Ward JS, Calligaro DO et al (1997) Xanomeline – a selective muscarinic agonist for the treatment of Alzheimer's disease. Drug Dev Res 40:158–170
21. Fisher A, Brandeis R (2002) AF150 (S) and AF267B M1 muscarinic agonists as innovative therapies for Alzheimer's disease. J Mol Neurosci 19:145–153
22. Foster DJ, Choi DL, Jeffrey Conn P, Rook JM (2014) Activation of M1 and M4 muscarinic

receptors as potential treatments for Alzheimer's disease and schizophrenia. Neuropsychiatr Dis Treat 10:183–191

23. Fisher A (2008) Cholinergic treatments with emphasis on M1 muscarinic agonists as potential disease-modifying agents for Alzheimer's disease. Neurotherapeutics 5:433–442
24. Johnson GVW, Stoothoff WH (2004) Tau phosphorylation in neuronal cell function and dysfunction. J Cell Sci 117(Pt 24):5721–5729
25. Hanger DP, Anderton BH, Noble W (2009) Tau phosphorylation : the therapeutic challenge for neurodegenerative disease. Trends Mol Med 15(3):112–119
26. Ghosh A, Osswald H (2014) BACE1 (β-Secretase) inhibitors for the treatment of Alzheimer's disease. Chem Soc Rev 43 (19):6765–6813
27. Vassar R (2014) BACE1 inhibitor drugs in clinical trials for Alzheimer's disease. Alzheimers Res Ther 6:89
28. Filser S, Ovsepian SV, Masana M, Blazquez-Llorca L, Elvang AB, Volbracht C et al (2015) Pharmacological inhibition of BACE1 impairs synaptic plasticity and cognitive functions. Biol Psychiatry 77:729–739
29. Nguyen L, Lucke-Wold BP, Mookerjee SA, Cavendish JZ, Robson MJ, Scandinaro AL et al (2015) Role of sigma-1 receptors in neurodegenerative diseases. J Pharmacol Sci 127:17–29
30. Hashimoto K (2015) Activation of sigma-1 receptor chaperone in the treatment of neuropsychiatric diseases and its clinical implication. J Pharmacol Sci 127:6–9
31. Kulshreshtha A, Piplani P (2016) Current pharmacotherapy and putative disease-modifying therapy for Alzheimer's disease. Neurol Sci 37:1403–1435
32. Mendiola-Precoma J, Berumen LC, Padilla K, Garcia-Alcocer G (2016) Therapies for prevention and treatment of Alzheimer's disease. Biomed Res Int 2016:2589276
33. Cole DC, Manas ES, Stock JR, Condon JS, Jennings LD, Aulabaugh A et al (2006) Acylguanidines as small-molecule secretase inhibitors. J Med Chem 49:6158–6161
34. Esbenshade TA, Browman KE, Bitner RS, Strakhova M, Cowart MD, Brioni JD (2008) The histamine H3 receptor: an attractive target for the treatment of cognitive disorders. Br J Pharmacol 154:1166–1181
35. Wu Z, Guo Z, Gearing M, Chen G (2014) Tonic inhibition in dentate gyrus impairs long-term potentiation and memory in an Alzheimer's [corrected] disease model. Nat Commun 5:4159
36. Fernandez-Enright F, Andrews JL (2016) Lingo-1: a novel target in therapy for Alzheimer's disease? Neural Regen Res 11:88–89
37. Ma T, Klann E (2014) PERK: a novel therapeutic target for neurodegenerative diseases? Alzheimers Res Ther 6:30
38. Gannon M, Peng Y, Jiao K, Qin W (2016) The α2 adrenergic receptor as a novel target for Alzheimer's disease. FASEB J 30(1):S707.2
39. Albani D, Polito L, Forloni G (2010) Sirtuins as novel targets for Alzheimer's disease and other neurodegenerative disorders: experimental and genetic evidence. J Alzheimer Dis 19:11–26
40. Cummings J, Morstorf T, Lee G (2016) Alzheimer's drug-development pipeline: 2016 Alzheimer's Dement. Transl Res Clin Interv 2:222–232
41. Kim J, Lee HJ, Lee KW (2010) Naturally occurring phytochemicals for the prevention of Alzheimer's disease. J Neurochem 112:1415–1430
42. Williams P, Sorribas A, Howes MJR (2011) Natural products as a source of Alzheimer's drug leads. Nat Prod Rep 28:48–77
43. Kumar A, Nisha CM, Kumar A, Bai BM, Vimal A (2016) Docking and ADMET prediction of few GSK-3 inhibitors divulges 6-bromoindirubin-3-oxime as a potential inhibitor. J Mol Graph Model 65:100–107
44. Airoldi C, Sironi E, Dias C, Marcelo F, Martins A, Rauter AP et al (2013) Natural compounds against Alzheimer's disease: molecular recognition of Ab1–42 peptide by Salvia sclareoides extract and its major component, rosmarinic acid, as investigated by NMR. Chem Asian J 8:596–602
45. Moniruzzaman M, Asaduzzaman M, Hossain MS, Sarker J, Rahman SMA, Rashid M et al (2015) In vitro antioxidant and cholinesterase inhibitory activities of methanolic fruit extract of Phyllanthus acidus. BMC Complement Altern Med 15:403
46. Saliu JA, Olabiyi AA (2016) Aqueous extract of Securidaca longipedunculata Oliv. and Olax subscorpioidea inhibits key enzymes (acetylcholinesterase and butyrylcholinesterase) linked with Alzheimer's disease in vitro. Pharm Biol 55:252–257

Part II

Computational Modeling Methods for Anti-Alzheimer Agents

Chapter 2

Recent Advances in Computational Approaches for Designing Potential Anti-Alzheimer's Agents

Sergi Gómez-Ganau, Jesús Vicente deJulián-Ortiz, and Rafael Gozalbes

Abstract

Alzheimer's disease (AD) is the leading cause of dementia in old people worldwide and one of the leading causes of death in developed countries. The current poor understanding of AD mechanisms makes it difficult to develop novel drugs that could be used to treat effectively this disease. Different enzymes are known to be crucial in the biochemical pathways involved in the development of AD and therefore can be considered as potential targets to design efficient drugs. Among them, three enzymes stand out: acetylcholinesterase (AChE), β-amyloid cleaving enzyme 1 (BACE1), and glycogen synthase kinase-3 (GSK-3).

Chemoinformatics and molecular modeling methodologies have been used for decades for the selection and optimization of new compounds with therapeutic properties in different areas. The initial works in computational drug discovery (CDD) were only considered as promising theoretical studies. However, currently, these in silico methodologies are part of the normal drug discovery process, and they are usually implemented in the search of novel drugs or for the optimization of therapeutic activity (or pharmacokinetic properties) of chemical series. In particular, the search for drugs in the field of neurodegenerative disorders is very active, though unfortunately no cure exists nowadays for AD.

In this review, we will present the advances on the use of computational techniques in the search for novel small molecules as effective therapeutic agents against AD. We will describe recent computational studies to find or design new anti-AD compounds applying different computational approaches, mainly quantitative structure-activity relationship (QSAR), molecular docking, pharmacophore development, and molecular dynamics simulations. Taking into account that there is a huge amount of literature on the field, we have chosen to focus on studies published in the last few years related to the main AD targets (AChE, BACE1, and GSK-3). We will conclude by providing a perspective on the future of this field.

Key words Alzheimer's disease (AD), QSAR, Docking, Molecular dynamics, Pharmacophores, New drugs

1 Introduction

Alzheimer's disease (AD), a progressive, irreversible neurological disorder among elderly people, is one of the leading causes of death in developed countries [1, 2]. AD is characterized by many symptoms such as steady cognitive impairment, memory loss, and a decline in the ability to use language. Different pathological

Kunal Roy (ed.), *Computational Modeling of Drugs Against Alzheimer's Disease*, Neuromethods, vol. 132, DOI 10.1007/978-1-4939-7404-7_2, © Springer Science+Business Media LLC 2018

alterations are the cause of AD, such as the amyloid-β cascade and the hyper-phosphorylated tau protein, the deficiency of central cholinergic neurotransmitter, inflammation, or mitochondrial dysfunction [3].The involvement of various enzymes and biochemical processes in the pathological conditions of AD reflects the complexity of the disease and the difficulties in finding adequate treatments. Several drugs have been commercialized to treat AD (Table 1), but unfortunately none have been able to stop the neurodegenerative process; they are simply useful for partially restoring memory and cognitive function in a moderate way [4]. Therefore, there is no current cure for AD patients, and effective therapies are needed.

The major hypothesis regarding the pathogenesis of AD is the pathogenic process of aggregation of amyloid-β (Aβ) peptide in the brain [5]. A transmembrane protein called amyloid precursor protein (APP) located in chromosome 21 is cleaved by the enzyme β-secretase (β-amyloid cleaving enzyme 1, BACE1) producing a 99 amino acid fragment (C99) that is further cleaved by the γ-secretase. This cleavage leads to Aβ peptides of 39–43 amino acids and the subsequent formation of insoluble Aβ fibrillary aggregates, which are the major constituents of senile plaques [5]. The enzyme BACE1 plays a critical role in A production as it cleaves APP at the extracellular space and initiates the formation of Aβ peptide(s). Therefore, the inhibition of this enzyme is widely accepted to regulate the production of Aβ peptide [6].

Alterations of tau protein have also been hypothesized as a potential cause of AD. Tau is a highly soluble microtubule-associated protein (MAP), which through its isoforms and phosphorylation interacts with tubulin to stabilize microtubule assembly [7]. Mutations that alter function and isoform expression of tau lead to hyper-phosphorylation and to the formation of neurofibrillary tangles. These tangles are insoluble structures that can damage cytoplasmic functions, interfere with axonal transport, and lead to cell death. This hyper-phosphorylation is a result of deregulation in several protein kinases, especially the glycogen synthase kinase-3 (GSK-3), which is the subject of much research and can be considered as a promising target for the treatment of AD [8].

Another important hypothesis of AD development is the reduction of the synthesis of acetylcholine [9], a neurotransmitter used by all cholinergic neurons, which has a very important role in the peripheral and central nervous systems and in cognitive processes. This neurotransmitter is released from nerve fibers during cholinergic transmission, and a deficit in this transmission can potentially influence all aspects of cognition and behavior, including cortical and hippocampal processing information. The cholinesterase enzyme AChE decreases the concentration of acetylcholine by hydrolyzing this molecule. Because of this, targeting AChE has

Table 1
US FDA-approved drugs to treat the symptoms of Alzheimer's disease

Generic name	Brand name	Year of approval	Chemical structure	Approval's stages
Tacrine	Cognex	1993	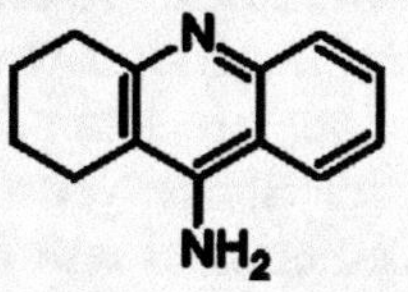 N, NH_2	Mild to moderate
Donepezil[a]	Aricept	1996	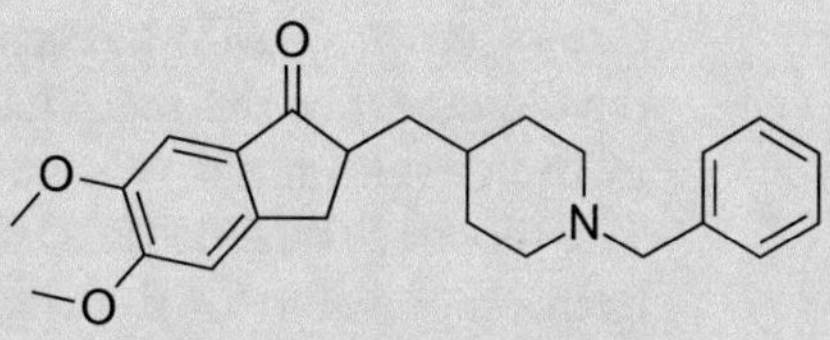 O, O, O, N	All stages
Rivastigmine	Exelon	2000	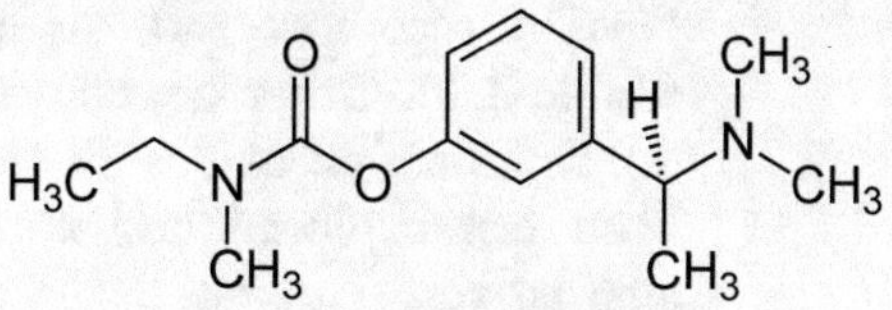 O, H_3C, N, CH_3, O, H, CH_3, N, CH_3, CH_3	Mild to moderate
Galantamine	Razadyne	2001	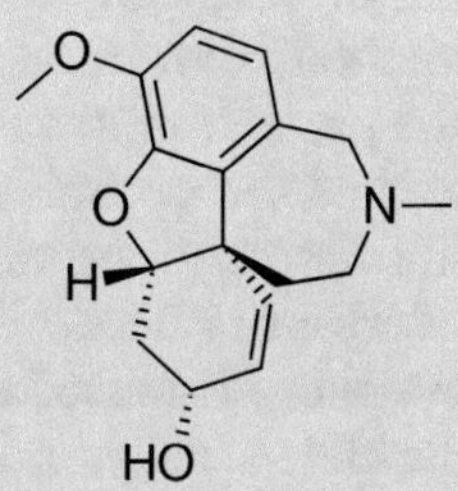 O, O, N—, H, HO	Mild to moderate
Memantine[a]	Namenda	2003	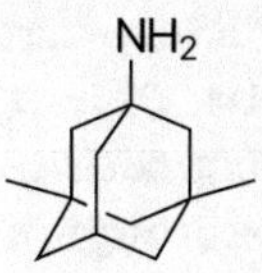 NH_2	Moderate to severe

[a]A combination of donepezil and memantine has been approved by FDA in 2014 under the brand name of Namzaric

been described as an important way to increase the concentration of acetylcholine in the synapses [10].

In addition to the mentioned enzymes, there are other targets associated with AD such as butyrylcholinesterase, cathepsin B, or the N-methyl-D-aspartate receptor. Butyrylcholinesterase (BChE) plays a secondary role in healthy brains, whereas it has been seen to increase its activity hydrolyzing acetylcholine in AD brains. A comparison between human AChE and BChE has revealed that aromatic residues Phe295 and Phe297 in the AChE are switched into the aliphatic residues Leu286 and Val288 in the BChE. These residues are in close proximity to the catalytic triad in both enzymes and contribute to their specificity and selectivity. Cathepsin B belongs to the peptidase family, and it has an important role in intracellular protein degradation through limited proteolysis. Cathepsin B reduces Aβ levels by inducing C-terminal truncation of A(1–42) which suggests that endogenous cathepsin B activity could protect against AD-related deficits. The N-methyl-D-aspartate receptor (NMDA receptor) has also an important role in the development of the mammalian brain, in the cognitive process and in the development of some neurodegenerative diseases like AD. The blockage of the NR2B subunits of NMDA receptors has shown to be responsible for the cognitive-stimulating and neuroprotective action of drugs, as well as to mitigate the neuropathic pain, and to prevent convulsions. Memantine is the first drug described to block NMDA receptors and has an effect in moderate-to-severe AD (Table 2.1). Other targets such as monoamine oxidase A (MAO-A), monoamine oxidase B (MAO-B), or caspase-3 [11, 12] have been less explored and currently are not considered as good alternatives.

Computational methods have demonstrated their great potential in designing hits/leads for complex diseases [13–15]. These methods allow the simulation of mechanisms of action of drugs and the prediction of values of therapeutic efficacy in humans [16], as well as the optimization of pharmacokinetic properties [17, 18], the discovery and validation of new targets [19], and the reduction of adverse effects of drugs [20, 21]. There are various computational or "in silico" approaches available; the use of one or another depends usually on the baseline information available.

Molecular docking is the most commonly used computational method involving the three-dimensional structure of the target ("structure-based drug design" methods, SBDD), since (i) the number of new therapeutic targets has increased greatly thanks to the sequencing of the human genome and (ii) new developments in high-throughput crystallography and nuclear magnetic resonance allow a very high level of detail when determining the atomic structures of proteins and protein–ligand complexes [22]. Docking simulations play an important role in predicting the preferred orientation of a stable ligand–protein complex, exploring the detailed interaction features, and investigating and interpreting the action mechanisms of biologically active compounds. A large number of structures have been studied with AD enzymes, and herein we will describe the more relevant recent docking studies performed in the last few years.

Chemoinformatics is an ensemble of methodologies for helping chemists make sense of the big (even massive) amounts of scientific data available [23]. Usually, the work of chemoinformaticians consists of the construction, organization, and analysis of information systems based on known scientific data, in order to extract new information, attempting to predict the properties of chemical substances from a sample of data and assist in the development of novel compounds [24]. Some chemoinformatic relevant tools include the following: chemical structure representation, data storage and registration on electronic databases, online chemical literature searches, diversity analysis and library design, and virtual screening for rational drug design. Among these, one of the most outstanding areas is the development of the so-called "Quantitative Structure-Activity Relationships (QSARs) for the prediction of biological/pharmacological activity based on chemical structure [25]. This technique involves the construction of a mathematical model linking the chemical structure of a series of molecules with a specific biological activity. The development of QSARs requires the previous characterization of the molecules by a set of numerical descriptors [26–28] and the application of statistical tools providing regression or classification models. Once a QSAR has been developed and validated [29], it can be used to predict the activity of new molecules whose chemical structure is known, and their numerical descriptors can be easily calculated [30]. QSAR studies for the discovery of new anti-AD drugs have been reported since 1996, and Ambure and Roy reviewed most of the more significant articles on this matter [31]. In this paper, we will discuss mainly the work published since then and address new advances and challenges in the field.

Pharmacophores are defined by IUPAC as "the ensemble of steric and electronic features that is necessary to ensure the optimal supramolecular interactions with a specific biological target structure and to trigger (or to block) its biological response" [32]. Therefore, pharmacophores describe the spatial arrangement of the key chemical features necessary for small compounds to interact with a target and induce a biochemical effect. Pharmacophores can be elucidated in two ways: (i) from the complementarities of a ligand interacting with a binding site (structure-based pharmacophores) or (ii), most frequently, from the alignment of a set of known active molecules (ligand-based pharmacophores). Pharmacophore modeling is a common method for describing the interactions of small molecules with macromolecular targets of therapeutic interest [33], and they can be used to screen virtual databases for searching for compounds that have the same or similar distribution of chemical features [34].

Molecular dynamics (MD) is another in silico method commonly used in drug design and drug discovery projects [35]. MD consists of the computational simulation of the physical movements of atoms and molecules, which are allowed to interact for a fixed period of time, giving a view of the dynamical evolution of the system. The calculations required to describe the motions and reactions of large molecular systems are very complex and computationally intensive; therefore, MD tries to overcome this limitation by using simple approximations based on Newtonian physics to simulate atomic motions. Some articles deal with molecular dynamic (MD) studies of the receptors alone in order to simulate their conformational behavior in diverse conditions and gain an insight into the mechanisms involved. Both atomistic and coarse-grained force fields have been applied to MD of AD-related molecules. Among the atomistic force fields employed, while computationally demanding but able to provide a greater level of detail, the most used in this field are AMBER [36, 37], CHARMM [38, 39], GROMOS [40, 41], and OPLS-AA [42, 43]. Coarse-grained force fields describe biomolecules as polyatomic fragments, such as functional groups, thus reducing the CPU calculation time. Coarse-grained force fields widely used include OPEP [44], MARTINI [45], and UNRES [46].

In this review, we will present several significant papers published in the last few years in the CDD field applied to AD. We will focus on studies describing the use of docking, QSAR, pharmacophores, and MD methods to develop novel small molecules which could inhibit the three main targets (BACE1, GSK-3, and AChE), since they are the most cited in the scientific literature and current efforts to find anti-AD drugs are mainly devoted to them. The chemical structures of a selection of compounds presented in this review are shown in Table 2.

2 BACE 1

The most important constituents of the senile plaques are Aβ peptides (Aβ1–40, Aβ1–42 and amyloid β-protein) which come from the cleavage of the APP located in chromosome 21 by the enzymes β-secretase and γ-secretase [56]. Importantly, β-secretase has been identified as the rate-limiting enzyme for Aβ production, and therefore, the inhibition of the β-site APP cleaving enzyme 1 (BACE1) represents a potential mechanism for treating AD [57].

Until now, more than 350 crystal structures of BACE1 have been deposited in the Protein Data Bank (PDB, http://www.rcsb.org/pdb/home/home.do). Due to its aspartyl protease nature, the active site of the enzyme has a catalytic dyad (Asp 32 and Asp 228) at its center [58] (Fig. 1). Moreover, it is covered by a flexible antiparallel β-hairpin (called a flap), which is believed to control substrate access to the active site and orientate it into the correct geometry for the catalytic process to begin. The flap of BACE1 is composed of 11 residues (Val67-Glu77) at the N-terminal domain.

Table 2
Selection of structures of chemicals reviewed in this article

Chemicals	Reference	Target	Chemical structure
Fucosterol	[47]	BACE1	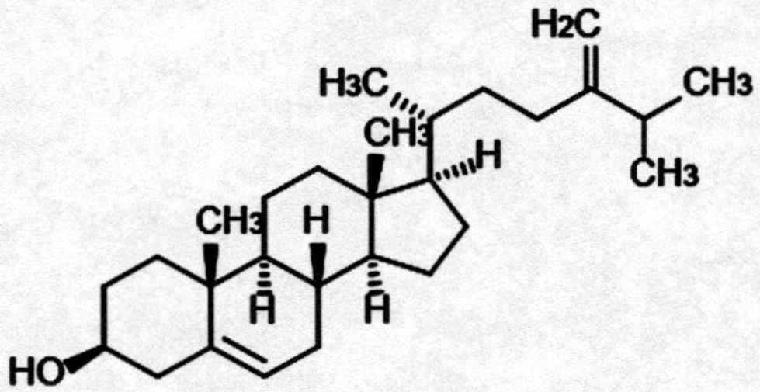
Fucoxanthin	[47]	BACE1	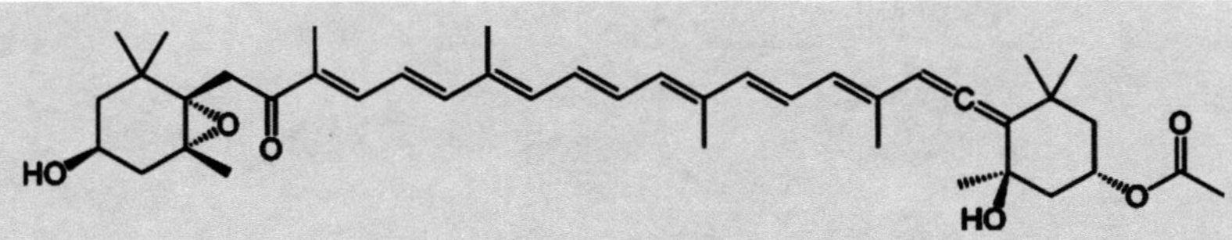
Cyclic sulfone hydroxyethylamines	[48]	BACE1	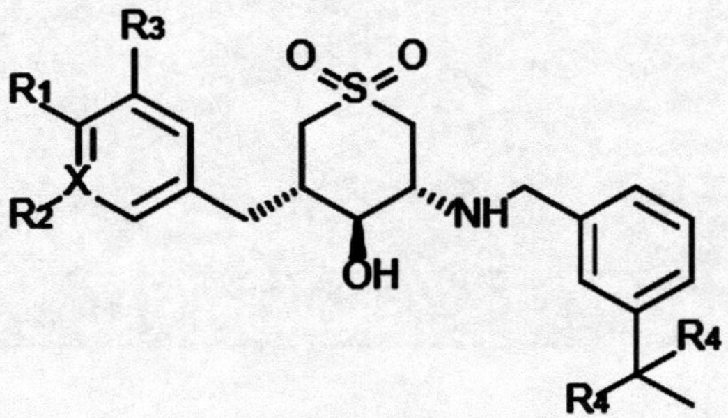

(continued)

Table 2
(continued)

Chemicals	Reference	Target	Chemical structure
Hydroxyethylamine (molecule 26)	[49]	BACE1	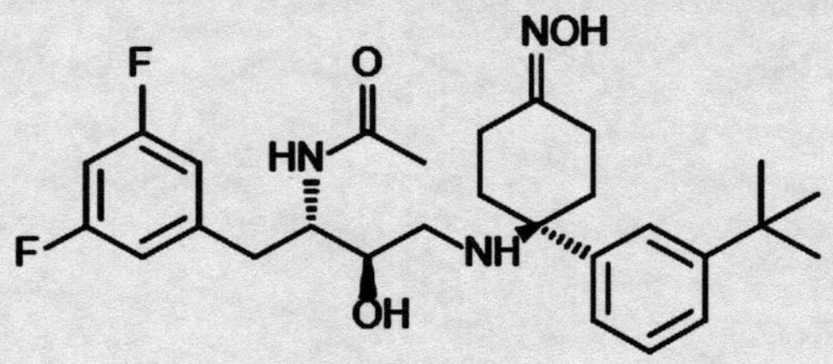
Oxadiazoles (derivatives series A)	[50]	GSK-3	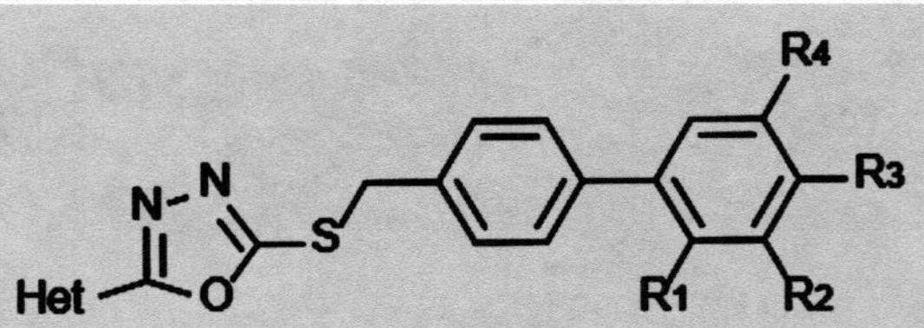
Oxadiazoles (derivatives series B)	[50]	GSK-3	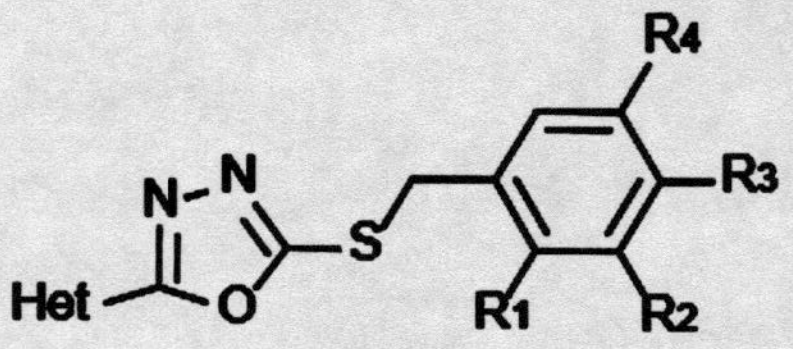

ChEMBL474807	[51]	GSK-3	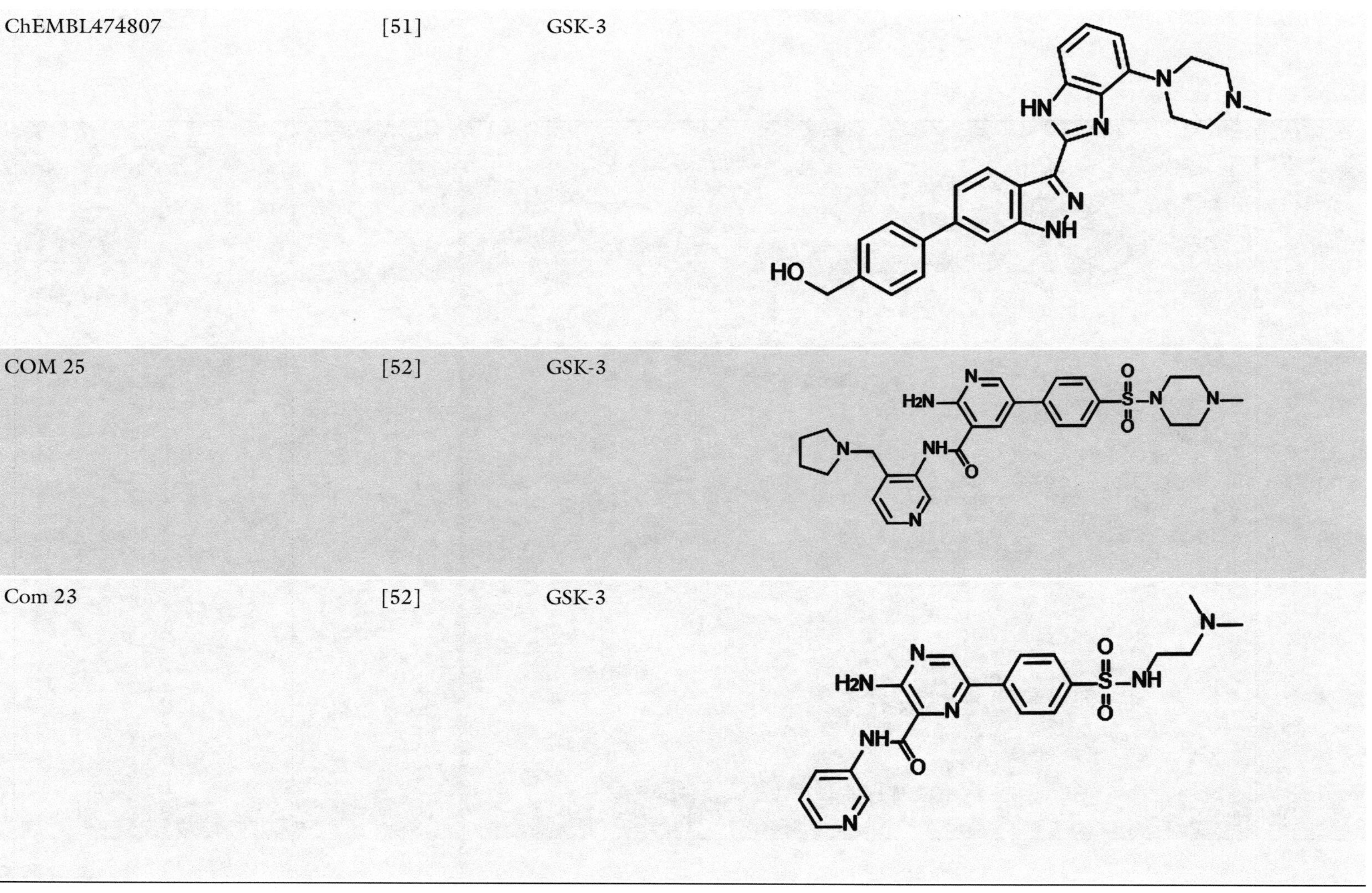
COM 25	[52]	GSK-3	
Com 23	[52]	GSK-3	

(continued)

Table 2
(continued)

Chemicals	Reference	Target	Chemical structure
1H-indazole-3-carboxamides	[53]	GSK-3	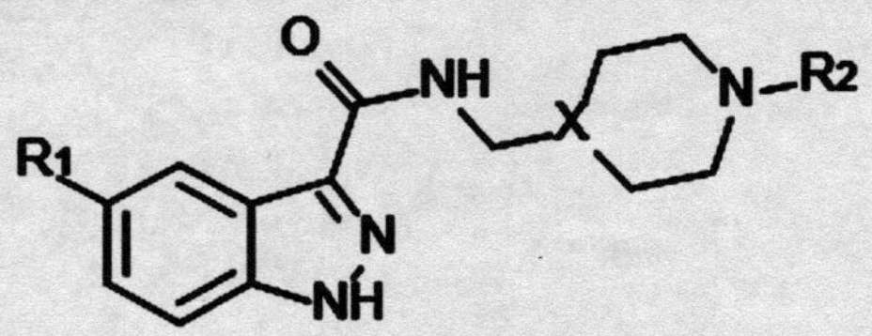
Benzothiazole piperazines	[54]	AChE	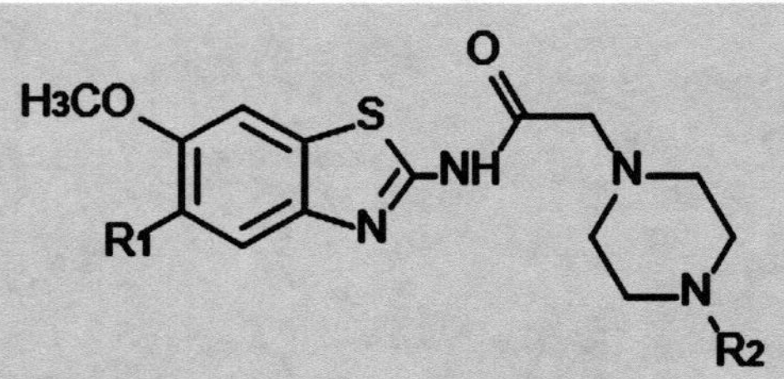
Isoalloxazines	[104]	AChE	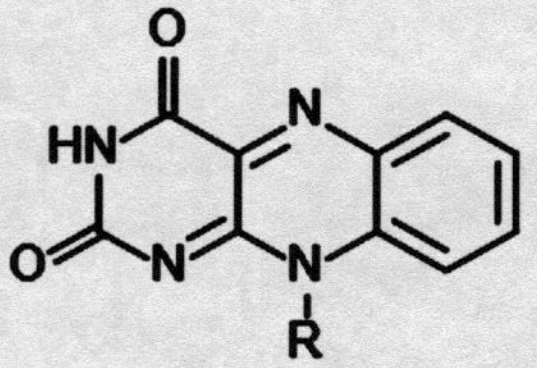

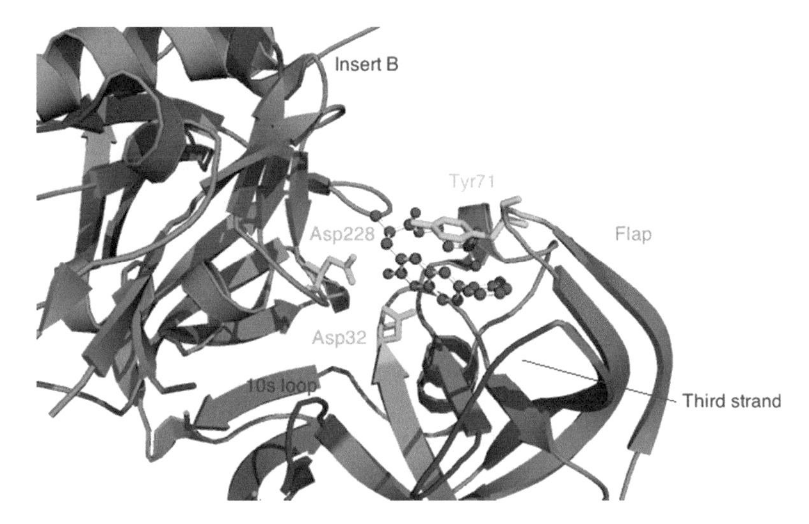

Fig. 1 Structure of human BACE1 in complex with an aminoquinoline (PDB ID: 5I3V) (Jordan et al. [59])

The flap of the inhibitor-bound form is tightly packed in a closed conformation; however, the substrate-free structure of BACE1 indicated that the flap was in an open conformation. Therefore, it can adopt different conformations in the presence and absence of inhibitors [60–62]. In addition, there are some key functional regions that lead to the binding of different substrates. These regions are 10s loops(Lys9-Tyr14), third strand (Lys107-Gly117), and insert A (Gly158- Leu167), which are present at the N-terminus, and insert B (Lys218-Asn221), insert C (Ala251-Pro258), insert D (Trp270-Thr274), insert E (Glu290-Ser295), and insert F (Asp311-Asp317) regions, which are located at the C-terminus [63]. Furthermore, two conserved water molecules (WAT1 and WAT2) are placed at the active site. It is thought that WAT1, which is closer to the catalytic dyad, is used in the hydrolytic cleavage of the peptide bond, whereas WAT2 participates in a continuous H-bonding network and stabilizes the flap in the closed conformation through structural organization [62, 63]. In summary, BACE1 cleaves certain peptide bonds of the substrates utilizing the catalytic Asp dyad. This Asp dyad interacts with the conserved water molecule (WAT1) present in the active site and utilizes it for peptide hydrolysis [64].

Since BACE1 has emerged as a promising target to reduce the Aβ levels in the brain, a large number of candidate compounds have been designed. These potential inhibitors are principally classified as peptidics and non-peptidics. Peptides are mainly designed based on substrate specificity and bear the transition-state isostere as a functional group that copies peptide bond hydrolysis (e.g., hydroxyethylene, statine, hydroxyethylamine, reduced amide) and usually binds to both Asp32 and Asp228. The first-generation studies on the protonation state of the Asp dyad focused on peptidomimetic hydroxyethylene-based inhibitors OM99-2 and OM00. They were highly peptidic with a high quantity of hydrogen bonds and large molecular size, thus making them pharmacologically unfavorable. Later generations are smaller and more potent, and some of them have advanced to clinical development [64]. Otherwise, promising non-peptidomimetic inhibitors have also shown the capability to inhibit BACE1, but unfortunately, to date, no BACE1 non-peptidomimetic inhibitor has passed to phase II/III of clinical trials.

Docking simulation studies have been performed by Ran Joo Choi et al. in order to predict binding energies of six ginsenosides isolated from *Panax ginseng* with BACE1 and to identify the interacting residues [65]. Ginsenosides have attracted attention because they show ability to pass the blood–brain barrier (BBB) and non-toxicity. Each of the ginsenosides investigated formed two or more hydrogen bonds with BACE1. The most active compound (binding affinity of −9.7 kcal/mol) was found to form single hydrogen bonds with the Tyr71, Asp228, and Gln326 residues of BACE1. In addition, van der Waals interactions were also present with Val332, Ile226, Ile118, Phe108, Lys107, Trp115, Trp76, Gly11, Val69, Asp32, Ile110, Asn233, Lys321, and Thr23.

Jung et al. have also applied docking to study the compounds fucosterol and fucoxanthin from some algae against BACE1 [47]. It was observed that one hydroxyl group of fucosterol interacted by hydrogen bonding with the Lys224 residue of the enzyme. Moreover, seven residues of BACE1 (viz., Ile118, Tyr71, Ile226, Thr231, Val332, Phe108, and Val69) were found to participate in hydrophobic interactions with the methyl group of fucosterol. Concerning fucoxanthin, two hydroxyl groups established two hydrogen-bonding interactions with the Gly11 and Ala127 residues of the enzyme. Furthermore, eight residues of BACE1 (Thr231, Tyr198, Thr232, Tyr71, Ile110, Ile118, Leu30, and Ile126) participated in hydrophobic interactions with the methyl group of fucoxanthin. Both compounds show a strong ability to inhibit BACE1 in experimental inhibition assays (IC_{50} values were 64.1 and 5.3 μM for fucosterol and fucoxanthin, respectively, whereas the positive control –quercetin – had an IC_{50} of 10.2 μM). The higher number of interactions of fucoxanthin with BACE1 could explain its higher inhibitory activity.

Several QSAR studies on BACE have been published since the end of the 1990s, and most have been compiled in the comprehensive review by Ambure and Roy [31]. In the last few years, several new studies in the field have been reported, such as the work of Cruz and Castilho, who have developed statistically sound descriptor-based and fragment-based QSAR models on a series of aminoimidazoles, aminohydantoins, and aminopyridines derivatives [66]. Furthermore, this work underlines polar interactions that are important for BACE-1 inhibition.

A chemometric study combining pharmacophore, 3D-QSAR, and docking was published by Hossain et al. [67]. The QSARs were developed from a database containing 980 structurally diverse potential BACE inhibitors assembled from different sources, and comparative molecular field analysis (CoMFA), comparative molecular similarity indices analysis (CoMSIA) and hologram QSAR (HQSAR) techniques were applied. The models were statistically robust and provide an explanation of the main features that play a role in the interaction with the receptor site cavity. Furthermore, a complementary docking study elucidated the more significant interactions at the catalytic site with residues Gly11, Thr72, Asp228, Gly230, Thr231, and Arg235, thus allowing de novo designed molecules to be conceived. Another study using the same 3D-QSAR methods was published by Zhang et al., which did not focus on a chemically diverse database but on a particular chemical series of BACE1 inhibitors sharing the same chemotype: cyclic sulfone hydroxyethylamines [48]. The CoMSIA model seemed to be better than the other two models for this kind of structures. Nevertheless, the three models yielded some important structural information that can provide guidelines for the design of new cyclic sulfone hydroxyethylamine inhibitors.

An ensemble of ligand-based drug design (LBDD) approaches has been reported by Subramanian et al., who compiled a data set of more than 1500 compounds together with their respective binding affinities (IC_{50} values) on BACE1 [68]. Different kinds of descriptors and chemoinformatic modeling approaches were applied on a chemically diverse data set of 205 chemicals (representing about 20% of the total database) and were tested on the remaining compounds. The machine-learning methods with 2D descriptors performed well in qualitative and quantitative models, thus providing a great way to further explore lead identification and optimization of BACE1 inhibitors.

MD has been used in the study of the amyloid-β (Aβ) peptide conformational ensemble, both by using partial sequences (peptide segments) and the whole, full-length protein [69, 70]. Due to the huge computational time necessary in MD simulations with proteins when the solvent is explicit, the Aβ peptide is usually divided into four fragments following functional criteria: N-terminal (res. 1–16), central hydrophobic core (17–21), fibril-turn region (22–29), and C-terminal (30–42). From these works, it can be deduced that the Aβ peptide shows intrinsic disorder and that several low-energy structures are possible. Depending on the calculation method employed, different transient secondary structures for the Aβ monomer in aqueous milieu have been obtained. Studies on the full-length peptide indicate that (i) the N-terminus region (1–10) shows a slight tendency to form beta-strands, probably in contact with most other regions; (ii) the central hydrophobic core (17–21) exhibits a tendency to form beta-rich structures; (iii) the fibril-turn region (22–29) is rich in turn motifs; (iv) and the C-terminal region (30–42) adopts a beta-hairpin structure.

A more recent study of the interaction between APP and BACE1 concluded that BACE1 is highly flexible in several loop regions that line the active site cavity of BACE1 along with the flap region that frames the cavity opening [61]. Concerted movements of all these regions lead to a conformational transition from open to closed form of free BACE1 and vice versa, as seen in the simulations. The BACE1-APP complex shows significant reduction of the BACE1 flexibility that affects the flap region. This region approaches to its opposite loop, while other loop regions move toward APP for effective recognition by the site cavity of BACE1. These actions cause the stabilization of the closed form of BACE1. The flap region interacts appropriately with APP to orient its cut-site toward the catalytic aspartic dyad. The identification of the interacting residues involved in APP recognition by the active site of BACE1 has been unveiled, and it is useful to design more potent inhibitors in AD.

Three recent MD studies have focused on the determination of the most feasible protonation states of the aspartic dyad (D32 and D228) in the BACE1 binding site. Depending on the competitive inhibitor that interacts with BACE1, the two catalytic aspartic acid residues and the aspartic dyad can show several protonation states. In the first article, quantum DFT, MD, and molecular docking calculations on the possible protonation states of the aspartic dyad were performed in the presence of an acyl guanidine inhibitor in the binding site that was anionic or as monoprotonated in different runs [71]. The DFT calculations in a simplified model have shown that the unprotonated dyad is about 14 kcal/mol more stable than the second best model, even when the inhibitor molecule is also unprotonated. The force field MD simulations also suggested a di-deprotonated state. However, the binding affinities obtained in docking studies were not conclusive. In the second study, MD simulations were performed for the four protonation states of BACE1 [72]. Then, the docking of 21 BACE1 inhibitors against the obtained MD conformations was studied. It was demonstrated that the BACE1-inhibitor complex remains stable during the simulation when only Asp32 has been protonated and Asp228 remains anionic. The model proposed, termed HD32D228, seemed to recognize the majority of the active compounds. In the third article, MD simulations were performed to identify the most feasible protonation state shown by the catalytic dyad in the presence of HEA transition state analogues [73]. The results indicate that the di-deprotonated state is preferred in the presence of HEA inhibitors. Nine protonation state configurations were evaluated. The HEA inhibitory mechanism was thus clarified, and a new pathway was opened for the design of new useful inhibitors. The apparent contradiction between the three results could indicate that the optimal protonation state (un-, mono-, or di-) can be different depending of the nature of the ligand.

As a continuation of the last cited article, the most probable initial configuration of the protonation states of BACE1 and HEA has been searched for, and the conformational change from the inactive to the closed form of the protein, the flap closure, has been observed [74]. Both by comparison with a crystal structure and by population density using different combinations of parameters, the HEA neutral and Asp289 protonated configuration was identified as the most probable. MD simulations with other HEA inhibitors with different structures suggested that ligand inhibitory activity was due to interactions with the Asp93, Gly95, Thr133, Gly291, and Asn294 residues.

The importance of Asp32 and Asp228 in the selective inhibition of BACE1 by four selective inhibitors has also been reported by studying their alignment, docking, and MD [75]. Electrostatic interactions with Asp32 and Asp228, hydrogen bonds, π–π stacking, and van der Waals interactions with the amino acid residues located inside the catalytic site have been revealed as important. This research also compares the catalytic sites of BACE1 and other aspartic proteases, such as BACE2 and cathepsin D (CTSD). Although the three aspartic proteases could show similar binding mode due to the residues present, the site in BACE2 is smaller, and steric effect made the cleavage of the inhibitors ineffective. In CTSD, by contrast, the catalytic site is a larger cavity, and the interactions are weaker. Although a pharmacophore model has not been proposed, selective aspartyl protease inhibitors could be designed with the aid of these results.

MD has frequently been used to analyze the stability, discard non-binders, optimize the best candidates, and provide a rationale for their potential as inhibitors, in putative BACE1-ligand complexes obtained by methods such as pharmacophore modeling or docking [76, 77]. Tandem studies with sequential application of other methods such as 3D-QSAR have been also published [78]. An extensive study of 128 hydroxyethylamine (HEA) BACE1 potential ligands using 3D-QSAR, molecular docking, and molecular dynamic (MD) approaches to explore the binding mode and structural requirements for high inhibitory potency has been recently published by Wu et al. [49]. A comparative molecular similarity indices analysis (CoMSIA) model displayed reasonable predictability. Docking and MD simulations have shown that these HEAs bind to BACE1 occupying three different pockets in a conformation significantly stabilized by H-bonds. The structural determinants of these HEA analogues were also determined.

3 GSK-3

Glycogen synthase kinase-3 (GSK-3) is a highly expressed and multifunctional serine/threonine protein kinase involved in a large number of cellular processes [79]. GSK-3 is contained in all eukaryotic organisms in very similar forms. In mammals, there are two isoforms encoded by separate genes: GSK-3α (51 kDa) and GSK-3β (47 kDa) [80]. Both proteins share 97% of sequence similarity in their kinase catalytic domains, but differ significantly from one another outside this region due to the presence of an additional 63 amino acid residues at the N-terminus in GSK-3α. Moreover, although they are structurally similar, their functions are different [80]. A neuron-specific splicing isoform (β2) having an insertion of 13 amino acids within the substrate-binding domain has also been described [81]. Though GSK-3 isoforms are regulated by phosphorylation, they are done so differently: phosphorylation of Ser9 (inhibitory) and Tyr216 (activating) in the case of GSK-3β and Ser21 (inhibitory) and Tyr279 (activating) in the case of GSK-3α.

In AD patients, GSK-3β abnormally hyper-phosphorylates the microtubule-associated tau protein leading to the detachment of tau from the microtubules in the neurons. An accumulation of hyper-phosphorylated tau protein gives rise to paired helical filaments, followed by neurofibrillary tangles, which first result in weak synapses between the neurons before neuronal death [82]. Furthermore, GSK-3β activity is believed to mediate Aβ production from its precursor and APP, and increased GSK-3β activity results in toxicity [83]. Taking into account all of the information available, GSK-3 has become one the preferred targets for novel compounds oriented to cure AD [84].

The structure of GSK-3β presents two major domains, a β-strand present at the N-terminus between amino acid residues 25 and 138 and an α-helical present at the C-terminus between amino acid residues 139 and 343. The ATP binding site is at the interface of the two domains and is bordered by a glycine-rich loop (residues 63–68) and a hinge region (residues 133–136) [85] (Fig. 2).

Before phosphorylation can be catalyzed by GSK-3, both the β-strand and α-helical domains must be aligned in a catalytically active conformation for effective binding of the substrate. Nearly all kinases use one or two phosphorylated residues on the activation loop for this purpose. Polar residues provided by the β-strand and α-helical domains bind the phosphate group of the phosphorylated residue on the activation loop, which allows the proper alignment of the two domains. The phosphorylation of the second residue (if present) opens the substrate-binding pocket and allows the substrate to bind [87]. GSK-3β phosphorylates multiple substrates but does not phosphorylate all targets in the same way and with the

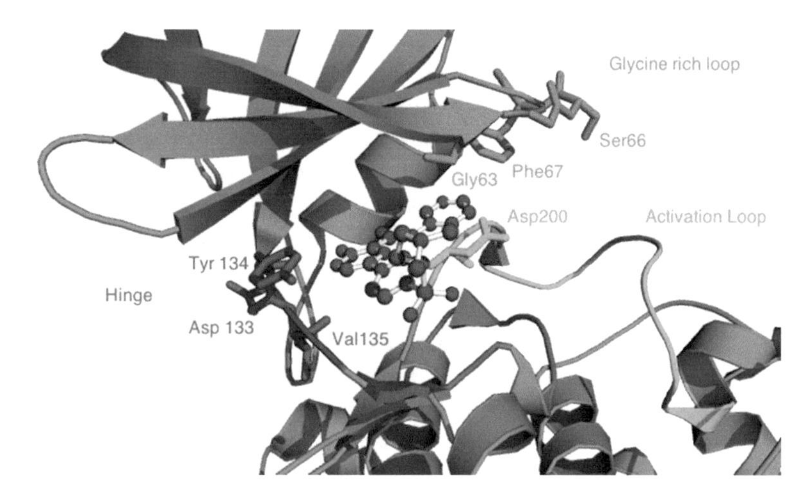

Fig. 2 Structure of human GSK-3β in complex with a pyrazine (PDB ID: 4ACC) (Berg et al. [86])

same efficiency. In some cases, GSK-3β directly phosphorylates the substrate, whereas in other circumstances, priming of the substrate by another kinase is required before phosphorylating the target. As a result, the function and mechanism of action of GSK-3β is substrate specific [88].

One of the first docking exercises with GSK-3 was reported by Polgar et al. [89]. In this study, a virtual screening protocol was developed from publicly available structural information (GSK-3 structures with PDB code 1Q4L, 1UV5, and 1Q3D), in order to distinguish GSK-3 inhibitors from a collection of virtual compounds. Experimental high-throughput screen (HTS) analysis was also performed in order to compare the virtual approach with a real screening situation and therefore to determine the effectiveness of the protocol and its limitations. A corporate collection of 16 299-membered diverse sub-libraries was screened. It was found that the hit rate in the virtual screening (12.9%) was significantly higher than the hit rate obtained by the HTS (0.55%). Nevertheless, the virtual screening protocol also picked a large number of false positives and false negatives, therefore suggesting that this method could be a good complementary approach to HTS that could be useful in the design of screening libraries, though not completely replace experimental screening.

Quesada-Romero and Caballero have reported two studies combining docking and QSAR for the discovery of new GSK-3-3β inhibitors for oxadiazole and maleimide derivatives, respectively. In the first study, 42 compounds were docked inside the GSK-3β structure (PDB code: 3F88) with similar results observed in all cases: all the oxadiazoles adopted the same mode of binding (a scorpion-shaped conformation inside the active site), and they accepted a hydrogen bond from the Val135 residue of the ATP binding site hinge region [50]. Different QSAR models were developed by using 3D vectors and SMILES-based optimal descriptors, and some key molecular features contributing to inhibitory activity were identified, such as the presence of a 2,3-dihydro-1,4-benzodioxine group. In the second study, the chemical structures and inhibitory activities of 77 maleimide derivatives were collected from the literature, and the binding modes inside the GSK-3β (PDB code: 2OW3) were studied by docking [90]. These compounds adopted propeller-like conformations, and their orientations prioritized the formation of hydrogen bonds with Asp133 and Val135 residues at the hinge region of GSK-3β. Several QSAR CoMSIA models were developed, and a robust model (Q^2 value of 0.54) was generated for non-cyclic maleimides (subset of 54 compounds). This model included steric, hydrophobic, and hydrogen-bond donor fields. Proper QSAR results were possible neither with the subseries of macrocyclic maleimide derivatives nor the whole data set of 77 chemicals.

Other studies with maleimides have been published such as that by Crisan et al. [91]. Dragon descriptors were calculated for a collection of 120 maleimide derivatives compiled from the literature and were used to generate two different kinds of "Projection to Latent Structures" (PLS) QSAR models: one model where the compounds from the data set were aligned with "Rapid Overlay of Chemical Structures" (ROCS, a robust 3D ligand-based virtual screening engine) using as a template the X-ray structure of one of the compounds and an "alignment-free" model including the lowest energy conformation of the compounds. Reliable models were obtained either for aligned or for alignment-free QSARs and suggested that the topology and conformational features of maleimides represent key factors which influence the inhibitory activity toward GSK-3.

Fu and coworkers developed a hierarchical QSAR analysis coupled with a combination of ligand-based and structure-based virtual screening to identify structurally novel GSK-3β inhibitors [92]. To this end, a collection of 728 chemicals with associated pIC_{50} values were compiled from the literature. These compounds came from seven different research groups using particular bioassays. Regression models were then built on the data collected by each group. Subsequently, in the upper level of the QSAR model, all the compounds were collected into a unique data set and classification models were generated to separate chemicals into different subclasses. Different approaches were implemented to construct the predictive learning models such as support vector machines (SVM), binary particle swarm (PS) algorithms, and random forests (RF). Virtual screening was done with a sample of more than 12,000 drug-like small molecules from ChemBridge's EXPRESS-Pick collection (http://www.chembridge.com), previously filtered by a "desirable chemical groups" query. Docking was also applied to further examine the predicted binding affinities of the hit compounds obtained from the hierarchical QSAR screening. At the end of this sequential process, five compounds with high predicted pIC_{50} values (>8.0) were tested. Two of them exhibited sub-micromolar inhibitory activity and therefore were considered as potential hits for discovering structurally novel GSK-3β inhibitors.

A similar approach (identification of new hits by a combination of LBDD and SBDD methods, followed by in vitro evaluation) was applied by Darshit and coworkers [93]. Interestingly, their work considered not only the inhibitory activity of GSK-3β but also the selectivity over CDK5, a structurally and biologically similar target. Initially, a pharmacophore model was identified by 3D-QSAR from a set of 135 molecules representing three different scaffolds. This pharmacophore was used to find hits from the roughly 200,000 molecules on the SPECS database (www.specs.net). The best primary hits (357 structures) were submitted for further docking analysis with crystal structures of GSK-3β and CDK5 (PDB codes 1Q5K and 1UNL, respectively). Finally, five hits having isoquinoline, thiazolidinedione, purine, quinazolinone, and benzothiophene scaffolds were selected as the best ones by the models. These compounds showed experimental IC_{50} values ranging from 0.82 μM to 1.12 μM. Furthermore, the experiments showed also that they were 23–57-fold selective with respect to CDK5. These compounds could then be used for further exploration and optimization, as well as in vitro and in vivo studies to evaluate their neuroprotective potential.

The question of selectivity has been raised by other authors, and the most usual way to solve this issue has been MDs. Several ATP-competitive inhibitor compounds have failed at the preclinical stage, mainly owing to the selectivity issues against cyclin-dependent kinase-2 (CDK-2), cyclin-dependent kinase-5 (CDK-5), and other phylogenetically related kinases, due to the close homology of the ATP binding sites of these enzymes. In order to provide a detailed understanding of the origin of the selectivity factors of ATP-competitive inhibitors, MD simulations with AMBER in combination with MM-PBSA free energy calculations were done using crystal structures of GSK-3β and CDK-2 in complex with 12 ATP-competitive inhibitors belonging to five different chemical classes [94]. The starting poses for the MD simulations were obtained from molecular docking on the PDB structures. Analyses of energy contributions indicate that electrostatic interaction energy dictates the selectivity of ATP-competitive inhibitors against CDK-2. Key interactions as well as residues that potentially make a major contribution to the binding free energy were identified at the ATP binding site. This analysis shows the need for the inhibitors to interact with Lys85, Thr138, and Arg141 in the binding site of GSK-3β to show selectivity. A more specific study was undertaken by Czeleń and Szefler [51], who focused on ChEMBL474807, an indirubin derivative, since indirubins are

ATP-competitive inhibitors. Docking simulations of ChEMBL474807 were done on the crystal structures of CDK-2 and GSK-3β (PDB codes: 1E9H and 1Q41, respectively), and were followed by a MDs procedure with AMBER. Important differences between the complexes were found, either in their structural or energetic properties. Conformations were stabilized by hydrogen bonds for both complexes, but differences related to the frequency of occurrence and strengths of particular hydrogen bonds and in binding affinities indicated a higher structural and energetic affinity of ChEMBL474807 for CKD-2 than for GSK-3β. More recently, Zhao et al. carried out a theoretical study on the selectivity of the same targets (CDK-2 and GSK-3β) by docking, MD, free energy calculations, and umbrella sampling simulations [52]. Two potent inhibitors of GSK-3β (K_i nanomolar values) with different selectivity with respect to CDK-2 (tenfold and 591-fold, respectively) were employed in this work. The four complexes (two inhibitors against the two targets) were used for the different computational approaches. It was observed that the Phe67 and Arg141 residues formed stronger binding with GSK-3β than the corresponding Tyr15 and Lys89 in CDK2. On the other hand, Leu132, Val135, Pro136, and Cys199 influenced the shape of the binding pocket of GSK-3β to accommodate a larger inhibitor than CDK2.

Ombrato et al. have performed a study on 1H-indazole-3-carboxamides in order to search new inhibitors of GSK-3 [53]. A virtual screening protocol was applied to a proprietary library of about 8,000 chemicals, and eight compounds were identified as hits (pIC_{50} values at low micro-molar level). All of the inhibitors showed some similarities: two hydrogen bond acceptors capable of interacting with the NH of Val135 and two hydrogen bond donors interacting with Val135 and Asp133. In addition, they interact with the side chains of the Ile62, Phe67, Leu188, and Cys199 residues and establish polar contacts with Thr138, Glu185, and Asp200. Specific crucial interactions were apparent with the ATP binding site residues Asp133, Val135, and Arg141; the hydrogen bonds with the hinge region, composed of Asp133 and Val135; and the main hydrophobic interactions with Ile62, Ala83, Lys85, Pro136, Glu137, Thr138, Leu188, and Arg141. The three-dimensional structure of GSK-3β complexed with one of the hits was solved by X-ray crystallography and confirmed that the interactions predicted by docking were correct.

Different GSK-3 virtual screening approaches have also been the subject of research of Zefirov and coworkers [95, 96]. These authors have applied two different strategies, namely, ligand-based and structure-based computational screenings. The ligand-based approach consisted in a one-class classification (OCC) system developed by using molecular fingerprints as descriptors of a series of 1226 inhibitors and 209 non-inhibitors of GSK-3β [95]. Only active samples and an auto-encoder neural network were used for model construction. Retrospective studies showed that the system was useful for the identification of new scaffolds and outperformed the results based on pharmacophore hypotheses and molecular docking. The structure-based approach consisted into the comparative assessment of nine different scoring functions for rigid docking implemented by FRED software (OpenEye, www.eyesopen.com) and SYBYL software (Certara, www.certara.com, former Tripos). A pre-generated conformations library of GSK-3β ATP-competitive inhibitors was used for this study. The best scoring function for this database screening was Chemgauss3 (OpenEye), since this function led to the best enrichment and elimination of true inactive compounds [96].

Another aspect to be considered in any drug discovery project is the prediction and optimization of the ADME-T properties of the candidates. In this sense, a recent study based purely on computational work reports a comparison of docking and computational ADME-T parameters of four GSK-3-targeted ligands (viz., indirubin, hymenialdisine, meridianin, and 6-bromoindirubin-3-oxime) against two control molecules (tideglusib and LY-2090314, two GSK-3 inhibitors under clinical trials and viewed as potential new medicines) to derive certain specifically tailored drug-like properties [97]. The best docking and ADME-T parameters were found to 6-bromoindirubin-3-oxime (6-BIO). This molecule showed LD_{50} values similar to those of the control ligands and a good ADME-T profile (considering parameters such as blood–brain barrier (BBB) passage, plasma protein binding (PPB) percentage, Caco-2 permeability, cytochrome P450 (CYP450) metabolism, and the absence of expected carcinogenesis/mutagenesis properties). Therefore, 6-BIO is a potential drug candidate to optimize, and considering that this molecule is derived from the basic molecule indirubin, it can be hypothesized that other variations around the indirubin chemical core could be explored.

4 AChE

Acetylcholinesterase (AChE) is expressed in cholinergic neurons and neuromuscular junctions, where its primary function is the rapid breakdown of the neurotransmitter acetylcholine that is released during cholinergic neurotransmission [9]. The structure of AChE has been extensively studied since the 1990s [98] (Fig. 3.). The active site of the enzyme is situated on a long and narrow gorge of 20 Å in deep, and it holds a catalytic triad (with amino acid residues Ser203, His447, and Glu334) catalyzing the hydrolysis of the ester bond. In addition, the active site contains a catalytic anionic site (CAS, where Trp86, Glu202, and Tyr337 are situated) which is responsible for the correct orientation and

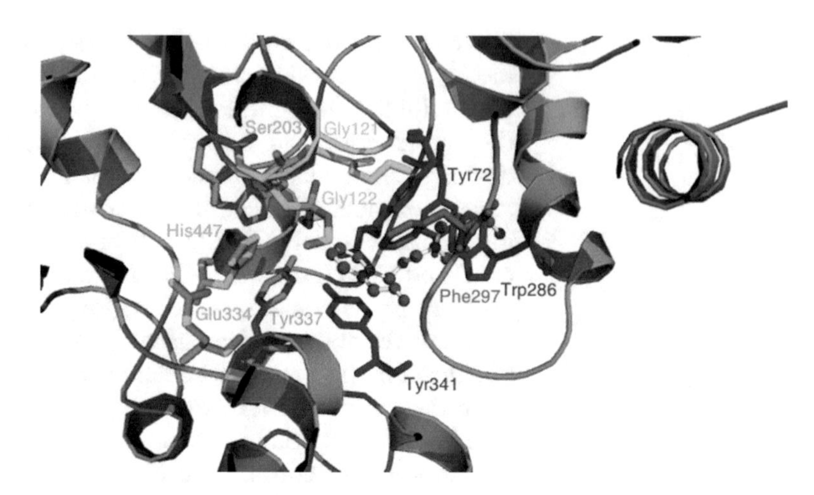

Fig. 3. Structure of human AChE in complex with HI6 (PDB ID: 5HF9) (Franklin et al. [99])

stabilization of the trimethylammonium group of acetylcholine. An oxygen anionic site (which contains Gly121, Gly122, and Ala204) has also been described as having a role in the arrangement of hydrogen bond donors that stabilize the tetrahedral enzyme substrate complex. The acyl pocket covers two phenylalanine residues, 295 and 297, which bind the acetyl group of acetylcholine. It is also known that at least a peripheral anionic site (PAS) exists, comprising residues Tyr72, Asp74, Tyr124, Trp286, and Tyr341. This PAS is placed on the enzyme surface around the cavity entrance, and it plays an important role in AD since its interaction with the Aβ peptide contributes to the formation of amyloid plaques by accelerating the aggregation [9]. In consequence, PAS has been recognized as a target for AChE, and, for example, tacrine and galantamine have been described to bind to PAS, whereas donepezil and some tacrine dimers bind simultaneously to both PAS and CAS [54]. It is thought that the ideal inhibitors of AChE should have several structural features such as a core ring system that interacts with PAS, a basic center that binds to CAS, and a linker such as –O–, CH_2, CONH, and $CONH(CH_2)n$ between the core ring system and the basic center.

Donepezil (DNP), an AChE inhibitor, is one of the preferred choices in AD therapy. Saglık et al. have recently studied new DNP analogues in order to explore their potential therapeutic activity against AChE [100]. A total of 38 analogues were synthesized, and their inhibitory activity on cholinesterase enzymes was determined. The two most relevant compounds were further investigated and revealed a good pharmacokinetic profile and non-toxicity. Finally, docking of these two compounds showed a lipophilic benzene and a 1-indanone ring system interacting with the PAS region, a piperazine moiety as a polar group, and a basic dimethylaminoalkyl side chain interacting with the CAS region. Methoxy or dimethoxy substituents at the fifth and/or sixth position of 1-indanone acted as linkers, with the oxygen being very important due to its polar interactions. The carbonyl substituents interacted with His287, Ser293, Gln291, Leu289, and Glu292 of AChE by the formation of hydrogen bonds. In addition, a hydrogen bond between 1-indanone carbonyl and the amino group of Phe295 of the acyl pocket was found in both compounds. Other important interactions observed were the cation–π interaction between the nitrogen atom of the dimethylamino group and Trp86 and a hydrogen bond interaction with Gly202 of the CAS region. Otherwise, a π–π interaction between the 1,4-disubstituted phenyl of the compounds and the Tyr341 of the PAS region was observed.

Benzothiazole–piperazines were designed by Özkay et al. in order to investigate their activity against AChE [101]. A series of 14 compounds were synthesized and their inhibitory potential determined. Complementary studies (cytotoxicity, genotoxicity, and pharmacokinetics) were also carried out. The two more active compounds showed interactions with Trp86, Tyr124, Ser203, Trp286, His287, Leu289, and Tyr341, belonging to both the CAS and PAS regions. In detail, the benzothiazole structure produced a π–π interaction with the Trp286 of the PAS. The methoxy substituents provided polar interactions with the amino groups of Trp286, His287, and Leu289 by forming hydrogen bonds. The carbonyl of the amide moiety generated a hydrogen bond with the amino group of the PAS of Tyr124. In the CAS region, a hydrogen bond was formed between the nitrogen of the dimethylamino group and the carbonyl of Ser203, as well as a cation–π interaction with Trp86. The ethyl/propyl group and terminal dimethyl group also induced van der Waals interactions with the active site.

Another interesting recent study includes the synthesis and in vitro evaluation of donepezil derivatives [102]. These derivatives showed moderate to good anti-aggregation activity, and three of them were better than the reference drug donepezil (IC_{50}s lower than 0.039 μM). The binding mode of these three molecules was elucidated by docking (using the X-ray crystal structure of human $A\beta_{1-42}$). Curiously, it was observed that the sitting pattern of the most active compound (IP-15) was very similar to that of curcumin. Moreover, their H-bond and hydrophobic interactions were also similar. A MD study proved also the structural stability of the binding of IP-15 with AChE and Aβ, as well as consistent interactions with the active site during 40 ns simulations.

Different chemical families have been studied in the last few years by computational means in the search for a series of compounds with potential interest as AChE inhibitors. Among the most studied are tacrine derivatives, since tacrine has already been shown to be efficient in delaying the symptoms of AD and in fact was the first drug to be approved by the US FDA for the treatment of AD [31]. Recently, tacrine analogues have been studied by Wong et al. [103]. Ten homogeneous groups of tacrine derivatives were extracted from different sources, with each class measured under different conditions. All the molecules were characterized by DRAGON descriptors [104], and multivariable linear regression (MLR) QSAR studies were specifically developed for each group. Satisfactory predictive models were obtained in this way, and no attempt to model the complete set of molecules was done, since different biochemical conditions were employed for measuring the inhibitory potency of the different groups. Another study on this structural family has been reported by Zhou and co-authors who combined 3D-QSAR with docking and MD to model the inhibitory activities of 60 tacrine derivatives for which the chemical structures and pIC_{50} values were known [78]. The main results of this study were two optimal QSAR models (a CoMFA and a CoMSIA model with $q^2 = 0.55$ and 0.58, respectively), as well as the identification of a few key residues at the binding site of AChE (mainly Tyr70, Trp84, Tyr121, Trp279, and Phe330). Furthermore, the models were used to predict the AChE inhibitory activities of five molecules synthesized by the authors. The results obtained were reasonably in agreement with expectations (the residual values obtained by calculating the difference between the predicted and actual pIC_{50} were below one logarithmic unit).

Huprines are potential AChE inhibitors combining the 4-aminoquinoline moiety of tacrine with the carbobicyclic substructure of (-)-huperzine A. In a recent study, Zhang et al. designed novel compounds of this family using 3D-QSAR methods [105]. Three models were developed from 35 huprines taken from the scientific literature and validated using an external set of six other huprines with high predictive r^2 values obtained (0.98, 0.97, and 0.78 for the CoMFA, CoMSIA, and HQSAR models, respectively). Based on this analysis, 14 new molecules of this family were designed as potent AChE inhibitors.

Isoalloxazine compounds have exhibited experimental inhibitory activities against both AChE and BuChE in micromolar ranges and, therefore, can be considered as promising anti-AD agents. Gurung et al. have built several QSAR models by different statistical approaches (e.g., multiple linear regression, MLR; partial least squares, PLS; and principal component regression, PCR) [55]. A 3D-pharmacophore model was also generated which revealed eight important pharmacophoric features: four hydrogen bond acceptors (the two O atoms and N atom of pyrimidine and the N atom of pyrazine), one hydrogen bond donor (one N atom of pyrimidine ring), one aromatic ring (the pyrazole ring), and two hydrophobic features (the benzene ring fused with pyrazine and side chain benzene ring attached to the N atom of pyrazine). Docking studies showed differences in the binding modes of the molecules and specially that the most active ones favorably bind to a peripheral binding subsite of AChE. Finally, the binding modes and stability of docked complexes were validated by MD simulations.

In a recent study, a series of functionalized piperidine scaffolds were synthesized, and their AChE inhibitory activity was evaluated in vitro and in silico [106]. Functionalized piperidine scaffolds are the core of numerous biologically active compounds, both natural and synthetic, including pharmaceuticals that have reached clinical trials. Brahmachari et al. synthesized 27 of these compounds following a diastereoselective multicomponent one-pot protocol under eco-friendly conditions. All of them were evaluated in vitro and IC_{50} values ranged from 0.09 to 5.39 mM. Halogenation, nitration, or 3,4-methylenedixoxy-substitution at the phenyl ring attached to the 2- and 6-positions of 1,2,5,6-tetrahydropyridine nucleus in several compounds greatly enhanced the AChE inhibition. A previously developed pharmacophore model against the AChE enzyme [107] was used for mapping the set of piperidines. This model consists of four distinct features: two hydrogen bond acceptors, one hydrophobic, and one aromatic feature. Most of the 27 compounds have three out of the four pharmacophores, and almost all the mapped compounds lacked one HBA feature. A docking analysis with the complete series in human AChE (PDB code: 4M0E) demonstrated that most of the compounds fitted well into the active site and showed good docking scores. Key interactions shown by compounds with the best scores were hydrogen bonding with Tyr72, Thr75, Phe295, and Arg296 and π–π/π–cation interactions with Trp286 and Tyr341, where Tyr72, Thr75, Trp286, and Tyr341 are active site residues present in peripheral anionic sites (PAS) and Phe295 and Arg296 are active site residues present in the acyl pocket. It should be emphasized that the PAS site is involved in accelerating the formation of amyloid-β deposits, and therefore these compounds might show dual action against AD. Finally, a QSAR study was also carried out with a set of 404 compounds [107] and the compounds synthesized for this study, therefore representing a total set of 430 compounds. The merger of the data set was reasonable since the same experimental method was applied to all the compounds to determine the activity values against AChE. A robust QSAR model was obtained by partial least squares (PLS) with five components. All the validation metrics showed acceptable values, confirming the reliability of the model.

Other computational studies are not oriented to the optimization of chemical series, but to the discovery of completely new AChE inhibitors as potential AD drugs. Martis et al. reported the development of a QSAR model by comparative residue interaction analysis (CoRIA), a QSAR formalism that yields detailed knowledge that can help to precisely alter a specific interaction, thus enhancing binding without drastically modifying the molecules [108]. A data set of AChE inhibitors was collected from literature reports and uniformly tested by the same in vitro assay method. The inhibitory activity ranged from $pIC_{50} = 4.1$ to 9.2 (where IC_{50} was expressed in moles). The CoRIA model was found to be statistically superior to other 3D-QSAR models such as CoMFA and CoMSIA and was also deemed to be efficient at identifying key residues involved in ligand recognition and binding. Furthermore, the model was developed using diverse classes of AChE inhibitors, and it was hypothesized that the model could be useful to design more novel inhibitors with new scaffolds. This hypothesis was confirmed since novel hits were retrieved by virtual screening of the model on public databases (ZINC, http://zinc.docking.org/) and their biological activities were experimentally measured and found in agreement with predictions. The most active molecule identified by this work, characterized by an imidazo[1,5-a] pyrimidine nucleus, could be considered as a completely novel lead with the potential to become a clinical candidate.

In another work, Simeon et al. proposed a large-scale QSAR investigation for predicting AChE inhibition with a large collection of 2,570 nonredundant compounds [109]. All of these compounds and their IC_{50} values against AChE were obtained from the ChEMBL 20 database [110]. The QSAR models were generated by a random forest (RF) considering a set of 12 fingerprint descriptors. The predictive performance of the QSAR models was verified by tenfold cross validation, external validation, and Y-scrambling test. The best model was selected according to the OECD guidelines [111] and a thorough analysis of the substructure fingerprint count was conducted to provide informative insights on the inhibitory activity of AChE inhibitors. A subset of 30 diverse compounds was selected from the full set of active AChE inhibitors ($IC_{50} < 1\ \mu M$) using the Kennard–Stone algorithm [112] and was employed for further molecular docking studies in order to gain structural insights on the AChE inhibition mechanisms. The co-crystal structure of human AChE with donepezil (PDB ID: 4EY7) was retrieved from the Protein Data Bank, and docking was performed after extraction of the co-crystal ligand, donepezil, from the PDB file. This study revealed that binding to human AChE is modulated by hydrogen bonding, π–π stacking and hydrophobic interactions inside the binding pocket.

5 Multi-target Approaches

The inhibition of individual targets seems a limited strategy since currently approved drugs cannot stop the progression of AD and can be considered as ineffective regarding its cure. It is thought that a possible alternative strategy would be to develop new inhibitors capable of simultaneously modulating several targets [77]. Some examples of this combined approach are discussed below.

Samadi et al. studied heterocyclic substituted alkyl and cycloalkyl propargyl amines for simultaneous MAO and cholinesterase enzymes inhibition [113]. The more active compound showed that AChE could accommodate its naphthyridine ring inside the aromatic gorge. This compound spans the narrow gorge, interacting with the anionic subsite halfway down the gorge and in the mouth of the aromatic gorge (PAS). The aromatic nitrogen at the cyano pyridine ring is hydrogen bonded to the hydroxyl group of Tyr124, and the cyano group is hydrogen bonded to the Phe295. In this orientation, the protonated pyrrolidine is buried inside the protein gorge and is involved in hydrophobic contacts with the central hydrophobic region (Trp86, Tyr124, and Tyr337). The results suggest that this compound does not compete for the same active site as the substrate.

Bhakta et al. evaluated the activity of loganin, morroniside, and 7-O-galloyl-D-sedoheptulose from *Corni fructus* as inhibitors of AChE and BACE1 [114]. Loganin forms hydrogen bonds with the His440, Try334, and Try 121 residues in the active site of AChE (*Tetronarce californica* numbering), which could explain its greater inhibitory activity with respect to the other compounds which do not establish hydrogen bonds with His440. Concerning BACE1, the most potent inhibitors were loganin and morroniside, which formed significant interactions with the enzyme's active site (five and six hydrogen bonds, respectively).

Panek et al. recently described a series of phthalimide and saccharin derivatives linked by different alicyclic fragments with phenylalkyl moieties attached to be candidates for efficient therapy [115]. In the case of AChE, all derivatives were arranged along the active gorge and bound to the CAS as well as the PAS. The benzyl moiety of the most active inhibitor was responsible for π–π stacking with Trp84 (*Tetronarce californica* numbering) in the CAS. The protonated piperazine nitrogen atom forms cation–π interactions with Phe330 and a hydrogen bond network with Tyr121 via a water molecule. The other part of the piperazine ring and the short two-carbon linker interacted with the aromatic side chains of Phe290, Phe331, and Tyr334 in the middle of the active gorge. The saccharin fragment was engaged in π–π stacking with Trp279 and CH-π interactions with Tyr70 in the PAS. Hydrogen bonds were formed by the carbonyl group with Tyr121 and one oxygen atom from sulfone with main chain of Phe288 via water molecules. Such a binding mode provided the highest activity against AChE for this compound. Phthalimide established weaker hydrogen bonds compared to saccharin and thus led to diminished activity. It is also worth noting that the linearity of the molecules provided by the disubstituted piperazine group is very important for binding to AChE. The investigated compounds revealed diverse binding modes within the active site of BACE1. The basic nitrogen atoms of alicyclic amines such as hexahydropyrimidine, piperazine, and 3-

aminopiperidine interacted with the active site aspartate residues Asp32 and Asp228 (catalytic dyad). The distance between these basic centers and catalytic aspartates was different when comparing the more and less active compounds. The remaining parts of the molecules were located in the neighboring pockets (S20, S2, and S3).

Makhaeva et al. [116] have synthesized and studied new multi-target neuroprotective drugs with inhibitory activity toward AChE, BchE, and structurally close carboxylesterase (CaE), as well as with binding affinity to NMDA receptors. These new compounds are conjugates based on latreperdine (Dimebon) and methylene blue (MB), both of which have demonstrated to protect neurons in different neurodegenerative models. The conjugates combined the γ-carboline fragment of latreperdine and the phenothiazine core of MB, which were linked by 1-oxo- and 2-hydroxypropylene spacers. These γ-carboline-phenothiazine conjugates showed high inhibitory activity against BChE (some of them at the low micromolar range), and very weakly inhibition of AChE and CaE (10–15 times lower). All of the conjugates were much more selective inhibitors of BChE than latreperdine. The activity and selectivity of the conjugates was assessed by docking to the BChE active site, and the results suggested that the interaction is characterized by very good geometrical complementarity rather than specific interactions. The most important interaction was a π–π stacking between the indole ring of Trp82 and the phenothiazine fragment, and π–π weak interactions exist between the γ-carboline fragment and Phe329. In addition, the interaction of the conjugated compounds with NMDA receptors was also assessed in the two main binding sites. Compounds with the 1-oxopropylene spacer increased their affinity in ten times toward NMDA-receptor binding sites compared to latreperdine, and compounds with the highest affinity to both NMDA-receptor sites were also significant inhibitors of BChE. Several studies suggested that the activity of AChE decreases in brains of AD people, while that of BChE gradually rises. In consequence, the remarkable activity and selectivity toward BChE showed by the γ-carboline-phenothiazine conjugates could contribute to develop more specific anti-AD therapies, without the adverse drug–drug interactions induced by AChE inhibition.

A related study has recently been published on novel conjugates to obtain multi-target compounds, by combining pharmacophore moieties of specific ligands whose biological targets are known to be involved in AD pathogenesis [117]. In this case, the conjugates were based on the combination of memantine and its aminoadamantane derivative amantadine (as the first pharmacophore) and carbazole and tetrahydrocarbazole derivatives (as the second pharmacophore). The inhibitory activity of conjugated compounds toward AChE, BChE, and CaE was investigated, as well as the binding to the NMDA receptor. Again, a very weak inhibition against AChE and CaE was found for all conjugates, whereas they showed a high inhibitory activity and selectivity toward BChE. Molecular docking analyses were performed to describe the nature of interactions with AutoDock and AutoDock Vina software (http://vina.scripps.edu/). These studies showed that both memantine and amantadine moieties occupied the cavity in the BChE active site gorge under the PAS and above the catalytic triad. The docked positions obtained for the compounds revealed a highly variable position for the linker while the position of the amantadine and memantine was very stable. The strongest interaction was obtained by the salt bridge between the positively charged amino group and Asp70 side chain of the BChE PAS. In addition, an important hydrogen bond was formed by the hydroxyl group of the 2-hydroxypropylene spacer with the BChE gorge. The positions of the amantadine/memantine and carbazole/tetrahydrocarbazole groups were generally very similar; thus, these substituents did not cause significant changes in the ligand position. Finally, several conjugated compounds showed to block the ifenprodil binding site of NMDA receptors, which coupled with the ability to selectively inhibit BChE could enhance the therapeutic effect in the treatment of AD.

6 Conclusions

Without doubt, AD is one of the most challenging human diseases, firstly because of its enormous incidence in older people worldwide and secondly because there is currently not an efficient treatment available. Several symptomatic drugs have been commercialized but there is a huge need to discover compounds able to target the complex mechanisms that affect the brain leading to AD.

Computational approaches have shown their usefulness when searching for new drugs, and the application of in silico methodologies is typical in many drug discovery programs, especially by companies searching for new products to reach the pharmaceutical market. In this review, we have reported a selection of studies related to the discovery or design of novel compounds with potential therapeutic efficacy in the treatment of AD through the application of computational approaches. We have focused on three enzymes that are potential targets for the treatment of AD, namely, acetylcholinesterase (AChE), glycogen synthase kinase-3 (GSK-3), and β-amyloid cleaving enzyme 1 (BACE1). Other targets have also been identified, though they are less explored in computational studies for different reasons, such as their uncertain role in AD or the lack of data known about them (which is essential in order to develop predictive models). Nevertheless, focus on these new targets is important in order to enlarge the opportunities to find new anti-AD drugs. A good example is the recent work of Dhanavade and coworkers [118], who have applied homology modeling, molecular docking, and MD simulation techniques to explore the potential role of cathepsin B and propose an efficient method to design alternative strategies for the treatment of AD.

In silico methods have been extensively used in AD, and a very large number of studies have been published in the field. Molecular docking, QSAR predictive models, pharmacophore approaches, and MD simulations have been successfully applied to find new hits in virtual screening campaigns or to optimize chemical series and propose new potential leads as drugs in AD treatment. A high percentage of these papers have coupled several computational techniques, and therefore, ligand-based and structure-based methods have been combined to gain information and improve the predictive ability of the models.

It is accepted that the best way to treat complex multifactorial diseases such as AD is the simultaneous modulation of multiple targets. In this review, some examples of this multi-target drug design strategy are shown, and in fact, it can be expected that further research efforts will be done in the next few years. In this sense, a particular area to exploit is the repurposing of current drugs for other therapeutic uses [119], for example, rasagiline and liraglutide, which are used, respectively, in Parkinson's disease and diabetes. These are currently among the most promising drugs being investigated in AD research.

Finally, it is really important to consider the role of the blood--brain barrier (BBB) in AD, which probably has been underestimated until now. First, multiple reports point out that BBB could play a significant role in the pathogenesis of AD, contributing to the disease through a number of different mechanisms, and vice versa, that AD risk factors could also contribute to different aspects of BBB dysfunction [120, 121]. Second, it is very well known that drug delivery to the central nervous system (CNS) is hindered by the BBB, and physicochemical and pharmacokinetic properties of compounds have to be optimized to reach the brain of AD patients. In this field, computational approaches have also an important role to play in order to generate good models for predicting the BBB permeability of compounds targeting the CNS, either for traditional small molecules or the more innovative nanotherapeutic molecules [122].

References

1. Winblad B, Prince M, Wimo A et al (2015) World Alzheimer report 2015. The global impact of dementia. An analysis of prevalence, incidence, cost & trends. Alzheimer's Disease International, London
2. Plassman BL, Langa KM, Fisher GG et al (2007) Prevalence of dementia in the United States: the aging, demographics, and memory study. Neuroepidemiology 29:125–132

3. Scheltens P, Blennow K, Breteler MM et al (2016) Alzheimer's disease. Lancet 388:505–517
4. Cummings JL, Morstorf T, Zhong K (2014) Alzheimer's disease drug development pipeline: few candidates, frequent failures. Alzheimers Res Ther 6:37
5. Karran E, Mercken M, De Strooper B (2011) The amyloid cascade hypothesis for Alzheimer's disease: an appraisal for the development of therapeutics. NatRevDrugDiscov 10:698–712
6. Knowles TP, Waudby CA, Devlin GL et al (2009) An analytical solution to the kinetics of breakable filament assembly. Science 326:1533–1537
7. Koss DJ, Jones G, Cranston A et al (2016) Soluble pre-fibrillar tau and β-amyloid species emerge in early human Alzheimer's disease and track disease progression and cognitive decline. Acta Neuropathol 132:875–895
8. Kumar A, Nisha CM, Silakari C et al (2016) Current and novel therapeutic molecules and targets in Alzheimer's disease. J Formos Med Assoc 115:3–10
9. Ferreira-Vieira TH, Guimaraes IM, Silva FR et al (2016) Alzheimer's disease: targeting the cholinergic system. Curr Neuropharmacol 14:101–115
10. Pinto T, Lanctôt KL, Herrmann N (2011) Revisiting the cholinergic hypothesis of behavioral and psychological symptoms in dementia of the Alzheimer's type. Ageing Res Rev 10:404–412
11. Bautista-Aguilera OM, Esteban G, Chioua M et al (2014) Multipotent cholinesterase/monoamine oxidase inhibitors for the treatment of Alzheimer's disease: design, synthesis, biochemical evaluation, ADMET, molecular modeling, and QSAR analysis of novel donepezil-pyridyl hybrids. Drug Des Devel Ther 8:1893–1910
12. Firoozpour L, Sadatnezhad K, Dehghani S et al (2012) An efficient piecewise linear model for predicting activity of caspase-3 inhibitors. DARU J Pharm Sci 20:31
13. Macalino SJ, Gosu V, Hong S et al (2015) Role of computer-aided drug design in modern drug discovery. Arch Pharm Res 38:1686–1701
14. Julián-Ortiz JV, Gozalbes R, Besalú E (2016) Discriminating drug-like compounds by partition trees with quantum similarity indices and graph invariants. Curr Pharm Des 22:5179–5195
15. Ramírez D (2016) Computational methods applied to rational drug design. Open Med Chem J 10:7–20
16. Nantasenamat C, Isarankura-Na-Ayudhya C, Prachayasittikul V (2010) Advances in computational methods to predict the biological activity of compounds. Expert Opin Drug Discov 5:633–654
17. Gozalbes R, Jacewicz M, Annand R et al (2011) QSAR-based permeability model for drug-like compounds. Bioorg Med Chem 19:2615–2624
18. Wang Y, Xing J, Xu Y et al (2015) In silico ADME/T modelling for rational drug design. Q Rev Biophys 48:488–515
19. Katsila T, Spyroulias GA, Patrinos GP et al (2016) Computational approaches in target identification and drug discovery. Comput Struct Biotechnol J 14:177–184
20. Saha A, Roy K (2012) In silico modeling for prediction of drug-induced adverse reactions and environmental hazards using QSAR tools. Curr Drug Saf 7:255–256
21. Chen YG, Wang YY, Zhao XM (2016) A survey on computational approaches to predicting adverse drug reactions. Curr Top Med Chem 16:3629–3635
22. de Ruyck J, Brysbaert G, Blossey R et al (2016) Molecular docking as a popular tool in drug design, an in silico travel. Adv Appl Bioinform Chem 9:1–11
23. Chen WL (2006) Chemoinformatics: past, present, and future. J Chem Inf Model 46:2230–2255
24. Lawless MS, Waldman M, Fraczkiewicz R et al (2016) Using cheminformatics in drug discovery. Handb Exp Pharmacol 232:139–168
25. Wang T, Wu MB, Lin JP et al (2015) Quantitative structure-activity relationship: promising advances in drug discovery platforms. Expert Opin Drug Discov 10:1283–1300
26. Gozalbes R, Doucet JP, Derouin F (2002) Application of topological descriptors in QSAR and drug design: history and new trends. Curr Drug Targets Infect Disord 2:93–102
27. Gozalbes R, Pineda-Lucena A (2011) Small molecule databases and chemical descriptors useful in chemoinformatics: an overview. Comb Chem High Throughput Screen 14:548–558
28. Danishuddin KAU (2016) Descriptors and their selection methods in QSAR analysis: paradigm for drug design. Drug Discov Today 21:1291–1302
29. Roy K, Mitra I (2011) On various metrics used for validation of predictive QSAR models with applications in virtual screening and focused library design. Comb Chem High Throughput Screen 14:450–474

30. Roy K, Das RN (2014) A review on principles, theory and practices of 2D-QSAR. Curr Drug Metab 15:346–379
31. Ambure P, Roy K (2014) Advances in quantitative structure–activity relationship models of anti-Alzheimer's agents. Expert Opin Drug Discov 9:697–723
32. Wermuth CG, Ganellin CR, Lindberg P et al (1998) Glossary of terms used in medicinal chemistry (IUPAC recommendations 1997). Annu Rep Med Chem 33:385–395
33. Güner OF (2002) History and evolution of the pharmacophore concept in computer-aided drug design. Curr Top Med Chem 2:1321–1332
34. Wolber G, Seidel T, Bendix F et al (2008) Molecule-pharmacophore superpositioning and pattern matching in computational drug design. Drug Discov Today 13:23–29
35. De Vivo M, Masetti M, Bottegoni G et al (2016) Role of molecular dynamics and related methods in drug discovery. J Med Chem 59:4035–4061
36. Cornell WD, Cieplak P, Bayly CI et al (1995) A second generation force field for the simulation of proteins, nucleic acids, and organic molecules. J Am Chem Soc 117:5179–5197
37. Lindorff-Larsen K, Piana S, Palmo K et al (2010) Improved side-chain torsion potentials for the Amber ff99SB protein force field. Proteins 78:1950–1958
38. MacKerell AD, Bashford D, Bellot M et al (1998) All-atom empirical potential for molecular modeling and dynamics studies of proteins. J Phys Chem B 102:3586–3616
39. Foloppe N, MacKerell AD (2000) All-atom empirical force field for nucleic acids: I. Parameter optimization based on small molecule and condensed phase macromolecular target data. J Comput Chem 21:86–104
40. Schuler LD, Daura X, van Gunsteren WF (2001) An improved GROMOS96 force field for aliphatic hydrocarbons in the condensed phase. J Comput Chem 22:1205–1218
41. Oostenbrink C, Villa A, Mark AE et al (2004) A biomolecular force field based on the free enthalpy of hydration and solvation: the GROMOS force-field parameter sets 53A5 and 53A6. J Comput Chem 25:1656–1676
42. Jorgensen WL, Maxwell DS, Tirado-Rives J (1996) Development and testing of the OPLS all-atom force field on conformational energetics and properties of organic liquids. J Am Chem Soc 118:11225–11236
43. Kaminski GA, Friesner RA, Tirado-Rives J et al (2001) Evaluation and reparameterization of the OPLS-AA force field for proteins via comparison with accurate quantum chemical calculations on peptides. J Phys Chem B 105:6474–6487
44. Maupetit J, Tuffery P, Derreumaux P (2007) A coarse- grained protein force field for folding and structure prediction. Proteins 69:394–408
45. Monticelli L, Kandasamy SK, Periole X et al (2008) The MARTINI coarse-grained force field: extension to proteins. J Chem Theory Comput 4:819–834
46. Nanias M, Czaplewski C, Scheraga HA (2006) Replica exchange and multicanonical algorithms with the coarse-grained united-residue UNRES force field. J Chem Theory Comput 2:513–528
47. Jung HA, Ali MY, Choi RJ et al (2016) Kinetics and molecular docking studies of fucosterol and fucoxanthin, BACE1 inhibitors from brown algae *Undaria pinnatifida* and *Ecklonia stolonifera*. Food Chem Toxicol 89:104–111
48. Zhang S, Lin Z, Pu Y et al (2016) Comparative QSAR studies using HQSAR, CoMFA, and CoMSIA methods on cyclic sulfone hydroxyethylamines as BACE1 inhibitors. Comput Biol Chem 67:38–47
49. Wu Q, Li X, Gao Q et al (2016) Interaction mechanism exploration of HEA derivatives as BACE1 inhibitors by in silico analysis. Mol BioSyst 12:1151–1165
50. Quesada-Romero L, Caballero J (2014) Docking and quantitative structure–activity relationship of oxadiazole derivatives as inhibitors of GSK3β. Mol Divers 18:149–159
51. Czeleń P, Szefler B (2015) Molecular dynamics study of the inhibitory effects of ChEMBL474807 on the enzymes GSK-3β and CDK-2. J Mol Model 21:74
52. Zhao S, Zhu J, Xu L et al (2016) Theoretical studies on the selective mechanisms of GSK3β and CDK2 by molecular dynamics simulations and free energy calculations. Chem Biol Drug Des 2016:1–10
53. Ombrato R, Cazzolla N, Mancini F et al (2015) Structure-based discovery of 1H-indazole-3-carboxamides as a novel structural class of human GSK-3 inhibitors. J Chem Inf Model 55:2540–2551
54. Unzeta M, Esteban G, Bolea I et al (2016) Multi-target directed donepezil-like ligands for Alzheimer's disease. Front Neurosci 10:205
55. Gurung AB, Aguan K, Mitra S et al (2016) Identification of molecular descriptors for design of novel Isoalloxazine derivatives as

potential Acetylcholinesterase inhibitors against Alzheimer's disease. J Biomol Struct Dyn 28:1–14
56. Selkoe DJ (2001) Alzheimer's disease: genes, proteins, and therapy. Physiol Rev 81:741–766
57. Ghosh AK, Osswald HL (2014) BACE1 (β-secretase) inhibitors for the treatment of Alzheimer's disease. Chem Soc Rev 43:6765–6813
58. Hong L, Koelsch G, Lin X et al (2000) Structure of the protease domain of memapsin 2 (β-secretase) complexed with inhibitor. Science 290:150–153
59. Jordan JB, Whittington DA, Bartberger MD et al (2016) Fragment-linking approach using (19)F NMR spectroscopy to obtain highly potent and selective inhibitors of beta-secretase. J Med Chem 59:3732–3749
60. Barman A, Schürer S, Prabhakar R (2011) Computational modeling of substrate specificity and catalysis of the β-secretase (BACE1) enzyme. Biochemistry 50:4337–4349
61. Chakraborty S, Basu S (2015) Structural insight into the mechanism of amyloid precursor protein recognition by β-secretase 1: a molecular dynamics study. Biophys Chem 202:1–12
62. Xu Y, Li MJ, Greenblatt H et al (2011) Flexibility of the flap in the active site of BACE1 as revealed by crystal structures and molecular dynamics simulations. Acta Crystallogr D Biol Crystallogr 68:13–25
63. Barman A, Prabhakar R (2014) Computational insights into substrate and site specificities, catalytic mechanism, and protonation states of the catalytic Asp dyad of β -secretase. Scientifica (Cairo) 2014:598728
64. Liu H, Wang L, Su W et al (2014) Advances in recent patent and clinical trial drug development for Alzheimer's disease. Pharm Pat Anal 3:429–447
65. Choi RJ, Roy A, Jung HJ et al (2016) BACE1 molecular docking and anti-Alzheimer's disease activities of ginsenosides. J Ethnopharmacol 190:219–230
66. Cruz DS, Castilho MS (2014) 2D QSAR studies on series of human beta-secretase (BACE-1) inhibitors. Med Chem 10:162–173
67. Hossain T, Mukherjee A, Saha A (2015) Chemometric design to explore pharmacophore features of BACE inhibitors for controlling Alzheimer's disease. Mol BioSyst 11:549–557
68. Subramanian G, Ramsundar B, Pande V et al (2016) Computational modeling of β-secretase 1 (BACE-1) inhibitors using ligand based approaches. J Chem Inf Model 56 (10):1936–1949. [Epub ahead of print]
69. Lemkul JA, Bevan DR (2012) The role of molecular simulations in the development of inhibitors of amyloid β-peptide aggregation for the treatment of Alzheimer's disease. ACS Chem Neurosci 3:845–856
70. Tran L, Ha-Duong T (2015) Exploring the Alzheimer amyloid-β peptide conformational ensemble: a review of molecular dynamics approaches. Peptides 69:86–91
71. Kocaka A, Erol I, Yildiz M et al (2016) Computational insights into the protonation states of catalytic dyad in BACE1–acyl guanidine based inhibitor complex. J Mol Graph Model 70:226–235
72. Sabbah DA, Zhong HA (2016) Modeling the protonation states of β-secretase binding pocket by molecular dynamics simulations and docking studies. J Mol Graph Model 68:206–215
73. Gueto-Tettay C, Pestana-Nobles R, Drosos-Ramirez JC (2016) Determination of the protonation state for the catalytic dyad in β-secretase when bound to hydroxyethylamine transition state analogue inhibitors: a molecular dynamics simulation study. J MolGraph Model 66:155–167
74. Gueto-Tettay C, Zuchniarz J, Fortich-Seca Y et al (2016) A molecular dynamics study of the BACE1 conformational change from Apo to closed form induced by hydroxyethylamine derived compounds. J Mol Graph Model 70:181–195
75. Hernández-Rodríguez M, Correa-Basurto J, Gutiérrez A et al (2016) Asp32 and Asp228 determine the selective inhibition of BACE1 as shown by docking and molecular dynamics simulations. Eur J Med Chem 124:1142–1154
76. Pradeep N, Munikumar M, Swargam S et al (2015) Combination of e-pharmacophore modeling, multiple docking strategies and molecular dynamic simulations to discover of novel antagonists of BACE1. J Biomol Struct Dyn 33(Suppl 1):129–130
77. Domínguez JL, Fernández-Nieto F, Castro M et al (2015) Computer-aided structure-based design of multitarget leads for Alzheimer's disease. J Chem Inf Model 55:135–148
78. Zhou A, Hu J, Wang L et al (2015) Combined 3D-QSAR, molecular docking, and molecular dynamics study of tacrine derivatives as potential acetylcholinesterase (AChE) inhibitors of Alzheimer's disease. J Mol Model 21:277

79. Medina M, Wandosell F (2011) Deconstructing GSK-3: the fine regulation of its activity. Int J Alzheimers Dis 2011:479249
80. Darrington RS, Campa VM, Walker MM et al (2012) Distinct expression and activity of GSK-3α and GSK- 3β in prostate cancer. Int J Cancer 131:E872–E883
81. Maqbool M, Mobashir M, Hoda N (2016) Pivotal role of glycogen synthase kinase-3: a therapeutic target for Alzheimer's disease. Eur J Med Chem 107:63–81
82. Aplin AE, Gibb GM, Jacobsen JS et al (1996) In vitro phosphorylation of the cytoplasmic domain of the amyloid precursor protein by glycogen synthase kinase-3beta. J Neurochem 67:699–707
83. Hanger DP, Hughes K, Woodgett JR et al (1992) Glycogen synthase kinase-3 induces Alzheimer's disease-like phosphorylation of tau: generation of paired helical filament epitopes and neuronal localization of the kinase. Neurosci Lett 147:58–62
84. Cohen P, Goedert M (2004) GSK3 inhibitors: development and therapeutic potential. Nat Rev Drug Discov 3:479–487
85. Dajani R, Fraser E, Roe SM et al (2001) Crystal structure of glycogen synthase kinase 3 beta: structural basis for phosphate-primed substrate specificity and autoinhibition. Cell 105:721–732
86. Berg S, Bergh M, Hellberg S et al (2012) Discovery of novel potent and highly selective glycogen synthase kinase-3β (Gsk3β) inhibitors for Alzheimer's disease: design, synthesis, and characterization of pyrazines. J Med Chem 55:9107–9119
87. Pandey MK, DeGrado TR (2016) Glycogen synthase kinase-3 (GSK-3)-targeted therapy and imaging. Theranostics 6:571–593
88. Thomas GM, Frame S, Goedert M et al (1999) A GSK3-binding peptide from FRAT1 selectively inhibits the GSK3-catalysed phosphorylation of axin and beta-catenin. FEBS Lett 458:247–251
89. Polgar T, Baki A, Szendrei GI et al (2005) Comparative virtual and experimental high-throughput screening for glycogen synthase kinase-3β inhibitors. J Med Chem 48:7946–7959
90. Quesada-Romero L, Mena-Ulecia K, Tiznado W et al (2014) Insights into the interactions between maleimide derivates and GSK3β combining molecular docking and QSAR. PLoS One 9:e102212
91. Crisan L, Pacureanu L, Avram S et al (2014) PLS and shape-based similarity analysis of maleimides – GSK-3 inhibitors. J Enzyme Inhib Med Chem 29:599–610
92. Fu G, Liu S, Nan X et al (2014) Quantitative structure-activity relationship analysis and a combined ligand-based/structure-based virtual screening study for glycogen synthase kinase-3. Mol Inf 2014(33):627–640
93. Darshit BS, Balaji B, Rani P et al (2014) Identification and in vitro evaluation of new leads as selective and competitive glycogen synthase kinase-3β inhibitors through ligand and structure based drug design. J Mol Graph Model 53:31–47
94. Arfeen M, Patel R, Khan T et al (2015) Molecular dynamics simulation studies of GSK-3β ATP competitive inhibitors: understanding the factors contributing to selectivity. J Biomol Struct Dyn 33:2578–2593
95. Karpov PV, Osolodkin DI, Baskin II et al (2011) One-class classification as a novel method of ligand-based virtual screening: the case of glycogen synthase kinase 3β inhibitors. Bioorg Med Chem Lett 21:6728–6731
96. Osolodkin DI, Palyulin VA, Zefirov NS (2011) Structure-based virtual screening of glycogen synthase kinase 3β inhibitors: analysis of scoring functions applied to large true actives and decoy sets. Chem Biol Drug Des 78:378–390
97. Nisha CM, Kumar A, Vimal A et al (2016) Docking and ADMET prediction of few GSK-3 inhibitors divulges 6-bromoindirubin-3-oxime as a potential inhibitor. J Mol Graph Model 65:100–107
98. Dvir H, Silman I, Harel M et al (2010) Acetylcholinesterase: from 3D structure to function. Chem Biol Interact 187:10–22
99. Franklin MC, Rudolph MJ, Ginter C et al (2016) Structures of paraoxon-inhibited human acetylcholinesterase reveal perturbations of the acyl loop and the dimer interface. Proteins 84:1246–1256
100. Sağlık BN, Ilgın S, Özkay Y (2016) Synthesis of new donepezil analogues and investigation of their effects on cholinesterase enzymes. Eur J Med Chem 124:1026–1040
101. Demir ÖÜ, Can ÖD, Sağlık BN et al (2016) Design, synthesis, and AChE inhibitory activity of new benzothiazole-piperazines. Bioorg Med Chem Lett 26:5387–5394
102. Mishra CB, Kumari S, Manral A et al (2017) Design, synthesis, in-silico and biological evaluation of novel donepezil derivatives as multi-target-directed ligands for the treatment of Alzheimer's disease. Eur J Med Chem 125:736–750

103. Wong KY, Mercader AG, Saavedra LM et al (2014) QSAR analysis on tacrine-related acetylcholinesterase inhibitors. J Biomed Sci 21:84
104. Todeschini R, Consonni V (2009) In: Mannhold R, Kubinyi H, Timmerman H (eds) Molecular descriptors for chemoinformatics, Methods and principles in medicinal chemistry, vol 41. Wiley - VCH, Weinheim
105. Zhang S, Hou B, Yang H (2016) Design and prediction of new acetylcholinesterase inhibitor via quantitative structure activity relationship of huprines derivatives. Arch Pharm Res 39:591–602
106. Brahmachari G, Choo C, Ambure P et al (2015) In vitro evaluation and in silico screening of synthetic acetylcholinesterase inhibitors bearing functionalized piperidine pharmacophores. Bioorg Med Chem 23:4567–4575
107. Ambure P, Kar S, Roy K (2014) Pharmacophore mapping-based virtual screening followed by molecular docking studies in search of potential acetylcholinesterase inhibitors as anti-Alzheimer's agents. Biosystems 116:10–20
108. Martis EA, Chandarana RC, Shaikh MS et al (2015) Quantifying ligand–receptor interactions for gorge-spanning acetylcholinesterase inhibitors for the treatment of Alzheimer's disease. J Biomol Struct Dyn 33:1107–1125
109. Simeon S, Anuwongcharoen N, Shoombuatong W et al (2016) Probing the origins of human acetylcholinesterase inhibition via QSAR modeling and molecular docking. PeerJ 4:e2322
110. Gaulton A, Bellis LJ, Bento AP et al (2012) ChEMBL: a large-scale bioactivity database for drug discovery. Nucleic Acids Res 40 (D1):D1100–D1107
111. The Organisation for Economic Co-operation and Development (OECD) (2007) Guidance document on the validation of (quantitative) structure-activity relationships [(Q)SAR] models, OECD environment health and safety publications, Series on testing and assessment no. 69. Organisation for Economic Co-operation and Development, Paris. (www.oecd.org/ehs/)
112. Kennard RW, Stone LA (1969) Computer aided design of experiments. Technometrics 11:137–148
113. Samadi A, de los Ríos C, Bolea I et al (2012) Multipotent MAO and cholinesterase inhibitors for the treatment of Alzheimer's disease: synthesis, pharmacological analysis and molecular modeling of heterocyclic substituted alkyl and cycloalkyl propargyl amine. Eur J Med Chem 52:251–262
114. Bhakta HK, Park CH, Yokozawa T et al (2016) Kinetics and molecular docking studies of loganin, morroniside and 7-O-galloyl-D-sedoheptulose derived from *Corni fructus* as cholinesterase and β-secretase 1 inhibitors. Arch Pharm Res 39:794–805
115. Panek D, Więckowska A, Wichur T et al (2017) Design, synthesis and biological evaluation of new phthalimide and saccharin derivatives with alicyclic amines targeting cholinesterases, beta-secretase and amyloid beta aggregation. Eur J Med Chem 125:676–695
116. Makhaeva GF, Lushchekina SV, Boltneva NP et al (2015) Conjugates of γ-Carbolines and Phenothiazine as new selective inhibitors of butyrylcholinesterase and blockers of NMDA receptors for Alzheimer disease. Sci Rep 18:13164
117. Bachurin SO, Shevtsova EF, Makhaeva GF et al (2017) Novel conjugates of aminoadamantanes with carbazole derivatives as potential multitarget agents for AD treatment. Sci Rep 7:45627
118. Dhanavade MJ, Parulekar RS, Kamble SA et al (2016) Molecular modeling approach to explore the role of cathepsin B from *Hordeum vulgare* in the degradation of Aβ peptides. Mol BioSyst 12:162–168
119. Hughes RE, Nikolic K, Ramsay RR (2016) One for all? Hitting multiple Alzheimer's disease targets with one drug. Front Neurosci 10:177
120. Gosselet F, Saint-Pol J, Candela P et al (2013) Amyloid-β peptides, Alzheimer's disease and the blood-brain barrier. Curr Alzheimer Res 10:1015–1033
121. Erickson MA, Banks WA (2013) Blood-brain barrier dysfunction as a cause and consequence of Alzheimer's disease. J Cereb Blood Flow Metab 33:1500–1513
122. Gu X, Chen H, Gao X (2015) Nanotherapeutic strategies for the treatment of Alzheimer's disease. Ther Deliv 6:177–195

Chapter 3

Computer-Aided Drug Design Approaches to Study Key Therapeutic Targets in Alzheimer's Disease

Agostinho Lemos, Rita Melo, Irina S. Moreira, and M. Natália D.S. Cordeiro

Abstract

Alzheimer's Disease (AD) is one of the most common and complex age-related neurodegenerative disorders in elderly people. Currently there is no cure for AD, and available therapeutic alternatives only improve both cognitive and behavioral functions. For that reason, the search for anti-AD therapeutic agents with neuroprotective properties is highly demanding. Several research studies have implicated the involvement of G-Protein-Coupled Receptors (GPCRs) in diverse neurotransmitter systems that are dysregulated in AD, mainly in modulation of amyloidogenic processing of Amyloid Precursor Protein (APP) and of microtubule-associated protein *tau* phosphorylation and in learning and memory activities in in vivo AD models subjected to numerous behavioral procedures. In this chapter, a special focus will be given to the structure- and ligand-based in silico approaches and their applicability on the development of small molecules that target various GPCRs potentially involved in AD such as 5-hydroxytryptamine receptors, adenosine receptors, adrenergic receptors, chemokine receptors, histamine receptors, metabotropic glutamate receptors, muscarinic acetylcholine receptors, and opioid receptors.

Key words Alzheimer's disease, GPCRs, G-proteins, Drug design, Docking, Pharmacophore, QSAR

1 Introduction

Alzheimer's disease (AD) is a neurodegenerative disorder clinically characterized by a progressive and irreversible loss of memory and impairment of other cognitive functions, which ultimately results in a complete degradation of intellectual and mental activities. Although age represents a critical risk factor, a combination of genetic, lifestyle, and environmental factors may contribute for the development of AD. Being the most common cause of dementia in elderly people, continuous research efforts have been devoted to unravel the etiology of AD with the objective of developing effective pharmacological treatments.

Although the underlying mechanism of AD is not yet well understood, several neuropathological hallmarks are thought to be

Kunal Roy (ed.), *Computational Modeling of Drugs Against Alzheimer's Disease*, Neuromethods,
vol. 132, DOI 10.1007/978-1-4939-7404-7_3, © Springer Science+Business Media LLC 2018

involved in the neurodegeneration in AD, including (i) deficiency on cholinergic transmission in the Central Nervous System (CNS) due to an extensive loss of cholinergic neurons which results in a deficit of AcetylCholine (ACh) in specific regions of the brain (cholinergic hypothesis) (reviewed in [1, 2]); (ii) abnormal clustering of neurotoxic β-amyloid (Aβ) fragments and formation of senile plaques that occur as a consequence of an imbalance between the amyloidogenic (mediated by β- and γ-secretases) and non-amyloidogenic (mediated by α- and γ-secretases) processing pathways of Amyloid Precursor Protein (APP) and an inefficient clearance of Aβ oligomers (amyloid hypothesis) (reviewed in [3, 4]); and (iii) hyperphosphorylation of Serine (Ser), Threonine (Thr), and Tyrosine (Tyr) sites in microtubule-associated *tau* proteins that leads to the destabilization of neuronal microtubules, the formation of *tau* aggregates and NeuroFibrillary Tangles (NFT), and the collapse of neuronal signaling (*tau* hypothesis) (reviewed in [5, 6]). With the increasing number of people suffering from age-related neurodegenerative disorders, particularly AD, effective therapeutic alternatives are highly demanding. Currently, pharmacological research has been focused on the discovery of drug candidates with neuroprotective properties, which target disease-modifying effects, contributing to the blockade of neuronal apoptosis and subsequent disease progression. These strategies are based on targeting key proteins involved in amyloidogenic processing of APP (activation of α-secretase, inhibition of β- and γ-secretases, prevention of Aβ aggregation, and promotion of Aβ clearance) and in *tau* pathology (inhibition of *tau*-phosphorylating kinases, prevention of *tau* aggregation, and promotion of *tau* aggregate disassembly). However, current clinically available AD therapies are essentially symptomatic and target mainly AcetylCholinEsterase (AChE) (donepezil, rivastigmine, and galantamine) and *N*-methyl-D-aspartate receptor (memantine), which lead to the reversion of dysfunctions on cholinergic and glutamatergic neurotransmission, respectively. Moreover, neurodegeneration is not restricted to a particular neurotransmitter system. Histaminergic, adenosinergic, adrenergic, and serotonergic, among other neurotransmitter systems, are also dysregulated in AD. Interestingly, numerous studies have implicated the role of G-Protein-Coupled Receptors (GPCRs) in the pathogenesis of AD, particularly in the modulation of the distinct therapeutic targets involved in amyloidogenic processing of APP and in microtubule-associated *tau* protein aggregation, and the influence of GPCR modulators in AD animal models subjected to various learning and memory paradigms. Potential GPCR-derived therapeutic targets for AD include 5-HydroxyTryptamine 2A, 2C, 4, and 6 Receptors (5-HT$_{2A}$R [7, 8], 5-HT$_{2C}$R [7, 9], 5-HT$_4$R [10, 11, 12, 13], and 5-HT$_6$R [14, 15, 16, 17]); Adenosine A$_1$ and A$_{2A}$ Receptors (A$_1$AR [18, 19, 20] and A$_{2A}$AR [18, 21, 22, 23, 24]); α_{2A}- and β_2-Adrenergic Receptors (α_{2A}-AR [25] and β_2-AR [26, 27, 28]); CC motif chemokine receptor 2 (CCR_2 [29, 30]); CXC motif chemokine receptor 2 ($CXCR_2$ [31, 32, 33]); corticotropin-releasing factor receptor 1 ($CRFR_1$ [34, 35, 36, 37]); δ-opioid receptors (DOR [38]); histamine H_3 receptor (H_3R [39, 40, 41]); metabotropic glutamate receptor types 1, 2, and 5 ($mGluR_1$ [42, 43, 44, 45], $mGluR_2$ [42, 46, 47], and $mGluR_5$ [42, 48, 49]); and M_1, M_2, and M_3 muscarinic acetylcholine receptors (M_1 mAChR [50, 51, 52, 53, 54], M_2 mAChR [54, 55], and M_3 mAChR [53, 54]), among others. In this chapter, we will provide an overview of the structure-based and ligand-based computational approaches widely employed in in silico medicinal chemistry to target the mentioned GPCRs potentially implicated in AD.

2 GPCRs: A Case Study of Potential Targets for AD

Being one of the most heavily investigated drug targets in the pharmaceutical industry, GPCR-targeting drugs represent about ~30–40% of the current market for human therapeutics and have been subjected to a considerable number of computational studies [56, 57]. They comprise a large family of membrane-embedded proteins that mediate important physiological functions through interaction with various endogenous ligands, including ions, proteins, peptides, amines, hormones, chemokines, and neurotransmitters [58, 59]. Structurally, a single polypeptide chain with a variable length that crosses the phospholipidic bilayer seven times adopting the typical structure of seven transmembrane (TM) α-helices connected to extracellular (ECL) and intracellular (ICL) loops characterizes the receptors belonging to this family [60]. Based on sequence homology and phylogenetic analysis, human GPCRs can be classified into five main families of receptors: *glutamate* (Class C, 22 members), *rhodopsin* (Class A, 672 members), *adhesion* (33 members), *frizzled/Taste2* (Class F, 36 members), and *secretin* (Class B, 15 members), which are usually shortened to the acronym *GRAFS* [60]. The complexity of GPCR-induced signaling is determined by their association with specific heterotrimeric guanine nucleotide-binding proteins (G-proteins) within the plasma membrane. Heterotrimeric G-proteins are composed of a guanine-binding α-subunit (G_α) and a dimer consisting of the β- and γ-subunits ($G_{\beta\gamma}$). In their inactive state, G_α is bound to guanosine diphosphate (GDP) and associated with $G_{\beta\gamma}$. In the extracellular site, the binding of an agonist stabilizes the active conformation of the receptor, which couples to heterotrimeric G-proteins, leading to GDP release and guanosine triphosphate (GTP) binding to the G_α subunit. Subsequently, the GTP binding induces a conformational switch on the G_α subunit, which promotes the release of G-proteins

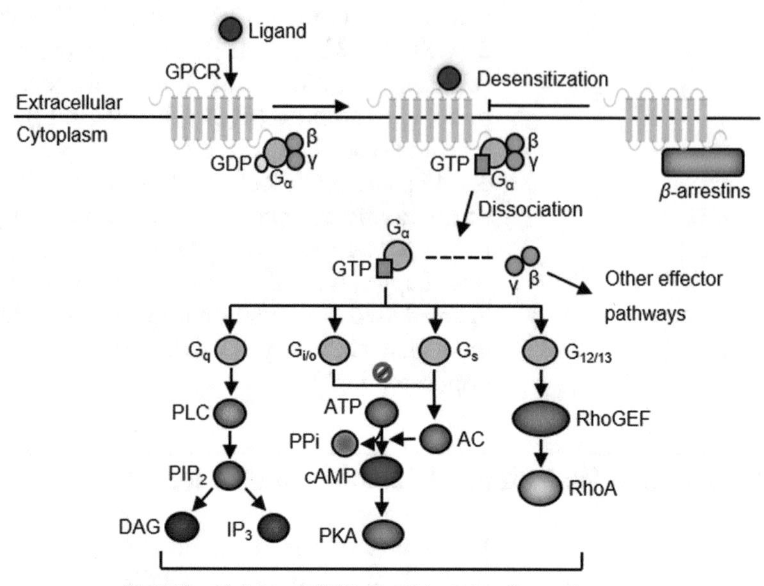

Fig. 1 General diagram of GPCR signaling mediated by activation of G_α subunit of heterotrimeric G-proteins. *AC* Adenylyl Cyclase, *ATP* Adenosine TriPhosphate, *cAMP* cyclic Adenosine MonoPhosphate, *DAG* DiAcylGlycerol, *GDP* Guanosine DiPhosphate, *GTP* Guanosine TriPhosphate, *IP_3* Inositol 1,4,5-trisPhosphate, *PIP_2* PhosphatidylInositol 4,5-bisPhosphate, *PKA* Protein Kinase A, *PLC* PhosphoLipase C, *PPi* inorganic PyroPhosphate, *RhoA* Ras homolog gene family, member A, *RhoGEF* Rho Guanine nucleotide Exchange Factor

from GPCR and the dissociation of heterotrimeric G-proteins into G_α and $G_{\beta\gamma}$ subunits [61, 62]. The G_α ($G_{\alpha s}$, $G_{\alpha i/o}$, $G_{\alpha q}$, $G_{\alpha 12/13}$) and $G_{\beta\gamma}$ subunits amplify and propagate their transduction signals by modulating the activity of distinct downstream cellular effectors, including adenylyl and guanylyl cyclases, phospholipases, phosphodiesterases, and phosphoinositide 3-kinases, that in turn induces an increasing or decreasing production of second messengers, such as Ca^{2+}, diAcylglycerol (DAG), inositol 1,4,5-trisphosphate (IP_3), cyclic adenosine monophosphate (cAMP), and cyclic guanosine monophosphate (cGMP) that triggers a wide range of cellular responses [63, 64] (Fig. 1).

Nevertheless, not all GPCR-dependent signaling pathways are mediated via heterotrimeric G-proteins. The persistent stimulation of a specific agonist may contribute to a decreasing responsiveness of GPCRs, eliciting a process of receptor desensitization, which terminates or attenuates the receptor signaling. Two families of regulatory proteins participate in the mechanism of GPCR desensitization, including second messenger-dependent protein kinases and G-protein-coupled receptor kinases (GRKs). Second messenger-dependent protein kinases, protein kinases A (PKA) and C (PKC), induce a conformational change in the receptor through GPCR phosphorylation, directly uncoupling GPCR to heterotrimeric G-proteins. This mechanism of receptor regulation can be mediated in the absence of GPCR occupancy by an agonist through a process of heterologous desensitization. In contrast, GPCR occupancy is required for the recruitment of GRKs on receptor desensitization (homologous desensitization). The GRKs preferentially induce the phosphorylation in an agonist-bound conformation, leading to a significant attenuation of receptor signaling [65]. GRK-dependent phosphorylation enables GPCRs to interact with high affinity to a class of multifunctional scaffold proteins called β-arrestins, which sterically blocks further interactions between the G-protein and the activated receptor, preventing GPCR signaling [66]. Additionally, receptor-bound β-arrestins can also promote different signaling pathways or act as adapter proteins, promoting receptor sequestration through interaction with components of the cellular machinery required for clathrin-mediated endocytosis [67]. This mechanism is critical not only for receptor signaling desensitization but also for receptor resensitization for a next round of GPCR activation. Other mechanisms of desensitization include the receptor proteolysis in lysosomes [68], dynamic regulation of receptor gene expression [69], and GTP hydrolysis by regulators of G-protein signaling (RGS) proteins [70, 71].

3 In Silico Approaches in the Discovery of New Modulators of GPCR-Derived Therapeutic Targets for AD

A wide array of Computer-Aided Drug Design (CADD) methodologies have been employed as a complementary tool to the high-throughput screening (HTS) approaches to identify new GPCR modulators with therapeutic potential for AD. One critical stage in in silico drug design of GPCR modulators is the discovery of novel lead compounds (or hit-to-lead optimization), which can be accomplished using different strategies such as virtual screening of large libraries of chemical compounds using structure-based or ligand-based drug design approaches (Fig. 2).

3.1 Structure-Based Drug Design Approaches

Over the last years, the progress of the structural biology on determination of accurate three-dimensional (3D) structures of GPCRs has furnished a valuable tool for drug design of GPCR modulators by structure-based drug design approaches, such as homology modeling, virtual screening, and fragment screening. In fact, X-ray crystallography and Nuclear Magnetic Resonance (NMR) studies provide detailed and atomic-level information of GPCR-drug interactions. As their function implicates, GPCRs are membrane-bound proteins, which make experimental 3D structure elucidation, by X-ray crystallography or NMR studies, an extremely

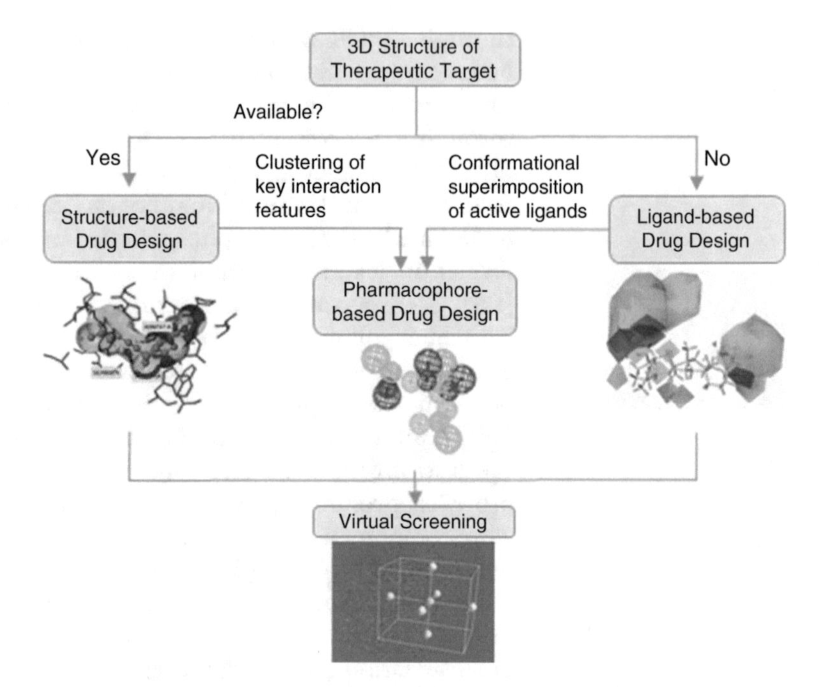

Fig. 2 General diagram of in silico drug design approaches based on the availability of 3D structural information of therapeutic targets (Representative images were extracted from [72, 73])

complex and challenging task compared to globular proteins (reviewed in [74]). Until the elucidation of the X-ray diffraction structure at 2.8 Å resolution of Class A GPCR bovine rhodopsin in 2000 [75], no X-ray structures of any GPCR were available. The high quality and detailed structure of bovine rhodopsin provided a huge progress of understanding of GPCRs at molecular level and paved the way for structure-based design approaches for GPCRs. Rhodopsin was chosen as the typical example for structural studies due to the fact that it is easy to obtain considerable quantities of functional protein with high stability under conditions that denature other GPCRs [76]. For many years, the structure of inactive state of rhodopsin provided the only template sequence for molecular modeling studies in homologous GPCRs (reviewed in [77]), which was a limitation for the study of other GPCR family members. Although rhodopsin-like or Class A GPCRs present similar structural features with the prototypical bovine rhodopsin, especially concerning the TM domain, they share a low overall homology. Moreover, other GPCRs belonging to *glutamate*, *adhesion*, *secretin*, and *frizzled/taste2* families have no homology with *rhodopsin*. Also, the distinct ligand binding and mechanism of activation of rhodopsin from other GPCRs make the understanding from rhodopsin structure how such a diverse plethora of ligands could activate the large family of GPCRs difficult. Additionally, X-ray structure of rhodopsin represents the inactive form of the receptor, while the active form would be much more suitable for rational drug design. The experimental progress in obtaining crystal GPCR structures was very slow. In fact, it took more than seven years until the 3D structures of β_2-AR complexed with carazolol [78, 79] and turkey β_1-AR complexed with cyanopindolol [80] were solved. With the development of receptor crystallization techniques, a number of technical issues derived from the low expression of GPCRs and their structural instability have been overcome, thereby resulting in an accelerated increase in solved GPCR structures. Currently, there are more than 150 3D structures of apo-, peptide-, natural ligand-, agonist-, and antagonist-bound GPCR complexes available within Protein Data Bank (PDB), in which the family A GPCR structures have been the most frequently reported ones. Only two family B, two family C, and one frizzled 3D GPCR structures have been published. Given the diverse physiological and pathological implications of their signaling, particularly in AD and other neurodegenerative disorders, GPCRs have been considered very promising therapeutic targets for pharmaceutical applications. Moreover, the identification of 3D GPCR structures provides a wealth of information to pharmaceutical researchers for drug design of GPCR modulators with neuroprotective properties for the treatment of AD.

Drug discovery efforts targeting GPCRs have been mainly focused on the development of ligands which interact with the orthosteric binding site for endogenous ligands, but a wide variety of GPCRs possess additional topographically distinct druggable sites (allosteric sites) (reviewed in [81, 82]). This allows the pharmacological modulation of particular GPCRs not only by conventional orthosteric agonists or antagonists but also by positive allosteric modulators (PAMs) or negative allosteric modulators (NAMs) with potentially high receptor subtype selectivity that either increase or reduce the receptor responsiveness, respectively (reviewed in [81, 82]). Since GPCRs interact with a plethora of intracellular signaling proteins, such as heterotrimeric G-proteins and β-arrestins, and modulate distinct intracellular pathways, distinct GPCR-targeted ligands are expected to stabilize various structural conformations and signaling states of GPCRs. In fact, specific GPCR-targeted ligands possess the ability to selectively evoke a particular stimulus-response, which results in a unique ligand-dependent signaling profile referred to as functional selectivity, biased signaling, or stimulus bias. The functional selectivity phenomenon has been explored in medicinal chemistry for the design of GPCR-targeted drugs with pathway selectivity (reviewed in [83, 84]).

In order to address how drug-dependent GPCR signaling relates to the concept of functional selectivity, atomistic-level information about the mode of ligand-GPCR interactions coupled with its two signaling partners, G-proteins and β-arrestins, is required [85]. However, relevant structure-function information is still scarce. The first X-ray crystal structure of a GPCR/G-protein complex only became available in 2011, in which the β_2-AR was complexed to $G_{\alpha s}$ protein [86]. Given the limitations of the crystallizable fragments and the static nature of this single available model, this important but restricted information is insufficient to understand the function of such a complex biological system. Nowadays, molecular dynamics (MD) simulations are a treasured resource for the study of GPCRs and can be applied to better understand their function. In fact, the usage of MD simulations has been extremely relevant to model the process of GPCR activation on an atomistic level [87, 88], to study ligand recognition or GPCR oligomerization [89] by generating ensembles of energetically accessible conformations [90, 91]. The overall construction of the membrane-protein systems is harder than for soluble proteins, but a few tools provide accurate and fast alternatives to step-by-step manual construction, such as Chemistry at HARvard Macromolecular Mechanics-Graphical User Interface (CHARMM-GUI) [92, 93], QwikMD [94], and high-throughput molecular dynamics (HTMD) [95]. The membrane environment can be explicitly (all atom) or implicitly (coarse grained (CG)) modeled. However, when a researcher aims to fully characterize the ligand-GPCR interactions, the explicit option should be chosen as it allows a detailed characterization of pairwise interactions and the measurement of a variety of chemical-physical features. While the dynamics of activation are beginning to be clarified for individual GPCRs, an increasingly important consideration pertains to the identity of the "signaling unit." Thus, for many years, the GPCRs were thought to function only as monomers, but increasing evidence suggests that they can form homodimers, heterodimers, or higher-order oligomers. It was already demonstrated that minimal functional signaling unit is a complex between a GPCR and heterotrimeric G proteins [56]. Various dimer interfaces have been proposed, but a rearrangement of the dimerization interface to form a TM4-TM4 interface is likely a critical component of activation [96]. Nonetheless, the mechanistic and structural details of the ligand-GPCR function are not known, either at the level of the receptor signaling unit or with regard to the functional epitope between GPCR/G-proteins and GPCR/β-arrestins. These aims could be also achieved upon long all-atom MD simulations of the complete systems and their subsequent analysis.

Another in silico approach widely employed in drug design is docking-based virtual screening, which consists of a wide range of computational methodologies that analyze the interaction of large databases of small-molecule drug candidates against a 3D representation of the structure of a therapeutic target protein of interest (reviewed in [97]). This approach is usually performed through molecular docking, in which each "virtual" drug candidate is docked into the X-ray crystallographic structure of the therapeutic target or, if 3D structure is not available, into a model of the target (homology model-based virtual screening), using algorithms that explore the multiple binding conformations of the ligand inside the binding cavity of a target protein. Subsequently, for each of the generated ligand conformation, the strength of their binding affinity to the target is predicted through the determination of a scoring function. In most of the automated molecular docking studies, a flexible ligand is docked in a rigid protein, since a flexible macromolecular target would demand a high cost of computational time (reviewed in [97]). Docking-based virtual screening can be applied to databases of commercially available compounds and *in-house* ligands that have been previously synthesized and tested in vitro or databases of virtual ligands that can be synthesized according to their calculated docking scores. Moreover, docking-based virtual screening may be also useful following in vitro studies for the interpretation of potential target-ligand interactions. Therefore, the main purpose of structure-based virtual screening is to select the ligand structures that are most likely to bind to a certain therapeutic target of interest, providing a library of the best scored ligands for experimental screening and, thereby, improving the overall efficacy of the drug screening process. Currently, there are a number of in silico tools widely employed in protein-ligand docking studies including automated docking (AutoDock) [98], AutoDock Vina [99], CHARMm-based DOCKER (CDOCKER) [100], FlexX [101], Genetic Optimization of Ligand Docking (GOLD) [102], Grid-based Ligand Docking with Energetics (GLIDE) [103], Internal Coordinate Mechanics (ICM) [104], molecular Interaction FingerPrints (IFP) [105], Induced-Fit Docking (IFD) [106], Library Docking (LibDock) [107], MolGridCal [108], and Protein-Ligand ANT System (PLANTS) [109], among others. Table 1 summarizes the most relevant structure-based studies performed by these docking programs for the GPCRs involved in AD.

3.2 Ligand-Based Drug Design Approaches

The GPCR ligand-based drug design useful for the identification of therapeutic agents for AD relies on knowledge of compounds that are recognized to modulate the activity of this family of TM proteins and represents a suitable in silico approach when the structural information of the therapeutic target is not available. In fact, the majority of potential drug candidates that act on GPCRs have been conceived from ligand-based methodologies, due to the restricted availability of 3D structural data on GPCRs. Various ligand-based drug design approaches have been used to better understand the mechanism of action of GPCR modulators and to screen for new bioactive molecules. Table 2 reports the applicability of ligand-based drug design approaches on the discovery of GPCR modulators with therapeutic potential for AD using large databases of

Table 1
Structure-based drug design techniques for the modulation of potential GPCR-derived therapeutic targets of AD

GPCR: Adenosine A_1 receptor (A_1AR)		
Ligands		
Adenosine		
Drug design technique(s)	**Computational tool(s)**	**References**
Docking into a human A_1AR model using the X-ray structure of bovine rhodopsin as template (PDBid 1F88)	AUTODOCK	[110]
Ligands		
Library of commercially available compounds (ZINC database) with molecular weight between 250 and 350 g/mol, less than 7 rotatable bonds, and a xlogP between 2.5 and 3.5		
Drug design technique(s)	**Computational tool(s)**	**References**
Docking into a human A_1AR model using the X-ray structure of $A_{2A}AR$ as template (PDBid 3EML)	DOCK	[111]
Ligands		
DPCPX, 52 active antagonists, and 1000 decoys		
Drug design technique(s)	**Computational tool(s)**	**References**
Docking into 12 models of A_1AR using the X-ray structure of $A_{2A}AR$ as template (PDBid 3EML)	DOCK, VINA, GOLD	[112]
GPCR: Adenosine A_{2A} receptor ($A_{2A}AR$)		
Ligands		
Library of 545,000 CNS drug-like compounds		
Drug design technique(s)	**Computational tool(s)**	**References**
Docking into a $A_{2A}AR$ model using the X-ray structure of turkey β_1-AR as template (PDBid 2VT4)	GLIDE	[113]
Ligands		
Library of 4,300,000 drug-like compounds		
Drug design technique(s)	**Computational tool(s)**	**References**
Docking into X-ray structure of $A_{2A}AR$ (PDBid 3EML)	ICM	[114]
Ligands		
ZM241385		

(continued)

Table 1
(continued)

Drug design technique(s)	Computational tool(s)	References
Docking into X-ray structure of A_{2A}AR (PDBid 3EML) and into a A_{2A}AR model using X-ray structure of β_2-AR as template (PDBid 2R4R)	GLIDE XP, InducedFit, MOE Tabu search	[115]
Ligands		
Library of commercially available compounds (ZINC database)		
Drug design technique(s)	**Computational tool(s)**	**References**
Docking into four X-ray structures of A_{2A}AR (PDBid 3QAK; PDBid 2YDO; PDBid 2YDV; PDBid 3EML)	DOCK	[116]
Ligands		
Library of commercially available compounds (ZINC database) with molecular weight less than 350 g/mol, less than seven rotatable bonds, and logP lower than 3.5		
Drug design technique(s)	**Computational tool(s)**	**References**
Docking into X-ray structure of A_{2A}AR (PDBid 3EML)	DOCK	[117]
GPCR: α_{2A}-Adrenergic Receptor (α_{2A}-AR)		
Ligands		
Library of WOMBAT 2007.1 compounds		
Drug design technique(s)	**Computational tool(s)**	**References**
Docking into a α_{2A}-AR model using the X-ray structure of human β_2-AR as template (PDBid 2RH1)	GLIDE	[118]
Ligands		
Chlorpromazine, spiperone, spiroxatrine, quinazolines, dopamine, adrenaline, clonidine, dexmedetomidine, BRL-44408, JP-1302, OPC-2836, ARC239, clozapine, WB4101		
Drug design technique(s)	**Computational tool(s)**	**References**
Docking into a α_{2A}-AR model using the X-ray structure of human dopamine D_3 receptor (D_3R) as template (PDBid 3PBL) as template	GLIDE	[119]
GPCR: β_2-Adrenergic receptor (β_2-AR)		
Ligands		
Library of commercially available compounds (ZINC database)		
Drug design technique(s)	**Computational tool(s)**	**References**
Docking into the X-ray structures of β_2-AR (PDBid 2RH1; PDBid 3P0G) and virtual screening	PLANTS, IFP	[120]

(continued)

Table 1
(continued)

Drug design technique(s)	Computational tool(s)	References
Ligands		
Library of commercially available compounds (ZINC database)		
Drug design technique(s)	**Computational tool(s)**	**References**
Docking into the X-ray structure of β_2-AR (PDBid 2RH1)	DOCK	[121]
Ligands		
Library of commercially available compounds (ZINC database)		
Drug design technique(s)	**Computational tool(s)**	**References**
Docking intothe X-ray structure of β_2-AR (PDBid 3SN6)	MolGridCal, AUTODOCK VINA, LibDock, CDOCKER, Discovery Studio 2.5, NAMD	[108]
GPCR: CC motif chemokine receptor 2 (CCR_2)		
Ligands		
Teijin, RS-504393, 2-amino-*N*-(2-(((1*S*,2*R*)-2-((4-(methylthio)benzyl)amino)cyclohexyl)amino)-2-oxoethyl)-5-(trifluoromethyl)benzamide, 2-amino-*N*-(2-(((3*S*,4*R*)-4-((4-(methylthio)benzyl)amino)piperidin-3-yl)amino)-2-oxoethyl)-5-(trifluoromethyl)benzamide		
Drug design technique(s)	**Computational tool(s)**	**References**
Docking into a CCR_2 model using the X-ray structure of CXC chemokine receptor 4 ($CXCR_4$) as template (PDBid 3ODU)	GROMACS, AUTODOCK	[122]
Ligands		
TAK779, Teijin-comp1, JnJ-comp1, Merck-comp55, INCB3344, and BMS-comp22		
Drug design technique(s)	**Computational tool(s)**	**References**
Docking into a CCR_2 model using the X-ray structure of $CXCR_4$ as template (PDBid 3ODU)	AMBER, GLIDE	[123]
GPCR: Corticotropin-releasing factor receptor 1 ($CRFR_1$)		
Ligands		
Dihydropyrrole[2,3-*d*]pyridines		
Drug design technique(s)	**Computational tool(s)**	**References**
Docking into a $CRFR_1$ model using the X-ray structure of glucagon and calcitonin receptors as templates	MacroModel/BatchMin	[124]
GPCR: δ-Opioid receptor (DOR)		
Ligands		
NTB, NTI, NTIR, BNTX, SNC80, SNC67, BW373U86, SIOM, TAN-67, SB219825, SB206848, SUPERFIT, *cis*-(+)-3-methylfentanyl		

(continued)

Table 1 (continued)

Drug design technique(s)	Computational tool(s)	References
Docking into three DOR models using the X-ray structure of bovine rhodopsin as template (PDBid 1F88)	AUTODOCK	[125]
Ligands		
Morphine		
Drug design technique(s)	**Computational tool(s)**	**References**
Docking into a DOR model using the X-ray structure of bovine rhodopsin (PDBid 1F88) and the theoretical model of bovine rhodopsin based on electron microscopy (PDBid 1B0J) as templates	MOE	[126]
GPCR: Histamine H_3 receptor (H_3R)		
Ligands		
Library of compounds derived from ChEMBL database and VU-MedChem fragment library		
Drug design technique(s)	**Computational tool(s)**	**References**
Virtual fragment screening into a H_3R model using the X-ray structure of histamine H_1 receptor (H_1R) as template (PDBid 3RZE)	PLANTS, GOLD	[127]
Ligands		
Library of 418 H_3R antagonists		
Drug design technique(s)	**Computational tool(s)**	**References**
Docking into a H_3R model using the X-ray structure of bivine rhodopsin (PDBid 1HZX) as template	GOLD	[128]
Ligands		
Library of non-imidazole H_3R antagonists		
Drug design technique(s)	**Computational tool(s)**	**References**
Docking into a H_3R model using the X-ray structure of bovine rhodopsin as template (PDBid 1L9H)	GOLD	[129]
GPCR: Metabotropic glutamate receptor type 1 ($mGluR_1$)		
Ligands		
L-Glu, QUIS, Ibo, (1*S*,3*R*)-ACPD, *S*-4CPG, *S*-4C3HPG, *S*-4H3CPG, M4CPG, *S*-3HPG, UPF523		
Drug design technique(s)	**Computational tool(s)**	**References**
Docking into a NH_2-terminal domain of $mGluR_1$ model using the X-ray structures of leucine/isoleucine/valine-binding protein	SYBYL	[130]

(continued)

Table 1
(continued)

(LIVBP) (PDBid 2LIV) and of leucine-binding protein (LBP) (PDBid 2LBP) as templates		
GPCR: M_1 muscarinic AcetylCholine Receptor (M_1 mAChR)		
Ligands		
L005771, L005772, L005773, L006454, L014151, pilocarpine, NCC11-1314, NCC11-1585, NCC11-1607, nebracetam, oxotremorine, oxotremorine-M, quinuclidine, RU47213, SDZ ENS 163, pilofrin, gliatilin (TN), sabcomeline, VU0357017, xanomeline, pentylthio-TZTP		
Drug design technique(s)	**Computational tool(s)**	**References**
Docking into a M_1 mAChR model using the X-ray structure of M_3 mAChR as template (PDBid 4DAJ)	GLIDE	[131]
Ligands		
Flavonoids		
Drug design technique(s)	**Computational tool(s)**	**References**
Docking into a M_1 mAChR model using the X-ray structure of M_3 mAChR as template (PDBid 4DAJ)	GLIDE, AUTODOCK	[132]
GPCR: M_2 muscarinic AcetylCholine Receptor (M_2 mAChR)		
Ligands		
Library of lead-like compounds and fragments derived from the ZINC database		
Drug design technique(s)	**Computational tool(s)**	**References**
Docking into the X-ray structure of M_2 mAChR (PDBid 3UON) and virtual screening	DOCK	[133]
GPCR: M_3 muscarinic AcetylCholine Receptor (M_3 mAChR)		
Ligands		
Library of lead-like compounds and fragments derived from the ZINC database		
Drug design technique(s)	**Computational tool(s)**	**References**
Docking into the X-ray structure of M_3 mAChR (PDBid 4DAJ) and virtual screening	DOCK	[133]
GPCR: 5-HydroxyTryptamine 2A Receptor (5-HT_{2A}R)		
Ligands		
Library of 5-HT_{2A}R agonists (serotonin, DOI, mescaline, LSD, 5-MeO-alpha-ET, psilocin, bufotenine, dimethyltryptamine) and antagonists (nefazodone, aripiprazole, haloperidol, cyproheptadine, trazodone, clozapine, ketanserin, spiperone, risperidone)		

(continued)

Table 1 (continued)

Drug design technique(s)	Computational tool(s)	References
Docking into human 5-HT$_{2A}$R model using the X-ray structure of β_2-AR (PDBid 3SN6) as template and virtual screening	GLIDE	[134]
Ligands		
Serotonin, dopamine, DOI, LSD, haloperidol, ketanserin, clozapine, risperidone		
Drug design technique(s)	**Computational tool(s)**	**References**
Docking into human 5-HT$_{2A}$R model using the X-ray structure of β_2-AR (PDBid 2RH1) as template	AUTODOCK	[135]
Ligands		
(Aminoalkyl)benzo and heterocycloalkanones		
Drug design technique(s)	**Computational tool(s)**	**References**
Docking into the transmembrane α-helices bundle of 5-HT$_{2A}$R model using the X-ray structure of bovine rhodopsin (PDBid 1F88) as template	AMBER	[136]
GPCR: 5-HydroxyTryptamine 2C Receptor (5-HT$_{2C}$R)		
Ligands		
(Aminoalkyl)benzo and heterocycloalkanones		
Drug design technique(s)	**Computational tool(s)**	**References**
Docking into the transmembrane α-helices bundle of 5-HT$_{2C}$R model using the X-ray structure of bovine rhodopsin (PDBid 1F88) as template	AMBER	[136]
Ligands		
11-Chloro-2,3,4,5-tetrahydro-1*H*-[1, 4]diazepino[1,7-*a*]indole, 8,9-dichloro-2,3,4,4*a*-tetrahydro-1*H*-pyrazino[1,2-*a*]quinoxalin-5(6*H*)-one, (*S*)-1-(2-aminopropyl)-7-fluoro-1*H*-indazol-6-ol, (*R*)-1-(7-(2-chlorophenyl)-5-fluoro-2,3-dihydrobenzofuran-2-yl)-*N*-methylmethanamine, *N*-(3-(4-methylimidazolidin-1-yl)phenyl)-5,6-dihydrobenzo[*h*]quinazolin-4-amine, *N*-(4-methoxy-3-(4-methylpiperazin-1-yl)phenyl)-1,2-dihydro-3*H*-benzo[*e*]indole-3-carboxamide, 1-(3,5-difluorophenyl)-3-(4-methoxy-3-(2-(piperidin-1-yl)ethoxy)phenyl)imidazolidin-2-one, *N*-(3-(2-((3-(piperazin-1-yl)pyrazin-2-yl)oxy)ethoxy)benzyl)propan-2-amine		
Drug design technique(s)	**Computational tool(s)**	**References**
Docking into 5-HT$_{2C}$R model using the X-ray structure of β_2-AR (PDBid 2RH1) as template	FlexX	[137]

Table 2
Ligand-based drug design techniques for the modulation of potential GPCR-derived therapeutic targets of AD

GPCR: Adenosine A_1 receptor (A_1AR)		
Ligands		
N^6-Substituted adenosines, 8-substituted xanthines		
Drug design technique(s)	**Computational tool(s)**	**References**
CoMFA	CHEM-X	[155]
GPCR: Adenosine A_{2A} receptor ($A_{2A}AR$)		
Ligands		
2-(Furan-2-yl)-[1, 2, 4]triazolo[1,5-*f*]pyrimidin-5-amines, 2-(furan-2-yl)-[1, 2, 4]triazolo[1,5-*a*] pyrazin-8-amine, and 2-(furan-2-yl)-[1, 2, 4]triazolo[1,5-*a*][1, 3, 5]triazin-7-amines		
Drug design technique(s)	**Computational tool(s)**	**References**
HQSAR	SYBYL	[154]
Ligands		
Thieno[3,2-*d*]pyrimidine-4-methanones, 4-arylthieno[3,2-*d*]pyrimidines, pyrazolo[3,4-*d*]pyrimidines, pyrrolo[2,3-*d*]pyrimidines, 6-arylpurines, pyrimidine-4-carboxamides, 7-aryltriazolo[4,5-*d*] pyrimidines		
Drug design technique(s)	**Computational tool(s)**	**References**
CoMFA, CoMSIA	SYBYL	[156]
Ligands		
Pyrimidines		
Drug design technique(s)	**Computational tool(s)**	**References**
CoMFA	SYBYL	[157]
Ligands		
2-Substituted adenosines, 2-substituted adenosine-5'uronamides, 2-substituted adenosine-5′ *N*-ethyluronamides		
Drug design technique(s)	**Computational tool(s)**	**References**
CoMFA	SYBYL	[158]
Ligands		
Triazolopyrimidines		
Drug design technique(s)	**Computational tool(s)**	**References**
CoMFA	SYBYL	[159]
Ligands		
2-Alkyloxy-, 2-aryloxy-, and 2-aralkyloxy-adenosines		

(continued)

Table 2 (continued)

Drug design technique(s)	Computational tool(s)	References
CoMFA	CHEM-X	[160]
GPCR: β_2-Adrenergic receptors (β_2-AR)		
Ligands		
Library of 94 β_2-AR agonists and antagonists		
Drug design technique(s)	**Computational tool(s)**	**References**
CoMFA, CoMSIA	SYBYL	[161]
Ligands		
Tryptamines		
Drug design technique(s)	**Computational tool(s)**	**References**
CoMFA	SYBYL	[162]
Ligands		
Fenoterol derivatives		
Drug design technique(s)	**Computational tool(s)**	**References**
CoMFA	SYBYL	[163, 164, 165]
GPCR: CXC motif chemokine receptor 2 ($CXCR_2$)		
Ligands		
N,N'-Diarylsquaramides, *N,N'*-diarylureas, diaminocyclobutenediones		
Drug design technique(s)	**Computational tool(s)**	**References**
CoMFA, CoMSIA	SYBYL	[166]
GPCR: δ-Opioid receptor (DOR)		
Ligands		
SNC80 analogs		
Drug design technique(s)	**Computational tool(s)**	**References**
CoMFA	SYBYL	[167]
GPCR: Histamine H_3 receptor (H_3R)		
Ligands		
Quinolines		
Drug design technique(s)	**Computational tool(s)**	**References**
CoMFA, CoMSIA	SYBYL	[168]
Ligands		

(continued)

Table 2
(continued)

4-(3-(Phenoxy)propyl)-1*H*-imidazoles, 4-aminoquinolines, 3-(1*H*-imidazol-4-yl)propanol derivatives, 1-(4-(phenoxymethyl)benzyl)piperidines		
Drug design technique(s)	**Computational tool(s)**	**References**
CoMFA and CoMSIA combined with the implementation of charged partial surface area and VolSurf descriptors, among others	SYBYL	[169]
Ligands		
Imidazoles, thiazoles		
Drug design technique(s)	**Computational tool(s)**	**References**
CoMFA, CoMSIA	SYBYL	[170]
GPCR: Metabotropic glutamate receptor type 1 ($mGluR_1$)		
Ligands		
Triazafluorenones		
Drug design technique(s)	**Computational tool(s)**	**References**
CoMFA	$CERIUS^2$	[171]
Ligands		
Quinolines		
Drug design technique(s)	**Computational tool(s)**	**References**
CoMFA	SYBYL	[172]
GPCR: Metabotropic glutamate receptor type 2 ($mGluR_2$)		
Ligands		
Triazolopyridines		
Drug design technique(s)	**Computational tool(s)**	**References**
CoMFA	SYBYL, PIPELINE PILOT	[173]
GPCR: Metabotropic glutamate Receptor type 5 ($mGluR_5$)		
Ligands		
N-(1,3-Diphenyl-1*H*-pyrazol-5-yl)benzamides		
Drug design technique(s)	**Computational tool(s)**	**References**
HQSAR	SYBYL	[174]
Ligands		
Benzodiazepines		
Drug design technique(s)	**Computational tool(s)**	**References**
CoMFA	SYBYL	[175]
Ligands		

(continued)

Table 2
(continued)

Aryl ethers		
Drug design technique(s)	**Computational tool(s)**	**References**
CoMFA	SYBYL	[176]
GPCR: M_2 muscarinic AcetylCholine Receptor (M_2 mAChR)		
Ligands		
Bisquaternary caracurine V derivatives		
Drug design technique(s)	**Computational tool(s)**	**References**
CoMSIA	SYBYL	[177]
Ligands		
Piperidinylpiperidines		
Drug design technique(s)	**Computational tool(s)**	**References**
CoMFA	SYBYL	[178]
GPCR: 5-HydroxyTryptamine 2A Receptor (5-HT_{2A}R)		
Ligands		
Tryptamines		
Drug design technique(s)	**Computational tool(s)**	**References**
HQSAR	SYBYL	[153]
Ligands		
Indoles, methoxybenzenes, quinazolinediones		
Drug design technique(s)	**Computational tool(s)**	**References**
CoMFA, CoMSIA	SYBYL	[179]
Ligands		
1,4-Disubstituted aromatic piperazines		
Drug design technique(s)	**Computational tool(s)**	**References**
CoMFA, CoMSIA	SYBYL	[180]
Ligands		
3-(Aminomethyl)tetralones, ketanserin analogs, 2-aminoethyl benzocyclanones, 2-(2-piperidinoethyl) benzocycloalkanones		
Drug design technique(s)	**Computational tool(s)**	**References**
CoMFA	SYBYL	[181]
Ligands		
Phenylalkylamines		

(continued)

Table 2
(continued)

Drug design technique(s)	Computational tool(s)	References
CoMFA	SYBYL	[182]
GPCR: 5-HydroxyTryptamine 2C Receptor (5-$HT_{2C}R$)		
Ligands		
1-(3-Pyridylcarbamoyl)indolines		
Drug design technique(s)	**Computational tool(s)**	**References**
CoMFA	SYBYL	[183]
GPCR: 5-HydroxyTryptamine 4 Receptor (5-HT_4R)		
Ligands		
Benzimidazoles		
Drug design technique(s)	**Computational tool(s)**	**References**
CoMFA	SYBYL	[184, 185]
Ligands		
Indole carbazimidamides, 5-hydroxytryptamine, 4-amino-5-chloro-2-methoxybenzoic acid esters, 4-amino-*N*-[2-(1-aminocycloalkan-1-yl)ethyl]-5-chloro-2-methoxybenzamides, (±)-1-hydroxy-3-aminopyrrolidones, 5-benzyloxytryptamines, 5-methoxytryptamines		
Drug design technique(s)	**Computational tool(s)**	**References**
CoMFA	SYBYL	[186]
Ligands		
Benzamides		
Drug design technique(s)	**Computational tool(s)**	**References**
CoMFA	SYBYL	[187]
GPCR: 5-HydroxyTryptamine 6 Receptor (5-HT_6R)		
Ligands		
Arylsulfonamides		
Drug design technique(s)	**Computational tool(s)**	**References**
HQSAR	HQSAR software	[152]
Ligands		
N_1-Arylsulfonylindoles		
Drug design technique(s)	**Computational tool(s)**	**References**
CoMFA, CoMSIA	SYBYL	[188]

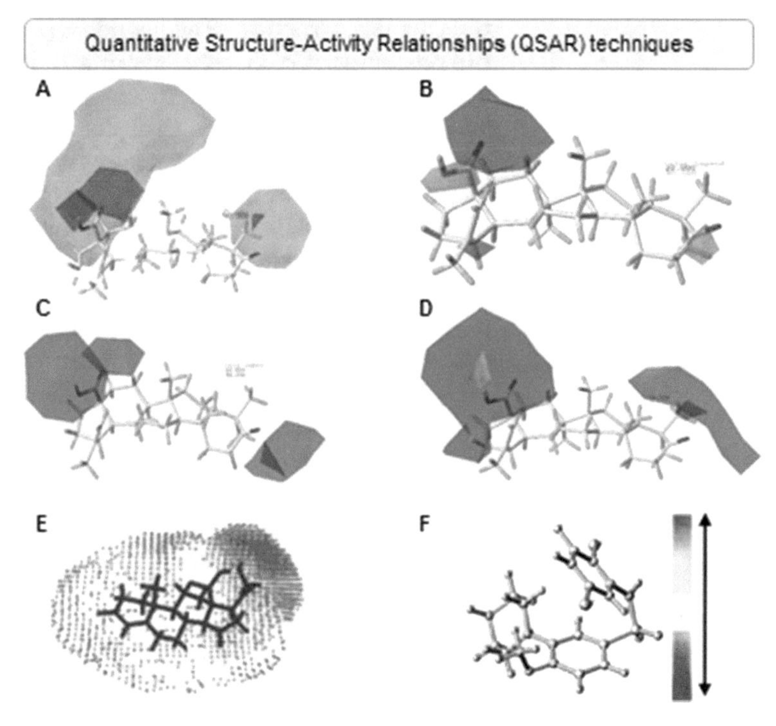

Fig. 3 Representative QSAR-based methodologies for drug design of modulators of potential GPCR-derived therapeutic targets of AD. Color-coded contour maps (**a**) and (**b**) Comparative Molecular Field Analysis (CoMFA) [72], color-coded contour maps (**c**) and (**d**) Comparative Molecular Similarity Index Analysis (CoMSIA) [72], color-coded contour map (**e**) Self-Organizing Molecular Field Analysis (SOMFA) [149], and color-coded contour map (**f**) Hologram Quantitative Structure-Activity Relationship (HQSAR) [152]

compounds with drug-like properties, including Quantitative Structure-Activity Relationship (QSAR) techniques such as Comparative Molecular Field Analysis (CoMFA) (Fig. 3a, b), Comparative Molecular Similarity Index Analysis (CoMSIA) (Fig. 3c, d), Self-Organizing Molecular Field Analysis (SOMFA) (Fig. 3e), and Hologram Quantitative Structure-Activity Relationships (HQSAR) (Fig. 3f).

The investigation of QSARs has been a ligand-based drug design approach of utmost importance for computational chemistry and has opened new perspectives on the drug discovery process. This useful computational methodology searches for mathematical models that explore the contribution of specific functional groups and moieties of the ligands (physicochemical parameters and/or theoretical molecular descriptors) to the experimental determined biological/pharmacological data for congeneric or non-congeneric series of chemical compounds (reviewed in [138, 139, 140]). The development of a robust and trustworthy QSAR model should take into account some considerations, particularly the guarantee that the chemical structure of all ligands is properly drawn or imported, the reliability of biological/pharmacological activity data, and the use of validated software to calculate the descriptor values. In addition, the biological/pharmacological activity data should possess a normal distribution pattern (reviewed in [138, 139, 140]). The main purposes of QSAR are focused on explaining the subtle differences in biological/pharmacological data, at the molecular level, of a statistical population of drug candidates (training set) through the use of appropriate and relevant molecular descriptors (e.g., topological descriptors, electronic descriptors, geometrical descriptors, constitutional descriptors, etc.) (reviewed in [138, 139, 140]). The construction of mathematical QSAR models usually employs a wide variety of statistical methods for linear modeling, such as multiple (or multivariate) linear regression (MLR) [141], partial least squares (PLS) regression [142], and linear discriminant analysis (LDA) [143], and nonlinear modeling, such as artificial neural networks (ANN) [144] or support vector machines (SVM) [145] to derive a robust mathematical correlation that explains the dependence of particular descriptor variables (independent variables) to the biological/pharmacological activity of a set of ligands (dependent variables). The choice of an appropriate statistical method is crucial especially when a large number of descriptors are calculated in order to neglect the least relevant or redundant descriptors and to select the other ones with the highest mutual intercorrelation with the activity data. The resulting QSAR model is subjected to several validation tests to verify the reliability of the correlation models. After its construction, a QSAR model is usually corroborated by applying multiple strategies of QSAR model validation, in particular the internal validation or cross-validation and the external validation, which provide information about its stability and predictivity (reviewed in [138, 139, 140]). Regarding internal validation or cross-validation, the training set is modified by deleting one (leave-one-out cross-validation, LOO) or more (leave-some-out cross-validation, LSO; leave-many-out cross-validation, LMO) ligands from the set. The QSAR model is reconstructed based on the remaining ligands using the combination of descriptors previously determined, and the biological/pharmacological activity of the eliminated ligand(s) is calculated from the developed QSAR equation. Subsequently, the same procedure is performed until all or a definite portion of the ligands of the training set have been eliminated once and the predictive activity values of the compounds are used for the calculation of several internal validation parameters, in particular the predictive correlation coefficient r^2_{cv} (reviewed in [138, 139, 140]). The external validation consists in the prediction of activity of a group of chemical compounds that are not included in the training set (i.e., test set) and the same parameters are used in the construction of QSAR model. The external predictive ability of the generated QSAR

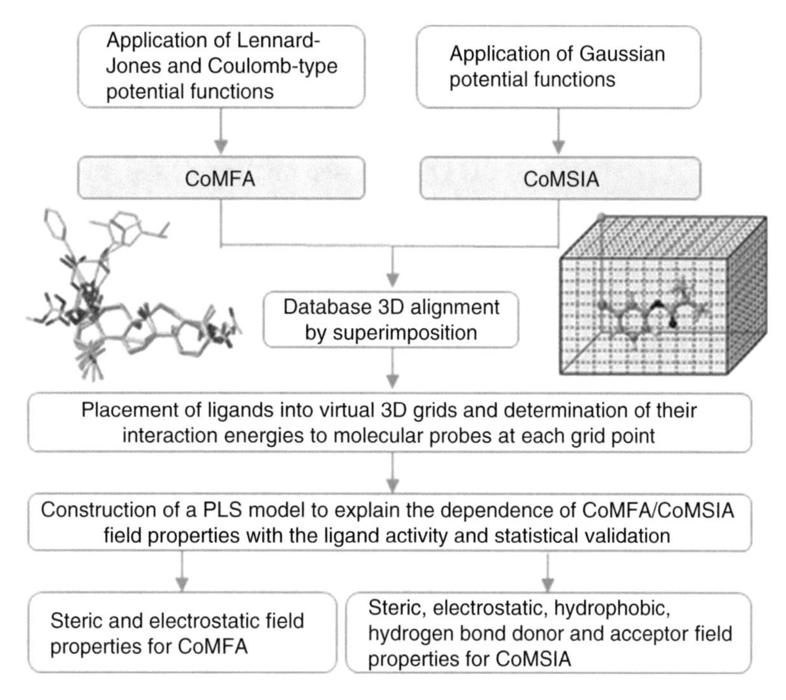

Fig. 4 General procedure for CoMFA and CoMSIA methodologies (Representative images were extracted from [72])

model is determined using the predictive correlation coefficient r^2_{pred} (reviewed in [138, 139, 140]).

The CoMFA and CoMSIA methodologies have been important ligand-based tools for the design and development of more potent drug candidates targeting GPCRs (Fig. 4). The basic concept of CoMFA methodology consists in finding differences in biological/pharmacological activity of a data set of ligands correlated to the differences in their 3D shape and the magnitude of molecular field properties. Particularly, CoMFA is restrained to steric (Lennard-Jones potential functions) and electrostatic components (Coulomb potential functions) for field calculation, and therefore these descriptors only take into account the ability of ligands to establish intermolecular interactions with a putative target protein (enthalpic contributions) [72, 146, 147]. A similar QSAR-based methodology, CoMSIA, was conceived based on arbitrary descriptors, so-called similarity indices. Unlike CoMFA, CoMSIA applies a smoother potential based on Gaussian-type distance-dependent functions, allowing the calculation of various similarity indices, in particular steric, electrostatic, hydrophobic, hydrogen-bond acceptor and donor properties, that were created to cover more broadly than the steric and electrostatic fields calculated by CoMFA, the possible major contributions for the binding free energy of ligands to a putative therapeutic target [148]. The 3D alignment of the chemical structures of ligands is required to perform both methodologies. An optimal structure alignment of a data set of molecules can be described as the alignment that reaches the maximum superimposition of steric, electrostatic, hydrophobic, hydrogen-bond acceptor and hydrogen-bond donor parameters. In 3D QSAR-based methodologies, the 3D alignment is a crucial step and should reveal the superimposition of molecular conformations that a data set of ligands adopt when interacting with a specific therapeutic target. Each member of the training set is aligned to a template molecule which shares a common molecular substructure, and the members of the aligned training sets are placed inside virtual 3D grid boxes with a default grid spacing in all Cartesian directions [72, 146, 147, 148]. Subsequently, the interaction energies are calculated between the ligands and molecular fragments (molecular probes) at each grid point. Using an appropriate method for regression analysis, usually by PLS, the 3D QSAR model is constructed to describe the variation of biological/pharmacological activity with the variation of CoMFA/CoMSIA interaction fields, and the predictive ability of 3D QSAR model is verified by cross-validation and prediction of activity of test set. The resulting QSAR model is usually interpreted in a graphic form as color-coded contour maps, which exhibit specific volumes of space where the magnitudes of the steric, electrostatic, hydrophobic, hydrogen-bond acceptor and hydrogen-bond donor parameters are positively or negatively correlated with the biological/pharmacological activity [72, 146, 147, 148]. This type of graphical representation can be assumed as a model of the binding site in which a training set of ligands is supposed to interact. While the colored contour maps relative to steric and electrostatic field contributions of CoMFA display the regions of space where the aligned ligands can favorably or unfavorably bind to a putative therapeutic target, the colored contour maps generated by CoMSIA-field contributions highlight the regions of the aligned molecules that can favor the presence of a moiety with a given physicochemical property [148]. From the information provided by these graphical representations of CoMFA/CoMSIA models, the activity of novel synthesized drug candidates can be predicted.

A grid-based 3D QSAR technique known as SOMFA (Fig. 5) was originally developed by *Robinson* et al. to estimate the binding affinity of steroid compounds with corticosteroid-binding globulin [149]. This methodology shares common characteristics with CoMFA, in which a grid-based approach is employed, and with molecular similarity methods, in which the intrinsic molecular properties, such as molecular shape and electrostatic potential, are used to construct SOMFA-based QSAR models [149, 150]. The first step in the SOMFA procedure consists in the determination of mean centered activities for a training set of known ligands. The mean centered activity for each ligand of the training set, which consists in the subtraction of mean activity of the training set from

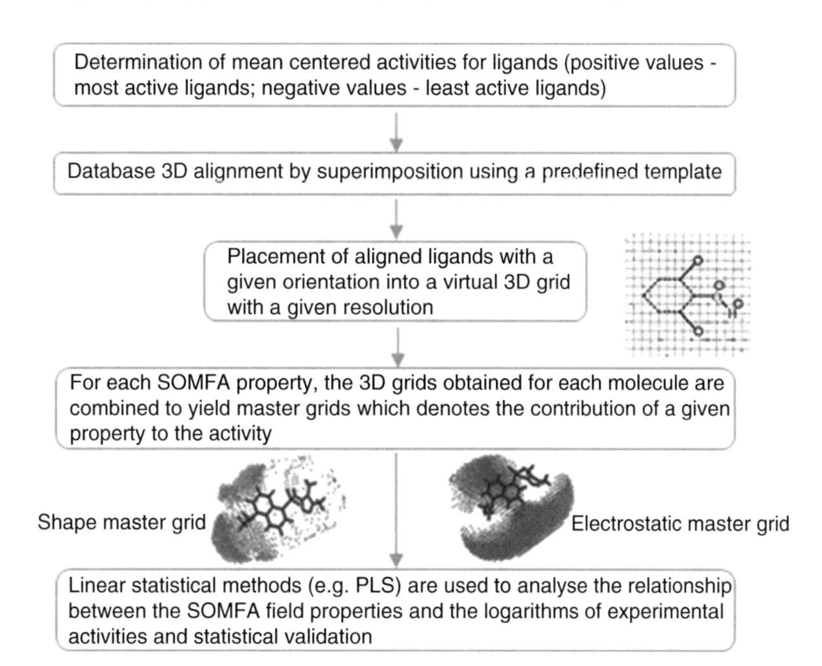

Fig. 5 General procedure for SOMFA methodology (Representative images were extracted from [149])

each ligand's activity, is calculated in order to obtain a scale of activity where the most and the least active ligands present positive and negative values, respectively. In general, the mean centered activity represents a form of descriptor filtering which denotes the structural features that discriminate high-activity from low-activity ligands [149]. Subsequently, the ligands are structurally aligned by superimposition using molecular alignment tools such as principal component analysis (PCA) method and placed on a given orientation into 3D grids with a given resolution, as in other QSAR methodologies, with values at each grid point representing the shape and electrostatic potential. Linear regression models are constructed to describe the dependence of a given SOMFA molecular property with the experimental training set activities represented on logarithmic scale. The calculation of correlation coefficient indicates the potential importance of a given property. The final result is a grid-based map representing each molecular property that can support the molecular design of novel compounds with improved activity (e.g., binding affinity, etc.) [149, 150]. In general, a SOMFA grid can be used to calculate any molecular property. For each molecular property, particularly for molecular shape and electrostatic properties, the grids for each ligand in the training set are combined to yield property master grids that highlight the regions of ligands where steric and electrostatic parameters might be expected to be correlated with the activity (e.g., binding affinity values, etc.) [149]. Highly active ligands sharing similar structural features superimpose these features at the same point on a master grid. The grid values for highly active ligands strengthen each other, resulting in a final master grid, in which the positive values are associated with common characteristics to these compounds. In a similar way, the least active compounds share common features that lead to a master grid of negative grid values. Since the grid values are assigned based on mean centered activity, moderately active ligands will have small effect on the final grid values. The quality of the model produced in SOMFA increases rapidly with the size of the training set data, and, for that reason, small data sets will not produce the overlapping features for SOMFA, contributing for a lower quality of correlation. The development of SOMFA has been revealed to be advantageous into the search for the best 3D QSAR model due to its speed and simplicity. Additionally, for the construction of a SOMFA model, the structural similarity between the suitably aligned ligands of a training set is not mandatory [149].

HQSAR has emerged as a novel 2D and fragment-based QSAR technique which employs fragment fingerprints as predictive variables of biological/pharmacological activity. The methodology employed in HQSAR procedure (Fig. 6) involves several steps, including the generation of structural fragments for each ligand in the training set and the encoding of these fragments in holograms [151]. Initially, the input molecules are broken into all possible structural fragments of atoms (e.g., linear, branched, cyclic, and

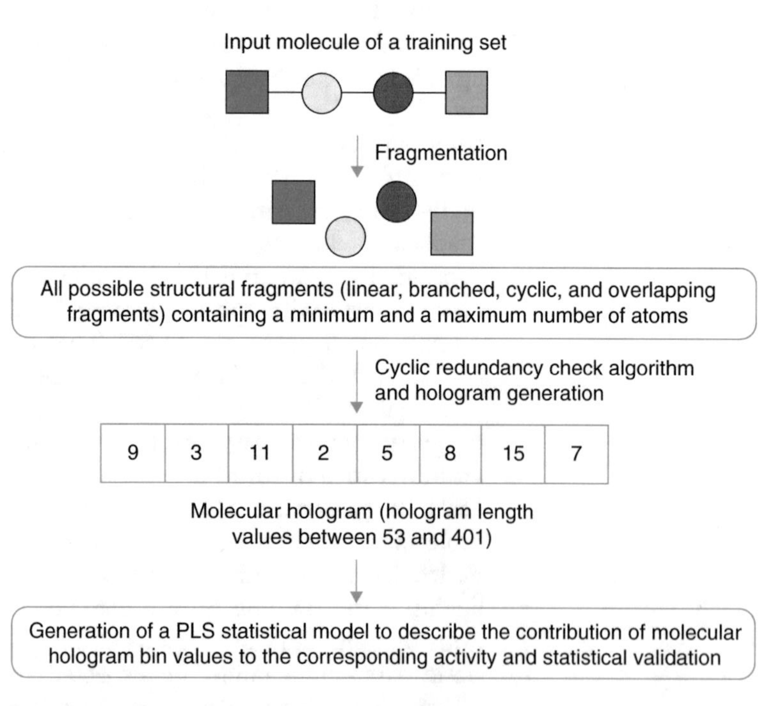

Fig. 6 General procedure for HQSAR methodology

overlapping fragments, etc.) connected in size between a minimum and a maximum number of atoms as defined by hologram length parameters [152, 153, 154]. Afterwards, each unique fragment in the data set is assigned a large positive integer by means of a cyclic redundancy check (CRC) algorithm. Each of these integers corresponds to a square array of integers in a specified hologram length L. The cell values (bin occupancies) are incremented according to the produced fragments. Therefore, all the generated fragments are hashed into boxes (array bins), composing a matrix in the range of 1 to L. The matrix now constitutes a molecular hologram, and the bin occupancies are the descriptor parameters [152, 153, 154]. These descriptors provide some information about chemical and topological features of ligands under study. The use of hashing significantly diminishes the size of the molecular hologram but induces a phenomenon of fragment collision. Upon production of molecular fragments, the identical ones are hashed to the same bin, and the respective bin occupancy is increased. With the objective of avoiding the occurrence of identical or similar fragment collisions between unique molecular fragments, the values of hologram length are often selected to be prime numbers (default hologram length values which are a set of 12 prime numbers ranging from 53 to 401) [152, 153, 154]. The development of HQSAR model is strongly correlated to a number of different parameters concerning hologram generation, in particular the fragment size, the hologram length, and the fragment distinction. Diverse patterns of following fragment distinction parameters, including atom types (A), bond types (B), connectivity (C), hydrogen atoms (H), chirality (Ch), and donors and acceptors (DA), are used for the generation of molecular fragments and for the construction of HQSAR models [152, 153, 154]. Once an optimal model is identified, linear statistical methods such as PLS yield a mathematical equation that explains the dependence of molecular hologram bin values to the corresponding biological/pharmacological activity of each ligand in the training set. The resulting HQSAR models can be graphically displayed as color-coded contribution maps in which the color of each molecular fragment exhibits the contribution (favorable contribution, intermediate contribution, or unfavorable contribution) of an atom or a small number of atoms to the overall activity of ligands under study [152, 153, 154]. As in other QSAR methodologies, the derived HQSAR models are validated, and the biological/pharmacological activities of external test sets are predicted from the generated models. The application of HQSAR as an alternative to the existing QSAR methodologies exhibited a plethora of potential advantages. It avoids the selection and calculation of the physicochemical descriptors by traditional QSAR, and no explicit 3D information for the ligands (e.g., determination of the 3D structure, putative binding conformation, and molecular alignment) is required for the generation of molecular holograms.

Additionally, HQSAR analyses could be easily and rapidly performed for both small and large data sets that are not analyzable by traditional QSAR techniques [152].

3.3 Pharmacophore-Based Drug Design

Pharmacophore modeling has been demonstrated to be a remarkably useful in silico approach for the discovery of potentially bioactive molecules acting on several therapeutic targets [189, 190]. A pharmacophore does not represent a real molecule or an association of functional groups, but it represents an ensemble of steric and electronic determining features that assure an optimal interaction toward a relevant biological/pharmacological target and trigger its biological/pharmacological activity. Therefore, a pharmacophore can be described as the highest common denominator shared by a set of active ligands with similar biological/pharmacological activity and which may interact to the same site of a protein (reviewed in [73]). A pharmacophore model can be developed either in the absence of therapeutic target structure (ligand-based pharmacophore modeling) or based on the 3D structure of a therapeutic target (structure-based pharmacophore modeling) (Fig. 7). The construction of receptor-based pharmacophore models implies the analysis of the pharmacophoric features (hydrogen-bond acceptors and donors, hydrophobic groups, aromatic rings, etc.) in the active site and their spatial relationships which are important for ligand binding (reviewed in [73]). Regarding ligand-based pharmacophore

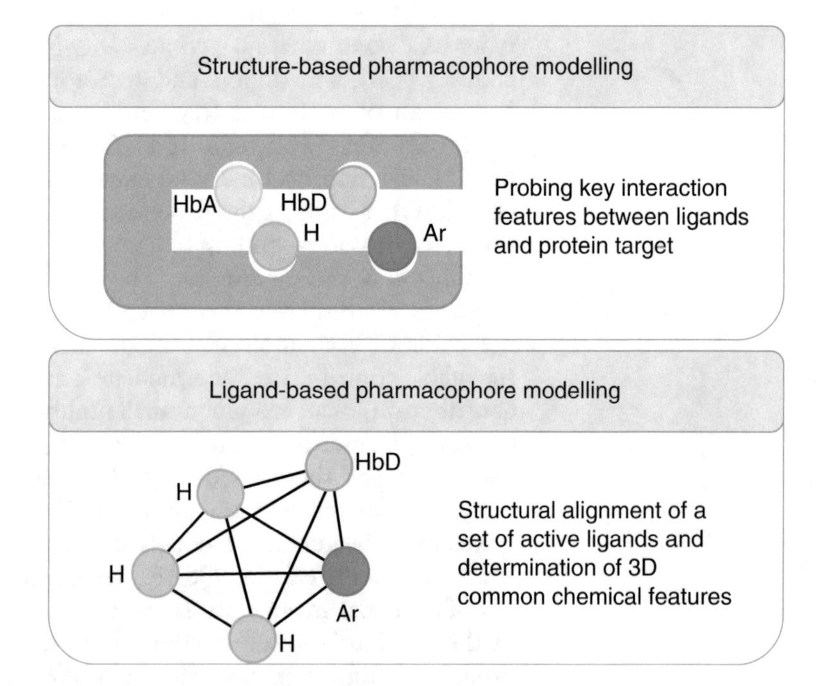

Fig. 7 Representative schemes of structure-based and ligand-based pharmacophore modeling techniques

modeling, the construction of a pharmacophore model involves initially the generation of a conformational space for each ligand of training set to represent their conformational flexibility. The major goal of conformation generation relies on the identification of bioactive conformation(s) of a set of ligands from conformational ensembles in the lowest amount of computational time. Various software tools and algorithms used for conformation generation possess the ability to calculate different conformational geometries containing the bioactive conformation and other similar geometries (reviewed in [73]). A suitable computational tool for conformational search needs to generate all conformational geometries that ligands adopt when they interact with protein targets, to select a short list of low-energy conformational geometries in order to avoid the excess of mass storage capacity and to calculate the conformational geometries in a lower computational time. Subsequently, the multiple ligands belonging to training set are superimposed, and the common 3D structural features crucial for biological/pharmacological activity are determined (reviewed in [73]). Currently, several computational functionalities for generation of pharmacophore models have been developed, including Pharmacophore Alignment and Scoring Engine (PHASE) [191], Activity Prediction Expert System-3D (Apex-3D) [192], MOLMOD [193], System Level Automation Tool for Engineers (SLATE) [194], LigandScout [195], distributed computing (DistComp) [196], SYBYL [197], CATALYST [198], discrete surface charge optimization (DISCO) [198], genetic algorithm for structure and phase production (GASP) [198], and molecular operating environment (MOE) [199], among others. Table 3 reports the most relevant examples of applicability of these software packages for the study of the most critical molecular and electronic features of ligand databases for the modulation of GPCRs with therapeutic potential for AD.

Once a pharmacophore model is created by either ligand-based or structure-based manner, it can be used as a query to perform a virtual screening of 3D chemical databases in the search for new therapeutic strategies for AD based on modulation of GPCRs (reviewed in [73]). In the pharmacophore-based virtual screening procedure, a pharmacophore hypothesis is considered as a template for the identification of hit ligands that present similar chemical features to those of the pharmacophoric template. Apart from the applicability of pharmacophore modeling for virtual screening, de novo drug design approaches have been explored specifically for the design of drug candidates with novel structures which cover the chemical features of a given pharmacophore hypothesis. The software programs of pharmacophore-based de novo drug design usually use as input a set of detached molecular fragments consistent with the pharmacophore hypothesis, and the pharmacophoric fragments are connected by using appropriate linkers (reviewed in [73]).

Table 3
Pharmacophore-based drug design approaches for the modulation of potential GPCR-derived therapeutic targets of AD

GPCR: Adenosine A_{2A} receptor ($A_{2A}AR$)		
Ligands		
1,2,4-Triazolo[5,1-*i*]purines, 2-*N*-butyl-9-methyl-8-[1–3]triazol-2-yl-9*H*-purin-6-ylamines, pyrazolo [4,3-*e*]-1,2,4-triazolo[1,5-*c*]pyrimidines, 2-amino-6-furan-2-yl-4-substituted nicotinonitriles, 4′-aza-carbocyclic nucleosides, 5,6-dihydro-(9*H*)-pyrazolo[3,4-*c*]-1,2,4-triazolo[4,3-*a*]pyridines, *N*-[6-amino-2-(heteroaryl)pyrimidin-4-yl]acetamides, 4-acetylamino-2-(3,5-dimethylpyrazol-1-yl)-6-pyridylpyrimidines		
Drug design technique(s)	**Computational tool(s)**	**References**
Pharmacophore modeling combined with QSAR	PHASE	[200]
Ligands		
Pyrazolo[4,3-*e*]-1,2,4-triazolo[1,5-*c*]pyrimidines, triazolopyridines, 4-amido-2-aryl-1,2,4-triazolo[4,3-*a*]quinoxalin-1-ones, 2-amino-5-benzoyl-4-(2- furyl)thiazoles, N^2-substituted pyrazolo[3,4-*d*] pyrimidines, 2-(benzimidazol-2-yl)quinoxalines, 5-amino-2-phenyl [1–3]triazolo[1,2-*a*] [1, 2, 4] benzotriazin-1-ones, 1,3-dipropyl-8-(1-heteroarylmethyl-1*H*-pyrazol-4-yl)-xanthines, 9-alkylpurines, pyrido[2,1-*f*]purine-2,4-diones, 1,3-dialkyl-8-*N*-substituted benzyloxycarbonylamino-9-deazaxanthines, 7-aryltriazolo[4,5-*d*]pyrimidines, 7-imino-2-thioxo-3,7-dihydro-2*H*-thiazolo [4,5-*d*] pyrimidines, 2-amino-6-furan-2-yl-4-substituted nicotinonitriles, 2-aminoimidazopyridines, 8-(furan-2-yl)-3-substituted thiazolo[5,4-*e*][1, 2, 4]triazolo-[1,5-*c*]pyrimidine-2(3*H*)-thiones, 2,6-diaryl-4-acylaminopyrimidines, 1,2,4-triazolo[1,5-*c*]pyrimidines, 1,2,4-triazolo[5,1-*i*]purines, *N*-1 monosubstituted 8-(pyrazol-4-yl)xanthenes, 1,3-dialkyl-8-(hetero)aryl-9-OH-9-deazaxanthines, pyrimidine-4-carboxamides, 4-acetylamino-2-(3,5-dimethylpyrazol-1-yl)-6-pyridylpyrimidines		
Drug design technique(s)	**Computational tool(s)**	**References**
Pharmacophore modeling combined with QSAR	PHASE	[201]
Ligands		
7-Substituted 5-amino-2-(2-furyl)pyrazolo[4,3-*e*]-1,2,4-triazolo[1,5-*c*]pyrimidines		
Drug design technique(s)	**Computational tool(s)**	**References**
Pharmacophore modeling	CATALYST	[202]
Ligands		
2,6-Diaryl-4-phenacylaminopyrimidines, 2-amino-*N*-pyrimidin-4-ylacetamides, 2-amino-*N*-pyrimidin-4-yl acetamides, *N*-pyrimidinyl-2-phenoxyacetamides, 4-acetylamino-2-(3,5-dimethylpyrazol-1-yl)-6-pyridylpyrimidines, *N*-[6-amino-2-(heteroaryl)pyrimidin-4-yl]acetamides, pyrazolo[4,3-*e*][1, 2, 4] triazolo[1,5-*c*]pyrimidin-5-amine, pyrazolo[4,3-*e*]-1,2,4-triazolo[1,5-*c*]pyrimidines, 6-(furanyl)-9*H*-purin-2-amines, 2-(2-furanyl)-7-phenyl[1, 2, 4]triazolo[1,5-*c*]pyrimidin-5-amines, 3*H*-[1, 2, 4]-triazolo[5,1-*i*]purin-5-amines, 1,2,4-triazolo[1,5-*c*]pyrimidines, biaryl, heteroaryl, and heterocyclic derivatives of SCH 58261		
Drug design technique(s)	**Computational tool(s)**	**References**
Pharmacophore modeling combined with QSAR based on GFA joined with *k*NN	CATALYST	[203]
GPCR: α_{2A}-Adrenergic receptor (α_{2A}-AR)		

(continued)

Table 3 (continued)

Drug design technique(s)	Computational tool(s)	References
Ligands		
Catecholamines, imidazolines, guanidines, structures possessing distinct scaffolds (rilmenidine, talipexole, xylazyne)		
Drug design technique(s)	**Computational tool(s)**	**References**
Pharmacophore modeling combined with CoMFA	DISCO, SYBYL	[204]
GPCR: CC motif chemokine receptor 2 (CCR_2)		
Ligands		
R-3-amino-pyrrolidines		
Drug design technique(s)	**Computational tool(s)**	**References**
Pharmacophore modeling combined with CoMFA and CoMSIA	SYBYL	[205]
Ligands		
Diaminopropionamide-glycine dipeptides, disubstituted and trisubstituted cyclohexanes		
Drug design technique(s)	**Computational tool(s)**	**References**
Pharmacophore modeling	CATALYST	[206]
GPCR: Corticotropin-releasing factor receptor 1 ($CRFR_1$)		
Ligands		
Arylquinolines, phenylpyrazolo[1,5-*a*]pyrimidines, benzoylpyrimidines, and arylpyrrolopyridines		
Drug design technique(s)	**Computational tool(s)**	**References**
Pharmacophore modeling	CATALYST	[207]
Ligands		
Anilinopyrimidines and triazines		
Drug design technique(s)	**Computational tool(s)**	**References**
Pharmacophore modeling	CATALYST	[208]
Ligands		
N^3-Phenylpyrazinones		
Drug design technique(s)	**Computational tool(s)**	**References**
Pharmacophore modeling	PHASE	[209]
GPCR: δ-Opioid receptor (DOR)		
Ligands		
SB219825, SIOM, (-) TAN-67, BNTX, naltriben, naltrindole, oxymorphindole		
Drug design technique(s)	**Computational tool(s)**	**References**
Pharmacophore modeling	SYBYL	[210]

(continued)

Table 3
(continued)

Ligands		
Non-peptides (xorphanol, naltrindole, BNTX, SIOM, Win44441, lofentanil, carfentanil, SNC80(+8)), cyclic peptides (DPDPE, DPLPE), linear peptides (TIPP, TIP, TI-NH_2)		
Drug design technique(s)	**Computational tool(s)**	**References**
Pharmacophore modeling	DistComp	[196]
Ligands		
(*E*)- and (*Z*)-arylidenenaltrexones		
Drug design technique(s)	**Computational tool(s)**	**References**
Pharmacophore modeling	SYBYL	[211]
Ligands		
DADLE, DPDPE, deltorphins, Leu- and Met-enkephalins, Dmt-Tic, ICI 174,864, TIPP		
Drug design technique(s)	**Computational tool(s)**	**References**
Pharmacophore modeling	SYBYL	[212]
GPCR: Histamine H_3 receptor (H_3R)		
Ligands		
Dibasic biphenyl derivatives, tetrahydroisoquinolines, tetrahydroquinolines, tetrahydroazepines, imidazolidinylidenepropanedinitriles		
Drug design technique(s)	**Computational tool(s)**	**References**
Pharmacophore modeling	CATALYST	[213]
Ligands		
Imidazoles		
Drug design technique(s)	**Computational tool(s)**	**References**
Pharmacophore modeling	SLATE	[214]
Ligands		
1-(4-(3-(Piperidin-1-yl)propoxy)benzyl)piperidine, 1-(4-chlorobenzyl)-1-(5-(pyrrolidin-1-yl)pentyl) guanidine, 3-(2,6-dibromo-4-(2-(dimethylamino)ethyl)phenoxy)-*N*,*N*-dimethylpropan-1-amine		
Drug design technique(s)	**Computational tool(s)**	**References**
Pharmacophore modeling	CATALYST	[215]
GPCR: Metabotropic glutamate receptor type 1 ($mGluR_1$)		
Ligands		
Methylglutamates		
Drug design technique(s)	**Computational tool(s)**	**References**
Pharmacophore modeling	APEX-3D	[216]

(continued)

Table 3 (continued)

Drug design technique(s)	Computational tool(s)	References
Ligands		
α-Substituted cyclobutylglycins, 4-carboxy phenylglycins, (*R*,*S*)-1-aminoindan-2,5-dicarboxylic acid, (±)-α-thioxanthylmethyl-3-carboxycyclobutylglycine		
Drug design technique(s)	**Computational tool(s)**	**References**
Pharmacophore modeling	MOLMOD	[217]
GPCR: Metabotropic glutamate receptor type 2 ($mGluR_2$)		
Ligands		
1,3-Dihydrobenzo[*b*][1, 4]diazepin-2-ones		
Drug design technique(s)	**Computational tool(s)**	**References**
Pharmacophore modeling combined with CoMFA and CoMSIA	DISCO, SYBYL	[218]
Ligands		
Methylglutamates		
Drug design technique(s)	**Computational tool(s)**	**References**
Pharmacophore modeling	APEX-3D	[216]
GPCR: 5-Hydroxytryptamine 2C Receptor (5-HT_{2C}R)		
Ligands		
RS-102221, SB240284, Haloperidol, S20098, 2-alkyl-4-aryl-pyrimidines, bisaryl imidazolidin-2-ones, 2-phenyl-dihydropyrrolones, *N*-substituted-pyridoindolines, *cis*-fused 2-*N*,*N*-dimethylaminomethyl-2,3,3*a*,12*b*-tetrahydrodibenzo[*b*,*f*]furo[2,3-*d*]oxepines, 1*H*-indole-3-carboxylic acid pyridine-3-ylamides, benzazepines		
Drug design technique(s)	**Computational tool(s)**	**References**
Pharmacophore modeling combined with CoMFA	CATALYST, SYBYL	[219]
Ligands		
Library of 16,560 ChemDiv GPCR compounds		
Drug design technique(s)	**Computational tool(s)**	**References**
Pharmacophore modeling	CATALYST	[220]
GPCR: 5-Hydroxytryptamine 4 Receptor (5-HT_4R)		
Ligands		
Indolecarbazimidamide, 3-*N*-isopropylbenzimidazolone amide, 3-*N*-ethylbenzimidazolone amide and benzamide, (*R*)-zacopride, 5-carbamoyltryptamine and metoclopramide		
Drug design technique(s)	**Computational tool(s)**	**References**
Pharmacophore modeling combined with CoMFA	SYBYL	[221]
Ligands		
Indolecarbazimidamides, azabicyclic indole esters, macrocyclic benzamides		
Drug design technique(s)	**Computational tool(s)**	**References**

(continued)

Table 3 (continued)

Pharmacophore modeling	CATALYST	[222]
GPCR: 5-Hydroxytryptamine 6 receptor (5-HT_6R)		
Ligands		
Arylsulfonamides, arylsulfonyl derivatives, *N*-arylsulfonylindoles, 2-substituted tryptamines		
Drug design technique(s)	**Computational tool(s)**	**References**
Pharmacophore modeling	CATALYST	[223]
Ligands		
Indoles; indole-like derivatives; monocyclic, bicyclic, and tricyclic aryl-piperazines; and miscellaneous derivatives		
Drug design technique(s)	**Computational tool(s)**	**References**
Pharmacophore modeling	CATALYST	[224]
Ligands		
2-Methylindoles, 2-phenylindoles		
Drug design technique(s)	**Computational tool(s)**	**References**
Pharmacophore modeling	CATALYST	[225]

GFA genetic function algorithm, *kNN k* nearest neighbor

4 Concluding Remarks

With the progress of the structural biology on elucidation of 3D crystal structures of GPCRs from X-ray crystallography and NMR techniques and of in silico-based drug design tools, a diverse plethora of GPCR modulators have been identified by structure- and ligand-based drug design strategies. Experimentally, the application of structure-based drug design methodologies allows the understanding of ligand-GPCR interactions at a molecular level, which is fundamental for the construction of reliable structure-based pharmacophores and generation of novel drugs. However, future drug candidates acting on GPCRs are likely to rely on ligand-based approaches because of limited structural data information for the majority of GPCRs. The present chapter provided a general overview of the structure- and ligand-based computational methodologies as well as their applicability on various potential GPCR-derived therapeutic targets for AD by small-molecule modulators. In fact, the pharmacological activation/inhibition of all the aforementioned GPCRs on Tables 1, 2, and 3 has provided therapeutic opportunities, and from the analysis of these tables, it has become evident that diverse chemical scaffolds of small molecules have been explored using structure-based, ligand-based, and pharmacophore-based methodologies in the search for anti-AD alternatives. Collectively, these in silico approaches have revealed to be of utmost importance in early stages of drug discovery, particularly in hit-to-lead optimization of drug candidates, in order to uncover the most favorable molecular modifications for the development of more potent and subtype-selective GPCR modulators targeting AD.

Apart from the extreme relevance of pharmacodynamic (PD) profile of GPCR modulators, pharmacokinetic (PK) properties, including absorption, distribution, metabolism, and excretion (ADME), and toxicology are vital features that should be taken into account in early phases of drug discovery since usually drug candidates with a promising PD profile may be failed at late stages of drug development due to unfavorable PK properties and toxicity. In silico structure- and ligand-based drug design approaches combined with in silico prediction of ADME properties are expected to contribute to the improvement of the computational methodologies used for drug discovery and be fundamental for the development of drugs targeting AD with enhanced PD and PK properties.

Acknowledgments and Funding

This work had the financial support of Fundação para a Ciência e a Tecnologia (FCT/MEC) through national funds and cofinanced by FEDER, under the Partnership Agreement PT2020 (projects UID/QUI/50006/2013 and POCI/01/0145/FEDER/007265). Irina S. Moreira acknowledges support by the FCT – Investigator Programme – IF/00578/2014 (cofinanced by European Social Fund and Programa Operacional Potencial Humano), a Marie Skłodowska-Curie Individual Fellowship MSCA-IF-2015 [MEMBRANEPROT 659826]. This work was also financed by the European Regional Development Fund (ERDF), through the Centro 2020 Regional Operational Programme under project CENTRO-01-0145-FEDER-000008, BrainHealth 2020, and through the COMPETE 2020 – Operational Programme for Competitiveness and Internationalisation and Portuguese national funds via FCT, under project POCI-01-0145-FEDER-007440. Rita Melo acknowledges support from the FCT (SFRH/BPD/97650/2013 and UID/Multi/04349/2013 project). MNDSC further acknowledges FCT for the sabbatical grant SFRH/BSAB/127789/2016.

References

1. Bartus RT (2000) On neurodegenerative diseases, models, and treatment strategies: lessons learned and lessons forgotten a generation following the cholinergic hypothesis. Exp Neurol 163(2):495–529
2. Craig LA, Hong NS, McDonald RJ (2011) Revisiting the cholinergic hypothesis in the development of Alzheimer's disease. Neurosci Biobehav Rev 35(6):1397–1409
3. Hardy J, Selkoe DJ (2002) The amyloid hypothesis of Alzheimer's disease: progress and problems on the road to therapeutics. Science 297(5580):353–356
4. Karran E, Mercken M, Strooper BD (2011) The amyloid cascade hypothesis for Alzheimer's disease: an appraisal for the development of therapeutics. Nat Rev Drug Discov 10(9):698–712
5. Tolnay M, Probst A (1999) Review: *tau* protein pathology in Alzheimer's disease and related disorders. Neuropathol Appl Neurobiol 25(3):171–187
6. Maccioni RB, Farias G, Morales I, Navarrete L (2010) The revitalized *tau* hypothesis on Alzheimer's disease. Arch Med Res 41 (3):226–231
7. Nitsch RM, Deng M, Growdon JH, Wurtman RJ (1996) Serotonin 5-HT_{2A} and 5-HT_{2C} receptors stimulate amyloid precursor protein ectodomain secretion. J Biol Chem 271 (8):4188–4194
8. Price DL, Bonhaus DW, McFarland K (2012) Pimavanserin, a 5-HT_{2A} receptor inverse agonist, reverses psychosis-like behaviors in a rodent model of Alzheimer's disease. Behav Pharmacol 23(4):426–433
9. Arjona AA, Pooler AM, Lee RK, Wurtman RJ (2002) Effect of a 5-HT_{2C} serotonin agonist, dexnorfenfluramine, on amyloid precursor protein metabolism in guinea pigs. Brain Res 951(1):135–140
10. Giannoni P, Gaven F, De Bundel D, Baranger K, Marchetti-Gauthier E, Roman FS, Valjent E, Marin P, Bockaert J, Rivera S (2013) Early administration of RS 67333, a specific 5-HT_4 receptor agonist, prevents amyloidogenesis and behavioral deficits in the 5xFAD mouse model of Alzheimer's disease. Front Aging Neurosci 5:96. doi:10.3389/fnagi.2013.00096. eCollection 2013
11. Pimenova AA, Thathiah A, De Strooper B, Tesseur I (2014) Regulation of amyloid precursor protein processing by serotonin signaling. PLoS One 9(1):e87014. doi:10.1371/journal.pone.0087014
12. Robert SJ, Zugaza JL, Fischmeister R, Gardier AM, Lezoualc'h F (2001) The human serotonin 5-HT_4 receptor regulates secretion of non-amyloidogenic precursor protein. J Biol Chem 276(48):44881–44888
13. Tesseur I, Pimenova AA, Lo AC, Ciesielska M, Lichtenthaler SF, De Maeyer JH, Schuurkes JA, D'Hooge R, De Strooper B (2013) Chronic 5-HT_4 receptor activation decreases $A\beta$ production and deposition in hAPP/PS1 mice. Neurobiol Aging 34 (7):1779–1789
14. Benhamú B, Martín-Fontecha M, Vázquez-Villa H, Pardo L, López-Rodríguez ML (2014) Serotonin 5-HT_6 receptor antagonists for the treatment of cognitive deficiency in Alzheimer's disease. J Med Chem 57 (17):7160–7181
15. Maher-Edwards G, Zvartau-Hind M, Hunter A, Gold M, Hopton G, Jacobs G, Davy M, Williams P (2010) Double-blind, controlled phase II study of a 5-HT_6 receptor antagonist, SB-742457, in Alzheimer's disease. Curr Alzheimer Res 7(5):374–385
16. Rosse G, Schaffhauser H (2010) 5-HT_6 receptor antagonists as potential therapeutics for cognitive impairment. Curr Top Med Chem 10(2):207–221
17. Upton N, Chuang TT, Hunter AJ, Virley DJ (2008) 5-HT_6 receptor antagonists as novel cognitive enhancing agents for Alzheimer's disease. Neurotherapeutics 5(3):458–469
18. Arendash G, Schleif W, Rezai-Zadeh K, Jackson E, Zacharia L, Cracchiolo J, Shippy D, Tan J (2006) Caffeine protects Alzheimer's mice against cognitive impairment and reduces brain β-amyloid production. Neuroscience 142(4):941–952
19. Giunta S, Andriolo V, Castorina A (2014) Dual blockade of the A_1 and A_{2A} adenosine receptor prevents amyloid β toxicity in neuroblastoma cells exposed to aluminum chloride. Int J Biochem Cell Biol 54:122–136
20. Angulo E, Casadó V, Mallol J, Canela EI, Viñals F, Ferrer I, Lluis C, Franco R (2003) A_1 adenosine receptors accumulate in neurodegenerative structures in Alzheimer's disease and mediate both amyloid precursor protein processing and *tau* phosphorylation and translocation. Brain Pathol 13(4):440–451
21. Canas PM, Porciúncula LO, Cunha GM, Silva CG, Machado NJ, Oliveira JM, Oliveira CR, Cunha RA (2009) Adenosine A_{2A} receptor blockade prevents synaptotoxicity and memory dysfunction caused by β-amyloid peptides

via p38 mitogen-activated protein kinase pathway. J Neurosci 29(47):14741–14751
22. Espinosa J, Rocha A, Nunes F, Costa MS, Schein V, Kazlauckas V, Kalinine E, Souza DO, Cunha RA, Porciúncula LO (2013) Caffeine consumption prevents memory impairment, neuronal damage, and adenosine A_{2A} receptors upregulation in the hippocampus of a rat model of sporadic dementia. J Alzheimers Dis 34(2):509–518
23. Nagpure BV, Bian JS (2014) Hydrogen sulfide inhibits A_{2A} adenosine receptor agonist induced β-amyloid production in SH-SY5Y neuroblastoma cells via a cAMP dependent pathway. PLoS One 9(2):e88508. doi:10.1371/journal.pone.0088508. eCollection 2014
24. Orr AG, Hsiao EC, Wang MM, Ho K, Kim DH, Wang X, Guo W, Kang J, Yu GQ, Adame A (2015) Astrocytic adenosine receptor A_{2A} and G_s-coupled signaling regulate memory. Nat Neurosci 18(3):423–434
25. Chen Y, Peng Y, Che P, Gannon M, Liu Y, Li L, Bu G, van Groen T, Jiao K, Wang Q (2014) α_{2A} adrenergic receptor promotes amyloidogenesis through disrupting APP-SorLA interaction. Proc Natl Acad Sci U S A 111 (48):17296–17301
26. Branca C, Wisely EV, Hartman LK, Caccamo A, Oddo S (2014) Administration of a selective β_2 adrenergic receptor antagonist exacerbates neuropathology and cognitive deficits in a mouse model of Alzheimer's disease. Neurobiol Aging 35(12):2726–2735
27. Ni Y, Zhao X, Bao G, Zou L, Teng L, Wang Z, Song M, Xiong J, Bai Y, Pei G (2006) Activation of β_2-adrenergic receptor stimulates γ-secretase activity and accelerates amyloid plaque formation. Nat Med 12 (12):1390–1396
28. Wisely EV, Xiang YK, Oddo S (2014) Genetic suppression of β_2-adrenergic receptors ameliorates *tau* pathology in a mouse model of tauopathies. Hum Mol Genet 23 (15):4024–4034
29. El Khoury J, Toft M, Hickman SE, Means TK, Terada K, Geula C, Luster AD (2007) CCR_2 deficiency impairs microglial accumulation and accelerates progression of Alzheimer-like disease. Nat Med 13(4):432–438
30. Westin K, Buchhave P, Nielsen H, Minthon L, Janciauskiene S, Hansson O (2012) CCL_2 is associated with a faster rate of cognitive decline during early stages of Alzheimer's disease. PLoS One 7(1):e30525. doi:10.1371/journal.pone.0030525
31. Bakshi P, Margenthaler E, Laporte V, Crawford F, Mullan M (2008) Novel role of $CXCR_2$ in regulation of γ-secretase activity. ACS Chem Biol 3(12):777–789
32. Bakshi P, Jin C, Broutin P, Berhane B, Reed J, Mullan M (2009) Structural optimization of a $CXCR_2$-directed antagonist that indirectly inhibits γ-secretase and reduces Aβ. Bioorg Med Chem 17(23):8102–8112
33. Bakshi P, Margenthaler E, Reed J, Crawford F, Mullan M (2011) Depletion of $CXCR_2$ inhibits γ-secretase activity and amyloid-β production in a murine model of Alzheimer's disease. Cytokine 53(2):163–169
34. Carroll JC, Iba M, Bangasser DA, Valentino RJ, James MJ, Brunden KR, Lee VMY, Trojanowski JQ (2011) Chronic stress exacerbates *tau* pathology, neurodegeneration, and cognitive performance through a corticotropin-releasing factor receptor-dependent mechanism in a transgenic mouse model of tauopathy. J Neurosci 31 (40):14436–14449
35. Justice NJ, Huang L, Tian JB, Cole A, Pruski M, Hunt AJ, Flores R, Zhu MX, Arenkiel BR, Zheng H (2015) Posttraumatic stress disorder-like induction elevates β-amyloid levels, which directly activates corticotropin-releasing factor neurons to exacerbate stress responses. J Neurosci 35(6):2612–2623
36. Rissman RA, Staup MA, Lee AR, Justice NJ, Rice KC, Vale W, Sawchenko PE (2012) Corticotropin-releasing factor receptor-dependent effects of repeated stress on *tau* phosphorylation, solubility, and aggregation. Proc Natl Acad Sci U S A 109 (16):6277–6282
37. Scullion GA, Hewitt KN, Pardon MC (2013) Corticotropin-releasing factor receptor 1 activation during exposure to novelty stress protects against Alzheimer's disease-like cognitive decline in AβPP/PS1 mice. J Alzheimers Dis 34(3):781–793
38. Cai Z, Ratka A (2012) Opioid system and Alzheimer's disease. NeuroMolecular Med 14(2):91–111
39. Medhurst AD, Atkins AR, Beresford IJ, Brackenborough K, Briggs MA, Calver AR, Cilia J, Cluderay JE, Crook B, Davis JB, Davis RK, Davis RP, Dawson LA, Foley AG, Gartlon J, Gonzalez MI, Heslop T, Hirst WD, Jennings C, Jones DNC, Lacroix LP, Martyn A, Ociepka S, Ray A, Regan CM, Roberts JC, Schogger J, Southam E, Stean TO, Trail BK, Upton N, Wadsworth G, Wald JA, White T, Witherington J, Woolley ML, Worby A,

Wilson DM (2007) GSK189254, a novel H_3 receptor antagonist that binds to histamine H_3 receptors in Alzheimer's disease brain and improves cognitive performance in preclinical models. J Pharm Exp Ther 321 (3):1032–1045
40. Nathan PJ, Boardley R, Scott N, Berges A, Maruff P, Sivananthan T, Upton N, Lowy MT, Nestor PJ, Lai R (2013) The safety, tolerability, pharmacokinetics and cognitive effects of GSK239512, a selective histamine H_3 receptor antagonist in patients with mild to moderate Alzheimer's disease: a preliminary investigation. Curr Alzheimer Res 10 (3):240–251
41. Haig GM, Pritchett Y, Meier A, Othman AA, Hall C, Gault LM, Lenz RA (2014) A randomized study of H_3 antagonist ABT-288 in mild-to-moderate Alzheimer's dementia. J Alzheimer Dis 42(3):959–971
42. Kim SH, Fraser PE, Westaway D, George-Hyslop PHS, Ehrlich ME, Gandy S (2010) Group II metabotropic glutamate receptor stimulation triggers production and release of Alzheimer's amyloid β42 from isolated intact nerve terminals. J Neurosci 30 (11):3870–3875
43. Kirazov L, Löffler T, Schliebs R, Bigl V (1997) Glutamate-stimulated secretion of amyloid precursor protein from cortical rat brain slices. Neurochem Int 30(6):557–563
44. Lee R, Wurtman RJ, Cox AJ, Nitsch RM (1995) Amyloid precursor protein processing is stimulated by metabotropic glutamate receptors. Proc Natl Acad Sci U S A 92 (17):8083–8087
45. Nitsch RM, Deng A, Wurtman RJ, Growdon JH (1997) Metabotropic glutamate receptor subtype $mGluR_{1\alpha}$ stimulates the secretion of the amyloid β-protein precursor ectodomain. J Neurochem 69(2):704–712
46. Lee HG, Zhu X, Casadesus G, Pallàs M, Camins A, O'Neill MJ, Nakanishi S, Perry G, Smith MA (2009) The effect of $mGluR_2$ activation on signal transduction pathways and neuronal cell survival. Brain Res 1249:244–250
47. Spinelli S, Ballard T, Gatti-McArthur S, Richards GJ, Kapps M, Woltering T, Wichmann J, Stadler H, Feldon J, Pryce CR (2005) Effects of the $mGluR_{2/3}$ agonist LY354740 on computerized tasks of attention and working memory in marmoset monkeys. Psychopharmacol 179(1):292–302
48. Um JW, Kaufman AC, Kostylev M, Heiss JK, Stagi M, Takahashi H, Kerrisk ME, Vortmeyer A, Wisniewski T, Koleske AJ, Gunther EC, Nygaard HB, Strittmatter SM (2013) Metabotropic glutamate receptor 5 is a coreceptor for Alzheimer Aβ oligomer bound to cellular prion protein. Neuron 79 (5):887–902
49. Kumar A, Dhull DK, Mishra PS (2015) Therapeutic potential of $mGluR_5$ targeting in Alzheimer's disease. Front Neurosci 9:215. doi:10.3389/fnins.2015.00215
50. Caccamo A, Oddo S, Billings LM, Green KN, Martinez-Coria H, Fisher A, LaFerla FM (2006) M_1 receptors play a central role in modulating AD-like pathology in transgenic mice. Neuron 49(5):671–682
51. Davis AA, Fritz JJ, Wess J, Lah JJ, Levey AI (2010) Deletion of M_1 muscarinic acetylcholine receptors increases amyloid pathology *in vitro* and *in vivo*. J Neurosci 30 (12):4190–4196
52. Jiang S, Wang Y, Ma Q, Zhou A, Zhang X, Zhang YW (2012) M_1 muscarinic acetylcholine receptor interacts with BACE1 and regulates its proteosomal degradation. Neurosci Lett 515(2):125–130
53. Nitsch RM, Slack BE, Wurtman RJ, Growdon JH (1992) Release of Alzheimer amyloid precursor derivatives stimulated by activation of muscarinic acetylcholine receptors. Science 258(5080):304–307
54. Züchner T, Perez-Polo JR, Schliebs R (2004) β-secretase BACE1 is differentially controlled through muscarinic acetylcholine receptor signaling. J Neurosci Res 77(2):250–257
55. Packard MG, Regenold W, Quirion R, White NM (1990) Post-training injection of the acetylcholine M_2 receptor antagonist AF-DX 116 improves memory. Brain Res 524(1):72–76
56. Han Y, Moreira IS, Urizar E, Weinstein H, Javitch JA (2009) Allosteric communication between protomers of dopamine class a GPCR dimers modulates activation. Nat Chem Biol 5(9):688–695
57. Moreira IS (2014) Structural features of the G-protein/GPCR interactions. Biochim Biophys Acta, Gen Subj 1840(1):16–33
58. Schioth HB, Fredriksson R (2005) The GRAFS classification system of G-protein coupled receptors in comparative perspective. Gen Comp Endocrinol 142(1–2):94–101
59. Ji TH, Grossmann M, Ji I (1998) G protein-coupled receptors. I. Diversity of receptor-ligand interactions. J Biol Chem 273 (28):17299–17302
60. Rosenbaum DM, Rasmussen SG, Kobilka BK (2009) The structure and function of G-protein-coupled receptors. Nature 459 (7245):356–363

61. Lang M, Beck-Sickinger AG (2006) Structure-activity relationship studies: methods and ligand design for G-protein coupled peptide receptors. Curr Protein Pept Sci 7 (4):335–353
62. Marinissen MJ, Gutkind JS (2001) G-protein-coupled receptors and signaling networks: emerging paradigms. Trends Pharmacol Sci 22(7):368–376
63. Birnbaumer L (2007) The discovery of signal transduction by G proteins. A personal account and an overview of the initial findings and contributions that led to our present understanding. Biochim Biophys Acta Biomembr 1768(4):756–771
64. Kontoyianni M, Liu Z (2012) Structure-based design in the GPCR target space. Curr Med Chem 19(4):544–556
65. Shukla AK, Xiao K, Lefkowitz RJ (2011) Emerging paradigms of β-arrestin-dependent seven transmembrane receptor signaling. Trends Biochem Sci 36(9):457–469
66. Lefkowitz RJ (1998) G protein-coupled receptors. III New roles for receptor kinases and β-arrestins in receptor signaling and desensitization. J Biol Chem 273 (30):18677–18680
67. Wolfe BL, Trejo J (2007) Clathrin-dependent mechanisms of G protein-coupled receptor endocytosis. Traffic 8(5):462–470
68. Tsao P, von Zastrow M (2000) Downregulation of G protein-coupled receptors. Curr Opin Neurobiol 10(3):365–369
69. Collins S, Caron MG, Lefkowitz RJ (1991) Regulation of adrenergic receptor responsiveness through modulation of receptor gene expression. Annu Rev Physiol 53:497–508
70. De Vries L, Zheng B, Fischer T, Elenko E, Farquhar MG (2000) The regulator of G protein signaling family. Annu Rev Pharmacol Toxicol 40:235–271
71. Ross EM, Wilkie TM (2000) GTPase-activating proteins for heterotrimeric G proteins: regulators of G protein signaling (RGS) and RGS-like proteins. Annu Rev Biochem 69:795–827
72. Ghemtio L, Zhang Y, Xhaard H (2012) CoMFA/CoMSIA and pharmacophore modelling as a powerful tools for efficient virtual screening: application to anti-leishmanial betulin derivatives. In: Taha MO (ed) Virtual screening. In Tech, Croatia, pp 55–82. doi:10.5772/36690
73. Yang SY (2010) Pharmacophore modeling and applications in drug discovery: challenges and recent advances. Drug Discov Today 15 (11–12):444–450
74. Ghosh E, Kumari P, Jaiman D, Shukla AK (2015) Methodological advances: the unsung heroes of the GPCR structural revolution. Nat Rev Mol Cell Biol 16(2):69–81
75. Palczewski K, Kumasaka T, Hori T, Behnke CA, Motoshima H, Fox BA, Le Trong I, Teller DC, Okada T, Stenkamp RE, Yamamoto M, Miyano M (2000) Crystal structure of rhodopsin: a G protein-coupled receptor. Science 289(5480):739–745
76. Costanzi S, Siegel J, Tikhonova IG, Jacobson KA (2009) Rhodopsin and the others: a historical perspective on structural studies of G protein-coupled receptors. Curr Pharm Des 15(35):3994–4002
77. Patny A, Desai PV, Avery MA (2006) Homology modeling of G-protein-coupled receptors and implications in drug design. Curr Med Chem 13(14):1667–1691
78. Cherezov V, Rosenbaum DM, Hanson MA, Rasmussen SG, Thian FS, Kobilka TS, Choi HJ, Kuhn P, Weis WI, Kobilka BK, Stevens RC (2007) High-resolution crystal structure of an engineered human β_2-adrenergic G protein-coupled receptor. Science 318 (5854):1258–1265
79. Rasmussen SG, Choi HJ, Rosenbaum DM, Kobilka TS, Thian FS, Edwards PC, Burghammer M, Ratnala VR, Sanishvili R, Fischetti RF, Schertler GF, Weis WI, Kobilka BK (2007) Crystal structure of the human β_2 adrenergic G-protein-coupled receptor. Nature 450(7168):383–387
80. Warne T, Serrano-Vega MJ, Baker JG, Moukhametzianov R, Edwards PC, Henderson R, Leslie AG, Tate CG, Schertler GF (2008) Structure of a β_1-adrenergic G-protein-coupled receptor. Nature 454(7203):486–491
81. Christopoulos A (2002) Allosteric binding sites on cell-surface receptors: novel targets for drug discovery. Nat Rev Drug Discov 1 (3):198–210
82. Conn PJ, Christopoulos A, Lindsley CW (2009) Allosteric modulators of GPCRs: a novel approach for the treatment of CNS disorders. Nat Rev Drug Discov 8(1):41–54
83. Chang SD, Bruchas MR (2014) Functional selectivity at GPCRs: new opportunities in psychiatric drug discovery. Neuropsychopharmacol 39(1):248–249
84. Schrage R, Kostenis E (2017) Functional selectivity and dualsteric/bitopic GPCR targeting. Curr Opin Pharmacol 32:85–90
85. Vishnivetskiy SA, Gimenez LE, Francis DJ, Hanson SM, Hubbell WL, Klug CS, Gurevich VV (2011) Few residues within an extensive binding interface drive receptor interaction

and determine the specificity of arrestin proteins. J Biol Chem 286(27):24288–24299
86. Rasmussen SG, DeVree BT, Zou Y, Kruse AC, Chung KY, Kobilka TS, Thian FS, Chae PS, Pardon E, Calinski D, Mathiesen JM, Shah ST, Lyons JA, Caffrey M, Gellman SH, Steyaert J, Skiniotis G, Weis WI, Sunahara RK, Kobilka BK (2011) Crystal structure of the β_2 adrenergic receptor-G_s protein complex. Nature 477(7366):549–555
87. Dror RO, Arlow DH, Maragakis P, Mildorf TJ, Pan AC, Xu H, Borhani DW, Shaw DE (2011) Activation mechanism of the β_2-adrenergic receptor. Proc Natl Acad Sci U S A 108(46):18684–18689
88. Kohlhoff KJ, Shukla D, Lawrenz M, Bowman GR, Konerding DE, Belov D, Altman RB, Pande VS (2014) Cloud-based simulations on Google Exacycle reveal ligand modulation of GPCR activation pathways. Nat Chem 6 (1):15–21
89. Bruno A, Costantino G (2012) Molecular dynamics simulations of G protein-coupled receptors. Mol Inform 31(3–4):222–230
90. Cozzini P, Kellogg GE, Spyrakis F, Abraham DJ, Costantino G, Emerson A, Fanelli F, Gohlke H, Kuhn LA, Morris GM, Orozco M, Pertinhez TA, Rizzi M, Sotriffer CA (2008) Target flexibility: an emerging consideration in drug discovery and design. J Med Chem 51(20):6237–6255
91. Feixas F, Lindert S, Sinko W, McCammon JA (2014) Exploring the role of receptor flexibility in structure-based drug discovery. Biophys Chem 186:31–45
92. Jo S, Lim JB, Klauda JB, Im W (2009) CHARMM-GUI membrane builder for mixed bilayers and its application to yeast membranes. Biophys J 97(1):50–58
93. Wu EL, Cheng X, Jo S, Rui H, Song KC, Dávila-Contreras EM, Qi Y, Lee J, Monje-Galvan V, Venable RM, Klauda JB, Im W (2014) CHARMM-GUI *Membrane Builder* toward realistic biological membrane simulations. J Comput Chem 35(27):1997–2004
94. Ribeiro JV, Bernardi RC, Rudack T, Stone JE, Phillips JC, Freddolino PL, Schulten K (2016) QwikMD - integrative molecular dynamics toolkit for novices and experts. Sci Rep 6:26536. doi:10.1038/srep26536
95. Doerr S, Harvey MJ, Noé F, De Fabritiis G (2016) HTMD: high-throughput molecular dynamics for molecular discovery. J Chem Theory Comput 12(4):1845–1852
96. Guo W, Shi L, Filizola M, Weinstein H, Javitch JA (2005) Crosstalk in G protein-coupled receptors: changes at the transmembrane homodimer interface determine activation. Proc Natl Acad Sci U S A 102(48):17495–17500
97. Kitchen DB, Decornez H, Furr JR, Bajorath J (2004) Docking and scoring in virtual screening for drug discovery: methods and applications. Nat Rev Drug Discov 3(11):935–949
98. Morris GM, Huey R, Olson AJ (2008) Using AutoDock for ligand-receptor docking. Curr Protoc Bioinformatics 24:8.14:8.14.1–8.14.40. doi:10.1002/0471250953.bi0814s24
99. Trott O, Olson AJ (2010) AutoDock Vina: improving the speed and accuracy of docking with a new scoring function, efficient optimization, and multithreading. J Comput Chem 31(2):455–461
100. Wu G, Robertson DH, Brooks CL 3rd, Vieth M (2003) Detailed analysis of grid-based molecular docking: a case study of CDOCKER-A CHARMm-based MD docking algorithm. J Comput Chem 24 (13):1549–1562
101. Rarey M, Kramer B, Lengauer T, Klebe G (1996) A fast flexible docking method using an incremental construction algorithm. J Mol Biol 261(3):470–489
102. Verdonk ML, Cole JC, Hartshorn MJ, Murray CW, Taylor RD (2003) Improved protein-ligand docking using GOLD. Proteins 52 (4):609–623
103. Repasky MP, Shelley M, Friesner RA (2007) Flexible ligand docking with GLIDE. Curr Protoc Bioinformatics 18:8.12:8.12.1–8.12.36. doi:10.1002/0471250953.bi0812s18
104. Neves MA, Totrov M, Abagyan R (2012) Docking and scoring with ICM: the benchmarking results and strategies for improvement. J Comput Aided Mol Des 26 (6):675–686
105. Marcou G, Rognan D (2007) Optimizing fragment and scaffold docking by use of molecular interaction fingerprints. J Chem Inf Model 47(1):195–207
106. Kalid O, Toledo Warshaviak D, Shechter S, Sherman W, Shacham S (2012) Consensus induced fit docking (cIFD): methodology, validation, and application to the discovery of novel CRM1 inhibitors. J Comput Aided Mol Des 26(11):1217–1228
107. Rao SN, Head MS, Kulkarni A, LaLonde JM (2007) Validation studies of the site-directed docking program LibDock. J Chem Inf Model 47(6):2159–2171
108. Bai Q, Shao Y, Pan D, Zhang Y, Liu H, Yao X (2014) Search for β_2 adrenergic receptor

ligands by virtual screening via grid computing and investigation of binding modes by docking and molecular dynamics simulations. PLoS One 9(9):e107837. doi:10.1371/journal.pone.0107837. eCollection 2014
109. Korb O, Stutzle T, Exner TE (2009) Empirical scoring functions for advanced protein-ligand docking with PLANTS. J Chem Inf Model 49(1):84–96
110. Gutiérrez-de-Terán H, Centeno NB, Pastor M, Sanz F (2004) Novel approaches for modeling of the A_1 adenosine receptor and its agonist binding site. Proteins: Struct Funct Bioinf 54(4):705–715
111. Kolb P, Phan K, Gao ZG, Marko AC, Sali A, Jacobson KA (2012) Limits of ligand selectivity from docking to models: in Silico screening for A_1 adenosine receptor antagonists. PLoS One 7(11):e49910. doi:10.1371/journal.pone.0049910
112. Ke YR, Jin HW, Liu ZM, Zhang LR (2010) Homology modeling and structure validation of the adenosine A_1 receptor. Acta Phys - Chim Sin 26(10):2833–2839
113. Langmead CJ, Andrews SP, Congreve M, Errey JC, Hurrell E, Marshall FH, Mason JS, Richardson CM, Robertson N, Zhukov A, Weir M (2012) Identification of novel adenosine A_{2A} receptor antagonists by virtual screening. J Med Chem 55(5):1904–1909
114. Katritch V, Jaakola VP, Lane J, Lin J, Ijzerman AP, Yeager M, Kufareva I, Stevens RC, Abagyan R (2010) Structure-based discovery of novel chemotypes for adenosine A_{2A} receptor antagonists. J Med Chem 53(4):1799–1809
115. Ivanov AA, Barak D, Jacobson KA (2009) Evaluation of homology modeling of G-protein-coupled receptors in light of the A_{2A} adenosine receptor crystallographic structure. J Med Chem 52(10):3284–3292
116. Rodríguez D, Gao ZG, Moss SM, Jacobson KA, Carlsson J (2015) Molecular docking screening using agonist-bound GPCR structures: probing the A_{2A} adenosine receptor. J Chem Inf Model 55(3):550–563
117. Carlsson J, Yoo L, Gao ZG, Irwin JJ, Shoichet BK, Jacobson KA (2010) Structure-based discovery of A_{2A} adenosine receptor ligands. J Med Chem 53(9):3748–3755
118. Ostopovici-Halip L, Curpăn R, Mracec M, Bologa CG (2011) Structural determinants of the α_2 adrenoceptor subtype selectivity. J Mol Graph Model 29(8):1030–1038
119. Jayaraman A, Jamil K, Kakarala KK (2013) Homology modelling and docking studies of human α_2-adrenergic receptor subtypes. J Comput Sci Syst Biol 6:136–149
120. Kooistra AJ, Vischer HF, McNaught-Flores D, Leurs R, de Esch IJP, de Graaf C (2016) Function-specific virtual screening for GPCR ligands using a combined scoring method. Sci Rep 6:28288. doi:10.1038/srep28288
121. Kolb P, Rosenbaum DM, Irwin JJ, Fung JJ, Kobilka BK, Shoichet BK (2009) Structure-based discovery of β_2-adrenergic receptor ligands. Proc Natl Acad Sci U S A 106 (16):6843–6848
122. Kothandan G, Gadhe CG, Cho SJ (2012) Structural insights from binding poses of CCR_2 and CCR_5 with clinically important antagonists: a combined in silico study. PLoS One 7(3):e32864. doi:10.1371/journal.pone.0032864
123. Singh R, Sobhia ME (2013) Structure prediction and molecular dynamics simulations of a G-protein coupled receptor: human CCR_2 receptor. J Biomol Struct Dyn 31 (7):694–715
124. Di Fabio R, Arban R, Bernasconi G, Braggio S, Blaney FE, Capelli AM, Castiglioni E, Donati D, Fazzolari E, Ratti E, Feriani A, Contini S, Gentile G, Ghirlanda D, Sabbatini FM, Andreotti D, Spada S, Marchioro C, Worby A, St-Denis Y (2008) Dihydropyrrole [2,3-*d*]pyridine derivatives as novel corticotropin-releasing factor-1 antagonists: mapping of the receptor binding pocket by in silico docking studies. J Med Chem 51 (22):7273–7286
125. Micovic V, Ivanovic MD, Dosen-Micovic L (2009) Docking studies suggest ligand-specific δ-opioid receptor conformations. J Mol Model 15(3):267–280
126. Bautista DL, Asher W, Carpenter L (2005) Development of the human μ-, κ-, and δ-opioid receptors and docking with morphine. J Ky Acad Sci 66(2):107–117
127. Sirci F, Istyastono EP, Vischer HF, Kooistra AJ, Nijmeijer S, Kuijer M, Wijtmans M, Mannhold R, Leurs R, de Esch IJP, de Graaf C (2012) Virtual fragment screening: discovery of histamine H_3 receptor ligands using ligand-based and protein-based molecular fingerprints. J Chem Inf Model 52 (12):3308–3324
128. Schlegel B, Laggner C, Meier R, Langer T, Schnell D, Seifert R, Stark H, Höltje HD, Sippl W (2007) Generation of a homology model of the human histamine H_3 receptor for ligand docking and pharmacophore-based screening. J Comput Aided Mol Des 21 (8):437–453
129. Levoin N, Calmels T, Poupardin-Olivier O, Labeeuw O, Danvy D, Robert P, Berrebi-Bertrand I, Ganellin CR, Schunack W, Stark

H, Capet M (2008) Refined docking as a valuable tool for lead optimization: application to histamine H_3 receptor antagonists. Arch Pharm (Weinheim) 341(10):610–623

130. Costantino G, Pellicciari R (1996) Homology modeling of metabotropic glutamate receptors. (mGluRs) structural motifs affecting binding modes and pharmacological profile of $mGluR_1$ agonists and competitive antagonists. J Med Chem 39(20):3998–4006
131. Chin SP, Buckle MJ, Chalmers DK, Yuriev E, Doughty SW (2014) Toward activated homology models of the human M_1 muscarinic acetylcholine receptor. J Mol Graph Model 49:91–98
132. Swaminathan M, Chee CF, Chin SP, Buckle MJ, Rahman NA, Doughty SW, Chung LY (2014) Flavonoids with M_1 muscarinic acetylcholine receptor binding activity. Molecules 19(7):8933–8948
133. Kruse AC, Weiss DR, Rossi M, Hu J, Hu K, Eitel K, Gmeiner P, Wess J, Kobilka BK, Shoichet BK (2013) Muscarinic receptors as model targets and antitargets for structure-based ligand discovery. Mol Pharmacol 84 (4):528–540
134. Gandhimathi A, Sowdhamini R (2016) Molecular modelling of human 5-hydroxytryptamine receptor (5-HT_{2A}) and virtual screening studies towards the identification of agonist and antagonist molecules. J Biomol Struct Dyn 34(5):952–970
135. Kanagarajadurai K, Malini M, Bhattacharya A, Panicker MM, Sowdhamini R (2009) Molecular modeling and docking studies of human 5-hydroxytryptamine 2A (5-HT_{2A}) receptor for the identification of hotspots for ligand binding. Mol BioSyst 5(12):1877–1888
136. Brea J, Rodrigo J, Carrieri A, Sanz F, Cadavid MI, Enguix MJ, Villazón M, Mengod G, Caro Y, Masaguer CF, Raviña E, Centeno NB, Carotti A, Loza MI (2002) New serotonin 5-HT_{2A}, 5-HT_{2B}, and 5-HT_{2C} receptor antagonists: synthesis, pharmacology, 3D-QSAR, and molecular modeling of (aminoalkyl)benzo and heterocycloalkanones. J Med Chem 45(1):54–71
137. Ahmed A, Nagarajan S, Doddareddy MR, Cho YS, Pae AN (2011) Binding mode prediction of 5-hydroxytryptamine 2C receptor ligands by homology modeling and molecular docking analysis. Bull Kor Chem Soc 32 (6):2008–2014
138. Perkins R, Fang H, Tong W, Welsh WJ (2003) Quantitative structure-activity relationship methods: perspectives on drug discovery and toxicology. Environ Toxicol Chem 22(8):1666–1679
139. Dudek AZ, Arodz T, Galvez J (2006) Computational methods in developing quantitative structure-activity relationships (QSAR): a review. Comb Chem High Throughput Screen 9(3):213–228
140. Roy K, Kar S, Das RN (2015) Statistical methods in QSAR/QSPR. In: A primer on QSAR/QSPR modeling. SpringerBriefs in Molecular Science, Cham, pp 37–59
141. Alexopoulos EC (2010) Introduction to multivariate regression analysis. Hippokratia 14 (1):23–28
142. Abdi H (2003) Partial least square regression (PLS regression). In: Lewis-Beck M et al (eds) Encyclopedia of social sciences research methods. Sage, Thousand Oaks, pp 792–795
143. Speck-Planche A, Kleandrova VV, Luan F, Cordeiro MN (2013) Multi-target inhibitors for proteins associated with Alzheimer: in silico discovery using fragment-based descriptors. Curr Alzheimer Res 10(2):117–124
144. Speck-Planche A, Kleandrova VV (2012) QSAR and molecular docking techniques for the discovery of potent monoamine oxidase B inhibitors: computer-aided generation of new rasagiline bioisosteres. Curr Top Med Chem 12(16):1734–1747
145. Chapelle O, Vapnik V, Bousquet O, Mukherjee S (2002) Choosing multiple parameters for support vector machines. Mach Learn 46 (1):131–159
146. Cramer RD, Patterson DE, Bunce JD (1988) Comparative molecular field analysis (CoMFA). 1. Effect of shape on binding of steroids to carrier proteins. J Am Chem Soc 110(18):5959–5967
147. Kubinyi H (2008) Comparative molecular field analysis (CoMFA). In: Gasteiger J (ed) Handbook of Chemoinformatics: from data to knowledge in 4 volumes. Wiley-VCH Verlag GmbH, Weinheim, pp 1555–1574. doi:10.1002/9783527618279.ch44d
148. Klebe G, Abraham U, Mietzner T (1994) Molecular similarity indices in a comparative analysis (CoMSIA) of drug molecules to correlate and predict their biological activity. J Med Chem 37(24):4130–4146
149. Robinson DD, Winn PJ, Lyne PD, Richards WG (1999) Self-organizing molecular field analysis: a tool for structure-activity studies. J Med Chem 42(4):573–583
150. Li M, Du L, Wu B, Xia L (2003) Self-organizing molecular field analysis on α_{1A}-adrenoreceptor dihydropyridine antagonists. Bioorg Med Chem 11(18):3945–3951
151. Moda TL, Montanari CA, Andricopulo AD (2007) Hologram QSAR model for the

prediction of human oral bioavailability. Bioorg Med Chem 15:7738–7745
152. Doddareddy MR, Lee YJ, Cho YS, Choi KI, KohHY PAN (2004) Hologram quantitative structure activity relationship studies on 5-HT_6 antagonists. Bioorg Med Chem 12 (14):3815–3824
153. Palangsuntikul R, Berner H, Berger ML, Wolschann P (2013) Holographic quantitative structure-activity relationships of tryptamine derivatives at NMDA, $5HT_{1A}$ and $5HT_{2A}$ receptors. Molecules 18 (8):8799–8811
154. Muñoz-Gutiérrez C, Caballero J, Morales-Bayuelo A (2016) HQSAR and molecular docking studies of furanyl derivatives as adenosine A_{2A} receptor antagonists. Med Chem Res 25(7):1316–1328
155. Doytchinova I (2001) CoMFA-based comparison of two models of binding site on adenosine A_1 receptor. J Comput Aided Mol Des 15(1):29–39
156. Lima E, Teixeira-Salmela LF, Simoes L, Guerra AC, Lemos A (2016) Assessment of the measurement properties of the post stroke motor function instruments available in Brazil: a systematic review. Braz J Phys Ther 20 (2):114–125
157. Pourbasheer E, Shokouhi Tabar S, Masand V, Aalizadeh R, Ganjali M (2015) 3D-QSAR and docking studies on adenosine A_{2A} receptor antagonists by the CoMFA method. SAR QSAR Environ Res 26(6):461–477
158. Rieger JM, Brown ML, Sullivan GW, Linden J, Macdonald TL (2001) Design, synthesis, and evaluation of novel A_{2A} adenosine receptor agonists. J Med Chem 44(4):531–539
159. Baraldi PG, Borea PA, Bergonzoni M, Cacciari B, Ongini E, Recanatini M, Spalluto G (1999) Comparative molecular field analysis (CoMFA) of a series of selective adenosine receptor A_{2A} antagonists. Drug. Dev Res 46 (2):126–133
160. Doytchinova I, Valkova I, Natcheva R (2001) CoMFA study on adenosine A_{2A} receptor agonists. Quant Struct-Act Relat 20 (2):124–129
161. Vilar S, Karpiak J, Costanzi S (2010) Ligand and structure-based models for the prediction of ligand-receptor affinities and virtual screenings: development and application to the β_2-adrenergic receptor. J Comput Chem 31 (4):707–720
162. Senthil Kumar P, Bharatam PV (2010) Comparative 3D QSAR study on β_1-, β_2-, and β_3-adrenoceptor agonists. Med Chem Res 19 (9):1121–1140
163. Jozwiak K, Khalid C, Tanga MJ, Berzetei-Gurske I, Jimenez L, Kozocas JA, Woo A, Zhu W, Xiao RP, Abernethy DR, Wainer IW (2007) Comparative molecular field analysis of the binding of the stereoisomers of fenoterol and fenoterol derivatives to the β_2 adrenergic receptor. J Med Chem 50 (12):2903–2915
164. Jozwiak K, Woo AY, Tanga MJ, Toll L, Jimenez L, Kozocas JA, Plazinska A, Xiao RP, Wainer IW (2010) Comparative molecular field analysis of fenoterol derivatives: a platform towards highly selective and effective β_2-adrenergic receptor agonists. Bioorg Med Chem 18(2):728–736
165. Plazinska A, Pajak K, Rutkowska E, Jimenez L, Kozocas J, Koolpe G, Tanga M, TollL WIW, Jozwiak K (2014) Comparative molecular field analysis of fenoterol derivatives interacting with an agonist-stabilized form of the β_2-adrenergic receptor. Bioorg Med Chem 22(1):234–246
166. Gunda SK, Anugolu RK, Tata SR, Mahmood S (2012) Structural investigations of $CXCR_2$ receptor antagonists by CoMFA, CoMSIA and flexible docking studies. Acta Pharma 62 (3):287–304
167. Peng Y, Keenan SM, Zhang Q, Welsh WJ (2005) 3D-QSAR comparative molecular field analysis on δ opioid receptor agonist SNC80 and its analogs. J Mol Graph Model 24(1):25–33
168. Ghasemi JB, Tavakoli H (2012) Improvement of the prediction power of the CoMFA and CoMSIA models on histamine H_3 antagonists by different variable selection methods. Sci Pharm 80(3):547–566
169. Chen HF (2008) Computational study of histamine H_3-receptor antagonist with support vector machines and three dimension quantitative structure activity relationship methods. Anal Chim Acta 624(2):203–209
170. Rivara S, Mor M, Bordi F, Silva C, Zuliani V, Vacondio F, Morini G, Plazzi P, Carrupt PA, Testa B (2003) Synthesis and three-dimensional quantitative structure-activity relationship analysis of H_3 receptor antagonists containing a neutral heterocyclic polar group. Drug Des Discov 18(2–3):65–79
171. Sekhar YN, Ravikumar M, Nayana MRS, Mallena SC, Kumar MK (2008) 3D-QSAR studies of triazafluorenone inhibitors of metabotropic glutamate receptor subtype 1. Eur J Med Chem 43(5):1025–1034
172. Sekhar YN, Nayana MRS, Ravikumar M, Mahmood S (2007) Comparative molecular field analysis of quinoline derivatives as

selective and noncompetitive $mGluR_1$ antagonists. Chem Biol Drug Des 70 (6):511–519

173. Tresadern G, Cid JM, Trabanco AA (2014) QSAR design of triazolopyridine $mGlu_2$ receptor positive allosteric modulators. J Mol Graph Model 53:82–91

174. de Paulis T, Hemstapat K, Chen Y, Zhang Y, Saleh S, Alagille D, Baldwin RM, Tamagnan GD, Conn PJ (2006) Substituent effects of *N*-(1,3-Diphenyl-1*H*-pyrazol-5-yl)benzamides on positive allosteric modulation of the metabotropic glutamate-5 receptor in rat cortical astrocytes. J Med Chem 49 (11):3332–3344

175. Lowe JEW, Ferrebee A, Rodriguez AL, Conn PJ, Meiler J (2010) 3D-QSAR CoMFA study of benzoxazepine derivatives as $mGluR_5$ positive allosteric modulators. Bioorg Med Chem Lett 20(19):5922–5924

176. Selvam C, Thilagavathi R, Narasimhan B, Kumar P, Jordan BC, Ranganna K (2016) Computer-aided design of negative allosteric modulators of metabotropic glutamate receptor 5 ($mGluR_5$): comparative molecular field analysis of aryl ether derivatives. Bioorg Med Chem Lett 26(4):1140–1144

177. Zlotos DP, Buller S, Stiefl N, Baumann K, Mohr K (2004) Probing the pharmacophore for allosteric ligands of muscarinic M_2 receptors: SAR and QSAR studies in a series of bisquaternary salts of caracurine V and related ring systems. J Med Chem 47 (14):3561–3571

178. Niu YY, Yang LM, Deng KM, Yao JH, Zhu L, Chen CY, Zhang M, Zhou JE, Shen TX, Chen HZ (2007) Quantitative structure--selectivity relationship for M_2 selectivity between M_1 and M_2 of piperidinyl piperidine derivatives as muscarinic antagonists. Bioorg Med Chem Lett 17(8):2260–2266

179. Silva ME, Heim R, Strasser A, Elz S, Dove S (2011) Theoretical studies on the interaction of partial agonists with the 5-HT_{2A} receptor. J Comput Aided Mol Des 25(1):51–66

180. Moeller D, Salama I, Kling RC, Hübner H, Gmeiner P (2015) 1,4-Disubstituted aromatic piperazines with high 5-HT_{2A}/D_2 selectivity: quantitative structure-selectivity investigations, docking, synthesis and biological evaluation. Bioorg Med Chem 23 (18):6195–6209

181. Raviña E, Negreira J, Cid J, Masaguer CF, Rosa E, Rivas ME, Fontenla JA, Loza MI, Tristán H, Cadavid MI, Sanz F, Lozoya E, Carotti A, Carrieri A (1999) Conformationally constrained butyrophenones with mixed dopaminergic (D_2) and serotoninergic (5-HT_{2A}, 5-HT_{2C}) affinities: synthesis, pharmacology, 3D-QSAR, and molecular modeling of (aminoalkyl)benzo- and -thienocycloalkanones as putative atypical antipsychotics. J Med Chem 42(15):2774–2797

182. Zhang Z, An L, Hu W, Xiang Y (2007) 3D-QSAR study of hallucinogenic phenylalkylamines by using CoMFA approach. J Comput Aided Mol Des 21(4):145–153

183. Bromidge SM, Dabbs S, Davies DT, Duckworth DM, Forbes IT, Ham P, Jones GE, King FD, Saunders DV, Starr S, Thewlis KM, Wyman PA, Blaney FE, Naylor CB, Bailey F, Blackburn TP, Holland V, Kennett GJ, Riley GJ, Wood MD (1998) Novel and selective 5-$HT_{2C/2B}$ receptor antagonists as potential anxiolytic agents: synthesis, quantitative structure–activity relationships, and molecular modeling of substituted 1-(3-pyridylcarbamoyl)indolines. J Med Chem 41 (10):1598–1612

184. Lopez-Rodriguez ML, Murcia M, Benhamú B, Viso A, Campillo M, Pardo L (2001) 3D-QSAR/CoMFA and recognition models of benzimidazole derivatives at the 5-HT_4 receptor. Bioorg Med Chem Lett 11 (21):2807–2811

185. López-Rodríguez ML, Murcia M, Benhamú B, Viso A, Campillo M, Pardo L (2002) Benzimidazole derivatives. 3. 3D-QSAR/CoMFA model and computational simulation for the recognition of 5-HT_4 receptor antagonists. J Med Chem 45(22):4806–4815

186. Iskander MN, Leung LM, Buley T, Ayad F, Di Iulio J, Tan YY, Coupar IM (2006) Optimization of a pharmacophore model for 5-HT_4 agonists using CoMFA and receptor based alignment. Eur J Med Chem 41 (1):16–26

187. Suzuki T, Imanishi N, Itahana H, Watanuki S, Miyata K, Ohta M, Nakahara H, Yamagiwa Y, Mase T (1998) Novel 5-hydroxytryptamine 4 (5-HT_4) receptor agonists. Synthesis and Gastroprokinetic activity of 4-amino-*N* (2-(1-aminocycloalkan-1-yl) ethyl)-5-chloro-2 methoxybenzamides. Chem Pharm Bull 46 (7):1116–1124

188. Doddareddy MR, Cho YS, Koh HY, Pae AN (2004) CoMFA and CoMSIA 3D QSAR analysis on N_1-arylsulfonylindole compounds as 5-HT_6 antagonists. Bioorg Med Chem 12 (15):3977–3985

189. Guner OF (2005) The impact of pharmacophore modeling in drug design. IDrugs 8 (7):567–572

190. Sun H (2008) Pharmacophore-based virtual screening. Curr Med Chem 15 (10):1018–1024

191. Dixon SL, Smondyrev AM, Knoll EH, Rao SN, Shaw DE, Friesner RA (2006) PHASE: a new engine for pharmacophore perception, 3D QSAR model development, and 3D database screening: 1. Methodology and preliminary results. J Comput Aided Mol Des 20 (10–11):647–671
192. Golender V, Vesterman B, Eliyahu O, Kardash A, Kletzin M, Shokhen M, Vorpagel E (1994) Knowledge engineering approach to drug design and its implementation in the APEX-3D Expert System. In: Proceedings of the 10th European Symposium on Structure-Activity Relationships, Barcelona: Prous Science, pp 249–254
193. Harris DL, Loew G (2008) Development and assessment of a 3D pharmacophore for ligand recognition of BDZR/GABAA receptors initiating the anxiolytic response. Bioorg Med Chem 8(11):2527–2538
194. Mills JEJ, de Esch IJP, Perkins TDJ, Dean PM (2001) SLATE: a method for the superposition of flexible ligands. J Comput Aided Mol Des 15(1):81–96
195. Wolber G, Langer T (2005) LigandScout: 3-D pharmacophores derived from protein-bound ligands and their use as virtual screening filters. J Chem Inf Model 45(1):160–169
196. Huang P, Kim S, Loew G (1997) Development of a common 3D pharmacophore for δ-opioid recognition from peptides and non-peptides using a novel computer program. J Comput Aided Mol Des 11(1):21–28
197. Clark M, Cramer RD, Van Opdenbosch N (1989) Validation of the general purpose Tripos 5.2 force field. J Comput Chem 10 (8):982–1012
198. Patel Y, Gillet VJ, Bravi G, Leach AR (2002) A comparison of the pharmacophore identification programs: CATALYST, DISCO and GASP. J Comput Aided Mol Des 16 (8–9):653–681
199. Chen IJ, Foloppe N (2008) Conformational sampling of druglike molecules with MOE and CATALYST: implications for pharmacophore modeling and virtual screening. J Chem Inf Model 48(9):1773–1791
200. Mustyala KK, Chitturi AR, Naikal James PS, Vuruputuri U (2012) Pharmacophore mapping and in silico screening to identify new potent leads for A_{2A} adenosine receptor as antagonists. J Recept Signal Transduct Res 32(2):102–113
201. Bacilieri M, Ciancetta A, Paoletta S, Federico S, Cosconati S, Cacciari B, Taliani S, Da Settimo F, Novellino E, Klotz KN, Spalluto G, Moro S (2013) Revisiting a receptor-based pharmacophore hypothesis for human A_{2A} adenosine receptor antagonists. J Chem Inf Model 53(7):1620–1637
202. Wei J, Wang S, Gao S, Dai X, Gao Q (2007) 3D-Pharmacophore models for selective A_{2A} and A_{2B} adenosine receptor antagonists. J Chem Inf Model 47(2):613–625
203. Khanfar MA, Al-Qtaishat S, Habash M, Taha MO (2016) Discovery of potent adenosine A_{2A} antagonists as potential anti-Parkinson disease agents. Non-linear QSAR analyses integrated with pharmacophore modeling. Chem Biol Interact 254:93–101
204. Balogh B, Jójárt B, Wágner Z, Kovács P, Máté G, Gyires K, Zádori Z, Falkay G, Márki Á, Viskolcz B, Mátyus P (2007) 3D QSAR models for α_{2A}-adrenoceptor agonists. Neurochem Int 51(5):268–276
205. Kothandan G, Gadhe CG, Madhavan T, Cho SJ (2011) Binding site analysis of CCR_2 through in silico methodologies: docking, CoMFA, and CoMSIA. Chem Biol Drug Des 78(1):161–174
206. Singh R, Balupuri A, Sobhia ME (2013) Development of 3D-pharmacophore model followed by successive virtual screening, molecular docking and ADME studies for the design of potent CCR_2 antagonists for inflammation-driven diseases. Mol Simul 39 (1):49–58
207. Ye Y, Liao Q, Wei J, Gao Q (2010) 3D-QSAR study of corticotropin-releasing factor 1 antagonists and pharmacophore-based drug design. Neurochem Int 56(5):107–117
208. Whitten JP, Xie YF, Erickson PE, Webb TR, De Souza EB, Grigoriadis DE, McCarthy JR (1996) Rapid microscale synthesis, a new method for lead optimization using robotics and solution phase chemistry: application to the synthesis and optimization of corticotropin-releasing factor 1 receptor antagonists. J Med Chem 39(22):4354–4357
209. Kaur P, Sharma V, Kumar V (2012) pharmacophore modelling and 3D-QSAR studies on N^3-phenylpyrazinones as corticotropin-releasing factor 1 receptor antagonists. Int J Med Chem 2012:452325. doi:10.1155/2012/452325
210. Bernard D, Coop A, MacKerell AD (2003) 2D conformationally sampled pharmacophore: a ligand-based pharmacophore to differentiate δ opioid agonists from antagonists. J Am Chem Soc 125(10):3101–3107
211. Coop A, Jacobson AE (1999) The LMC δ opioid recognition pharmacophore: comparison of SNC80 and oxymorphindole. Bioorg Med Chem Lett 9(3):357–362

212. Bernard D, Coop A, MacKerell AD (2005) Conformationally sampled pharmacophore for peptidic δ opioid ligands. J Med Chem 48(24):7773–7780

213. Levoin N, Labeeuw O, Krief S, Calmels T, Poupardin-Olivier O, Berrebi-Bertrand I, Lecomte JM, Schwartz JC, Capet M (2013) Determination of the binding mode and interacting amino-acids for dibasic H_3 receptor antagonists. Bioorg Med Chem 21 (15):4526–4529

214. De Esch IJP, Mills JEJ, Perkins TDJ, Romeo G, Hoffmann M, Wieland K, Leurs R, WMPB M, PHJ N, Dean PM, Timmerman H (2001) Development of a pharmacophore model for histamine H_3 receptor antagonists, using the newly developed molecular modeling program SLATE. J Med Chem 44 (11):1666–1674

215. Axe FU, Bembenek SD, Szalma S (2006) Three-dimensional models of histamine H_3 receptor antagonist complexes and their pharmacophore. J Mol Graph Model 24 (6):456–464

216. Jullian N, Brabet I, Pin JP, Acher FC (1999) Agonist selectivity of $mGluR_1$ and $mGluR_2$ metabotropic receptors: a different environment but similar recognition of an extended glutamate conformation. J Med Chem 42 (9):1546–1555

217. Filizola M, Tasso SM, Loew GH, Villar HO (2001) Global physicochemical properties as activity discriminants for the $mGluR_1$ subtype of metabotropic glutamate receptors. J Comput Chem 22(16):2018–2027

218. Zhang MQ, Zhang XL, Li Y, Fan WJ, Wang YH, Hao M, Zhang SW, Ai CZ (2011) Investigation on quantitative structure activity relationships and pharmacophore modeling of a series of $mGluR_2$ antagonists. Int J Mol Sci 12 (9):5999–6023

219. Lu C, Jin F, Li C, Li W, Liu G, Tang Y (2011) Insights into binding modes of 5-HT_{2C} receptor antagonists with ligand-based and receptor-based methods. J Mol Model 17 (10):2513–2523

220. Ahmed A, Choo H, Cho YS, Park WK, Pae AN (2009) Identification of novel serotonin 2C receptor ligands by sequential virtual screening. Bioorg Med Chem 17 (13):4559–4568

221. Iskander MN, Coupar IM, Winkler DA (1999) Investigation of 5-HT_4 agonist activities using molecular field analysis. J Chem Soc Perkin Trans 2(2):153–158

222. Bureau R, Daveu C, Lemaître S, Dauphin F, Landelle H, Lancelot JC, Rault S (2002) Molecular design based on 3D-pharmacophore. Application to 5-HT_4 receptor. J Chem Inf Comput Sci 42(4):962–967

223. López-Rodríguez ML, Benhamú B, de la Fuente T, Sanz A, Pardo L, Campillo M (2005) A three-dimensional pharmacophore model for 5-hydroxytryptamine 6 (5-HT_6) receptor antagonists. J Med Chem 48 (13):4216–4219

224. Kim HJ, Doddareddy MR, Choo H, Cho YS, No KT, Park WK, Pae AN (2008) New serotonin 5-HT_6 ligands from common feature pharmacophore hypotheses. J Chem Inf Model 48(1):197–206

225. Hayat F, Cho S, Rhim H, Indu Viswanath AN, Pae AN, Lee JY, Choo DJ, Choo HY (2013) Design and synthesis of novel series of 5-HT_6 receptor ligands having indole, a central aromatic core and 1-amino-4-methyl piperazine as a positive ionizable group. Bioorg Med Chem 21(17):5573–5582

Chapter 4

Virtual Screening in the Search of New and Potent Anti-Alzheimer Agents

Livia Basile

Abstract

Alzheimer's disease (AD) is a multifaceted neurodegenerative disorder for which there is no cure, but only symptomatic drugs are available. Most of the scientific efforts are addressed toward the employment of computational approaches able to speed up the discovery of putative anti-AD drugs. Among these, virtual screening allows to select very quickly and at extremely low cost compounds endowed with optimum physicochemical, pharmacokinetic, and biological properties to develop new potential drugs. This chapter presents recent works on the use of virtual screening for the design of specific ligands of targets related with AD, most of which were subsequently validated by experimental assays.

Key words Alzheimer/Drug design/Virtual screening/Similarity search/Acetylcholinesterase inhibitors/ BACE-1 inhibitors/γ-Secretase inhibitors

1 Introduction

Alzheimer's disease (AD) is a progressive neurodegenerative brain disorder, known as the most common cause of dementia in the Western world. Clinically AD leads to a gradual cognitive impairment that in advanced stages becomes strongly disabling. At the central level, AD causes a progressive loss of cortical neurons, especially pyramidal cells involved into the higher cognitive functions. From a molecular point of view, AD is considered a multifactorial disease characterized by various molecular and cellular processes, among which causes of protein aggregation, oxidative stress, cell cycle deregulation, neuroinflammation, and decreased acetylcholine levels are still unknown [1, 2]. As a result, up to now no drugs able to modify the progression of the disease are available, and current therapeutic strategies cure only the symptomatic aspect. An example of symptomatic drugs is given by acetylcholinesterase (AChE) inhibitors, namely, donepezil, rivastigmine, and galantamine. Low levels of the neurotransmitter acetylcholine (ACh) in fact lead to the death of cholinergic neurons underlying

Kunal Roy (ed.), *Computational Modeling of Drugs Against Alzheimer's Disease*, Neuromethods, vol. 132, DOI 10.1007/978-1-4939-7404-7_4,

© Springer Science+Business Media LLC 2018

synaptic failure and cognitive dysfunction [3]. Experimental evidences demonstrated that the key event involved in the pathogenesis of AD is the increased production of amyloid beta protein (Aβ), a 39–42 amino acid peptide that comes from the enzymatic cleavage of the membrane-embedded amyloid precursor protein (APP) [4] by the transmembrane aspartyl protease, β-secretase (BACE-1) [5, 6], followed by presenilin-dependent γ-secretase cleavage. It is a hypothesis that high levels of Aβ(1–42) allow the aggregation of monomeric Aβ(1–42) into neurotoxic oligomeric species, leading to neuronal death [7]. Unfortunately, even though it is accepted that Aβ is involved in the pathogenesis of AD, at present no inhibitors of Aβ aggregation or BACE-1 activity are in the market for AD therapy. A large number of potential Aβ fibrillogenesis inhibitors have been investigated as polyamines, metal chelators, carbohydrate-containing compounds, polyphenols, osmolytes, short peptides, RNA aptamers, etc. [8, 9]. It is also believed that DNA polymerase-β might have a causal role in Aβ-induced DNA replication and neuronal death. This DNA repair enzyme intervenes in Aβ-induced DNA replication representing an intriguing target in the treatment of AD [10, 11].

To the best of our knowledge, all the efforts of scientific community aim to the discovery of more efficient anti-AD therapeutic strategies.

Computer-aided drug design approaches are frequently used in drug development [12], since they allow to improve effectiveness and efficiency minimizing time and costs of chemical synthesis and biological testing [13]. Large compound libraries can be screened computationally to reduce the number of compounds for bioassays screen, thus decreasing time and resources [14, 15]. Furthermore, the identified hits can be optimized to lead compounds by means of these in silico approaches. Some of these "rational" drug design strategies are based on the molecular similarity principle, which states that "structurally similar compounds are more likely to exhibit similar biological activity" [16]. Molecular similarity concept takes hold of a wide range of applications: ligand-based (2D fingerprints, shape based) virtual screening technologies (LBVS); 3D pharmacophore modeling; estimation of absorption, distribution, metabolism, excretion, and toxicity (ADMET); and prediction of physicochemical properties (solubility, pKa, etc.). These similarity-based approaches aim to find bioactive molecules by measuring their similarity to a known active compound or their complementarity to a protein binding site.

This chapter focuses on the role of virtual screening approaches to discover new drug candidates for different targets associated with AD. Recent works will be presented (Table 1) and discussed.

Table 1
Recent works based on virtual screening in the search for new AD agents

Target	Approach	Database	N/leads obtained	Authors	Ref.
AChE	LBVS[a]	*In-house* 3D database, NPD	2	Rollinger et al.	[91]
AChE	LBVS, SBVS[b]	CERMN	2	Sopkova-de Oliveira Santos et al.	[92]
AChE	LBVS	NCI, Specs, IBScreen	2	Guptaet al.	[93]
AChE	LBVS	NCI		Lu et al.	[95]
AChE	SBVS	Specs	2	Chen et al.	[96]
AChE	LBVS	*In-house* database	2	Chaudhaeryet al.	[98]
AChE	LBVS, SBVS	DrugBank database	10	Bag et al.	[99]
AChE	LBVS	InterBioScreen NC	4	Ambure et al.	[100]
AChE, BChE	LBVS	ZINC	2	Nogara et al.	[101]
AChE, BChE	LBVS	ZINC	2	Bajda et al.	[102]
AChE	LBVS, SBVS	ZINC	2	Dhanjal et al.	[104]
AChE, BChE	LBVS, SBVS	Chemdiv	5	Chen et al.	[105]
γ-Secretase	LBVS	*In-house* database	10	Gundersen et al.	[117]
γ-Secretase	SVM[c]-RF[d]	ZINC	6	Yang et al.	[120]
BACE-1	SBVS	–	–	Polgár et al.	[121]
BACE-1	LBVS	Chemdiv, Chemnavigator	4	Huang et al.	[122]
BACE-1	LBVS, SBVS	Maybridge database, LeadQuest	2	Vijayan et al.	[124]
BACE-1	SBVS	Specs		Xu et al.	[129]
BACE-1	LBDD[e]	–	3	Hossain et al.	[130]
BACE-1	LBVS	Chemdiv, Zinc	1	Ju et al.	[133]
Tau protein	LBVS	Maybridge database	2	Larbig et al.	[134]
p38-alpha MAPK	LBVS, SBVS	CNS, Kinacore KINASet, ZINC	13	Pinsetta et al.	[136]
$A\beta_{40}$, $A\beta_{42}$	LBVS, SBVS	TCM database	1	Viet et al.	[139]
ABAD	LBVS, SBVS	Local database		Valasani et al.	[143]

(continued)

Table 1 (continued)

Target	Approach	Database	N/leads obtained	Authors	Ref.
Mt-iii	LBVS, SBVS	InterBioScreen	3	Roy et al.	[144]
DNA pol-β	LBVS	ZINC	1	Merlo et al.	[148]

[a]Ligand-based virtual screening
[b]Structure-based virtual screening
[c]Support vector machine
[d]Random forest
[e]Ligand-based drug design

2 Virtual Screening

In virtual screening, a large database of molecules is screened for their similarity to known active molecules or their capability to bind a target molecule. This computational approach has been proposed as an alternative or complementary to high-throughput screening (HTS) [17], since it allows to prioritize the selection and testing of large molecule data sets. Based on the similar property principle [16], which states that *molecules with similar structure have similar biological profile*, virtual screening can be carried out employing 2D [18] or 3D descriptors [19, 20], experimental or computed physicochemical properties, and cheminformatic fingerprints of the molecules. The approach can also help in lead discovery to find hits through library enrichment [21]. At last, those molecules that have the largest a priori probabilities of activity in a lead discovery perspective will be assayed. Two approaches are parts of virtual screening: (1) the structure-based approach, namely, docking and de novo design, and (2) the ligand-based approach used when there are no structural information on the target protein (Fig. 1). Ligand-based methods are based on structural information of biologically active small molecules. Further, this approach offers the advantage of using pre-calculated descriptors finally being faster than molecular docking in screening large databases. It is often used as a first filter to slim down the size of molecular libraries before docking [22–24]. Ligand-based methods include similarity methods, pharmacophore methods that are essentially made up of a pharmacophore model from a set of known actives to perform a 3D substructure search, and machine learning methods used to develop a clustering rule from a training set with known active and known inactive compounds. By contrast, structure-based approaches are usually more computationally expensive but help to derive 3D structural hypotheses on the putative binding mode of a ligand with its target [25]. However, when both receptor and ligand 3D structures are available, combined ligand- and structure-based approaches might be a more advisable way toward more reliable results. The most common ligand- and structure-based virtual screening techniques will be described in the following paragraphs.

2.1 Ligand-Based Virtual Screening

2.1.1 Similarity Search

Molecular similarity provides an accessible and well-known method for virtual screening based on the *similar property principle*. A database of molecules with unknown activity but structurally similar to an active biologically molecule (reference or target structure) is also *likely* to be active. This capability strongly depends on both the likelihood of ligands to assume conformers that fit into the active site of the target protein and their physicochemical properties. Molecular similarity measures take into account both molecules' shapes and their physicochemical properties. Based on the fact that if two molecules have a similar scaffold, they most probably also have a similar surface, hence, similar binding properties; molecular similarity measures consider molecular scaffold rather than molecular shape. However, two molecules endowed with a similar surface do not always possess a similar scaffold. Consequently, unlike scaffold-based approaches, surface-based approaches are able to identify molecules with similar binding properties. The similarity concept can be applied globally between two whole objects, for instance, small molecule, protein, etc., or locally to an object such as atom, substituents, DNA strand, etc. Though, the use of a local and global similarity analysis on the same sample can lead finally to contrasting results.

According to what is reported by Sheridan and Kearsley [26], similarity methods are useful in the pharmaceutical field firstly because they are computationally unexpensive and, therefore, can be routinely employed; no external information is needed. Moreover, the target may not be known, and the analysis starts from one or two known *actives*; finally, in accordance with the *similar property principle*, the molecules show "neighborhood behavior" [27]. There are many different modes to measure molecular similarity, and as a rule they involve three principal components: the descriptors, the coefficients, and the weighting scheme. The descriptors are representations used to characterize the molecules to compare which arise from the structure (constitutional, configurational, and conformational descriptors) or the physicochemical and biological properties of the molecules [28]. Preferably, the descriptors used to build the model should be rapidly calculated and accessible to the computers and users. The number and the complexity of the molecular descriptors are remarkably increased in the last few years. The software CODESSA calculates about 400 molecular descriptors among which topological, topographical, and quantum chemical descriptors [29]. DRAGON software provides more than

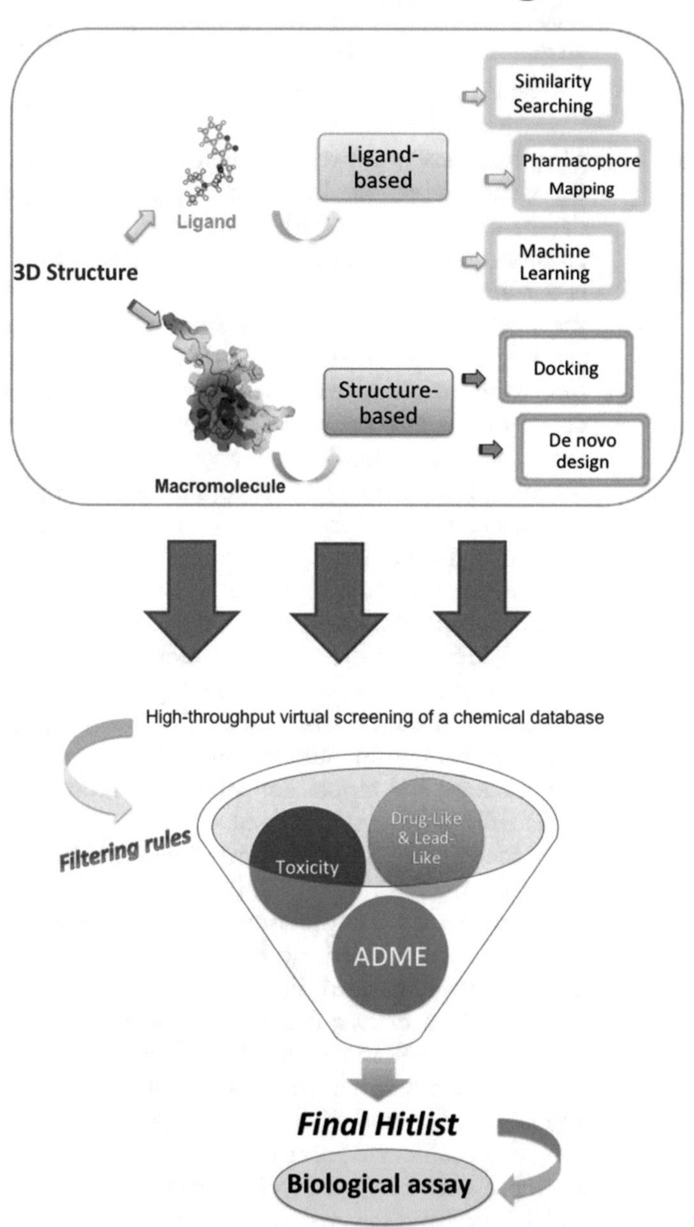

Fig. 1 Ligand-based and structure-based approaches are used as virtual screening methods. They mainly differ for the input to query large compound databases

1600 molecular descriptors that are divided into 20 logical blocks including topological and geometrical descriptors [30]. The descriptor-based similarity method represents a molecule as a set of descriptors in a multidimensional descriptor space, and the similarity measure of compounds is related to their intermolecular distance in that space. One example is given by constitutional descriptors that express the order of bonds and atoms, presence or absence of substituents, and other 2D elements. Configurational and conformational descriptors express the 3D arrangements of atoms which are thermodynamically stable [28].

The similarity coefficients, such as index and distance, are functions that transform molecular representations into real numbers over any ranges, and they may include basic property, or they can be dichotomous values, for instance, binary values, where the numbers 0 and 1 indicate the absence or presence of some particular features of the object. They are the quantitative expression of a measure of the chemical resemblance degree [31–34]. The most popular coefficient for comparison fingerprints is the Tanimoto coefficient, also known as Jaccard coefficient. It is the most commonly used coefficient in both *in-house* and commercial software for chemoinformatics [18]. The Tanimoto coefficient is expressed by the formula:

$$\frac{c}{a + b - c}$$

where a and b are bits in the fingerprints of two molecules and c represents the bits in common in both of the fragment bit strings. It can assume values from zero (none bit in common) to one (all bits in common). Despite its large use, the Tanimoto coefficient has the limitation to be affected by the size of the molecules in analysis. It has been noted, in fact, that it gives low similarity scores when the query consists of a few bit strings in its fingerprints [35] showing a characteristic bias to particular similarity values [36]. A solution to this problem has been suggested by Holiday [37] and Salim [38] which noted that some of the similarity coefficients express a different type of structural resemblance. According to these authors, the use of multiple similarity coefficients in a virtual screening procedure should improve screening performance rather than make a search using a single coefficient. These *data fusion* methods elaborate outcomes from database screening carried out using the same reference structure but different similar coefficients. Unfortunately, there does not exist a single combination (the best one) that is appropriate to properly screen all active classes.

Values from 0 to 1 offer a simple way for moving between a similarity coefficient and a complementary distance coefficient. The similarity coefficients can also be classified into correlation, probabilistic, and distance coefficients [39]. The latter corresponds to distances in a multidimensional geometric space.

The third distinctive component of molecular similarity is the weighting scheme, which indicates the level of importance of the various components of these representations.

In molecular similarity approaches, a very large number of descriptors can be used. Overall there are many types of descriptors that come from the molecular graph or are associated to molecular shape or are based on calculations of the molecular wave functions.

There are also descriptors that represent a change in the molecular structure such as ionization and pKa or extracted from experimental measurements, for instance, the logP [40]. One-dimensional descriptors describe physicochemical properties, such as solubility, polarizability, volume, and molecular weight, useful for the prediction of physical or key properties for receptor binding like the miscibility. Topological indices represent molecular connectivity and do not consider conformational flexibility and three-dimensional structures [41, 42]. Yet they are fragment- or substructure-based descriptors, often used in the search of substructure "similar" to a reference substructure [43]. Another group of descriptors comes from the combination of descriptors through correlation or principal component methods [44, 45]. Most popular are field-based descriptors based on the surface potential of molecules calculated by a probe such as positive charge, water molecule, or methyl group on a regular grid [46]. These last descriptors consider the three-dimensionality and require alignment of the molecules to compare. The quantum mechanical and non-quantum mechanical methods belong to this group, where quantum similarity takes into account electron distribution [33] and electron density [47], the latter simplified through a Gaussian approximation [48, 49] that increases the performance. One example of grid-based descriptors associated with a robust statistical method to obtain predictive models is the comparative molecular field method, comparative molecular field analysis (CoMFA) [46]. Another class of descriptors arise from pharmacophore hypothesis that represents *the geometrical arrangement of key functional groups responsible of the biological activity* or according to the IUPAC definition *the ensemble of steric and electrostatic features that is necessary to ensure the optimal supramolecular interactions with a specific biological target structure and to trigger (or block) its biological response.* These approaches go beyond molecular alignment and are often known as two-point pharmacophores (2PP) or atom pairs [50] since they correspond to atom pairs in the molecule. Three-point pharmacophores (3PP) and four-point pharmacophores (4PP) [51, 52] have been also reported.

Hence, the choice to use a certain descriptor is determinant and puts up with certain conditions. Despite 2D descriptors reveal the physical properties and reactivity of molecules, 3D descriptors are equally important since they are associated with biological activity. However, when the choice falls on 3D descriptors, the flexibility question has to be considered, and it is not so simple determining the energy for each conformer. Another issue is the way to depict the structure: by patterns of atoms or by its mode of interaction with a receptor. On the whole, the opportunity to choose one particular descriptor is not mutually exclusive; however, the redundancy phenomena by excluding correlated descriptors should be avoided. In addition, the selection of descriptors for screening of large database is affected by the speed of calculation. More generally, there are no "right" or "wrong" descriptors, and the choice for a type of descriptor is strictly linked to the investigation scope and the study context [53].

2.1.2 Pharmacophore-Based Virtual Screening

In the pharmacophore-based virtual screening, a pharmacophore hypothesis is screened on 3D chemical databases with the aim of finding hits having chemical features similar to those of the pharmacophore query. Hence, pharmacophore-based virtual screening provides molecules that match with the chemical features of the template. However, some of these molecules might be different for the scaffold but share the same biological activity as in the case of scaffold hopping [54]. The concept of pharmacophore was first formulated by Ehrlich in 1909, who described it as "a molecular framework that carries (phoros) the essential features responsible for a drug's (pharmacon) biological activity" [55]. A century later, the basic pharmacophore concept was modified in part according to the most recent definition of IUPAC, namely, a pharmacophore model is "an ensemble of steric and electronic features that is necessary to ensure the optimal supramolecular interactions with a specific biological target and to trigger (or block) its biological response" [56]. A pharmacophore model can be obtained either by a ligand-based approach by superimposing a series of active molecules and selecting common substituents essential for their bioactivity or by a structure-based approach. The latter is accomplished by searching putative key groups in the macromolecular target [57]. The process of deriving pharmacophores, named pharmacophore mapping, foresees three main actions: identifying common chemical determinants that affect the biological activity, calculating putative conformations that the active compound can assume, and establishing 3D relationships between pharmacophoric points in the calculated conformations. Conformational analysis and manual alignment are very time-consuming steps, especially when the number of active ligands is high and the key chemical features of the pharmacophore model are difficult to locate. Actually, several automated pharmacophore generators are available as commercial software, such as HipHop [58], Hypogen [59], DISCO [60], GASP [61], and PHASE [15]. Each of these programs uses several algorithms that differently compute the flexibility of the ligands and perform the alignment of molecules.

In the receptor-based pharmacophore modeling, the pharmacophore hypothesis can be derived from the active site of the 3D structure of a target macromolecule by analyzing the complementarity and the 3D rearrangement of the side chains in the binding site. This procedure is more accurate than the ligand-based approach since the search is constrained to a region of the receptor where the interactions with ligands take place. Unlike its counterpart docking-based virtual screening, pharmacophore-based virtual

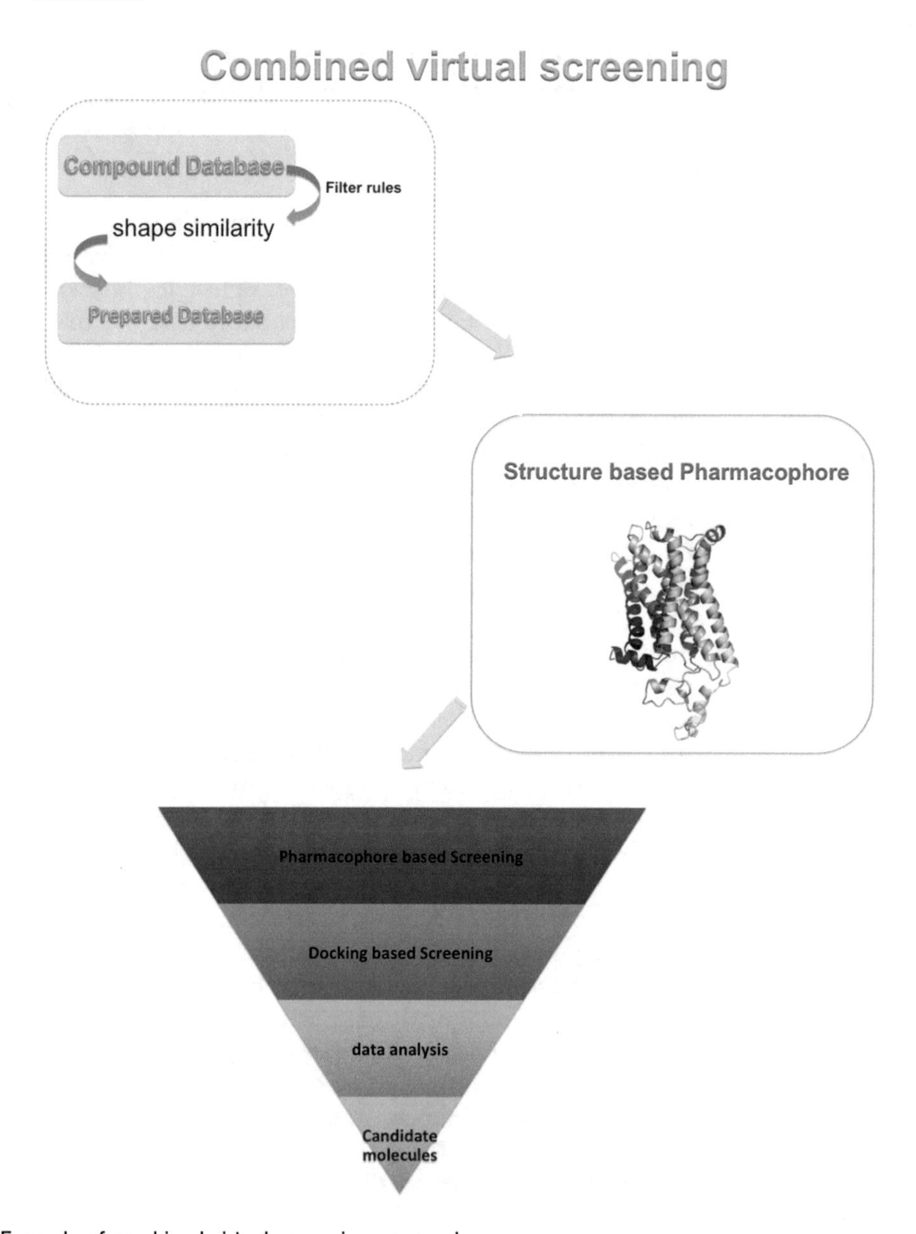

Fig. 2 Example of combined virtual screening approach

screening overcomes the issue related to the protein flexibility by considering a tolerance radius for each pharmacophoric feature. Furthermore, pre-calculating multiple conformations for each ligand or conformational sampling is used to handle the protein flexibility [57]. Beyond these solutions, the screening of large chemical databases with flexible molecules remains a key challenge for the pharmacophore-based approach. A key strategy that in part allows limiting the search time, known as multilevel searching approach, uses screening filters in an increasing order of complexity able to slow down the entire database to a small amount of molecules [62]. Furthermore, the quality and composition of the pharmacophore model are considered to be another limit of this virtual screening approach that can provide a more restricted number of actives. Even though the problem of flexibility is taken into account by using a tolerance radius for each pharmacophoric feature, not always it is able to completely remedy for the receptor flexibility. For this reason, recent efforts are focused toward the introduction of MD simulations in pharmacophore models that will have an adequate accounting of flexibility [63]. The combined use of pharmacophore-based VS and docking-based VS in a hybrid approach can help to fill these gaps and find putative hits (Fig. 2).

2.1.3 Machine Learning

Machine learning techniques are useful to build and optimize predictive models by using active and inactive molecules [64]. The classification or regression models obtained from the training sets allow to estimate the relationship between compound structural features and biological activity. Machine learning techniques can be categorized into unsupervised (training the machine learning task in an input to target way) and supervised methods (training the machine learning task only with a set of inputs) [65]. The first one includes self-organizing maps and principal component analysis (PCA) and considers only the descriptor contents to represent the data set [66]. In general, this technique reduces high-dimensional descriptor data sets to few (two or three) dimensions, clustering compounds based on the original descriptors. Hence active or inactive compounds can be found by detecting biological activities onto the data space. In the case of novel molecules, they are predicted to be active according to their closeness with the active region. It is interesting that unsupervised methods do not suffer from overfitting for the very low number of the tunable descriptors. Supervised methods [67] where the biological activity (dependent variable) is considered to set up the model use regression or classification techniques for model validation. A test set from the initial data set containing active and inactive molecules is employed to evaluate the predictive ability of the model. Therefore, models that are based on adjustable parameters have predictive power for the data set on which the model was built, but they could not predict the activity for novel compounds, a phenomenon called overfitting. It follows that the model quality is really accomplished when the model is able to predict the activity of not parameterized compounds. This model validation is often made by using the model to screen compound databases. Support vector machine (SVM) and random forest (RF) are actually among machine learning techniques/methods of choice in virtual screening. In SVM techniques, the descriptors are projected on to a higher-dimensional space through a classifying function named "kernel" [68]. The virtual screening search is carried out by projecting the descriptors of the database molecules into their high-dimensional representation by means of the kernel. RF methods are obtained from the combination of decision tree techniques coming from different subsets of the training data [69]. It is generally employed when the use of single decision tree gives rise to over-fitting. In the decision tree methods, the training data set is distributed at each node of the tree into active and inactive molecules according to the presence of a particular substructure or a particular descriptor [70]. Finally, the best tree has leaves full of either just active or inactive molecules. In the case of a new molecule, according to the decision procedure at each node of the tree, it reaches the leaf with only active or inactive molecules. The applications of machine learning for virtual screening are widely discussed in literature [71, 72]. In particular, these studies show that the predictive capability of several different machine learning methods for virtual screening, when applied on a set of different biological targets, is strongly affected by the data set, and the obtained results demonstrated that SVMs and RFs are the best approaches in relation to their predictive ability.

2.2 Structure-Based Virtual Screening

In structure-based virtual screening (SBVS), high-throughput docking is used to virtually screen a compound database into a target binding site. In modern drug design, molecular docking explores the various conformations that a ligand can assume within the binding sites of macromolecular target [73]. By using the 3D structural information of the target protein, docking analysis allows to investigate molecular interactions needed for the binding between ligand and protein. Docking and scoring provide a ranking of different molecules that reflects the binding affinity to the protein. The combination of different scoring algorithms (*consensus scoring*) might provide better results than the use of an individual one [74]. Performing SBVS includes protein preparation, compound database selection, molecular docking, and post-docking analysis. A careful analysis of both the target and known ligand structures along with an adequate choice of docking algorithms are needed for successful strategies [75]. Structural information of the receptor can be obtained from X-ray crystallography, NMR, or homology modeling. This latter technique allows the building of the target structure from the 3D structure of a known homologous protein. When several structures of a particular receptor are known, information as conformational changes and structural resolution should be taken into account for the selection of the most appropriate structure. Furthermore, being the crystal structure representative of a single conformation of the protein without any information on possible conformational changes, the attempt to include receptor flexibility in docking program with the aim to gain a more realistic protein model is still a challenge. The most common approach used to handle the flexibility issue foresees the use of an ensemble of receptor conformations (ensemble docking) [76]. The preparation steps for proteins include adding hydrogen atoms, eliminating water molecules when they are not involved in important interactions, correcting protonation and tautomerization states of the binding site residues, and calculating partial charges. The content and quality of a compound collection affect the success of a docking-based virtual screening. These libraries contain a wide amount of chemical data, collected in a specific format, such as SMILES, SMARTS, and InChI [77], which are then converted into 3D structures. Moreover they provide information on the chemical suppliers. Each compound in the library is virtually docked into the binding site of the protein in order to predict the most stable ligand-target interactions and obtain an optimal fit of steric and physicochemical properties. The search algorithm then explores the energy landscape of each complex. Top-scoring compounds by a so-called scoring function [78–80] are then chosen as potential ligands. Since thousands or millions of molecules are screened, post-docking analysis is conducted to choose which compound to prioritize. Unfortunately false positives and/or false negatives can be found because of the many times poor performance of the scoring functions in estimating binding affinity [81]. Therefore visual inspection of the predicted poses can be a good way to integrate the analysis and have more sound results. The validation of docking methods can be assessed by their ability in reproducing the binding mode of co-crystals. It consists in redocking the ligand into the protein and comparing the predicted conformation with the experimental coordinates by means measure of the rootmean-square deviation (RMSD) [82].

3 Cases Studies

3.1 Anticholinesterase Inhibitors

It is a well-accepted hypothesis that both acetylcholinesterase (AChE) and butyrylcholinesterase (BChE) are involved in the pathogenesis of AD [83]. In fact, the reduction of acetylcholinesterase levels is believed to be one of the possible causes for this cognitive disorder. Hence, maintaining the concentration of acetylcholine (Ach) at physiological level represents a determinant for treating AD [84]. The two cholinesterases (ChEs) are responsible for the hydrolytic cleavage of acetylcholine at central level. For this reason, the use of cholinesterase inhibitors would regulate the acetylcholine levels in the brains of people affected by AD [85]. Since butyrylcholinesterase differs from acetylcholinesterase for sensitivity to substrates and inhibitors [86], it is believed that acetylcholinesterase inhibitors (AChEIs) can provide more specific therapeutic effects than BChE inhibitors (BChEIs). However, in the last decade, a functional role of BChE in the late stage of the disease has been recognized. Thus, searching for selective BChE inhibitors is also an important task of drug developers. As anticipated before, to date three ChE inhibitors have been used in the therapy of AD, acting only on symptomatic effects because they are unable to cure or prevent neurodegeneration. Among these, donepezil and galantamine are selective AChE inhibitors, while rivastigmine is a dual inhibitor for both AChE and BChE. The analysis of the 3D structure of AChE revealed the presence of a deep and narrow gorge at the active site mainly composed of dual binding sites, namely, the catalytic anionic site (CAS) at the bottom of the gorge and the peripheral anionic site (PAS) near the entrance of the gorge [83]. The interaction with both CAS and PAS activates multiple therapeutic effects, because they are functionally different. CAS is responsible for the hydrolysis of Ach, whereas PAS acts on both Ach hydrolysis and $A\beta$ aggregation [87]. According to these insights, numerous efforts were addressed to the design of ligands as dual binding site inhibitors (Fig. 3). Furthermore, structural differences in the active site of both cholinesterases have prompted the research in this field toward the design of selective BChE inhibitors. The application of computational tools in drug discovery to find out novel potential cholinesterase inhibitors is most recurring in literature [88, 89]. Both ligand-based [90] and structure-based [25] approaches could help to interpret ligand-receptor interactions. A 3D query in screening database for potential biologically active compounds has been often used.

For example, a structure-based virtual screening of a 3D multiconformational database, consisting of more than 110,000 natural products, led to the discovery of scopoletin and its glucoside scopolin as potential AChE inhibitors. Both the molecules demonstrated to increase the extracellular acetylcholine concentration in rat brain compared to basal release. This was the first report on the investigation of the increase in the AChE in vivo levels after administration of coumarins and shows the reliability of the models that correctly predicts a distinct interaction with all the generated features for both the ligands [91].

In the search for new dual binding site inhibitors of AChE able to bind both the catalytic and the peripheral anionic sites, two kinds of virtual screening of the CERMN chemolibrary were done by Sopkova-de Oliveira Santos et al. [92]. A receptor-based screening using the 3D structure of AChE as a template for docking (Gold software) was performed firstly followed by a ligand-based screening based on a 3D pharmacophore models using catalyst. The best hypothesis provided 40 compounds with ten derivatives showing an IC_{50} in the range of μM–nM against AChE. Interestingly, the authors reported on a higher efficiency of the ligand-based screening compared to that of the structure-based (docking) approach by virtue of the availability of the conformation of the ligands donepezil and (S,S)-(-)-bis(10)-hupyridone in the active site used to generate the models. Ligand- and structure-based approaches are to be considered in a complementary fashion. This is because pharmacophore approaches provide correct insights on the ability of the ligand to bind certain regions of the AChE whether docking gives information related to the putative orientations of the ligands in the AChE binding pocket.

The outcome of a pharmacophore-based virtual screening approach especially depends on the accuracy and specificity of the pharmacophore query employed as demonstrated in the work of Gupta et al. for discovering dual binding site and selective AChE inhibitors using pharmacophore modeling and sequential virtual screening techniques [93]. The 3D models were built basing on a data set containing potent and selective AChE inhibitors obtained from a previous high-throughput in vitro screening study that involved a collection of about 56,000 compounds [94]. These compounds were selected for their ability to interact at the dual binding site of the targeted AChEs. Sequential virtual screening techniques were carried out on three different small compound databases, namely, National Cancer Institute (NCI), Specs, and InterBioScreen (IBScreen) using the pharmacophore model as a 3D query. Lipinski's rule of five, molecular docking, and absorption, distribution, metabolism, excretion, and toxicity (ADMET) were some of the criteria employed to refine the retrieved hits. In order to avoid false positives during the virtual screening, the efficiency of the model was evaluated by measuring sensitivity (%S = 73.5), goodness of fit (G = 0.75), and enrichment (E = 7.3). The nine putative dual binding site ligands were purchased and tested by an in vitro AChE inhibition assay. Finally, by using an integrated in vitro–in silico approach, two potent and selective dual binding site AChE inhibitors were obtained that might be used for AD.

A combination of pharmacophore modeling, virtual screening, and molecular docking studies were performed by Lu et al. [95]. Therein ten pharmacophore models were built and validated by statistical analysis. The best one ($R^2 = 0.830$) was used as a 3D query in a virtual screening process on the NCI database. The top 252 hits were subsequently filtered using docking methods that provided successfully putative novel AChE inhibitors.

In another interesting study, on pharmacophore-based virtual screening [96], the authors combined the 3D model with docking analysis and molecular dynamics to gain insights into the binding mode of tacrine hybrids and AChE. The structure-based pharmacophore (SBP) model was developed from the binding models and employed to screen the commercial Specs database for new compounds that satisfy the wanted chemical features. A focused hit list of 162 compounds was screened out from 263,148 entries based on the "best flexible search." Finally, the hit compounds were ranked on the basis of their molecular binding energy calculated by MM/Poisson–Boltzmann surface area (PBSA) continuum

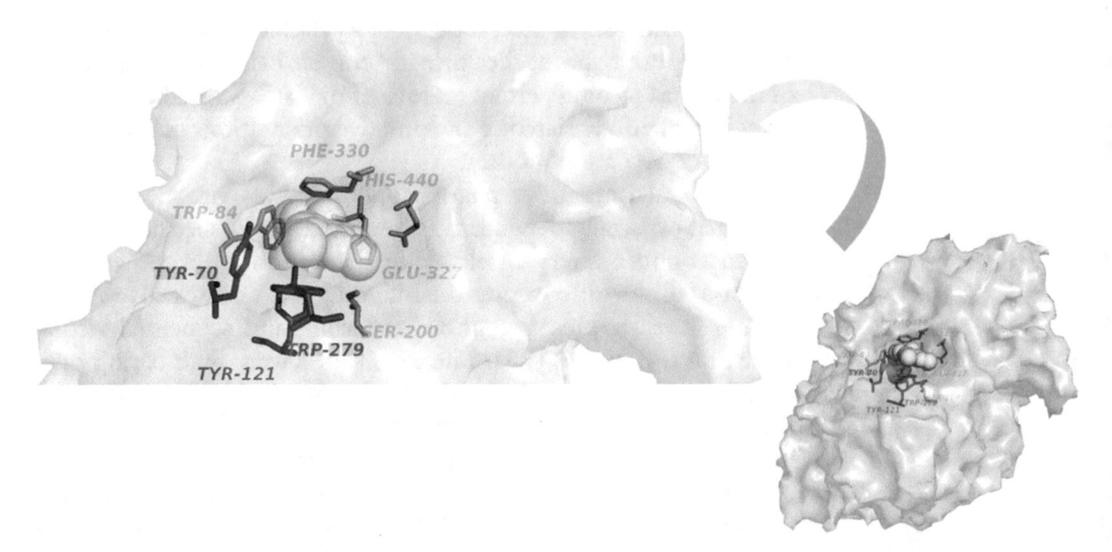

Fig. 3 Structure of *Torpedo californica* AChE (TcAChE; PDB code: 1EVE) complexed with E2020 (Aricept). The structure of TcAChE is depicted as a surface representation. Bound E2020 molecule lying into the gorge of AChE is shown as spheres. Side chains of catalytic anionic site (CAS: *Trp84* and *Phe330*), peripheral anionic site (PAS: *Tyr70*, *Tyr121*, and *Trp279*), and catalytic triad (CT: *Ser200*, *Glu237*, and *His440*) are represented in stick format

solvation methods [97]. Fifteen compounds were selected and purchased to test their in vitro activity against AChE. Among these, two compounds showed potent activity against AChE and good *drug-like* properties.

Recently, Chaudhaery et al. reported a simple pharmacophore-based virtual screening using Hypogen [59] starting from a set of 24 carbamates as AChE inhibitors [98]. The best model ($R^2 = 0.844$) finally find out two novel lead compounds that were submitted to pharmacological evaluation and proposed as candidate molecules for further modifications.

It is interesting to note that several commercially available marketed drugs can also act as potential AChE inhibitors. This hypothesis was considered by Bag et al. in their attempt to discover acetylcholinesterase inhibitory activity among already marketed drugs; the DrugBank database and natural alkaloids were screened by using a four-point pharmacophore model [99]. Therein both the ligand-based and structure-based mining strategies were applied to find new potential AChE inhibitors. The four-point pharmacophore model showed good predictive ability ($R^2 = 0.94$; $Q^2 = 0.73$) and was used to screen the DrugBank database along with ten different alkaloids that are believed to have anti-AChE activity. The selected molecules were further screened by docking analysis. The most promising ligands were evaluated by in vitro analysis, and many of them showed good inhibitory activity and, indeed, the reliability of the model.

Hence, the combined use of pharmacophore model and docking analysis successfully allows to identify putative novel AChE inhibitors from a database. The Ambure research group utilized known AChE inhibitors to develop a broad spectrum pharmacophore model as query to screen InterBioScreen NC database natural compounds [100]. The data set considered in that work spanned a wide range of activities range and was comprehensive of diverse structures. A first screening was carried out for drug-like molecules by using different rational filters such as the Lipinski's rule of five and drug-like ADME properties. Next, pharmacophore-based virtual screening and docking analysis further reduced the number of molecules retrieved from the data bank. Among the top hits, coumarin, furanocoumarin, pyranocoumarin, pyridocoumarin, and non-coumarin classes (like indole) of natural compounds were found as potential AChE inhibitors candidates.

A simple virtual screening procedure of the ZINC database using the Lipinski's rule of five and docking leads to find compounds similar to AChE with quaternary ammonium moiety in their structure [101]. The top-scoring compounds were further screened by in vitro assays, and relevant anti-AChE and anti-BChE activities were found. The results suggest further that using together pharmacophoric properties and molecular docking is a promising approach that might be employed for high-throughput screening with the aim to detect new bioactive compounds.

Recently, another work concerning the identification of novel cholinesterase inhibitors was proposed by Bajda et al. [102]. In that work the authors carried out a pharmacophore screening of the ZINC database, containing over 35 million of commercially available compounds. Two different pharmacophore models were generated for both cholinesterases. Crystal structures of AChE complexes with four inhibitors (donepezil, bis-tacrine, decamethonium, and tacrine) were employed. This showed that hydrophobic and electrostatic interaction properties play major role in the ligand binding process for the presence of several aromatic residues and cationic centers at the active gorge of the enzyme. Due to the lack of crystal structures of BChE complexes with inhibitors, the pharmacophore model was obtained on the basis of bis-nor-meptazinol, which is a potent inhibitor of both cholinesterases. This pharmacophore model was interpreted based on the putative interactions described in literature [103]. The screened compounds were further selected by docking to open, half-open, and closed gate structures of AChE (PDB codes: 1EVE, 1VOT, 1ACJ, respectively). The seven selected compounds were tested against both cholinesterases, and the theoretical prediction was compared with the experimental activity data. Two of them showed, respectively, 50.1% and 79.5% of inhibition against butyrylcholinesterase at the concentration of 100 μM, representing a starting point for further optimization in a lead discovery process.

The same workflow was followed by Dhanjal et al. [104] which proposed a combined strategy of ligand-based pharmacophore modeling high-throughput virtual screening and docking analysis to find natural products that could have anti-acetylcholinesterase activity for Alzheimer's disease. For this strategy, 16 known AChE inhibitors were downloaded from PubChem, and for each ligand several conformers were generated. Then, pharmacophore hypotheses for all the ligands were created by a set of pharmacophore features of the 3D structures. The common pharmacophores were ranked by a scoring function to find the best hypothesis that gave the best alignment of the chosen ligands. The highest-scoring hypothesis was screened against a data set of known natural drug-like compounds from ZINC obtaining finally several hundred hits that matched the selected hypothesis. These compounds were further filtered by docking to AChE using the software Glide. The two best-scored compounds which based on their binding affinity can be considered as potential drugs in the symptomatic treatment of the Alzheimer's disease were further analyzed for their binding mode in the AChE active site.

A relevant study was very recently described by Chen et al. [105]. Three-dimensional shape-based similarity screening combined with structure-based pharmacophore modeling and molecular docking was applied to identify new scaffolds for the design of AChE inhibitors. The screening was performed on the Chemdiv library containing 1,293,896 molecules. Donepezil was used as the template to build a 3D-shaped model. Thus, the collection of molecules was selected by rapid overlay of chemical structures (ROCS module of the OpenEye suite). The software uses a Gaussian function to depict molecular volumes of chemical structures to screen [106] and then calculates the shape similarity of a compound to the model, scoring the results in the range from 0 to 1. At the same time, a structure-based pharmacophore model was generated using the structure of donepezil co-crystalized with AChE (PDB code: 4EY7). This model was used to screen the Chemdiv collection providing 12,939 as primary hits. The selected hits were subjected to a second run of virtual screening, and 1089 compounds with fit values above 3.0 predicted by the structure-based pharmacophore model were removed. A further selection was done by molecular docking (CDOCKER). The obtained 24 hits were subjected to biological validation against AChE, leading to 5 active hits with IC_{50} values in the μM–nM range against AChE. The interesting results coming from this work provide good starting points for the discovery of highly selective AChE inhibitors.

3.2 BACE-1 and γ-Secretase Inhibitors

It is well accepted that the key player in the pathogenesis of AD is recognized to be the amyloid beta protein (Aβ), which is a proteolytic cleavage product of the Aβ-amyloid precursor protein (APP). Two enzymes, the isoform β-secretase (β-site APP cleaving enzyme or BACE-1) and γ-secretase, are responsible of the conversion of APP to Aβ, the accumulation of which into neurotoxic oligomeric species leads to neural apoptosis [7]. Hence the inhibition of these two enzymes BACE-1and γ-secretase, with the aim to reduce the production of Aβ or increasing of its clearance, represents another strategy under investigation for drug development that would aid in the management and delay the progression of AD. Major interest in this field was aimed at the production of peptidomimetic transition state isostere-based inhibitors, such as statine, homostatine, norstatine, and hydroxyethylamine. Some inhibitors of γ-secretase have been designed using the amino acid sequence of the APP cleavage site as a reference [107] and were submitted to preclinical and clinical studies [108–113]. However none of the already reported BACE-1 inhibitors has been marketed due to their inability to cross the blood–brain barrier (BBB) [114–116]. For that, finding selective non-peptidic BACE-1 or γ-secretase inhibitors with desirable pharmacokinetic features for CNS penetration is of current interest.

In a recent report, Gundersen et al. [117] described the design of a series of γ-secretase inhibitor through shape-based virtual screening using the software ROCS. In that study, a sulfonamide [118], known as potent inhibitor of Aβ production, was used as 3D query of the similarity search of an *in-house* database. Almost 500 hits were selected using a Tanimoto coefficient cutoff of 0.7. The analysis of the retrieved structures provided the chemical features considered important for γ-secretase activity. Two hydrophobes about 4 Å apart and a hydrogen bond acceptor about 6–8 Å from the hydrophobes were recognized as determinants for γ-secretase inhibition. After biological evaluation, the obtained data suggested that the compounds act via γ-secretase inhibition. Although the activity exhibited by these compounds is only moderate, this approach was proposed by the authors to find new scaffold for lead optimization for AD therapy.

Support vector machine (SVM) [119] and random forest (RF) [69] approaches were used to generate predictive models and perform screening of γ-secretase inhibitors [120]. For the analysis, 675 very structurally different γ-secretase inhibitors were downloaded from literature. Furthermore, a total of 189 molecular descriptors were calculated based on the 3D structure of each compound. Among these descriptors, a total of 18 constitutional descriptors, 124 topological descriptors, 22 quantum chemical descriptors, and 25 geometrical descriptors were considered. The screening was carried out employing RF method against 319,729 molecules in the "fragment-like" subset of the ZINC database. The selected 368 hits were compared with the known γ-secretase inhibitors by means of the Tanimoto coefficient measure. Low values of Tanimoto coefficient suggested that one or more hit candidates may be proposed as novel γ-secretase inhibitors, because they are broadly dissimilar to the known ones. Finally, it is interesting to note that all of these hit candidates have in common substituents like SO_2 or Ph-SO_2 that were accurately revealed by the RF model, demonstrating itself to be able to screen large compound libraries for aiding the discovery of lead compounds in the development of γ-secretase inhibitors.

Some works have been reported for the design of low-molecular-weight inhibitors of BACE-1 using in silico screening approaches based on docking analysis. For instance, the comparative virtual screening for BACE-1 inhibitors uses different docking methods (FlexX and FlexX-Pharm), scoring functions (Dock, Gold, Chem, PMF, FlexX), protonation states, and protein conformations (apo "unbound" and holo *ligand bound*) [121]. Interestingly, the authors demonstrated that the calculated enrichments, namely, the number of the ligands among the top-scoring docking hits in comparison to their number throughout the entire database, are affected by the protonation states of Asp32 and Asp228 residues of BACE-1 and improve with pharmacophore constraints such as Vertex pharmacophore using FlexX-Pharm/D-Score.

Huang et al. [122] applied a virtual high-throughput approach consisting of fragment-based docking, ligand conformational search by a genetic algorithm, and evaluation of free energy of binding to find non-peptidic BACE-1 inhibitors. Two unrelated compound libraries, Chemdiv (10,067 compounds) and Chemnavigator (306,022 compounds), were screened in silico based on the size and physicochemical character of the substrate binding site. The four-step library-docking procedure foresaw (1) decomposition of library compounds into fragments, (2) fragment docking with calculation of electrostatic solvation, (3) flexible docking of library compounds using a genetic algorithm, and (4) pose clustering binding energy evaluation of the top-ranked poses by Linear Interaction Energy with Continuum Electrostatics (LIECE) approach [123]. The top ten binders showing a common triazine scaffold of the retrieved 88 compounds were tested in vitro in a BACE-1 enzymatic test, showing a binding affinity lower than 100 μM. Four compounds had activity in a mammalian cell assay ($EC_{50} = 20$ μM), indicating that they are able to cross the biomembranes. According to these results, the combination of in silico screening with validation by biological assays appears to be a useful strategy to identify several molecules with excellent potential as lead compounds against BACE-1.

Vijayan et al. proposed a hybrid structure-based virtual screening protocol that merges features of both ligand-based and structure-based techniques for the identification of potential small molecule inhibitors [124]. On one hand, pharmacophore-based virtual screening provides hits more reliable because they define the determinants responsible of the interaction and selectivity. However, docking analysis is used for further screening of hits that lack complementarities with the target protein outside the pharmacophoric scheme [125]. Both the methods could be used independently, being able to successful screen library of ligands. In order to make the most of advantages of these two independent techniques, the authors employed a modified docking methodology that incorporates the structural and similarity features of the bound ligand and the protein to accelerate hit selection in a virtual screening program overcoming the structural constraints of the binding site typical of the docking and scoring. None of the already reported BACE-1 inhibitors has been marketed due to their inability to cross the blood–brain barrier [6, 114–116, 126]. ROCS-based GOLD docking [127] was carried out using a weighted [128] consensus scoring to improve the predictability of the analysis due to high flexibility of the known inhibitors that makes more difficult, reproducing the right binding mode. Finally, to validate the ability of the enrichment, receiver operator characteristic (ROC) curves (measure the performance of any approach across a reference data set endowed with both actives and decoys) and enrichment factor plots were used. The results obtained demonstrated the efficacy of this modified strategy over conventional docking methods (GOLD and LigandFit) in identifying true positives in a virtual screening procedure.

Xu et al. introduced alternative scaffolds as novel pharmacophores of BACE-1 inhibitors applying an integrated in silico/ in vitro approach to find potential inhibitors of BACE-1 [129]. Specs database containing ≈280.000 chemicals was screened using a co-crystalized structure of BACE-1 with an inhibitor as reference for receptor-based virtual screening. Based on the mode of interaction, 42 compounds were selected and submitted to biological assays. Although the tested compounds have proven to be weaker inhibitors than the positive control, the authors suggested them as potential key pharmacophore groups for further designing of non-peptide compounds as more powerful inhibitors against BACE-1.

Recently, Hossain et al. used a combined ligand-based pharmacophore mapping with 3D QSAR modeling approach that consists of CoMFA, CoMSIA, and HQSAR methods to inspect structural and physicochemical determinants for anti-BACE1 activity [130]. For this purpose a data set of 98 structurally different compounds with inhibitory activity against BACE-1 was employed. Molecular docking was carried out to clarify the binding mode of the key substituents coming from the ligand-based models, which were also employed to screen the fragment library created through the de novo procedure [131] HQSAR technique [132], that correlates predictive variables (descriptors) with biological activity using fragment fingerprints or molecular holograms. It was shown that the most important contribution is associated to macrocyclic substituents and phenyl rings. Furthermore, the backbone alkyl chain affects positively the activity, like H bond donor, acceptor, and chiral atoms. The benzimidazole scaffold favors the biological activity. The reported de novo approach provided three seed structures: benzene (S1), pyperidine-1-carbaldehyde (S2), and 3-(1, 2 Dihydro-pyridin-3-yl)-4,5-dihydro-1H-pyrrolo[3,2-c]pyridine (S3), being considered in the context of drug design as prime molecular scaffolds. After docking analysis, these fragment showed good interactions profile with BACE-1 binding site (PDB code: 3OHH) and gained satisfactory ADMET properties. Thus, the reported approach could be a successful process to obtain new hit molecules to propose as BACE-1 inhibitors.

One of the latest attempts for discovering novel BACE-1 inhibitors by combining pharmacophore modeling, structure-based design, and in vitro assay was accomplished by Ju et al. [133]. Here, the 3D determinants for the interaction between BACE-1 and highly active inhibitors were defined in pharmacophore models. The best model (Pearson correlation = −0.719) that consists of 1 H bond acceptor, 1 H bond donor, 1 hydrophobic group, 1 positive ionization, and 25 excluded volumes was used to query the Chemdiv database and ZINC database. The 11,019 hits reflecting the criteria of the best pharmacophore model referred as Hypo18 were further screened by docking and drug-likeness analysis that allowed the final identification of 3 hits. Among these three compounds, only one demonstrated in vitro activity against the target enzyme and was considered as a potential lead for the design of more potent inhibitors.

3.3 Inhibitors of Other Targets

Due to the complexity and multifactorial nature of AD, other biological targets have been the focus of in silico screening approaches. For instance, the screening for inhibitors of tau protein phosphorylation and aggregation represents one of the hallmarks of AD [134]. For this purpose, 10 pharmacophore hypotheses were generated from a high-throughput screening of 200,000 compounds for tau aggregation [135]. About 1266 compounds were active, with 77 in the low micromolar concentration. At the end, three hypotheses were considered for the analysis and used as 3D query for the Maybridge database (Accelrys) containing 59,676 commercial available molecules. All three pharmacophore models that consist of two aromatic rings, one hydrophobic region, and one hydrogen bond acceptor provided a hit list of 136 potential tau aggregation inhibitors. In vitro evaluation of the 19 hits selected from the 3D searching provided two novel leads with an $IC_{50} < 13$ μM that might be employed as starting points for the development of tau aggregation inhibitors.

Another example is reported by Pinsetta et al. [136] which focused on novel p38-alpha mitogen-activated protein kinase (MAPK) inhibitors for the therapy of AD. The p38-alpha MAPK is believed to be involved in the regulation of two events at the basis of AD, namely, tau protein phosphorylation and consequent formation of neurofibrillary tangles (NFTs) and inflammation [137]. For these reasons it is considered an interesting target for treatment strategy. Unfortunately, only few inhibitors of p38-alpha MAPK inhibitors have reached the clinical phase III because of notable side effects and inadequate pharmacokinetic properties [138]. Virtual screening was carried out including more than 500,000 molecules from Chembridge [(subcollections Diverset, central nervous system (CNS)], Kinacore KINASet, and ZINC (subcollection CNS) databases, using the search algorithm for generation of conformers. The binding mode of the 100 best fitted ligands for each database was analyzed by docking analysis. A library containing up to 100 conformers per molecule, from the best fit of pharmacophore-based virtual screening, was generated and used as a new database of molecules for 3D virtual screening experiments. Furthermore, shape (ROCS) and electrostatic similarity (EON of OpenEye Scientific Software) were used as filters in a further virtual screening cycle. The selected compounds by pharmacophore similarity and docking were analyzed for their activity (PharmMapper), toxicity (Deductive Estimation of Risk from Existing Knowledge DEREK), and physicochemical (QikProp) and pharmacokinetic properties. Thirteen compounds satisfactorily showed requirements of low or no toxicity, good pharmacotherapeutic profile, and putative activities, along with the main interactions pattern observed for the most potent inhibitors.

Natural products even provide another potential source of possible AD therapeutic agents [8]. Screening of Eastern herbs and plants is one of the cases reported in the literature to find promising ligands for $A\beta_{40}$ and $A\beta_{42}$ [139]. The 3699 ligands that satisfied the Lipinski's rule were validated by docking for their ability to interact with $6A\beta_{9-40}$ and $5A\beta_{17-42}$. Ligands were accommodated inside the fibrils yielding more stable complex in $6A\beta_{9-40}$ which were mainly located outside of fibrils in $5A\beta_{17-42}$. The ten top-scoring compounds have in common two rings that are critical for binding affinity. Among them dihydrochalcone which is derived from *Daemonorops draco* tree was purchased and tested in vitro ($IC_{50} = 2{,}46$ μM). It showed good ability to prevent $A\beta$ precipitation. Interestingly, the dihydrochalcone has a similar shape of curcumin that is in the second phase of clinical trials. Molecular mechanic/Poisson–Boltzmann surface area (MM/PBSA) methods and MD correctly predict the binding to the peptide target, showing that the van der Waals interaction dominates over the electrostatic interaction. Docking analysis on $6A\beta_{9-40}$ revealed the importance of aromatic rings for ligand binding, whereas H bond network did not play a critical role in the stabilization of the complex. Dihydrochalcone shows all the criteria of AD drugs such as the binding affinity, blood–brain barrier (BBB) crossing capability, and nontoxicity. It is the subject for further in vivo studies and possible clinical trials [139].

Amyloid binding alcohol dehydrogenase (ABAD), a mitochondrial protein, is believed to be a plausible target for the therapy of AD, since it is able to interact with Aβ causing mitochondrial and synaptic dysfunction [140, 141]. Several molecules were identified as ABAD inhibitors [142], and the current efforts are addressed toward the design of ligands with the right ADME properties. According to that, Valasani et al. [143] designed and tested novel small drug molecules as benzothiazole amine and frentizole phosphonate derivatives as promising ABAD inhibitors. A total of 20 compounds and their anti-ABAD activity were predicted by applying QSAR, 3D pharmacophore methods, preADME properties, docking studies, and drug-like analysis. The preliminary results obtained from docking analysis revealed that the benzothiazole amine moiety is critical for inhibition of Aβ–ABAD interaction. Substitutions on the benzothiazole ring and phenyl rings negatively influence the potency. Small electron-withdrawing groups at the benzothiazole ring, especially F, have a positive impact. Furthermore, the presence of hydroxyl substituents in para position to the phenyl ring might have a positive effect. A local database of novel leads was generated from the equilibrated conformations and used to calculate 2D and 3D descriptors, including Lipinski's rules. Overall, analyses of the models based on these methods provide reliable hypotheses, since most of these compounds are reasonable ABAD inhibitors, and the most active one that has also shown the best drug-like properties could be the starting point for the discover of more potent analogues for the treatment of the Alzheimer's disease.

Recently, virtual screening and molecular modeling approaches were applied to find inhibitors against metallothionein-III (MT-III) [144]. This target belongs to a superfamily of intracellular metal-binding proteins considered responsible of copper sequestration and protection from reactive oxygen species. Furthermore, it is believed to be a growth inhibitory factor (GIF) [145], mainly located in the astrocytes of the human brain. In people affected by AD, there is a reduction of its expression levels [146], whereas an overproduction impedes neuronal cell death in animal models of brain damage [147]. In order to find natural molecule-based MT-III inhibitors, a database of natural compounds (InterBioScreen) was screened considering drug-like ADMET properties and Lipinski's rule of five. The 50,536 natural compounds retrieved from the InterBioScreen database were further screened by docking analysis, providing 156 best-scoring compounds that were submitted to in silico ADMET profiling. Among the 156 tested compounds, 41 have satisfactory ADMET properties, but only 3 compounds showed high BBB permeability and a plausible interaction pattern in the active site of MT-III.

Recent evidences indicate the pol-β-mediated cell cycle induction as pathogenic mechanism underlying different neurodegenerative diseases, comprehensive of AD [10, 11]. In this regard, nine putative DNA pol-β inhibitors were identified in silico by querying the ZINC database. In this study, a field-based virtual screening was carried out, and the 3D reference template to query the database was built by aligning the putative bioactive conformations of three known DNA pol-β inhibitors. Among the screened ligands, only 5-methoxyflavone was identified as the only candidate compound with inhibitory activity against DNA pol-β in multiple in vitro assays and able to prevent cell cycle initiation and subsequent neuronal apoptosis in Aβ challenged primary neuronal cultures. The three-ring skeleton of flavones offers a suitable starting point for development of nature-based compounds more stable and safer, with the advantage to be more accessible from large-scale chemical synthesis [148]. Furthermore, methoxylated flavones have chemopreventive properties superior to their nonmethylated counterparts along with the advantage of a potentially greater of blood–brain barrier permeability because of their more lipophilic character.

4 Conclusions

Many works regarding the use of different in silico approaches have been reported for the design of specific ligands of targets related to AD. It is evident from the literature that in silico approaches offer a viable solution to drug molecules with acceptable ADME, safety, and efficacy properties. In particular, virtual screening has emerged as a reliable, cost-effective, and timesaving approach for the discovery of lead compounds for new and more effective therapies in AD. The current status of virtual screening comprehensive of similarity searching in 2D and 3D databases of chemical structures has been reviewed here highlighting the great potential of both ligand-based virtual screening (LBVS) and structure-based virtual screening (SBVS) approaches, the progress of which has really been great over the last years, to discover new AD drugs.

Some of the presented works describe combinations of ligand-based and structure-based drug approaches for the fast screening as well as scanning more efficiently the structural diversity of chemical compounds in search of lead candidates followed by experimental validation. The LBVS studies are based on pharmacophore models to screen existing compound libraries with the aim of generating hits against multiple targets. The interest toward a multi-target drug design was raised by the understanding that AD is a complex and multifactorial disease underlying many interlinked pathological pathways [2]. Overall, there is still the need to extend the existing techniques to explore a major chemical diversity and complexity in order to design more potent and versatile target inhibitors associated with AD; therefore, considerable efforts are being made to yield new and better algorithms and selection methods.

References

1. Roberson E, Mucke L (2006) 100 years and counting: prospects for defeating Alzheimer's disease. Science 314:781–784
2. Cavalli A, Bolognesi ML, Minarini A et al (2008) Multi-target-directed ligands to combat neurodegenerative diseases. J Med Chem 51:347–372
3. Stahl M, Markowitz JS, Gutterman EM et al (2003) Co-use of donepezil and hypnotics among Alzheimer's disease patients living in the community. J Clin Psychiatry 64:466–467
4. Selkoe DJ (2001) Alzheimer's disease: genes, proteins, and therapy. Physiol Rev 81:741–766
5. Hussain I (2004) The potential for BACE-1 inhibitors in the treatment of Alzheimer's disease. IDrugs 7:653–658
6. Hills ID, Vacca JP (2007) Progress toward a practical BACE-1 inhibitor. Curr Opin Drug Discov Devel 10:383–391
7. Walsh DM, Selkoe DJ (2007) A beta oligomers – a decade of discovery. J Neurochem 101:1172–1184
8. Cummings JL (2004) Alzheimer's disease. N Engl J Med 351:56–67
9. Hawkes CA, Ng V, MacLaurin J (2009) Small molecule inhibitors of aβ-aggregation and neurotoxity. Drug Dev Res 70:111–124
10. Copani A, Sortino MA, Caricasole A et al (2002) Erratic expression of DNA polymerases by beta-amyloid causes neuronal death. FASEB J 16:2006–2008
11. Copani A, Hoozemans JJ, Caraci F et al (2006) DNA polymerase-beta is expressed early in neurons of Alzheimer's disease brain and is loaded into DNA replication forks in neurons challenged with beta-amyloid. J Neurosci 26:10949–10957
12. Anderson AC, Wright DL (2005) The design and docking of virtual compound libraries to structures of drug target. Curr Comput Aided Drug Des 1:103–127
13. Jorgensen WL (2004) The many role of computation in drug discovery. Science 303:1813–1818
14. Schoichet BK (2004) Virtual screening of chemical libraries. Nature 432:862–865
15. Dixon SL, Smondyrev AM, Rao SN (2006) PHASE: a novel approach to pharmacophore modelling and 3D database searching. Chem Biol Drug Des 67:370–372
16. Johnson MA, Maggiora GM (1990) Concept and applications of molecular similarity. John Wiley, New York
17. Stahura FL, Bajorath J (2004) Virtual screening methods that complement high-throughput screening. Comb Chem High Throughput Screen 7:259–269
18. Willett P (2006) Similarity-based virtual screening using 2D fingerprints. Drug Discov Today 11:1046–1053
19. Cheeseright T, Mackey M, Rose S et al (2006) Molecular field extrema as descriptors of biological activity: definition and validation. J Chem Inf Model 46:665–676
20. Nikolic K, Mavridis L, Djikic T et al (2016) Drug design for CNS diseases: polypharmacological profiling of compounds using cheminformatic, 3D-QSAR and virtual screening methodologies. Front Neurosci 10:265. doi:10.3389/fnins.2016.00265
21. Pozzan A (2006) Molecular descriptors and methods for ligand based virtual high throughput screening in drug discovery. Curr Pharm Des 12:2099–2110
22. Walters WP, Stahl MT, Murcko MA (1998) Virtual screening – an overview. Drug Discov Today 3:160–178
23. Böhm HJ, Schneider G (2000) Virtual screening for bioactive molecules. Wiley-VCH, Weinheim
24. Klebe G (2000) Virtual screening: an alternative or complement to high throughput screening. Kluwer, Drordrecht
25. McGaughey GB, Sheridan RP, Bayly CI et al (2007) Comparison of topological, shape and docking methods in virtual screening. J Chem Inf Model 2007(47):1504–1519
26. Sheridan RP, Kearsley SK (2002) Why do we need so many chemical similarity search methods? Drug Discov Today 7:903–911
27. Patterson DE, Cramer RD, Ferguson AM et al (1996) Neighborhood behavior: a useful concept for validation of 'molecular diversity' descriptors. J Med Chem 39:3049–3059
28. Nikolova N, Jaworska J (2003) Approaches to measure chemical similarity – a review. QSAR Comb Sci 22:1006–1026
29. Katritzky AR, Lobanov VS, Karelson M (1996) CODESSA reference manual, Version 2.0, Gainville
30. Todeschini R, Consonni V, Mannhold R et al (2009) Molecular descriptors for chemoinformatics. WILEY-VCH, Weinheim
31. Good AC, Mason JS (1995) Three-dimensional structure database searches. Rev Comput Chem 7:67–117

32. Cheng C, Maggiora G, Lajiness M et al (1996) Four association coefficients for relating molecular similarity measures. J Chem Inf Comput Sci 36:909–915
33. Carbo R, Arnau M, Leyda L (1980) How similar is a molecule to another? An electron density measure of similarity between two molecular structures. Int J Quantum Chem 17:1185–1189
34. Reynolds CA, Burt C, Richards WG (1992) A linear molecular similarity index. Quant Struct-Activ Relat 11:34–35
35. Flower DR (1988) On the properties of bit sting based measure of chemical similarity. J Chem Inf Comput Sci 38:379–386
36. Godden JW (2000) Combinatorial preferences affect molecular similarity/diversity calculations using binary fingerprints and Tanimoto coefficients. J Chem Inf Comput Sci 40:163–166
37. Holiday JD, Hu CY, Willett P (2002) Grouping of coefficients for the calculation of intermolecular similarity and dissimilarity using 2D fragment bit string. Comb Chem High Throughput Screen 5:155–166
38. Salim N, Holliday J, Willett P (2003) Combination of fingerprint based similarity coefficient using data fusion. J Chem Inf Comput Sci 43:435–442
39. Willet P (1998) Chemical similarity searching. J Chem Inf Comput Sci 38:983–996
40. Bender A, Glen RC (2004) Molecular similarity: a key technique in molecular informatics. Org BiomolChem 2:3204–3218
41. Estrada E, Uriarte E (2001) Recent advances on the role of topological indices in drug discovery research. Curr Med Chem 8:1573–1588
42. Balaban AT, Basak SC, Colburn T et al (1994) Correlation between structure and normal boiling points of haloalkanes C1-c4 using neural networks. J Chem Inf Comput Sci 34:1118–1121
43. Cone MM, Venkataraghavan R, McLafferty FW (1977) Molecular structure comparison program for the identification of maximal common substructures. J Am Chem Soc 99:7668–7671
44. Burden FR, Winkler DA (1999) Robust QSAR models using Bayesian regularized artificial neural networks. J Med Chem 42:3183–3187
45. Pearlman RS, Smith KM (1998) Novel software tools for chemical diversity. Perspect Drug Discov Des 9-11:339–353
46. Cramer RD, Patterson DE, Bunce JD (1988) Comparative molecular field analysis (CoMFA). 1. Effect of shape on binding of steroids to carrier proteins. J Am Chem Soc 110:5959–5967
47. Hodgkin EE, Richards WG (1987) Molecular similarity based on electrostatic potential and electric field. Int J Quantum Chem 87:105–110
48. Walker PD, Arteca GA, Mezey PG (1991) A complete shape characterization for molecular charge densities represented by Gaussian-type functions. J Comput Chem 12:220–230
49. Good AC, Hodgkin EE, Richards WG (1993) The utilisation of Gaussian functions for the rapid evaluation of molecular similarity. J Chem Inf Comput Sci 32:188–191
50. Schneider G, Neidhart W, Giller T et al (1999) "Scaffold-hopping" by topological pharmacophore search: a contribution to virtual screening. Angew Chem Int Ed Engl 38:2894–2896
51. Mason JS, Good AC, Martin EJ (2001) 3-D pharmacophores in drug discovery. Curr Pharm Des 7:567–597
52. Mason JS, Morize I, Menard PR et al (1999) New 4-point pharmacophore method for molecular similarity and diversity applications: overview of the method and applications, including a novel approach to the design of combinatorial libraries containing privileged substructures. J Med Chem 42:3251–3264
53. Maldonado AG, Doucet JP, Petitjean M et al (2006) Molecular similarity and diversity in chemoinformatics: from theory to applications. Mol Divers 10:39–79
54. Zhao H (2007) Scaffold selection and scaffold hopping in lead generation: a medicinal chemistry perspective. Drug Discov Today 12:149–155
55. Ehrlich P (1909) Ueber den jetzigen Stand der Chemotherapie. Ber Dtsch Chem Ges 42:17–47
56. Wermuth CG, Ganellin CR, Lindberg P et al (1998) Glossary of terms used in medicinal chemistry (IUPAC recommendations 1997). Annu Rep Med Chem 33:385–395
57. Yang SH (2010) Pharmacophore modeling and applications in drug discovery: challenges and recent advances. Drug Discov Today 15:444–450
58. Barnum D, Greene J, Smellie A et al (1996) Identification of common functional configurations among molecules. J Chem Inf Comput Sci 36:563–571
59. Li H, Sutter JM, Hoffmann R (2000) HypoGen: an automated system for generating 3D predictive pharmacophore models. In: Güner OF (ed) Pharmacophore perception,

development, and use in drug design. International University Line, La Jolla, pp 171–189
60. Martin YC (2000) DISCO: what we did right and what we missed. In: Güner OF (ed) Pharmacophore perception, development, and use in drug design. International University Line, La Jolla, pp 49–68
61. Jones G, Willet P (2000) GASP: genetic algorithm superimposition program. In: Güner OF (ed) Pharmacophore perception, development, and use in drug design. International University Line, La Jolla, pp 85–106
62. Dror O, Shulman-Peleg A, Nussinov R et al (2006) Predicting molecular interactions in silico. I. An updated guide to pharmacophore identification and its applications to drug design. Front Med Chem 3:551–584
63. Deng J, Sanchez T, Neamati N et al (2006) Dynamic pharmacophore model optimization: identification of novel HIV-1 integrase inhibitors. J Med Chem 49:1684–1692
64. Antonio L (2015) Machine-learning approaches in drug discovery: methods and applications. Drug Discov Today 20:318–331
65. Ding H, Takigawa I, Mamitsuka H et al (2014) Similarity-based machine learning methods for predicting drug-target interactions: a brief review. Brief Bioinform 15:734–747
66. Ivanenkov YA, Savchuk NP, Ekins S et al (2009) Computational mapping tools for drug discovery. Drug Discov Today 14:767–775
67. Hastie T, Tibshirami R, Friedman J (2001) The elements of statistical learning. Springer, New York
68. Schölkopf B, Smola AJ (2002) Learning with kernels. MIT Press, London
69. Svetnik V, Liaw A, Tong C et al (2003) Random forest: a classification and regression tool for compound classification and QSAR modeling. J Chem Inf Comput Sci 43:1947–1958
70. Rusinko A, Farmen MW, Lambert CG et al (1999) Analysis of a large structure/biological activity data set using recursive partitioning. J Chem Inf Comput Sci 39:1017–1026
71. Ma XH, Jia J, Zhu F et al (2009) Comparative analysis of machine learning methods in ligand-based virtual screening of large compound libraries. Comb Chem High Throughput Screen 12:344–357
72. Plewczynski D, Spieser SAH, Koch U (2009) Performance of machine learning methods for ligand-based virtual screening. Comb Chem High Throughput Screen 12(4):358–368
73. Lionta E, Spyrou G, Vassilatis DK et al (2014) Structure-based virtual screening for drug discovery: principles, applications and recent advances. Curr Top Med Chem 14:1923–1938
74. Charifson PS, Corkery JJ, Murcko MA et al (1999) Consensus scoring: a method for obtaining improved hit rates from docking databases of three-dimensional structures into proteins. J Med Chem 42:5100–5109
75. Kitchen DB, Decornez H, Furr JR et al (2004) Docking and scoring in virtual screening for drug discovery: methods and applications. Nat Rev Drug Discov 3:935–949
76. Craig IR, Essex JW, Spiegel K (2010) Ensemble docking into multiple crystallographically derived protein structure: an evaluation based on statistical analysis of enrichments. J Chem Inf Model 50:511–524
77. Heller S, McNaught A, Stein S et al (2013) InChI – the worldwide chemical structure identifier standard. J Cheminform 5:7. doi:10.1186/1758-2946-5-7
78. Spyrakis F, Cavasotto CN (2015) Open challenges in structure-based virtual screening: receptor modeling, target flexibility consideration and active site water molecules description. Arch Biochem Biophys 583:105–119
79. Ke YY, Coumar MS, Shiao HY et al (2014) Ligand efficiency based approach for efficient virtual screening of compound libraries. Eur J Med Chem 83:226–235
80. Liu J, Wang R (2015) Classification of current scoring functions. J Chem Inf Model 55:475–482
81. Korb O, Brink TT, Raj FRVP et al (2012) Are predefined decoy sets of ligand poses able to quantify scoring function accuracy? J Comput Aided Mol Des 26:185–197
82. Kufareva I, Abagyan R (2012) Methods of protein structure comparison. Methods Mol Biol 857:231–257
83. Sonmez F, Kurt BZ, Gazioglu I et al (2016) Design, synthesis and docking study of novel coumarin ligands as potential selective acetylcholinesterase inhibitors. J Enzyme Inhib Med Chem 32:285–297
84. Holzgrabe U, Kapkova P, Alptuzun V et al (2007) Targeting acetylcholinesterase to treat neurodegeneration. Expert OpinTher Targets 11:161–179
85. Giacobini E, Spiegel R, Enz A et al (2002) Inhibition of acetyl- and butyrylcholinesterase in the cerebrospinal fluid of patients with Alzheimer's disease by rivastigmine: correlation with cognitive benefit. J Neural Transm 109:1053–1065

86. Catto M, Pisani L, Leonetti F et al (2013) Design, synthesis and biological evaluation of coumarin alkylamines as potent and selective dual binding site inhibitors of acetylcholinesterase. Bioorg Med Chem 21:146–152
87. Bajda M, Guzior N, Ignasik M et al (2011) Multi-target-directed ligands in Alzheimer's disease treatment. Curr Med Chem 18:4949–49759
88. Kaur J, Zhang MQ (2000) Molecular modelling and QSAR of reversible acetylcholinesterase inhibitors. Curr Med Chem 7:273–294
89. Bermudez-Lugo JA, Rosales-Hernandez MC, Deeb O et al (2011) In silico methods to assist drug developers in acetylcholinesterase inhibitor design. Curr Med Chem 18:1122–1136
90. Speck-Planche A, Luan F, Cordeiro MNDS (2012) Role of ligand-based drug design methodologies toward the discovery of new anti-Alzheimer agents: futures perspectives in fragment-based ligand design. Curr Med Chem 19:1635–1645
91. Rollinger JM, Hornick A, Langer T et al (2004) Acetylcholinesterase inhibitory activity of scopolin and scopoletin discovered by virtual screening of natural products. J Med Chem 47:6248–6254
92. Sopkova-de Oliveira Santos J, Lesnard A, Agondanou JH et al (2010) Virtual screening discovery of new acetylcholinesterase inhibitors issued from CERMN chemical library. J Chem Inf Model 50:422–428
93. Gupta S, Fallarero A, Järvinen P et al (2011) Discovery of dual binding site acetylcholinesterase inhibitors identified by pharmacophore modeling and sequential virtual screening techniques. Bioorg Med Chem Lett 21:1105–1112
94. Jarvinen P, Fallarero A, Gupta S et al (2010) Miniaturization and validation of the Ellman's reaction based acetylcholinesterase inhibitory assay into 384-well plate format and screening of a chemical library. Comb Chem High Throughput Screen 13:278–284
95. Lu SH, Wu JW, Liu HL et al (2011) The discovery of potential acetylcholinesterase inhibitors: a combination of pharmacophore modeling, virtual screening, and molecular docking studies. J Biomed Sci 18:8. doi:10.1186/1423-0127-18-8
96. Chen Y, Fang L, Peng S et al (2012) Discovery of a novel acetylcholinesterase inhibitor by structure-based virtual screening techniques. Bioorg Med Chem Lett 22:3181–3187
97. Genheden S, Ryde U (2015) The MM/PBSA and MM/GBSA methods to estimate ligand-binding affinities. Expert Opin Drug Discov 10:449–461
98. Chaudhaery SS, Roy KK, Shakya N et al (2010) Novel carbamates as orally active acetylcholinesterase inhibitors found to improve scopolamine-induced cognition impairment: pharmacophore-based virtual screening, synthesis, and pharmacology. J Med Chem 53:6490–6505
99. Bag S, Tulsan R, Sood A et al (2013) Pharmacophore modeling, virtual and in vitro screening for acetylcholinesterase inhibitors and their effects on amyloid-β self- assembly. Curr Comput Aided Drug Des 9:2–14
100. Ambure P, Kar S, Roy K (2014) Pharmacophore mapping-based virtual screening followed by molecular docking studies in search of potential acetylcholinesterase inhibitors as anti-Alzheimer's agents. Biosystems 116:10–20
101. Nogara PA, SaraivaRde A, CaeranBueno D et al (2015) Virtual screening of acetylcholinesterase inhibitors using the Lipinski's rule of five and ZINC databank. Biomed Res Int 2015:870389. doi:10.1155/2015/870389
102. Bajda A, Panek D, Hebda M et al (2015) Search for potential cholinesterase inhibitors from the ZINC database by virtual screening method. Acta Pol Pharm 72:737–745
103. Xie Q, Tang Y, Li W et al (2006) Investigation of the binding mode of (−)-meptazinol and bis-meptazinol derivatives on acetylcholinesterase using a molecular docking method. J Mol Model 12:390–397
104. Dhanjal JK, Sharma S, Grover A et al (2015) Use of ligand-based pharmacophore modeling and docking approach to find novel acetylcholinesterase inhibitors for treating Alzheimer's. Biomed Pharmacother 71:146–152
105. Chen Y, Lin H, Yang H et al (2017) Discovery of new acetylcholinesterase and butyrylcholinesterase inhibitors through structure-based virtual screening. RSC Adv 7:3429–3438
106. Kirchmair J, Distinto S, Markt P et al (2009) How to optimize shape-based virtual screening: choosing the right query and including chemical information. J Chem Inf Model 49:678–692
107. Josien H (2002) Recent advances in the development of gamma-secretase inhibitors. Curr Opin Drug Discov Devel 5:513–525
108. Peters JU, Galley G, Jacobsen H et al (2007) Novel orally active, dibenzazepinone-based gamma-secretase inhibitors. Bioorg Med Chem Lett 17:5918–5923

109. Josien H, Bara T, Rajagopalan M et al (2007) Small conformationally restricted piperidine N-arylsulfonamides as orally active gamma-secretase inhibitors. Bioorg Med Chem Lett 17:5330–5335
110. Shaw D, Best J, Dinnell K et al (2006) 3,4-Fused cyclohexyl sulfones as gamma-secretase inhibitors. Bioorg Med Chem Lett 16:3073–3076
111. Thompson LA, Liauw AY, Ramanjulu MM et al (2006) Synthesis and evaluation of succinoyl-caprolactam gamma-secretase inhibitors. Bioorg Med Chem Lett 16:2357–2363
112. Prasad CVC, Noonan JW, Sloan CP et al (2004) Hydroxytriamides as potent gamma-secretase inhibitors. Bioorg Med Chem Lett 14:1917–1921
113. Best JD, Smith DW, Reilly MA et al (2007) The novel gamma secretase inhibitor N-[cis-4-[(4-chlorophenyl)sulfonyl]-4-(2,5-difluorophenyl)cyclohexyl]-1,1,1-trifluoromethanesulfonamide (MRK-560) reduces amyloid plaque deposition without evidence of notch-related pathology in the Tg2576 mouse. J Pharmacol Exp Ther 320:552–558
114. Siemers ER, Quinn JF, Kaye J et al (2006) Effects of a gamma-secretase inhibitor in a randomized study of patients with Alzheimer disease. Neurology 66:602–604
115. Congreve M, Aharony D, Albert J et al (2007) Application of fragment screening by X-ray crystallography to the discovery of aminopyridines as inhibitors of beta-secretase. J Med Chem 50:1124–1132
116. Barrow JC, Stauffer SR, Rittle KE et al (2008) Discovery and X-ray crystallographic analysis of a spiro piperidine iminohydantoin inhibitor of beta-secretase. J Med Chem 51:6259–6262
117. Gundersen E, Fan K, Haas K et al (2005) Molecular-modeling based design, synthesis, and activity of substituted piperidines as γ-secretase inhibitors. Bioorg Med Chem Lett 15:1891–1894
118. Smith DW, Munoz B, Srinivasan K et al (2000) Preparation of sulfonamide derivatives as amyloid production inhibitors useful in treating or preventing disease related to. WO 0050391, 2000. Chem Abstr 2000 133:207678
119. Vapnik VN (1995) The nature of statistical learning theory. Springer, New York
120. Yang XG, Lv W, Zong Y et al (2009) In silico prediction and screening of γ-secretase inhibitors by molecular descriptors and machine learning methods. J Comput Chem 31:1249–1258
121. Polgár T, Keserü GM (2005) Virtual screening for beta-secretase (BACE1) inhibitors reveals the importance of protonation states at Asp32 and Asp228. J Med Chem 48:3749–3755
122. Huang D, Lüthi U, Kolb P et al (2006) In silico discovery of Β-secretase inhibitors. J Am Chem Soc 128:5436–5443
123. Huang D, Caflisch A (2004) Efficient evaluation of binding free energy using continuum electrostatics solvation. J Med Chem 47:5791–5797
124. Vijayan RSK, Prabu M, Mascarenhas NM et al (2009) Hybrid structure-based virtual screening protocol for the identification of novel BACE1 inhibitors. J Chem Inf Model 49:647–657
125. Schuster D, Nashev LG, Kirchmair J et al (2008) Discovery of nonsteroidal 17 Aβ-hydroxysteroid dehydrogenase 1 inhibitors by pharmacophore-based screening of virtual compound libraries. J Med Chem 51:4188–4199
126. Xiao K, Li X, Li J et al (2006) Design, synthesis, and evaluation of Leu*Ala hydroxyethylene-based non-peptide beta-secretase (BACE) inhibitors. Bioorg Med Chem 14:4535–4541
127. Grant JA, Mosyak L, Nicholls A (2005) A shape-based 3-D scaffold hopping method and its application to a bacterial protein-protein interaction. J Med Chem 48:1489–1495
128. Jacobsson M, Karlén A (2006) Ligand bias of scoring functions in structure-based virtual screening. J Chem Inf Model 46:1334–1343
129. Xu W, Chen G, Liew OW et al (2009) Novel non-peptide beta-secretase inhibitors derived from structure-based virtual screening and bioassay. Bioorg Med Chem Lett 19:3188–3192
130. Hossain T, Mukherjee A, Saha A (2015) Chemometric design to explore pharmacophore features of BACE inhibitors for controlling Alzheimer's disease. Biomed Res Int 11:549–557
131. Yang Y, Pei J, Lai L (2011) LigBuilder 2: a practical de novo drug design approach. J Chem Inf Model 51:1083–1091
132. Doddareddy MR, Lee YJ, Cho YS et al (2004) Hologram quantitative structure activity relationship studies on 5-HT6 antagonists. Bioorg Med Chem 12:3815–3824
133. Ju Y, Li Z, Deng Y et al (2016) Identification of novel BACE1 inhibitors by combination of pharmacophore modeling, structure-based design and in vitro assay. Curr Comput Aided Drug Des 12:73–82

134. Larbig G, Pickhardt M, Lloyd DG et al (2007) Screening for inhibitors of tau protein aggregation into Alzheimer paired helical filaments: a ligand based approach results in successful scaffold hopping. Curr Alzheimer Res 4:315–323
135. Pickhardt M, von Bergen M, Gazova Z et al (2005) Screening for inhibitors of tau polymerization. Curr Alzheimer Res 2:219–226
136. Pinsetta FR, Taft CA, de Paula da Silva CH (2014) Structure- and ligand-based drug design of novel p38-alpha MAPK inhibitors in the fight against the Alzheimer's disease. J Biomol Struct Dyn 32:1047–1063
137. Banerjee A, Koziol-White C, Panettieri R Jr (2012) p38 MAPK inhibitors, IKK2 inhibitors, and TNFα inhibitors in COPD. Curr Opin Pharmacol 12:287–292
138. Lengauer T (2002) Bioinformatics: from genomes to drugs. Methods and principles in medicinal chemistry. Wiley-VHC Verlag, Weinheim
139. Viet MH, Chen CY, Hu CK et al (2013) Discovery of dihydrochalcone as potential lead for Alzheimer's disease: in silico and in vitro study. PLoS One 8(11):e79151
140. Lustbader JW, Cirilli M, Lin C et al (2004) ABAD directly links Abeta to mitochondrial toxicity in Alzheimer's disease. Science 304:448–452
141. Takuma K, Yao J, Huang J et al (2005) ABAD enhances Abeta-induced cell stress via mitochondrial dysfunction. FASEB J 19:597–598
142. Marques AT, Fernandes PA, Ramos MJ (2008) Molecular dynamics simulations of the amyloid-beta binding alcohol dehydrogenase (ABAD) enzyme. Bioorg Med Chem 16:9511–9518
143. Valasani KR, Hu G, Chaney MO et al (2013) Structure-based design and synthesis of benzothiazole phosphonate analogues with inhibitors of human ABAD-ab for treatment of Alzheimer's disease. Chem Biol Drug Des 81:238–249
144. Roy S, Kumar A, Baig MH et al (2015) Virtual screening, ADMET profiling, molecular docking and dynamics approaches to search for potent selective natural molecules based inhibitors against metallothionein-III to study Alzheimer's disease. Methods 15:105–110
145. Coyle P, Phylcox JC, Carey LC et al (2002) Metallothionein: the multipurpose protein. Cell Mol Life Sci 59:627–647
146. Uchida Y, Takio K, Titani K et al (1991) The growth inhibitory factor that is deficient in the Alzheimer's disease brain is a 68 amino acid metallothionein-like protein. Neuron 7:337–347
147. Hozumi I, Inuzuka T, Hiraiwa M et al (1995) Changes of growth inhibitory factor after stab wounds in rat brain. Brain Res 688:143–148
148. Merlo S, Basile L, Giuffrida ML et al (2015) Identification of 5-methoxyflavone as a novel DNA polymerase-beta inhibitor and neuroprotective agent against beta-amyloid toxicity. J Nat Prod 78:2704–2711

Chapter 5

Molecular Field Topology Analysis (MFTA) in the Design of Neuroprotective Compounds

Eugene V. Radchenko, Vladimir A. Palyulin, and Nikolay S. Zefirov

Abstract

The Molecular Field Topology Analysis (MFTA) is a QSAR method designed to model the activities mediated by small molecule binding to biotargets using the local physicochemical descriptors reflecting the major types of ligand–target interactions. A molecular supergraph provides a common frame of reference for the meaningful comparison of the properties of atoms in different structures and the visualization of their effects. The MFTA method has been successfully used in the activity and selectivity modeling, design, and virtual screening of promising ligands of various enzymes and receptors (the NMDA and AMPA receptor antagonists and modulators as well as the acetyl- and butyrylcholinesterase inhibitors are of particular interest in the design of the anti-Alzheimer and other neuroprotective compounds).

The design of potential anti-Alzheimer and other neuroprotective compounds based on the MFTA structure–activity models involves the following basic steps: (1) preparation of a training set containing structures of the compounds and the experimental activity and/or selectivity values, (2) generation of a series of MFTA models using various descriptor combinations, (3) evaluation of model quality and selection of the optimal model, (4) interpretation of the model, (5) preparation of the virtual screening library, (6) prediction of the activity and/or selectivity endpoints, (7) selection of promising compounds, and (8) prediction of the relevant ADMET properties. As a result, a focused library of potential neuroprotective compounds having the desired activity/selectivity profile and acceptable ADMET properties is obtained. The workflow presented in this chapter using a case study involving the inhibitors of glycogen synthase kinase 3β is applicable to many other relevant targets and ligand classes. In addition, some of its elements may be incorporated into other virtual screening and design workflows.

Key words Molecular Field Topology Analysis (MFTA), QSAR, Neuroprotective compounds, Virtual screening, Molecular design, ADMET, Glycogen synthase kinase 3β

1 Introduction

The quantitative structure–activity relationships (QSAR) analysis aims to derive predictive models capturing the patterns in the experimental activity data for a series of chemical compounds. Their structures are usually characterized by a suitable set of numerical molecular descriptors representing the relevant facets of a

Kunal Roy (ed.), *Computational Modeling of Drugs Against Alzheimer's Disease*, Neuromethods, vol. 132, DOI 10.1007/978-1-4939-7404-7_5, © Springer Science+Business Media LLC 2018

structure, and the descriptor values are analyzed by applicable machine learning techniques. The benefits of a QSAR model are twofold. First, it can be used to predict the activity values for "new" compounds of interest, in particular, in the molecular design and virtual screening context. Second, the analysis and interpretation of a model can provide insights into the mechanisms of action of the compounds and their structural features critical for the activity.

The *Molecular Field Topology Analysis (MFTA)* [1, 2] is a QSAR method designed primarily to model the activities mediated by binding of small molecule ligands to biotargets such as receptors and enzymes. The molecular descriptors in this approach are the local physicochemical parameters (properties of atoms) reflecting major types of interactions that may be involved in such binding. In particular, the possible electrostatic interactions can be characterized by the effective atomic charges, the steric interactions by the van der Waals radii of atoms and groups, the hydrophobic interactions by the local lipophilicity contributions, and the hydrogen bonding interactions by the hydrogen bond donor and acceptor abilities. A common frame of reference for the meaningful comparison of atom properties in different structures is provided by a so-called molecular supergraph, i.e., a kind of topological network such that all structures of interest can be superimposed onto it. Thus, by considering the topological rather than geometrical alignment, the MFTA method solves the problems of conformational flexibility and molecular structure alignment typical in the 3D QSAR approaches. However, this may overestimate the structure flexibility and limits the training set and the intended applicability domain of a model to reasonably congeneric compounds with a common basic scaffold. The partial least squares regression (PLSR) is commonly used to build a predictive model from a supergraph-based uniform descriptor set. The effects of various local descriptors identified by the model can be visualized on the supergraph, providing graphical activity and selectivity maps for convenient analysis. They highlight the important structural features of the compounds and can help in detecting significant ligand–target interactions and in designing the novel promising structures in a chemically intuitive way. In addition, the supergraph captures the structural variability of a training set that can be used to generate a representative structure library for virtual screening within the model applicability domain [3].

The MFTA method has been successfully used by us [4–13] and other groups [14–17] in the activity and selectivity modeling, design, and virtual screening of promising ligands of various enzymes and receptors as well as viral fusion inhibitors. In the design of the anti-Alzheimer and other neuroprotective compounds, the studies on the NMDA and AMPA receptor antagonists and modulators [2, 7] and on the covalent [8–12] and reversible [13, 17] inhibitors of acetyl- and butyrylcholinesterase are of particular interest. As an example, let us consider the model of inhibitory activity of organophosphate compounds toward butyrylcholinesterase [12]. The molecular supergraph for this series of compounds and the example of the superimposition of the training set structure are shown in Fig. 1a. The activity maps for the butyrylcholinesterase inhibition shown in Fig. 1b indicate that the activity is increased by longer-chain alkylphosphate groups (easily explained by larger size of the butyrylcholinesterase active site [10]) as well as by the halomethyl and ester moieties in the leaving group, while the benzyl and trifluoromethyl moieties in the leaving group tend to decrease the activity. These structural requirements are evident in the training set structure with the highest anti-butyrylcholinesterase activity (Fig. 1a); however, similar activity can be expected for compounds containing various other groups with similar charge, steric, and lipophilicity properties.

The design of potential neuroprotective compounds based on the MFTA structure–activity models involves the following basic steps: (1) preparation of a training set containing structures of the compounds and the experimental activity and/or selectivity values, (2) generation of a series of MFTA models using various descriptor combinations, (3) evaluation of model quality and selection of the optimal model, (4) interpretation of the model, (5) preparation of the virtual screening library, (6) prediction of the activity and/or selectivity endpoints, (7) selection of promising compounds, and (8) prediction of the relevant ADMET properties. The proposed workflow is illustrated in Fig. 2. In addition, some of these elements may be incorporated into other virtual screening and design workflows.

2 Materials

Preparation and Manipulation of Structure–Activity Databases: The ChemAxon Instant JChem [18] software is recommended for the structure database management tasks in the proposed workflow, although other solutions supporting the SDF format can also be used and some simple tasks can be done in the MFTA software itself (see below). Instant JChem supports Windows and Linux platforms. Free academic licenses are available. Detailed documentation for the software is provided at the ChemAxon site.

Molecular Field Topology Analysis (MFTA): The MFTAWin [19] software implements the MFTA method, providing the model generation and analysis, activity prediction, and related additional functions. The software (including a detailed reference manual) is available on request from the authors. Currently only Windows platform is supported; however, the Windows version can be used on Linux in an emulated environment.

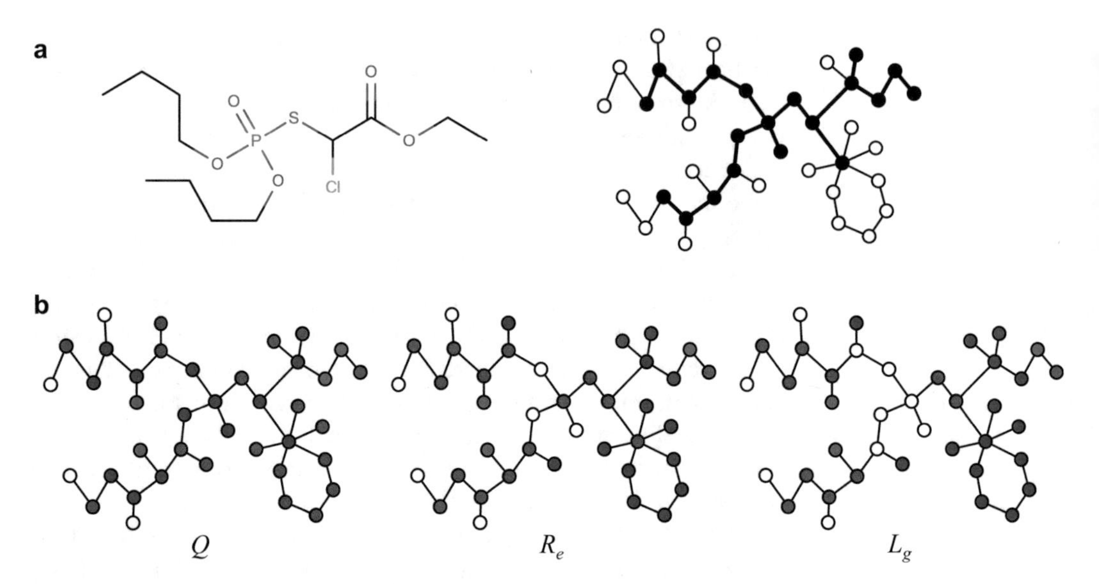

Fig. 1 Molecular Field Topology Analysis (MFTA) modeling of the butyrylcholinesterase inhibition by organophosphate compounds. (**a**) Example of the training set structure superimposed onto the molecular supergraph for the series of compounds. (**b**) Activity maps illustrating the effect of local molecular descriptors on the inhibitory activity. In the positions of a molecular supergraph marked with the *red circles*, an increase in the descriptor values is favorable for the activity; conversely, in positions marked with the *blue circles*, an increase in the descriptor tends to decrease the activity. Intensity of colors reflects the magnitude of the influence. Local descriptors: Q effective atomic charge, R_e effective van der Waals radius taking into account the steric requirements of the central non-hydrogen atom and other attached atoms, L_g group lipophilicity taking into account the contributions of the central non-hydrogen atom and the attached hydrogen atoms

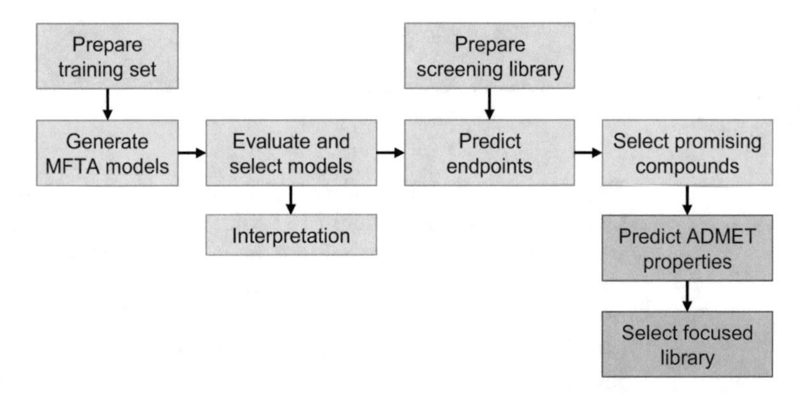

Fig. 2 The MFTA-based design workflow is centered around the construction of the MFTA models and the endpoint prediction (*cyan*). In addition, it involves data preparation (*mint*), model interpretation (*green*), prediction of the ADMET properties (*lavender*), and finally selection of the focused library of promising compounds (*pink*)

Prediction of ADMET Properties: The integrated online service for ADMET properties prediction [20] developed at the Laboratory of Medicinal Chemistry of the Department of Chemistry, Lomonosov Moscow State University, can be used to estimate a number of important pharmacokinetic and toxicity endpoints for the compounds identified by virtual screening, including the blood–brain barrier permeability (*LogBB*) [21], human intestinal absorption (HIA) [22], hERG-mediated cardiac toxicity [23], etc. The models implemented in this service are built from broad, diverse, and accurate training sets using fragmental structural descriptors and artificial neural networks, thus ensuring good prediction accuracy and applicability domain. Convenient tools for the visualization and analysis of the results are available.

3 Methods

In this section, the MFTA workflow for the design of potential neuroprotective compounds is presented. A case study involving the inhibitors of glycogen synthase kinase 3β (GSK-3β) based on the indirubin scaffold will be used as an illustration (the case study results are discussed in Sect. 5).

3.1 Training Set Preparation

1. Open the Instant JChem software and create a new project.
2. Create a new data table to store the structures and activity values for the training set compounds (*New Data Tree and structure entity*).
3. Add the data fields for the measured activity values, compound name/ID, and other necessary information. Take care to select the appropriate field types.
4. Enter the structures and other data for the training set compounds.
5. Add the calculated fields to convert the measured activities to the logarithmic quantities (pK_i, pIC_{50}, etc.) and calculate the necessary selectivity and multi-target activity endpoint values (see Note 1). For convenience, convert the calculated field values to the standard data fields using the *Convert type* feature.
6. Export the training set to the SDF format file.
7. Open the MFTAWin software and import the training set SDF file by selecting *File | Open*. If necessary, use the substructure search feature to label the atoms in the training set structures as explained in the reference manual in order to explicitly specify the researcher's view of a central core/scaffold and the allowed atom correspondence (see Note 2). To save the dataset in the internal MDS (MFTA DataSet) format for further analysis, click the right mouse button and choose *File | Save As* from the pop-up menu, or simply close the dataset window and confirm saving changes.

3.2 Generation of MFTA Models

The generation of an MFTA model consists of two basic procedures. The first one involves the calculation of local descriptors, the construction of a molecular supergraph, and the formation of a uniform MFTA descriptor set. In the second procedure, the predictive model is built from this descriptor set using the partial least squares regression (PLSR).

3.2.1 MFTA Descriptor Calculation

1. Open the MFTAWin software. Choose *Modeling | MFTA descriptors*. The *MFTA Calibration Parameters* dialog box appears containing the parameters for the first procedure.
2. Using the *Database* edit box and browse button, specify the dataset (MDS) file containing the training set.
3. Select the necessary local descriptors in the *Descriptors* box. As a rule of thumb, it is usually advisable to start with the descriptor combinations listed in Tables 1 and 2 that cover major interaction types (see Note 3).
4. If desired, clear the *Bond match | Cyclic* flag and set the *Penalty for Cyclic to Acyclic mapping* to 0 in order to allow superimposing the cyclic and acyclic fragments of the molecules (see Note 4).
5. For other parameters keep the default values. In the *Job* box, enter some name or identifier for the calculation job (the dataset name and the job name will be used to generate the names of the resulting files). It is advisable to specify the selected descriptors and other important parameters in the job name. Press the *OK* button to start the calculation.
6. When the calculation is complete, the job results are opened in the *MFTA Descriptor Values* window (Fig. 3a). The results can also be reopened later by selecting *Tools | Analyze | MFTA*. In the left pane, the selected dataset structure is shown schematically. The right pane displays the molecular supergraph, the mapping of the selected structure (bold lines), and the distribution of a selected local descriptor. If desired, adjust the supergraph layout by dragging the atoms (vertices) with the mouse. (Click the right mouse button and select *Marker | None* from the pop-up menu to turn off the display of atom numbers.) The displayed supergraph view can be copied to the clipboard using the *Edit* options in the pop-up menu.
7. Inspect the resulting molecular supergraph. If a supergraph (or the mappings of individual structures) seems too complicated or chemically undesirable, modify the calculation setup and repeat the procedure (see Note 5).

3.2.2 Construction of a Predictive Model

1. In the MFTAWin software, choose *Modeling | PLS Calibrate*. The *PLS Calibration Parameters* dialog box appears containing the parameters for the second procedure.

Table 1
Recommended local descriptor combinations for MFTA modeling

N	Descriptor codes	Meaning
1	Charge.Q, VdW.Re	Electrostatic and steric interactions
2	Charge.Q, VdW.Re, HBond.Hd, HBond.Ha	Electrostatic, steric, hydrogen bonding interactions
3	Charge.Q, VdW.Re, Lipo.Lg	Electrostatic, steric, hydrophobic interactions
4	Charge.Q, VdW.Re, HBond.Hd, HBond.Ha, Lipo.Lg	Electrostatic, steric, hydrogen bonding, hydrophobic interactions

Individual descriptors are explained in Table 2

Table 2
Recommended local descriptors for MFTA modeling [1, 2]

N	Descriptor code	Descriptor	Meaning
1	Charge.Q	Q	Effective atomic charge calculated by the electronegativity equalization method
2	VdW.Re	R_e	Effective van der Waals radius of atom environment taking into account the steric requirements of the central non-hydrogen atom and other attached atoms
3	HBond.Hd	H_d	Ability of an atom to act as a hydrogen bond donor
4	HBond.Ha	H_a	Ability of an atom to act as a hydrogen bond acceptor
5	Lipo.Lg	L_g	Group lipophilicity taking into account the contributions of the central non-hydrogen atom and the attached hydrogen atoms

2. Using the *Descr file* edit box and browse button, specify the MFTA descriptor (DSC) file created during the first procedure.
3. Using the *Activity file* edit box and browse button, specify the dataset (MDS) file containing the training set. Then select the desired endpoint from the *Activity name* box.
4. In the *Stop | N factors* box, specify 6 as the maximum number of factors (latent variables) in the PLSR model.
5. If necessary, set the *Genetic selection of descriptors* flag in order to use the Q^2-guided optimization of a descriptor subset (see Note 6).
6. For other parameters keep the default values. In the *Job* box, enter some name or identifier for the calculation job (the descriptor set name and the job name will be used to generate the names of the resulting files). It is advisable to specify the modeled endpoint and other important parameters in the job name. Press the *OK* button to start the calculation.

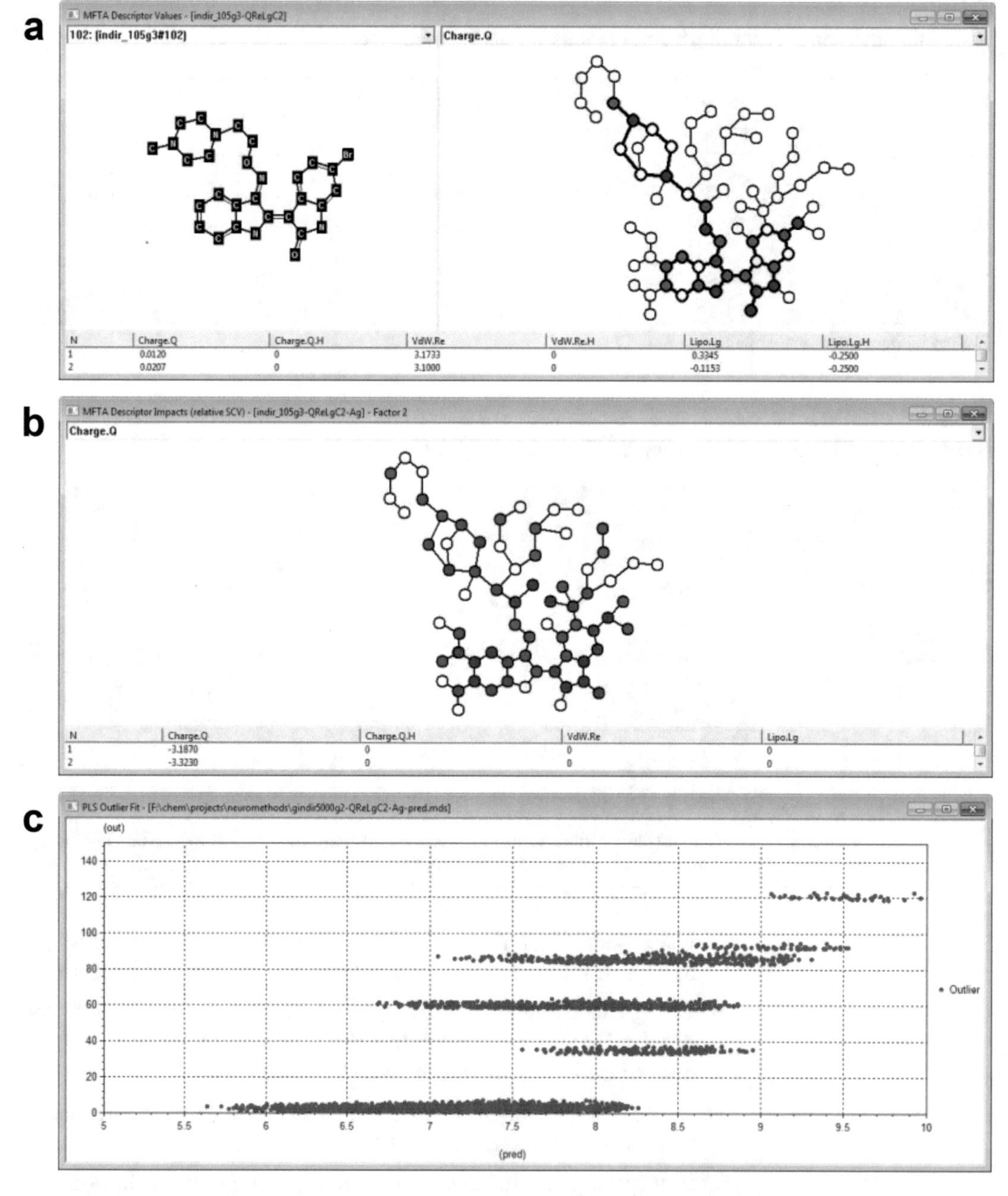

Fig. 3 Some key steps of the MFTA-based design workflow in the MFTAWin software. (**a**) Inspection of a molecular supergraph providing a uniform frame of reference for local atomic descriptors. (**b**) Inspection of the local descriptor influence on activity. (**c**) Inspection of the prediction results

7. When the calculation is complete, the job results are opened in the *PLS Calibration Results* window (they can also be reopened later by selecting *Tools* | *Analyze* | *PLS*). The statistical quality parameters of the models with different number of factors are shown in the displayed table. The model with the highest Q^2 value is selected by default.

3.3 Evaluation of Model Quality and Selection of the Optimal Model

1. For the generated series of models based on different descriptor sets and parameters for a given endpoint, inspect the statistical quality parameters as well as the correspondence plots between the experimental endpoint values and the values predicted during the cross-validation (in the *PLS Calibration Results* window, click the right mouse button and choose *Plot | Prediction fit | Multiple CV* from the pop-up menu). The displayed plot view can be copied to the clipboard using the *Edit* options in the pop-up menu.
2. Select the optimal model taking into account the model predictivity, number of descriptors, and interpretability (see Note 7).
3. Save the optimal model parameters for future use during prediction by clicking the right mouse button and selecting *File | Save model.*

3.4 Model Interpretation

1. In the *PLS Calibration Results* window for the optimal model, click the right mouse button and choose *Graph | Variable impact | Relative SCV* from the pop-up menu. The *MFTA Descriptor Impacts* window appears (Fig. 3b).
2. Inspect and analyze the activity/selectivity maps for the model by selecting different local descriptors from a drop-down list (see Note 8). Click the right mouse button and select *Marker | None* from the pop-up menu to turn off the display of atom numbers. The displayed activity/selectivity map view can be copied to the clipboard using the *Edit* options in the pop-up menu.

3.5 Preparation of Virtual Screening Library

1. Identify the desired scaffold, substitution positions, and substituents for the virtual screening library taking into account the structural diversity of the training set as well as the structural features affecting the activity.
2. Generate the structure library using the Markush Enumeration feature in the Marvin Sketch or Instant JChem software (refer to the documentation if necessary) (see also Note 9).
3. Export the library to the SDF format file.

3.6 Activity and/or Selectivity Prediction

1. Open the MFTAWin software and import the generated library (prediction set) SDF file by selecting *File | Open.*
2. Label the atoms in the prediction set structures, and/or split it into subsets in exactly the same way as it was done during the model generation (see Note 2 and Note 5).
3. Save the dataset in the MDS format for further processing by clicking the right mouse button and choosing *File | Save As* from the pop-up menu, or simply close the dataset window and confirm saving changes.

4. Choose *Prediction | MFTA descriptors*. The *MFTA Prediction Parameters* dialog box appears containing the parameters for the descriptor generation procedure.
5. Using the *Database* edit box and browse button, specify the dataset (MDS) file containing the prediction set.
6. Using the *Prediction dictionary* edit box and browse button, specify the prediction dictionary (DCT) file created during the model generation. The descriptor generation parameters are loaded automatically. No modifications are usually needed.
7. In the *Job* box, enter some name or identifier for the calculation job (the dataset name and the job name will be used to generate the names of the resulting files). It is advisable to specify the selected descriptors and other important parameters in the job name. Press the *OK* button to start the calculation.
8. When the calculation is complete, the job results are opened in the *MFTA Descriptor Values* window (they can also be reopened later by selecting *Tools | Analyze | MFTA*). If desired, inspect the mappings of the dataset structures to the molecular supergraph (see Sect. 3.2.1 for more details). Then close the window.
9. Choose *Prediction | PLS Predict*. The *PLS Prediction Parameters* dialog box appears containing the parameters for the prediction procedure.
10. Using the *Descr file* edit box and browse button, specify the MFTA descriptor (DSC) file created during the descriptor generation procedure for the prediction set.
11. Using the *Model file* edit box and browse button, specify the model parameter file saved previously (see Sect. 3.3 step 3).
12. In the *Job* box, enter some name or identifier for the calculation job (the descriptor set name and the job name will be used to generate the names of the resulting files). It is advisable to specify the modeled endpoint and other important parameters in the job name. Press the *OK* button to start the calculation.
13. When the calculation is complete, the job results are opened in the *PLS Prediction Dataset* window (they can also be reopened later by selecting *Tools | Analyze | PLS*). If necessary, adjust the generated field names to reflect the predicted endpoint (click the respective table column, then click the right mouse button and choose *Field | Rename* from the pop-up menu).
14. If additional endpoints have been predicted, export their prediction results by choosing *File | Export | Table* from the pop-up menu. Then import them into the first prediction dataset by choosing *File | Import*. See also Note 10.

3.7 Selection of Promising Compounds

1. Using the options in the *Plot* pop-up menu, check the model applicability using the outlier rating plots (Fig. 3c) as well as inspect the endpoint distribution and define the desirable screening criteria. Remove the compounds falling outside of the applicability domains of the models [2] using the prediction outlier ratings marked as "(out)" in the prediction datasets. The outlier rating values below 40 are generally acceptable, while higher values indicate unreliable predictions. From the pop-up menu, choose *Structure* | *Select* | *Filter by Field*, select the outlier rating field, specify "< 40" as the condition, and press *OK*. The compounds violating the condition are unmarked. Then choose *Structure* | *Delete* | *Unused*. (See also Note 10.)
2. The virtual screening of promising compounds should take into account the features of a specific problem, relevant endpoints, their desired values, as well as the distribution of the endpoint values in the prediction set. Identify the required conditions and generate a focused library using the *Structure* | *Select* | *Filter by Field* and *Structure* | *Delete* | *Unused* options. Then export the focused library in SDF format (or just the prediction table) using the *File* | *Export* menu. (See also Note 10.)

3.8 Prediction of ADMET Properties

1. To predict the relevant ADMET properties such as the blood–brain barrier permeability, human intestinal absorption, or hERG-mediated cardiac toxicity for the promising compounds of a focused library, open the integrated online service for ADMET properties prediction [20] in a Web browser.
2. Identify the structure of interest in the Instant JChem software (either by checking the structure number in a generated library or directly in the filtering results), click it with the right mouse button, and choose *Copy* in order to copy the SMILES representation of a structure to the clipboard.
3. On the prediction service page, click the structure editor applet with the right mouse button and choose *Paste MOL or SDF or SMILES* from the pop-up menu. Once the structure appears in the editor, press the *Calculate!* button.
4. Evaluate the acceptability or desirability of a structure from the ADMET perspective using the predicted parameter values as well as the color codes and the graphical representations of the compound position with respect to known experimental parameter measurements.
5. Prepare a final focused library of potential neuroprotective compounds having the desired activity/selectivity profile and acceptable ADMET properties.

4 Notes

1. For the pairwise selectivities, a simple difference between the respective activity values (in the logarithmic scale) can be used. If a molecular design problem requires a multi-target/multi-objective optimization of the activity profile, it may be useful to introduce additional integral parameters of positive and negative effects that are based on domain knowledge and easily interpretable [11, 12]. They can be constructed as generalized mean values of the individual activities. From our experience, the power-3 (cubic) mean is a simple function suitable for summarizing negative effects (all of which are undesirable and should be avoided) as it is monotonically increasing with increasing individual values while being partially dominated by larger values. For the positive multi-target activities (all of which are desirable), the geometric mean is suitable as an integral parameter [24].

   From the statistical uncertainty perspective, direct modeling of the structure–selectivity relationships provides more reliable selectivity estimates. However, sometimes the experimental activity measurements for some of the training set compounds may be missing, making the calculation of their selectivity values impossible. In such cases it may be useful to model only the individual activities and calculate the estimated selectivity for the virtual screening purposes from the predicted activity values (provided that the predictivities and the applicability domains of the individual activity models are adequate) [13].

   As an alternative to using Instant JChem, the logarithmic endpoints and selectivity parameters can be calculated in the MFTAWin software itself after the dataset is imported (click the right mouse button and choose *Field | Add* from the pop-up menu).

2. The labeling of atoms in the training set structures allows the researcher to explicitly specify the central core/scaffold and the allowed atom correspondence, as well as significantly improves the performance at the supergraph construction and descriptor calculation steps (see also Note 5).

3. The descriptor combinations listed in Tables 1 and 2 cover the major interaction types that may be important for the ligand-biotarget binding. All the descriptors are explained in the MFTAWin reference manual. In some cases, additional or different combinations may be tried in order to improve the model quality and/or test the mechanism of action hypotheses. However, usually the recommended combinations are sufficient. Some of the descriptors available in the software require external utilities and/or are experimental and should be used with care.

4. The default parameters do not allow superimposing the cyclic and acyclic fragments of the molecules, reflecting differences in their conformational behavior. However, this requirement may be too restrictive, especially for peripheral parts of a structure, causing unnecessary separation of the substituents. If such a problem is detected during the inspection of a resulting molecular supergraph (see Sect. 3.2 step 7), repeat the analysis after allowing the cyclic to acyclic mapping as explained.
5. Overly complicated or chemically undesirable molecular supergraphs and structure mappings may significantly degrade the quality of the resulting models and complicate their interpretation. To avoid this, several aspects of the calculation setup can be adjusted. (1) The cyclic to acyclic mapping can be allowed (see Sect. 3.2 step 4 and Note 4). (2) The atoms in the training set structures can be labeled in order to explicitly specify the researcher's view of a central core/scaffold and the allowed atom correspondence. Such labeling is performed using the substructure search feature in the dataset file window (see Sect. 3.1 step 7) as explained in the reference manual. The approaches to automated detection of scaffolds in the MFTA analysis are currently in development. (3) If the training set structures are substantially different in their topology, mechanism of action, and/or binding mode, adequately covering them with one supergraph and model can be problematic. In such cases it may be useful to split the dataset into subsets containing related structures [13]. If only a few structures are different from the rest, it is better to exclude them from the analysis in order to obtain simpler model with better predictivity and more clear applicability domain. Such splitting or exclusion can be done in Instant JChem (see Sect. 3.1 step 6) or directly in MFTAWin by marking/unmarking and deleting the structures in the dataset file window (see Sect. 3.1 step 7) as explained in the reference manual.
6. The Q^2-guided descriptor selection by means of a genetic algorithm allows one to improve the model predictivity by optimizing the descriptor subset. However, doing this may entail significant trade-offs [2]. The PLSR implementation for the MFTA method uses a so-called stable cross-validation (Stable-CV) procedure to provide a more objective and reliable estimate of the model predictivity by averaging the Q^2 values over random reshufflings of the compounds between the cross-validation groups until the resulting Q^2 is stabilized within a specified precision (0.001 by default). This prevents obtaining chance correlations with overestimated predictivity that are often associated with simple Q^2-guided descriptor selection. Nevertheless, the model based on a full descriptor set, despite a lower Q^2 value, may provide easier interpretation as well as

more reliable predictions in a wider applicability domain by taking into account the influence and intercorrelations of all descriptors. Thus, it is recommended to try the Q^2-guided descriptor selection only if the predictivity of a full model is not acceptable.

7. When selecting the model with optimal predictivity, number of descriptors, and interpretability, the following features are desirable: (1) higher Q^2 values, preferably above 0.7 (the values above 0.5 are acceptable); (2) smaller number of local descriptors (see Note 3 and Table 1); (3) the Q^2-guided descriptor selection not used; (4) acceptable scatter and distribution of data points, no signs of a systematic bias in the prediction fit plot; (5) the molecular supergraph and structure mappings that are clear and chemically meaningful.
8. During the analysis and interpretation of the activity/selectivity maps, keep in mind that in the positions of molecular supergraph marked with the red circles, an increase in the descriptor values is favorable for the modeled endpoint; conversely, in positions marked with the blue circles, an increase in the descriptor tends to decrease the endpoint. Intensity of colors reflects the magnitude of the influence. To get a clearer picture of the structural features affecting the activity, it may be useful to check the structure mappings to the supergraph in the *MFTA Descriptor Values* window. Refer to Sect. 1 for a general discussion of the MFTA activity maps.
9. Alternatively, the structure generator software can be used to build a representative library of compounds in a comprehensive or stochastic mode based on an MFTA molecular supergraph [3] or on a central fragment (scaffold) and the substituents formed by combining various microfragments [25].
10. Alternatively, all exported prediction results may then be imported into the Instant JChem software for further processing and analysis, including filtering and selection of promising compounds.

5 Case Study: Indirubin Inhibitors of Glycogen Synthase Kinase GSK-3β

Glycogen synthase kinase 3β (GSK-3β) is a constitutively active serine/threonine protein kinase that plays a pivotal regulatory role in many signaling pathways associated with a number of serious diseases, including diabetes mellitus type II, inflammation, and cancer, as well as schizophrenia, bipolar disorder, Alzheimer's disease, Parkinson's disease, and other neurodegenerative disorders [26–29]. Thus, inhibition of GSK-3β is viewed as a promising therapeutic approach, especially for cancer, diabetes mellitus, and Alzheimer's disease. The search for potential drugs having the required activity and selectivity profiles [26] usually employs a combination of synthetic studies and the structure-based [30–32] and ligand-based (including QSAR) [32–33] drug design methods.

The indirubin derivatives are among the more thoroughly studied ATP-competitive GSK-3β inhibitors [26, 28, 29]. Based on the available body of experimental data and our earlier results (E. V. Radchenko, A. Ya. Safanyaev, V. A. Palyulin, unpublished data), in this case study, we use the MFTA-based design workflow to design indirubin compounds with high anti-GSK activity and good pharmacokinetics and toxicity profiles.

The training set containing the structure and activity data for 105 indirubin derivatives with diverse substituents was compiled from the original publications [34–39] using the Instant JChem software [18]. The inhibitory activity was represented by IC_{50} concentration values; in further analysis, the logarithmic quantities $pIC_{50} = \log(1/IC_{50})$ were used. Then the training set was imported into the MFTAWin software for the QSAR analysis. The labeling of the indirubin scaffold atoms specified in order to ensure more reliable and fast supergraph mapping, as well as an example of the training set structure and its superimposition onto the supergraph, are shown in Fig. 4a. The MFTA QSAR models obtained for the local descriptor combinations listed in Table 1 had similar statistical quality and consistent picture of the local descriptor influence on activity. The Q^2-guided descriptor selection was not used. The optimal model containing the atomic charge, steric, and lipophilicity descriptors had the number of PLSR factors $N_f = 2$, squared correlation coefficient $R^2 = 0.65$, root mean square error RMSE = 0.73, cross-validation parameter $Q^2 = 0.52$, and the root mean square error for cross-validation $RMSE_{cv} = 0.86$. The stable cross-validation procedure was used that yields more reliable predictivity estimates; however, they are usually lower and not directly comparable to ordinary cross-validation parameters. The predictivity of the model is sufficient for virtual screening of promising inhibitors. This is confirmed by the comparison of the experimental activity values and the values predicted during the cross-validation (Fig. 4b). As can be seen, for the medium- and high-activity compounds, good prediction accuracy is achieved.

The activity maps illustrating the effect of local molecular descriptors on the inhibitory activity (see Fig. 4c) indicate a preference for the larger, moderately polar, hydrophobic substituents, especially in the 3′, 5, and 6 positions of the indirubin scaffold. This agrees well with the X-ray data on the structure of the indirubin binding site (Fig. 5). Indeed, the ligand is located in a deep but narrow hydrophobic or slightly polar cleft/pocket between the two kinase lobes. The 3′ and 5 positions are oriented toward the cleft entrance, explaining the preference for long flexible substituents. The 6 position substituent is located in a small hydrophobic

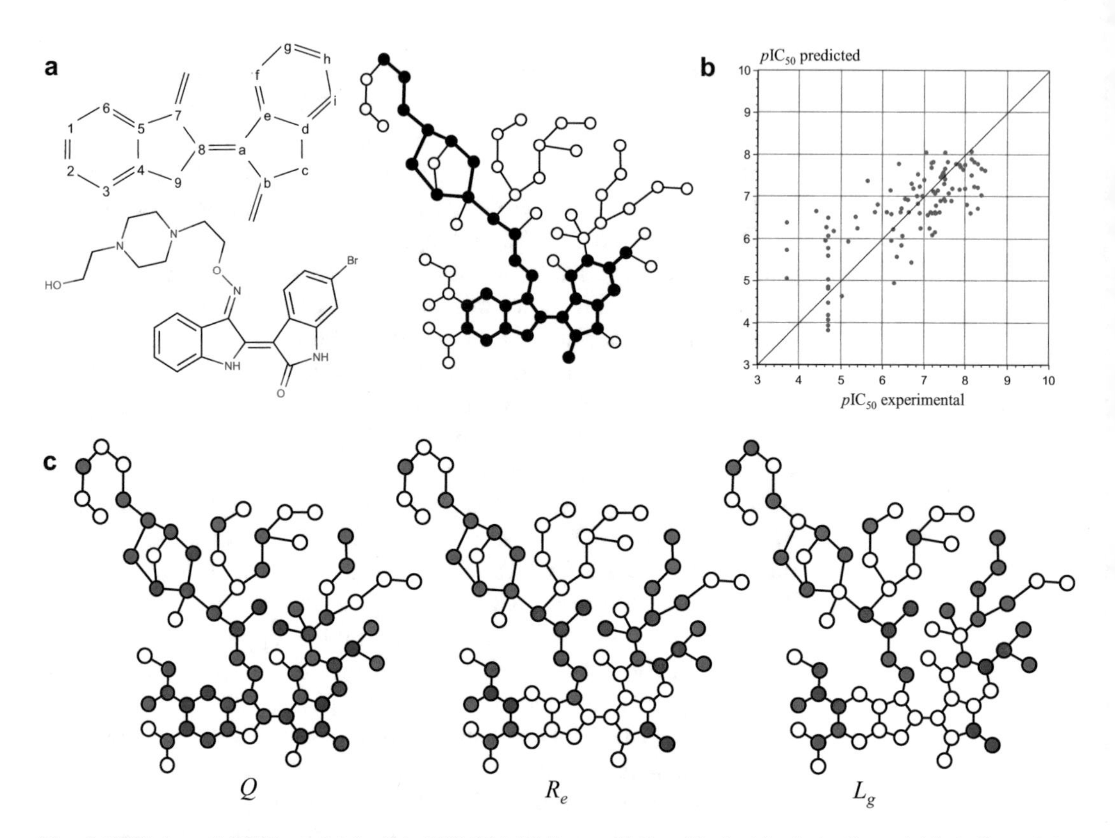

Fig. 4 MFTA-based QSAR model for the GSK-3β inhibitory activity of indirubin derivatives. (**a**) Labeling of the indirubin scaffold atoms (*left top*), example of the training set structure (*left bottom*) and its superimposition onto the molecular supergraph shown in *bold lines* (*right*). (**b**) Correspondence of the experimental inhibitory activity values and the values predicted during the cross-validation procedure. (**c**) Activity maps illustrating the effect of local molecular descriptors on the inhibitory activity (for notation see Fig. 1b)

pocket. Although such interpretation is useful, it should be stressed that the MFTA model includes the contributions from the properties in many positions of the molecular supergraph, and the overall picture of their effects may be complicated. Therefore, a virtual screening-based approach should be used to design structures with good predicted activity.

Taking into account the structural features affecting the activity as well as the structural diversity of the training set compounds (limiting the applicability domain of the model), the scaffold (including the substitution positions) and the set of substituents shown in Fig. 6a were selected. The virtual screening library containing 5000 structures randomly drawn out of 10,648 possible structures was generated using the Markush Enumeration feature in the Instant JChem software. After importing it into the MFTA-Win software, the indirubin scaffold atoms were labeled in the same way as in the training set (Fig. 4a). Then the activity values were predicted for the library structures using the previously saved MFTA model. The inspection of the outlier rating *vs* predicted

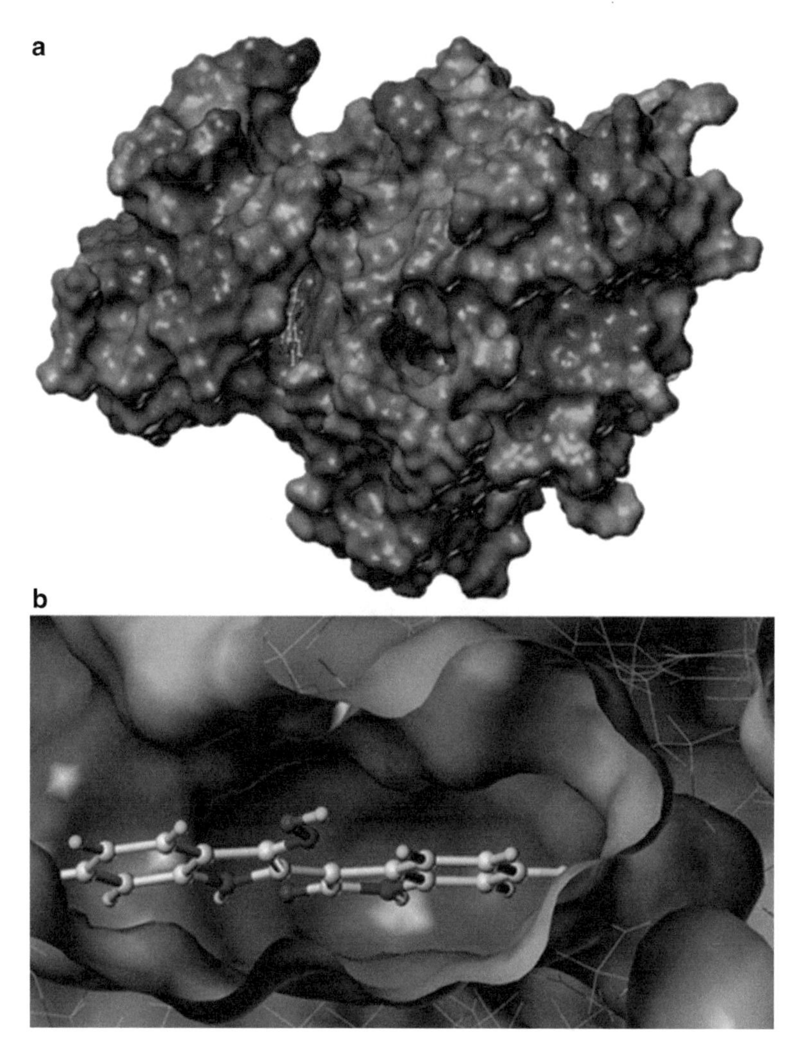

Fig. 5 Indirubin binding site in the GSK-3β structure. (**a**) General view of the GSK-3β complexed with indirubin-3′-oxime (PDB: 1Q41) [40] and the location of the binding site. The protein Connolly surface is colored according to the molecular lipophilic potential from *brown* (hydrophobic) to *blue* (hydrophilic) regions. The ligand molecule is represented by ball-and-stick model. The Sybyl-X 2.1 software [41] was used for the visualization. (**b**) Detailed view of the inhibitor position in the binding site

value plot (Fig. 6b) shows that safe predictions (OR2X < 40) cover a reasonable range of activity values approximately up to $pIC_{50} = 9$, while higher predicted values are less reliable. Out of 3140 structures falling within the model applicability domain, the 95 structures having the predicted pIC_{50} value above 8.5 were selected and exported for further analysis.

The structures with higher predicted activity were inspected visually, and their pharmacokinetic parameters and toxicity were predicted using the integrated online ADMET prediction service [20]. For a potential anti-Alzheimer lead candidate, high GSK-3β inhibitory activity ($pIC_{50} > 8.5$), reasonable blood–brain barrier

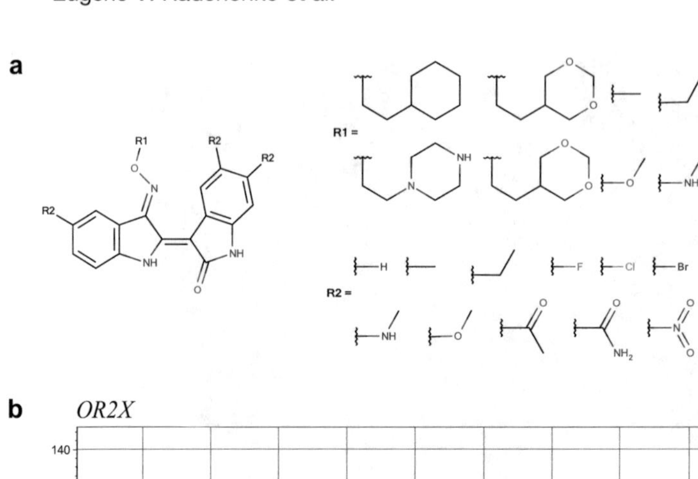

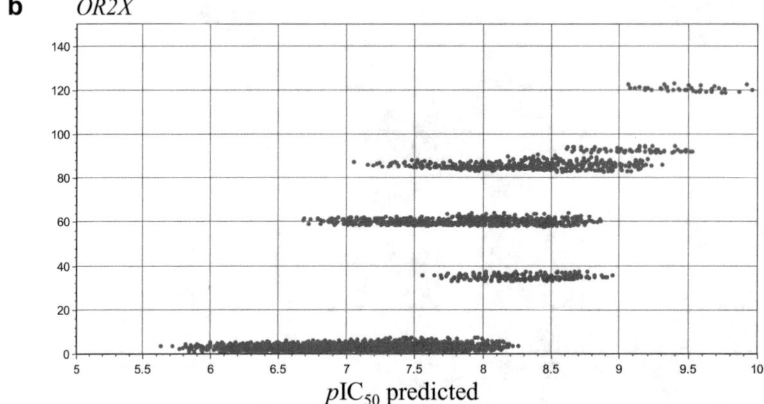

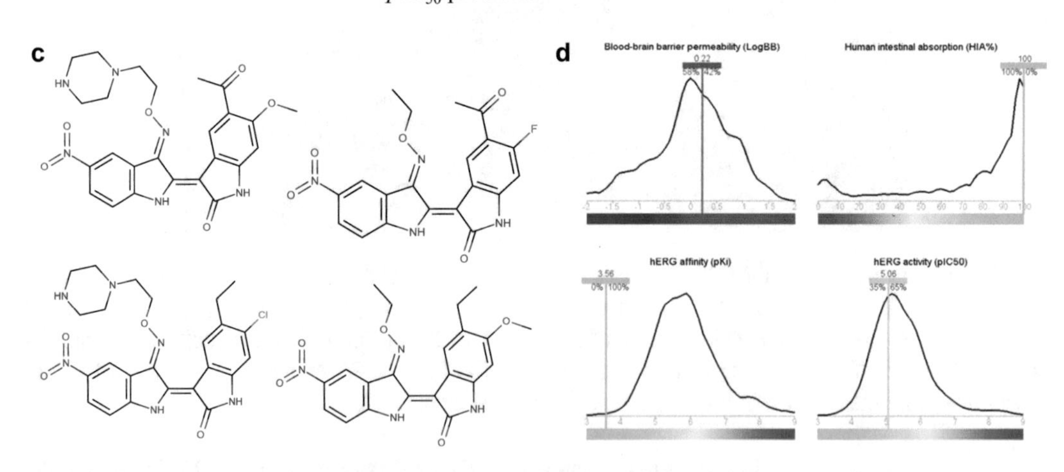

Fig. 6 MFTA-based design of potential indirubin inhibitors of GSK-3β. (**a**) Scaffold and substituents for the generation of the virtual screening library. (**b**) Plot of the outlier rating *vs* predicted activity values for the virtual screening library. (**c**) Examples of the focused library compounds having good predicted activity and ADMET properties. (**d**) ADMET predictions for the first compound in (**c**) The predicted property value (*top*), compound position in comparison to known compounds (percentage and plot), as well as the color code of the property magnitude (*blue/red scale*) or desirability (*red/green scale*) are shown

permeability (*LogBB* about zero or higher), good human intestinal absorption (*HIA* close to 100%), and low hERG-mediated cardiac toxicity risk (pK_i and $pIC_{50} < 5.5$) are desirable. Some of the focused library compounds meeting these criteria are shown in

Fig. 6c. The ADMET predictions for the first compound are illustrated in Fig. 6d. If necessary, the virtual screening of a larger and more diverse structure library can be performed in order to identify more promising structures.

6 Conclusion

The Molecular Field Topology Analysis (MFTA) is a QSAR method designed to model the activities mediated by small molecule binding to biotargets using the local physicochemical descriptors reflecting the major types of ligand–target interactions. It has been successfully used in the activity and selectivity modeling, design, and virtual screening of promising ligands of various enzymes and receptors (the NMDA and AMPA receptor antagonists and modulators as well as the acetyl- and butyrylcholinesterase inhibitors are of particular interest in the design of the anti-Alzheimer and other neuroprotective compounds).

The MFTA-based workflow for the design of potential anti-Alzheimer and other neuroprotective compounds presented in this chapter using a case study involving the inhibitors of glycogen synthase kinase 3β (GSK-3β) is also applicable to many other relevant targets and ligand classes. As a result, a focused library of compounds having the desired activity/selectivity profile and acceptable ADMET properties is obtained. In addition, some elements of this workflow may be incorporated into other virtual screening and design workflows.

Acknowledgments

This work was supported by the Russian Foundation for Basic Research (project 15–03-09084). The authors are grateful to the ChemAxon company for kindly providing the academic licenses for the structural data management software.

References

1. Palyulin VA, Radchenko EV, Zefirov NS (2000) Molecular Field Topology Analysis method in QSAR studies of organic compounds. J Chem Inf Comput Sci 40 (3):659–667. doi:10.1021/ci980114i
2. Radchenko EV, Palyulin VA, Zefirov NS (2008) Molecular Field Topology Analysis in drug design and virtual screening. In: Varnek A, Tropsha A (eds) Chemoinformatics approaches to virtual screening. RSC, Cambridge, pp 150–181. doi:10.1039/9781847558879-00150
3. Mel'nikov AA, Palyulin VA, Radchenko EV, Zefirov NS (2007) Generation of chemical structures on the basis of QSAR models of Molecular Field Topology analysis. Dokl Chem 415(2):196–199. doi:10.1134/S0012500807080058
4. Sun A, Prussia A, Zhan W, Murray EE, Doyle J, Cheng L-T, Yoon J-J, Radchenko EV, Palyulin VA, Compans RW, Liotta DC, Plemper RK, Snyder JP (2006) Nonpeptide inhibitors of measles virus entry. J Med Chem 49 (17):5080–5092. doi:10.1021/jm0602559

5. Chupakhin VI, Bobrov SV, Radchenko EV, Palyulin VA, Zefirov NS (2008) Computer-aided design of selective ligands of the benzodiazepine-binding site of the $GABA_A$ receptor. Dokl Chem 422(1):227–230. doi:10.1134/S0012500808090073
6. Radchenko EV, Koshelev SO, Tsareva DA, Voronkov AE, Palyulin VA, Zefirov NS (2012) Computer-aided design of arylphthalazines as potential Smoothened receptor antagonists. Dokl Chem 443(2):97–100. doi:10.1134/S0012500812040027
7. Radchenko EV, Karlov DS, Palyulin VA, Zefirov NS, Pentkovski VM (2012) Computer-aided modeling of activity and selectivity of quinazolinones as noncompetitive NMDA receptor antagonists. Dokl Biochem Biophys 443(1):118–122. doi:10.1134/S1607672912020159
8. Radchenko EV, Makhaeva GF, Malygin VV, Sokolov VB, Palyulin VA, Zefirov NS (2008) Modeling of the relationships between the structure of O-phosphorylated oximes and their anticholinesterase activity and selectivity using Molecular Field Topology Analysis (MFTA). Dokl Biochem Biophys 418 (1):47–51. doi:10.1134/S1607672908010122
9. Radchenko EV, Makhaeva GF, Sokolov VB, Palyulin VA, Zefirov NS (2009) Study of the structural determinants of acute and delayed neurotoxicity of O-phosphorylated oximes by Molecular Field Topology Analysis (MFTA). Dokl Biochem Biophys 429(1):309–314. doi:10.1134/S1607672909060064
10. Makhaeva GF, Radchenko EV, Baskin II, Palyulin VA, Richardson RJ, Zefirov NS (2012) Combined QSAR studies of inhibitor properties of O-phosphorylated oximes toward serine esterases involved in neurotoxicity, drug metabolism and Alzheimer's disease. SAR QSAR Environ Res 23(7–8):627–647. doi:10.1080/1062936X.2012.679690
11. Radchenko EV, Mel'nikov AA, Makhaeva GF, Palyulin VA, Zefirov NS (2012) Molecular design of O-phosphorylated oximes – selective inhibitors of butyrylcholinesterase. Dokl Biochem Biophys 443(1):91–95. doi:10.1134/S1607672912020093
12. Makhaeva GF, Radchenko EV, Palyulin VA, Rudakova EV, Aksinenko AY, Sokolov VB, Zefirov NS, Richardson RJ (2013) Organophosphorus compound esterase profiles as predictors of therapeutic and toxic effects. Chem Biol Interact 203(1):231–237. doi:10.1016/j.cbi.2012.10.012
13. Radchenko EV, Makhaeva GF, Boltneva NP, Serebryakova OG, Serkov IV, Proshin AN, Palyulin VA, Zefirov NS (2016) Molecular design of N,N-disubstituted 2-aminothiazolines as selective inhibitors of carboxylesterase. Russ Chem Bull 65 (2):570–575. doi:10.1007/s11172-016-1339-6
14. Girgis AS, Tala SR, Oliferenko PV, Oliferenko AA, Katritzky AR (2012) Computer-assisted rational design, synthesis, and bioassay of non-steroidal anti-inflammatory agents. Eur J Med Chem 50:1–8. doi:10.1016/j.ejmech.2011.11.034
15. Oliferenko PV, Oliferenko AA, Poda GI, Osolodkin DI, Pillai GG, Bernier UR, Tsikolia M, Agramonte NM, Clark GG, Linthicum KJ, Katritzky AR (2013) Promising *Aedes aegypti* repellent chemotypes identified through integrated QSAR, virtual screening, synthesis, and bioassay. PLoS One 8(9):e64547. doi:10.1371/journal.pone.0064547
16. Oliferenko PV, Oliferenko AA, Girgis AS, Saleh DO, Srour AM, George RF, Pillai GG, Panda CS, Hall CD, Katritzky AR (2014) Synthesis, bioassay, and Molecular Field Topology Analysis of diverse vasodilatory heterocycles. J Chem Inf Model 54(4):1103–1116. doi:10.1021/ci400723m
17. Jabeen F, Oliferenko PV, Oliferenko AA, Pillai GG, Ansari FL, Hall CD, Katritzky AR (2014) Dual inhibition of the α-glucosidase and butyrylcholinesterase studied by Molecular Field Topology Analysis. Eur J Med Chem 80:228–242. doi:10.1016/j.ejmech.2014.04.018
18. Instant JChem 17.1.9 (2017) ChemAxon Ltd., http://www.chemaxon.com
19. MFTAWin 3.5 (2016) Department of Chemistry, Lomonosov Moscow State University, http://qsar.chem.msu.ru
20. ADMET Prediction Service. http://qsar.chem.msu.ru/admet/. Accessed 01 Jan 2017
21. Dyabina AS, Radchenko EV, Palyulin VA, Zefirov NS (2016) Prediction of blood-brain barrier permeability of organic compounds. Dokl Biochem Biophys 470(1):371–374. doi:10.1134/S1607672916050173
22. Radchenko EV, Dyabina AS, Palyulin VA, Zefirov NS (2016) Prediction of human intestinal absorption of drug compounds. Russ Chem Bull 65(2):576–580. doi:10.1007/s11172-016-1340-0
23. Radchenko EV, Rulev YA, Safanyaev AY, Palyulin VA, Zefirov NS (2017) Computer-aided estimation of the hERG-mediated cardiotoxicity risk of potential drug compounds. Dokl Biochem Biophys 473(2):128–131

24. Palyulin VA, Radchenko EV, Zefirov NS (2014) Molecular Field Topology Analysis (MFTA) as a tool for multi-target QSAR. In: 20th European symposium on quantitative structure-activity relationships Euro-QSAR2014, St. Petersburg, August 31–September 4, p 46
25. Mel'nikov AA, Palyulin VA, Zefirov NS (2007) Generation of molecular graphs for QSAR studies: an approach based on supergraphs. J Chem Inf Model 47(6):2077–2088. doi:10.1021/ci700156f
26. Osolodkin DI, Palyulin VA, Zefirov NS (2013) Glycogen synthase kinase 3 as an anticancer drug target: novel experimental findings and trends in the design of inhibitors. Curr Pharm Des 19(4):665–679. doi:10.2174/1381612811306040665
27. Llorens-Martín M, Jurado J, Hernández F, Avila J (2014) GSK-3β, a pivotal kinase in Alzheimer disease. Front Mol Neurosci 7:46. doi:10.3389/fnmol.2014.00046
28. Maqbool M, Mobashir M, Hoda N (2016) Pivotal role of glycogen synthase kinase-3: a therapeutic target for Alzheimer's disease. Eur J Med Chem 107:63–81. doi:10.1016/j.ejmech.2015.10.018
29. Khan I, Tantray MA, Alam MS, Hamid H (2017) Natural and synthetic bioactive inhibitors of glycogen synthase kinase. Eur J Med Chem 125:464–477. doi:10.1016/j.ejmech.2016.09.058
30. Osolodkin DI, Palyulin VA, Zefirov NS (2011) Structure-based virtual screening of glycogen synthase kinase 3β inhibitors: analysis of scoring functions applied to large true actives and decoy sets. Chem Biol Drug Des 78(3):378–390. doi:10.1111/j.1747-0285.2011.01159.x
31. Shulga DA, Osolodkin DI, Palyulin VA, Zefirov NS (2012) Simulation of intramolecular hydrogen bond dynamics in manzamine A as a sensitive test for charge distribution quality. Nat Prod Commun 7(3):295–299
32. Karpov PV, Osolodkin DI, Baskin II, Palyulin VA, Zefirov NS (2011) One-class classification as a novel method of ligand-based virtual screening: the case of glycogen synthase kinase 3β inhibitors. Bioorg Med Chem Lett 21(22):6728–6731. doi:10.1016/j.bmcl.2011.09.051
33. Tsareva DA, Osolodkin DI, Shulga DA, Oliferenko AA, Pisarev SA, Palyulin VA, Zefirov NS (2011) General purpose electronegativity relaxation charge models applied to CoMFA and CoMSIA study of GSK-3 inhibitors. Mol Inform 30(2–3):169–180. doi:10.1002/minf.201000141
34. Leclerc S, Garnier M, Hoessel R, Marko D, Bibb JA, Snyder GL, Greengard P, Biernat J, Wu YZ, Mandelkow EM, Eisenbrand G, Meijer L (2001) Indirubins inhibit glycogen synthase kinase-3 beta and CDK5/p25, two protein kinases involved in abnormal tau phosphorylation in Alzheimer's disease. A property common to most cyclin-dependent kinase inhibitors? J Biol Chem 276(1):251–260. doi:10.1074/jbc.M002466200
35. Guengerich FP, Sorrells JL, Schmitt S, Krauser JA, Aryal P, Meijer L (2004) Generation of new protein kinase inhibitors utilizing cytochrome p450 mutant enzymes for indigoid synthesis. J Med Chem 47(12):3236–3241. doi:10.1021/jm030561b
36. Polychronopoulos P, Magiatis P, Skaltsounis AL, Myrianthopoulos V, Mikros E, Tarricone A, Musacchio A, Roe SM, Pearl L, Leost M, Greengard P, Meijer L (2004) Structural basis for the synthesis of indirubins as potent and selective inhibitors of glycogen synthase kinase-3 and cyclin-dependent kinases. J Med Chem 47(4):935–946. doi:10.1021/jm031016d
37. Wu ZL, Aryal P, Lozach O, Meijer L, Guengerich FP (2005) Biosynthesis of new indigoid inhibitors of protein kinases using recombinant cytochrome P450 2A6. Chem Biodivers 2(1):51–65. doi:10.1002/cbdv.200490166
38. Beauchard A, Ferandin Y, Frère S, Lozach O, Blairvacq M, Meijer L, Thiéry V, Besson T (2006) Synthesis of novel 5-substituted indirubins as protein kinases inhibitors. Bioorg Med Chem 14(18):6434–6443. doi:10.1016/j.bmc.2006.05.036
39. Vougogiannopoulou K, Ferandin Y, Bettayeb K, Myrianthopoulos V, Lozach O, Fan Y, Johnson CH, Magiatis P, Skaltsounis AL, Mikros E, Meijer L (2008) Soluble 3′,6-substituted indirubins with enhanced selectivity toward glycogen synthase kinase-3 alter circadian period. J Med Chem 51(20):6421–6431. doi:10.1021/jm800648y
40. Bertrand JA, Thieffine S, Vulpetti A, Cristiani C, Valsasina B, Knapp S, Kalisz HM, Flocco M (2003) Structural characterization of the GSK-3β active site using selective and non-selective ATP-mimetic inhibitors. J Mol Biol 333(2):393–407. doi:10.1016/j.jmb.2003.08.031
41. Sybyl-X 2.1, Certara L.P., St. Louis, Mo., 2013. www.certara.com

Part III

Modeling of Ligands Acting Against Specific Anti-Alzheimer Drug Targets

Chapter 6

Galantamine Derivatives as Acetylcholinesterase Inhibitors: Docking, Design, Synthesis, and Inhibitory Activity

Irini Doytchinova, Mariyana Atanasova, Georgi Stavrakov, Irena Philipova, and Dimitrina Zheleva-Dimitrova

Abstract

Galantamine (GAL) is a well-known acetylcholinesterase (AChE) inhibitor, and it is widely used for treatment of Alzheimer's disease. GAL fits well in the catalytic site of AChE, but it is too short to block the peripheral anionic site (PAS) of the enzyme, where the amyloid beta (Aβ) peptide binds and initiates the Aβ aggregation. Here, we describe a docking-based technique for designing of GAL derivatives with dual-site binding fragments – one blocking the catalytic site and another blocking the PAS. The highly scored compounds are synthesized and tested. Protocols for docking, design, synthesis, and AChE inhibitory test are given.

Key words Galantamine, Acetylcholinesterase inhibition, Docking, Drug design, Dual-site binding

1 Introduction

The binding site of recombinant human AChE (*rh*AChE) consists of several domains [1–5]. The catalytic anionic site (CAS) lies on the bottom of 20 Å deep and narrow binding gorge. It consists of the catalytic triad: Ser203, Glu334, and His447. The anionic domain binds the quaternary trimethylammonium choline moiety of acetylcholine (ACh). Despite its name, it does not contain any anionic residues but only aromatic ones: Trp86, Tyr130, Tyr337, and Phe338. They are involved in cation-pi interactions with the protonated head of ACh. The acyl pocket consists of two bulky residues, Phe295 and Phe297, and it determines the selective binding of ACh by preventing the access of larger choline esters. The oxyanion hole hosts one molecule of structural water and consists of Gly121, Gly122, and Ala204. The water molecule bridges the binding between enzyme and substrate by hydrogen-bond networking and stabilizes the substrate tetrahedral transition state.

Kunal Roy (ed.), *Computational Modeling of Drugs Against Alzheimer's Disease*, Neuromethods, vol. 132, DOI 10.1007/978-1-4939-7404-7_6, © Springer Science+Business Media LLC 2018

Finally, the peripheral anionic site (PAS) lies at the entrance to binding gorge. It is composed of five residues: Tyr72, Asp74, Tyr124, Trp286, and Tyr341. PAS allosterically modulates catalysis [6] and is implicated in non-cholinergic functions such as amyloid deposition [7], cell adhesion, and neurite outgrowth [8, 9]. It was found that the Aβ peptide binds close to PAS, interacting hydrophobically with residues 275–305 [7]. This interaction promotes the formation of amyloid fibrils characteristic of Alzheimer's disease [10, 11]. The blockade of PAS prevents the AChE-induced Aβ aggregation [7]. This pivotal finding prompted the design of novel AChEIs with dual binding moieties – one blocking the CAS at the bottom of the gorge and one blocking the PAS [12–17].

Galantamine (GAL) is an alkaloid initially isolated from the bulbs and flowers of *Galanthus caucasicus*, *Galanthus woronowii* (Amaryllidaceae), and related genera [18]. Paskov first developed GAL as an industrial drug under the trade name Nivalin (Sopharma, Bulgaria) [19]. It has been used for treatment of myasthenia gravis, myopathy, residual poliomyelitis paralysis syndromes, sensory and motor disorders of CNS, and decurarization [20–24]. Because of its ability to cross the blood-brain barrier and to affect the central cholinergic function, in the 1980s, GAL was investigated for treatment of Alzheimer's disease (AD) and in 2000 was approved for use in Europe, the United States, and Asia [25–27].

In addition to its acetylcholinesterase (AChE)-inhibiting ability, GAL has been identified as an allosteric modulator of nicotinic acetylcholine receptors (nAChRs) [28–31]. The stimulation of nAChRs can increase intracellular Ca^{2+} levels and facilitate noradrenaline release; both effects enhance the cognitive brain function [32]. Recently, Takata et al. have demonstrated that treatment of rat microglia with GAL significantly enhanced microglial amyloid-β (Aβ) phagocytosis and facilitated Aβ clearance in brains of rodent AD models [33]. This multiple-target action of GAL makes it a most valuable drug for AD treatment and prompts the synthesis of novel GAL derivatives with improved binding to AChE [33–39]. Several series of GAL derivatives with dual-site binding to the enzyme have been prepared and tested [16, 36, 37, 40–42]. All of them showed good AChE inhibitory activities.

In the present paper, we summarize our experience in developing of galantamine derivatives as acetylcholinesterase inhibitors with dual-site binding [43–45]. We used docking-based design in a stepwise manner to develop derivatives with binding affinity more than 1000 times higher than that of GAL.

2 Materials

2.1 Docking Software

GOLD [46] is a docking tool based on a genetic algorithm, and it has proven successful in virtual screening, lead optimization, and identification of the correct binding mode of active molecules [47, 48]. GOLD takes into account the flexibility of the ligand as well as the flexibility of the residues in the binding site. GOLD v. 5.1 was used in our studies.

2.2 Reactants for Synthesis

Fine chemicals: 5-hydroxyindole, 97%; 5-aminoindole, 97%; 4-hydroxybenzaldehyde, 99%; 1,6-dibromohexane, 98%; 7-bromoheptanoic acid, 97%; 6-bromohexanoic acid, 98%; 5-bromovaleric acid, 97%; 4-bromobutyric acid, 98%; aniline, 99.8%; benzylamine, 99%; phenethylamine, 99%; (S)-(+)-2-phenylglycine methyl ester hydrochloride, 97%; L-phenylalanine methyl ester hydrochloride, 98%; L-tryptophan methyl ester hydrochloride, 98%; galantamine hydrobromide, 95%.

Reagents: potassium carbonate, 99 + %; sodium triacetoxyborohydride, 97%; *N*-(3-dimethylaminopropyl)-*N*′-ethylcarbodiimide hydrochloride (EDC), 98%; 1-hydroxybenzotriazole hydrate, 99%; *N*,*N*-diisopropylethylamine, for synthesis; 3-chloroperoxybenzoic acid, 70–75%; ammonia water, 25%, pure for analysis; iron(II) sulfate heptahydrate, 99 + %; sodium sulfate, anhydrous, 99 + %; sodium bicarbonate, 99 + %; triethylamine, 99%; hydrochloric acid, ca. 37% solution in water.

Solvents: 1,2-dichloroethane, extra pure; acetonitrile, 99.9%, extra dry over molecular sieve; methylene chloride, extra pure; petroleum ether 40–60 °C, extra pure; ethyl acetate, for analysis; methanol for HPLC.

2.3 Reactants for AChE Inhibition Assay

Electrophorus electricus AChE (Sigma-Aldrich), acetylthiocholine (ATCl), 5,5′-dithio-bis-(2-nitrobenzoic acid) (DTNB), phosphate buffer (pH 7.6), UV-Vis Spectrophotometer Shimadzu UV-1203 at 405 nm.

3 Methods

3.1 Design

The GAL derivatives are N-substituted products (Fig. 1). Linkers of different type and length connect the GAL core to aromatic fragments. The GAL core fits into the catalytic site, while the aromatic fragment aims to block the PAS (Fig. 2).

3.2 Molecular Docking Protocol

The optimized docking protocol includes the following settings:

1. Protein: X-ray structure of rhAChE in complex with galantamine (GAL) (pdb id: 4EY6, R = 2.15 Å) [49]. The ligand is removed.

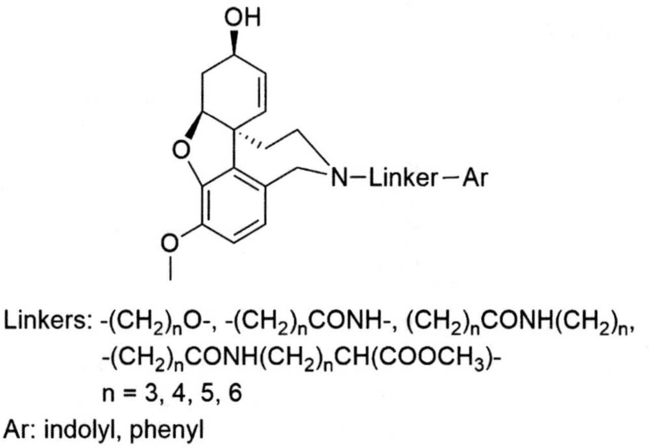

OH
O
N—Linker—Ar
O
Linkers: $-(CH_2)_nO-$, $-(CH_2)_nCONH-$, $(CH_2)_nCONH(CH_2)_n$,
$-(CH_2)_nCONH(CH_2)_nCH(COOCH_3)-$
n = 3, 4, 5, 6
Ar: indolyl, phenyl

Fig. 1 N-substituted GAL derivatives

Peripheral binding site (PAS)
Catalytic anionic site (CAS)

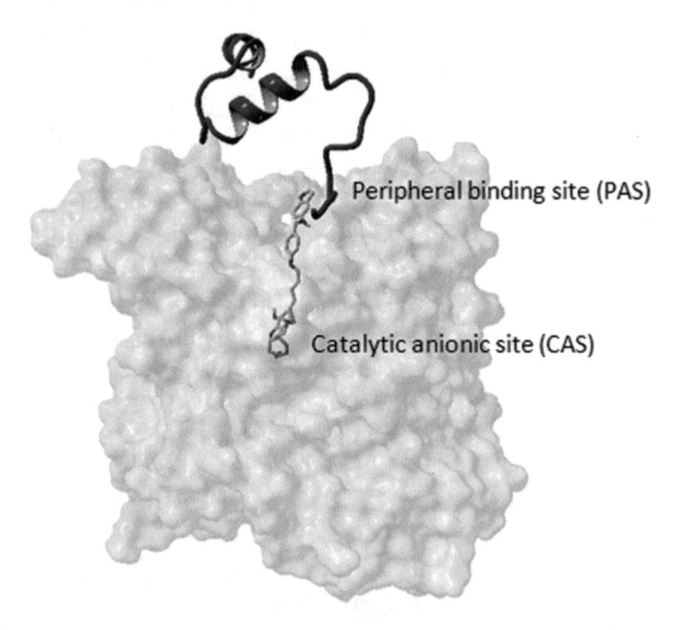

Fig. 2 The GAL core fits into the catalytic site, while the aromatic fragment blocks the PAS. AChE given as *gray* surface, inhibitor in *cyan*, and amyloid beta peptide in *blue*

2. Ligands: The structures of the designed molecules are entered in .mol2 format.
3. Scoring function: GoldScore or ChemPLP.
4. Radius of the binding site: 10Å.
5. One water molecule (HOH846) is kept within the binding site.
6. Flexible side chains in the binding site: Tyr72, Asp74, Trp86, Tyr124, Ser125, Trp286, Phe297, Tyr337, Phe338, and Tyr341.

The compounds with the highest GoldScores or ChemPLP scores are selected for synthesis.

3.3 Synthesis

3.3.1 Synthesis of Bromo-Containing Linkers of Different Type and Length Connected to Aromatic Fragments: Br-Linker-Ar

Protocol 1: *Synthesis of 4-(6-bromohexyloxy)benzaldehyde*

HO–C6H4–CHO → ($Br(CH_2)_6Br$, K_2CO_3, CH_3CN, 5 h, reflux, 83%) → $Br(CH_2)_6O$–C6H4–CHO

1. *Load a 100 ml round bottom flask with 4-hydroxybenzaldehyde (0.5 g, 4.1 mmol), K_2CO_3 (2.83 g, 20.5 mmol), and CH_3CN (20 ml).*
2. *Add to the stirred suspension 1,6-dibromohexane (12.3 mmol, 1.86 ml), equip the flask with a condenser, and reflux for 5 hours.*
3. *Stop the heating, remove the condenser, and concentrate till dry under reduced pressure.*
4. *Separate the residue between ethyl acetate and water, extract the aqueous phase with ethyl acetate, and dry the organic extracts over Na_2SO_4. Filter from the solids and concentrate.*
5. *Purify* via *flash column chromatography using silica gel as stationary phase and a mixture of ethyl acetate/petroleum ether (1:9) as mobile phase.*

Protocol 2: *Synthesis of* N*-(4-(6-bromohexyloxy)benzyl)-1*H*-indol-5-amine*

1H-indol-5-amine (NH_2) → (OHC–C6H4–$O(CH_2)_6Br$, $NaBH(OAc)_3$, DCE, 24 h, r.t., 99%) → N-(4-($O(CH_2)_6Br$)benzyl)-1H-indol-5-amine

1. *Load a 50 ml round bottom flask with 1*H*-indol-5-amine (0.100 g, 0.757 mmol), 4-(6-bromohexyloxy)benzaldehyde (0.216 g, 0.757 mmol), and 1,2-dichloroethane (8 ml).*
2. *Stir the mixture for 30 min at room temperature, add NaBH $(OAc)_3$ (0.240 g, 1.136 mmol), and continue the stirring for 24 h.*
3. *Quench the reaction with sat.aq.$NaHCO_3$ (50 ml), extract the product with CH_2Cl_2, and dry the organic extracts over Na_2SO_4. Filter from the solids and concentrate.*
4. *Purify* via *flash column chromatography using silica gel as stationary phase and a mixture of petroleum ether/ethyl acetate/triethylamine (2:1:0.5) as mobile phase.*

Protocol 3: *Synthesis of 5-(6-bromohexyloxy)-1H-indole*

Reaction scheme: 1H-indol-5-ol → ($Br(CH_2)_6Br$, K_2CO_3, CH_3CN, 24 h, 60 °C, 68%) → 5-($O(CH_2)_6Br$)-1H-indole

1. *Load a 50 ml round bottom flask with 1H-indol-5-ol (0.100 g, 0.750 mmol), CH_3CN (15 ml), and K_2CO_3 (0.310 g, 2.250 mmol).*
2. *Add to the stirred suspension 1,6-dibromohexane (0.220 g, 0.900 mmol) and heat at 60 °C for 24 h.*
3. *Cool to room temperature, filter through a pad of Celite, and concentrate till dry under reduced pressure.*
4. *Purify* via *flash column chromatography using silica gel as stationary phase and a mixture of petroleum ether/ethyl acetate (4:1) as mobile phase.*

Protocol 4: *Synthesis of bromo-amide intermediates*

Reaction scheme: $Br-(CH_2)_n-C(O)OH$ → (H_2N-R, EDC/HOBT, CH_2Cl_2, 1-24 h, r.t.) → $Br-(CH_2)_n-C(O)NHR$

Bromo carboxylic acid	Appropriate amine	Product (bromo-amide intermediate)	Yield %
$Br-(CH_2)_n-C(O)OH$	H_2N-R	$Br-(CH_2)_n-C(O)NHR$	
$n = 5$	H_2N-(1H-indol-5-yl)	$Br-(CH_2)_5-C(O)NH$-(1H-indol-5-yl)	84
$n = 6$	H_2N-Ph	$Br-(CH_2)_6-C(O)NHPh$	88
$n = 4$	H_2N-Bn	$Br-(CH_2)_4-C(O)NHBn$	90
$n = 4$	$H_2N-(CH_2)_2Ph$	$Br-(CH_2)_4-C(O)NH(CH_2)_2Ph$	67
$n = 5$	$H_2N-(CH_2)_2Ph$	$Br-(CH_2)_5-C(O)NH(CH_2)_2Ph$	89

(continued)

Bromo carboxylic acid	Appropriate amine	Product (bromo-amide intermediate)	Yield %
$Br-(CH_2)_n-C(=O)OH$	H_2N-R	$Br-(CH_2)_n-C(=O)NHR$	
$n = 6$	$H_2N-(CH_2)_2Ph$	$Br-(CH_2)_6-C(=O)NH(CH_2)_2Ph$	94
$n = 5$	HCl · H_2N–CH(CO_2Me)(Ph)	$Br-(CH_2)_5-C(=O)$NH–CH(CO_2Me)(Ph)	78
$n = 6$	HCl · H_2N–CH(CO_2Me)(Ph)	$Br-(CH_2)_6-C(=O)$NH–CH(CO_2Me)(Ph)	80
$n = 3$	HCl · H_2N–CH(CO_2Me)(Bn)	$Br-(CH_2)_3-C(=O)$NH–CH(CO_2Me)(Bn)	88
$n = 4$	HCl · H_2N–CH(CO_2Me)(Bn)	$Br-(CH_2)_4-C(=O)$NH–CH(CO_2Me)(Bn)	90
$n = 5$	HCl · H_2N–CH(CO_2Me)(Bn)	$Br-(CH_2)_5-C(=O)$NH–CH(CO_2Me)(Bn)	86
$n = 6$	HCl · H_2N–CH(CO_2Me)(Bn)	$Br-(CH_2)_6-C(=O)$NH–CH(CO_2Me)(Bn)	89
$n = 5$	HCl · H_2N–CH(CO_2Me)(CH_2-indol-2-yl, N–H)	$Br-(CH_2)_5-C(=O)$NH–CH(CO_2Me)(CH_2-indol-2-yl, N–H)	99

1. *Load a 50 ml round bottom flask with bromo carboxylic acid (1.1 mmol), CH_2Cl_2(20 ml),* N*-[3-(dimethylamino)propyl]-*N*-ethylcarbodiimide (0.210 g, 1.1 mmol), 1-hydroxybenzotriazole (0.148 g, 1.1 mmol), and the appropriate amine (1 mmol).*
2. *In the cases of amino acid methyl ester hydrochlorides, add also* N, N*-diisopropylethylamine (1.1 mmol).*
3. *Stir the mixture at room temperature. Follow the reaction development by TLC.*
4. *Concentrate under reduced pressure till dry.*
5. *Purify* via *flash column chromatography using silica gel as stationary phase and a mixture of dichloromethane/ethyl acetate (20:1) as mobile phase.*

3.3.2 Synthesis of the Target Compounds

Protocol 5: *Synthesis of norgalantamine and its subsequent alkylation with Br-Linker-Ar:*

OH, O, O, HBr, N–CH_3 — 1. NH_4OH; CH_2Cl_2 2. MCPBA; CH_2Cl_2 3. $FeSO_4.7H_2O$; MeOH 78% → OH, O, O, NH — Br-Linker-Ar, K_2CO_3, CH_3CN, 24 h, 60 °C → OH, O, O, N–Linker-Ar

1. *To a stirred suspension of galantamine hydrobromide (0.734 g, 2 mmol) in dichloromethane (10 ml), add 25% aqueous NH_4OH (1 ml) and stir till clear. Separate the organic layer, dry it over Na_2SO_4, and concentrate under reduced pressure to isolate free galantamine base.*
2. *Dissolve the galantamine (2 mmol) into dichloromethane (10 ml), add portionwise 3-chloroperoxybenzoic acid (0.542 g, 2.2 mmol), and stir at room temperature for 1 h.*
3. *Cool the mixture to 0 °C and add methanol (20 ml) followed by $FeSO_4.7H_2O$ (0.556 g, 2 mmol). Stir for ½ h at 0 °C and for 3 h at room temperature.*
4. *Quench the reaction with 5 N HCl (40 ml), remove the organic solvents under reduced pressure, and wash the aqueous residue twice with ether.*
5. *Cool the aqueous layer to 0 °C and basify it with 25% aqueous NH_4OH.*
6. *Extract with dichloromethane, dry over Na_2SO_4, and concentrate.*

7. *Purify* via *flash column chromatography using silica gel as stationary phase and a mixture of dichloromethane/methanol (5:1) as mobile phase to isolate norgalantamine.*
8. *Prepare under argon atmosphere a solution of norgalantamine (0.200 g, 0.732 mmol) in anhydrous acetonitrile (25 mL).*
9. *Add the appropriate Br-Linker-Arom (0.951 mmol) and anhydrous* K_2CO_3 *(0.303 g, 2.2 mmol).*
10. *Stir at 60 °C for 24 h, cool to room temperature, filter through a pad of Celite, and remove the solvent under reduced pressure.*
11. *Purify* via *flash column chromatography using silica gel as stationary phase and a mixture of dichloromethane/methanol/ammonia (20:1:0.05) to isolate the desired product.*

Structure elucidation: *The structures of the newly synthesized compounds were confirmed by 1D and 2D NMR spectra. Their purity was proven by elemental analysis or HRMS. The spectral analyses were in accordance with the assigned structures. Additionally, the compounds were characterized by melting points and specific rotation for the chiral compounds.*

3.4 AChE Inhibition Assay

The AChE activity is assayed as described by Ellman et al. [50] with some modifications [51]. 50 μL of *E. electricus* AChE (Sigma-Aldrich) in buffer phosphate (pH 7.6) and 50 μL of the test compounds (4–500 μM in methanol) dissolved in 700 μL in the same buffer are mixed. The mixtures are incubated for 30 min at room temperature. 100 μL of the substrate solution (0.5 M DTNB, 0.6 mM ATCI in buffer, pH 7.6) is added. The absorbance is read in Shimadzu spectrophotometer at 405 nm after 3 min. Enzyme activity is calculated as a percentage compared to an assay using a buffer without any inhibitor using nonlinear regression. IC_{50} values are means ± SD of three individual determinations each performed in triplicate. GAL is used as a positive control.

4 Some Results Illustrating the Docking-Based Design Efficiency

In order to illustrate the efficiency of the proposed docking-based design of novel dual-site binding GAL derivatives, in Table 1., we present some results from our in-house database of GAL derivatives designed to block simultaneously the CAS and PAS [44, 45]. More than hundred compounds were designed and docked into AChE and the best scored of them were synthesized and tested.

Most of the compounds were 10–100 times more active than GAL. However, two of them – compounds *9* and *15* – were more than 1000 times more active. They are subject to further research.

Table 1
Docking scores and experimental IC_{50} values of the designed GAL derivatives

Comp.	Linker	Ar	Docking score[a]	IC_{50} (exp) μM	Times more active than GAL
1	$(CH_2)_6O$	Phenyl-CH_2NH-indolyl	119.15	0.011	95
2	$(CH_2)_6O$	Indolyl	112.32	0.012	93
3	$(CH_2)_5CONH$	Indolyl	109.06	0.015	72
4	$(CH_2)_5CONHCH(COOCH_3)CH_2$	Indolyl	113.89	0.094	11
5	$(CH_2)_6CONH$	Phenyl	100.87	0.0169	63
6	$(CH_2)_4CONHCH_2$	Phenyl	99.82	0.0308	35
7	$(CH_2)_4CONH(CH_2)_2$	Phenyl	101.79	0.0308	35
8	$(CH_2)_5CONH(CH_2)_2$	Phenyl	106.09	0.021	51
9	$(CH_2)_6CONH(CH_2)_2$	Phenyl	109.94	0.0008	1338
10	$(CH_2)_5CONHCH(COOCH_3)$	Phenyl	107.52	0.0527	20
11	$(CH_2)_6CONHCH(COOCH_3)$	Phenyl	111.86	0.0211	51
12	$(CH_2)_3CONHCH(COOCH_3)CH_2$	Phenyl	97.53	0.0958	11
13	$(CH_2)_4CONHCH(COOCH_3)CH_2$	Phenyl	104.38	0.0264	40
14	$(CH_2)_5CONHCH(COOCH_3)CH_2$	Phenyl	108.24	0.0246	43
15	$(CH_2)_6CONHCH(COOCH_3)CH_2$	Phenyl	116.95	0.0011	1008
GAL HBr			74.56	1.07	1

Data are taken from Refs. [44, 45]
[a]GoldScores for compounds *1–4*
GAL HBr ChemPLP scores for compounds *5–15*

5 Conclusions

Molecular docking is a widely used structure-based method for virtual screening of huge databases of compounds as well as for binding prediction of newly designed ligands. It was used solely or in combination with 2D- and 3D-QSAR or machine learning methods [52–64]. Here, we described a protocol for molecular docking of GAL derivatives into the enzyme AChE. The protocol was optimized stepwise to find the settings (scoring function, flexible binding site, radius of the binding site, presence/absence of a water molecule) correlating best with the binding affinities of compounds. The optimized protocol was used to predict the affinities of newly designed GAL derivatives with dual-site binding to AChE. The best scored compounds were synthesized and tested. All of them are 10–100 times more active than GAL. Even more, two of them are more than 1000 times more active.

6 Note

The BLAST alignment of AChEs from human, rat, rabbit, and *E. electricus* shows that the main residues forming the binding site are conserved.

Acknowledgments

This work has been supported by the Council on Medical Science at the Medical University of Sofia, Bulgaria (Grants 2-S/2013, 1-S/2014, 8-S/2015, and 10-S/2016).

References

1. Harel M, Schalk I, Ehret-Sabatier L, Bouet F, Goeldner M, Hirth C, Axelsen PH, Silman I, Sussman JL (1993) Quaternary ligand binding to aromatic residues in the active-site gorge of acetylcholinesterase. Proc Natl Acad Sci USA 90:9031–9035
2. Ordentlich A, Barak D, Kronman C, Flashner Y, Leitner M, Segall Y, Ariel N, Cohen S, Velan B, Shafferman A (1993) Dissection of the human acetylcholinesterase active center determinants of substrate specificity. Identification of residues constituting the anionic site, the hydrophobic site, and the acyl pocket. J Biol Chem 268:17083–17095
3. Ordentlich A, Barak D, Kronman C, Ariel N, Segall Y, VelanB SA (1998) Functional characteristics of the oxyanion hole in human acetylcholinesterase. J Biol Chem 273:19509–19517
4. Radic Z, Pickering NA, Vellom DC, Camp S, Taylor P (1993) Three distinct domains in the cholinesterase molecule confer selectivity for acetyl- and butyrylcholinesterase inhibitors. Biochemistry 32:12074–12084
5. Sussman JL, Harel M, Frolow F, Oefner C, Goldman A, Toker L, Silman I (1991) Atomic structure of acetylcholinesterase from Torpedo californica: a prototypic acetylcholine-binding protein. Science 253:872–879
6. Kitz RJ, Braswell LM, Ginsburg S (1970) On the question: is acetylcholinesterase an allosteric protein? Mol Pharmacol 6:108–121
7. De Ferrari GV, Canales MA, Shin I, Weiner LM, Silman I, Inestrosa NC (2001) A structural motif of acetylcholinesterase that promotes amyloid beta-peptide fibril formation. Biochemistry 40:10447–10457
8. Johnson G, Moore SW (2004) Identification of a structural site on acetylcholinesterase that promotes neurite outgrowth and binds laminin-1 and collagen IV. Biochem Biophys Res Commun 319:448–455
9. Johnson G, Moore SW (2006) The peripheral anionic site of acetylcholinesterase: structure, functions and potential role in rational drug design. Curr Pharm Des 12:217–225
10. Inestrosa NC, Alvarez A, Perez CA, Moreno RD, Vicente M, Linker C, Casanueva OI, Soto C, Garrido J (1996) Acetylcholinesterase accelerates assembly by amyloid beta-peptides into Alzheimer's fibrils; possible role pf the peripheral site of the enzyme. Neuron 16:881–891

11. Inestrosa NC, Alarcon R (1998) Molecular interactions of acetylcholinesterase with senile plaques. J Physiol Paris 92:341–344
12. Viayna E, Sabate R, Munoz-Torrero D (2006) Dual inhibitors of β-amyloid aggregation and acetylcholinesterase as multi-target anti-Alzheimer drug candidates. Curr Top Med Chem 13:1820–1842
13. Camps P, Formosa X, Galdeano C, Munoz-Torrero D, Ramirez L, Gomez E, Isambert N, Lavilla R, Badia A, Clos MV, Bartolini M, Mancini F, Andrisano V, Arce MP, Rodrigues-Franco MI, Huertas O, Dafni T, Luque FJ (2009) Pyrano[3,2-c]quinoline-6-chlorotacrine hybrids as a novel family of acetylcholinesterase- and beta-amyloid-directed anti-Alzheimer compounds. J Med Chem 52:5365–5379
14. Yu L, Cao R, Yi W, Yan Q, Chen Z, Ma L, Peng W, Song H (2010) Synthesis of 4-[(diethylamino)methyl]-phenol derivatives as novel cholinesterase inhibitors with selectivity towards butyrylcholinesterase. Bioorg Med Chem Lett 20:3254–3258
15. Leon R, Marco-Contelles J (2011) A step further towards multitarget drugs for Alzheimer and neuronal vascular diseases: targeting the cholinergic system, amyloid-β aggregation and Ca(2+) dyshomeostasis. Curr Med Chem 18:552–576
16. Simoni E, Daniele S, Bottegoni G, Pizzirani D, Trincavelli ML, Goldoni L, Tarozzo G, Reggiani A, Martini C, Piomelli D, Melchiorre C, Rosini M, Cavalli A (2012) Combining galantamine and memantine in multitargeted, new chemical entities potentially useful in Alzheimer's disease. J Med Chem 55:9708–9721
17. Nepovimova E, Uliassi E, Korabechy J, Pena-Altamira LE, Samez S, Pesaresi A, Garcia GE, Bartolini M, Andrisano V, Bergamini C, Fato R, Lamba D, Roberti M, Kuca K, Monti B, Bolognesi ML (2014) Multitarget drug design strategy: quinone-tacrine hybrids designed to block amyloid-β aggregation and to exert anticholinesterase and antioxidant effects. J Med Chem 57:8576–8589
18. Heinrich M, Teoh HL (2004) Galanthamine from snowdrop–the development of a modern drug against Alzheimer's disease from local Caucasian knowledge. J Ethnopharmacol 92:147–162
19. Paskov DS (1959) Nivalin: pharmacology and clinical application. MedicinaiFizkultura Sofia Bulgaria
20. Paskov DS (1957) Effect of galanthamine on the skeletal muscles. Proceedings of the department of biological and medical sciences at the Bulgarian academy of sciences. Ser Exp Biol Med, Sofia 1:29–35
21. Bubeva-Ivanova L (1957) Phytochemical study on *Galanthus nivalis* var. *gracilis*. Pharmacia (Sofia) 2:23–26
22. Nastev G (1960) Nivalin treatment of patients with diseases of nervous system. Cult Med (Roma) 15:87–97
23. Stoyanov EA (1964) Galanthaminum hydrobromicum "Nivalin", einneues Antidot der nichtdepolarisierenden Muskelrelaxantien. Pharmakologie und klinische Anwendung. Anaesthesist 13:217–220
24. Paskov DS (1986) Galanthamine. In: Kharkevich DA (ed) Handbook of experimental pharmacology. Springer, Berlin/Heidelberg/New York/Tokyo, pp 653–672
25. Parys W (1998) Development of Reminyl (galantamine), a novel acetylcholinesterase inhibitor, for the treatment of Alzheimer's disease. Alzheimer's Reports 53:S19–S20
26. Farlow MR (2001) Pharmacokinetics profiles of current therapies for AD: implications for switching to galantamine. Clin Ther 23: A13–A24
27. Marco-Contelles J, Rodríguez C, Carreiras MC, Villarroya M, García AG (2006) Synthesis and pharmacology of galantamine. Chem Rev 106:116–133
28. Pereira EF, Reinhardt-Maelicke S, Schrattenholz A, Maelicke A, Albuquerque EX (1993) Identification and functional characterization of a new agonist site on nicotinic acetylcholine receptors of cultured hippocampal neurons. J Pharmacol Exp Ther 265:1474–1491
29. Storch A, Schrattenholz A, Cooper JC, Abdel Ghani EM, Gutbrod O, Weber KH, Reinhardt S, Lobron C, Hermsen B, Soskic V, Pereira EFR, Albuquerque EX, Methfessel C, Maelicke A (1995) Physostigmine, galanthamine and codeine act as 'noncompetitive nicotinic receptor agonists' on clonal rat pheochromocytoma cells. Eur J Pharm 290:207–219
30. Schrattenholz A, Pereira EF, Roth U, Weber KH, Albuquerque EX, Maelicke A (1996) Agonist responses of neuronal nicotinic acetylcholine receptors are potentiated by a novel class of allosterically acting ligands. Mol Pharmacol 49:1–6
31. Albuquerque EX, Alkondon M, Pereira EF, Castro CF, Schrattenholz A, Barbosa CT, Bonfante-Canarcas R, Aracava Y, Eisenberg HM, Maelicke A (1997) Properties of neuronal nicotinic acetylcholine receptors: pharmacological characterization and modulation of synaptic function. J Pharmacol Exp Ther 280:1117–1136

32. Dajas-Bailador FA, Heimala K, Wonnacott S (2003) The allosteric potentiation of nicotinic acetylcholine receptors by galantamine is transduced into cellular responses in neurons: Ca2+ signals and neurotransmitter release. Mol Pharmacol 64:1217–1226
33. Takata K, Kitamura Y, Saeki M, Terada M, Kagitani S, Kitamura R, Fujikawa Y, Maelicke A, Tomimoto H, Taniguchi T, Shimohama S (2010) Galantamine-induced amyloid-{beta} clearance mediated via stimulation of microglial nicotinic acetylcholine receptors. J Biol Chem 285:40180–40191
34. Han SY, Sweeney JE, Bachman ES, Schweiger EJ, Forloni G, Coyle JT, Davis BM, Joullié MM (1992) Chemical and pharmacological characterization of galanthamine, an acetylcholinesterase inhibitor, and its derivatives. A potential application in Alzheimer's disease? Eur J Med Chem 27:673–687
35. Bores GM, Kosley RW (1996) Galanthamine derivatives for the treatment of Alzheimer's disease. Drugs Future 21:621–635
36. Mary A, Renko DZ, Guillou C, Thal C (1998) Potent acetylcholinesterase inhibitors: design, synthesis, and structure-activity relationships of bis-interacting ligands in the galanthamine series. Bioorg Med Chem 6:1835–1850
37. Guillou C, Mary A, Renko DZ, Thal C (2000) Potent acetylcholinesterase inhibitors: design, synthesis and structure-activity relationships of alkylene linked bis-galanthamine and galanthamine-galanthaminium salts. Bioorg Med Chem Lett 10:637–639
38. Treu M, Jordis U, Mereiter K (2001) 12H-[2]-Benzothiepino[6,5a,5-bc]benzofuran: synthesis of a sulfuranalog of galanthamine. Heterocycles 55:1727–1735
39. Poschalko A, Welzig S, Treu M, Nerdinger S, Mereiter K, Jordis U (2002) Synthesis of (±)-6-H-benzofuro[3a,3,2,ef][3]benzazepine: anunnaturalanalog of (−)-galanthamine. Tetrahedron 58:1513–1518
40. Herlem D, Martin MT, Thal C, Guillou C (2003) Synthesis and structure-activity relationships of open D-ring galanthamine analogues. Bioorg Med Chem Lett 13:2389–2391
41. Greenblatt HM, Guillou C, Guenard D, Argaman A, Botti S, Badet B, Thal C, Silman I, Sussman JL (2004) The complex of a bivalent derivative of galanthamine with torpedo acetylcholinesterase displays drastic deformation of the active-site gorge: implications for structure-based drug design. J Am Chem Soc 126:15405–15411
42. Bartolucci C, Haller LA, Jordis U, Fels G, Lamba D (2010) Probing Torpedo californica acetylcholinesterase catalytic gorge with two novel bis-functional galanthamine derivatives. J Med Chem 53:745–751
43. Atanasova M, Yordanov N, Dimitrov I, Berkov S, Doytchinova I (2015) Molecular docking study on galantamine derivatives as cholinesterase inhibitors. Mol Inf 34:394–403
44. Atanasova M, Stavrakov G, Philipova I, Zheleva D, Yordanov N, Doytchinova I (2015) Galantamine derivatives with indole moiety: docking, design, synthesis and acetylcholinesterase inhibitory activity. Bioorg Med Chem 23:5382–5389
45. Stavrakov G, Philipova I, Zheleva D, Atanasova M, Konstantinov S, Doytchinova I (2016) Docking-based design of galantamine derivatives with dual-site binding to acetylcholinesterase. Mol Inf 35:278–285
46. Jones G, Willett P, Glen RC, Leach AR, Taylor R (1997) Development and validation of a genetic algorithm for flexible docking. J Mol Biol 267:727–748
47. Kellenberger E, Rodrigo J, Muller P, Rognan D (2004) Comparative evaluation of eight docking tools for docking and virtual screening accuracy. Proteins 57:225–242
48. Perola E, Walters WP, Charifson PS (2004) A detailed comparison of current docking and scoring methods on systems of pharmaceutical relevance. Proteins 56:235–249
49. Cheung J, Rudolph MJ, Burshteyn F, Cassidy MS, Gary EN, Love J, Franklin MC, Height JJ (2012) Structures of human acetylcholinesterase in complex with pharmacologically important ligands. J Med Chem 55:10282–10286
50. Ellman GL, Courtney KD, Andreas V Jr, Featherstone RM (1961) A new and rapid colorimetric determination of acetylcholinesterase activity. Biochem Pharmacol 7:88–95
51. Ortiz J, Berkov S, Pigni N, Theoduloz C, Roitman G, Tapia A, Bastida J, Feresin GE (2012) Wild Argentinian Amaryllidaceae, a new renewable source of the acetylcholinesterase inhibitor galanthamine and other alkaloids. Molecules 17:13473–13482
52. Jansen JJ, Martin EJ (2004) Target-biased scoring approaches and expert systems in structure-based virtual screening. Curr Opin Chem Biol 8:359–364
53. Jain AN (2004) Virtual screening in lead discovery and optimization. Curr Opin Drug Discov Devel 7:396–403
54. Kishan KV (2007) Structural biology, protein conformations and drug designing. Curr Protein Pept Sci 8:376–380
55. Villoutreix BO, Renaut N, Lagorce D, Sperandio O, Montes M, Miteva MA (2007) Free

resources to assist structure-based virtual ligand screening experiments. Curr Protein Pept Sci 8:381–411

56. Fukunishi Y (2009) Structure-based drug screening and ligand-based drug screening with machine learning. Comb Chem High Throughput Screen 12:397–408
57. Sobhia ME, Singh R, Kare P, Chavan S (2010) Rational design of CCR2 antagonists: a survey of computational studies. Expert Opin Drug Discov 5:543–557
58. Kirchmair J, Distinto S, Liedl KR, Markt P, Rollinger JM, Schuster D, Spitzer GM, Wolber G (2011) Development of anti-viral agents using molecular modeling and virtual screening techniques. Infect Disord Drug Targets 11:64–93
59. Scotti L, BezerraMendonca FJ Jr, Magalhaes Moreira DR, da Silva MS, Pitta IR, Scotti MT (2012) SAR, QSAR and docking of anticancer flavonoids and variants: a review. Curr Top Med Chem 12:2785–2809
60. Debnath AK (2013) Rational design of HIV-1 entry inhibitors. Methods Mol Biol 993:185–204
61. Safavi M, Baeeri M, Abdollahi M (2013) New methods for the discovery and synthesis of PDE7 inhibitors as new drugs for neurological and inflammatory disorders. Expert Opin Drug Discov 8:733–751
62. Kumar V, Chandra S, Siddiqi MI (2014) Recent advances in the development of anti-viral agents using computer-aided structure based approaches. Curr Pharm Des 20:3488–3499
63. Tomioka H (2014) Current status and perspective on drug targets in tubercle bacilli and drug design of antituberculous agents based on structure-activity relationship. Curr Pharm Des 20:4305–4306
64. Siddiqi NI, Siddiqi MI (2014) Recent advances in QSAR-based identification and design of anti-tubercular agents. Curr Pharm Des 20:4418–4426

Chapter 7

Modeling of BACE-1 Inhibitors as Anti-Alzheimer's Agents

Odailson Santos Paz, Thamires Quadros Froes, Franco Henrique Leite, and Marcelo Santos Castilho

Abstract

Although Alzheimer's disease (AD) is the most common cause of dementia worldwide, the drugs available to treat it do not cure or prevent the disease progression. Aiming at circumventing this drawback, beta-secretase (BACE-1) has been targeted for new drug development. This chapter focuses on the importance of computational tools in the hit identification and lead optimization steps with reference to anti-Alzheimer drug development. The influence of BACE-1 flexibility and aspartic acid protonation state regarding inhibitors' identification is described by molecular dynamic studies, whereas the role of pharmacophore models toward the identification of structurally diverse BACE-1 inhibitors is highlighted. Last, but not least, the contributions of QSAR models to improve the potency of BACE-1 inhibitors are described. The tools and strategies employed in each step are thoroughly discussed so that the reader can grasp the importance of in silico tools to the development of novel BACE-1 inhibitors.

Key words BACE-1 inhibitors, QSAR, Pharmacophore, Molecular dynamics, Beta-secretase

1 Introduction

Alzheimer's disease (AD) is the most common cause of dementia affecting 47 million people worldwide [1]. Currently, the treatment of AD costs approximately $10.2 billion to the USA healthcare system, but the amount of money required to treat AD patients is expected to escalate to a trillion dollar by 2018 [2]. On top of that, the number of AD patients might reach 131 million people by 2050 [2]. These figures show the economic and social impact of AD in the near future. The currently available drugs are prescribed to improve AD symptoms such as cognitive dysfunction, psychiatric and behavioral disturbances, and difficulties in performing daily activities [3, 4]. In order to achieve these goals, traditional AD treatments focus on inhibition of cholinesterase enzymes [5, 6]. Although such approach has a positive impact on the patients' quality of life, they do not cure or prevent the disease progression [7].

Kunal Roy (ed.), *Computational Modeling of Drugs Against Alzheimer's Disease*, Neuromethods, vol. 132, DOI 10.1007/978-1-4939-7404-7_7, © Springer Science+Business Media LLC 2018

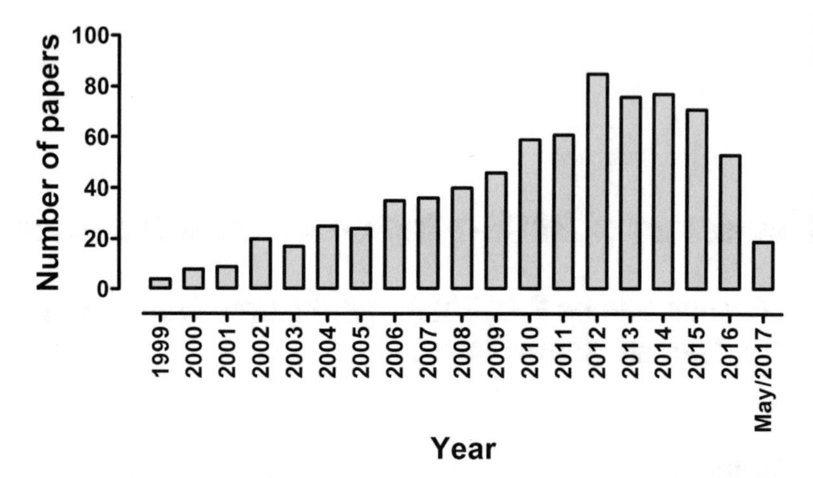

Fig. 1 Number of PubMED (https:/www.ncbi.nlm.nih.gov/pubmed/) indexed papers describing BACE-1 drug development efforts. Searching strategy ("amyloid precursor protein secretases" [MeSH Terms] OR ("amyloid"[All Fields] AND "precursor"[All Fields] AND "protein"[All Fields] AND "secretases"[All Fields]) OR "amyloid precursor protein secretases"[All Fields] OR ("beta"[All Fields] AND "secretase"[All Fields]) OR "beta secretase"[All Fields]) AND drug[All Fields] AND ("growth and development"[Subheading] OR ("growth"[All Fields] AND "development"[All Fields]) OR "growth and development"[All Fields] OR "development"[All Fields])

Aiming at circumventing this drawback, much research has been directed toward understanding AD etiology at a molecular level. One of the pathological hallmarks of AD is the presence of amyloid plaques which are produced by the proteolytic cleavage of amyloid precursor protein (APP) into β-amyloid peptide (A-β) [8]. As beta-secretase (BACE-1) is the main protease involved in this step, it is considered a promising therapeutic target for AD treatment [9–13]. This hypothesis is supported by the fact that BACE-knockout mice show no phenotypic effects, suggesting that no major side effects would be associated with BACE-1 inhibition [1, 14].

This background information helps to explain the great number of papers that describe drug development efforts toward this target (Fig. 1).

As a consequence, it would be impossible to summarize all the knowledge provided by several groundbreaking papers and seminal reviews regarding this subject. Instead, this chapter aims at highlighting the importance of computational tools in the hit identification and lead optimization steps, so that the reader can grasp the contributions of several techniques and strategies toward the development of BACE-1 inhibitors that shall reach the market in the near future.

2 Methods

2.1 Molecular Dynamics Simulations

Molecular dynamics (MD) is an approach that calculates the motion of atoms in a molecule, in accordance with Newton's laws of motion [15, 16]. Hence, molecules are treated as a collection of atoms which are described as particles bound together by harmonic or elastic forces that move under the influence of classical mechanics (Eq. 1) [17].

$$\mathrm{F}i(t) = \mathrm{m}i.\mathrm{a}i \tag{1}$$

where $\mathrm{F}i$ is the force acting on each particle system at a time t and $\mathrm{a}i$ is the acceleration of atom with mass $\mathrm{m}i$.

Among the several software available to carry out MD simulations, Kocak and workers (2016) employed Gromacs 5.0 [18] and 5.1 [19], which employs Gromos96 (54A7) force field to study BACE-1. Although amino acid parameters are available in most MD software, ligand atom types generally are determined using either PRODRG or ATB servers [20, 21]. Kocak and coworkers (2016) [18] followed this trend. Besides this, they employed PropKa 3.1 [22] to predict the protonation state of each amino acid at pH 4.5, except for catalytic aspartic residues, whose protonation was manually set. On the other hand, Marvin [23] and Propka 3.1 were used to calculate the ionization state (pKa) of the free and bound ligand, respectively.

Then, the protein-ligand complex was placed in the center of a cubic box with dimensions 9.0 × 9.0 × 9.0 nm and surrounded by SPC-model water molecules [24], so that a 10 Å margin was kept in each dimension. In order to avoid artifacts from periodic boundary conditions, it is necessary to neutralize the system using either Na+ or Cl− ions to replace water molecules. This precaution was undertaken in all MD studies reported in this chapter.

Once minor steric clashes would result in great acceleration and system instability, it is usual to carry out minimization steps before the actual MD simulations [25]. Kocak and coworkers (2016) [18], for instance, employed 500 cycles of steepest descent minimization and another 500 cycles of conjugate gradient minimization, using Verlet cutoff scheme [26], to prepare their MD input structures. During this step, electrostatic and van der Waals interaction cutoff of 14 Å was set, along with the Particle Mesh Ewald (PME) method. After this minimization step, a 200 ps MD simulation with canonical (NVT) ensemble was carried out, and only after the system had reached equilibrium, data from MD simulations, at the NPT ensemble, were considered.

In contrast to the previous approach, Ellis and coworkers (2016) [27] employed CHARMM (version C37b) to study the effect of constant pH over the residues' ionization state. The CHARMM22/CMAP all-atom force field was used to represent

BACE-1 atoms, whereas the force field parameters for the inhibitors were obtained from the CHARMM general force field (CGenFF) protocol. Using replicas 0.3 pH units apart, the effect of pH (1.3–8.0) over titratable amino acids (Asp, Glu, and His), as well as the amine group on the pyridine/pyrimidine ring of the inhibitor, was investigated under NPT conditions.

2.2 Virtual Screening

Molecular docking strives to predict the best conformation of a ligand within the active site of its target molecule [28]. In order to accomplish this task, conformational flexibility is given to the putative ligands during the virtual screening and the proposed binding modes, also known as poses [29], are scored according to the compounds' binding free energy to the molecular target. Three types of scoring functions are available: (1) based on force fields, (2) empirical and (3) methods based on knowledge [30]. In fact, the effect of different scoring functions on BACE-1 inhibitors hit enrichment was investigated by Dominguez and workers [31].

Pharmacophore-based virtual screening also relies on a conformational search and an evaluation protocol, but the theoretical background for each step is quite different from docking. GALAHAD (Genetic Algorithm with Linear Assignment for Hypermolecular Alignment of Datasets), for instance, employs a proprietary technology to generate molecular alignments that share common binding interactions [32]. Next, the aligned molecules have their pharmacophore features' superposition ranked [33].

Catalyst on the other hand has a built-in conformational analysis tool (catalyst/best or catalyst/fast) that affords low-energy conformers (i.e., 100 conformers within 10Kcal/mol from the lowest energy conformer) and then strives to build the simplest pharmacophore model that satisfies the SAR of training set compounds. In order to achieve this goal, pharmacophore elements that are present in the highly active compounds, but missing from the least active ones, are identified by HypoGen [34].[1] During the pharmacophore model generation, three phases take place: First, pharmacophores that are common among the most active compounds are identified (additive phase). Then next are those pharmacophore features that are also found in weak active compounds removed from the pharmacophore models (subtractive phase). Finally, small adjustments are made to the remaining models (i.e., random translations of features, rotations of vectored features, etc.) in order to maximize its fit (Eq. 2) [35]

$$\text{Fit} = \sum_{\text{maped Hypofunctions}} w \left[1 - \sum_{\text{spheres}} \left(\frac{D}{T} \right)^2 \right] \tag{2}$$

[1] In case only potent inhibitors are employed, HipHop instead of HypoGen protocol is employed.

where w is the weight of the pharmacophore functions, D represents the distance between the pharmacophore features in the molecule, and T stands for the tolerance radius.

The correlation between fit values and the biological property, obtained by linear regression methods, can be employed as a QSAR model to predict the activity of each compound from the training set [36]. This strategy along with QSAR models was used to explain the SAR of 68 BACE inhibitors by Al-Nadaf and Taha (2013) [37].

Pharmacophore models can be evaluated by their usefulness to enlighten the structure-activity relationship for known ligands [38], but if virtual screening is the main goal, their ability to differentiate true ligands from decoys should also be taken into consideration [39]. For that reason, Leite and coworkers claim that a combination of these strategies have higher chance of selecting bioactive compounds [40]. Another major issue in virtual screening campaigns is the time required to evaluate the compounds in silico. Then, several authors employ the best pharmacophore model as a query for UNITY 3D [41], which ranks molecule according to the QFIT value [42].

2.3 QSAR Model Development

Quantitative structure-activity relationship (QSAR) modeling is an important approach in drug discovery once it correlates the molecular structure with the biological activity (i.e., IC50, Ki, etc.). Therefore, QSAR is one of the major computational tools employed in medicinal chemistry [43]. QSAR models can be classified according to the type of the descriptor employed to capture the chemical features that influence the biological activity. For instance, 2D QSAR methods rely on topological descriptors that are not influenced by the conformation of the ligands [44]. Similarly, hologram QSAR (HQSAR) employs hashed fingerprints to create molecular hologram that can be related to the biological activity. Molecular holograms are influenced by the hologram length (HL), the fragment distinction (atoms (A), bonds (B), connections (C), hydrogen atoms (H), chirality (Ch), and donor and acceptor (DA)), and the fragment size [45]. For that reason, most HQSAR studies report the influence of such parameters over the predictive ability of the models [46]. Accordingly, Cruz and Castilho reported the influence of such parameters over the predictive ability of HQSAR models built with BACE-1 inhibitors. One major advantage of HQSAR models over other 2D QSAR methods is the possibility to build contribution maps that provide a visual guide to analyze the contribution of atoms/groups for the biological property under investigation. In these maps, colors in the red-orange end of the spectrum indicate unfavorable contributions, whereas those in yellow-green end of the spectrum reflect as favorable contributions. Atoms and fragments colored white/gray indicates neutral or intermediate contributions, either because these fragments are present in all molecules of the series or because the sum of the positive and negative atomic contributions render these fragments as "neutral" [45, 47]. Despite being useful, these contribution maps do not explain the underlying reasons for the positive or negative contributions. Hence, the synergic interpretation of descriptor-based (classical 2D QSAR models) and fragment-based QSAR (HQSAR) models helps to shed some light on the structural and chemical features that are essential for the biological activity [48], as demonstrated in the analysis of BACE-1 inhibitors by Cruz and Castilho (2014).

Another very popular QSAR approach relies on molecular interaction field descriptors (3D descriptors) that are deeply influenced by the molecular alignment of the ligands as well as their spatial arrangement [49]. A proper molecular alignment would have all ligands in their bioactive conformation [50]. Whenever crystallographic data is available for most ligands, this information guides the molecular alignment rules. The great number of BACE-1 crystal structures available allowed Liu and coworkers (2012) [51] to follow this approach.[2] However, this situation is far from the real for most macromolecular targets. Alternatively, low-energy, stable conformations are assumed to be good surrogates for the bioactive conformation [49], and for that reason they can be superposed and employed to build 3D QSAR models. Last but not least, docked poses can be employed to build reasonable molecular alignments [52]. In fact, most of the examples described in this chapter rely on these two last approaches.

Comparative molecular field analysis (CoMFA) is the most well-known representative of this 3D QSAR approach [53]. Once all molecules are already in the "bioactive conformation," entropic contributions to the free energy of binding are ignored, and the compounds' affinity toward their target is described only by steric and electrostatic interactions. In order to simplify the calculation, a probe atom is positioned in equally spaced positions (grid) around the ligands, and Lennard-Jones and Coulomb potentials are employed to calculate steric (S) and electrostatic (E) interactions that are either favorable or unfavorable to the biological property. Visual representation of these data leads to the construction of polyhedral regions (contour maps) that are located in the regions with the greatest variance of the steric and electrostatic values. Thus, green polyhedra suggest that bulky substituents at that position increase the biological property, whereas those colored yellow indicate a negative steric contribution; blue polyhedra suggest that positively charged moieties would favor the biological property; and red polyhedrons indicate that electron-rich groups would improve the biological property [54].

[2] The comparative binding energy analysis (COMBINE) employed by these authors, to build their QSAR models, is not discussed in this chapter.

Although intuitive, the contour maps from CoMFA have sharp edges that are not chemically meaningful. In order to overcome this limitation as well as the bias introduced by alignment rules, comparative molecular similarity analysis (CoMSIA) was developed [55]. This approach allows the user to investigate not only the steric and electrostatic contributions to potency but also to probe how hydrophobic and hydrogen donor groups influence it. CoMSIA model analysis is similar to the one carried out for CoMFA, but the maps have blunt shapes and indicate positions within the ligands that should be modified (not favorable/unfavorable regions in their vicinity) [56].

3 Results

3.1 Flexibility Issues Regarding BACE-1 Inhibitors Development

Molecular docking helps to predict protein-ligand binding profiles [57]. Thus, it has been used routinely for virtual screening or lead optimization steps in drug development campaigns [2], as will be presented in Sect. 3.1 of this chapter. However, this technique has some limitations such as disregard for the conformational flexibility of the protein and inadequate description of water participation in complex stabilization [9].

Binding site water molecules play a crucial role in protein-ligand recognition, either being displaced upon ligand binding or interacting with it. However, discriminating water molecules that are crucial for binding from those that should be removed before docking simulations depend on the analysis of ensemble structures [8]. Moreover, water molecules are involved in a range of interactions (i.e., H-bonding, van der Waals contacts, etc.), and for that reason, their displacement seldom increases ligand affinity. Nevertheless, a suitable description of hydration shells provides important insights into the properties of the binding site as well as the intermolecular forces that guide drug-like molecules binding toward the protein [10, 11], none of which are considered in most docking routines [15].

In order to circumvent these limitations, molecular dynamics (MD) have been applied not only to explicitly account for water molecules within the binding site but also to explore how fast internal motions (i.e., side-chain rotamers shift) or even domain movements impact ligand binding [13, 14, 57]. This in silico approach calculates the atom trajectory according to Newton's laws of motion [15]. Hence, MD evaluates the conformational flexibility of the whole system (protein-protein or ligand-protein complexes) during a certain amount of time [16]. MD has significantly contributed to the development of selective and potent BACE-1 inhibitors [8, 11, 58], and some emblematic examples of its utility are presented in the next section.

3.2 Molecular Dynamic Studies on BACE-1

It has been proposed that BACE-1 catalytic mechanism involves two aspartic residues (Asp^{32} and Asp^{228}) and one water molecule, which acts as base during the hydrolysis of the substrate. Although it is well known that the mutation of either aspartate residues abolishes BACE-1 activity [59] and the general catalytic mechanism has been largely explained and supported by structural and kinetic data [60, 61], the details regarding the protonation of the aspartic acid residues and their reorientation during the catalysis remained largely unexplored until recently. The precise location of protons on the two aspartic acid residues cannot be determined by X-ray crystallography data, due to limited resolution of structures [62]. However, knowledge about the protonation state is crucial to design BACE-1 inhibitors that have low affinity to other human proteases. In fact, this matter has already been discussed in the literature [11, 18, 63], but only recently MD simulations have provided further details on the aspartic residues' protonation state [64]: whereas ionized Asp^{32} and protonated Asp^{228} are the preferred state in the presence of ligand, the di-deprotonated state seems to be more stable when the ligand is absent. This hypothesis is supported by QM/MD studies carried out by Liu and coworkers [58] that show ionized Asp^{32} and neutral Asp^{228} are the preferred state for the free enzyme (APO-BACE), but neutral $Asp^{32}$and ionized Asp^{228} are predominant states at the transition state and product generation steps. This information is quite different from the one provided by QM/MM studies, when no time-dependent effects were accounted for, which support neutral Asp^{32} and ionized Asp^{228} as the favored states [65]. This apparent contradiction might be due to the lack of parameters or force field standardization in in silico studies (i.e., differences in the pH employed during the simulations) or simply due to differences in inhibitors' scaffold [18, 66]. Park and Lee [67], for instance, carried out MD simulations of BACE-1 with the peptidomimetic hydroxyethylene-based inhibitor (OM99–2) considering either the neutral Asp^{32} and deprotonated Asp^{228} (state 1) or deprotonated Asp^{32} and neutral Asp^{228} (state 2). The RMSD (all heavy atoms) values for states 1 and 2 are similar (state 1, 1.50 Å; state 2, 1.56 Å), but the hydrogen bonds between the ligand and state 1 (Asp^{32} HD2) is retained during 99% of the simulation productive phase, whereas its H-bonds to state 2 (Asp^{228} HD2) lasts 4% of the time. This result supports state 1 as the major protonation state for peptidomimetic drug development campaigns. Despite its importance, such data has not been taken into consideration during the QSAR model development described in Sect. 3.4 of this chapter.

Kocak and coworkers (2016) [18] carried out a similar study for non-peptidomimetic inhibitors considering all possible protonation states of Asp residues at pH of 4.5. MD simulations suggest that the deprotonated Asp^{32}/Asp^{228} is the major protonation state when the non-peptide inhibitor is found within the BACE-1 active

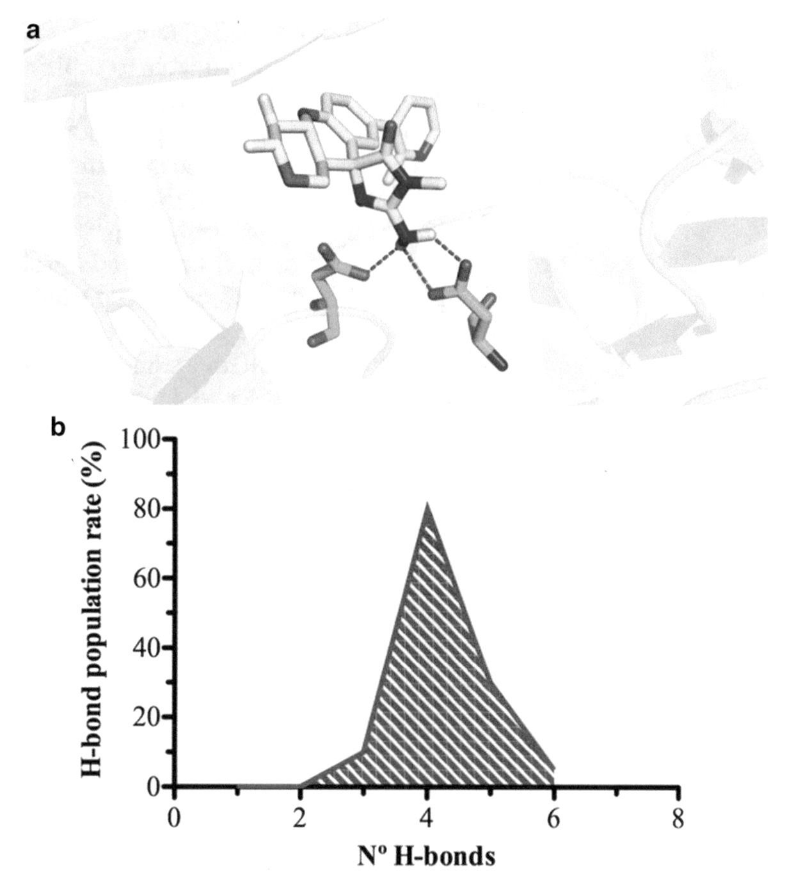

Fig. 2 BACE-1 protonation state in the presence of non-peptidomimetic inhibitor. (**a**) Di-deprotonated state of the aspartic residues in the presence of (4R,4a'S,10a'S)-2-amino-8′-(2-fluoropyridin- 3-yl)-1-methyl-3′,4′,4a',10a'-tetrahydro-1'H-spiro[imidazole-4,10′-pyrano[4,3-b]chromen]-5(1H)-one (PDB, 4 N00). H-bonds are depicted by *red dashed* lines. (**b**) H-bonds formed between the aspartic residues during MD simulation. Color code: *red*, oxygen; blue, nitrogen; green, protein carbon; yellow, inhibitor carbon (Adapted from Kocak et al. [18])

site (Fig. 2), as this is the only protonation state that keeps the dihedral angle required for ligand binding. Unfortunately, this information has not influenced yet the development of QSAR models that aim at improving non-peptidomimetic inhibitors' potency. Frequently, protonation states of residues, substrates, and ligands are kept at experimentally supported, fixed values during MD simulations. However, chemical environment changes along the binding process and so does the pKa value of ionizable residues. The importance of this issue can be grasped from the analysis of iminopyrimidinone and aminothiazine compounds that inhibit BACE-1 [29, 30]. Despite having a similar structure and nearly identical binding modes, these inhibitors have different potency against BACE-1. Hence, the analysis of static X-ray crystallographic structure is unable to explain the potency variation. In order to further investigate this matters, Ellis and coworkers (2016) [30] employ the constant pH molecular dynamics (CpHMD) approach that allows for the dynamic change of protonation states of ionizable groups during the simulation [68–70].

According to this study, Asp^{228} protonation state (deprotonated) is not influenced by the inhibitor's binding; on the other hand, when BACE-1 is bound to the iminopyrimidinone inhibitor, the Asp^{32} pKa is significantly lower than that in its absence (2.5 *vs* 4.1, respectively) and the aminothiazine inhibitor causes only a minor shift in the pKa (4.1 to 3.7) (Fig. 3). As a consequence, Asp^{32} binding profile toward these compounds is expected to differ: at pH 3.5, **2** participates in H-bonding to BACE-1 less than 20% of the time, whereas at the same condition approximately

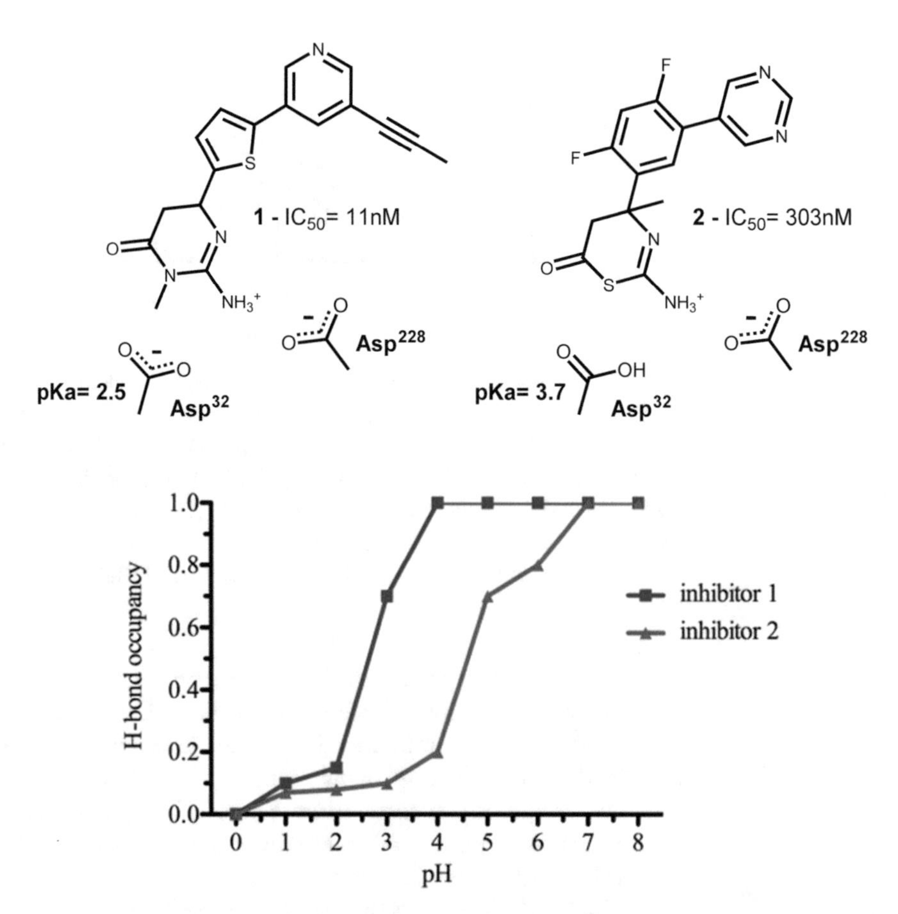

Fig. 3 Schematic representation of iminopyrimidinone (1) and aminothiazine (2) inhibitors within BACE-1 active site. *Upper panel*, pKa shift due to inhibitor binding; *lower panel*, the fraction of time that hydrogen bonds exist between inhibitors **1** (*blue*) and **2** (*red*); and the protein residues, during MD simulation, are represented as H-bond occupancy. *Highlighted area* corresponds to the BACE-1 active pH range (Adapted from Ellis et al. [27])

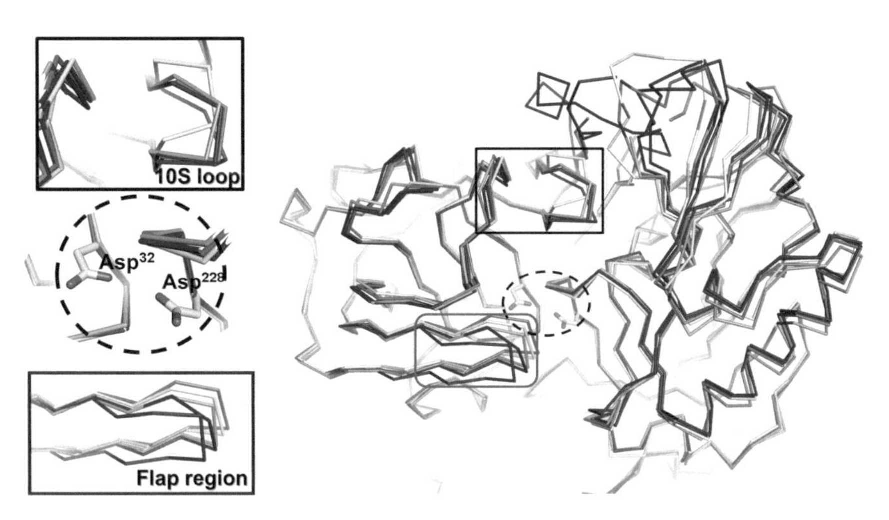

Fig. 4 Flexibility within BACE-1 crystallographic structures. Inserts highlight flexible regions (10s loop and FLAP), as well as the catalytic residues. PDB color code: 4 N00, *pink*; 2BL8, *light gray*; 1 W51, *salmon*; 2WEZ, *slate blue*; 2HM1, *dark gray*; 3OHH, *purple*. Figure generate in Pymol V. 0.99

70% of **1** is H-bonded (Fig. 3, lower panel). Accordingly, the lower potency of inhibitor **2** might be explained by the presence of the partially protonated Asp^{32}: MD simulations support this hypothesis once the aminothiazine inhibitor hydrogen bonds to Thr232 instead of the catalytic aspartic acid residues, at pH 3.5. This interaction is made possible by the lack of propyne tail (decreased steric clash), which allows the pyridine ring to lie within H-bonding distance.

Another crucial information that might be provided by MD simulation concerns induced fit seen in BACE-1 X-ray structures upon ligand binding (Fig. 4). In fact, several regions from BACE-1 (10s loop, flap, insert-A, insert-D, and insert-F) are recognized to control either access or substrate/inhibitor binding [59], despite the fact that 10s loop and aspartic residues (Asp^{32}/Asp^{228}) are the most important moieties responsible for the catalytic mechanism [71]. The induced fit is expected to change the volume available for ligand binding, and this issue can be explored to design selective BACE-1 inhibitors.

For instance, structural comparison of BACE-1 to BACE-2- (global identity, 64%) and cathepsin D (CTSD, global identity, 42%) shows that BACE-1 active site volume is approximately equal to BACE-2 (354.34 $Å^3$vs 349.50 $Å^3$, respectively), whereas CTSD has a larger binding site (723.456 $Å^3$). Although BACE-1 and BACE-2 have leucine (Leu^{30} and Leu^{46}, respectively) lining the S1 pocket, whereas CTSD has a smaller substituent (Val^{46}) in the equivalent position, this structural difference is not sufficient to explain the active-site volume differences. A more convincing argument comes from the electrostatic interaction of His^{77} and Asp^{75} in the CTSD FLAP region that stabilizes its open conformation.

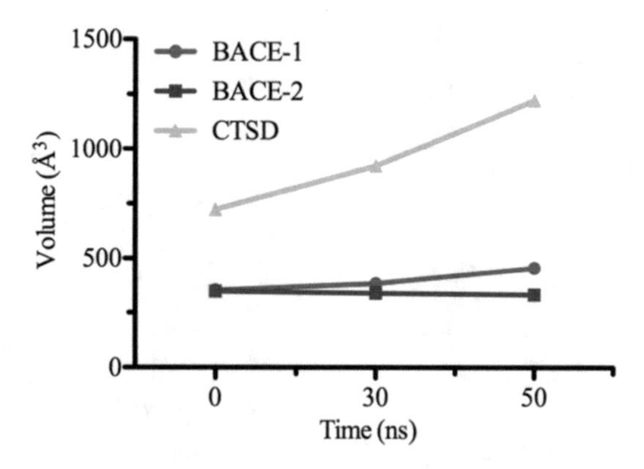

Fig. 5 Binding site volume evolution during MD simulations (Adapted from Hernández-Rodríguez et al. [73])

Indeed, MD studies suggest that the available volume of the active site in CTSD increases during the simulation (Fig. 5) [72]. On the other hand, the available volume in BACE-1 and BACE-2 does not change significantly. This result may be explained by a weaker ion-dipole interaction in BACE-2 (Lys^{86} and Thr^{84} are found in the equivalent positions to His^{77}and Asp^{75} in CTSD), which is not sufficient to stabilize the FLAP open conformation and the lack of interaction between Pro^{70} and Tyr^{68} in BACE-1 (positions equivalent to His^{77} and Asp^{75}in CTSD) [27, 28].

Last but not least, Halima and coworkers (2016) [74] employed MD simulations to shed some light on the BACE-1 higher affinity and catalytic efficiency toward non-amyloid substrate, in comparison to APP. According to them, NRG (Gly-Ile-Glu-Phe-Met-Glu-Ala-Glu) displays favorable electrostatic interaction (salt bridges) among NRG-GLU Arg^{235} and Arg^{128}of BACE-1. As APP has a positively charged residue (lysine) in the equivalent position (underlined in the NRG sequence), an electrostatic repulsion to Arg^{235} would explain why it is a poor substrate. Such information highlights the critical role of GLU at P2 and shall be useful to design novel, selective and potent BACE-1 inhibitors.

Despite the advantages of this technique, computational costs limit its widespread use in virtual screening campaigns or to study large systems [75]. As a consequence, simplified models and indirect representations of the binding profile of inhibitors are commonly used in drug design efforts, whereas MD has been largely employed to investigate fundamental questions that have a critical role in BACE-1 inhibition.

3.3 Virtual Screening of BACE-1 Inhibitors

Virtual screening has become a promising tool for the discovery of lead compounds, and for that reason it is considered as one of the main computational tools to prioritize compounds for biological evaluation [76]. Despite the fact that the crystallographic structure of BACE-1 was solved in 1999, several papers still describe ligand-based strategies as an alternative to identify hits that lie outside the chemical space of previously known BACE-1 inhibitors [37, 77–80]. The underlying reasons to choose a ligand-based approach are as follows: (A) the scoring functions employed in most docking tools having little, if any, correlation to the biological property [81]; (B) the macromolecular induced-fit caused by the ligand being overlooked or restricted to side-chain flexibility [82]; and (C) limited performance to identify protonated compounds [83]. Of course, these limitations could be overcome by a rigorous treatment of the solvation-desolvation effect [9] as well as accounting for the entropic loss due to the ligand binding [84] and proper description of the protonation states of the Asp residues. Besides, macromolecular flexibility can be represented by an ensemble of structures derived from X-ray structures or molecular dynamic simulations [85]. However, all these strategies significantly increase the time required (computational cost) for virtual screening. Instead, pharmacophore models rely on a simple description of molecular interactions that should be found at specific distances and angles [81]. As a matter of fact, IUPAC's definition of pharmacophore "the ensemble of steric and electronic features that is necessary to ensure the optimal supra-molecular interactions with a specific biological target structure and to trigger (or to block) its biological response" emphasizes that steric and electronic features are required for the biological activity. However, it neglects to highlight that compounds might fulfill the pharmacophore requirements and still not be active due to steric clashes related to auxophore moieties. In order to overcome this sort of limitation, structure-based pharmacophore models have been built or docking tools have been employed in tandem to filter out those compounds that do not fit into the active site of BACE-1 [82, 85]. Further details on how pharmacophore models and docking tools have been explored to develop novel BACE-1 inhibitors, as well as the insights gained from representative papers, are described in the following section.

3.3.1 Contributions and Limitations of Structure-Based Virtual Screening Toward BACE-1 Inhibitors' Identification

Several effective methods are currently available to dock a flexible ligand to a rigid protein [86–88]. On the other hand, considering protein flexibility during docking remains a challenge, in terms of both computational cost and efficiency. BACE-1 can be considered as an outstanding example of the task to be solved, once this enzyme is known to undergo a massive rearrangement of the FLAP region (residues 68–74) upon ligand binding, not to mention the 10S loop (residues 9–14) flexibility [88].

In fact, Chirapu and workers (2009) [89] have underscored that docking software provides no near-native pose when crystallographic structures are considered as rigid (PDB ID, 2B8L, and

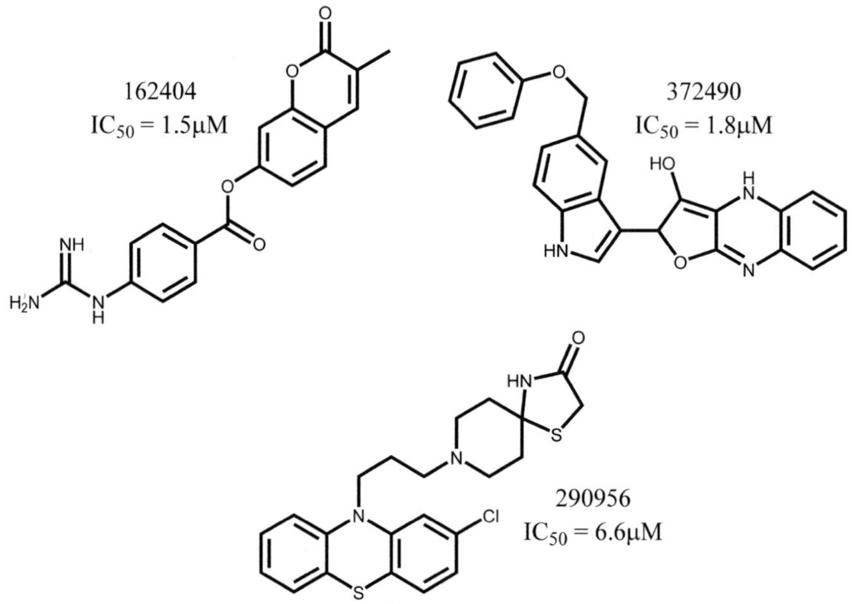

Fig. 6 Example of BACE-1 inhibitors identified by structure-based virtual screening approach developed by Cosconati and workers (2012) [88]

PDB ID, 1 W51). In order to circumvent this limitation, LIU and coworkers (2012) [51] investigated whether one BACE-1 crystallographic structure would perform better than the others for virtual screening purposes. Although they claim that 1 W51 affords better enrichment factor (EF) than 1SGZ, 1FKN, 1XS7, and 1M4H, this conclusion may be highly biased by database properties (fragment-like, lead-like, or drug-like). Furthermore, it offers no biophysical-based solution to the BACE-1 flexibility issue. In order to deal with this problem, Cosconati and workers (2012) [88] incorporated protein flexibility in their docking protocol by using an ensemble of X-ray enzyme structures (PDB ID, 1FKN, 1 W51, 1TQF, 1XN3, 2G94) to screen the National Cancer Institute database. A simplistic description of the protein motion was also generated by computing and combining a set of grid maps using an energy-weighting scheme. Assessment of the enrichment factors from these two virtual screening approaches demonstrated comparable predictive power to the energy-weighted method being faster than the ensemble method. After in vitro evaluation of 32 ligands selected by this combined method, 17 bioactive compounds were identified (Fig. 6). Such an impressive success rate (53%) suggests that this approach is a promising alternative to overcome BACE-1 flexibility issues in the search for BACE-1 inhibitors.

A concern for structure-based virtual screening is the protonation state of the ligands within the database. FLEXX ranks neutral compounds better than charged ones (at pH 4.5). On the other hand, SURFLEX shows no significant dependence on the protonation of the ligands [90]. This information is rather important as EF comparison of these docking software would hint that FLEXX performs better than SURFLEX (EF 69 vs EF 58, respectively). To make matters worse, the protonation state of the residues also plays a major influence in docking results. Surprisingly, Domínguez and workers (2013) [31] have shown that GOLD affords higher success rate for near-native poses at near neutral pH. This result is quite disturbing if one considers that most inhibition assays are carried out at acid pH values and MD studies reported in the previous section support either di-deprotonated Asp^{32}/Asp^{228} (non-peptidomimetic inhibitors) or neutral Asp^{32} and deprotonated Asp^{228} (peptidomimetic inhibitors) protonation states. Hence, the predicted binding profile shown by docking might not depict what really happens within the binding site.

3.3.2 Contributions and Limitations of Ligand-Based Virtual Screening Toward BACE-1 Inhibitors' Identification

An indirect strategy to overcome the problem with the protonation of BACE residues is through pharmacophore-based approaches. John and coworkers (2011) [91], for instance, developed a pharmacophore hypothesis, using 20 BACE-1 inhibitors, that helps to shed some light on the structure-activity relationships of congeneric inhibitors (Fig. 7a). In fact, RMSD deviation from pharmacophore features (donor hydrogen bonding group, an ionizable

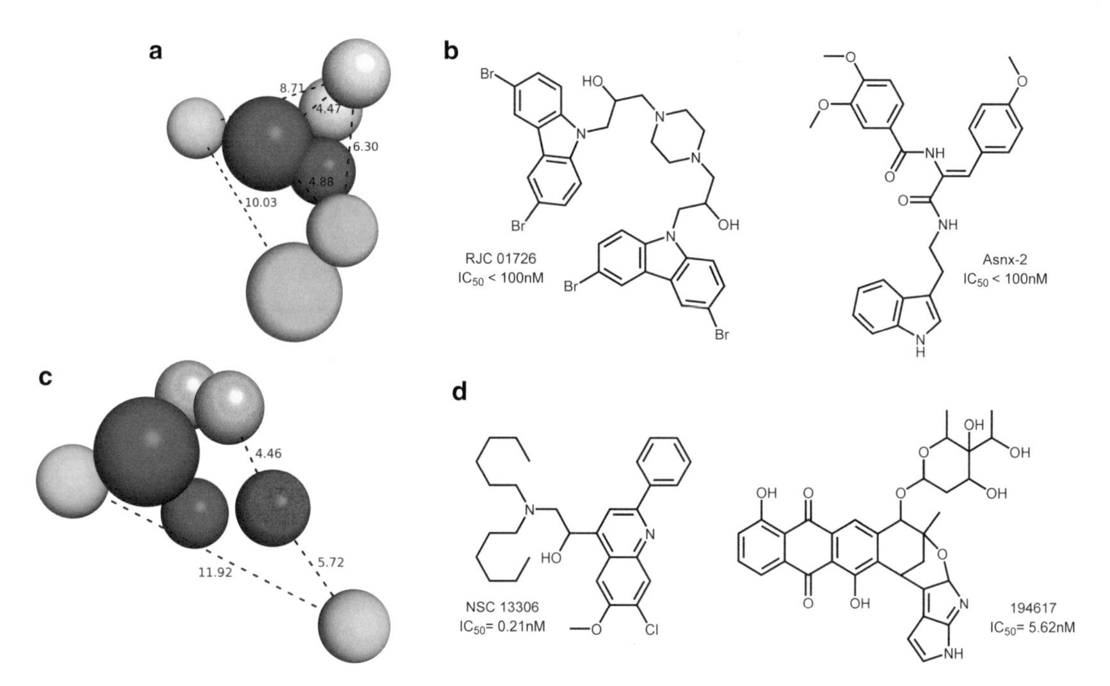

Fig. 7 Pharmacophore hypothesis (**a** = adapted from John and coworkers (2011) [91] and **c** = adapted from AL-NADAF and TAHA (2013) [37]) generated from different BACE-1 inhibitor datasets. Distances are presented in angstroms. **b** = Hits identified with the pharmacophore model shown in (**a**). **d** = Hits identified with the pharmacophore model shown in (**c**). Color code: *magenta*, hydrogen bond donor; *green*, hydrogen bond acceptor; *orange*, aromatic rings; *cyan*, hydrophobic center

positive, an aromatic ring, and two hydrophobic features) shows high correlation coefficient ($r^2 = 0.98$) to compounds potency. Then, it is reasonable to assume that this model captures the essential chemical features for BACE-1 inhibition. According to this model, 793 compounds from MayBridge, Chembridge, NCI, and Asinex databases might inhibit BACE-1 with $IC_{50} < 100$ nM (Fig. 7b). Using a similar approach, AL-NADAF and TAHA (2013) [37] employed a ligand-based virtual screening approach to identify pharmacophore features among 68 BACE-1 inhibitors (Fig. 7c). As the chemical diversity covered by this model is larger than the one employed by John and coworkers [91], it is expected to provide additional information on BACE-1 inhibition requirements. The best pharmacophore model is capable of discriminating active compounds from inactive ones, as evaluated by a ROC curve analysis, and allowed the identification of BACE-1 inhibitors with novel scaffolds (Fig. 7d).

Both pharmacophore models agree that a hydrophobic center, along with two hydrogen bond donor groups and two aromatic rings, are crucial for BACE-1 inhibition. However, the pharmacophore model built by John and coworkers [54] hints that one additional hydrophobic center is also required, whereas Al-Nadaf and Taha [37] claim that additional H-bond donors contribute to BACE-1 inhibitors potency. One way to unravel which one is correct would be to rely on the bioactive conformations of inhibitors to build the pharmacophore models. Then, it is appropriate to analyze the work carried out by SEMIGHINI and coworkers (2013) [78] which took advantage of the crystallographic structures of BACE-1 in complex with different inhibitors (PDB 2VJ7, 2VNM, 2WF1, 2WF2, 2WF3, 2WF0, 2VNN, 2WEZ, and 2VIJ) to build a structure-based pharmacophore model. This pharmacophore model partially matches the previous ones once it underscores that two hydrophobic centers, one H-bonding donor and two H-bond acceptor groups, are required for the BACE-1 inhibition. It might be argued though that the differences among the pharmacophore models are due to the structural diversity among dataset employed to build each model and that they represent partial views of the binding interactions available in BACE-1 active site.

The model built by Semighini and coworkers (2013) [78] led to the selection of 98 compounds from CNS ZINC database as putative BACE-1 inhibitors. Further investigation with GRID software showed that nine of them do not fit the active due to either poor electrostatic or steric complementarity. The synergic use of these tools afforded the identification of micromolar BACE-1 inhibitors. Then, it is tentative to assume that Al-Nadaf and Taha pharmacophore model is the most appropriate to identify potent BACE-1 inhibitors.

Despite the advantages of pharmacophore-based virtual screening approaches, this tool also has its flaws. For instance, most BACE-1 pharmacophore models were built within HypoGen [37, 77–80], which generates a discrete set of low-energy conformations for each input ligand [92]. Although this approach has several merits, peptidomimetic compounds, such as several BACE-1 inhibitors, can adopt dozens, sometimes hundreds, of low-energy conformations that might not be properly sampled by HypoGen. Such under-sampled conformational space search can lead to the selection of pharmacophore features that do not really correspond to the expected interaction profile within the BACE-1 active site.

However, several software SCAMPI [93], GAMMA [94] (MATSUSHITA, MURAKAMI, IWAMOTO, 2012), GASP [95], and GALAHAD [33] carry out a conformational search during the pharmacophore feature selection, thus minimizing the limitation reported above.

3.4 Optimization of Lead Compounds that Inhibit BACE-1

Both virtual and high-throughput screening campaigns afforded a great number of hits against BACE-1 that require further potency or selectivity optimization to be considered as good lead compounds for drug development efforts. Despite the fact that docking provides several hints on how to increase the steric and electronic complementarity, some compounds adopt different binding profile after their groups are modified [96]. Besides, docking affords only a qualitative standpoint of view over the optimization of structure-activity relationships. On the other hand, quantitative structure-activity relationship (QSAR) models strive to depict the chemical features that contribute to potency or selectivity through a mathematical equation [97], using different types of descriptors (1D, 2D, 3D, 4D). Simply put, quantitative structure-activity relationship (QSAR) models allow the further exploration of biological data from previously evaluated compounds to design more potent and selective inhibitors in a fast and predictable way.

From a historical point of view, QSAR modeling is a ligand-based approach that does not require structural information from the target, which was very scarce in the 1960s (when QSAR was born). The evolution X-ray crystallography along with QSAR modeling techniques overcomes this barrier. As a consequence, most QSAR models described in the following section take advantage of crystallographic structure of BACE-1 in complex with a hit compound. For example, 3D QSAR methods (e.g., CoMFA) employ molecular interaction fields (MIFs) to account for electronic (E) and steric (S) features that influence the biological activity [49]. For MIFs to have any biological significance, compounds must be in the bioactive conformation and aligned in the space. As this kind of information rarely is available for all the compounds employed for QSAR model development, this is one of the most challenging and sensitive steps in 3D QSAR model development [98]. However, the great number of BACE-1 crystallographic structures available in the PDB databank (>335 on 2 February 2017, http://www.rcsb.org/pdb/home/home.do) makes it easier to build acceptable 3D QSAR models.

As mentioned before, the QSAR models reported here do not capture all the contributions that QSAR modeling has made toward the design of potent BACE-1 inhibitors. Instead, examples that help to illustrate the pros and cons of QSAR models were selected. Following this approach, it is crucial to remind the reader that BACE-1 inhibitors can be broadly classified into peptidomimetics and non-peptidomimetics [99, 100]. QSAR models reported in the next section are grouped according to this classification.

3.4.1 QSAR Models for Peptidomimetic BACE-1 Inhibitors

Peptidomimetic inhibitors are transition state bioisosteres that are stable to peptide-bond hydrolysis [99]. Hence, these compounds bind to Asp^{32} and Asp^{228} residues [101], and their optimization relies on increasing steric and electronic complementarity toward BACE-1 subsites S1 (Leu30, Phe108, Typ115, and Ile118), S2 (Arg235, Gln12, and Asn233), S3 (Ala335, Ile110, and Ser113), S1' (Lys224 and Thr329), and S2' (Ile126 and Arg128) (Fig. 8). This goal can be treated as a multidimensional problem whose solution requires a quantitative description of the chemical features that are responsible for the structure- activity relationship of each class of compounds.

Pandey and coworkers (2010) [102], for instance, employed comparative molecular field analysis (CoMFA) to investigate the steric and electronic requirements of 43 hydroxyethylamine BACE-1 inhibitors, whose potency ranges from 1 to 7.950 nM. The crystallographic conformation of the inhibitor found in PDB ID

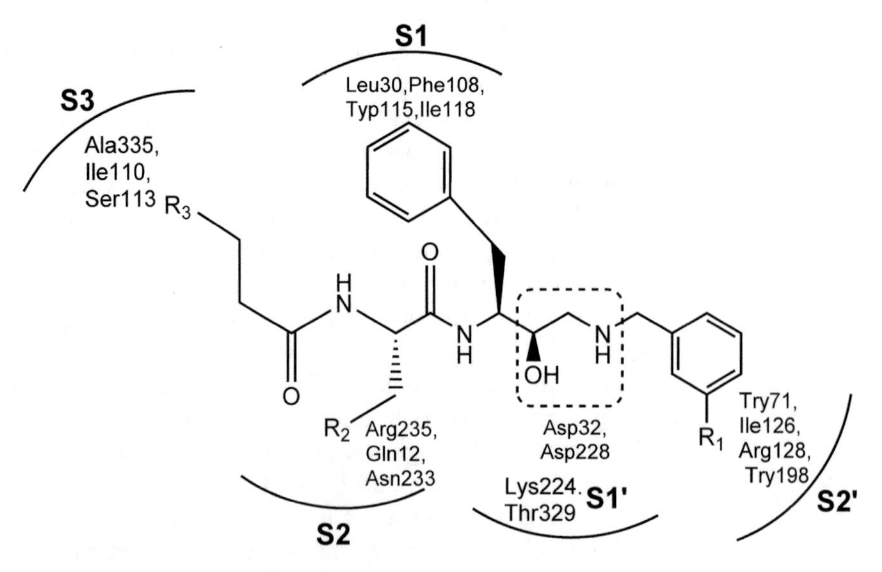

Fig. 8 Peptidomimetic scaffold within the BACE-1 binding site

2HM1 was considered as a template, upon which all other compounds were superposed using maximum common substructure alignment option, available in SYBYL platform. This simplistic approach afforded CoMFA models with excellent statistical parameters ($n = 34$, $r^2 = 0.99$, $q^2 = 0.81$) which suggest that electrostatic fields (E) have a slightly larger contribution to explain the potency of the inhibitors than the steric fields (S) ($E = 52.6\%$ vs $S = 47.4\%$). Although informative, CoMFA maps have sharp edges that lack physical meaning [55, 103]. Another concern comes from the fact that r^2_{pred} value (0.93, $n = 9$) is higher than cross-validated ($q^2 = 0.81$), suggesting that overfitting might have occurred. These authors also employed CoMSIA to further investigate the steric, electrostatic, hydrophobic, and hydrogen bond donor features that contribute to hydroxyethylamine potency. CoMSIA models not only proved to have comparable statistical parameters ($n = 34$, $r^2 = 0.98$, $q^2 = 0.75$) but also show predictive power within the expected range ($r^2_{pred} = 0.75$, $n = 9$). In contrast to CoMFA, CoMSIA similarity maps follow a Gaussian-type distance-dependent function that avoids sharp edges seen in CoMFA contour maps [69]. Then, they can be easily interpreted. Similarity maps support that electronic (E) is the major feature for BACE-1 inhibition (34.0%). On the other hand, steric (S) seems to be as important as hydrophobicity (H) ($S = 24.8\%$ vs $H = 26.3\%$) toward potency, and hydrogen bond donor (HD) plays a minor role (14.9%). This result emphasizes the importance of the protonation state of aspartic (Asp) residues (implicitly captured by electrostatic feature) in the binding of inhibitors. Although this sort of information is not taken into account during QSAR model development, CoMSIA and CoMFA models agree that positively charged groups nearby the S1 subsite improve potency, whereas negatively charged moieties should be placed within S2 subsite proximity. CoMSIA similarity maps further underscore that bulky moieties nearby the S1, S2, S2', and S3 subsites would increase potency, as long as hydrophilic groups are placed in the S2 and S3 pocket and hydrophobic groups are located at S1, S1, and S2' pockets. This information is particularly interesting because S3 subsite can influence BACE-1/BACE-2 selectivity [103].

Salum and Valadares (2010) [104] investigated the influence of molecular alignment over CoMFA models built from a dataset of 128 hydroxyethylamines whose potency ranges from 2 nM to 5.89 μM. Their work proposes a novel alignment strategy (fragment-guided alignment) (Table 1) that employs molecular holograms to cluster the inhibitors into families, which have at least one compound whose bioactive conformation (from X-ray crystallography) is known. Using this information, they conclude that benzyl substituted rings in the S2' subpocket, commonly found in this position, are detrimental to the potency due to steric

Table 1
Statistical parameters of CoMFA models[a] for BACE-1 peptidomimetic inhibitors, according to their molecular alignment

Alignment	r^2	r^2 pred	S	E
Fragment-guided	0.88	0.72	0.66	0.34
Fragment guided + rigid fit	0.87	0.69	0.66	0.34
Docked poses	0.88	0.40	0.64	0.36
Docked poses + visual analysis[b]	0.91	0.48	0.71	0.29

[a]MMFF94 charges were employed
[b]This is the only CoMFA model that uses five principal component (PCs) instead of six

clash. In contrast to the previous example, this CoMFA model suggests that steric complementarity explains 66% of inhibitors' potency. This apparent contradiction might be explained by the differences in the datasets or the type of charge employed in each work.

More recently, Zhang and coworkers (2017) [105] compared the results from hologram QSAR (2D QSAR) and CoMFA/CoMSIA models built with 55 cyclic sulfone hydroxyethylamine inhibitors whose potency (IC_{50}) ranges from 0.002 to 2.75 μM. Hologram QSAR has proved to be as robust and predictive as 3D QSAR techniques and not be influenced by molecular alignment rules. Furthermore, this method allows an easy interpretation of the results according to contribution maps that colors atoms whose contribution to potency is positive in green/yellow, whereas atoms that disfavor potency are colored in orange/red. The remaining atoms are colored in white. Accordingly, the best HQSAR model ($q^2 = 0.69$, $r^2 = 0.98$, $r^2_{pred} = 0.69$) indicates that the cyclic sulfone moiety has a positive contribution to potency. On the other hand, the t-butyl group and the benzene ring favor the potency in some compounds and disfavors in others (Fig. 9). This result may be related to different binding profile (orientation) of these compounds within the BACE-1 active site.

CoMFA ($r^2 = 0.91$, $q^2 = 0.53$, $r^2_{pred} = 0.54$) models suggests that electrostatic field descriptors have a major contribution to the model ($E = 57.3\%$ vs $S = 42.7\%$), whereas CoMSIA model suggests that electrostatic similarity explains just 21.4% of inhibitors' potency, while hydrogen bond donating ability accounts for 32.3% and hydrophobicity for 46.3%.

For a matter of clarity, the main results from QSAR models presented so far are summarized in Fig. 10.

3.4.2 QSAR Models for Non-peptidomimetic BACE-1 Inhibitors

Although the peptidomimetic inhibitors are potent and effective in in vitro assays, their low oral bioavailability, metabolic instability, and poor central nervous system (CNS) penetration features have limited their advance into the clinical phases [107]. Then, non-

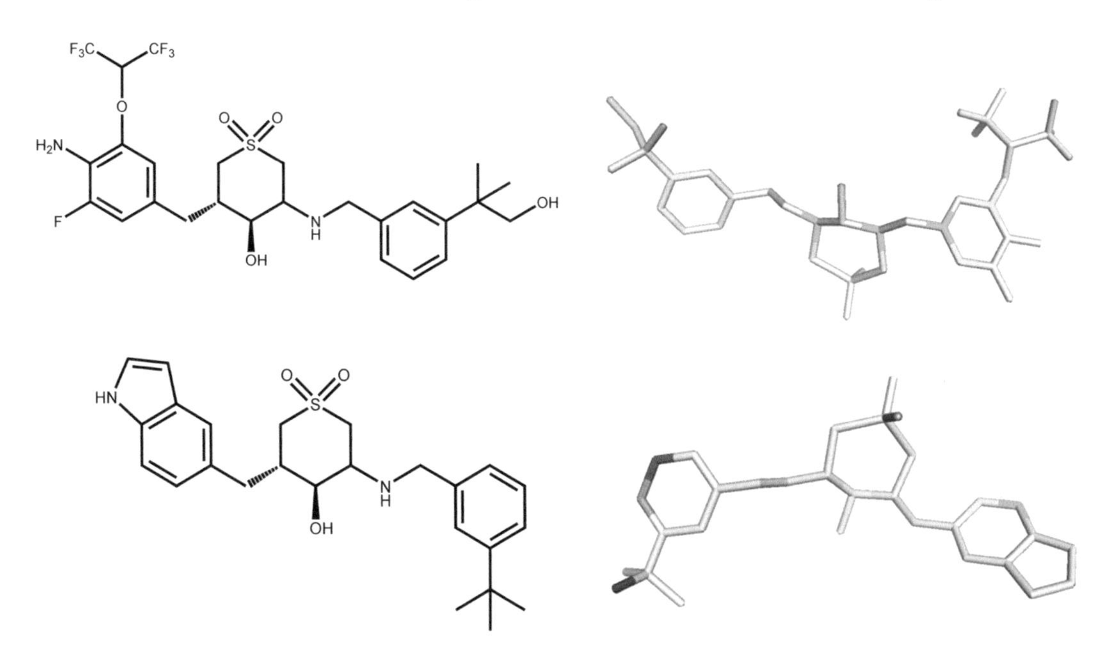

Fig. 9 HQSAR contribution map from HQSAR model for cyclic-peptidomimetic derivatives that inhibit BACE-1 (Adapted from Zhang and coworkers (2017) [105])

peptidomimetic inhibitors pose as a promising alternative due to their smaller size, better metabolic stability, and pharmacokinetic profile [99].

Accordingly, Malamas and coworkers have shown that aminoimidazole, aminohydantoin, and aminopyridine derivatives not only selectively inhibit BACE-1 but also display improved pharmacokinetic profile. Therefore, Cruz and Castilho (2014) [108] employed 2D QSAR approach to investigate 102 aminoimidazole, aminohydantoin, and aminopyridine derivatives, whose potency (IC_{50}) ranges from 0.01 to 45 μM. Their HQSAR models ($r^2 = 0.85$ / $q^2 = 0.84$ and $r^2_{Pred} = 0.70$) suggest that the amino group attached to the central aminoimidazole and aminohydantoin ring, as well as the nitrogen from the pyridine ring, has positive contribution to the potency (Fig. 11).

However, contribution maps do not enlighten whether this result is due to steric or electrostatic reasons. To clarify that matter, topological descriptors were employed to build statistically sound QSAR models ($r^2 = 0.87$, $q^2 = 0.85$, and $r^2_{Pred} = 0.84$, grid spacing = 1.0 and CoMFA focusing with Stdev_Coefficient = 0.6) that point out the importance of electrostatic features toward the potency. Further investigation of this dataset with CoMFA ($r^2 = 0.89$ / $q^2 = 0.82$ and $r^2_{Pred} = 0.76$) (Fig. 12) [108] confirms that electron-rich groups (i.e., H-bond acceptor), in the pyridine ring, might favor the potency (red volume in Fig. 12b). Furthermore, this 3D QSAR model suggests that additional substituents on this ring are allowed (green surface in Fig. 12c).

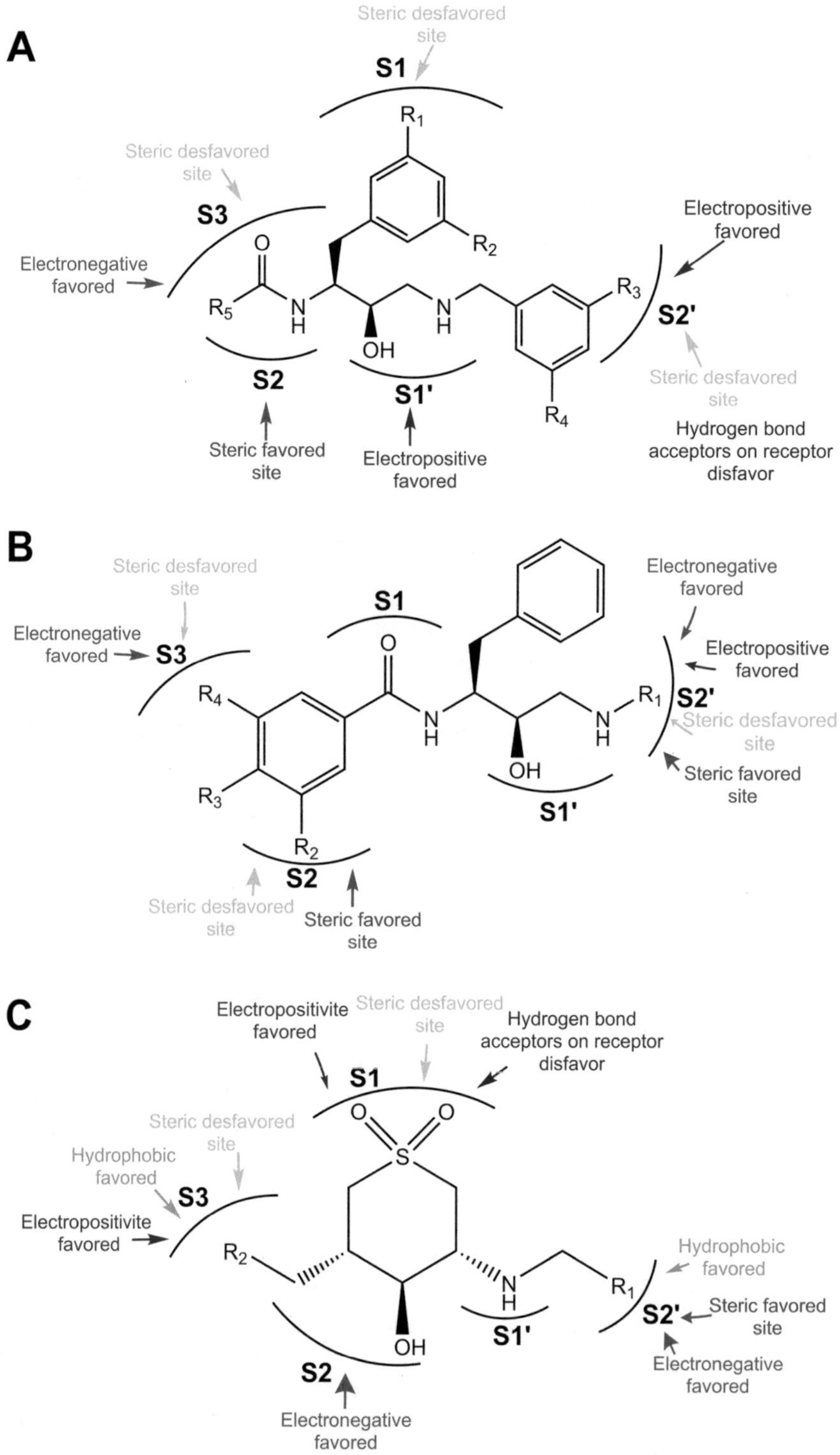

Fig. 10 Optimization guidelines for BACE-1 inhibitors according to 3D QSAR studies: (a) Pandey and coworkers (2010) [106], (b) Salum and Valadares (2010) [104], and (c) Zhang and coworkers (2017) [105]

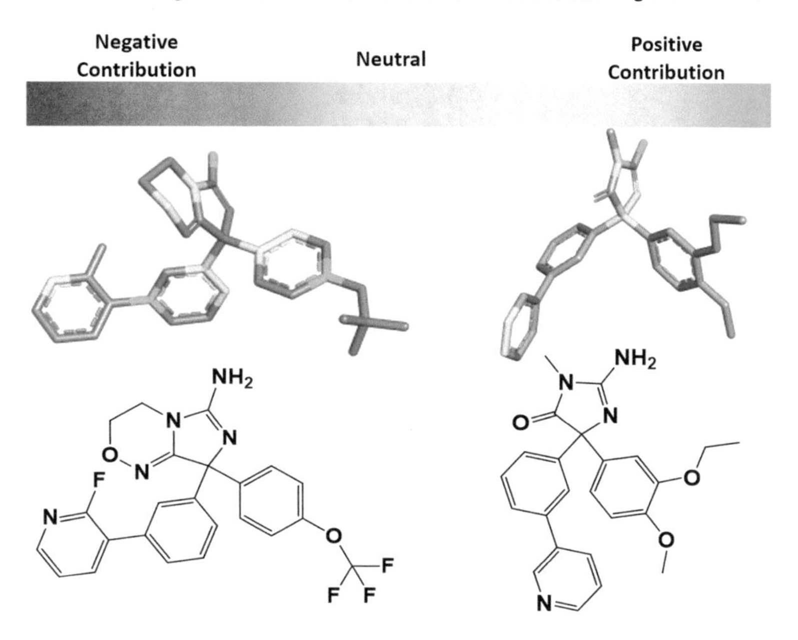

Fig. 11 HQSAR contribution maps for aminoimidazole and aminohydantoin derivatives that inhibit BACE-1

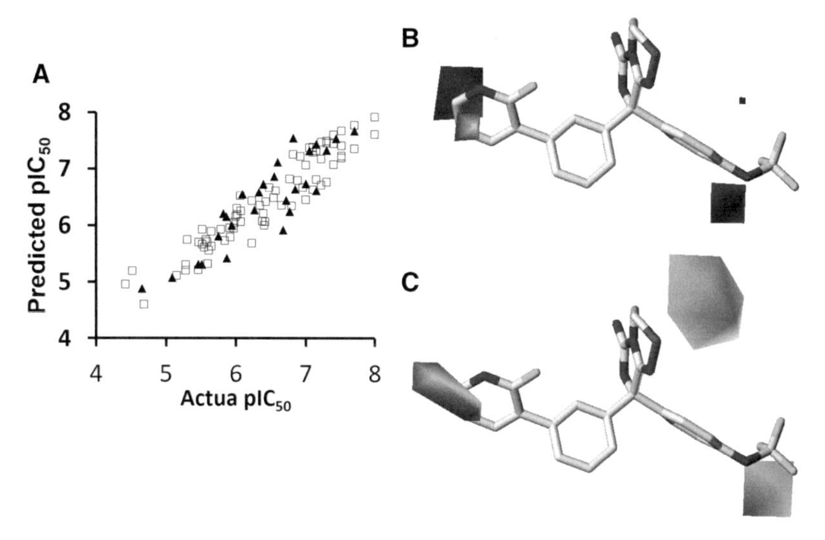

Fig. 12 CoMFA results for the dataset reported in Cruz (2012) [108]. (**a**) Predicted pIC50 values of aminoimidazole, aminohydantoin, and aminopyridine derivatives, according to CoMFA model (molecular alignment = docking + rigid fit). □ Training-set compounds, ▲ test-set compounds. (**b**) Steric contour maps surrounding a potent BACE-1 inhibitor (*pIC50* = 7.5). Map levels, *green* = 0.05 Kcal/mol and *yellow* = −0.017 Kcal/mol. (**c**) Electrostatic contour maps surrounding a potent BACE-1 inhibitor (pIC50 = 7.5). Map levels, *red* = 0.078 Kcal/mol and *blue* = −0.065 Kcal/mol. color code: *red*, oxygen; *blue*, nitrogen; *white*, inhibitor carbon

Hossain and coworkers (2015) developed HQSAR and 3D QSAR models (CoMFA and CoMSIA) from 980 structurally diverse BACE-1 inhibitors, whose potency ranges from 1 nM to 2.8 M. Such large dataset would provide massive insights into the non-peptidomimetic SAR and further speed up the development of BACE-1 inhibitors. All compounds were docked to BACE-1 (PDB ID, 3OHH) using the software LigandFit and then aligned to each other based on their shape. This strategy afforded a robust CoMFA model ($r^2 = 0.88$, $q^2 = 0.60$, $r^2_{pred} = 0.71$) that suggests that steric (S) and electrostatic (E) fields have a balanced contribution to explain the potency of the inhibitors (S = 46.70% and E = 53.30%). According to the authors, these S contour maps hint that positive inductive effect of alkyl chain increases the potency of the inhibitors. Besides, S contour maps suggest that macrocyclic ring in position S1 (Fig. 8) improves the biological activity, but bulkier groups at this position are not allowed.

The best CoMSIA model ($r^2 = 0.90$, $q^2 = 0.58$, $r^2_{pred} = 0.72$) further suggests that electron-rich moieties (i.e., hydroxyl, amine, or sulfonyl) in S2 have favorable electrostatic contribution to potency. However, sulfonyl group has a negative contribution according to the H-bonding similarity map toward S2 subsite.

4 Conclusion and Perspectives

Although in silico tools have contributed to identify BACE-1 inhibitors and highlight how their structure should be modified to increase potency or selectivity, there are several shortcomings for the successful development of BACE-1 inhibitors. For instance, much of the knowledge gained from MD studies has not been employed to set the protonation state of aspartic residues during virtual screening, or has it been considered in the interpretation of CoMFA contour maps. Moreover, many QSAR model predictions have not been proved right or wrong. In fact, most computational efforts are dissociated from following-up experiments that evaluate their usefulness. The synergic combination of efforts between organic and computational chemists is paramount to identify errors and discrepancies in the pharmacophore/QSAR models, so that new paradigms are established, innovative methods developed, and BACE-1 inhibitors become approved drugs for AD patient treatment.

Finally, it is important to underscore that in silico cryptic and allosteric sites prediction [109–111] may revolutionize BACE-1 drug development efforts in the near future. However, after these new binding pockets are found and proved druggable [112], traditional virtual screening tools (docking and pharmacophore models) and QSAR models shall pave the way, once more, to rationalize the structure-activity relationships of yet undiscovered noncompetitive BACE-1 inhibitors.

References

1. Wilson RS, Segawa E, Boyle PA et al (2012) The natural history of cognitive decline in Alzheimer â€™ s disease. Psychol Aging 27:1008–1017. doi:10.1037/a0029857.The
2. Prince M, Comas-Herrera A, Knapp M et al (2016) World Alzheimer Report 2016 Improving healthcare for people living with dementia. Coverage, Quality and costs now and in the future 1–140
3. Alzheimer's Association (2016) 2016 Alzheimer's Disease facts and figures. 2016 http://www.alz.org/facts/
4. Brookmeyer R, Johnson E, Ziegler-Graham K, Arrighi HM (2007) Forecasting the global burden of Alzheimer's disease. Alzheimers Dement 3:186–191. doi:10.1016/j.jalz.2007.04.381
5. Deardorff WJ, Feen E, Grossberg GT (2015) The use of cholinesterase inhibitors across all stages of Alzheimer???s disease. Drugs Aging 32:537–547. doi:10.1007/s40266-015-0273-x
6. Ferreira-vieira TH, Guimaraes IM, Silva FR, Ribeiro FM (2016) Alzheimer's Disease : targeting the cholinergic system. Curr Neuropharmacol 14:101–115. doi:10.1186/1750-1326-4-48
7. Chen Y-CC (2015) Beware of docking! Trends Pharmacol Sci 36:78–95. doi:10.1016/j.tips.2014.12.001
8. Alonso H, Bliznyuk AA, Gready JE (2006) Combining docking and molecular dynamic simulations in drug design. Med Res Rev 26:531–568. doi:10.1002/med.20067
9. Elokely KM, Doerksen RJ (2013) Docking challenge: protein sampling and molecular docking performance. J Chem Inf Model 53:1934–1945. doi:10.1021/ci400040d
10. Ganesan A, Coote ML, Barakat K (2016) Molecular dynamics-driven drug discovery: leaping forward with confidence. Drug Discov Today 22(2):249–269. doi:10.1016/j.drudis.2016.11.001
11. Kim MO, Blachly PG, McCammon JA (2015) Conformational dynamics and binding free energies of inhibitors of BACE-1: from the perspective of protonation equilibria. PLoS Comput Biol 11:1–28. doi:10.1371/journal.pcbi.1004341
12. Kitchen D, Decornez H, Furr J, Bajorath J (2004) Docking and scoring in virtual screening for drug discovery: methods and applications. Nat Rev Drug Discov 3:935–949. doi:10.1038/nrd1549
13. Durrant JD, McCammon JA (2011) Molecular dynamics simulations and drug discovery. BMC Biol 9:71. doi:10.1186/1741-7007-9-71
14. Mortier J, Rakers C, Bermudez M et al (2015) The impact of molecular dynamics on drug design: applications for the characterization of ligand-macromolecule complexes. Drug Discov Today 20:686–702. doi:10.1016/j.drudis.2015.01.003
15. Verli H (2014) Dinâmica Molecular. In: Bioinformática da Biol. à Flexibilidade Mol, 1st edn, Porto Alegre, pp 1–291 e-book: https://www.ufrgs.br/bioinfo/ebook/
16. Namba AM, Da Silva VB, Da Silva CHTP (2008) Dinâmica molecular: Teoria e aplicações em planejamento de fármacos. Ecletica Quim 33:13–24. doi:10.1590/S0100-46702008000400002
17. Maginn EJ, Elliott JR (2010) Historical perspective and current outlook for molecular dynamics as a chemical engineering tool. Ind Eng Chem Res 49:3059–3078
18. Kocak A, Erol I, Yildiz M, Can H (2016) Computational insights into the protonation states of catalytic dyad in BACE1???Acyl guanidine based inhibitor complex. J Mol Graph Model 70:226–235. doi:10.1016/j.jmgm.2016.10.013
19. Abraham MJ, Murtola T, Schulz R et al (2015) Gromacs: high performance molecular simulations through multi-level parallelism from laptops to supercomputers. SoftwareX 1:19–25. doi:10.1016/j.softx.2015.06.001
20. Malde AK, Zuo L, Breeze M et al (2011) An automated force field topology builder (ATB) and repository: version 1.0. J Chem Theory Comput 7:4026–4037. doi:10.1021/ct200196m
21. Schüttelkopf AW, Van Aalten DMF (2004) PRODRG: a tool for high-throughput crystallography of protein-ligand complexes. Acta Crystallogr Sect D Biol Crystallogr

60:1355–1363. doi:10.1107/S0907444904011679

22. Rostkowski M, Olsson MHM, Soendergaard CR, Jensen JH (2011) Graphical analysis of pH-dependent properties of proteins predicted using PROPKA. BMC Struct Biol 11:1–6. doi:10.1186/1472-6807-11-6
23. ChemAxon (2015) Marvin Sketch version 15.4.20: ChemAxon
24. Berendsen HJC, Grigera JR, Straatsma TP (1987) The missing term in effective pair potentials. J Phys Chem 91:6269–6271. doi:10.1021/j100308a038
25. Lindahl E, Hess B, Spoel D (2001) GROMACS 3.0: a package for molecular simulation and trajectory analysis. J Mol Model 7:306–317. doi:10.1007/s008940100045
26. Verlet L (1967) Computer Experiments on classical fluids. I. Thermodynamical properties of Lennard, –Jones molecules. Phys Rev 159:98–103. doi:10.1103/PhysRev.159.98
27. Ellis CR, Tsai CC, Hou X, Shen J (2016) Constant pH molecular dynamics reveals pH-modulated binding of two small-molecule BACE1 inhibitors. J Phys Chem Lett 7:944–949. doi:10.1021/acs.jpclett.6b00137
28. Nicolaou CA, Brown N (2013) Multi-objective optimization methods in drug design. Drug Discov Today Technol 10:1–9. doi:10.1016/j.ddtec.2013.02.001
29. Thangapandian S, John S, Sakkiah S, Lee KW (2011) Molecular docking and pharmacophore filtering in the discovery of dual-inhibitors for human leukotriene A4 hydrolase and leukotriene C4 synthase. J Chem Inf Model 51:33–44. doi:10.1021/ci1002813
30. Sliwoski G, Kothiwale S, Meiler J, Lowe EW (2014) Computational methods in drug discovery. Pharmacol Rev 66:334–395. doi:10.1124/pr.112.007336
31. Domínguez JL, Villaverde MC, Sussman F (2013) Effect of pH and ligand charge state on BACE-1 fragment docking performance. J Comput Aided Mol Des 27:403–417. doi:10.1007/s10822-013-9653-7
32. Avakul P, Nishiyama H, Kato N et al (2012) Benefit of selecting number of active Mesh routers in disaster oriented wireless Mesh network. J Softw Eng Appl 2012:36–41. doi:10.4236/jsea.2012.512b
33. Liu M, Sun Z, Hu W (2012) Three-dimensional pharmacophore screening for fentanyl derivatives. Neural Regen Res 7:1398–1405. doi:10.3969/j.issn.1673-5374.2012.18.006
34. Sliwoski G, Kothiwale S, Meiler J, Lowe EW (2013) Computational methods in drug discovery. Pharmacol Rev 66:334–395. doi:10.1124/pr.112.007336
35. Xiao Z, Varma S, De XY, Tropsha A (2004) Modeling of p38 mitogen-activated protein kinase inhibitors using the catalyst??? HypoGen and k-nearest neighbor QSAR methods. J Mol Graph Model 23:129–138. doi:10.1016/j.jmgm.2004.05.001
36. Qing X, Lee XY, De Raeymaeker J et al (2014) Pharmacophore modeling: advances, limitations, and current utility in drug discovery. J Receptor Ligand Channel Res 7:81–92. doi:10.2147/JRLCR.S46843
37. Al-Nadaf A, Taha MO (2013) Ligand-based pharmacophore exploration and QSAR analysis of transition state analogues followed by in silico screening guide the discovery of new sub-micromolar β-secreatase inhibitors. Med Chem Res 22:1979–1997. doi:10.1007/s00044-012-0204-x
38. Amaravadhi H, Baek K, Yoon HS (2014) Revisiting de novo drug design: receptor based pharmacophore screening. Curr Top Med Chem 14:1890–1898
39. Taha MO, Habash M, Hatmal MM et al (2015) Ligand-based modeling followed by in vitro bioassay yielded new potent glucokinase activators. J Mol Graph Model 56:91–102. doi:10.1016/j.jmgm.2014.12.003
40. Leite FHA, Froes TQ, da Silva SG et al (2017) An integrated approach towards the discovery of novel non-nucleoside Leishmania major pteridine reductase 1 inhibitors. Eur J Med Chem 132:322–332. doi:10.1016/j.ejmech.2017.03.043
41. Kaserer T, Beck KR, Akram M et al (2015) Pharmacophore models and pharmacophore-based virtual screening: concepts and applications exemplified on hydroxysteroid dehydrogenases. Molecules 20:22799–22832. doi:10.3390/molecules201219880
42. Jiang L, Zhang X, Chen X et al (2015) Virtual screening and molecular dynamics study of potential negative allosteric modulators of mGluR1 from Chinese herbs. Molecules 20:12769–12786. doi:10.3390/molecules200712769
43. Danishuddin KAU (2016) Descriptors and their selection methods in QSAR analysis: paradigm for drug design. Drug Discov Today 21:1291–1302. doi:10.1016/j.drudis.2016.06.013
44. Cherkasov A, Muratov EN, Fourches D et al (2015) QSAR modeling: where have you been? Where are you going to? J Med Chem 57:4977–5010. doi:10.1021/jm4004285. QSAR

45. Heritage TW, Lowis DR (1999) Molecular Hologram QSAR. In: Parrill A (ed) Ration. Drug Des. American Chemical Society, Washington, pp 212–225
46. Hurst T, Heritage T (1997) HQSAR – a highly predictive QSAR technique based on molecular holograms, 213th edn. ACS Natl. Meeting. CINF 019, San Francisco
47. Andrade CH, Salum LDB, Castilho MS et al (2008) Fragment-based and classical quantitative structure-activity relationships for a series of hydrazides as antituberculosis agents. Mol Divers 12:47–59. doi:10.1007/s11030-008-9074-z
48. Castilho MS, RV CG, Andricopulo AD (2007) Classical and hologram QSAR studies on a series of Tacrine derivatives as Butyrylcholinesterase inhibitors. Lett Drug Des Discov 4:106–113. doi:10.2174/157018007779422505
49. Kubinyi H (1993) 3D QSAR in drug design: theory, methods and applications. ESCOM, Leiden
50. Cramer RD, Patterson DE, Bunce JD (1988) {C}omparative {M}olecular {F}ield {a}nalysis ({C}o{MFA}). 1. {E}ffect of {S}hape on {B}inding of {S}teroids to {C}arrier {P}roteins. J Am Chem Soc 110:5959–5967. doi:10.1021/ja00226a005
51. Liu S, Fu R, Cheng X et al (2012) Exploring the binding of BACE-1 inhibitors using comparative binding energy analysis (COMBINE). BMC Struct Biol 12:21. doi:10.1186/1472-6807-12-21
52. Langham JJ, Cleves AE, Spitzer R et al (2010) Physical binding pocket induction for affinity prediction. J Med Chem 52:6107–6125. doi:10.1021/jm901096y
53. Kubinyi H (1997) QSAR and 3D QSAR in drug design part 1: methodology. DDT 2:457–467
54. Rodrigues CR, Castro HC, Brito MA (2011) Métodos de Correlação Quantitativa entre Estrutura Molecular Tridimensional e Atividade Farmacológica (3D-QSAR). In: Química Med. Métodos e Fundam. em Planej. Fármacos. Edusp, São Paulo, pp 455–484
55. Klebe G, Abraham U (1999) Comparative molecular similarity index analysis (CoMSIA) to study hydrogen bonding properties and to score combinatorial libraries. J Comput Aided Mol Des 13:1–10
56. Klebe G, Abraham U, Mietzner T (1994) Molecular similarity indexes in a comparative-analysis (Comsia) of drug molecules to correlate and predict their biological-activity. J Med Chem 37:4130–4146
57. Perilla JR, Goh BC, Cassidy CK et al (2015) Molecular dynamics simulations of large macromolecular complexes. Curr Opin Struct Biol 31:64–74. doi:10.1016/j.sbi.2015.03.007
58. Liu S, Fu R, Cheng X et al (2012) Exploring the binding of BACE-1 inhibitors using comparative binding energy analysis (COMBINE) exploring the binding of BACE-1 inhibitors using comparative binding energy analysis (COMBINE). BMC Struct Biol 12:1–20
59. Barman A, Prabhakar R (2013) Elucidating the catalytic mechanism of beta secretase (BACE1): a quantum mechanics / molecular mechanics (QM / MM) approach. J Mol Graph Model 40:1–9. doi:10.1016/j.jmgm.2012.12.010
60. Paul TJ, Barman A, Ozbil M, Ram Prasad Bora TZ (2016) Mechanisms of peptide hydrolysis by aspartyl and metalloproteases Thomas. Phys Chem Chem Phys 18:24790–24801. doi:10.1039/C6CP02097F
61. Kennedy ME, Wang W, Song L et al (2003) Measuring human ??-secretase (BACE1) activity using homogeneous time-resolved fluorescence. Anal Biochem 319:49–55. doi:10.1016/S0003-2697(03)00253-7
62. Tounge BA, Reynolds CH (2003) Calculation of the binding affinity of beta-secretase inhibitors using the linear interaction energy method. J Med Chem 46:2074–2082. doi:10.1021/jm020513b
63. Sabbah DA, Zhong HA (2016) Modeling the protonation S tates of β -secretase binding pocket by molecular dynamics simulations and docking studies. J Mol Graph Model 68:206–215. doi:10.1016/j.jmgm.2016.07.005
64. Rajamani R, Reynolds CH (2004) Modeling the protonation states of the catalytic aspartates in β-secretase. J Med Chem 47:5159–5166. doi:10.1021/jm049817j
65. Yu N, Hayik SA, Wang B et al (2006) Assigning the protonation states of the key aspartates in β-secretase using QM/MM X-ray structure refinement. J Chem Theory Comput 2:1057–1069. doi:10.1021/ct0600060
66. Domı L, Christopeit T, Villaverde MC et al (2010) Effect of the protonation state of the titratable residues on the inhibitor affinity. Biochemical 49:7255–7263. doi:10.1021/bi100637n
67. Park H, Lee S (2003) Determination of the active site protonation state of β-secretase from molecular dynamics simulation and docking experiment: implications for

structure-based inhibitor design. J Am Chem Soc 125:16416–16422. doi:10.1021/ja0304493

68. Lee J, Miller BT, Brooks BR (2016) Computational scheme for pH-dependent binding free energy calculation with explicit solvent. Protein Sci 25:231–243. doi:10.1002/pro.2755
69. Swails JM, Roitberg AE (2012) Enhancing conformation and protonation state sampling of hen egg white lysozyme using pH replica exchange molecular dynamics. J Chem Theory Comput 8:4393–4404. doi:10.1021/ct300512h
70. Lee J, Miller BT, Damjanović A, Brooks BR (2014) Constant pH molecular dynamics in explicit solvent with enveloping distribution sampling and hamiltonian exchange. J Chem Theory Comput 10:2738–2750. doi:10.1021/ct500175m
71. McGaughey GB, Colussi D, Graham SL et al (2007) ??-secretase (BACE-1) inhibitors: accounting for 10s loop flexibility using rigid active sites. Bioorg Med Chem Lett 17:1117–1121. doi:10.1016/j.bmcl.2006.11.003
72. Mirsafian H, Mat Ripen A, Merican AF, Bin Mohamad S (2014) Amino acid sequence and structural comparison of BACE1 and BACE2 using evolutionary trace method. Sci World J 2014:482463. doi:10.1155/2014/482463
73. Hernández-Rodríguez M, Correa-Basurto J, Gutiérrez A et al (2016) Asp32 and Asp228 determine the selective inhibition of BACE1 as shown by docking and molecular dynamics simulations. Eur J Med Chem 124:1142–1154. doi:10.1016/j.ejmech.2016.08.028
74. Ben Halima S, Mishra S, Raja KMP et al (2016) Specific inhibition of beta-secretase processing of the Alzheimer disease amyloid precursor protein. Cell Rep 14:2127–2141. doi:10.1016/j.celrep.2016.01.076
75. Perego C, Salvalaglio M, Parrinello M (2015) Molecular dynamics simulations of solutions at constant chemical potential. J Chem Phys 142:144113. doi:10.1063/1.4917200
76. Nikolic K, Mavridis L, Djikic T et al (2016) Drug design for CNS diseases: Polypharmacological profiling of compounds using Cheminformatic, 3D-QSAR and virtual screening methodologies. Front Neurosci 10:1–21. doi:10.3389/fnins.2016.00265
77. Kumar PS, Singh A, Sharma S et al (2015) Ligand based pharmacophore modeling virtual screening and molecular docking for identification of novel CYP51 inhibitors abstract dataset collection. J Med Chem 1:1–10
78. Semighini EP, A.Taft C, CHTP S (2013) Structure and ligand based rational drug Design for Bace-1 inhibitors. Curr Bioact Compd 9:14–20. doi:10.2174/157340721130901000З
79. Wei HY, Chen GJ, Chen CL, Lin TH (2012) Developing consensus 3D-QSAR and pharmacophore models for several beta-secretase, farnesyl transferase and histone deacetylase inhibitors. J Mol Model 18:675–692. doi:10.1007/s00894-011-1094-4
80. Zhang WT, Jiang FC, Liu D, Du G (2012) Pharmacopore hypothesis generation of BACE-1 inhibitors and pharmacophore-driven identification of potent multi-target neuroprotective agents. Med Chem Res 21:3656–3668. doi:10.1007/s00044-011-9885-9
81. Palakurti R, Sriram D, Yogeeswari P, Vadrevu R (2013) Multiple e-pharmacophore modeling combined with high-throughput virtual screening and docking to identify potential inhibitors of beta- secretase(BACE1). Mol Inform 32:385–398. doi:10.1002/minf.201200169
82. Razzaghi-Asl N, Sepehri S, Ebadi A et al (2015) Molecular docking and quantum mechanical studies on biflavonoid structures as BACE-1 inhibitors. Struct Chem 26:607–621. doi:10.1007/s11224-014-0523-2
83. Liu S, Fu R, Zhou LH, Chen SP (2012) Application of consensus scoring and principal component analysis for virtual screening against ??-secretase (BACE-1). PLoS One 7 (6):e38086. doi:10.1371/journal.pone.0038086
84. Manoharan P, Chennoju K, Ghoshal N (2015) Target specific proteochemometric model development for BACE1 – protein flexibility and structural water are critical in virtual screening. Mol BioSyst 11 (7):1955–1972. doi:10.1039/c5mb00088b
85. Kumar A, Roy S, Tripathi S, Sharma A (2015) Molecular docking based virtual screening of natural compounds as potential BACE1 inhibitors: 3D -QSAR pharmacophore mapping and molecular dynamics analysis. J Biomol Struct Dyn 1102:1–46. doi:10.1080/07391102.2015.1022603
86. Pradeep N, Munikumar M, Swargam S et al (2015) 197 combination of e-pharmacophore modeling, multiple docking strategies and molecular dynamic simulations to discover of

novel antagonists of BACE1. J Biomol Struct Dyn 33(Suppl 1):129–130. doi:10.1080/07391102.2015.1032834

87. Thiyagarajan C, Shanthi S, Karthikeyan M, Thiruneelakan-Dan G (2013) Structure-based virtual screening for identification of novel inhibitors against BACE1 from selective medicinal plant compounds. Int J Curr Res 5:4097–4101
88. Cosconati S, Marinelli L, Di Leva FS et al (2012) Protein flexibility in virtual screening: the BACE-1 case study. J Chem Inf Model 52:2697–2704. doi:10.1021/ci300390h
89. Chirapu SR, Pachaiyappan B, Nural HF et al (2009) Molecular modeling, synthesis, and activity studies of novel biaryl and fused-ring BACE1 inhibitors. Bio Med Chem Let 19 (1):264–274. doi: 10.1016/j.bmcl.2008.10.096
90. Danishuddin M, Khan AU (2015) Structure based virtual screening to discover putative drug candidates: necessary considerations and successful case studies. Methods 71:135–145. doi:10.1016/j.ymeth.2014.10.019
91. John S, Thangapandian S, Sakkiah S, Lee KW (2011) Potent BACE-1 inhibitor design using pharmacophore modeling, in silico screening and molecular docking studies. BMC Bioinformatics 12(Suppl 1):S28. doi:10.1186/1471-2105-12-S1-S28
92. Wong Y-H, Lin C-L, Chen T-S et al (2015) Multiple target drug cocktail design for attacking the core network markers of four cancers using ligand-based and structure-based virtual screening methods. BMC Med Genet 8(Suppl 4):S4. doi:10.1186/1755-8794-8-S4-S4
93. Chang C, Ekins S, Bahadduri P, Swaan PW (2006) Pharmacophore-based discovery of ligands for drug transporters. Adv Drug Deliv Rev 58:1431–1450. doi:10.1016/j.addr.2006.09.006
94. Matsushita K, Murakami C, Iwamoto S (2012) Outage planning method for electrical power facilities using MOGA. 257–262
95. Nicolotti O, Giangreco I, Introcaso A et al (2011) Strategies of multi-objective optimization in drug discovery and development. Expert Opin Drug Discov 6(9):1–14
96. Kozakov D, Hall DR, Jehle SS et al (2015) Ligand deconstruction: why some fragment binding positions are conserved and others are not. Proc Natl Acad Sci U S A 112:E2585–E2594. doi:10.1073/pnas.1501567112
97. Cherkasov A, Muratov EN, Fourches D et al (2015) NIH Public Access 57:4977–5010. doi:10.1021/jm4004285.QSAR
98. Böhm M, Stürzebecher J, Klebe G (1999) Three-dimensional quantitative structure–activity relationship analyses using comparative molecular field analysis and comparative molecular similarity indices analysis to elucidate selectivity differences of inhibitors binding to trypsin, thrombin, and factor Xa. J Med Chem 42:458–477. doi:10.1021/jm981062r
99. Ghosh AK, Osswald HL (2014) BACE1 (β-secretase) inhibitors for the treatment of Alzheimer's disease. Chem Soc Rev 43:6765–6813. doi:10.1039/c3cs60460h
100. Nguyen J-T, Hamada Y, Kimura T, Kiso Y (2008) Design of potent aspartic protease inhibitors to treat various diseases. Arch Pharm (Weinheim) 341:523–535. doi:10.1002/ardp.200700267
101. Dash C, Kulkarni A, Dunn B, Rao M (2003) Aspartic peptidase inhibitors: implications in drug development. Crit Rev Biochem Mol Biol 38:89–119. doi:10.1080/713609213
102. Pandey A, Mungalpara J, Mohan CG (2010) Comparative molecular field analysis and comparative molecular similarity indices analysis of hydroxyethylamine derivatives as selective human BACE-1 inhibitor. Mol Divers 14:39–49. doi:10.1007/s11030-009-9139-7
103. Freskos JN, Fobian YM, Benson TE et al (2007) Design of potent inhibitors of human beta-secretase. Part 2. Bioorg Med Chem Lett 17:78–81. doi:10.1016/j.bmcl.2006.09.091
104. Salum LB, Valadares NF (2010) Fragment-guided approach to incorporating structural information into a CoMFA study: BACE-1 as an example. J Comput Aided Mol Des 24:803–817. doi:10.1007/s10822-010-9375-z
105. Zhang S, Lin Z, Pu Y et al (2017) Comparative QSAR studies using HQSAR, CoMFA, and CoMSIA methods on cyclic sulfone hydroxyethylamines as BACE1 inhibitors. Comput Biol Chem 67:38–47
106. Tyangi A, Shikhar Gupta AP, GM (2010) Alzheimer's disease multi-target directed inhibitor design sequential virtual screening techniques. Pharm Sci 11:29–32
107. Silvestri R (2009) Boom in the development of non-peptidic β-secretase (BACE1) inhibitors for the treatment of Alzheimer's disease. Med Res Rev 29:295–338. doi:10.1002/med.20132

108. Cruz D (2012) Estudos de QSAR 2D e 3D para derivados de aminoimidazóis, aminohidantoínas e aminipiridinas com actividade inibitória sobre a enzima beta-secretase humana. Univ. Estadual da Feira Santana – Pós Grad. em Biotecnol.

109. Guarnera E, Berezovsky IN (2016) Allosteric sites: remote control in regulation of protein activity. Curr Opin Struct Biol 37:1–8. doi:10.1016/j.sbi.2015.10.004

110. Lu S, Huang W, Zhang J (2014) Recent computational advances in the identification of allosteric sites in proteins. Drug Discov Today 19:1595–1600. doi:10.1016/j.drudis.2014.07.012

111. Oleinikovas V, Saladino G, Cossins BP, Gervasio FL (2016) Understanding cryptic pocket formation in protein targets by enhanced sampling simulations. J Am Chem Soc 138:14257–14263. doi:10.1021/jacs.6b05425

112. Surade S, Blundell TL (2012) Structural biology and drug discovery of difficult targets: the limits of ligandability. Chem Biol 19:42–50. doi:10.1016/j.chembiol.2011.12.013

Chapter 8

Design of Anti-Alzheimer's Disease Agents Focusing on a Specific Interaction with Target Biomolecules

Yoshio Hamada and Kenji Usui

Abstract

Alzheimer's disease (AD) is the most common cause of dementia, characterized by progressive intellectual deterioration. Amyloid β peptide (Aβ), the main component of senile plaques in the brains of patients with AD, is formed from amyloid precursor protein (APP) by two processing enzymes. According to the amyloid hypothesis, a processing enzyme β-secretase (BACE1; β-site APP cleaving enzyme) that triggers Aβ formation in the rate-limiting first step of Aβ processing appears to be a promising molecular target for therapeutic intervention in AD. Many researchers have revealed BACE1 inhibitors for the AD treatment. Early BACE1 inhibitors were designed based on the first reported X-ray crystal structure, 1FKN, of a complex between recombinant BACE1 and inhibitor OM99-2. Although OM99-2 seemed to interact with BACE1-Arg235 side chain by hydrogen bonding, we found that a quantum chemical interaction, such as σ-π interaction or π-π stacking, plays a critical role in BACE1 inhibition mechanism. Moreover, we proposed a novel "electron-donor bioisostere" concept in drug discovery study and designed potent BACE1 inhibitors using this concept.

Key words BACE-1, Amyloid β peptide, Amyloid precursor protein, Electron-donor bioisostere

1 Introduction

1.1 Pathology of Alzheimer's Disease

Alzheimer's disease (AD), which is the most common cause of dementia, is characterized by progressive intellectual deterioration. In 1901, Alois Alzheimer, a psychiatrist and neuropathologist, observed a 51-year-old female patient at Frankfurt Asylum and reported on the case in 1906. The patient showed strange behavioral symptoms, including loss of short-term memory, which was later called "AD." Unfortunately, AD's cause has been unclear until now, and there have been no treatment approaches since that first report by Dr. Alzheimer over 100 years ago. A breakthrough was made through the genetic study of some patients with familial AD (FAD) with a mutation of the gene coding for amyloid precursor protein (*APP*) or presenilin gene. As these mutations caused an increase in amyloid β peptides (Aβs) that are the main components

Kunal Roy (ed.), *Computational Modeling of Drugs Against Alzheimer's Disease*, Neuromethods, vol. 132, DOI 10.1007/978-1-4939-7404-7_8, © Springer Science+Business Media LLC 2018

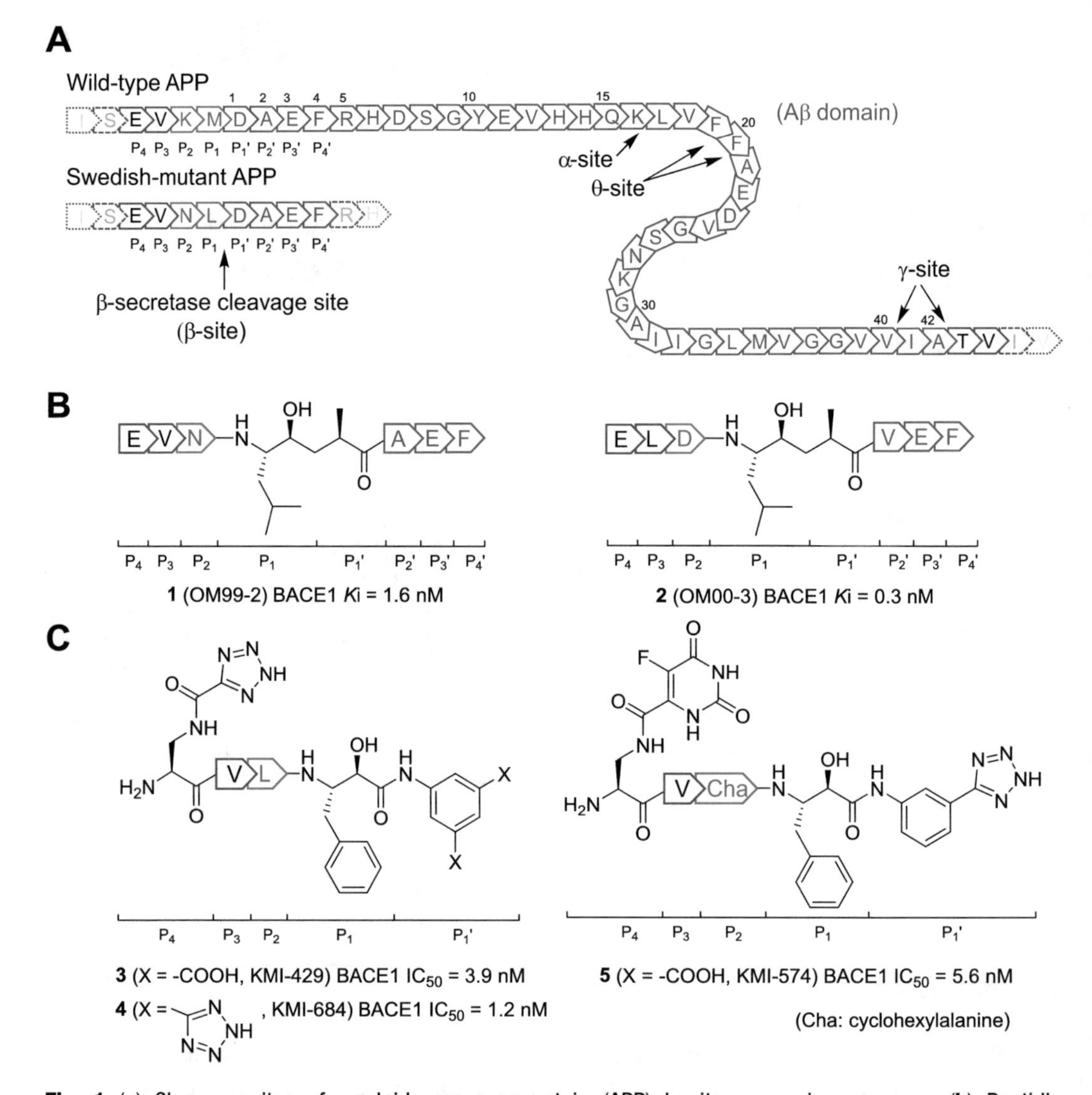

Fig. 1 (**a**) Cleavage sites of amyloid precursor protein (APP) by its processing enzymes. (**b**) Peptidic β-secretase (BACE1) inhibitors by Ghosh et al. (**c**) Peptidic BACE1 inhibitors prepared by us

of senile plaques in the brain of patients with AD, it indicates their involvement in AD's pathogenesis [1–4]. Aβ is produced from APP by two processing enzymes, β-secretase (BACE1; β-site APP cleaving enzyme) and γ-secretase, which are potential molecular targets for the development of anti-AD drugs [5–8]. The cleavage sites of APP are shown in Fig. 1a. There are the full-length APP (APP_{770}) and its isoforms, APP_{695} and APP_{751}, resulting from the alternative splicing of its mRNA. BACE1, one of the processing enzymes of APP, triggers Aβ formation in the rate-limiting first step by cleaving APP at the Aβ domain *N*-terminus (β-site). BACE1 is a type I transmembrane aspartic protease with 501 amino acids. At the next stage in the amyloidogenic pathway of APP, an aspartic protease, γ-secretase, cleaves at the *C*-terminus of the Aβ domain, releasing Aβ that consists mainly of two molecular species, $A\beta_{1-42}$ and $A\beta_{1-40}$. γ-Secretase cleaves two cleavage sites "γ-sites" forming two species of Aβs, $A\beta_{1-40}$ and $A\beta_{1-42}$ (Fig. 1a). APP, BACE1, and γ-secretase are located in the same intracellular granules, including endosomes and the trans-Golgi network, which have an acidic environment, suggesting that Aβs are produced in these locations [9]. $A\beta_{1-42}$ shows greater neurotoxicity and aggregability than $A\beta_{1-40}$ and appears to be a key biomolecular marker of AD pathogenesis. In contrast, α-secretase is a disintegrin and metalloprotease (ADAM) family metalloprotease, e.g., ADAM9, ADAM10, and TNF-α-converting enzyme (TACE, also known as ADAM17), which cleaves APP at the α-site between Lys16 and Leu17 in the Aβ domain [9]. BACE2, a homolog enzyme of BACE1, cleaves at two sites, namely, θ-sites, between Phe19 and Phe20, and between Phe20 and Ala21 in the Aβ domain [10]. Since the α-site and θ-sites are located in the center of the Aβ domain, their cleavage involves no Aβ production, which leads to the non-amyloidogenic pathway of APP. Based on the amyloid hypothesis, BACE1 and γ-secretase appear to be the molecular targets for developing anti-AD drugs. However, because γ-secretase can cleave other single-pass transmembrane proteins in vivo, such as Notch that plays a critical role in cell differentiation, its inhibition appeared to lead to serious side effects. The fact that BACE1 knockout transgenic mice could survive normally gave us a promising road map, in which BACE1 is a molecular target for developing AD drugs [11]. Many BACE1 inhibitors have been revealed until now. We have also reported a series of peptidic and non-peptidic BACE1 inhibitors [12–18].

1.2 Early Peptidic BACE1 Inhibitors

Because BACE1 is an aspartic protease, early BACE1 inhibitors with a substrate transition state analogue have been reported on the basis of inhibitor design approach as well as other aspartic proteases such as renin and HIV protease [12–18]. Many mutations in the *APP* gene that affect Aβ formation, $A\beta_{1-42}/A\beta_{1-40}$ ratio, or Aβ toxicity have been reported. The Swedish mutation (K670N, M671L double mutation; Fig. 1a) around the β-site induces β-cleavage by BACE1, increasing the $A\beta_{1-42}$ and $A\beta_{1-40}$ levels. Given the fact that Swedish-mutant APP is cleaved faster than wild-type APP, early BACE1 inhibitors were designed on the basis of the Swedish-mutant APP amino acid sequence. In 1999, Sinha et al. at Elan Pharmaceuticals succeeded in purifying BACE1 from the human brain using a transition state analogue based on the Swedish-mutant sequence and cloning the BACE1 enzyme [8]. The results indicated the possibility of substrate-based BACE1 inhibitors as AD drugs. In 2000 and 2001, Ghosh and Tang et al. reported on the potent inhibitors **1** (OM99-2, $Ki = 1.6$ nM) and **2** (OM00-3, $Ki = 0.3$ nM) with a hydroxyethylene unit (corresponding to a dipeptide unit at the P_1-P_1' positions) as a substrate transition state analogue (Fig. 1b) and the first X-ray crystal structure (PDB ID: 1FKN) of a complex between recombinant BACE1 and the inhibitor OM99-2 [19–22]. We have also reported a series of peptidic BACE1 inhibitors possessing a norstatine-type transition state analogue, phenyl norstatine [Pns: (2*R*,3*S*)-3-amino-2-hydroxy-4-phenylbutyric acid], at the P_1 position as shown in Fig. 1c [23–29]. These inhibitors have a Glu bioisostere at the P_4 position and a substituted anilide at the C-terminus. Of these inhibitors, **3** (KMI-429, $IC_{50} = 3.9$ nM) exhibited effective inhibition of BACE1 activity in cultured cells and significant reduction of Aβ production in vivo (*via* direct administration into the hippocampi of APP transgenic and wild-type mice) [23, 24]. Inhibitor **4** (KMI-684, $IC_{50} = 1.2$ nM), in which two carboxylic acid residues of **3** at the P_1' position were replaced with their bioisostere, a tetrazolyl ring, showed the most potent BACE1 inhibitory activity [25]. Inhibitor **5** (KMI-574, $IC_{50} = 5.6$ nM), possessing a 5-fluoroortyl group in the *N*-terminus residue, showed improved inhibition in cultured cells [26].

1.3 Prologue

Early BACE1 inhibitors were designed using the coordinates of the first reported X-ray crystal structure 1FKN. OM99-2 has an Asn residue at the P_2 position, which corresponds to the P_2 residue of Swedish-mutant sequence. Since OM99-2 in 1FKN appeared to interact with the BACE1-Arg235 side chain *via* a hydrogen bonding, many researchers have designed BACE1 inhibitors that possess a hydrogen bond receptor in their P_2 moieties, using the 1FKN coordinates. However, most of our previously reported peptidic BACE1 inhibitors have a hydrophobic amino acid residue, such as a Leu or cyclohexylalanine (Cha) residue at the P_2 position as shown in Fig. 1c. Interestingly, the replacement of the P_2 residue with an Asn residue corresponding to the Swedish-mutant APP sequence eliminated its BACE1 inhibitory activity. Thus, our design strategy seemed to require a fundamental review. In this study, we hypothesized that a quantum chemical interaction between the inhibitor's P_2 moiety and BACE1-Arg235 side chain played a critical role for BACE1 inhibition and designed small-sized potent BACE1 inhibitors using our hypothesis and computational approaches.

2 Materials

2.1 Docking Simulation and Steric Energy Calculation

The docking simulation was performed using software Molecular Operating Environment (MOE, Chemical Computing Group, Inc., Canada) under the MMFF94x force field. Before the docking simulation, the addition of hydrogen atoms and deletion of crystal water to the original coordinate set of X-ray crystal structure from PDB were performed. Optimized partial charges were assigned to all atoms in the MMFF94x force field, and the energy minimization against hydrogen atoms was performed after the fixing of heavy atoms. Although 1FKN was used as coordinates for the docking simulation at the early stage, 1W51, 2IQG, 1M4H, and 2B8L were latterly used depending on the inhibitor's size. Steric energies of virtual inhibitors with a respective dihedral angle around the bond of the P_3 amide and P_2 aromatic scaffold in the MM2 force field were calculated using the "dihedral energy plot" and the "dihedral contour plot" functions in MOE software or the "dihedral driver" in ChemBio3D ultra 14.0 (PerkinElmer, USA). After the energy minimization of the virtual inhibitors, the inhibitors with the same dihedral angle to the inhibitor bound in the BACE1 active site were screened and synthesized. These inhibitors were able to bind to the active site of BACE1 in unchanged configuration.

2.2 Synthesis of BACE1 Inhibitors

Peptidic inhibitors were synthesized by traditional 9-fluorenylmethoxycarbonyl (Fmoc)-based solid-phase or *tert*-butoxycarbonyl (Boc)-based solution-phase peptide synthetic methods. Non-peptidic BACE1 inhibitors were synthesized to connect in tandem the building blocks corresponding to the P_3-P_2 residues and the P_1-P_1' residues. The amide bonds were formed using 1-ethyl-3-(3-dimethylaminopropyl)carbodiimide·HCl (EDC·HCl) in the presence of 1-hydroxybenzotriazole (HOBt) as coupling agents, and the Boc deprotection was performed by 4 N HCl/dioxane. In the case of peptidic BACE1 inhibitors, amide bonds were formed using diisopropylcarbodiimide (DIPCDI) in the presence of HOBt as coupling agents, and Fmoc deprotection was performed by 20% piperidine/*N*,*N*-dimethylformamide (DMF). The final deprotection of peptidic inhibitors was performed using trifluoroacetic acid (TFA) in the presence of *m*-cresol, thioanisole, and water. All compounds were purified by preparative reverse-phase high-performance liquid chromatography (RP-HPLC) and were stored in a refrigerator after freeze drying. For instance, the synthetic scheme of non-peptidic inhibitor **23** (KMI-1303) and BACE1 inhibitor peptide **25** are shown in Scheme 1.

2.3 BACE1 Inhibition Assay

BACE1 inhibitory activities of the synthetic inhibitors were determined by enzymatic assay using a recombinant human BACE1 and fluorescence resonance energy transfer (FRET) substrate.After the enzymatic reaction with 7 nM BACE1 (R&D System, Inc., USA) and 25 μM FRET substrate, (7-methoxycoumarin-4-yl)-acetyl-Ser-Glu-Val-Asn-Leu-Asp-Ala-Glu-Phe-Arg-Lys(2,4-dinitrophenyl)-Arg-Arg-NH_2 (Peptide Institute, Inc., Japan), with 2 or 0.2 μM inhibitors in incubation buffer [2-morpholinoethanesulfonic acid (MES), 62.5 mM, pH 5.0] at 37 °C, the N-terminal cleavage fragment of the substrate was analyzed by RP-HPLC using a fluorescence detector.

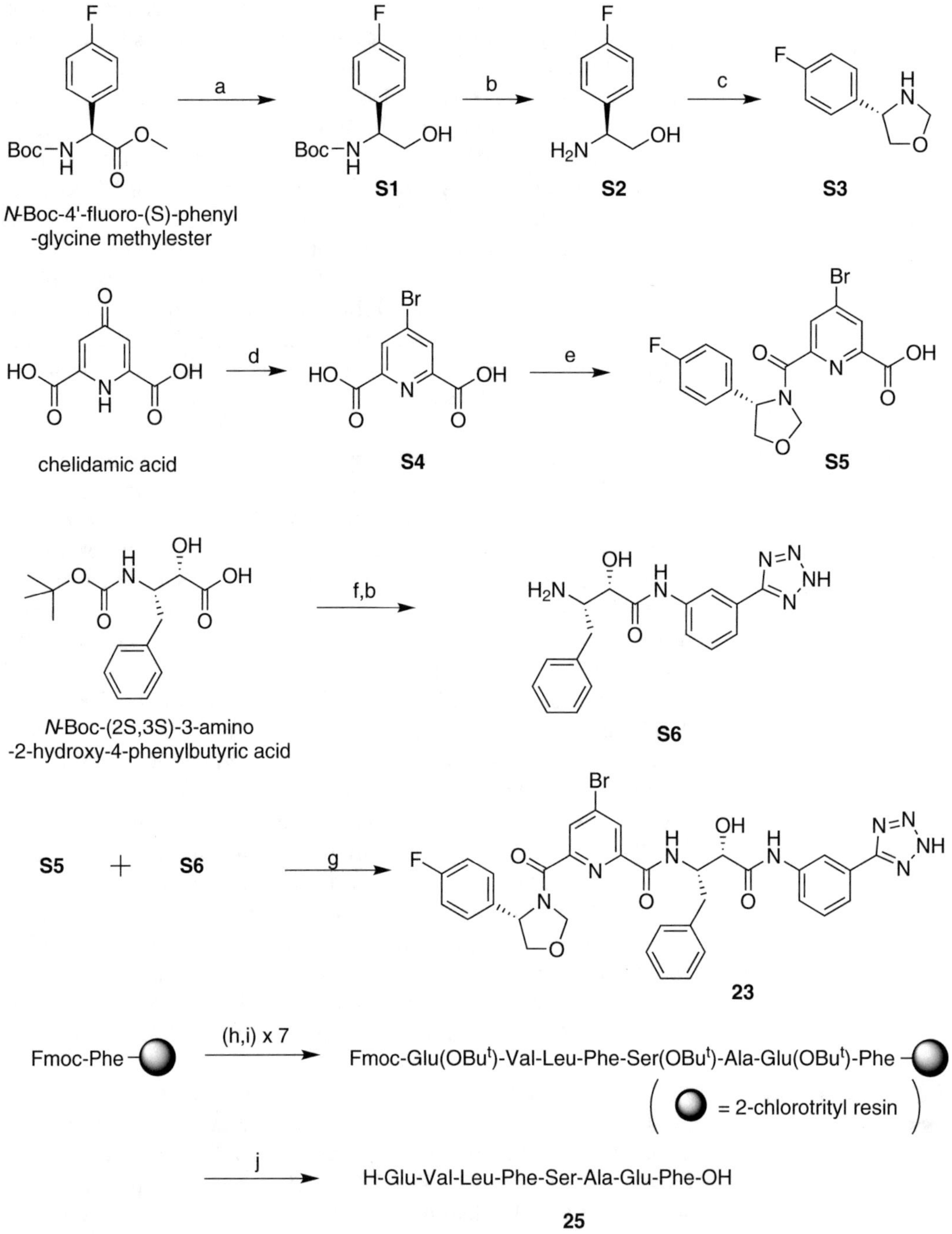

N-Boc-4'-fluoro-(S)-phenyl-glycine methylester →(a) **S1** →(b) **S2** →(c) **S3**

chelidamic acid →(d) **S4** →(e) **S5**

N-Boc-(2S,3S)-3-amino-2-hydroxy-4-phenylbutyric acid →(f,b) **S6**

S5 + **S6** →(g) **23**

Fmoc-Phe-● →(h,i) x 7 Fmoc-Glu(OBu^t)-Val-Leu-Phe-Ser(OBu^t)-Ala-Glu(OBu^t)-Phe-●

(● = 2-chlorotrityl resin)

→(j) H-Glu-Val-Leu-Phe-Ser-Ala-Glu-Phe-OH

25

Scheme 1 Reagents and condition: (**a**) $LiBH_4$/MeOH, rt; (**b**) 4 N HCl/dioxane, rt, 2 h; (**c**) HCHO, water, rt, 1 h; (**d**) tetra-*n*-butylammonium bromide, P_2O_5, THF, reflux; (**e**) **S3**, HOBt, EDC·HCl/DMF, 4 h, rt; (**f**) 5-(3-aminophenyl) tetrazole, HOBt, EDC·HCl/DMF, 1 day, rt; (**g**) HOBt, EDC·HCl/DMF, 1 day, rt; (**h**) 20% piperidine/DMF, 20 min, rt; (**i**) Fmoc-amino acid, DIPCDI, HOBt, DMF, 2 h, rt; and (**j**) TFA, *m*-cresol, thioanisole, H_2O, rt

3 Methods

3.1 Interactions Between Inhibitors and BACE1-Arg235 Side Chain

OM99-2 is the first reported X-ray crystal structure 1FKN reported by Ghosh et al. appeared to interact with BACE1-Arg235 side chain by hydrogen bonding. Whereas many early BACE1 inhibitors, containing OM99-2, have a hydrogen bond receptor, such as Asn side chain, at the P_2 position, our inhibitors have a hydrophobic Leu or Cha residue at the same position, as stated previously. We predicted the importance of the interaction between P_2 hydrophobic amino acid residues and BACE1-Arg235 side chain. The Arg235 side chain is only found outside the opening of the active site that is formed by the flap domain of BACE1. Considering that other amino acid side chains of BACE1, which can interact with inhibitors, are found inside the active site between the flap domain and cleft region of BACE1, the interaction with the S_2 pocket that consists of the Arg235 side chain might play a key role in the BACE1 inhibitory mechanism. To understand the role of this interaction, we compared the publicly available X-ray crystal structures of BACE1-inhibitor complexes. Surprisingly, the guanidine group of BACE1-Arg235 in most of the crystal structures, except 1FKN, showed similar figures flopping over the P_2 region of the inhibitors, and the nearest distances between the guanidino-plane of Arg235 and the inhibitor's P_2 region show similar values of about 3 Å, as shown in Table 1 [30]. The P_2 functional group in most of the crystal structures found to interact with the BACE1-Arg235 side chain was the methyl group that was bound to the Arg235 side chain by a CH-π interaction. Subsequently, an O-π interaction with an amide oxygen atom, S-π interaction with a methionine sulfur atom, CH-π stacking with a pyridine proton, and π-π stacking with an aromatic ring were observed. These facts suggest that π-orbital on the guanidino-plane interacts with the P_2 region by a weak quantum force such as stacking or σ-π interaction. The only exception is the interaction in the crystal structure 1FKM. The P_2 moiety of OM99-2 in the crystal structure 1FKN seemed to interact with BACE1-Arg235 side chain *via* a hydrogen bonding (Fig. 2a). However, there is a different structure from the former in the same X-ray crystal data 1FKN (Fig. 2b). BACE1-Arg235 side chain in another crystal structure interacted with the π-orbital on the amide plane of inhibitor's P_2-Asn side chain *via* a σ-π interaction. OM00-3 reported by the same researchers [22] is an inhibitor that is structurally similar to OM99-2, and surprisingly the P_2-Asp side chain of OM00-3 interacted with the π-orbital on the guanidino-plane of BACE1-Arg235 side chain *via* O-π interaction (Fig. 2c; PDB ID: 1M4H). Although many early BACE1 inhibitors possessing a hydrogen bond receptor at the P_2 position have been designed on the basis of the first reported crystal structure 1FKN of the complex between OM99-2 and BACE1, the

Table 1
The distance of BACE1-Arg235 from the P_2 parts of inhibitors

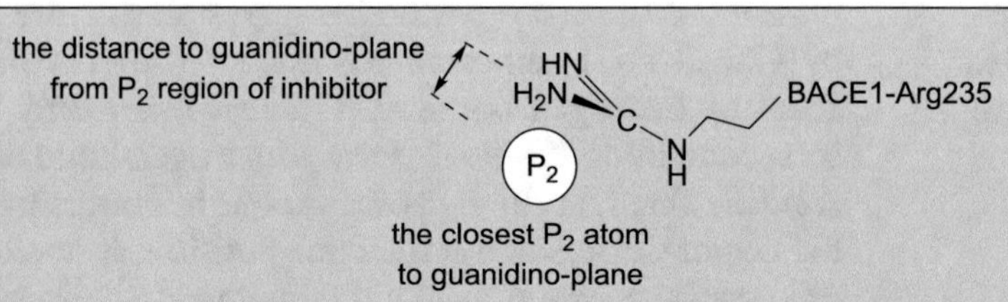

the distance to guanidino-plane from P_2 region of inhibitor
HN
H_2N
C
N
H
BACE1-Arg235
P_2
the closest P_2 atom to guanidino-plane

PDB ID	The distance (Å) from P_2 part	The closest P_2 atom to guanidino-plane
2P83	2.7	-H (*N*-methyl)
2B8L	2.8	-H (*N*-methyl)
2P8H	3.0	-H (*N*-methyl)
2PH6	2.8	-H (*N*-methyl)
20AH	2.8	-H (*N*-methyl)
2QZL	2.8	-H (*N*-methyl)
2P4J	2.7	-H (*N*-methyl)l
2IRZ	2.9	-H (*N*-methyl)
2IS0	2.8	-H (*N*-methyl)
2QK5	3.0	-H (*N*-methyl)
2B8V	2.9	-H (*N*-methyl)
1M4H	2.9	-COOH
2HM1	2.4	-H (pyridine ring)
1TQF	2.7	=O ($C_6H_5CH_2SO_2$-)
1YM2	3.5	-S- (methionine)
2G94	3.3	=O ($CH_3SO_2CH_2$-)
2HIZ	2.5	-H (methylene)
1XS7	2.6	=O (amide)
2IQG	3.0	-H (*N*-methyl)
2QK5	3.1	-H (*N*-methyl)
2QP8	2.8	-H (*N*-methyl)
3CIB	3.0	-H (*N*-methyl)

(continued)

Table 1 (continued)

PDB ID	The distance (Å) from P_2 part	The closest P_2 atom to guanidino-plane
3CIC	2.9	-H (*N*-methyl)
3CID	2.9	-H (*N*-methyl)
2QMD	2.9	-H (*N*-methyl)l
2QMF	2.8	-H (*N*-methyl)
2QMG	2.7	-H (*N*-methyl)
1W51	3.2	-H (isophthalic ring)
2FDP	3.1	-H (isophthalic ring)
2VIE	3.0	-H (pyrrlidone)
2VJ9	3.7	-H (butanesultan)
2VIZ	2.6	-H (butanesultan)
2VNM	3.3	-H (butanesultan)
2VIJ	3.6	-H (butanesultan)
2VNN	2.7	-H (*N*-methyl)
1PH8	3.3	-H (*N*-methyl)

hydrogen-bonding interaction between these inhibitors and the BACE1-Arg235 side chain was not shown in their crystal structures. For instance, the X-ray crystal structure of the complex between Merck's (MSD) inhibitor and BACE1 was shown in Fig. 2d. The researchers at MSD probably appear to design their inhibitor possessing an *N*-methyl-*N*-methanesulfonyl group on the P_2 isophthalic scaffold in anticipation of the hydrogen-bonding interaction between the sulfonyl oxygen atom and a BACE1-Arg235 side chain. However, the *N*-methyl group of MSD's inhibitor interacted with the π-orbital on the guanidino-plane of the BACE1-Arg235 side chain at a distance of 2.8 Å, meaning that it appears to interact with the BACE1-Arg235 side chain *via* a CH-π interaction. The inhibitors reported by Elan (Fig. 2e, PDB ID: 2IQG) and Pfizer (Fig. 2f, PDB ID: 2P83) appear to interact with BACE1-Arg235 side chain *via* CH-π and O-π interactions, respectively. As seen above, most of the BACE1 inhibitors, except OM99-2 in crystal structure 1FKN, interact with BACE1-Atg235 side chain by a weak quantum force such as stacking or σ-π interaction. The Arg235 side chain of the BACE1-OM99-2 complex

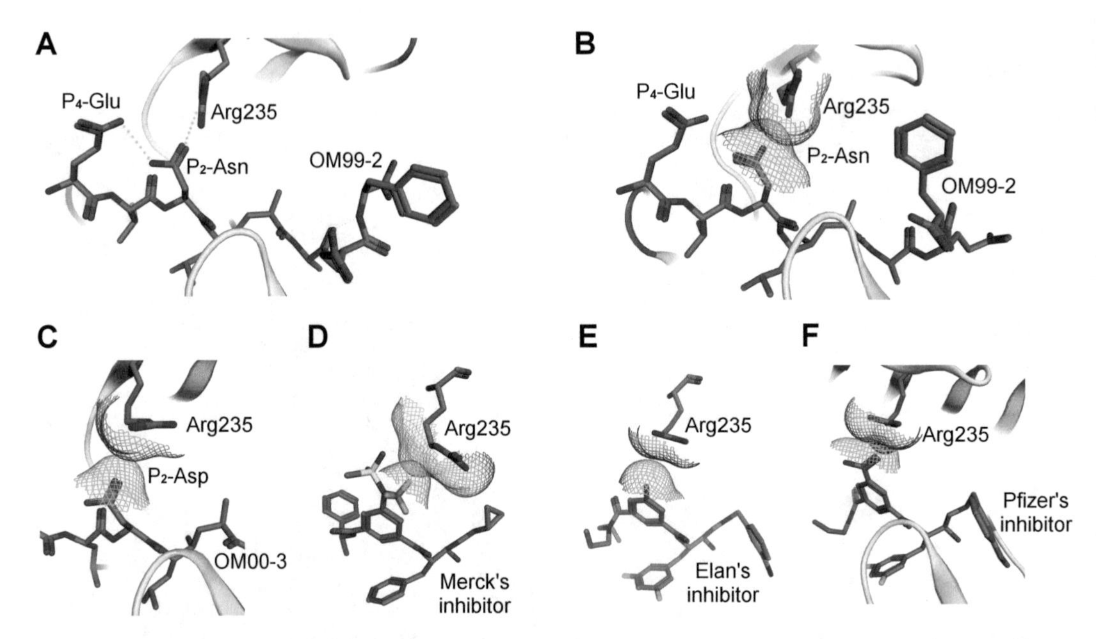

Fig. 2 Interaction of BACE1-Arg235 with BACE1 inhibitors in X-ray crystal structures (**a**) PDB ID: 1FKN. (**b**) Another binding mode in the crystal structure, 1FKN. (**c**) PDB ID: 1M4H. (**d**) PDB ID: 2B8L. (**e**) PDB ID: 2IQG (**f**) PDB ID: 2P83

(1FKN) assumed an exceptionally different pose to the other crystal structures because the BACE1 complex seems to stabilize by intramolecular hydrogen-bonding interaction between the P_4-Glu and P_2-Asn side chains of OM99-2. By a carefully observing of the crystal structure (Fig. 2a), the P_2-Asn's carbonyl oxygen atom is non-coplanar to the guanidino-plane of BACE1-Arg235, indicating that the contribution of a hydrogen bonding to the stabilization of the inhibitor-BACE1 complex seems to be a little. OM00-3 that interacts with BACE1-Arg235 side chain by a quantum chemical interaction has not such an intramolecular hydrogen bonding. Because many researchers have designed BACE1 inhibitors with a hydrogen bond receptor on the basis of the first reported crystal structure 1FKN, their results seem to require the review as a drug discovery approach. Furthermore, we found a serious issue for the drug discovery of BACE1 inhibitors using an in silico approach [30]. The guanidino-planes of BACE1-Arg235 in the crystal structures of most BACE1 complexes showed similar distances from the P_2 regions of the inhibitors regardless of their variety and size (Table 1). This indicates that the side chain of BACE1-Arg235 can move in concert with the inhibitor's size (Fig. 3a). The superimposed figure of three crystal structures (PDB ID: 2B8L, 2IQG, and 2P83) of the complex with MSD's, Elan's, and Pfizer's inhibitors, respectively, was depicted in Fig. 3b. The side chain of BACE1-Arg235 seems to move in concert with inhibitor's size in these crystal structures. Moreover, we found that

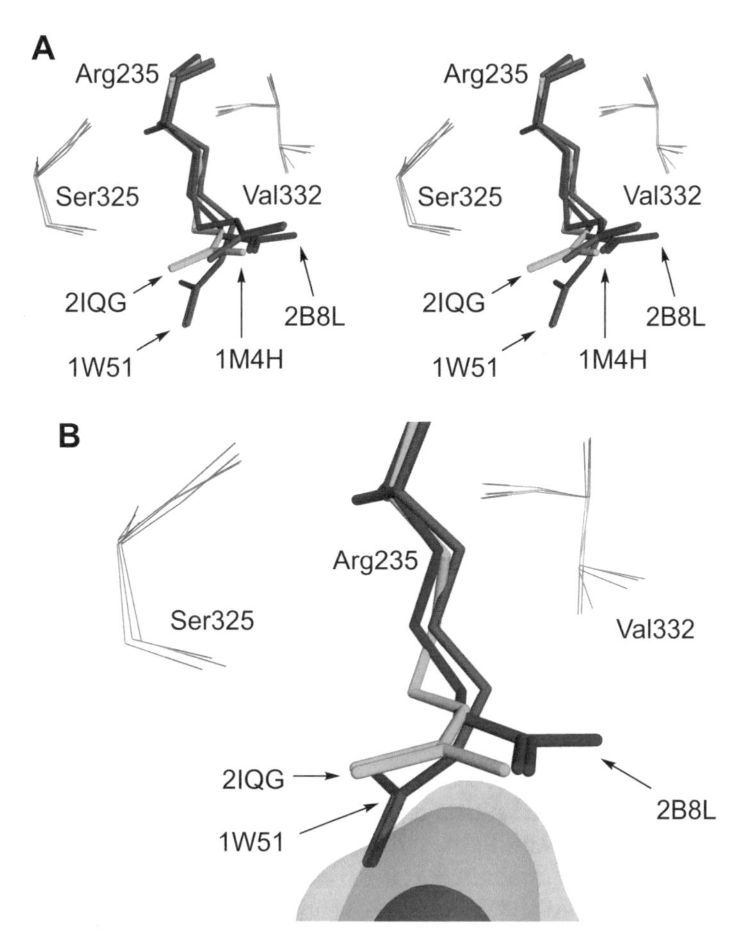

Fig. 3 Location of BACE1-Arg235s in the respective crystal structures (**a**) Stereo view of Arg235 side chains from the four crystal structures. (**b**) Interaction of Arg235s with the respective inhibitors. The *blue*, *green*, and *red* transparent surface models represent the inhibitors in the crystal structures of BACE1-inhibitor complexes, 1 W51, 2IQG, and 2B8L, respectively

the BACE1-Arg235 side chain has a restricted range of motion. The BACE1-Arg235 side chain slides sideways, not up and down, along the wall of the β-sheet structure that consists of four peptide strands behind the flap domain of BACE1. Probably, the location of BACE1-Arg235 side chain might be predicted by the inhibitor's size, because it follows the same pattern. We hypothesized an important key for considering the role of interactions with a BACE1-Arg235 side chain. The guanidino-plane of Arg235 that can move in concert with the inhibitor's size pushes down on the P_2 region of the inhibitors, causing them to be affixed to the active site of BACE1. The inhibitors that can interact with BACE1-Arg235 side chain *via* a quantum chemical interaction appear to indicate the potent BACE1 inhibitory activities because of this "flop-over" mechanism by BACE1-Arg235 side chain. Although a quantum chemical force such as σ-π interaction has a weaker binding energy than a hydrogen-bonding interaction, this "flop-over" mechanism allows the strong binding mode with the active site of BACE1.

3.2 Significance of Quantum Chemical Interactions for Molecular Recognition and "Electron-Donor Bioisostere" in Drug Design

In silico approaches, such as a docking simulation between a target biomolecule and drugs, have greatly contributed to drug discovery research. However, most docking simulation softwares adopt molecular mechanics/molecular dynamics (MM/MD) calculations based on classical Newtonian mechanics, and docking simulations using these calculations do not appear to estimate a weak quantum chemical interaction, such as stacking or σ-π interaction between drugs and the Arg235 side chain of BACE1. The quantum chemical interactions that also involve other aromatic amino acids, such as Phe, Tyr, and Trp side chains, seem to be approximately optimized using several descriptors based on classical mechanics in the docking simulation software that is based on MM/MD calculations. However, this software recognizes Arg as one of the charged amino acids, and the quantum chemical interactions involving an Arg side chain are unlikely to yield a reasonable output. It is well known that quantum chemical interactions involving a π-orbital of a guanidine group are found commonly in proteins and play an important role in molecular recognition by proteins. Crowley et al. surveyed cation-π interactions in protein interfaces using the Protein Data Bank and the Protein Quaternary Structure server [31]. They evaluated the cation-π interactions using a variant of the optimized potentials for liquid simulations (OPLS) force field and found that approximately half of the protein-protein complexes and one-third of the homodimers contained at least one intermolecular cation-π pair. This finding indicates the significance of these interactions in molecular recognition because the occurrence rate of cation-π pairs in protein-protein interfaces is higher than that in homodimer interfaces, which are similar to the protein interior. Among them, the interactions between an Arg and a Tyr were found to be the most abundant. Moreover, 53% of them involved planar π-π stacking by the quantum chemical interaction between the guanidine group of an Arg and the aromatic ring of a Tyr. Persson et al. reported the "arginine switch mechanism" for the molecular recognition of the CD46 receptor by an adenovirus [32]. A group of highly pathogenic species B adenoviruses can infect a host using CD46 as a cellular receptor. The knob domain of the fiber protein on the capsid of an adenovirus mediates interactions with CD46. The knob domain of adenovirus type 11 (Ad11 knob) has two Arg residues (Arg279 and Arg280) at the recognition site for CD46. Since the Agr280 side chain interacts with the Arg279 side chain by a π-π stacking interaction (turn on), the Arg280 side chain can stably bind to the Phe35 side chain of CD46 *via* a π-π stacking. Thus, the quantum chemical interactions play an important role in the molecular recognition by the Ad11 knob of the CD46 receptor. The Arg279 of Ad11 is mutated into a Gln residue in the knobs of adenovirus type 7 (Ad7) and adenovirus type 14 (Ad14); the affinities of the Ad7/Ad14 knobs against CD46 are known to be low. The quantum chemical interactions of Arg side chains and aromatic rings of biomolecules are also shown in the thrombin-RNA aptamer complex [33]. The π-π stacking interaction of some Arg side chains with nucleobase aromatic rings plays an important role in the docking between thrombin and RNA. Considering the evolution of life, "from ribozyme to enzyme," such quantum chemical interaction of Arg235 side chain might be one of the oldest interactions for a biomolecular recognition. As observed above, the quantum chemical interactions appear to be significant in medicinal science; however, their treatment using the docking simulation software based on an MM/MD approach requires particular attention.

The researchers at Bristol-Myers Squibb (BMS) Research reported a series of BACE1 inhibitors that can interact with BACE1-Arg235 side chain by a π-π stacking as shown in Fig. 4a. According to their structure-activity relationship (SAR) study as shown in Fig. 4b, the introduction of an electron-donating methoxy group on the *p* position of phenyl ring that interacts with BACE1-Arg235 side chain enhanced BACE1 inhibitory activity. This indicated that an inhibitor possessing a P_2 aromatic ring with higher electron density could strongly dock to the active site of BACE1 that has an electron-poor π-orbital on the guanidino-plane of the BACE1-Arg235 side chain. An important concept "bioisostere" for drug design is known. Bioisosteres are functional or atomic groups that display a similar physiochemical property. A functional or atomic group of compounds often replaced with its bioisostere, which exhibits similar biological properties to the parent drugs, have been designed to develop more practical drugs. However, in the case of BACE1 inhibitor design, the bioisostere of the P_2 moiety according to the Swedish-mutant APP is an Asn or an amide residue based on a classical bioisostere concept that does not assume quantum chemical interactions, and inhibitors that can interact with the Arg235 side chain on the basis of a quantum chemical interaction could never be designed using such a classical concept. Hence, we proposed the new "electron-donor bioisostere," concept, which can interact with an electron-poor π-orbital, such as the guanidine group of Arg235 by quantum chemical interactions.

3.3 Design of Non-peptidic BACE1 Inhibitor Based on "Electron-Donor Bioisostere"

We have designed and synthesized non-peptidic BACE1 inhibitors from our peptidic BACE1 inhibitors **3–5** as lead compounds [34–37]. The researchers at MSD, Elan, and Pfizer had reported many BACE1 inhibitors possessing an isophthalic scaffold at the P_2 position as shown in Fig. 2d–f [15, 17]. Because the distance between the flap domain and the cleft domain that form the S_2

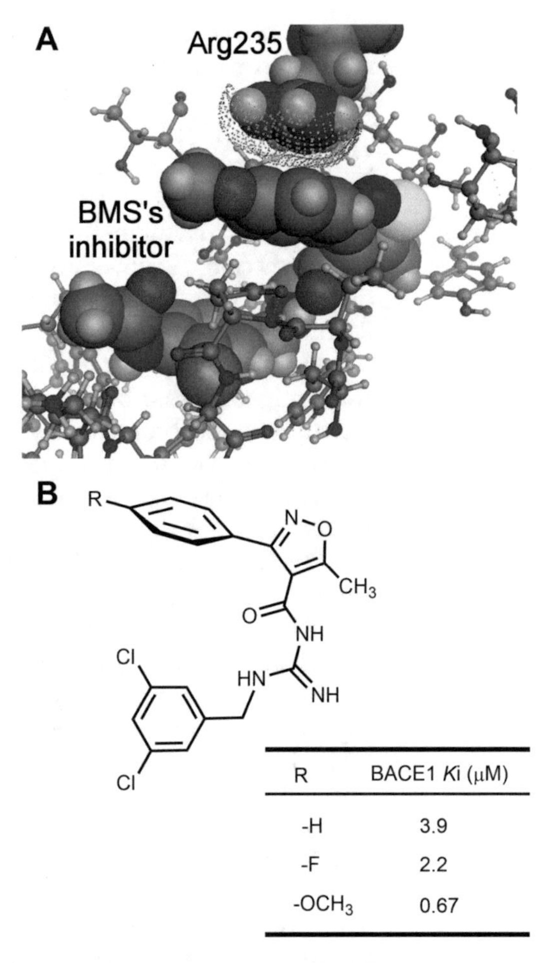

R	BACE1 *Ki* (μM)
-H	3.9
-F	2.2
$-OCH_3$	0.67

Fig. 4 BMS's BACE1 inhibitors. (**a**) X-ray crystal structure of complex between BACE1 and BMS's inhibitor (PDB ID: 4FSL). (**b**) SAR study by the researchers at BMS

pocket of BACE1 is narrow, a planar aromatic ring such as an isophthalic scaffold can closely dock in the S_2 pocket of BACE1. Hence, we designed a series of BACE1 inhibitors from a virtual inhibitor, in which the P_2 moiety of our peptidic inhibitors was replaced with an isophthalic scaffold. First, we focused the steric-hindered interaction between the P_3 amide and a proton on the P_2 isophthalic ring of the virtual inhibitor, which seems to restrict its configuration. Using an approach, "in silico conformational structure-based design," based on a conformer of the docked inhibitor in BACE1 [34, 38], we adopted a pyridinedicarboxylic scaffold as a P_2 moiety, which lacked the 2-proton from the isophthalic ring. We calculated the steric energies in the respective conformers around the bond of the P_3 amide and P_2 aromatic ring of the virtual inhibitors with P_2 isophthalic and P_2 pyridinedicarboxylic scaffolds, respectively (Fig. 5a). Whereas the stable conformer of the virtual

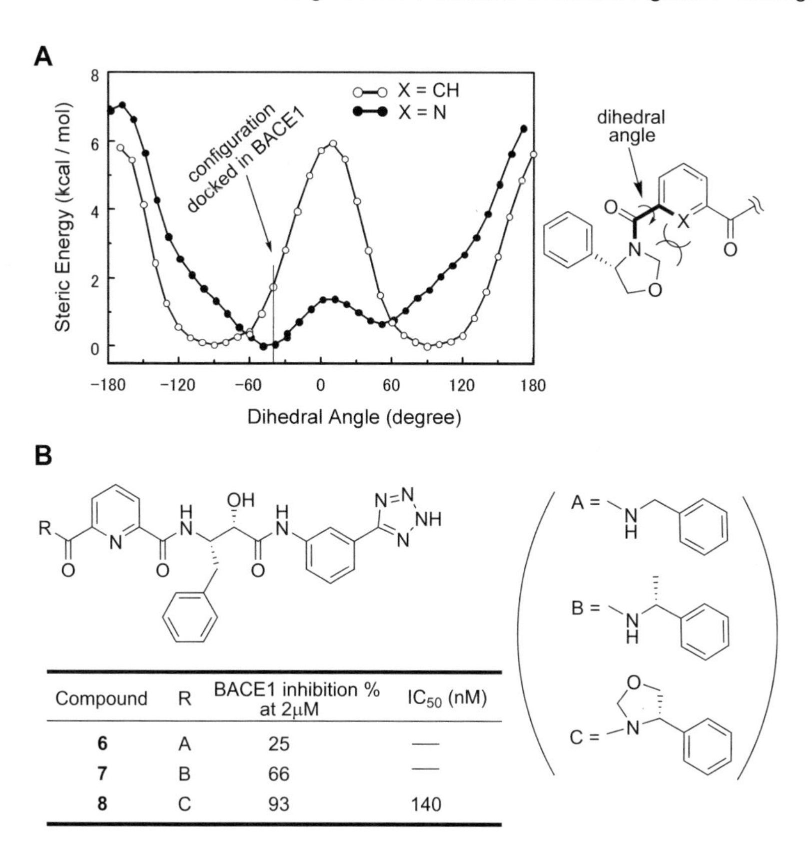

Compound	R	BACE1 inhibition % at 2μM	IC_{50} (nM)
6	A	25	—
7	B	66	—
8	C	93	140

Fig. 5 Optimization of isophthalic-type BACE1 inhibitors on the basis of a conformational structure docked in BACE. (**a**) The design of BACE1 inhibitors with a pyridinedicarboxylic scaffold. (**b**) Fixing of a folding pose by cyclization of the inhibitor's P_3 moiety

inhibitor with a P_2 pyridinedicarboxylic scaffold showed the same dihedral angle to the conformer docked in BACE1, the conformer of the P_2 isophthalic virtual inhibitor with the same dihedral angle to the conformer docked in BACE1 showed a high steric energy. Next, we optimized the P_3 region of inhibitors with a P_2 pyridine-dicarboxylic scaffold (Fig. 5b). Inhibitor 7 with an (R)-α-methyl-benzylamine at the P_3 position, which was reported by the researchers at MSD, exhibited a stronger inhibitory activity than inhibitor **6** with a benzylamide group at the same position. There is the S_3 sub-pocket behind the active site of BACE1, and the P_3 phenyl group of **6** and **7** appears to interact with the S_3 sub-pocket. We envisioned that inhibitors with a P_3 benzylamide group assumed a folding pose between the P_2 aromatic ring and the P_3 benzylamide, and the α-methyl group on the P_3 benzylamide stabilized this folding structure. Hence, we designed inhibitor **8** introducing a five-membered ring, oxazolidine, at the P_3 region to fix the folding structure. The oxazolidine ring fixes the direction of the phenyl ring at the P_3 position, so the P_3 phenyl ring might be able to bind closely to the S_3 sub-pocket of BACE1. Thus, we could design potent BACE1 inhibitor **8** using a computational approach based on the conformer docked in BACE1. However, **8** still showed a lower inhibitory activity (IC_{50} = 140 nM) than its lead compounds **3–5** (IC_{50} = 1.2–5.6 nM), and there is room for further optimization of these inhibitors.

We hypothesized that quantum chemical interaction of an inhibitor with the side chain of BACE1-Arg235 plays a critical role in the inhibition mechanism, as previously mentioned. Therefore, we focused on the optimization around the P_2 region. The SAR study around the inhibitor's P_2 region was shown in Table 2. Inhibitors **9**, **10**, **13**, **15**, and **18** with a hydrophobic and small-sized functional groups, such as methanesulfonyl, methoxy, methyl mercaptan, azide, and methyl groups, on the P_2 aromatic ring, tended to show higher inhibitory activity than inhibitors with a bulky or hydrophilic group. On the basis of the "electron-donor bioisostere" concept, we speculated that an electron-rich halogen atom could interact with the electron-poor guanidine π-orbital by Coulomb's force. Using the ab initio molecular orbital approach, Imai et al. indicated that the calculated Cl-π interaction energy is slightly stronger than that of CH-π interaction, and its energy is affected by π-electron density [39]. Inhibitors **19–21** possessing a halogen atom on the P_2 aromatic ring exhibited more potent inhibitory activities (IC_{50}: 22, 15, and 24, respectively), and additionally inhibitors **22–24** possessing a fluorine atom on the *p* position of P_3 phenyl group exhibited the most potent inhibitory activities (IC_{50}: 13, 9, and 10, respectively). Among them, inhibitor **23** (KMI-1303) exhibited the most potent inhibitory activity and is available from Wako Pure Chemical Industries (Japan) as a reagent for biological research. The fact that the introduction of a halogen atom into the P_2 position of our compounds drastically improved BACE1 inhibitory activity appears to support our hypothesis, namely, that the quantum chemical interactions between BACE1 and its inhibitors play a critical role in the mechanism of BACE1 inhibition, and the P_2 part of a potent BACE1 inhibitor can bind to the Arg235 side chain of BACE1, not by a hydrogen-bonding interaction but by a quantum chemical interaction.

3.4 Design of the First BACE1 Inhibitor Peptides: Can an Infinitely Small k_{cat} Value Turn the Substrate of an Enzyme into Its Inhibitor?

Although most BACE1 inhibitors interacted with the Arg235 side chain by a quantum chemical interaction, it is known that the Swedish-mutant APP possessing an hydrogen bond acceptor, such as an Asn residue, at the P_2 position, elicits higher catalytic efficiency by BACE1 than the wild-type APP. Whereas the k_{cat} value of Swedish-mutant APP is higher than that of wild-type APP, the K_m values of both types of substrate are similar, suggesting that the affinities for both substrates against BACE1 seem to be similar [40]. We hypothesized that the hydrogen-bonding interaction between the BACE1-Arg235 side chain and the P_2-Asn side chain

Table 2
BACE1inhibitors with a pyridinedicarboxylic scaffold

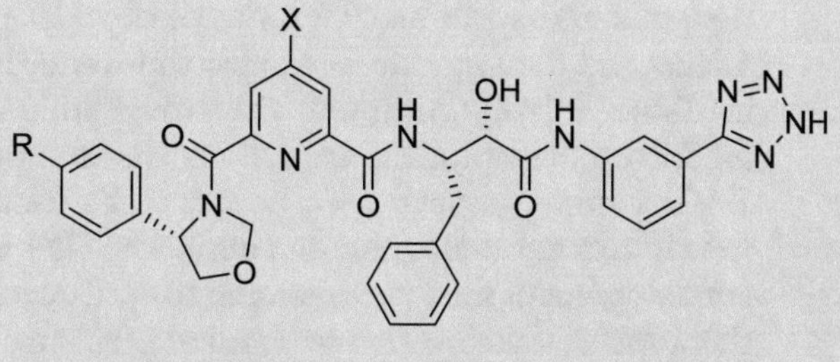

Compound	R	X	BACE1 inhibition % at 2 μM	at 0.2 μM	IC_{50} (nM)
8	H	-H	89		140
9	H	$-SO_2CH_3$	96	73	96
10	H	$-OCH_3$	91	53	151
11	H	-OEt	79		
12	H	-OPr	64		
13	H	$-SCH_3$	95	66	89
14	H	$-SCH_2(CH_3)$	72		
15	H	$-N_3$	95	68	79
16	H	$-NH_2$	78	53	
17	H	$-N(CH_3)SO_2CH_3$	75		
18	H	$-CH_3$	95	68	
19	H	-Cl	98	87	22
20	H	-Br	99	88	15
21	H	-I	99	86	24
22	F	-Cl	99	91	13
23 (KMI-1303)	F	-Br	99	93	9
24	F	-I	99	92	10

of the Swedish-mutant-type substrates activates the "turnover" required for enzymatic catalysis, thereby largely improving the k_{cat} value. However, such interaction appears to be unfavorable for BACE1 inhibition. In fact, most BACE1 inhibitors interacted with the Arg235 side chain by a quantum chemical interaction, not a hydrogen bonding. Our potent peptidic BACE1 inhibitors such as **3–5** also have a hydrophobic amino acid, such as Leu or Cha, at the P_2 position, and their side chains appear to behave as an electron-donor bioisostere. Therefore, we proposed a novel hypothesis—if the P_2 moiety of a substrate for BACE1 is replaced with an electron-donor bioisostere that can reduce the "turnover" of BACE1, it would turn into a BACE1 inhibitor. Hence, we designed and synthesized a series of peptides possessing a natural amino acid, Leu residue, as an electron-donor bioisostere as shown in Table 3 [41]. Although the octapeptide **25** exhibited weak inhibitory activities, **25** crucially is the first peptide that showed BACE1 inhibitory activity. The fact that a usual peptide without a transition state analogue could inhibit BACE1 enzyme activity is an encouragement for future research toward developing a gene-based therapy for Alzheimer's disease using DNA that codes for a peptide sequence with BACE1 inhibitory activity. As octapeptide **26** possessing a D-amino acid residue at the P_1' position exhibited slightly higher inhibitory activity than **25**, pentapeptides **27** and **28** possessing a D-amino acid were synthesized, and they replicated the inhibitory activity of **26**. Previously, we reported that replacing the Glu residue at the P_4 position of our inhibitors with its bioisostere, an N^{β}-(5-fluoroorotyl)-L-2,3-diaminopropionic acid (5FO) residue, improved their BACE1 inhibitory activity [26]. Hence, peptides **29–34** in which the P_4 amino acid residue was replaced with a 5FO residue were designed. Peptides **29**, **30**, and **32–34** exhibited moderate BACE1 inhibitory activity. In contrast, peptide **31** possessing a P_2-Asn residue corresponding to the Swedish-mutant APP sequence exhibited no inhibitory activity, as predicted. Since tetrapeptide **34** also showed a moderate BACE1 inhibitory activity, we designed small-sized tripeptides **35** and **36** possessing an α,α-dimethylphenethylamine at the C-terminus using an in silico conformational structure-based drug design approach. The *gem*-dimethyl group on the C-terminal moiety of tripeptides **35** and **36** was introduced to fix the folding structure of the P_1-Phe side chain of peptide **34** (Fig. 6). Among them, tripeptide **34** ($IC_{50} = 50$ nM) exhibited the highest BACE1 inhibitory activity. Moreover, we designed the smallest BACE1 inhibitor peptide **37** ($IC_{50} = 168$ nM) with a hydrophobic cyclic amine at the C-terminus.

4 Conclusion

We found that a specific interaction, a quantum chemical interaction between Arg235 side chain and inhibitor's P_2 region, played a critical role for BACE1 inhibition mechanism. Whereas most BACE1 inhibitors, except OM99-2, interact with BACE1-Arg235 by a quantum chemical interaction, such as stacking and σ-π interaction, many researchers had used the first reported X-ray crystal structure 1FKN for early BACE1 inhibitor design. As the crystal structure 1FKN has a hydrogen bonding between the BACE1-Arg235 side chain and OM99-2, the early studies on

Table 3
BACE1 inhibitor peptides with no transition state analogue

Compound	P_4 P_3 P_2 P_1 P_1' P_2' P_3' P_4'	BACE1 inhibition % at 2 μM	IC_{50} (nM)
25		6	
26		28	
27		23	
28		29	
29		85	263
30		94	138
31		< 5	
32		89	173
33		95	86
34		90	184
35	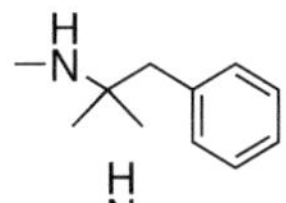	90	
36	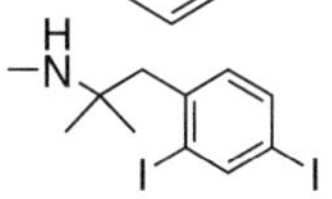	96	50
37	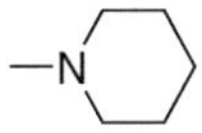	91	168

: D-serine

: D-asparagine

: cyclohexylalanine

: N^{β}-(5-fluoroorotyl)-L-2,3-diaminopropionic acid

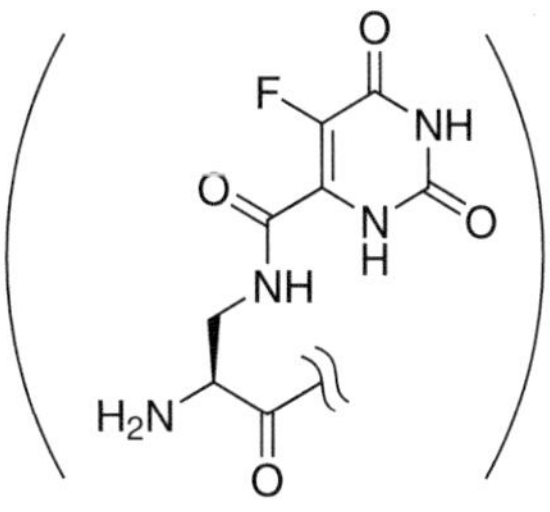

BACE1 inhibitor design might be misled, and a docking simulation using 1FKN seems to be meaningless. In fact, there is no hydrogen-bonding interaction in most X-ray crystal structures, except 1FKN. We focused on a quantum chemical interaction and designed potent non-peptidic BACE1 inhibitor, KMI-1303, using the "electron-donor bioisostere concept" that was proposed by us.

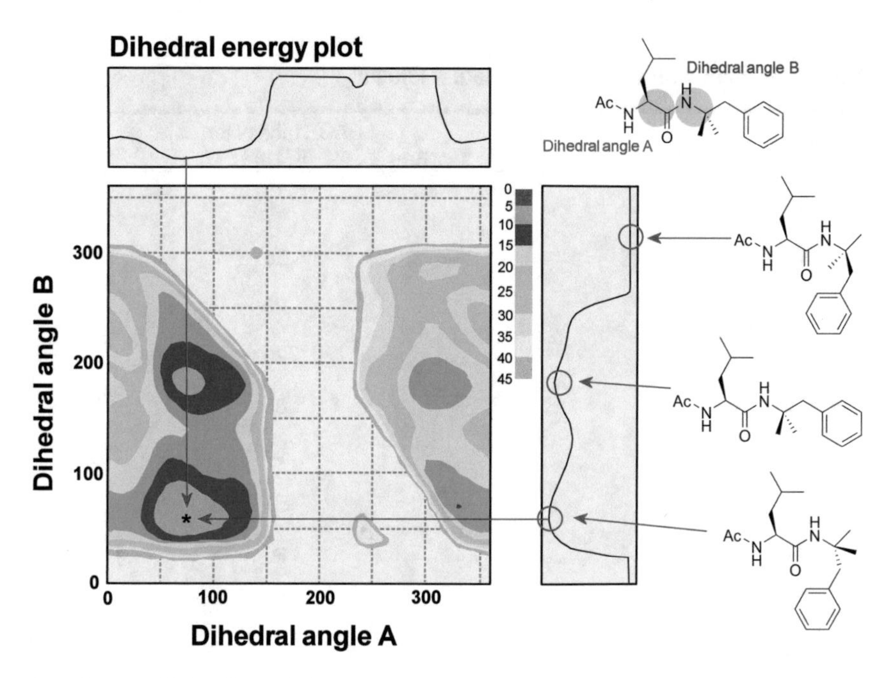

Fig. 6 Design of tripeptidic BACE1 inhibitors from a virtual inhibitor using in silico conformational structure-based design. The *black* star indicates configuration of inhibitor docked in BACE1

Moreover, we found the first BACE1 inhibitor peptides possessing P_2-Leu residue as an electron-donor bioisostere. Because these peptides have no substrate transition state analogue, we anticipate its application in a gene-based therapy for Alzheimer's disease using DNA that codes for peptide sequences with BACE1 inhibitory activity. Although the inhibitory activities of our peptides are low at this time, gene-based therapy is a worthwhile avenue for future investigation.

Acknowledgment

The study was supported in part by the Grants-in-Aid for Scientific Research from MEXT (Ministry of Education, Culture, Sports, Science and Technology), Japan (KAKENHI No. 23590137 and No. 26460163), and the donation from Professor Emeritus Tetsuro Fujita of Kyoto University. At the time writing, we received word that Prof. Fujita had passed away on January 1, 2017. Prof. Fujita was the teacher of one of the authors, Y. Hamada, and was known as the inventor of a treatment agent for multiple sclerosis. We dedicate this article to Prof. Fujita.

References

1. Selkoe DJ (2000) Toward a comprehensive theory for Alzheimer's disease. Hypothesis: Alzheimer's disease is caused by the cerebral accumulation and cytotoxicity of amyloid β-protein. Ann N Y Acad Sci 924:17–25
2. Selkoe DJ (1989) The deposition of amyloid proteins in the aging mammalian brain: implications for Alzheimer's disease. Ann Med 21:73–76
3. Selkoe DJ (1999) Translating cell biology into therapeutic advances in Alzheimer's disease. Nature 399:A23–A31
4. Sinha S, Lieberburg I (1999) Cellular mechanisms of β-amyloid production and secretion. Proc Natl Acad Sci U S A 96:11049–11053
5. Vassar R, Bennett BD, Babu-Khan S et al (1999) β-Secretase cleavage of Alzheimer's amyloid precursor protein by the transmembrane aspartic protease BACE. Science 286:735–741
6. Hussain I, Powell D, Howlett DR et al (1999) Identification of a novel aspartic protease (asp 2) as β-Secretase. Neuroscience 14:419–427
7. Yan R, Bienkowski MJ, Shuck ME et al (1999) Membrane-anchored aspartyl protease with Alzheimer's disease β-secretase activity. Nature 402:533–537
8. Sinha S, Anderson JP, Barbour R et al (1999) Purification and cloning of amyloid precursor protein β-secretase from human brain. Nature 402:537–540
9. Turner PR, O'Connor K, Tate WP, Abraham WC (2003) Roles of amyloid precursor protein and its fragment in regulating neural activity, plasticity and memory. Prog Neurobiol 70:1–32
10. Fluhrer R, Capell A, Westmeyer G et al (2002) A non-amyloidogenic function of BACE-2 in the secretory pathway. J Neurochem 81:1011–1020
11. Roberds SL, Anderson J, Basi G et al (2001) BACE knockout mice are healthy despite lacking the primary β-secretase activity in brain: implications for Alzheimer's disease therapeutics. Hum Mol Genet 10:1317–1324
12. Hamada Y, Kiso Y (2016) New directions for protease inhibitors directed drug discovery. Biopolymers 106:563–579
13. Hamada Y, Kiso Y (2015) Aspartic protease inhibitors as drug candidates for treating various difficult-to-treat diseases. Amino Acids, Peptides and Proteins 39. Royal Society of Chemistry, London:114–147
14. Hamada Y (2014) Drug discovery of β-secretase inhibitors based on quantum chemical interactions for the treatment of Alzheimer's disease. SOJ Pharmacy & Pharmaceutical Sci 1(3):1–8
15. Hamada Y, Kiso Y (2009) Recent progress in the drug discovery of non-peptidic BACE1 inhibitors. Expert Opin Drug Discov 4:391–416
16. Hamada Y, Kiso Y (2012) The application of bioisosteres in drug design for novel drug discovery: focusing on acid protease inhibitors. Expert Opin Drug Discov 7:903–922
17. Hamada Y, Kiso Y (2013) Advances in the identification of β-secretase inhibitors. Expert Opin Drug Discov 8:709–731
18. Nguyen J-T, Hamada Y, Kimura T, Kiso Y (2008) Design of potent aspartic protease inhibitors to treat various diseases. Arch Pharm Chem Life Sci 341:523–535
19. Ghosh AK, Shin D, Downs D et al (2000) Design of potent inhibitors for human brain memapsin 2 (β-secretase). J Am Chem Soc 122:3522–3523
20. Hong L, Koelsch G, Lin X et al (2000) Structure of the protease domain of memapsin 2 (β-secretase) complexed with inhibitor. Science 290:150–153
21. Ghosh AK, Bilcer G, Harwood C et al (2001) Structure-based design: potent inhibitors of human brain memapsin 2 (β-secretase). J Med Chem 44:2865–2868
22. Hong L, Turner RT, Koelsch G, Shin D, Ghosh AK, Tang J (2002) Crystal structure of memapsin 2 (β-secretase) in complex with an inhibitor OM00–3. Biochemistry 41:10963–10967
23. Kimura T, Shuto D, Hamada Y et al (2005) Design and synthesis of highly active Alzheimer's β-secretase (BACE1) inhibitors, KMI-420 and KMI-429, with enhanced chemical stability. Bioorg Med Chem Lett 15:211–215
24. Asai M, Hattori C, Iwata N et al (2006) The novel β-secretase inhibitor KMI-429 reduces amyloid beta peptide production in amyloid precursor protein transgenic and wild-type mice. J Neurochem 96:533–540
25. Kimura T, Hamada Y, Stochaj M et al (2006) Design and synthesis of potent β-secretase (BACE1) inhibitors with P1' carboxylic acid bioisostere. Bioorg Med Chem Lett 16:2380–2386
26. Hamada Y, Igawa N, Ikari H et al (2006) β-Secretase inhibitors: modification at the P_4 position and improvement of inhibitory activity in cultured cells. Bioorg Med Chem Lett 16:4354–4359

27. Hamada Y, Abdel-Rahman H, Yamani A et al (2008) BACE1 inhibitors: optimization by replacing the P_1' residue with non-acidic moiety. Bioorg Med Chem Lett 18:1649–1653
28. Tagad HD, Hamada Y, Nguyen J-T et al (2010) Design of pentapeptidic BACE1 inhibitors with carboxylic acid bioisosteres at P_1' and P_4 positions. Bioorg Med Chem 18:3175–3186
29. Tagad HD, Hamada Y, Nguyen J-T et al (2011) Structure-guided design and synthesis of P_1' position 1-phenylcycloalkylamine-derived pentapeptidic BACE1 inhibitors. Bioorg Med Chem 19:5238–5246
30. Hamada Y, Ohta H, Miyamoto N et al (2009) Significance of interaction of BACE1-Arg235 with its ligands and design of BACE1 inhibitors with P_2 pyridine scaffold. Bioorg Med Chem Lett 19:2435–2439
31. Crowley PB, Golovin A (2005) Cation-π interactions in protein-protein interfaces. Proteins 59:231–239
32. Persson BD, Müller S, Reiter DM (2009) An arginine switch in the species B adenovirus knob determines high-affinity engagement of cellular receptor CD46. J Virology 83:673–686
33. Long SB, Long MB, White RR (2008) Crystal structure of an RNA aptamer bound to thrombin. RNA 14:2504–2512
34. Hamada Y, Ohta H, Miyamoto N et al (2008) Novel non-peptidic and small-sized BACE1 inhibitors. Bioorg Med Chem Lett 18:1654–1658
35. Hamada Y, Nakanishi T, Suzuki K et al (2012) Novel BACE1 inhibitors possessing a 5-nitroisophthalic scaffold at the P_2 position. Bioorg Med Chem Lett 22:4640–4644
36. Suzuki K, Hamada Y, Nguyen J-T et al (2013) Novel BACE1 inhibitors with a non-acidic heterocycle at the P_1' position. Bioorg Med Chem 21:6665–6673
37. Hamada Y, Suzuki K, Nakanishi T et al (2014) Structure-activity relationship study of BACE1 inhibitors possessing a chelidonic or 2,6-pyridinedicarboxylic scaffold at the P_2 position. Bioorg Med Chem Lett 24:618–623
38. Hamada Y, Tagad HD, Nishimura Y et al (2014) Tripeptidic BACE1 inhibitors devised by in-silico conformational structure-based design. Bioorg Med Chem Lett 22:1130–1135
39. Imai YN, Inoue Y, Nakanishi I, Kitaura K (2008) Cl–π interactions in protein–ligand complexes. Protein Sci 17:1129–1137
40. Grüninger-Leitch F, Schlatter D, Küng E, Nelböck P, Döbeli H (2002) Substrate and inhibitor profile of BACE (β-Secretase) and comparison with other mammalian aspartic proteases. J Biol Chem 277:4687–4693
41. Hamada Y, Ishiura S, Kiso Y (2013) BACE1 inhibitor peptides: can an infinitely small k_{cat} value turn the substrate of an enzyme into its inhibitor? ACS Med Chem Lett 3:193–197

Chapter 9

Molecular Docking and Molecular Dynamics Simulation to Evaluate Compounds That Avoid the Amyloid Beta 1-42 Aggregation

Maricarmen Hernández Rodríguez, Leticia Guadalupe Fragoso Morales, José Correa Basurto, and Martha Cecilia Rosales Hernández

Abstract

The aggregation of amyloid beta (Aβ) is one of the principal hallmarks related to Alzheimer's disease (AD). Therefore, avoiding Aβ production and aggregation could be an important therapeutic strategy to design compounds against AD. In silico studies have introduced ways to explore and analyze the structure of biomolecules, and they are also of great help in designing new compounds based on the 3D structure of proteins. This chapter analyzes the applications of molecular docking and molecular dynamics (MD) simulations, two in silico methods employed to estimate the binding energy, and binding pose of compounds that could avoid Aβ42 aggregation.

Key words Alzheimer treatment, Docking, Molecular dynamics simulations, Amyloid beta

1 Introduction

Senile plaques are one of the principal neuropathological features related to the development and progression of Alzheimer's disease (AD). AD is the principal form of dementia that currently has no cure. The amyloid beta (Aβ) is produced by the proteolytic cleavage of the amyloid precursor protein (APP) by the action of the beta secretase protein. The Aβ can be produced with different lengths of amino acids, giving Aβ of 40 amino acid residues (Aβ40) and Aβ of 42 amino acid residues (Aβ42), principally. Although Aβ40 and Aβ42 differ only in terms of two amino acids (Ile41 and Ala42), Aβ42 is more neurotoxic and characterized by faster aggregation, as Aβ42 is involved in the development and progression of Alzheimer's disease (AD). During Aβ40 and Aβ42 aggregation, a conformational change has been identified, whereby a salt bridge is formed between Lys28 and Asp23 or Glu22 [1–3]. In addition, several structural regions of Aβ42 have been identified as important

Kunal Roy (ed.), *Computational Modeling of Drugs Against Alzheimer's Disease*, Neuromethods, vol. 132, DOI 10.1007/978-1-4939-7404-7_9, © Springer Science+Business Media LLC 2018

during aggregation, allowing the design and evaluation of small molecules. Some of these regions are the N-terminus [4, 5], hydrophobic core [6, 7], hinge or turn regions [8, 9], and C-terminus [10, 11]. Although all of these regions have been identified, some of the data have been considered important from the Aβ40 aggregation; however, it has been difficult to obtain details about Aβ42 aggregation and fibril formation. Although, it was recently reported that the process of Aβ42 aggregation is different to that reported for Aβ40; a salt bridge was identified between Lys28 and the C-terminal from Ala42, which was reported as a unique salt bridge in the Aβ42 fibrils [12]. In addition, the known structural detail of Aβ42 is necessary to design small molecules and to avoid Aβ42 aggregation to be used for the treatment of AD in the future.

There are presently no small molecules that avoid the conformational change or the self-assembly of Aβ42 in the clinical treatment [13]; this could be due to the fact that the small molecules that have been evaluated have not been designed with the perturbation conformational changes during Aβ42 aggregation taken into account. Several small molecules that were evaluated as Aβ42 aggregation inhibitors in clinical trials emerged from natural products or as observations from known drugs such as acetylcholinesterase inhibitors (galantamine) or from drugs repurposed from a database of small molecules approved by the Food and Drug Administration (FDA) such as bexarotene, which is still in phase II of AD clinical trials, and tramiprosate, which failed in phase III (Table 1) [14].

Thus, designing new molecules that avoid Aβ42 changing its conformation when it is released from the membrane and prevent Aβ42 from acquiring a beta sheet conformation is of great interest. In this sense, computer-aided drug design (CADD) could be of great help given that this has been employed in three principal areas: (1) select promisor compounds from large libraries of possible actives compounds; (2) improve lead compounds, increasing its affinity through the addition of a functional group; or (3) design new molecules from a target structure by getting pharmacophore features.

Furthermore, CADD methods can be employed in two different ways, one target structure based and the other ligand based. The first could be due to the fact that the structure of Aβ42 is known and because several regions of Aβ42 have been implicated in its aggregation mechanism. The second ligand-based approach is when an active ligand is enhanced to have a better ligand to inhibit Aβ42 aggregation.

Therefore, in order to obtain small molecules with Aβ42 antiaggregation activity, several works have focused on the design of new molecules from one starting molecule, improving it by adding different functional groups based on the knowledge of the structure target, in this case Aβ42 employing molecular docking and

Table 1
Small molecules evaluated as Aβ42 aggregation inhibitor in clinical trials and some patented lead compounds

Compound	Clinical phase	Designed by in silico studies	Chemical structure	In silico studies	Ref.
Tramiprosate (3APS)	Phase III (was terminated by poor clinical efficacy)	NO	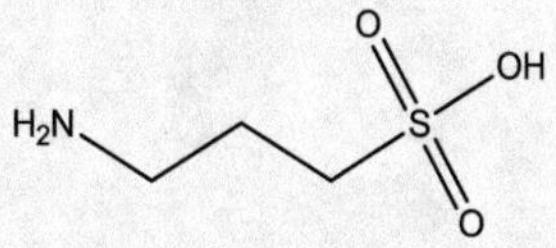	[15]	[16] [17]
Scyllo-inositol	Is in phase II	NO	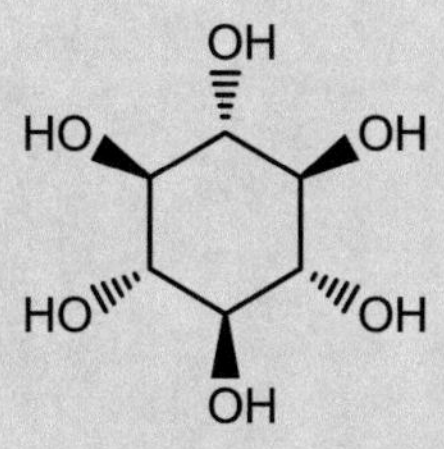	[18]	[19]
Epigallo-catechin-3-gallate (EGCG)	Is in phase II/III	NO	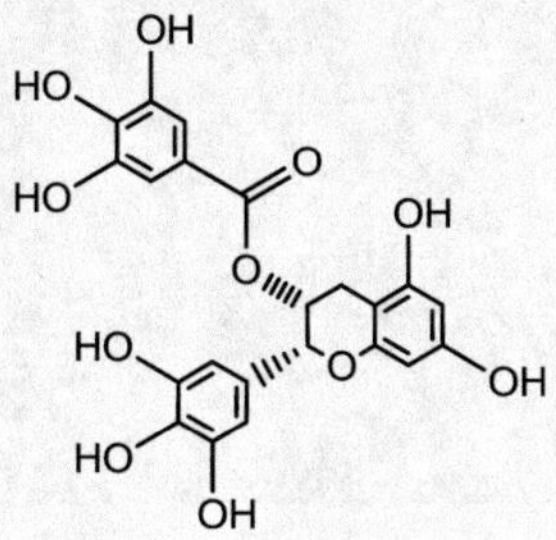	[20]	[21]
PBT1 (clioquinol)	Phase II (was terminated because of its toxicity)	NO	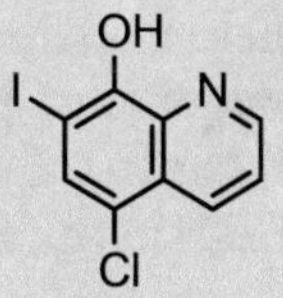	[22]	[23]
PBT2 (8-hydroxy quinoline analog) PBT2	Is in phase II	NO	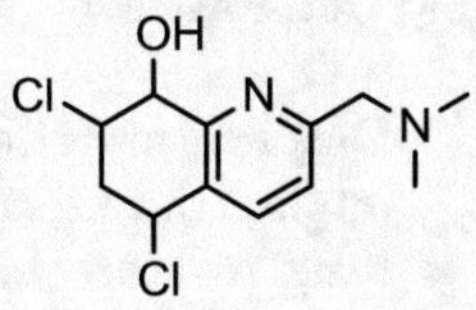	NO	[24]

(continued)

Table 1 (continued)

Compound	Clinical phase	Designed by in silico studies	Chemical structure	In silico studies	Ref.
Apomorphine	NO	NO	HO, HO, NH, H	NO	[25]
Peptidomimetic derivatives compound 8a	NO	NO	O, N, N, H, S, N, O	NO	[26]
Heterocyclic compounds such as compound 9a	NO	NO	N, NH, NH, N	NO	[27]

molecular dynamics (MD) simulation as computational tools that help to estimate the binding affinity and binding mode between the target and the new designed compound. Then, this chapter is focused on reviewing the application of docking and MD simulations in the evaluation of new Aβ42 aggregation molecules (Fig. 1).

2 Molecular Docking Simulations

Despite great efforts to elucidate the neuropathological mechanism implied during AD progression in order to develop a more effective drug to stop the AD, currently, the principal neuropathological feature is Aβ42 aggregation; several strategies can be adopted that yield Aβ target therapies, including the design and evaluation of drugs that (a) avoid Aβ production by the inhibition of β-secretase, (b) inhibit Aβ aggregation, and (c) promote Aβ clearance [28]. Therefore, in silico methods have been employed to design and

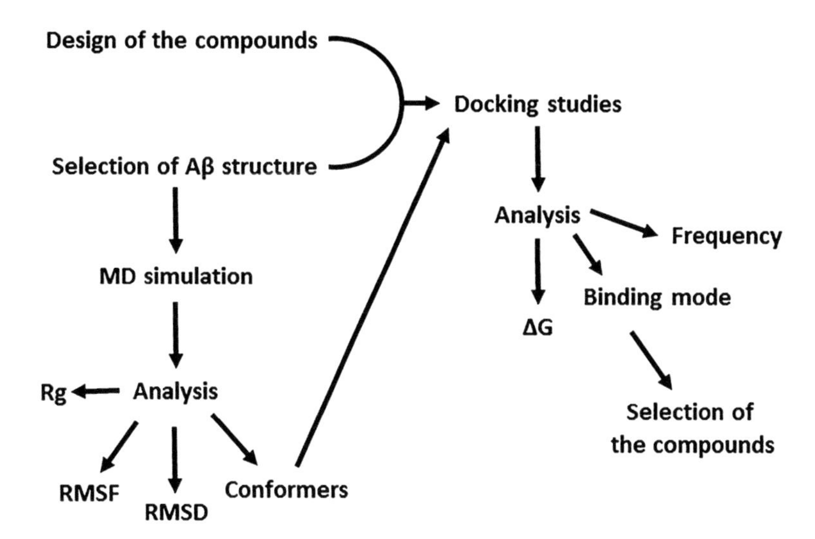

Fig. 1 Use of docking studies and MD simulations in the design of novel compounds to prevent Aβ42 aggregation

evaluate new molecules and peptides that could modify the above-mentioned events related to Aβ42 production and aggregation; the focus is to avoid Aβ42 aggregation. For these purposes, the three-dimensional (3D) structure of Aβ in a monomer, oligomer (dimer, tetramer, pentamer, etc.), or steric zipper assembly has been employed to perform docking studies and MD simulations to evaluate the affinity of novel compounds and thus know more of the structural features of Aβ conformational changes [29].

Molecular docking is a computational technique employed to estimate the binding affinity between two molecules such as the protein-protein and ligands-protein among others; prediction is based on different scoring (sampling and function); the score function is the sum functions referring to the mathematical method based on the strength of non-covalent interaction. For instance, AutoDock is a software which uses the AMBER force field in which the affinities are estimated by summing the strength of intermolecular van der Waals, hydrogen bonding, electrostatic interactions, conformational entropy, and desolvation terms between all atoms of the two molecules involved in a complex ligand-target.

As is known elsewhere, the conformational changes during the binding between the ligand and the protein are of great importance because the recognition between both does not consider a rigid mechanism; the ligand and the protein can adopt different conformations until they reach a conformation of binding with the lowest free energy. Therefore, there are three different forms to perform the docking methodology; the first is to consider both the ligand and the protein as rigid molecules; the second is to consider the ligand flexible and the protein rigid; and the third is to consider that both the ligand and protein are flexible. The most common procedure is to consider the protein rigid and the ligand flexible, as is done in the AutoDock, although the most recent version AutoDock 4 also considers the side chain of residues as having protein flexibility [30]. Because docking studies are very useful for the prediction of protein-ligand interactions, they provide an opportunity to explore recognition properties and identify potential pharmacophores to Aβ [31, 32]. Currently, there are several software programs that perform docking studies, including DOCK 6.1 [33], FlexX (v2.0.3) [34], GLIDE v4.5 [35], PhDOCK [36], Surflex (v2.1) [37], AutoDock 4.2 [38], and AutoDock Vina [39]. AutoDock 4.2 has been widely employed in the design of novel Aβ aggregation inhibitors [40–42].

2.1 Aβ42 Structures Employed in Molecular Docking and Molecular Dynamics Simulations

Molecular docking and MD simulation have been performed employing several Aβ structures based on the goal of the study; three principal features are taken into account: first, the length of Aβ sequence, i.e., Aβ (16–21) [43] and Aβ (18–41) [44] (to compare the length of the sequences, multiple alignments have been performed employing several programs such as the STRAP program [(http://www.bioinformatics.org/strap/)]); second, considering the number of Aβ monomers employed in the study, i.e., a monomer, dimer, trimer, pentamer, etc.; third, the conformation of Aβ employed in the study, as it has been demonstrated that the conditions of crystallization affect the proportion of secondary structures of Aβ. Once the researchers have selected the conditions of study, the Aβ tridimensional structure can be obtained from repositories such as the Protein Data Bank (PDB) or obtained by previous studies of MD simulations, or one can inclusively build the Aβ tridimensional structure under homology modeling [43]. The PDB shows that there are 11 structures obtained from Aβ42: 2NAO [45], 5KK3 [46], 5AEF [47], 2MXU [12], 2M5K, 2M5M, 3ZPK [48], 2M4J [49], 1Z0Q [50], 2BEG [51], and 1IYT [52]; some of these are in monomeric or fibrillary form; some are with the most complete sequence which are shown in Fig. 2. Mohamed T et al. reported that many small molecules are not able to prevent all forms of Aβ42 aggregates and they exhibit different inhibitory activity depending on the Aβ42 structure employed [53].

To study or analyze the interaction between Aβ42 monomers and other self-monomers or molecules, it is recommended to use during docking studies several secondary conformations of Aβ42 in order to enhance the screening results. For example, α-helix, random coil, and β-sheet due to the Aβ42 present several conformational changes during aggregation, acquiring principally these structures [54]. The Aβ42 in α-helix PDB, 1Z0Q, and Aβ17–42 in a β-sheet PDB, 2BEG – both structures have been employed in several research studies [55–58]. Also, the structures obtained from MD simulations such as Aβ42 in the random coil (RC)

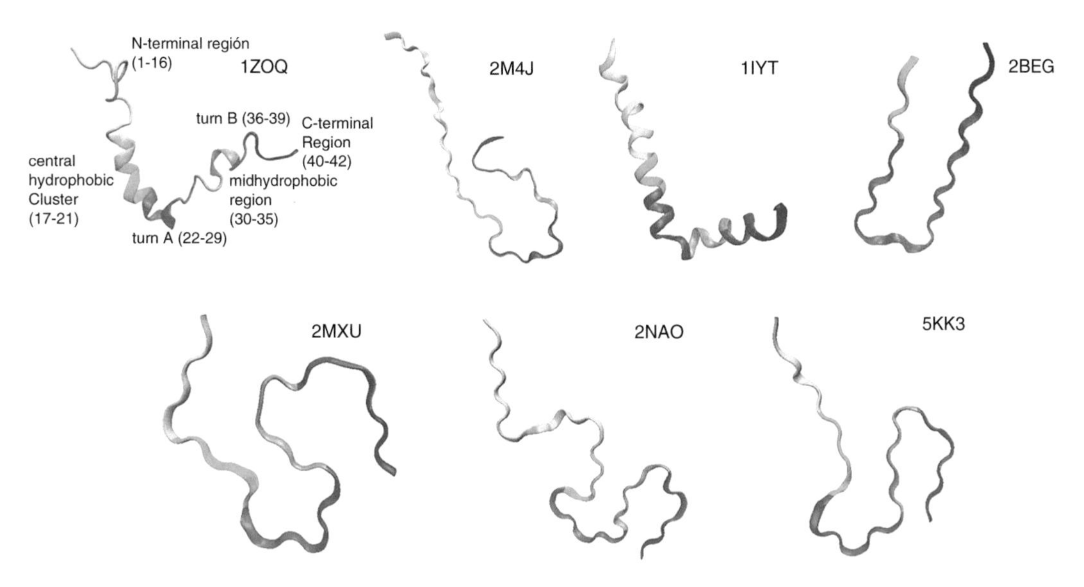

Fig. 2 Representation of one monomer from different Aβ42 structures obtained from PDB. The primary structure of Aβ42 is DAEFRHDSGYEVHHQKLVFFAEDVGSNKGAIIGLMVGGVVIA. Above each structure is the PDB code and in the 1ZOQ is indicated each Aβ42 region [50]

conformation obtained at 10 ns through MD simulations from the 1ZOQ structure have been employed [59]. Then, the results of designing compounds that interact with the Aβ42 monomer to avoid oligomer formations are interesting [60]. Developing compounds with high affinity to intermediaries of Aβ42 fibrils present problems, as it has been reported that oligomer Aβ42 species are highly toxic [53].

2.2 Ligands Evaluated by Molecular Docking to Inhibit Aβ42 Aggregation

The design of novel Aβ42 aggregation inhibitors could be performed in different ways. The compounds could be selected (a) by virtual screening employing databases of chemical compounds (Table 2), (b) by chemical modifications of compounds that have demonstrated activity as Aβ42 aggregation inhibitors, (c) or by designing a compound according to pharmacophore models against Aβ42.

2.2.1 Ligands Selected by Virtual Screening

Bansode SB. et al., performed a virtual screening employing 140 FDA-approved drugs used to treat pathologies of the nervous system using docking studies on human AChE, β-secretase, and Aβ42 [55]. Their results suggested the viability of the tricyclic group of antidepressants against three major AD targets, which allows selecting five compounds to perform in vitro studies. Among the drugs selected, protriptyline (Fig. 3) showed multifunctional activity against protein targets [61]. In this way, recently Das and Smid reported the use of the ZINC chemical database to select five compounds with anti-aggregatory and neuroprotective effects against Aβ42 [62].

Table 2
Databases of chemical compounds employed in docking studies to find and select several molecules

Database	Webpage
Zinc	http://zinc.docking.org/
PubChem	https://pubchem.ncbi.nlm.nih.gov/
ChemSpider	http://www.chemspider.com/
ChEMBL	https://www.ebi.ac.uk/chembl/
NuBBE DB	http://nubbe.iq.unesp.br/portal/nubbedb.html
ChemBank	http://chembank.broadinstitute.org/
eMolecules	https://www.emolecules.com/
DrugBank	https://www.drugbank.ca/
Binding DB	https://www.bindingdb.org/bind/index.jsp
DrugBank	http://www.drugbank.ca

A study of the molecular topology model reported for AD to find molecules with anti-Aβ aggregation activity from the database of the drugs approved by FDA shows that eight molecules (from A2641934 to A5261232; Fig. 3) were identified as lead compounds; despite the fact that the authors do not explain the interactions between the molecules with Aβ42 structures, it is possible that these molecules interact with the region where the salt bridge is formed and the hydrophobic region by π-π interaction with Phe, as several of the compounds have a carboxylic acids and all have aromatics groups [61].

2.2.2 Chemical Modifications on Known Ligands as Aβ42 Aggregation Inhibitors

Several works have employed molecules known elsewhere, such as polyphenols, to analyze their anti-aggregative properties and neuronal toxicity against Aβ42. Among the compounds evaluated are flavonoids such as luteolin and myricetin, the lignin honokiol, and the ellagitannin punicalagin (Fig. 3). These compounds showed neuroprotective effects in pheochromocytoma (PC12) cells exposed to native non-fibrillar Aβ42, and in silico studies showed that the compounds evaluated interact principally by hydrophobic interaction and by the hydrogen bond in the monomer and pentamer Aβ42 structures [63].

α-Mangostin is a polyphenolic xanthone derivative that inhibits Aβ42 aggregation by interacting with Phe19, Asp23, Glu22, and Lys16; these interactions were analyzed by in silico methods employing the PDB: 1BA4 structure and evaluated by in vitro studies using ThT fluorescence, finding that Aβ42 fibril formation was inhibited at a rate close to 72.32%. In addition, the Aβ42 fibrils

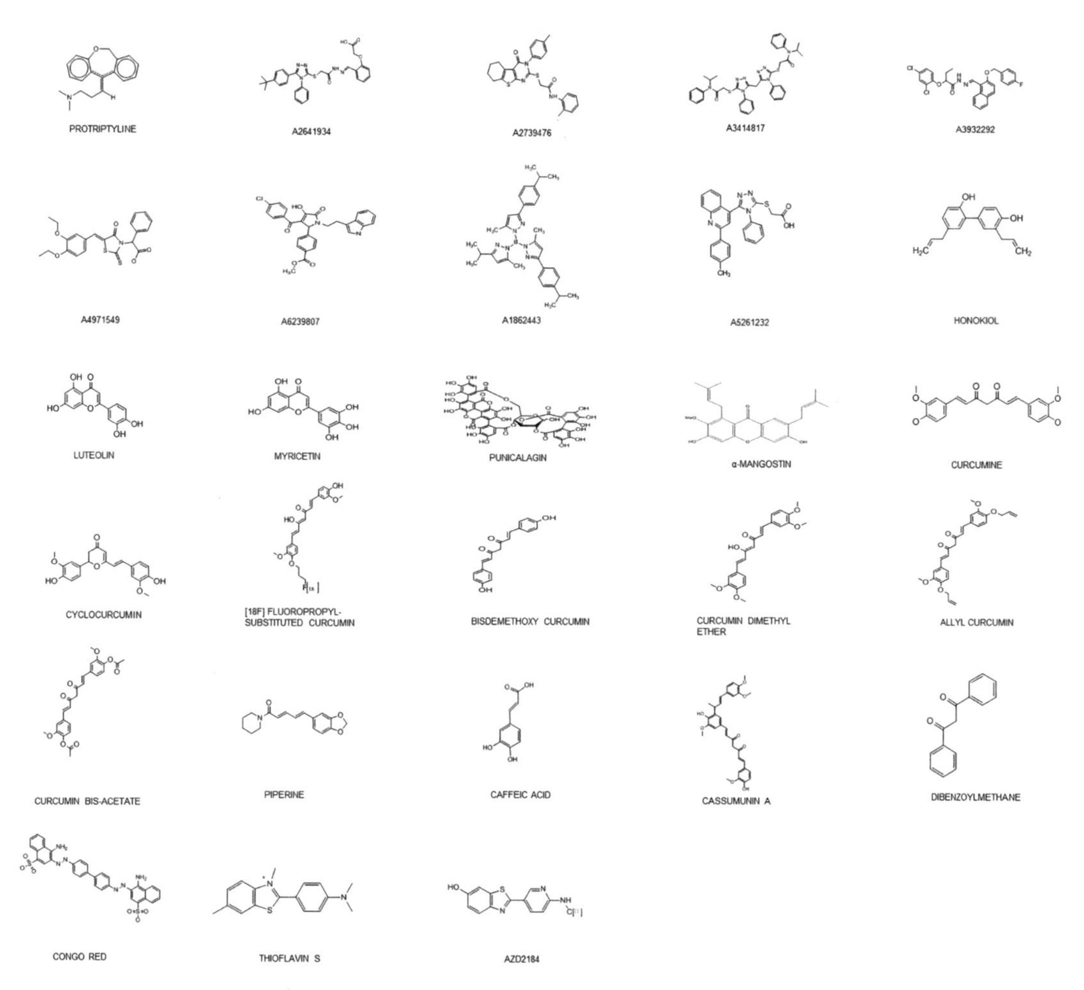

Fig. 3 Chemical structure of known ligands as Aβ42 aggregation inhibitors

were visualized by transmission electron microscopy (TEM), which were prevented by the incubation of Aβ42 with Mangostin observing only amorphous deposits [64].

Other works have reported several chemical modifications of the known molecule such as curcumin, where some derivatives of curcumin were evaluated such as cyclocurcumin, bisdemethoxycurcumin, curcumin dimethyl ether, allyl curcumin, curcumin bis-acetate, and [18F] fluoropropyl-substituted curcumin or analogs such as piperine, caffeic acid, cassumunin A, cassumunin B, chlorogenic acid, dehydrozingerone, dibenzoylmethane, ferulic acid, chalconeonoid, and yakuchinone-A; in this work, the Aβ42 structures PDB: 1IYT and 2BEG were employed to estimate the binding affinity of the compound identifying that hydrophobic interactions and the hydrogen bond are important as well as the interactions in the region of Lys16 and Phe20 [15, 65]. Figure 4a shows the interaction of curcumin with Lys 16 from Aβ42 structure (PDB: 1IYT).

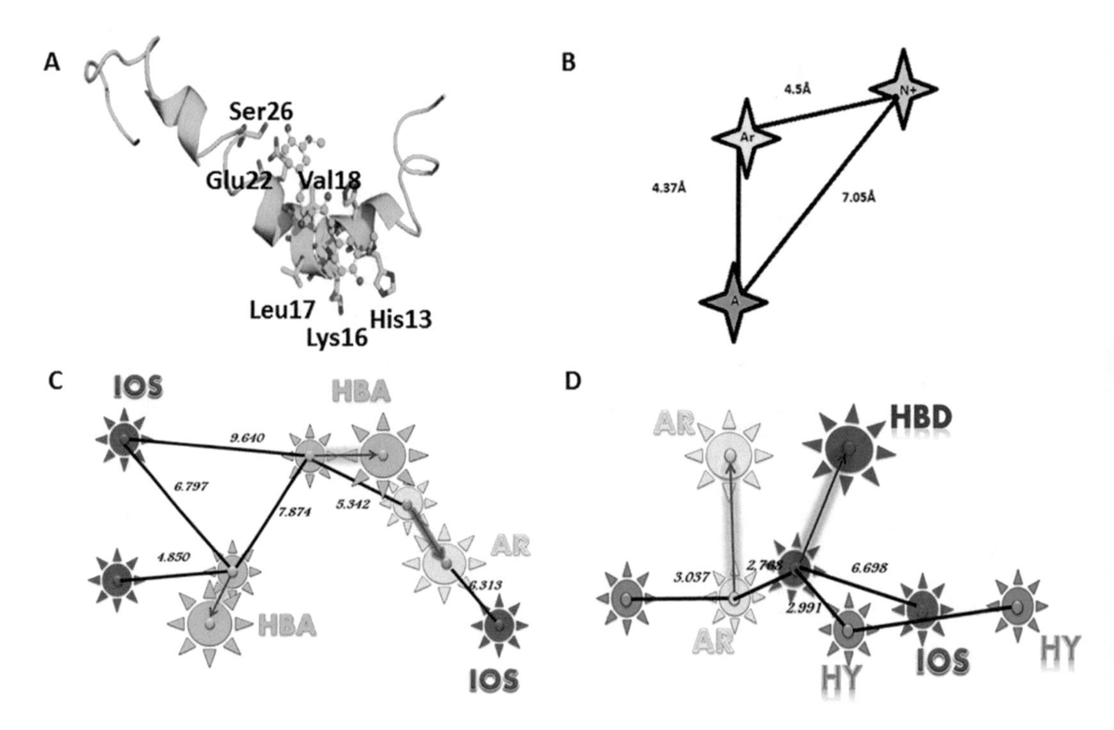

Fig. 4 Chemical interacction and pharmacophore models for Aβ. A) Interactions of curcumin with Aβ42. Proposed pharmacophore model for Aβ in different structures: B) monomer structure in α-helix. C) dimer structre and D) steric zipper assembly

It is important to mention that despite the existence of known molecules that bind to Aβ42 with great affinity, such as Congo red, thioflavin S (fluorescent probe), and AZD2184 (PET tracer), molecules such as Congo red and thioflavin S (Fig. 3) have not been used in AD treatment because they are not able to cross the blood-brain barrier; however, with in silico methods, it is possible to know the principal interaction with Aβ42 in order to design a molecule that has the pharmacophore to inhibit the Aβ42 aggregation [66]. In addition, some molecules act in a specific region of Aβ42 such as scyllo-inositol (Table 1), which interacts with the C-terminal [67]. This region plays a key role in controlling the oligomerization process, which could be due to the C-terminus of Aβ42, which is more rigid than the C-terminus of Aβ40. This increased rigidity has been attributed to interactions involving C-terminal residues Ile41and Ala42, which stabilize a putative turn conformation.

After evaluating the interaction of known molecules or new molecules to avoid Aβ42 aggregation, several chemical characteristics of anti-aggregation Aβ42 compounds have emerged; for instance, it is important that the molecule has two aromatics rings, as is observed in the curcumin structure (Fig. 3), and a linker between the aromatic rings. According to this, Reinke AA. and Gestwicki JE. published a study that argued the length between the two aromatic rings should be between 8 and 16 Å [68].

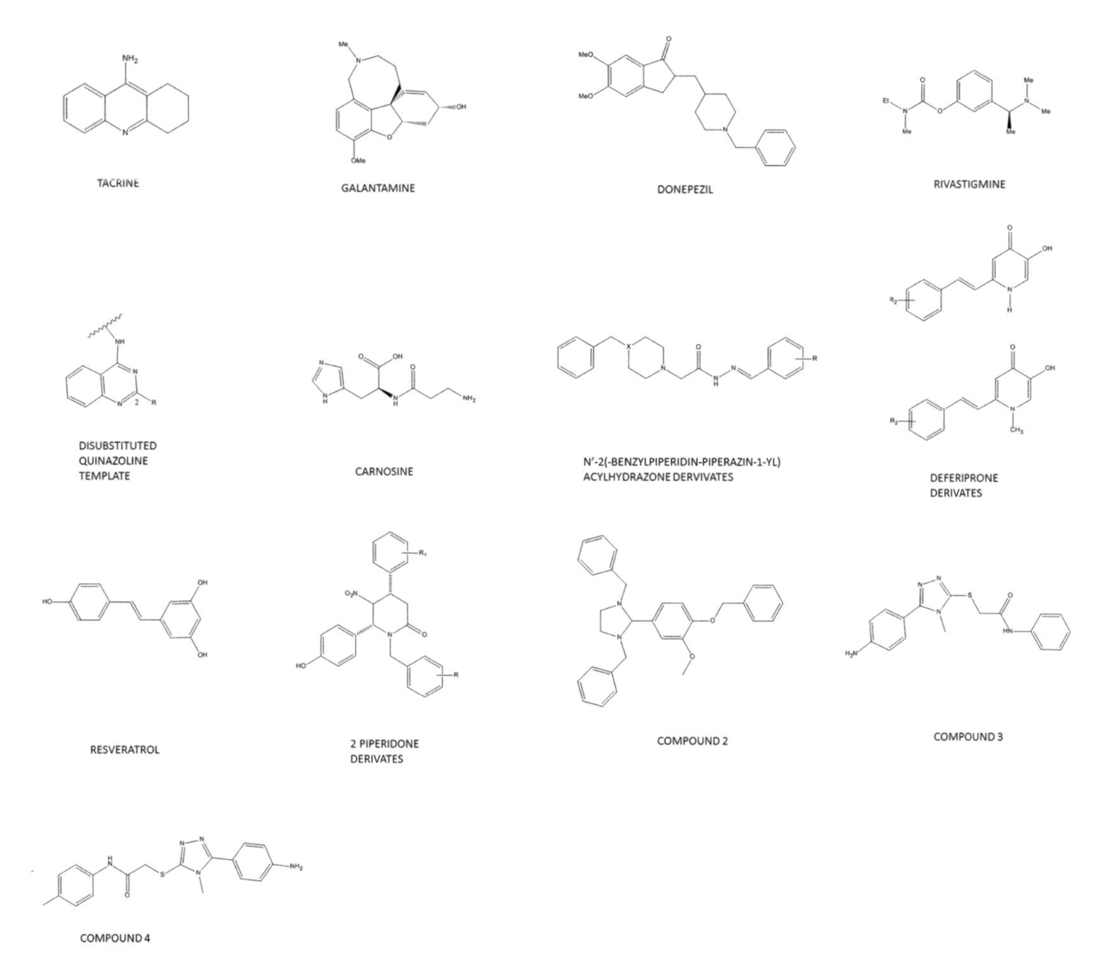

Fig. 5 Chemical structure of ligands designed according to Aβ42 pharmacophore models and ligands evaluated according to designed Aβ42 pharmacophore models

However, it was recently proposed that the ligands should not necessarily contain two aromatic rings. Some important molecular features are the presence of a tertiary amine in a ring of five carbons, which is able to establish electrostatic interactions with Glu22 and Asp23, the presence of one aromatic ring establish the π-π interactions with Phe19 and Phe20, and the presence of aliphatic substituents in the aromatic rings establish hydrophobic interactions with the methylene side chain of Lys16 and π-cation with the NH_3 group. Also, some known compounds showing anti-aggregations of Aβ42 possess a positive charge, such as nicotine, melatonin, and thioflavin S [32]. In addition, there are other compounds that have been employed as acetylcholinesterase inhibitors (AChEi) that have a tertiary amine, such as tacrine, rivastigmine, donepezil, and galantamine, and several of these show anti-Aβ42 aggregations (Fig. 5).

2.2.3 Ligands Designed According to Aβ42 Pharmacophore Models and Ligands Evaluated to Design Aβ42 Pharmacophore Models

Recent works indicated that compounds with a heterocycle ring containing one, two, or three N atoms are good small molecules for evaluation as Aβ42 aggregation inhibitors. Interesting works on this were published by Mohamed T. et al., and Hernández M. et al.,who reported pharmacophore models depending on the Aβ structure used(Fig.4b-d). The pharmacophore models are different as it uses the Aβ structure as monomer, dimer or Aβ steric zipper assembly. In the Aβ dimer model, compound 14 exhibited the best binding mode and interaction energy; then the pharmacophore model was proposed, taking into account its chemical structure; as shown in Fig. 4c-d, compound 14 has a large volume and therefore was able to have several contacts with the Aβ dimer; despite the fact that compound 14 does not have any hydrogen bond donators (HBD), it has three ionizable groups (IOS). However, the pharmacophore model for the Aβ steric zipper assembly showed that the presence of hydrophobic groups (HY), either five- or six-membered aromatic rings (AR), HBD, and IOS is necessary [53]. In accordance with the observations of Mohamed T. et al., a pharmacophore model proposed to avoid the Aβ aggregation in α-helix is different from that described for the Aβ dimer and Aβ steric zipper assembly [32, 53]. Despite the fact that the presence of an aromatic ring (AR) and five- or six-membered rings with a tertiary amine (IOS) is necessary for docking, studies have shown that molecules with a distance of 4.3–5 Å between the amine and aromatic groups have selectivity for Aβ in the α-helix [32].

The compounds evaluated in the work of Mohamed T. et al., were tacrine hybrids, piperidine derivatives, benzyl phenyl pyridine, pyrimidines, isoindoline-1,3-dione, 3-aminopyrazole derivatives, and triazine derivatives (Fig. 5). Several works have also demonstrated the importance of tertiary amine in the inhibitors of Aβ42 aggregation; for instance, the small natural dipeptide carnosine (Fig. 5), which has a histidine residue with an imidazole group, was evaluated as an anti-Aβ42 aggregation molecule employing in silico studies on the PDB:2BGE Aβ42 structure, showing that carnosine interacts in the region where Lys28 and Asp 23 are located. The principal interactions occurred between the imidazole ring of carnosine and Asp 23 and between the β-alanine and Lys28, explaining these interactions of anti-Aβ42 aggregation activity and of carnosine; in addition, ThT fluorescence and atomic force microscopy (AFM) assays showed that carnosine has anti-Aβ42 aggregation that is dose dependent [68]. Also, compound derivatives of piperidine and piperazine were evaluated as inhibitors of Aβ42 aggregation (Fig. 5) [69].

Most recently, deferiprone derivatives have been reported as Aβ42 aggregation inhibitors; the compounds were designed as multitarget compounds for AD. The chemical molecular structures of the compounds show a combination of a deferiprone molecule to resveratrol giving several derivatives (Fig. 5). The results for the molecular docking of the compounds with better inhibitory potency against Aβ42 aggregation employing the PDB: 1IYT structure show binding to Lys28 by the hydrogen bond, indicating that this interaction avoids the salt bridge formation between Lys28 and Asp23, and consequently, the structure of Aβ42 does not form a β-pleated-hairpin structure, which is stabilized principally by hydrophobic interactions and by the salt bridge formation [70].

In addition, a series of 2,4-disubstituted quinazoline derivatives has been evaluated as anti-aggregation molecules for Aβ42; these compounds were also designed as multitarget compounds for AD. The molecular docking of the best compound named *14c* showed that the compound binds in the loop where Asp 23 is located [71]. The inhibitory activity on Aβ42 aggregation of the compounds was evaluated in vitro by ThT fluorescence and obtained an IC_{50} of 45%.

Das S. and Smid SD. also reported molecules with anti-Aβ42 properties containing a five-membered heterocycle ring scaffold with nitrogen heteroatom, two dibenzyl phenyl imidazolines, and two triazole sulfanyl acetamides and one with benzoxazole ring. The molecular docking employing the PDB: 1IYT Aβ42 structure showed that compounds 2 and 3 (Fig. 5) interact with Phe19 and Glu22. In vitro studies of ThT fluorescence showed that compound 2 inhibits Aβ42 aggregation and it is less neurotoxic [62].

2-piperidone derivatives (Fig. 5) also act as inhibitors of Aβ42 aggregation, and specifically, the compound *7q* had the best inhibition on Aβ42 self-aggregation (59.11%) evaluated by ThT fluorescence, despite the fact that the authors do not make an in silico study of the compound which shares chemical characteristics with some of the compounds mentioned before, which has a heterocycle ring with one N atom, and it also has a phenolic group [72].

3 Molecular Dynamics Simulations

As mentioned before, the Aβ42 conformational changes during the ligand recognition are of great interest when one wants to estimate the binding affinity to consider the best ligand to be evaluated. Then, relevant results are yielded when the flexibility of the protein is considered; MD simulation is a technique that makes it possible to consider protein flexibility, and despite the fact that it has some disadvantage such as inadequate sampling and high computational expense, its combination with the docking approach is a great help in evaluating the recognition and binding energies of new ligands. By MD simulations, one can obtain different conformers of the protein, which can be employed to perform docking analyses; it is also possible to do MD simulations between the ligand-proteins complex.

Inhibiting the Aβ aggregation process by employing small molecules represents a considerable challenge. As has been mentioned before, Aβ undergoes large and continual conformational changes. Therefore, the evaluation of novel compounds appears to be complicated. This type of information is difficult to gather using traditional experimental techniques and thus represents an area well suited to computational methods [73]. Information obtained from MD simulations will likely accelerate the process of novel Alzheimer's drug development, and MDsimulation has recently been successfully employed in designing Aβ aggregation inhibitors [74, 75].

MD simulations have been used to study the conformational change of Aβ42 alone or to analyze the interaction between monomers to study the dimer formation, which is the first step during the Aβ42 assembly; in this way, with MD simulations, it has been possible to elucidate the role of structural disorder, water solvation, and salt bridge formation upon the Aβ42 monomer to dimer conversion [76]. In addition, MD simulations make it possible to analyze the interaction of small molecules with Aβ42. In this case, the SwissParam server (http://www.swissparam.ch/) [77] can be employed to obtain the topology and parameters for small organic molecules compatible with the CHARMM all atoms force field, for use with CHARMM and GROMACS. There are a great number of MD simulation programs. However, the most popular simulation suites are CHARMM (Chemistry at Harvard Macromolecular Mechanics) [78], AMBER [79], GROMACS [80], and NAMD [81]. The enlisted programs share features; however, these present several differences, such as capacity, performance, processing of data, and adaptability to new hardware, such as graphic processing units (GPUs) that are capable of yielding MD simulation results in milliseconds (ms) [82].

3.1 Molecular Dynamics Simulations Between Small Molecules with Aβ42

Due to the tendency of Aβ42 to establish conformational changes, MD simulations can be employed to analyze the nonbonding interactions between small molecules and peptides with Aβ42 (oligomers and fibrils).

Bajda M. and Filipek S. perform 100 ns of MD simulations using a Aβ42 pentamer (PDB code: 2BEG) in the absence and presence of curcumin and other compounds with and without Aβ42 aggregation inhibitor activity to decipher the changes of these compounds on Aβ42 conformation, showing that the stability of amyloid aggregates was well correlated with the number of intermolecular backbone hydrogen bonds (HB) between adjacent Aβ42 monomers. It was observed that in case of inactive compounds, the number of HB was comparable to that in the case of Aβ42 pentamer alone. On the other hand, when MD simulation was performed in the presence of Aβ42 pentamer and Aβ42 aggregation inhibitors, the number of hydrogen bonds decreased significantly [83].

Kundaikar HS. and Degani MS. combined computational studies to decipher the binding site and the interactions, which drives the recognition of the Aβ42 pentamer structure (PDB code: 2BEG) by curcumin thioflavin T (ThT) and florbetapir. Among these studies, they performed 150 ns of MD simulations of Aβ42 pentamer in the presence and absence of the mentioned compounds, where they observed that curcumin destabilizes the peripheral Aβ42 monomers in comparison with the central ones during the course of MD simulations by altering the salt bridge formation between Asp23 and Lys28, which may further dissociate from the protofibril, leaving the inner chains exposed to dissociation or to binding by the other molecules present in the system. In contrast, florbetapir and ThT do not destabilize the peripheral Aβ42 monomers [84]. A study of docking and MD simulations of thioflavin S and AZD2184 (Fig. 3) using the ten models of Aβ42 structures of 2BEG showed that the principal interaction with the oligomers is with Met 35 on the surface, and in a similar way as in the two works mentioned previously, they also perform an interaction with Phe19 and Leu17 [66].

Wang Q. et al., design, characterize, and identify a series of hexapeptides as potential Aβ inhibitors by combining a quantitative structure-activity relationship (QSAR) model, MD simulations, and biophysical experiments. The computational and experimental results demonstrate the proof of concept of the "like interacts like" design principle, whereby self-assembling peptides with amyloidogenic properties can interact strongly with conformationally similar motifs of Aβ peptides and competitively reduce Aβ-Aβ interactions, thus preventing Aβ aggregation and Aβ-induced toxicity [85].

In a posterior study, Quiao Y. et al., regarding the mentioned peptides employing MD simulations, showed that three of the designed peptides mentioned before show that the three peptide inhibitors tend to bind strongly to the two Aβ fragments of 16KLVFFA21 and 27NKGAII32, suggesting binding modes (i.e., affinities, forces, and residues) via electrostatic and VDW interactions. Since Aβ oligomers/fibrils grow via monomer addition to its edges, these peptides binding to the edges of the oligomers apparently block the interactive sites for fibril elongation [86].

Zhang T. et al., perform aMD simulation study employing two molecules of 1,4-naphthoquinone-2-yl-L-tryptophan (NQ-trp) and Aβ42 dimer structure in which were observed interactions with amino acid residues involved during the conformational change, such as hydrophobic residues Phe19, Phe20, and Leu34 and Met35 and hydrophilic residues such as Arg5, Tyr10, Lys16, Asp7, and Lys18; these facts explain the activity of NQ-trp as an Aβ42 aggregation inhibitor; it was also possible to identify that NQ-trp has better activity than carnosine due to the fact that this was only in contact with Aβ42 in 20% of the MD simulations; therefore, NQ-trp has better affinity for Aβ42 than carnosine [87].

An interesting MD simulation study was done employing the PDB: 1BA4 structure of Aβ, at which two amino acids were added to obtain Aβ42. The structure was subjected to microsecond levels of MD simulations in aqueous solution to identify the most stable conformations to carry out virtual screening and select an inhibitor for Aβ42 aggregation. The most stable Aβ42 conformation comprised of β-sheets and coils structure and AE-848 was the best compounds with dual activity for Aβ42 aggregation and BACE-1 inhibition. It is interesting to observe that this compound has several nitrogen atoms, and the interactions reported between the compound and Aβ42 are observed with V40, I41 near to the C-terminal instead of the region where a salt bridge is formed [88].

Furthermore, Barjad M. and Filipek S. reported the MD simulations between compound derivatives of both phthalimide and indole, which contain an extra amine group incorporated into the benzylpiperidine amine, benzylamine, or diethylamine moiety with an Aβ42 structure the initial position of ligands was determined by docking, and after the complex ligand Aβ42 was submitted to 100 ns of MD simulations, curcumin was employed as a control; the structure of Aβ42 started changing near the ligand. The results showed that at the end of MD simulations, curcumin was out of the fibril interior. Compound 5, which was the best active inhibitor, changed the secondary structure of the Aβ42 fibril; at the end of MD simulation, the fifth chains of the Aβ42 fibril were completely unorganized; however, with inactive compounds and the Aβ42 fibril alone, the changes to the secondary structure were smaller [83].

4 Conclusion

Given that to date there are no drugs for treatment patients with Alzheimer's disease obtained by modifying the production and aggregation of Aβ42, the design of new molecules that can alter this process represents a beneficial goal. In order to decrease time and cost, the design and evaluation of these potential drugs can be done by employing in silico tools, where molecule docking and MD simulations play an important role because they allow one to know the affinity and interaction between the Aβ42 and its ligands. This allows one to identify the chemical groups that are necessary to avoid Aβ42 aggregation in which it has been demonstrated that the molecules should have two aromatics rings with a linker of 8–16 Å or that the molecule possesses one aromatic ring and a ring of five or six members with an amine group able to protonate at pH 7.4 with a linker between the two rings.

References

1. Melquiond A, Dong X, Mousseau N et al (2008) Role of the region 23–28 in Abeta fibril formation: insights from simulations of the monomers and dimers of Alzheimer's peptides Abeta40 and Abeta42. Curr Alzheimer Res 5:244–250
2. Buchete NV, Hummer G (2007) Structure and dynamics of parallel beta sheets, hydrophobic core, and loops in Alzheimer's A beta fibrils. Biophys J 92:3032–3039
3. Sciarretta KL, Gordon DJ, Petkova AT et al (2005)Abeta40-Lactam(D23/K28) models a conformation highly favorable for nucleation of amyloid. Biochemistry 44:6003–6014
4. Gardberg AS, Dice LT, Ou S et al (2007) Molecular basis for passive immunotherapy of Alzheimer's disease. Proc Natl Acad Sci U S A 104:15659–15664
5. McLaurin J, Cecal R, Kierstead ME et al (2002) Therapeutically effective antibodies against amyloid-beta peptide target amyloid-beta residues 4-10 and inhibit cytotoxicity and fibrillogenesis. Nat Med 8:1263–1269
6. Wasmer C, Lange A, Van Melckebeke H et al (2002) Amyloid fibrils of the HET-s (218–289) prion form a beta solenoid with a triangular hydrophobic core. Science 319:1523–1526
7. Permanne B, Adessi C, Saborio GP et al (2002) Reduction of amyloid load and cerebral damage in a transgenic mouse model of Alzheimer's disease by treatment with a beta-sheet breaker peptide. FASEB J 16:860–862
8. Hoyer W, Gronwall C, Jonsson A et al (2008) Stabilization of a beta-hairpin in monomeric Alzheimer's amyloid-beta peptide inhibits amyloid formation. Proc Natl Acad Sci U S A 105:5099–5104
9. Fawzi NL, Phillips AH, Ruscio JZ et al (2008) Structure and dynamics of the Abeta (21–30) peptide from the interplay of NMR and molecular simulations. J Am Chem Soc 130:6145–6158
10. Fradinger EA, Monien BH, Urbanc B et al (2008) C-terminal peptides coassemble into Abeta42 oligomers and protect neurons against Abeta42-induced neurotoxicity. Proc Natl Acad Sci U S A 105:14175–14180
11. McLaurin J, Kierstead ME, Brown ME et al (2006) Cyclohexanehexol inhibitors of Abeta aggregation prevent and reverse Alzheimer phenotype in a mouse model. Nat Med 12:801–808
12. Xiao Y, Ma B, McElheny D et al (2015) Aβ (1–42) fibril structure illuminatesself-recognitionand replication of amyloid in Alzheimer's disease. Nat Struct Mol Biol 22:499–505
13. Cummings JL, Morstorf T, Zhong K (2014) Alzheimer's disease drug-development pipeline: few candidates, frequent failures. Alzheimers Res Ther 6:37
14. Habchi J, Arioso P, Perni M et al (2016) An anticancer drug suppresses the primary nucleation reaction that initiates the production of the toxic Aβ42 aggregates linked with Alzheimer's disease. Sci Adv 2:e1501244
15. Singh DB, Gupta MK, Kesharwani RK et al (2013) Comparative docking and ADMET study of some curcumin derivatives and herbal congeners targeting β-amyloid.Netw Model Anal Health Inform Bioinforma 2:13–27
16. Gervais F, Paquette J, Morissette C et al (2007) Targeting soluble Abeta peptide with Tramiprosate for the treatment of brain amyloidosis. Neurobiol Aging 28:537–547
17. Aisen PS, Gauthier S, Ferris SH et al (2011) Tramiprosate in mild-to-moderate Alzheimer's disease – a randomized, double-blind, placebo-controlled, multi-centre study (the Alphase study). Arch Med Sci 7:102–111
18. Kadam PD, Chuan HH (2016) Erratum to: Rectocutaneous fistula with transmigration of the suture: a rare delayed complication of vault fixation with the sacrospinous ligament. Int Urogynecol J 27:505
19. Ma K, Thomason LA, Mclaurin J (2012) Scyllo-inositol, preclinical, and clinical data for Alzheimer's disease. Adv Pharmacol 64:177–212
20. Liu FF, Dong XY, He L et al (2011) Molecular insight into conformational transition of amyloid β-peptide 42 inhibited by (−)-epigallocatechin-3-gallate probed by molecular simulations. J Phys Chem B 115:11879–11887
21. Mandel SA, Amit T, Kalfon L et al (2008) Cell signaling pathways and iron chelation in the neurorestorative activity of green tea polyphenols: special reference to epigallocatechin gallate (EGCG). J Alzheimers Dis 15:211–222
22. Hauser-Davis RA, de Freitas LV, Cukierman DS et al (2015) Disruption of zinc and copper interactions with Aβ(1-40) by a non-toxic, isoniazid-derived, hydrazone: a novel biometal homeostasis restoring agent in Alzheimer's disease therapy? Metallomics 7:743–747
23. Adlard PA, Cherny RA, Finkelstein DI et al (2008) Rapid restoration of cognition in Alzheimer's transgenic mice with 8-hydroxy

quinoline analogs is associated with decreased interstitial Abeta. Neuron 59:43–55
24. Lannfelt L, Blennow K, Zetterberg H et al (2008) Safety, efficacy, and biomarker findings of PBT2 in targeting Aβ as a modifying therapy for Alzheimer's disease: a phase IIa, double-blind, randomised, placebo-controlled trial. Lancet Neurol 7:779–786
25. Cytokine Pharmasciences, Inc. (2008) US20080 $32#096909 A1
26. Almqvist Fredrik, Sweden (2009) WO20091 $32#34203A1
27. Neuropore Therapies, Inc., USA (2013) WO2013134371A1
28. Liu H, Wang L, Su W et al (2014) Advances in recent patent and clinical trial drug development for Alzheimer's disease. Pharm Pat Anal 3:429–447
29. Awasthi M, Singh S, Pandey VP et al (2016) Alzheimer's disease: an overview of amyloid beta dependent pathogenesis and its therapeutic implications along with in silico approaches emphasizing the role of natural products. J Neurol Sci 361:256–271
30. Dhanik A, Kavraki L (2012) Protein–Ligand Interactions: Computational Docking. Encyclopedia of Life Science. John Wiley Sons, Ltd, Chichester
31. Arulalapperumal V, Sakkiah S, Thangapandian S (2012) Ligand base pharmacophore identification and molecular docking studies for Grb2 inhibitors. Korean Chem Soc 33:1707
32. Hernández M, Correa J, Nicolás MI et al (2015) Virtual and in vitro screens reveal a potential Pharmacophore that avoids the fibrillization of Aβ1-42. PLoS One 10:e0130263
33. Moustakas DT, Lang PT, Pegg S et al (2006) Development and validation of a modular, extensible docking program: DOCK 5. J Comput Aided Mol Des 20:601–619
34. Kramer B, Rarey M, Lengauer T (1999) Evaluation of the FLEXX incremental construction algorithm for protein-ligand docking. Proteins Struct Funct Bioinf 37:228–241
35. Repasky MP, Shelley M, Friesner RA (2007) Flexible ligand docking with Glide. Curr Protoc Bioinformatics. doi:10.1002/0471250953.bi0812s18
36. Eldridge MD, Murray CW, Auton TR et al (1997) Empirical scoring functions: I. The development of a fast empirical scoring function to estimate the binding affinity of ligands in receptor complexes. J Comput Aided Mol Des 11:425–445
37. Jain AN (2003) Surflex: fully automatic flexible molecular docking using a molecular similarity-based search engine. J Med Chem 46:499–511
38. Morris GM, Huey R, Lindstrom W et al (2009) Autodock4 and AutoDockTools4: automated docking with selective receptor flexibility. J Comput Chem 30:2785–2791
39. Trott O, Olson AJ (2010) AutoDockVina: improving the speed and accuracy of docking with a new scoring function, efficient optimization, and multithreading. J Comput Chem 31:455–461
40. Chen D, Martin ZS, Soto C et al (2009) Computational selection of inhibitors of Abeta aggregation and neuronal toxicity. Bioorg Med Chem 17:5189–5197
41. Chebaro Y, Jiang P, Zang T et al (2012) Structures of Aβ17-42 trimers in isolation and with five small-molecule drugs using a hierarchical computational procedure. J Phys Chem B 116:8412–8422
42. Hernández M, Correa J, Martínez F et al (2014) Design of multi-target compounds as AChE, BACE1, and amyloid-β (1-42) oligomerization inhibitors: in silico and in vitro studies. J Alzheimers Dis 41:1073–1085
43. Rao PP, Mohamed T, Teckwani K et al (2015) Curcumin binding to beta amyloid: a computational study. Chem Biol Drug Des 86:813–820
44. Churches QI, Caine J, Cavanagh K et al (2014) Naturally occurring polyphenolic inhibitors of amyloid beta aggregation. Bioorg Med Chem Lett 24:3108–3112
45. Wälti MA, Ravotti F, Arai H et al (2016) Atomic-resolution structure of a disease-relevant Aβ(1-42) amyloid fibril. Proc Natl Acad Sci U S A 113:E4976–E4984
46. Colvin MT, Silvers R, Ni QZ et al (2016) Atomic resolution structure of monomorphic $A\beta_{42}$ amyloid fibrils. J Am Chem Soc 138:9663–9674
47. Schmidt M, Rohou A, Lasker K et al (2015) Peptide dimer structure in an Aβ(1-42) fibril visualized with cryo-EM. Proc Natl Acad Sci U S A 112:11858–11863
48. Fitzpatrick AW, Debelouchina GT, Bayro MJ et al (2013) Atomic structure and hierarchical assembly of a cross-β amyloid fibril. Proc Natl Acad Sci U S A 110:5468–5473
49. Lu JX, Qiang W, Yau WM et al (2013) Molecular structure of β-amyloid fibrils in Alzheimer's disease brain tissue. Cell 154:1257–1268
50. Tomaselli S, Esposito V, Vangone P et al (2006) The alpha-to-beta conformational transition of Alzheimer's Abeta-(1-42) peptide in aqueous media is reversible: a step by step conformational analysis suggests the location of beta conformation seeding. Chembiochem 7:257–267

51. Lührs T, Ritter C, Adrian M et al (2005) 3D structure of Alzheimer's amyloid-beta (1-42) fibrils. Proc Natl Acad Sci U S A 102:17342–17347
52. Crescenzi O, Tomaselli S, Guerrini R et al (2002) Solution structure of the Alzheimer amyloid beta-peptide (1-42) in an apolar microenvironment. Similarity with a virus fusion domain. Eur J Biochem 269:5642–5648
53. Mohamed T, Shakeri A, Rao PP (2016) Amyloid cascade in Alzheimer's disease: recent advances in medicinal chemistry. Eur J Med Chem 113:258–272
54. Fezoui Y, Teplow DB (2002) Kinetic studies of amyloid beta-protein fibril assembly. Differential effects of alpha-helix stabilization. J Biol Chem 277:36948–36954
55. Bansode SB, Jana AK, Batkulwar KB et al (2014) Molecular investigations of protriptyline as a multi-target directed ligand in Alzheimer's disease. PLoS One 9:e105196
56. Cui L, Zhang Y, Cao H et al (2013) Ferulic acid inhibits the transition of amyloid-β42 monomers to oligomers but accelerates the transition from oligomers to fibrils. J Alzheimers Dis 37:19–28
57. Aloisi A, Barca A, Romano A et al (2013) Anti-aggregatingeffect of the naturally occurring dipeptide carnosine on Aβ1–42 fibril formation. PLoS One 8:e68159
58. Singh SK, Gaur R, Kumar A et al (2014) The flavonoid derivative 2-(4′ Benzyloxyphenyl)-3-hydroxy-chromen-4-one protects against Aβ42-induced neurodegeneration in transgenic Drosophila: insights from in silico and in vivo studies. Neurotox Res 26:331–350
59. Hernández M, Correa J, Benitez CG et al (2013) In silico and in vitro studies to elucidate the role of Cu2+ and galanthamine as the limiting step in the amyloid beta (1-42) fibrillation process. Protein Sci 22:1320–1335
60. Manassero G, Guglielmotto M, Zamfir R et al (2016) Beta-amyloid1-42monomers, but not oligomers, producePHF-likeconformation of Tau protein. Aging Cell 15:914–923
61. Wang J, Land D, Ono K et al (2014) Molecular topology as novel strategy for discovery of drugs with Aβ lowering andanti-aggregation-dual activities for Alzheimer's disease. PLoS One 9:e92750
62. Das S, Smid SD (2017) Identification of dibenzyl imidazolidine and triazole acetamide derivatives through virtual screening targeting amyloid beta aggregation and neurotoxicity in PC12 cells. Eur J Med Chem 130:354–364
63. Das S, Stark L, Musgrave IF et al (2016) Bioactive polyphenol interactions with β amyloid: a comparison of binding modelling, effects on fibril and aggregate formation and neuroprotective capacity. Food Funct 7:1138–1146
64. Wang Y, Xia Z, Xu JR, Wang YX, Hou LN, Qiu Y, Chen HZ (2012) ς-mangostin, a polyphenolic xanthone derivative from mangosteen, attenuates β-amyloid oligomers-induced neurotoxicity by inhibiting amyloid aggregation. Neuropharmacology 62(2):871–881
65. Orlando RA, Gonzales AM, Royer RE et al (2012) A chemical analog of curcumin as an improved inhibitor of amyloid Abeta oligomerization. PLoS One 7:e31869
66. Kuang G, Murugan NA, Tu Y et al (2015) Investigation of the binding profiles of AZD2184 and Thioflavin T with Amyloid-β (1–42) fibril by molecular docking and molecular dynamics methods. J Phys Chem B 119:11560–11567
67. Nie Q, Du XG, Geng MY (2011) Small molecule inhibitors of amyloid β peptide aggregation as a potential therapeutic strategy for Alzheimer's disease. Acta Pharmacol Sin 32:545–551
68. Reinke AA, Gestwicki JE (2007) Structure–activity relationships of amyloid Beta-aggregation inhibitors based on Curcumin: influence of linker length and flexibility. Chem Biol Drug Des 70:206–215
69. ÖzturanÖzer E, Tan OU, Ozadali K et al (2013) Synthesis, molecular modeling and evaluation of novel N'-2-(4-benzylpiperidin–/piperazin-1-yl) acylhydrazone derivatives as dual inhibitors for cholinesterases and Aβ aggregation. Bioorg Med Chem Lett 23:440–443
70. Xu P, Zhang M, Sheng R, Ma Y (2017) Synthesis and biological evaluation of deferiprone-resveratrol hybrids as antioxidants, Aβ1-42 aggregation inhibitors and metal-chelating agents for Alzheimer's disease. Eur J Med Chem 127:174–186
71. Mohamed T, Rao PP (2017) 2,4-Disubstituted quinazolines as amyloid-β aggregation inhibitors with dual cholinesterase inhibition and antioxidant properties: development and structure-activity relationship (SAR) studies. Eur J Med Chem 126:823–843
72. Li L, Chen M, Jiang FC (2016) Design, synthesis, and evaluation of2-piperidone-derivatives for the inhibition of β-amyloid aggregation and inflammation mediated neurotoxicity. Bioorg Med Chem 24:1853–1865
73. Teplow DB, Lazio ND, Bitan G et al (2006) Elucidating amyloid beta-protein folding and

assembly: a multidisciplinary approach. Acc Chem Res 39:635–645

74. Novick PA, Lopes DH, Branson KM et al (2012) Design of β-amyloid aggregation inhibitors from a predicted structural motif. J Med Chem 55:3002–3010
75. Lemkul JA, Bevan DR (2012) The role of molecular simulations in the development of inhibitors of amyloid β-peptide aggregation for the treatment of Alzheimer's disease. ACS Chem Neurosci 3:845–856
76. Barz B, Urbanc B (2012) Dimer formation enhances structural differences between amyloid β-protein(1–40) and (1–42): anexplicit-solventmolecular dynamics study. PLoS One 7:e34345
77. Zoete V, Cuendet MA, Grosdidier A et al (2011) SwissParam: a fast force field generation tool for small organic molecules. J Comput Chem 32:2359–2368
78. Case DA, Cheatham TE, Darden T et al (2005) The Amber biomolecular simulation programs. J Comput Chem 26:1668–1688
79. Gotz AW, Williamson MJ, Xu D et al (2012) Routine microsecond molecular dynamics simulations with AMBER on GPUs. 1. Generalized born. J Chem Theory Comput 8:1542–1555
80. Oostenbrink C, Villa A, Mark AE et al (2004) A biomolecular force field based on the free enthalpy of hydration and solvation: the GROMOS force-field parameter sets 53A5 and 53A6. J Comput Chem 25:1656–1676
81. Phillips JC, Braun R, Wang W et al (2005) Scalable molecular dynamics with NAMD. J Comput Chem 26:1781–1802
82. Hernández M, Rosales MC, Mendieta JE et al (2016) Current tools and methods in molecular dynamics (MD) simulations for drug design. Curr Med Chem 23:3909–3924
83. Bajda M, Filipek S (2017) Computational approach for the assessment of inhibitory potency againstbeta-amyloidaggregation. Bioorg Med Chem Lett. 27:212–216
84. Kundaikar HS, Degani MS (2015) Insights into the interaction mechanism of ligands with Aβ42 based on molecular dynamics simulations and mechanics: implications of role of common binding site in drug Design for Alzheimer's disease. Chem Biol Drug Des 86:805–812
85. Wang Q, Liang G, Zhang M (2014) De novo design of self-assembled hexapeptides as β-amyloid (Aβ) peptide inhibitors. ACS Chem Neurosci 5:972–981
86. Qiao Y, Zhang M, Liang Y et al (2016) A computational study of self-assembled hexapeptide inhibitors against amyloid-β (Aβ) aggregation. Phys Chem Chem Phys 19:155–166
87. Zhang T, Xu W, Mu Y, Derreumaux P (2014) Atomic and dynamic insights into the beneficial effect of the 1,4-naphthoquinone-2-yl-L-tryptophan inhibitor on Alzheimer's Aβ1-42 dimer in terms of aggregation and toxicity. ACS Chem Neurosci 5:148–159
88. Wang YY, Li L, Chen TT, Chen WY, Xu YC (2013) Microsecond molecular dynamics simulation of A_{42} and identification of a novel dual inhibitor of $Aβ_{42}$ aggregation and BACE1 activity. Acta Pharmacol Sin 34:1243–1250

Chapter 10

In Silico Strategies to Design Small Molecules to Study Beta-Amyloid Aggregation

Praveen P.N. Rao and Deguo Du

Abstract

The amyloid cascade hypothesis of Alzheimer's disease is an intriguing and complex pathway which alludes to the accumulation of amyloid (Aβ) aggregates as a key toxic event. In this regard, computational tools can be effectively used to study and design amyloid aggregation inhibitors and modulators. A number of in silico methods have been described in the literature, which use full-length Aβ proteins. However, these methods are computationally expensive. Herein, we describe an in silico method that uses the Aβ hexapeptide-derived steric-zipper octamer assembly as an alternative and effective model to predict the binding interactions of planar small molecule libraries with complex Aβ structures including oligomers, protofibrils, and fibrils. The method provides detailed steps involved in conducting molecular docking, the interpretation of results obtained including ligand-Aβ hexapeptide interactions, calculation of ligand binding energies, and its correlation with the ligand binding affinity. This octamer steric-zipper model represents a relevant prototype to explore and study the mechanisms of Aβ aggregation using small molecules.

Key words Amyloid-beta, Alzheimer's disease, Tricyclics, KLVFFA, Steric-zipper, Molecular docking, CDOCKER, CHARMm, Binding energy

1 Introduction: Beta-Amyloid

The term "beta-amyloid" or "amyloid-beta (Aβ)" refers to a group of peptides ranging in length from 39 to 43 amino acids. They are implicated in the pathogenesis of Alzheimer's disease (AD) [1–5]. Particularly, the two isoforms Aβ40 and Aβ42 have the tendency to aggregate into higher-order species including dimers, trimers, oligomers, protofibrils, and fibrils that exhibit different degrees of solubility and neurotoxicity. These observations constitute the basis of the "amyloid cascade hypothesis" of AD [5–8]. Preventing the accumulation of Aβ aggregates in the brain represents one of the key strategies aimed at preventing AD and associated dementia [9]. This quest has led to the development of a large number of

Kunal Roy (ed.), *Computational Modeling of Drugs Against Alzheimer's Disease*, Neuromethods, vol. 132, DOI 10.1007/978-1-4939-7404-7_10, © Springer Science+Business Media LLC 2018

novel synthetic small molecules that are known to prevent the formation of Aβ aggregates by direct binding. In addition, several molecules present in nature are known to exhibit anti-Aβ activity [10, 11].

1.1 Computational Methods to Study and Predict Aβ Binding

Despite these efforts, understanding the binding mechanism of small molecules to Aβ aggregates continues to be challenging while at the same time intriguing due to the fact that unlike enzymes, Aβ aggregates do not have well-defined grooves or binding sites. The complexity is increased further as Aβ aggregates can exist in multiple toxic forms with some molecules exhibiting selective or nonselective binding toward either lower- or higher-order Aβ aggregates. The mechanism of action of molecules that prevent Aβ aggregation by direct binding can be determined experimentally by fluorescence spectroscopy, thioflavin T (ThT)-based aggregation kinetics, atomic force of microscopy (AFM), transmission electron microscopy (TEM), and Western blotting [12]. All these methods have their benefits and drawbacks. In this regard, computational methods are invaluable tools to understand and predict the binding potential of direct Aβ inhibitors. These in silico methods can assist in understanding the mechanisms of Aβ aggregation. The availability of solved structures of Aβ is extremely useful to study and investigate small molecule-Aβ binding interactions. For example, the binding interactions of potent direct Aβ inhibitors can be studied by a combination of molecular docking and molecular dynamics using Aβ monomer and dimer models [13]. However, predicting the binding of small molecules toward other Aβ structures especially oligomers and fibrils is difficult as it correlates with the conversion from toxic to less toxic or nontoxic forms after binding [14, 15]. Moreover, current literature shows that some molecules can inhibit oligomerization by promoting the fibrillization process instead of direct binding to lower-order Aβ aggregates [16, 17]. Modeling and predicting these outcomes is challenging and can be computationally expensive. An alternative approach is to model short segments of aggregation-prone regions of Aβ instead of using the full-length Aβ sequence. In this approach, we used the solved structure of Aβ-derived hexapeptide ^{16}KLVFFA^{21} steric-zipper assembly [18] to model the binding interactions of methylene blue (MB), a tricyclic phenothiazine-based small molecule (Fig. 1). MB represents an interesting molecule in the literature as it does not prevent Aβ aggregation [16, 19, 20]. Remarkably it can promote the formation of less toxic fibrils [16]. MB can undergo reduction in vivo to form leucomethylene blue (leucoMB) shown in Fig. 1. Other MB metabolites include the demethylated forms azure A and azure B (Fig. 1). We used the Aβ-derived hexapeptide ^{16}KLVFFA^{21} steric-zipper assembly to understand the binding interactions of MB and its derivatives as inhibitors of Aβ oligomerization. An octamer steric-zipper model of Aβ-derived hexapeptide

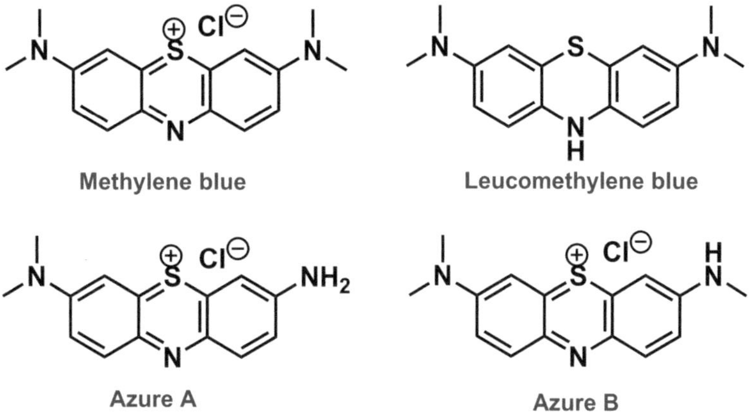

Fig. 1 Chemical structures of phenothiazine-based tricyclics

was built and was used to conduct molecular docking experiments of fused tricyclics, and their binding interactions were investigated by qualitative and quantitative methods.

2 Materials

The computational simulations were carried out using the Discovery Studio 2017 Client – Structure-Based Design (SBD) software, version v17.1.0.16143 from Dassault Systemes Biovia Corp, France. The SBD suite is capable of preparing 3D structures of macromolecules and small molecule ligands/libraries, generate conformations, perform molecular docking and molecular dynamics, generate pharmacophores, and calculate 2D and 3D molecular properties. The CHARMm-based DOCKER (CDOCKER) algorithm in the SBD is capable of performing flexible ligand-based molecular docking. CDOCKER is a grid-based computational algorithm that incorporates molecular dynamics (MD) in docking and offers complete ligand flexibility [21]. The SBD software suite also includes CHARMm molecular mechanics simulation program. Molecular visualization was carried out using the BIOVIA Discovery Studio Visualizer. The computational software was run using a Dell Optiplex 3040 PC with an Intel(R) Core (TM) i5-6500 CPU at 3.2 GHz processor.

3 Methods

3.1 Preparation of Small Molecule Ligands

Small molecules MB, leucoMB, azure A, and azure B were built using the *small molecules* module in the SBD suite. The small molecules can be initially built as 2D diagrams which can be converted to 3D structures. Then the 3D structures were assigned

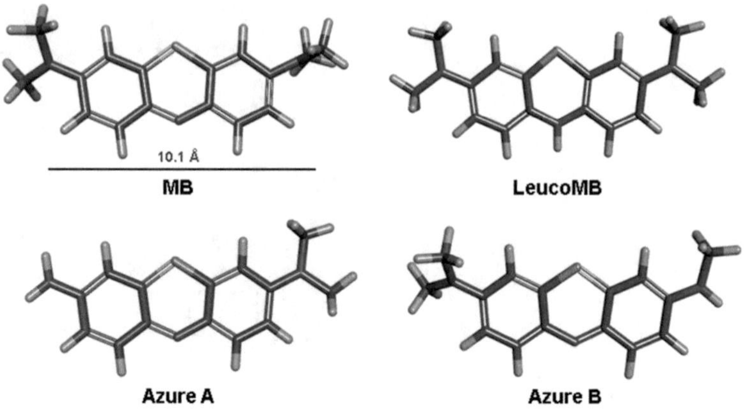

Fig. 2 Energy-minimized 3D structures of MB, leucoMB, azure A, and azure B. The distance along the long axis of MB is shown

proper ionization status at pH 7.4 using the *Prepare Ligands* tool in the *Small Molecules* module. This method can also generate tautomers and isomers depending on the chemical structures of the small molecules. In our case, the four compounds described (MB, leucoMB, azure A, and azure B) do not exhibit any tautomerism or have isomeric forms. Among them, MB, azure A and azure B possess a positively charged sulfur atom (Fig. 1). The prepared ligands were then subjected to minimization by 500 steps each of steepest descent and conjugate gradient minimization, respectively (RMS gradient 0.1 kcal/mol), using the CHARMm force field and SHAKE algorithm to constrain bond lengths. A distance-dependent dielectric constant was used as an implicit solvent model during this minimization step. From our experience, we have seen that using an implicit solvent function provides a better starting conformation of small molecule ligands for molecular docking studies compared to running the minimization in vacuum. The energy minimized structures were saved as .dsv files (Fig. 2). The energy minimization was carried out using the *Simulation* module in the SBD suite.

3.2 Aβ-Derived ^{16}KLVFFA^{21}-Hexapeptide Steric-Zipper Model

The solved ^{16}KLVFFA^{21} steric-zipper tetramer assembly available in the protein data bank (pdb id: 3OVJ, resolution 1.8 Å) is an x-ray crystal structure (Fig. 3). This structure represents the interface or cross-β spine of the Aβ protofibrils [22] with side chains of polar lysine (^{16}KLVFFA), nonpolar amino acids leucine (K^{17}LVFFA), valine (KL^{18}VFFA), and phenylalanine (KLVF^{20}FA) representing the binding site where small molecules can interact and bind. This x-ray structure contains the known Aβ aggregation inhibitor orange G bound which undergoes polar and nonpolar contacts with amino acids valine, phenylalanine, and lysine [18]. This model represents a simple system to dock small molecule libraries

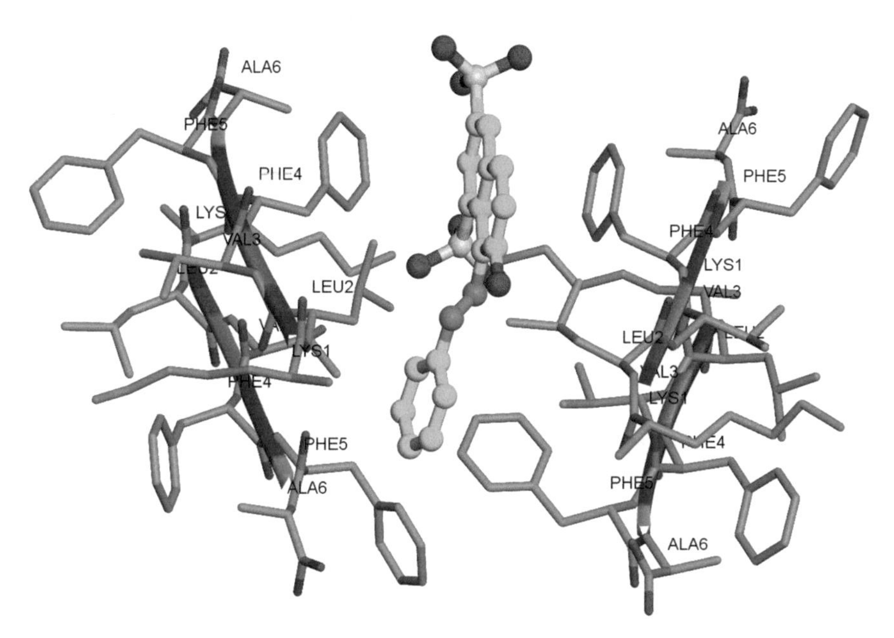

Fig. 3 X-ray crystal structure of orange G bound to the Aβ-derived [16]KLVFFA[21] tetramer steric-zipper assembly (PDB ID: 3OVJ)

to predict their Aβ-binding potential and prevent Aβ aggregation by direct binding. Using the full-length Aβ40 or Aβ42 solution structures to study and design small molecules as potential tools or therapeutic agents by computational methods is exhaustive and time-consuming. Compared to the Aβ40 fibril solution structure which has a U shape, the recently solved Aβ42 fibril structure turns out to be more complex due to its double-horse shoe-like shape further highlighting the complexity involved in studying these full-length proteins [23–25]. In contrast, the steric-zipper assembly of Aβ is a simple and efficient model to screen large compound libraries as putative inhibitors or modulators of Aβ aggregation [13, 26]. Moreover, this can be used as a general model to study the binding potential of small molecules toward both Aβ40 and Aβ42 at the hydrophobic KLVFFA interface as this region is implicated in the initiation of Aβ aggregation [27, 28]. We believe that this hexapeptide steric-zipper represents a suitable in silico model to study the mechanisms of Aβ aggregation using small molecules.

While using this steric-zipper tetramer model to conduct molecular docking studies, it is important to consider the length of small molecules along their long axis. As such, planar naphthalene containing orange G's length spans approximately 9.1 Å [26] and still can be accommodated in the cross-β spine of Aβ steric-zipper tetramer assembly due to the presence of the sp2 hybridized azo group substituted with a planar benzene ring (Fig. 3). In contrast, this tetramer steric-zipper assembly is not suitable for small molecules that possess a longer axis length > 9.0 Å, especially for fused planar ring systems such as tricyclics and planar small molecules that possess conjugated systems as exemplified by curcumin. These small molecules would not be accommodated in the tetramer assembly, and molecular docking with this setup can provide incorrect and inefficient binding orientations. These shortcomings can be overcome by designing alternative steric-zipper models to accommodate small molecules with longer axis lengths. Our research group has developed octamer assemblies of Aβ steric-zipper that can accommodate and predict the binding modes of small molecules that possess longer axis length in the range of 9–13 Å [26]. For small molecules with axis length beyond 13 Å, we recommend building a dodecamer steric-zipper assembly consisting of 12 individual KLVFFA hexapeptide segments arranged in antiparallel fashion.

3.3 Preparation of Aβ-Derived ^{16}KLVFFA^{21}-Hexapeptide Steric-Zipper Octamer Model

The x-ray structure of orange G bound to ^{16}KLVFFA^{21} steric-zipper tetramer assembly was obtained (pdb id: 3OVJ). The steric-zipper octamer assembly was constructed using the *Macromolecules* module by copying the steric-zipper tetramer assembly and orienting it close to the other tetramer assembly manually using the select/rotate/translate tool in SBD such that the intrastrand and interstrand distances between individual strands were 5 Å and 10 Å, respectively (Fig. 4), while monitoring the distance parameters. The entire octamer assembly was subjected to minimization protocol in the *Simulation* module using CHARMm force field with SHAKE constraints using 2000 steps each of steepest descent followed by conjugate gradient minimization, respectively, to a convergence of 0.1 kcal/mol to remove any bump or bad contact. In the next step, the entire octamer assembly was selected as the receptor, and one of the orange G molecules was deleted, whereas the other orange G was selected and used to create a sphere of 15 Å radius as the ligand binding site using the *Ligand-Receptor Interactions* module. Then orange G was deleted. It should be noted that this 15 Å sphere covers the entire span of the cross-β spine binding site in the octamer assembly (Fig. 4). Then the steric-zipper octamer assembly was assigned proper ionization state at pH 7.4 using the *Prepare Protein* tool in *Macromolecules* module.

3.4 Molecular Docking of MB, LeucoMB, Azure B, and Azure A in the Aβ-Derived ^{16}KLVFFA^{21}-Hexapeptide Steric-Zipper Octamer Model

The molecular docking studies were carried out using the CDOCKER algorithm available in the *Receptor-Ligand Interactions* module in the SBD suite. The CDOCKER algorithm provides receptor-ligand poses based on a simulated annealing protocol, and the poses are ranked using energy parameters, CDOCKER energy, and CDOCKER interaction energy in kcal/mol, respectively. The docking steps include 2000 heating steps, with a 700 K target temperature, and 5000 cooling steps with a target temperature of 300 K to generate 10 docked ligand poses. The CHARMm force field was used during the docking simulation. The quality of

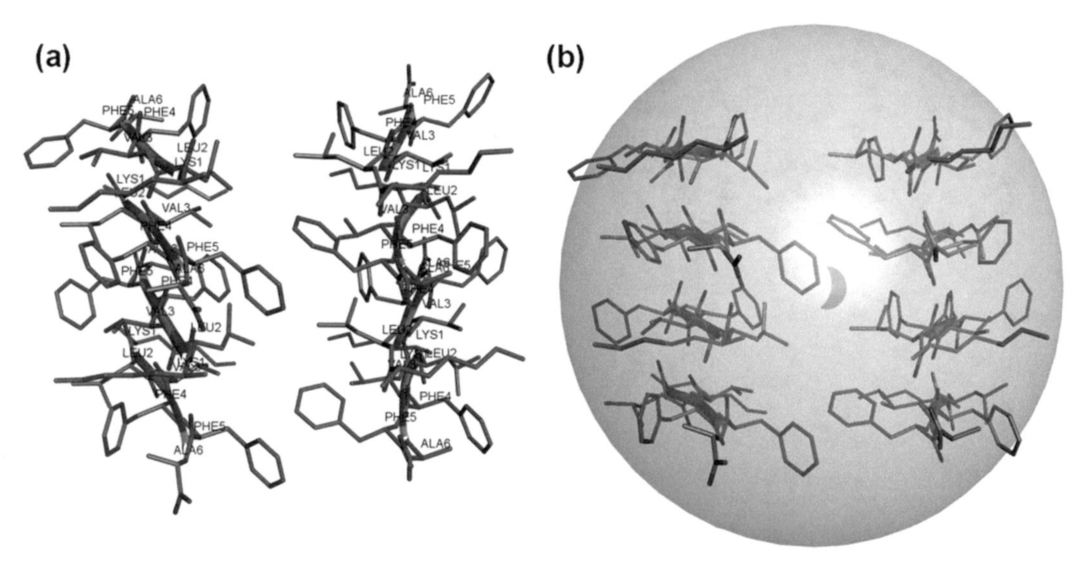

Fig. 4 (**a**) Octamer steric-zipper model for docking studies derived from the [16]KLVFFA[21] tetramer steric-zipper assembly; (**b**) a 15 Å sphere in the Aβ hexapeptide octamer steric-zipper model used for molecular docking studies of small molecules

docking was evaluated by considering the energy parameters such as CDOCKER energy and CDOCKER interaction energy between the ligand and the receptor. If the energy parameters are in the positive range, they represent high energy complex which may not be stable, whereas negative numbers indicate reasonable structures which can be further validated by calculating the ligand-receptor binding energy using the equation E_{binding} = Energy of complex ($E_{\text{ligand-receptor}}$) –Energy of ligand ($E_{\text{ligand}}$)– Energy of receptor ($E_{\text{receptor}}$). We used the implicit solvent model Generalized Born with a simple SWitching (GBSW) to calculate the energy of binding in kcal/mol. Positive values indicate high energy complex, whereas negative values indicate a stable ligand-receptor complex. In addition, monitoring the type and number of polar and nonpolar contacts of the ligand with the receptor assists in determining the best binding mode of the ligand.

The best docked poses of MB, leucoMB, azure A, and azure B are shown in Figs. 5 and 6. For all the small molecule ligands docked, pose 1 was identified as the best binding mode among the list of ten poses obtained. This was determined based on the stability of the ligand-Aβ steric-zipper octamer complex energy represented by the CDOCKER energy in kcal/mol (Table 1). The other energy parameter CDOCKER interaction energy compares the quality of ligand binding. Analysis of the best binding mode of MB in the steric-zipper assembly shows that the planar tricyclic MB was oriented parallel to the octamer fiber axis (Fig. 5) and the positively charged sulfur atom of the central thiazinium ring underwent a stabilizing cation-π interaction with the aromatic

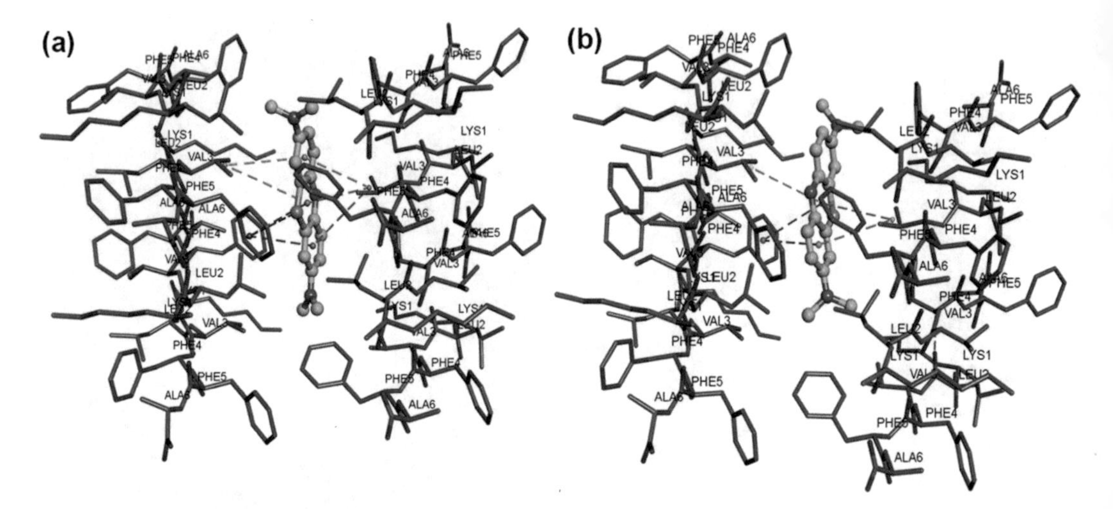

Fig. 5 (**a**) The tricyclic compound MB docked in the ^{16}KLVFFA^{21}octamer steric-zipper model; (**b**) the tricyclic compound leucoMB docked in the ^{16}KLVFFA^{21}octamer steric-zipper model. Small molecules are shown as ball and stick cartoon. Polar and nonpolar contacts of the ligand with the steric-zipper amino acids are shown. Hydrogen atoms are not shown to enhance clarity

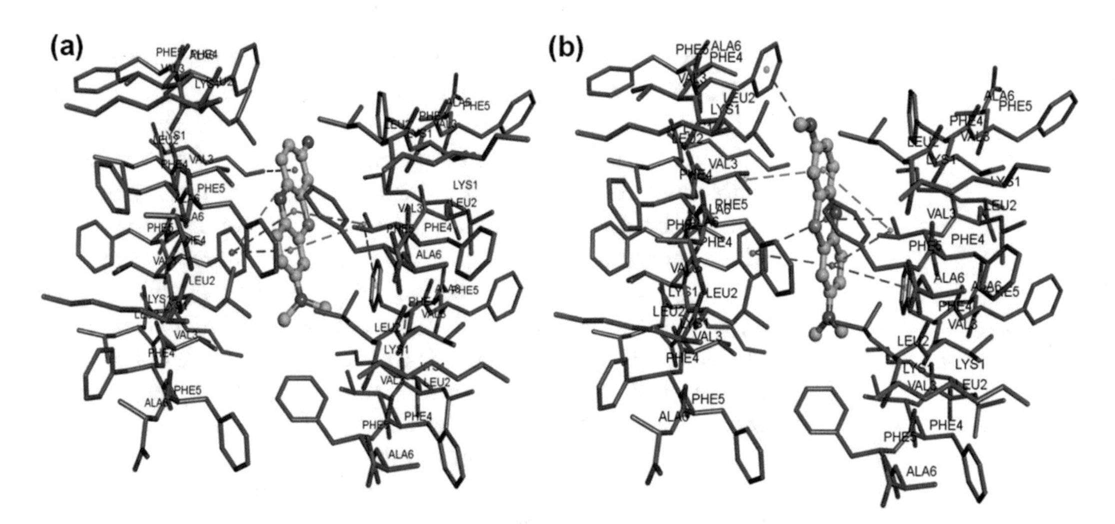

Fig. 6 (**a**) The tricyclic compound azure A docked in the ^{16}KLVFFA^{21}octamer steric-zipper model; (**b**) the tricyclic compound azure B docked in the ^{16}KLVFFA^{21}octamer steric-zipper model. Polar and nonpolar contacts of the ligand with the steric-zipper amino acids are shown. The small molecules are shown as ball and stick cartoon. Hydrogen atoms are not shown to enhance clarity

ring of phenylalanine (distance ~5.0 Å). The central thiazinium ring also underwent π-alkyl interaction with side chains of two valine residues (distance <5.5 Å). The two substituted benzene rings on either side of the central thiazinium ring were in van der Waal's contact with side chains of phenylalanines and valines at the steric-zipper interface, whereas the two dimethylamino substituents on either benzene rings at C3 and C7 position were accommodated in

Table 1
Ligand-receptor complex energy parameters for small molecules docked in the Aβ hexapeptide octamer steric-zipper model using the CDOCKER algorithm

Energy (kcal/mol)			
Compd	CDOCKER energy[a]	CDOCKER interaction energy[a]	Binding energy[b]
MB	−10.01	−27.93	−20.31
LeucoMB	−18.71	−27.72	−14.21
Azure A	−09.19	−24.49	−11.69
Azure B	−07.66	−24.82	−15.50

[a]The energy terms CDOCKER energy and CDOCKER interaction energy were calculated using the CDOCKER molecular docking algorithm in the Discovery Studio Client – Structure-Based Design software (Dassault Systemes Biovia Corp, France)
[b]Binding energy was calculated using the Receptor-Ligand Interactions module based on the equation $E_{binding}$ = Energy of complex ($E_{ligand\text{-}receptor}$) − Energy of ligand ($E_{ligand}$) − Energy of receptor ($E_{receptor}$) using the implicit solvent function GBSW

hydrophobic pockets consisting of phenylalanines, valines, and leucines (distance <5.5 Å). The docking pose of leucoMB which contains a central thiazine ring instead of an aromatic thiazinium exhibits an almost planar orientation parallel to the fiber axis. The tricyclic ring was flipped by 180° angle compared to the binding mode of MB, and the central thiazine ring underwent van der Waal's contacts (π-π and alkyl-π interactions) with phenylalanine and valine side chains (distance <5.3 Å). The flanking benzene rings fused to the thiazine ring on either side underwent van der Waal's contact with phenylalanine and valine side chains (distance <5.8 Å). The dimethylamino substituents were closer to hydrophobic regions made up of phenylalanine, leucine, and valine (Fig. 5). The binding modes of azure A which contains a primary amine instead of a dimethylamino substituent and azure B which contains a secondary amine (methylamino) instead of a dimethylamino substituent (Fig. 1) are shown in Fig. 6. These two MB metabolites also exhibited planar conformation similar to MB and were oriented parallel to the fiber axis. The docking studies show that these planar tricyclics are able to stabilize the cross-β spine assembly by binding in a parallel fashion across the fiber axis which can prevent their dissociation into lower-order, more toxic structures. The next question to ask would be among these four tricyclics, which would exhibit a better binding affinity in the octamer steric-zipper model? This can be answered by considering the CDOCKER interaction energy which suggests that the binding affinity would be of the order MB > leucoMB > azure B > azure A (Table 1). However, using scoring function such as CDOCKER interaction energy to predict binding affinity could be misleading as it doesn't consider all the ligand and receptor energy parameters

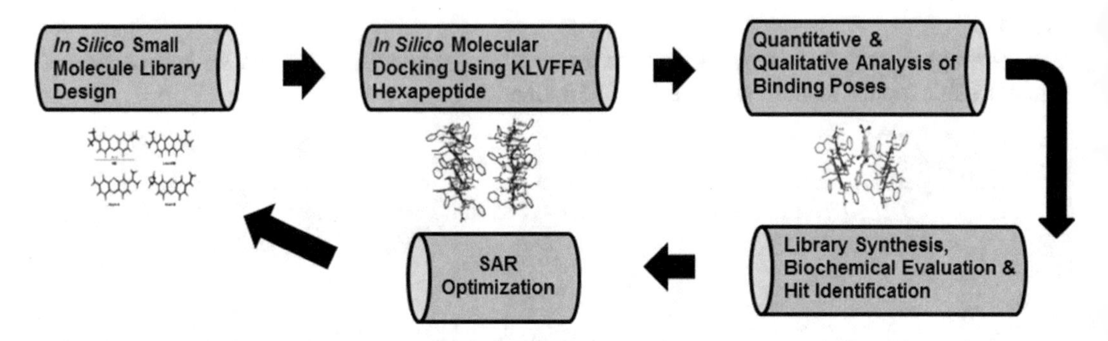

Fig. 7 A flow chart which shows the steps involved in anti-Aβ screening of small molecule libraries using the Aβ-derived ^{16}KLVFFA^{21} hexapeptide

such as ligand energy and protein energy. In addition, solvent effects are not considered. In contrast, the calculation of binding energy provides a more accurate prediction of ligand binding affinity. This parameter was calculated using the *Receptor-Ligand Interactions* module in the SBD suite and by selecting *Calculate Binding Energies* tool. This scoring function considers ligand entropic energies, ligand conformations as per the equation E_{binding} = Energy of complex ($E_{\text{ligand-receptor}}$) – Energy of ligand ($E_{\text{ligand}}$) – Energy of receptor ($E_{\text{receptor}}$), and implicit solvent effects. We used GBSW implicit solvent functionality to obtain binding energies (Table 1) which show that MB exhibited greater binding affinity (−20.31 kcal/mol) compared to other MB derivatives. The binding affinity trend was of the order MB (−20.31 kcal/mol) > azure B (−15.50 kcal/mol) > leucoMB (−14.21 kcal/mol) > azure A (−11.69 kcal/mol). This suggests that efficiency of steric-zipper octamer stabilization was mainly dependent on cation-π and alkyl-π interactions as opposed to hydrogen bonding interactions.

The protocol described here using the Aβ-derived KLVFFA hexapeptide steric-zipper octamer model is applicable to investigate the binding modes and mechanisms of planar small molecule libraries to uncover the potential mechanisms of Aβ oligomer, protofibril, and fibril aggregation. A flow chart is provided which gives a summary on the approach used (Fig. 7). It should be noted that similar in silico approaches using other steric-zipper assembly polymorphs of aggregation-prone Aβ or disease-related amyloid proteins would provide a good understanding on selective and nonselective binding exhibited by diverse class of small molecules. The molecular docking protocol described here is open to further refinement. However, we feel that it may not change the qualitative output significantly. Instead it might affect the quantitative data generated such as energy parameters based on the minimization steps, molecular docking algorithm, and explicit or implicit solvent functions used.

4 Conclusions

A versatile in silico modeling technique based on molecular docking was described for the rapid screening and investigation of binding interactions of small molecule libraries as potential Aβ aggregation modulators. The Aβ hexapeptide KLVFFA sequence-derived octamer steric-zipper model developed can be used as a valuable computational tool to study the binding interactions of small molecules with Aβ oligomers, protofibrils, and fibrils to gain insight into their potential binding modes and mechanisms of binding. This method is particularly useful to study the binding interactions of small molecule libraries with planar geometry such as fused bicyclics, tricyclics, as well as other planar conjugated systems. The in silico approach described is ideally suited for small molecule design and optimization and to elucidate mechanisms of Aβ aggregation.

5 Notes

1. The choice of using either the tetramer or octamer steric-zipper assembly of KLVFFA fragment is dependent on the length of the small molecule across its longer axis. This is especially true for small molecules with planar geometry. For small molecules that have an axis length less than 9 Å, KLVFFA tetramer assembly provides reliable data, whereas for small molecules with greater axis length (9–13 Å range), the KLVFFA octamer assembly is appropriate.
2. The molecular docking protocol described using the KLVFFA steric-zipper model is primarily designed to explore the binding interactions of planar ring systems and compound libraries with higher-order Aβ aggregates such as oligomers, protofibrils, and fibrils.

Acknowledgments

The PPNR would like to thank the University of Waterloo, NSERC-Discovery RGPIN 03830-2014, and Ministry of Research and Innovation, Government of Ontario, Canada, for an Early Researcher Award (ERA) for the financial support. DD also thanks the National Institutes of Health (R15GM116006) for the financial support.

References

1. LaFerla FM, Green KN, Oddo S (2007) Intracellular amyloid-beta in Alzheimer's disease. Nat Rev Neurosci 8:499–509
2. Maji SK, Wang L, Greenwald J, Riek R (2009) Structure-activity relationship of amyloid fibrils. FEBS Lett 583:2610–2617
3. Murphy MP, LeVine H (2010) Alzheimer's disease and the β-amyloid peptide. J Alzheimers Dis 19:311–323
4. Masters CL, Selkoe DJ (2012) Biochemistry of amyloid β-protein and amyloid deposits in Alzheimer's disease. Cold Spring Harb Perspect Med 2:a006262
5. Selkoe DJ, Hardy J (2016) The amyloid hypothesis of Alzheimer's disease at 25 years. EMBO Mol Med 8:595–608
6. Kevin WT, Oosten-Hawle PV, Hewitt EW, Radford SE (2015) Amyloid fibres: inert end-stage aggregates or key players in disease? Trends Biochem Sci 40:719–727
7. Ries HM, Nussbaum-Krammer C (2016) Shape matters: the complex relationship between aggregation and toxicity in protein-misfolding diseases. Essays Biochem 60:181–190
8. Hamley IW (2012) The amyloid beta peptide. Chem Rev 112:5147–5192
9. Villemagne VL, Burnham S, Bourgeat P, Brown B, Ellis KA, Salvado O, Szoeke C, Macaulay SL, Martins R, Maruff P, Ames D, Rowe CC, Master CL (2013) Amyloid β deposition, neurodegeneration, and cognitive decline in sporadic Alzheimer's disease: a prospective cohort study. Lancet Neurol 12:357–367
10. Belluti F, Rampa A, Gobbi S, Bisi A (2013) Small-molecule inhibitors/modulators of amyloid-β peptide aggregation and toxicity for the treatment of Alzheimer's disease: a patent review (2010 –2012). Expert Opin Ther Pat 23:581–596
11. Bu XL, Rao PPN, Wang YJ (2016) Anti-amyloid aggregation activity of natural compounds: implications for Alzheimer's drug discovery. Mol Neurobiol 53:3565–3575
12. Bruggink KA, Muller M, Kuiperij HB, Verbeek MM (2012) Methods for analysis of amyloid-β aggregates. J Alzheimers Dis 28:735–758
13. Mohamed T, Shakeri A, Rao PPN (2016) Amyloid cascade in Alzheimer's disease: recent advances in medicinal chemistry. Eur J Med Chem 113:258–272
14. Nguyen P, Derreumaux P (2014) Understanding amyloid fibril nucleation and Aβ oligomer/drug interactions from computer simulations. Acc Chem Res 47:603–611
15. Jiang L, Liu C, Leibly D, Landau M, Zhao M, Hughes MP, Eisenberg DS (2013) Structure-based discovery of fiber-binding compounds that reduce the cytotoxicity of amyloid beta. elife 2:e00857
16. Necula M, Breydo L, Milton S, Kayed R, van der Veer WE, Tone P, Glabe CG (2007) Methylene blue inhibits amyloid Aβ oligomerization by promoting fibrillization. Biochemistry 46:8850–8860
17. Bieschke J, Herbst M, Wiglenda T, Friederich RP, Boeddrich A, Schiele F, Kleckers D, Lopez del Amo JM, Gruning BA, Wang Q, Schmidt MR, Lurz R, Anwyl R, Schnoegl S, Fandrich M, Frank RF, Reif B, Gunther S, Walsh DM, Wanker EE (2011) Small-molecule conversion of toxic oligomers to nontoxic sheet-rich amyloid fibrils. Nat Chem Biol 8:93–101
18. Landau M, Sawaya MR, Faull KF, Laganowsky A, Jiang L, Sievers SA, Liu J, Barrio JR, Eisenberg D (2011) Towards a pharmacophore for amyloid. PLoS Biol 9:e1001080
19. Schirmer RH, Adler H, Pickhardt M, Mandelkow E (2011) Lest we forget you—methylene blue. Neurobiol Aging 32:2325.e7–2325.16
20. Petzer A, Harvey BH, Wegener G, Petzer JP (2012) Azure B, a metabolite of methylene blue, is a high-potency, reversible inhibitor of monoamine oxidase. Toxicol Appl Pharmacol 258:403–409
21. Wu G, Roberson DH, Brooks CL III, Vieth M (2003) Detailed analysis of grid-based molecular docking: a case study of CDOCKER- A CHARMm-based MD docking algorithm. J Comput Chem 24:1549–1562
22. Colletier JP, Laganowsky A, Landau M, Zhao M, Soriaga AB, Goldschmidt L, Flot D, Cascio D, Sawaya MR, Eisenberg D (2011) Molecular basis for amyloid-β polymorphism. Proc Natl Acad Sci U S A 108:16938–16943
23. Petkova A, Yau W, Tycko R (2006) Experimental constraints on the quaternary structure in Alzheimer's beta-amyloid fibrils. Biochemistry 45:498–512
24. Walti MA, Ravotti F, Arai H, Glabe CG, Wall JS, Bockmann A, Guntert P, Meier BH, Riek R (2016) Atomic-resolution structure of a disease-relevant Aβ(1–42) amyloid fibril. Proc Natl Acad Sci U S A 113:E4976–E4984
25. Colvin MT, Silvers R, Frohm B, Su Y, Linse S, Griffin RG (2015) High resolution structural characterization of Aβ42 amyloid fibrils by

magic angle spinning NMR. J Am Chem Soc 137:7509–7518

26. Rao PPN, Mohamed T, Teckwani K, Tin G (2015) Curcumin binding to beta amyloid: a computational study. Chem Biol Drug Des 86:813–820

27. Lopez de la Paz M, Serrano L (2004) Sequence determinants of amyloid fibril formation. Proc Natl Acad Sci U S A 101:87–92

28. Pastor MT, Esteras-Chopo A, Serrano L (2007) Hacking the code of amyloid formation. Prion 1:9–14

Chapter 11

Computational Approaches to Understand Cleavage Mechanism of Amyloid Beta (Aβ) Peptide

Kailas Dashrath Sonawane and Maruti Jayaram Dhanavade

Abstract

Molecular modeling methods are being used routinely for the prediction and analysis of three-dimensional structures of biological macromolecules. Generally, prediction of three-dimensional structures of macromolecules is done by X-ray crystallography or NMR spectroscopy. But these methods are time-consuming and not always suitable for all types of proteins, especially for larger proteins. For some proteins, it is difficult to obtain the crystals; therefore, 3D structure determination of those proteins by X-ray crystallography is not possible. Limitations for NMR spectroscopy are the size of the systems, as NMR spectroscopy can be applied only to smaller protein systems. Therefore, prediction of three-dimensional structures of proteins/ enzymes from their amino acid sequence can be achieved by using different types of approaches such as "comparative" or "homology" modeling. Thus, computational methods have been used to investigate the three-dimensional structures of amyloid beta peptide-degrading enzymes. Molecular modeling methods like energy minimization, molecular docking, molecular dynamics simulation, virtual screening, binding free energy, and solvent-accessible surface calculations are widely used to understand molecular interactions between Aβ-degrading enzymes and amyloid beta peptides. Thus, the present chapter deals with the applications of computational methods to investigate structure-function relationship of enzyme Aβ complex and to design new lead molecules to control amyloid beta peptide levels in Alzheimer's disease.

Key words Alzheimer's disease, Aβ peptides, Molecular modeling, Molecular docking, MD simulation

1 Introduction

Alzheimer's disease (AD) is the most common form of dementia in elderly people. AD is mainly characterized by neurofibrillary tangles (NFTs) and senile plaques. These senile plaques are extracellular deposits, composed of aggregated peptides called amyloid beta (Aβ) [1]. The Aβ peptides are generated by sequential cleavage of the amyloid precursor protein (APP), which is a transmembrane glycoprotein [2]. The fibrillary forms of Aβ peptide clusters outside the neurons in dense formation known as senile plaques [3]. The Aβ peptide accumulation is caused due to overproduction or inefficient clearance and defects in the proteolytic degradation [4].

Kunal Roy (ed.), *Computational Modeling of Drugs Against Alzheimer's Disease*, Neuromethods, vol. 132, DOI 10.1007/978-1-4939-7404-7_11,

© Springer Science+Business Media LLC 2018

Therefore, Aβ degradation and clearance could be a promising therapeutic approach in the AD treatment. The regulation of Aβ peptides by proteolysis has significant implications in pathogenesis and AD treatment [5]. Decreasing production and enhancing the clearance of Aβ peptides are currently being targeted as potential therapeutic strategies for AD treatment [6]. The potential anti-amyloid approaches include inhibitors of γ- and β-secretases, enhancers of α-secretase, inhibitors of Aβ aggregation, de-aggregators of amyloid plaques, inhibitors of receptor for advanced glycation end products (RAGE), and antibodies directed against Aβ [7].

The current research is focused on Aβ peptide degradation which would be useful to reduce Aβ levels in AD brain [8, 9]. Various enzymes such as proteases or peptidases are reported having capability to degrade Aβ either in vitro or in vivo. These enzymes like neprilysin (NEP) [10–12], endothelin-converting enzyme-1 (ECE-1) [13], insulin-degrading enzyme (IDE) [14–16], angiotensin-converting enzyme (ACE) [17], uPA/tPA-plasmin system [18, 19], cathepsin D [20, 21], gelatinase A [20], gelatinase B [22], matrix metalloendopeptidase-9 [23], aminopeptidase A [24], cathepsin B [25], coagulation factor XIa [26], antibody light chain c23.5 and hk14 [27], and α2-macroglobulin complexes [28] contribute to intracellular or extracellular degradation of Aβ peptides in AD brain. Recently, nattokinase from *Bacillus subtilis Natto* have been shown to degrade Aβ peptides in vitro [29, 30]. Similarly, aminopeptidase from *Streptomyces griseus* KK565 also showed Aβ peptide degradation activity [30], which suggest that the bacterial enzymes may be used for Aβ peptide clearance. The enzymes having potential to degrade amyloid beta peptide have been studied experimentally [29, 30].

Computational techniques have been proved useful to understand the mechanism of Aβ peptide degradation in details at the molecular level [31–34]. Different computational techniques such as homology modeling, energy minimization, molecular dynamics (MD), etc. have been used to build three-dimensional structures of Aβ peptide-degrading enzymes [35–38]. These reports showed that the interactions between active-site residues of enzymes and Aβ peptide can be studied using molecular docking and molecular dynamics simulation techniques. Finally, computational methods have been found reliable to understand the mechanism of Aβ peptide degradation and could be helpful to design novel therapeutic approaches in the treatment of Alzheimer's disease. These computational methods are briefly explained in the following sections.

2 Computational Methods

2.1 Basic Concepts in Homology Search

To understand molecular machinery of the cell, it is essential to investigate the functions of each and every biomolecule. The functions of desired proteins can be studied directly through wet-lab experiments or indirectly by predicting their three-dimensional structures. The functions of targeted proteins can be identified by searching one or more homologous proteins through sequence similarity. The identical proteins with known functions are being used to determine the functions of desired proteins. For this purpose, homology search is an important option to find similar proteins present in databases. Such homologous proteins can be identified using sequence analysis studies. An important goal of sequence analysis is to get information about the probable functions of the targeted protein depending upon its primary amino acid sequence. Sequence analysis studies have been found useful to compare human enzyme sequences with other enzymes from different sources having ability to degrade amyloid beta peptide (31, 35–36, 38).

2.2 Sequence Similarity, Homology, and Alignment

The basic concept of sequence analysis is to find sequences having homology with the target sequence. The proteins having common evolutionary ancestors are said to be homologous to each other. The protein subfamily contains the set of homologous proteins descended from a common ancestor. These proteins resemble in overall three-dimensional structures or folds, and similarities remain fairly constant over evolutionary time. It has been identified that various members of a protein family typically share a common three-dimensional fold. This similarity of fold within the homologous proteins suggests similar functions. To trace the descent one from a common ancestral sequence is the only way to conclusively prove that the two sequences are homologous to each other [39]. Sometimes, it is very hard to find common ancestor. So homology can be deduced from restricted sequences with an increase in degree of similarity. The sequence similarity can therefore be investigated. The goal of sequence analysis is to investigate, deduce, and quantify homology between two sequences.

Efforts have been made to find out the sequence similarity between human enzymes such as aminopeptidases, cathepsin B having ability to degrade amyloid beta peptide with bacterial enzymes [31, 36]. In sequence alignment, conserved domains and active-site residues of the target sequence can be found by comparing with the template sequence. In one of the studies, sequence analysis has been done on cysteine protease from *Xanthomonas campestris* having potential to degrade amyloid beta peptide [36]. This sequence analysis studies revealed that the residues Cys17, His87, and Gln88 of cysteine protease from *Xanthomonas campestris* are similar to the active-site residues (Cys29, His110, and His111) of human cathepsin B. Hence, these residues might play an important role in the function of cysteine protease [36].

2.3 Homology Modeling

Homology modeling, also known as comparative modeling, generally means to predict 3D structures of target proteins using a template protein, which is an experimentally solved structure. The main basis for the prediction of 3D structure of target protein is the alignment of its amino acid sequence with a related protein (template) having known 3D structure [40, 41]. In certain conditions, more than one template can be considered. Homology modeling protocol is currently believed to be the most reliable protein structure prediction method [42]. Homology modeling is based on two different approaches, de novo and comparative modeling. In de novo method, structure of a protein can be predicted from the sequence alone, without relying on similarity between the target sequence and any known structure. On the other hand, in comparative modeling, 3D model for a target protein is built using similarity with the template protein of a known structure.

Certain enzymes from human as well as bacterial sources have been found to degrade amyloid beta peptide. The three-dimensional structural information of such enzymes could be useful in the drug discovery process. Hence, homology modeling technique is applicable to build 3D structures of new enzymes which had showed the ability to degrade amyloid beta peptide. The structure of aminopeptidase from *Streptomyces griseus* KK565 was built by homology modeling, and its role is studied further [31]. The homology modeling protocol contains four steps: template finding, sequence alignment, model building, and model assessment.

2.3.1 Template Finding

Homology modeling usually starts with a template selection process against a protein structure database like Protein Data Bank (PDB) [43]. The sequence alignment method is used to identify one or more templates having significant alignment with the target sequence. Several approaches have been known for homologous sequence search. Out of these, the standard sequence searching tools are BLAST [44] or PSI-BLAST [45] and FASTA [46]. The sequence alignment programs like Basic Local Alignment Search Tool (BLAST) [44] and FASTA [46] are mostly used to compare target sequence with all sequences present in PBD which are having known 3D structures. A great deal of progress has been made to develop sensitive alignment methods based on integrative searches such as PSI-BLAST [45], hidden Markov models (HMM), e.g., SAM [47], and HMMER [48]. These methods are used to generate a large sequence alignment of protein that bears a clear evolutionary relationship. These results are useful to find better template from a number of potential templates available in respective structure database. Finally, the template which shows highest homology with target sequence is selected for further studies.

2.3.2 Sequence Alignment

The pairwise sequence alignment between the target sequence and template sequence has been widely used routinely to find homology between two sequences to build three-dimensional model of a target. The pairwise sequence comparison programs can provide a fast initial estimate of the difficulty level of homology modeling process. These sequence comparison programs are adequate to detect homologous proteins which share 25–30% sequence similarity. However, in many cases, if alignment score is low, the alignment needs some improvement. When aligned sequences are over 40–50% identical to each other and have few or no gaps, it can be expected that alignments may be accurate in a structural sense and accurate 3D structure can be predicted for the target sequence. If there are errors in sequence alignment, this may cause deviations in the homology model even though the correct template is chosen. During sequence alignment corrections, the features such as secondary structural elements, hydrophobic and hydrophilic regions, conservation of active-site residues, etc. should be considered. The information about the similar structural domains and conserved residues among the related sequences can be obtained by multiple sequence alignment (MSA). If the sequences are diverged, the sequence alignment allows insertion and deletion of certain residues [49]. This insertion and deletion of residues are implemented in the most multiple sequence alignment programs such as ClustalW2 and BioEdit [50, 51]. In more difficult alignment cases, it is wise to depend on multiple structures and sequence information [52].

2.3.3 Model Building

Once a statistically significant alignment is obtained between target and template sequence, homology model construction of the query sequence is done by taking template structure. There are algorithms through which homology models are constructed. The rigid body assembly method involves two steps, where at first the structures for all sequence conserved regions are directly taken from the templates. These regions are then kept rigid, and additional sampling is performed to build structures for gaps and mismatched regions. This type of procedure is implemented in program like SWISS-MODEL [53]. The second algorithm used is called "satisfaction of spatial restraints," implemented in MODELLER, a homology modeling software [54]. In this method, many types of spatial restraints are generated by matching the target sequence with the template structure. These restraints are used to create a probability density function (PDF) for the coordinates of each atom, and then a homology model is built by optimizing the molecular PDF. In addition to these two methods, several other model building protocols are available [55] like segment matching [56] and artificial evolution model building [57]. The alignment and template selection are important to predict an accurate 3D structure of the target sequence. However, the modeling method offers a degree of flexibility and automation, making it easier and faster to obtain good models. Also, the quality of model built by different homology modeling programs has been evaluated by different servers such as PROCHECK [58], Verify3D [59], and PROSA [60].

2.3.4 Model Refinement

The accuracy of a model may be reduced due to deviations in the structural geometry between target and template. Such structural geometry differences result from relative orientations and a number of secondary structural elements and large indel (insertions and deletions) in different regions of the structure [49]. The studies done by van Gelder and coworkers suggest that model refinement is an essential part, which requires accurate sampling for conformational space and identifies near native structure [61]. Energy minimization using molecular mechanics force field is the best approach to refine the starting model [62, 63]. Further, refinement can be done by using different techniques such as molecular dynamics, Monte Carlo, and genetic algorithm-based techniques [64, 65]. These techniques are helpful to improve the accuracy of the predicted model significantly by optimization of backbone conformation and placement of the core side chains at proper places.

2.3.5 Model Assessment

As we know that every step in homology modeling process is dependent on the previous task. Therefore, certain errors may be accidentally incorporated and proliferated in predicted homology model. The predicted model can be evaluated as a whole or in the individual regions to rectify such errors. There are different programs and servers available for evaluation of predicted model [66, 67]. When several models are built for the same target, the "best" model can be selected on the basis of having lowest value of the MODELLER objective function, which is reported in the second line of the model PDB file. Similarly, other programs are being used to determine whether the predicted model satisfies standard steric and geometric criteria [68, 69]. Once a final model is selected, then initial assessment of this selected model is done based on fold comparison with a template structure that reflects conservation of functional and structural residues in target sequence. This fold comparison is evaluated by energy-based Z-score in PROSAII analysis [68]. The stereochemistry of the selected model investigated to predict its quality using useful programs such as PROCHECK [58], AQUA [70], and WHATCHECK [71]. Both PROSAII [68] and PROCHECK [58] programs confirms suitability of the model, with a Z-score comparable to that of the template. Scoring functions implementing programs like Verify3D [72], PROSAII [68], and ANOLEA [73] are also used to assess the model quality.

The programs like PROCHECK [58] and PROSA [68] have been used to analyze the quality of three-dimensional structure of aminopeptidase from *Streptomyces griseus* KK565 built by homology modeling technique [31]. Hence, model validation has been found useful to check the model quality which is very important to carry out further computational studies.

2.4 Molecular Docking

Molecular docking study mainly aims to predict the structure of intramolecular complexes formed between two or more molecules. Taylor and coworkers in the year 2006 described that molecular docking is a useful tool in structure-based drug design and discovery [74]. With an increasing number of known protein crystal structures, the interest in molecular docking has also been increased. In the recent years, more progress has been made in the computational biology field [75–79]. Molecular docking is a method to predict preferred orientation of the receptor and a ligand while interacting to each other to form a stable complex [80]. Earlier the docking method considered only translational and orientational degrees of freedom of the ligand with respect to the receptor [81]. However, the molecule's geometry may change upon binding into the active site of an enzyme; therefore, now most of the methods treat the ligand as a flexible molecule. Proteins are not rigid structures; therefore, ideally it should be flexible during the docking simulations. However, introducing flexibility in the macromolecular proteins during docking is difficult, as the computational workload will be high. Hence, molecular docking studies are performed using a flexible ligand and a rigid receptor protein. Docking is a multistep process which starts with molecular representation followed by docking algorithms to identify best-matched conformation to receptor structure. For the evaluation of interactions between ligand and receptor, the docking algorithms complemented by scoring functions are used based on shape and electrostatic complementarities [82]. The AutoDock [83] software can be used to predict protein-protein and protein-ligand interactions with appropriate docking algorithms and scoring functions. Molecular docking techniques have been found useful to understand interactions between various enzymes from different sources with amyloid beta peptides [31–36]. Molecular docking studies are important to find the actual interactions between the active-site residues of enzymes with the various ligands. Dhanavade and coworkers performed molecular docking study between aminopeptidase from *Streptomyces griseus* KK565 with amyloid beta peptide (31). Earlier study showed that some active-site residues such as Glu140, Asp168, and His255 of aminopeptidase from *Streptomyces griseus* KK565 form hydrogen bonding interactions with the amyloid beta peptide suggesting their role in the degradation of amyloid beta peptide [31].

2.5 Molecular Dynamics (MD) Simulation

In recent years, theoretical methods, especially molecular dynamics (MD) simulations, have been applied increasingly to study structure-function relationships of proteins at atomic level. MD simulation is a valuable tool to study the dynamic behavior of system by considering the positions of every atom as a function of time in an iterative fashion using Newton's classical equation of motion:

$$F_i = m_i a_i$$

where

F_i – force exerted on particle i
m_i – mass of particle i
a_i – acceleration of particle i

Newton's classical equations of motions are solved simultaneously by considering small-time steps. The system automatically implies the use of classical mechanics for the description of motions of atoms at every time steps by maintaining temperature and pressure at desired values. The output file containing the coordinate as a function of time generated at regular intervals represents trajectory of the system. MD simulation method has been found useful to correlate theory with experimentation, and for this there are three main states. In the first state, MD-simulated properties are compared with experimental results, and when both of these agree, then it is reasonable to claim that the experimental results can be explained by the simulation model. In the second scenario, MD simulations are used to interpret experimental results. In the third state, MD simulations are used as an exploratory tool to help gain an initial understanding of a problem by providing guidance among possible lines of investigation, be it theoretical or experimental. In all these scenarios, it is often the case that larger simulations are necessary to get a more realistic model.

During MD simulation protocol, the atoms and molecules are allowed to interact with each other over a time period at a specified temperature following the laws of classical mechanics. Molecular dynamics simulations provide the crucial information considering individual atomic motion as a function of time [84]. The ensemble refinement produces better representations of the distributions of atomic positions in the simulated structures rather than the single conformer refinements [85]. MD simulation methods are used to reveal the role of molecular motions in enzymatic activity, signaling mechanisms, binding properties, and molecular assemblies within their biological context. MD simulation methodology is being widely used to investigate several biochemical and biophysical problems, such as conformational analysis of proteins, and ligand-receptor interactions along with drug design. Molecular dynamics simulation technique has been applied to investigate structural stability of enzymes involved in the degradation of amyloid beta peptide [31–36].

In one of the reports, MD simulation is used to confirm the stability of docked complex of hECE-1 with amyloid beta peptide, suggesting capabilities of hECE-1 to cleave amyloid beta peptides potentially to reduce risk of AD [32]. Similarly, other enzymes, viz., cathepsin B, ACE-1, cysteine protease, aminopeptidase, and gelatinase, have been studied to understand their role in the degradation of amyloid beta peptide in detail at the atomic level using molecular dynamics simulation technique [31, 33–36, 38].

2.5.1 MD Simulation Algorithms

To integrate Newton's equations of motions, several algorithms have been developed. Out of these, the most frequently used algorithms are Verlet algorithm [86], Gear predictor-corrector algorithm [87], Beeman algorithm [88], and leapfrog algorithm [89], which is a modified version of the Verlet algorithm. To calculate the interactions during the simulation, simple MD simulation algorithms are used. By considering this, MD simulation is usually divided into two classes such as short-range interactions and long-range electrostatic interactions depending on the range of interactions and cutoff radius [90]. The cell list [91] and Verlet list [92] algorithmic approaches are used to evaluate the short-range interactions, whereas particle Mesh Ewald (PME) and particle-particle particle-mesh (P3M) methods are used to evaluate long-range electrostatic interactions in the macromolecular system [93, 94]. Many algorithms have been proposed for solving constraint problem, which is considered as a physical behavior of bond vibrations. SHAKE algorithm is mostly used for larger molecules [95]. For small molecules, SETTLE is a faster algorithm which is an analytical solution of SHAKE algorithm [96]. SHAKE algorithm varies a set of unconstrained coordinates to the set of coordinates that fulfills list of distance constraints with solving a set of Lagrange multipliers in the constrained equations of motion [97]. LINear Constraint Solver (LINCS) algorithm resets bonds to their correct lengths after an unconstrained update. LINCS algorithm is used for large molecules, and it is three to four times faster than the SHAKE algorithm [98]. All these algorithms have been applied successfully in the studies done by several authors [31–36].

2.5.2 Topology

There are several atom types, so only atom types occurring in biological systems are parameterized in the force field. The topology file represents the atomic position, and their interactions such as bonds, angles, and dihedrals are determined as fixed list that are included in the topology file [99]. Topology files are essential for nonstandard atoms, ions, and molecules. During MD simulation, the potential functions act on atoms according to parameters defined in the topology file. Therefore, for the MD simulation of nonstandard molecules like ligands, lipids, and ions, additional molecular topology files are required. The topology files for such nonstandard molecules were obtained directly from online servers and included in respective topology file.

2.5.3 Force Fields

Force field refers to the combination of parameters and mathematical equations used to describe the properties of atoms and their bonded and nonbonded interactions. The force field parameters include the definitions of the atomic masses and charges for different atoms as well as the bond lengths, bond angles, and dihedral angles. The definitions of the parameters and the equations collectively describe the behavior and potential energy of the system. According to the mathematical point of view, a force field is a function of potential energy which exclusively depends on the positions of the nuclei. However, the potential functions are divided into three types: bonded, nonbonded, and restrains. The bonded interactions can be further subdivided with regard to the number of particles involved, which resulted into the terms describing bond stretching, angle bending, and torsion angle. However, nonbonded interactions are calculated between all pairs of atoms, which are in the same or different molecules but separated by at least three bonds. Therefore, these nonbonded interactions comprise electrostatic interactions and van der Waals interactions. Nonbonded potentials are described by Lennard-Jones potential and Coulomb interaction [99]. The force fields like CHARMM [100], AMBER [62], GROMOS [101], and OPLS [102] are most widely used force fields to simulate the biomolecules since the past several years. Currently, the optimized potentials for liquid simulations-all atom (OPLS-AA) force field has been developed which includes the parameter sets for every types of atoms in the system along with improved set of potential functions. Hence, this improved OPLS-AA reproduces well conformational energetic in agreement with experimental values and ab initio calculation [103].

2.5.4 Periodic Boundary Condition

The periodic boundary conditions are used to minimize the edge effect of cubic and finite systems. MD simulations of proteins are carried out in which proteins are solvated by water molecules under periodic boundary conditions to provide the optimum environment. The simulation takes place in a computational box, which is virtually surrounded by an infinite number of identical replica boxes, stacked in a space-filling way, all having exactly the same contents. Only performance of "central box" has to be simulated, while other boxes behave in a similar manner. Particles may freely cross box boundaries when periodic boundary conditions are applied. For every particle which is leaving the box, at the same time, an identical particle from an adjacent replica box enters the box at the opposite side. In an MD simulation system with periodic boundary conditions, particles are influenced by another particles in their own box and particles in surrounding boxes. The computational box shape should be such that it can be stacked in a space-filling way. For efficiency only convex boxes are used. There are five types of boxes such as the triclinic box, the hexagonal prism, two types of dodecahedrons, and the truncated octahedron. Out of which mostly cubic box is used for simulation. The classical way to minimize edge effect of cubic and finite systems is to implement periodic boundary conditions. But for long-range electrostatic interactions, this is not always accurate enough; therefore, GROMACS included lattice sum methods like Ewald Sum, PME, and PPPM [93, 104]. All atoms present in primary cell are replicated in all simulated directions as image cell having the same number, size, shape, position, and momentum of atom to form an infinite lattice. Within the box, the interactions are calculated between a given atom and any other atom using the image that provides shortest distance between two atoms. Therefore, a molecule behaves as an infinite system and is free to move from one side of the box to the other side [99, 105]. The main consideration during simulation of macromolecules is that length of each box vector should be larger than the length of macromolecule and twice the cutoff of non-bonded interaction [106, 107].

2.5.5 Thermodynamic Ensembles and Water Model

An ensemble is nothing but the collection of all possible systems which have different microscopic states but have an identical thermodynamic state. The parameters such as temperature (T), pressure (P), volume (V), energy (E), and the number of particles (N) are mainly important to define the thermodynamic state of a system [108]. There are different thermodynamic ensembles available like microcanonical (NVE), canonical ensembles (NVT), and isothermal-isobaric (Gibbs) ensemble (NPT) [109]. For the simulation of macromolecules, an isothermal-isobaric and constant pressure and temperature (NPT) ensemble is frequently used because it reproduces experimental conditions. In this ensemble, the number of molecules, temperature, and pressure are constant. Therefore, it is necessary to ensure that both the temperature and pressure remain constant throughout the entire MD simulation period [110, 111]. The system temperature is related to the kinetic energy of the system which can be kept constant by adjusting the velocities. The temperature and pressure coupling is implemented using Berendsen weak coupling scheme. This coupling method is having advantage that the strength of the coupling can be varied and adapted depending upon the user requirement [99, 112].

In nature, water is considered as the most important solvent. During MD simulation, biomolecules are solvated in water to investigate a variety of perceptions ranging from solvent dynamic behavior at the protein surfaces to solvent effect associated with the conformational behavior of biomolecules [113, 114]. The different water models are developed for molecular simulations, such as transferable intermolecular potential-3 point (TIP3P), transferable intermolecular potential-4 point (TIP4P) [115], simple point charge (SPC) [116], and extended simple point charge (SPC/E)

[117]. These water models have mainly three interaction sites as observed exactly in nature, but they differ in Lennard-Jones (LJ) and Coulombic terms in quality of reproducing the experimentally observed bulk properties for liquid water [118]. It was found that simpler water models are computationally more efficient and allow longer simulations of large systems [119].

2.6 Energy Minimization

The energy minimization methods are used efficiently to refine molecular structures, which are totally inadequate for sampling conformational space. The complex like amyloid beta-degrading enzymes and different amyloid beta peptides can be used to remove steric clashes before and after MD simulations similar to earlier reports [31, 36]. For an unrefined molecular structure with bond angles and lengths distorted from their respective minima or with steric clashes between atoms, energy minimization methods are found to be very useful for correcting these flaws and are therefore routinely applied to protein systems. The most popular energy minimization methods include those that use derivatives of various orders, including the first-order steepest-descent and conjugate gradient methods (which utilizes first-order derivatives) and the second order (i.e., utilizes second-order derivatives), Newton-Raphson method [120, 121].

The steepest-descent (SD) method uses first-order derivative scheme for locating minima on the molecular potential energy surface [97, 122]. The SD method uses first derivative to determine the direction toward the minimum and the direction in which the geometry is first minimized opposite to the direction in which the gradient is largest (i.e., steepest) at the initial point. This SD method is found to be robust and easy to implement. The energy minimization is carried out mainly to remove steric clashes and relax the system [121, 123]. The conjugate gradient and Newton-Raphson methods keep track of earlier steps used in minimization in contrast to steepest-descent method.

2.7 Molecular Dynamics Simulation Setup

The MD simulation can be performed by using various input parameters like starting structure, system preparation, thermodynamic ensembles, position restrained, and fully unrestrained MD simulation.

2.7.1 The Starting Structure

The selection of the starting structure is the first step. For MD simulation, the initial structure is very important. In order to understand the cleavage mechanism of amyloid beta peptide degradation by amyloid beta peptide-degrading enzymes such as aminopeptidase, cathepsin B, ECE-1, ACE-1, etc., the initial structure of either individual enzyme or a complex is very essential. The starting structure is taken preferably from the X-ray or NMR data; if not available, then a homology model can be used. Prior to MD simulation, the starting structure is subjected to energy minimization to remove steric clashes.

2.7.2 System Preparation

System preparation is nothing but the generation of topologies for standard as well as nonstandard molecules associated with starting structure. For MD simulation, it is necessary to construct periodic box for the whole system, followed by solvation with appropriate water model. Then to neutralize the system, various monovalent ions must be added in the simulation box by replacing the water molecules. The energy minimization of the full system should be done by taking care as the initial arrangements to remove steric conflicts, which can often determine the success or failure of a simulation.

2.7.3 Position Restrained MD

In order to bring the system (complex of enzyme and amyloid beta peptide or any drug) to a more stable state, artificial restraint has to be added to specific groups of atoms in a method called position-restrained (PR) MD. The position-restrained trajectories were checked to ensure system stability.

2.7.4 Fully Unrestrained MD

The simulation with no restrains begins once the position-restrained MD provides a stable simulation system. The MD simulation software alternates between calculating the forces on each atom at each time step and updating the atom positions due to those forces. The molecular structure is adjusted in the potential energy field and changes in acceleration. Therefore, the time step in MD must be very small. It was found that the smaller is the time step, the better the approximation. Larger time steps result into unreasonably large movements of the atoms. For most situations, particularly larger biomolecules, the time step of 2 femtoseconds (fs) is recommended. Long MD simulations are necessary to sample all conformational transitions occurring in a biomolecular system. The length of MD simulation depends upon what type of information is needed.

2.8 Virtual Screening Method

Computational methodologies have played a vital role in many drug discovery programs, from hit identification to lead optimization. Virtual screening technique used routinely in all aspects of drug discovery today. This is more direct and rational drug discovery approach and has the advantage of low-cost and effective screening as compared with traditional experimental high-throughput screening (HTS) [82, 124]. Virtual screening is mainly divided into two classes, namely, ligand-based and structure-based approaches. In the ligand-based methods, one or more ligands are used to screen the ligand database. After screening a similarity score between known ligand and each database entry is calculated, whereas in the structure-based method, structural information is used to identify the ligand which will bind to a target protein [125]. Different approaches have been found useful in the absence of three-dimensional structural information of potential targets. These approaches include quantitative structure-activity relationship (QSAR), pharmacophore modeling, molecular field analysis, and 2D or 3D similarity assessment. All these methods can give important insights about the interactions between drug targets and ligands.

In virtual screening, selection of an appropriate set of compounds carried out by applying adequate filtering strategies such as Lipinski's rule of five and the ADME restrictions [126, 127]. The interaction between the ligand and the protein at the atomic level can be studied by molecular docking approach. There are several docking approaches; DOCK Blaster is a fully automated docking system composed of six modules, and screening results are judged using pose fidelity and enrichment [128]. The AutoDock tool is being routinely used for virtual screening approach [129]. Different search strategies and scoring algorithms are used to identify best-matched conformation and ranking of ligand, respectively [83, 131].

The potential inhibitors of ACE and ECE were identified using a virtual screening method [129, 130]. These inhibitors might be potent therapeutic molecules in the treatment of Alzheimer's disease [129, 130]. This technique can also be used to search new drugs which might inhibit the aggregation of amyloid beta peptide a causative agent of Alzheimer's disease.

2.9 The Calculation of Binding Free Energy

The binding free energies of macromolecules such as enzymes complexed with amyloid beta peptide of a virtually screened drug can be calculated by using the Molecular Mechanics/Poisson-Boltzmann Surface Area (MM/PBSA) method [132–135]. In this calculation, a number of snapshots of docked complex of amyloid beta-degrading enzymes and amyloid beta peptide as a ligand can be taken evenly. Different energetic parameters using the MM/PBSA method were calculated by using the same snapshots which were taken earlier. The binding energy in three terms such as van der Waals contribution (ΔEvdw), electrostatic contribution (ΔEele), and solvation contribution (ΔEsol) was calculated for each receptor (eq. aminopeptidase, cathepsin B, ECE-1, ACE-1, etc.) and ligand ($A\beta_{1-42}$ Peptide). In the studies performed by Barage et al., the binding free energy calculations of MD-simulated hECE-1 and $A\beta_{1-42}$ peptide complex revealed higher binding affinity of $A\beta_{1-42}$ peptide toward hECE-1 residues. This binding energy calculation revealed that His 143, Arg 145, Arg 324, Arg 325, Glu 327, Val 565, Asn 566, Thr 572, Thr 729, Asp 730, and His 732 are the key residues of hECE-1 which are involved in binding with Aβ1–42 peptide [132]. Similarly binding free energy calculations can also be used for other biomolecules [136]. Hence, binding energy calculations are important to understand the key residues and to check the structural stability of docked complexes [132].

3 Conclusion

Sequence analysis and homology modeling studies have been found useful to determine the sequence and structural similarities. The molecular docking and MD simulation studies are used to get exact information about the active-site residues involved in the degradation of Aβ peptide. Additionally, the binding free energy calculations could be helpful to understand the stability of docked complex of amyloid beta peptide-degrading enzymes and Aβ peptides/virtually screened drugs. Thus, these computational techniques are useful to understand the role of various enzymes in the degradation of amyloid beta peptides and can be implemented to design new therapeutic approaches in treatment of Alzheimer's disease.

Acknowledgment

KDS is thankful to the University Grants Commission, New Delhi, for providing financial support. KDS gratefully acknowledges the Department of Science and Technology, New Delhi, for providing financial grants. MJD is thankful to DST for providing fellowship as a research assistance under the scheme DST-PURSE. The authors are thankful to Computer Centre, Shivaji University, Kolhapur, for providing the computational facilities.

References

1. Selkoe DJ (1994) Normal and abnormal biology of the beta-amyloid precursor protein. Annu Rev Neurosci 17:489–517
2. Korenberg JR, Pulst SM, Neve RL, West R (1989) The Alzheimer amyloid precursor protein maps to human chromosome 21 bands q21.105-q21.05. Genomics 5:124–127
3. Tiraboschi P, Hansen LA, Thal LJ, Corey-Bloom J (2004) The importance of neuritic plaques and tangles to the development and evolution of AD. Neurology 62:1984–1989
4. Tanzi RE, Moir RD, Wagner SL (2004) Clearance of Alzheimer's Aβ peptide: the many roads to perdition. Neuron 43:605–608
5. Sarah MS, Yungui Z, Hideaki A, Roberson ED, Sun B, Chen J, Wang X, Yu G, Esposito L, Lennart M, Li G (2006) Antiamyloidogenic and neuroprotective functions of cathepsin B: implications for Alzheimer's disease. Neuron 51:703–714
6. Evin G, Weidemann A (2002) Biogenesis and metabolism of Alzheimer's disease A beta amyloid peptides. Peptides 23:1285–1297
7. Hardy J, Selkoe DJ (2002) The amyloid hypothesis of Alzheimer's disease: progress and problems on the road to therapeutics. Science 297:353–356
8. Wang DS, Dickson DW, Malter J (2006) β-Amyloid degradation and Alzheimer's disease. J Biomed Biotechnol 3:1–12
9. Saido T, Leissring MA (2012) Proteolytic degradation of amyloid β-protein. Cold Spring Harb Perspect Med 2: a006379–a006397
10. Howell S, Nalbantoglu J, Crine P (1995) Neutral endopeptidase can hydrolyze beta-amyloid (1-40) but shows no effect on betaamyloid precursor protein metabolism. Peptides 16:647–652
11. Iwata N, Tsubuki S, Takaki Y, Shirotani K, Lu B, Gerard NP, Gerard C, Hama E, Lee HJ, Saido TC (2001) Metabolic regulation of brain Aβ by neprilysin. Science 292:1550–1552
12. Iwata N, Tsubuki S, Takaki Y, Watanabe K, Sekiguchi M, Hosoki E, Kawashima-Morishima M, Lee HJ, Hama E, Sekine-

Aizawa Y, Saido TC (2000) Identification of the major Abeta1-42-degrading catabolic pathway in brain parenchyma: suppression leads to biochemical and pathological deposition. Nat Med 6:143–150

13. Eckman EA, Reed DK, Eckman CB (2001) Degradation of the Alzheimer's amyloid beta peptide by endothelin-converting enzyme. J Biol Chem 276:24540–24548
14. Kurochkin IV, Goto S (1994) Alzheimer's beta-amyloid peptide specifically interacts with and is degraded by insulin degrading enzyme. FEBS Lett 345:33–37
15. McDermott JR, Gibson AM (1997) Degradation of Alzheimer's beta-amyloid protein by human and rat brain peptidases: involvement of insulin-degrading enzyme. Neurochem Res 22:49–56
16. Qiu WQ, Walsh DM, Ye Z, Vekrellis K, Zhang J, Podlisny MB, Rosner MR, Safavi A, Hersh LB, Selkoe DJ (1998) Insulin-degrading enzyme regulates extracellular levels of amyloid beta-protein by degradation. J Biol Chem 273:32730–32738
17. Hu J, Igarashi A, Kamata M, Nakagawa H (2001) Angiotensin converting enzyme degrades Alzheimer amyloid beta-peptide (A beta); retards A beta aggregation, deposition, fibril formation; and inhibits cytotoxicity. J Biol Chem 276:47863–47868
18. Sasaki H, Saito Y, Hayashi M, Otsuka K, Niwa M (1988) Nucleotide sequence of the tissue-type plasminogen activator cDNA from human fetal lung cells. Nucleic Acids Res 16:5692–5695
19. Verde P, Boast S, Franze A, Robbiati F, Blasi F (1988) An upstream enhancer and a negative element in the 5_ flanking region of the human urokinase plasminogen activator gene. Nucleic Acids Res 16:10699–10716
20. Yamada T, Kluve-Beckerman B, Liepnieks JJ, Benson MD (1995) In vitro degradation of serum amyloid A by cathepsin D and other acid proteases: possible protection against amyloid fibril formation. Scand J Immunol 41:570–574
21. Hamazaki H (1996) Cathepsin D is involved in the clearance of Alzheimer's beta-amyloid protein. FEBS Lett 396:139–142
22. Backstrom JR, Lim GP, Cullen MJ, Tokes ZA (1996) Matrix metalloproteinase-9 (MMP-9) is synthesized in neurons of the human hippocampus and is capable of degrading the amyloid-beta peptide (1-40). J Neurosci 16:7910–7919
23. Carvalho KM, Franca MS, Camarao GC, Ruchon AF (1997) A new brain metalloendopeptidase which degrades the Alzheimer beta-amyloid 1-40 peptide producing soluble fragments without neurotoxic effects. Braz J Med Biol Res 30:1153–1156
24. Sevalle J, Amoyel A, Robert P, Fournie´-Zaluski MC, Roques B, Checler F (2009) Aminopeptidase A contributes to the N-terminal truncation of amyloid β-peptide. J Neurochem 109:248–256
25. Mueller-Steiner S, Zhou Y, Arai H, Roberson ED, Sun B, Chen J, Wang X, Yu G, Esposito L, Mucke L, Gan L (2006) Antiamyloidogenic and neuroprotective functions of cathepsinB: implications for Alzheimer's disease. Neuron 51:703–714
26. Saporito-Irwin SM, Van Nostrand WE (1995) Coagulation factor XIa cleaves the RHDS sequence and abolishes the cell adhesive properties of the amyloid beta-protein. J Biol Chem 270:26265–26269
27. Rangan SK, Liu R, Brune D, Planque S, Paul S, Sierks MR (2003) Degradation of beta-amyloid by proteolytic antibody light chains. J Biochem 42:14328–14334
28. Qiu WQ, Borth W, Ye Z, Haass C, Teplow DB, Selkoe DJ (1996) Degradation of amyloid beta-protein by a serine proteasealpha2-macroglobulin complex. J Biol Chem 271:8443–8451
29. Hsu RL, Lee KT, Wang JH, Lily Y, Lee L, Rita P, Chen Y (2009) Amyloid-degrading ability of nattokinase from *Bacillus subtilis* Natto. J Agric Food Chem 57:503–508
30. Yoo C, Ahn K, Park JE, Kim MJ, Jo SA (2010) An aminopeptidase from Streptomyces sp. KK565 degrades beta amyloid monomers, oligomers and fibrils. FEBS Lett 584:4157–4162
31. Dhanavade MJ, Sonawane KD (2014) Insights into the molecular interactions between aminopeptidase and amyloid beta peptide using molecular modeling techniques. Amino Acids 46:1853–1866
32. Barage SH, Sonawane KD (2013) Exploring mode of phosphoramidon and Aβ peptide binding to hECE-1 by molecular dynamics and docking studies. Protein Pept Lett 21:140–152
33. Barage SH, Jalkute CB, Dhanavade MJ, Sonawane KD (2014) Simulated interactions between endothelin converting enzyme and Aβ peptide: insights into subsite recognition and cleavage mechanism. Int J Pept Res Ther 20:409–420
34. Jalkute CB, Barage SH, Dhanavade MJ, Sonawane KD (2013) Molecular dynamics simulation and molecular docking studies of

angiotensin converting enzyme with inhibitor lisinopril and amyloid beta peptide. Protein J 32:356–364

35. Dhanavade MJ, Parulekar RS, Kamble SA, Sonawane KD (2016) Molecular modeling approach to explore the role of cathepsin B from Hordeum vulgare in the degradation of Aβ peptides. Mol BioSyst 12:162–168
36. Dhanavade MJ, Jalkute CB, Barage SH, Sonawane KD (2013) Homology modeling, molecular docking and MD simulation studies to investigate role of cysteine protease from *Xanthomonas campestris* in amyloid beta degradation. Comput Biol Med 43:2063–2070
37. Thakar SB, Dhanavade MJ, Sonawane KD (2016) Phylogenetic, sequence analysis and structural studies of Maturase K proteins from mangroves. Curr Chem Biol 10:135–141
38. Jalkute CB, Barage SH, Sonawane KD (2015) Insight into molecular interactions of Aβ peptide and gelatinase from enterococcus faecalis: a molecular modeling approach. RSC Adv 5:10488–10496
39. Grundy WN (1998) A Bayesian Approach to Motif-based Protein Modeling, A PhD Thesis submitted to the University of California, San Diego
40. Sanchez R, Sali A (1998) Large-scale protein structure modeling of the Saccharomyces cerevisiae genome. Proc Natl Acad Sci U S A 95:13597–13602
41. Marti-Renom MA, Stuart AC, Fiser A (2000) Comparative protein structure modeling of genes and genomes. Annu Rev Biophys Biomol Struct 29:291–325
42. Lounnas V, Ritschel T, Kelder J, McGuire R, Bywater RP, Foloppe N (2013) Current progress in structure-based rational drug design marks a new mindset in drug discovery. Comput Struct Biotechnol J 5:e201302011
43. Berman HM, Westbrook J, Feng Z, Gilliland G, Bhat TN, Weissig H, Shindyalov IN, Bourne PE (2000) The Protein Data Bank. Nucleic Acids Res 28:235–242
44. Altschul SF, Gish W, Miller W, Myers EW, Lipman DJ (1990) Basic local alignment search tool. J Mol Biol 215:403–410
45. Altschul SF, Madden TL, Schaer AA, Zhang J, Zhang Z, Miller W (1997) Gapped BLAST and PSI-BLAST: a new generation of protein. Nucleic Acids Res 25:3389–33402
46. Pearson WR, Lipman DJ (1988) Improved tools for biological sequence comparison. Proc Natl Acad Sci U S A 85:2444–2448
47. Karplus K, Barrett C, Hughey R (1998) Hidden Markov models for detecting remote protein homologies. Bioinformatics 14:846–856
48. Eddy SR (1998) Profile hidden Markov models. Bioinformatics 14:755–763
49. Petrey D, Honig B (2005) Protein structure prediction: inroads to biology. Mol Cell 20:811–819
50. Thompson JD, Higgins DG, Gibson TJ (1994) CLUSTAL W: improving the sensitivity of progressive multiple sequence alignment through sequence weighting, position-specific gap penalties and weight matrix choice. Nucleic Acids Res 22:4673–4680
51. Hall TA (1999) BioEdit: a user-friendly biological sequence alignment editor and analysis program for Windows 95/98/NT. Nucleic Acids Symp Ser 41:95–98
52. Taylor WR, Flores TP, Orengo CA (1994) Multiple protein structure alignment. Protein Sci 3:1858–1870
53. Schwede T, Kopp J, Guex N, Peitsch MC (2003) SWISS-MODEL: an automated protein homology modeling server. Nucleic Acids Res 31:3381–3385
54. Sali A, Blundell TL (1993) Comparative protein modeling by satisfaction of spatial restraints. J Mol Biol 234:779–815
55. Xiang Z (2007) Homology based modeling of protein structure. In: Xu Y, Xu D, Liang J (eds) Computational methods for protein structure prediction and modeling: basic characterization. Springer, New York, pp 319–357
56. Levitt M (1992) Accurate modeling of protein conformation by automatic segment matching. J Mol Biol 226:507–533
57. Xiang Z, Honig B (2001) Extending the accuracy limits of prediction for side-chain conformations. J Mol Biol 311:421–430
58. Laskowaski RA, McArther MW, Moss DS, Thornton JM (1993) PROCHECK a program to check sterio-chemical quality of a protein structures. J Appl Crystallogr 26:283–291
59. Eisenberg D, Luthy R, Bowie JU (1997) VERIFY3D: assessment of protein models with three-dimensional profiles. Methods Enzymol 277:396–404
60. Wiederstein M, Sippl MJ (2007) ProSA-web: interactive web service for the recognition of errors in three-dimensional structures of proteins. Nucleic Acids Res 35:W407–W410
61. van Gelder CW, Leusen FJ, Leunissen JA, Noordik JH (1994) A molecular dynamics

approach for the generation of complete protein structures from limited coordinate data. Proteins 18:174–185
62. Cornell WD, Cieplak P, Bayly CI, Gould IR, Merz KM, Ferguson DM, Spellmeyer DC, Fox T, Caldwell JW, Kollman PA (1995) A second generation force field for the simulation of proteins, nucleic acids, and organic molecules. J Am Chem Soc 117:5179–5197
63. Zhu J, Fan H, Periole X, Honig B, Mark AE (2008) Refining homology models by combining replica-exchange molecular dynamics and statistical potentials. Proteins 72:1171–1188
64. Das R, Qian B, Raman S, Vernon R, Thompson J, Bradley P, Khare S, Tyka MD, Bhat D, Chivian D, Kim DE, Sheffler WH, Malmström L, Wollacott AM, Wang C, Andre I, Baker D (2007) Structure prediction for CASP7 targets using extensive all-atom refinement with Rosetta@home. Proteins 69:118–128
65. Han R, Leo-Macias A, Zerbino D, Bastolla U, Contreras-Moreira B, Ortiz AR (2008) An efficient conformational sampling method for homology modeling. Proteins 71:175–188
66. Laskowski RA, MacArthur MW, Thornton JM (1998) Validation of protein models derived from experiment. Curr Opin Struct Biol 8:631–639
67. Wilson C, Gregoret LM, Agard DA (1993) Modeling side-chain conformation for homologous proteins using an energy-based rotamer search. J Mol Biol 229:996–1006
68. Sippl MJ (1993) Recognition of errors in three-dimensional structures of proteins. Proteins 17:355–362
69. Vyas VK, Ukawala RD, Ghate M, Chintha C (2012) Homology modeling a fast tool for drug discovery: current perspectives. Indian J Pharm Sci 74:1–17
70. Laskowski RA, Rullmannn JA, MacArthur MW, Kaptein R, Thornton JM (1996) AQUA and PROCHECK-NMR: programs for checking the quality of protein structures solved by NMR. J Biomol NMR 8:477–486
71. Hooft RW, Vriend G, Sander C, Abola EE (1996) Errors in protein structures. Nature 381:272
72. Luthy R, Bowie JU, Eisenberg D (1992) Assessment of protein models with three-dimensional profiles. Nature 356:83–85
73. Melo F, Feytmans E (1998) Assessing protein structures with a non-local atomic interaction energy. J Mol Biol 277:1141–1152
74. Sliwoski G, Kothiwale S, Meiler J, Lowe EW (2014) Computational methods in drug discovery. Pharmacol Rev 66:334–395
75. Blaney JM, Dixon JS (1993) A good ligand is hard to find: automatic docking methods. Perspect Drug Discovery Des 15:301–319
76. Jones G, Willet P (1995) Docking small-molecule ligands into active sites. Curr Opin Biotechnol 6:652–656
77. Lybrand TP (1995) Ligand-protein docking and rational drug design. Curr Opin Biotechnol 5:224–228
78. Rosenfeld R, Vajda S, Delisi C (1995) Flexible docking and design. Annu Rev Biophys Biomol Struct 24:677–700
79. Gschwend DA, Good AC, Kuntz ID (1996) Molecular docking towards drug discovery. J Mol Recognit 8:175–186
80. Lengauer T, Rarey M (1996) Computational methods for biomolecular docking. Curr Opin Struct Biol 6:402–406
81. Kuntz ID, Blaney JM, Oatley SJ, Langridge R, Ferrin TE (1982) A geometric approach to macromolecule-ligand interactions. J Mol Biol 161:269–288
82. Kitchen DB, Decornez H, Furr JR, Bajorath J (2004) Docking and scoring in virtual screening for drug discovery: methods and applications. Nat Rev Drug Discov 3:935–949
83. Morris GM, Huey R, Lindstrom W, Sanner MF, Belew RK, Goodsell DS, Olson AJ (2009) AutoDock4 and AutoDockTools4: automated docking with selective receptor flexibility. J Comput Chem 30:2785–2791
84. Dodson GG, Lane DP, Verma CS (2008) Molecular simulations of protein dynamics: new windows on mechanisms in biology. EMBO Rep 9:144–150
85. Levin EJ, Kondrashov DA, Wesenberg GE, Phillips GN Jr (2007) Ensemble refinement of protein crystal structures: validation and application. Structure 15:1040–1052
86. Verlet L (1967) Computer experiments on classical fluids. I. Thermodynamical properties of Lenard-Jones molecules. Phys Rev 159:98–103
87. Gear CW (1971) Numerical initial value problems in ordinary differential equations. Prentice- Hall, Inc., Englewood Cliffs, pp 17–253
88. Beemann D (1976) Some multistep methods for use in molecular dynamics simulations. J Comput Phys 20:130–139
89. Hockney RW (1970) The potential calculation and some applications. In: Alder B, Fernbach S, Rotenberg M (eds) Methods in

computational physics, vol 9. Plasma Physics Academic Press, New York/London, p 136. ISBN: 1439810958

90. Zeyao M, Jinglin Z, Qingdong C (2002) Dynamic load balancing for short-range parallel molecular dynamics simulations. Int J Comput Math 79:165–177
91. Mattson W, Rice BM (1999) Near-neighbor calculations using a modified cell linked list method. Comput Phys Commun 119:135–148
92. Hairer E, Lubich C, Wanner G (2003) Geometric numerical integration illustrated by the Störmer/Verlet method. Acta Numerica 12:399–450
93. Darden T, York D, Pedersen L (1993) Particle mesh Ewald: an N -log(N) method for Ewald sums in large systems. J Chem Phys 98:10089–10092
94. Luty BA, Tironi IG, van Gunsteren WF (1995) Lattice-sum methods for calculating electrostatic interactions in molecular simulations. J Chem Phys 103:3014–3021
95. Ryckaert JP, Ciccotti G, Berendsen HJC (1977) Numerical integration of the cartesian equations of motion of a system with constraints; molecular dynamics of n-alkanes. J Comput Phys 23:327–341
96. Miyamoto S, Kollman PA (1992) Settle: an analytical version of the SHAKE and RATTLE algorithm for rigid water models. J Comput Chem 13:952–962
97. Van Der Spoel D, Lindahl E, Hess B, Groenhof G, Mark AE, Berendsen HJ (2005) b GROMACS: fast, flexible, and free. J Comput Chem 26:1701–1718
98. Hess B, Bekker H, Berendsen HJC, Fraaije JGEM (1997) LINCS: a linear constraint solver for molecular simulations. J Comput Chem 18:1463–1472
99. Van Der Spoel D, Lindahl E, Hess B, Groenhof G, Mark AE, Berendsen HJ (2005) a Gromacs user manual version 3.3, www.gromacs.org
100. MacKerell AD, Bashford D, Bellott M, Dunbrack RL, Evanseck JD, Field MJ, Fischer S, Gao J, Guo H, Ha S, Joseph-McCarthy D, Kuchnir L, Kuczera K, Lau FT, Mattos C, Michnick S, Ngo T, Nguyen DT, Prodhom B, Reiher WE, Roux B, Schlenkrich M, Smith JC, Stote R, Straub J, Watanabe M, Wiórkiewicz-Kuczera J, Yin D, Karplus M (1998) All-atom empirical potential for molecular modeling and dynamics studies of proteins. J Phys Chem B 102:3586–3616
101. Oostenbrink C, Villa A, Mark AE, van Gunsteren WF (2004) A biomolecular force field based on the free enthalpy of hydration and solvation: the GROMOS force-field parameter sets 53A5 and 53A6. J Comput Chem 25:1656–1676
102. Jorgensen WL, Maxwell DS, Tirado-Rives J (1996) Development and testing of the OPLS all-atom force field on conformational energetics and properties of organic liquids. J Am Chem Soc 118:11225–11236
103. Kaminski GA, Friesner RA (2001) Evaluation and reparametrization of the opls-aa force field for proteins via comparison with accurate quantum chemical calculations on peptides. J Phys Chem B 105:6474–6487
104. Berendsen HJC, van der Spoel D, van Drunen R (1995) Gromacs: a message-passing parallel molecular dynamics implementation. Comput Phys Commun 91:43–56
105. Hansson T, Oostenbrink C, van Gunsteren W (2002) Molecular dynamics simulations. Curr Opin Struct Biol 12:190–196
106. Weber W, Hünenberger PH, McCammon JA (2000) Molecular dynamics simulations of a polyalanineoctapeptide under Ewald boundary conditions: influence of artificial periodicity on peptide conformation. J Phys Chem B 104:3668–3675
107. Fonseca JE (2008) Temporal and steric analysis of ionic permeation and binding in NA+, K+ -ATPase *via* molecular dynamic simulations, Ph.D. Thesis, Department of Electrical Engineering and Computer Science, The Russ College of Engineering and Technology of Ohio University, Athens
108. Brooks CL 3rd (1995) Methodological advances in molecular dynamics simulations of biological systems. Curr Opin Struct Biol 5:211–215
109. Hunenberger PH (2005) Thermostat algorithms for molecular dynamics simulations. Adv Polym Sci 173:105–149
110. Evans DJ, Morriss GP (1983) Isothermal/isobaric molecular dynamics ensemble. Phys Lett A 98:433–436
111. Eslami H, Muller-Plathe F (2007) Molecular dynamics simulation in the grand canonical ensemble. J Comput Chem 28:1763–1773
112. Berendsen HJC, Postma JPM, Vangunsteren WF, Dinola A, Haak JR (1984) Molecular dynamics with coupling to an external bath. J Phys Chem 81:3684–3690
113. Marechal Y (2004) Water and biomolecules: an introduction. J Mol Struct 70:207–210
114. Fornili A, Autore F, Chakroun N, Martinez P, Fraternali F (2012) Protein-water interactions in MD simulations: POPS/POPS-COMP solvent accessibility analysis,

solvation forces and hydration sites. Methods Mol Biol 819:375–392

115. Jorgensen WL, Chandrasekhar J, Madura JD, Impey RW, Klein ML (1983) Comparison of simple potential functions for simulating liquid water. J Chem Phys 79:926–935
116. Berendsen HJC (1981) Interaction models for water in relation to protein hydration. Intermolecular forces, Jerusalem symposia on quantum chemistry and biochemistry, 14:331–342
117. Berendsen HJC, Griegera GR, Straatsma TP (1987) The missing term in effective pair potentials. J Phys Chem 91:6269–6271
118. Mark P, Nilsson L (2001) Structure and dynamics of the TIP3P, SPC, and SPC/E water models 298 K. J Phys Chem A 105:9954–9960
119. Levitt M, Hirshberg M, Sharon R, Laidig KE, Daggett V (1997) Calibration and testing of a water model for simulation of the molecular dynamics of proteins and nucleic acids in solution. J Phys Chem B 101:5051–5061
120. Hestenes MR, Eduard S (1952) Methods of conjugate gradients for solving linear systems. J Res Natl Bur Stand 49:410–436
121. Leach AR (2001) Molecular modelling: principles and applications. Prentice Hall Pearson Education, New York, p 784. ISBN: 978-0582382107
122. Wiberg KB (1965) A scheme for strain energy minimization application to the cycloalkanes. J Am Chem Soc 87:1070–1078
123. Kini RM, Evan HJ (1991) Molecular modeling of proteins: a strategy for energy minimization by molecular mechanics in the AMBER force field. J Biomol Struct Dyn 9:475–488
124. Meng XY, Zhang HX, Mezei M, Cui M (2011) Molecular Docking: a powerful approach for structure-based drug discovery. Curr Comput Aided Drug Des 7:146–157
125. Ou-Yang SS, Lu JY, Kong XQ, Liang ZJ, Luo C, Jiang H (2012) Computational drug discovery. Acta Pharmacol Sin 33:1131–1140
126. Lipinski CA, Christopher AL (1997) Experimental and computational approaches to estimate solubility and permeability in drug discovery and development settings. Adv Drug Deliv Rev 23:3–25
127. Bielska E et al (2011) Virtual screening strategies in drug design-methods and applications. Comput Biol Bionanotech 92:249–264
128. Irwin JJ et al (2009) Automated docking screens: a feasibility study. J Med Chem 52:5712–5720
129. Jalkute CB, Barage SH, Dhanavade MJ, Sonawane KD (2014) Identification of angiotensin converting enzyme inhibitor: an in Silico perspective. Int J Pept Res Ther 32:356–364
130. Barage SH, Jalkute CB, Dhanavade MJ, Sonawane KD (2013) Virtual screening and molecular dynamics simulation study of hECE-1 protease inhibitors. Res J Pharm Biol Chem Sci 4:1279–1291
131. Trott O, Olson AJ (2010) AutoDock Vina: improving the speed and accuracy of docking with a new scoring function, efficient optimization and multithreading. J Comput Chem 31:455–461
132. Sonawane KD, Barage SH (2014) Structural analysis of membrane-bound hECE-1 dimer using molecular modeling techniques: insights into conformational changes and Aβ1–42 peptide binding. Amino Acids 47:543–559
133. Hou T, Wang J, Li Y, Wang W (2011) Assessing the performance of the MM/PBSA and MM/GBSA methods. 1. The accuracy of binding free energy calculations based on molecular dynamics simulations. J Chem Inf Model 51:69–82
134. Genheden S, Ryde U (2010) How to obtain statistically converged MM/GBSA results. J Comput Chem 31:837–846
135. Vorontsov II, Miyashita O (2011) Crystal molecular dynamics simulations to speed up MM/PB(GB) SA evaluation of binding free energies of di-mannose deoxy analogs with P51G-m4-Cyanovirin-N. J Comput Chem 32:1043–1053
136. Sonawane KD, Sambhare SB (2015) The influence of hypermodified nucleosides lysidine and t^6A to recognize the AUA codon instead of AUG: a molecular dynamics simulation study. Integr Biol 7:1387–1395

Chapter 12

Computational Modeling of Gamma-Secretase Inhibitors as Anti-Alzheimer Agents

Prabu Manoharan and Nanda Ghoshal

Abstract

γ-Secretase (gamma secretase) is a complex unusual aspartyl protease, responsible for the production of amyloid-β peptides (Aβ) involved in Alzheimer's disease (AD). Inhibition of gamma secretase (GS) is an attractive therapeutic strategy to slow down the pathological progression of AD. For a long time, GS-targeted structure-based drug designing remained unrealistic without the 3D structural knowledge of GS. Hence, to meet the prevailing urgent need for AD drugs, several groups individually tried to develop GS inhibitors, with the aid of computational drug designing methods. This chapter mainly provides with a detailed discussion on a QSAR-guided fragment-based virtual screening method for GS inhibitor design and identification. In this study, we took advantage of the wealth of available known small molecular GS inhibitors and applied in this drug designing program. Here, the non-transition state small molecular GS inhibitors with corresponding affinity values were used to develop 2D- and 3D-QSAR models investigating alternative site-binding GS inhibitors. HipHop pharmacophore-based alignment-dependent (CoMFA and CoMSIA) and GRIND-based alignment-independent 3D-QSAR models were developed to elucidate the potential 3D features involved in GS inhibition. Consensus of QSAR results from this study underscores the reliability and accuracy of the results and provides a rationale for the design of novel potent GS inhibitors that can have AD therapeutic application.

Key words Gamma secretase, Bioisosteric replacement, 3D-QSAR, CoMFA, GRIND, Inverse QSAR

1 Introduction

1.1 Alzheimer's Disease: An Overview

Alzheimer's disease is the most common cause of dementia and has emerged as the most prevalent form of late-life mental failure in humans. AD was first described by the German psychiatrist and neuropathologist Alois Alzheimer [1] in 1906 from whom it takes its name. He reported the existence of two abnormal structures, senile plaques and neurofibrillary tangles, in the brain of his patient, a woman referred to as Auguste D. The symptoms, observed in his patient, exemplified several cardinal features of the disorder that is observed in AD patients even today, viz., progressive memory impairment, disordered cognitive function, altered behavior including paranoia, delusions, loss of social appropriateness, and a

Kunal Roy (ed.), *Computational Modeling of Drugs Against Alzheimer's Disease*, Neuromethods, vol. 132, DOI 10.1007/978-1-4939-7404-7_12, © Springer Science+Business Media LLC 2018

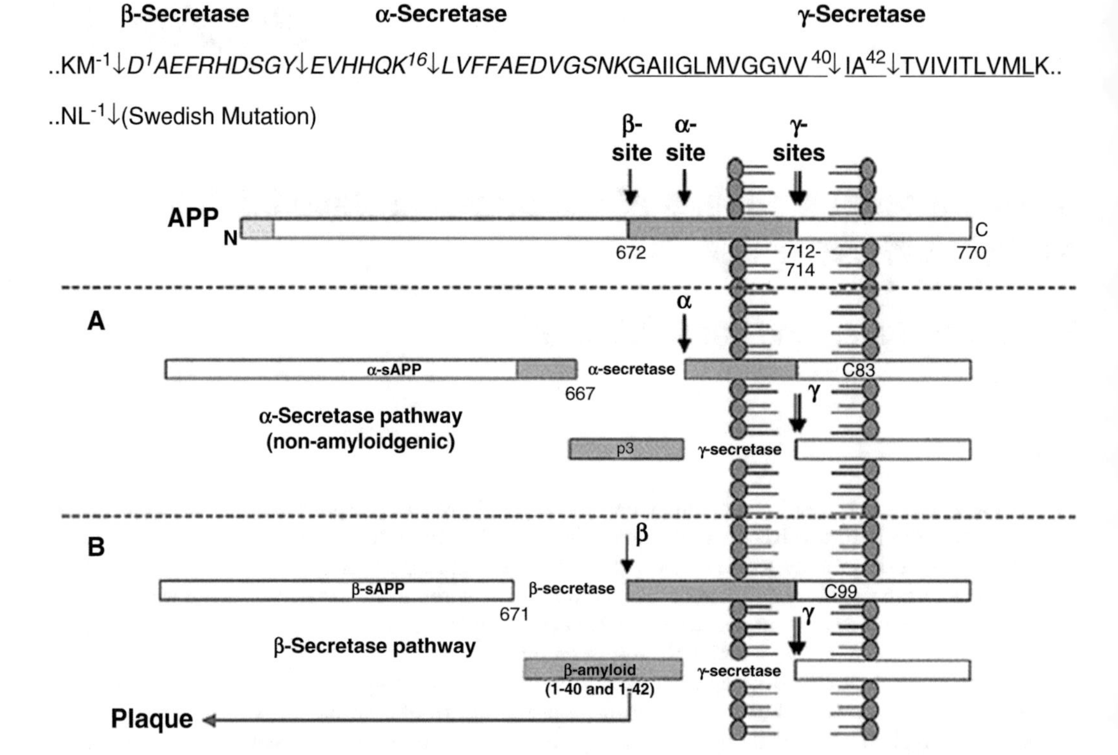

Fig. 1 (**a**) shows the α-secretase-mediated non-amyloidogenic pathway. (**b**) Shows the amyloidogenic pathway: production of pathogenic Aβ by sequential proteolytic cleavage of APP by BACE1 and γ-secretase [9] (Reproduced from Journal of Medicinal Chemistry, 2003 with permission from American Chemical Society)

progressive decline in language function [2]. AD is a devastating and ultimately fatal neurodegenerative disease, characterized by progressive cognitive and memory deterioration that leads to difficulty in carrying out everyday activities [1, 3, 4].

At present, more than 26 million people worldwide suffer from AD, and the annual total costs have been estimated to exceed US $200 billion. A recent estimate predicts that this number will quadruple by 2050 to more than 106 million [5]. This projection is set to double every 20 years [6]. Hence, AD is a disease of great socioeconomical concern and a burgeoning epidemic that exerts a tremendous toll on the individuals it affects, along with their families and caregivers.

The amyloid cascade hypothesis states that amyloid is at the center of the pathophysiology of AD. Amyloid deposits in the central nervous system (CNS) are the primary cause and instigate the process that drives a pathological cascade, which eventually culminates in the manifestation of the disease [7]. The pathogenic Aβ peptides are formed by sequential proteolytic cleavage of the amyloid precursor protein (APP) by β- and γ-secretases [2]. The cleavage of APP, by BACE1 (β-secretase) and γ-secretase, results in the pathogenic Aβ peptides. In the event of Aβ generation, the membrane-anchored APP is primarily cleaved by the transmembrane protease BACE1 at the N-terminus and then by the enzyme γ-secretase at the C-terminus. BACE1 cleavage of APP results in the production of soluble β-APP fragment and a membrane-bound C-99 fragment. This C-99 fragment then serves as the substrate for subsequent cleavage by γ-secretase, resulting in the formation of APP intracellular domain (AICD) and various Aβ species of differing lengths ranging from 37 to 49 amino acids. A schematic representation of the Aβ region of APP showing the amino acid sequence of Aβ and the major sites of processing by the α -, β -, and γ -secretases [8, 9] is presented in Fig. 1.

The processing of APP by γ-secretase also generates other Aβ isoforms with various lengths. Of these, Aβ_{42} is more pathogenic than the other isoforms, and it constitutes between 5% and 10% of the overall Aβ population and is more hydrophobic and prone to aggregation than the shorter isoforms. It is believed to be neurotoxic when aggregated, resulting in the formation of amyloid plaques. Investigations have revealed a link between these plaques and the pathogenesis of AD [10–12]. Recent studies show that γ-secretase cleaves APP transmembrane domain in a progressive, stepwise manner at the ε, ζ, and γ sites, resulting in Aβ species of varying length [12, 13]. The pathway from the monomeric Aβ to the plaques progresses through many stages of aggregation where non-fibrillary and fibrillary aggregates of different dimensions are formed. Aβ_{42} is the key to these intermediates because it has a high propensity to form these aggregates with itself and other proteins. It is because of the two additional hydrophobic amino acids which facilitate to form more aggregates with itself and also with other Aβ species [14].

1.2 γ-Secretase as AD Therapeutic Target

γ-Secretase is the other enzyme that cleaves the APP after BACE1. It is only because of the non-specific cleavage by γ-secretase that releases the amyloidogenic Aβ_{42}. Hence, inhibition or modulation of γ-secretase is an obvious therapeutic strategy with the goal of decreasing the concentration of Aβ and in particular Aβ_{42}. One of the difficulties in targeting γ-secretase is that it also cleaves other substrates within the transmembrane region [15]. One of the substrates is Notch, which is cleaved by γ-secretase to release a smaller cytosolic fragment NICD (Notch intracellular domain), important in signal transduction pathways [16]. However, research has shown that ~15-fold Notch sparing in gamma-secretase inhibitors (GSI) is sufficient to avoid Notch-related side effects (NRSEs) [17]. Also ~30% inhibition of Aβ synthesis is sufficient to reverse cognitive impairment [18]. Hence, pharmacological inhibition of both β-secretase and γ-secretase may still prove to be therapeutically beneficial. Partial inhibition of their proteolytic activities may be sufficient to lower the amount of Aβ to a level that may delay the onset and progression of the disease without producing intolerable side effects. Going by this approach, one could promise the delivery of disease-modifying anti-Alzheimer's drug (DMAAD).

1.3 γ-Secretase: Structural Biology

Unlike BACE1, which is a small protein, γ-secretase is a large protein complex. γ-Secretase belongs to a diverse family of intramembrane-cleaving proteases (I-CliPs) [19] and composed of four integral membrane proteins named as presenilin (PSEN), nicastrin (Nct), anterior pharynx-defective 1 (Aph-1), and presenilin enhancer protein 2 (Pen-2) [20]. Among these, the presenilins constitute the aspartic protease catalytic subunit of γ-secretase [21], and they belong to a family of related multipass transmembrane proteins that function as a part of the γ-secretase protease complex. In mammals, there are two presenilins (PS), expressed by two genes PSEN1 and PSEN2, that encode for PS-1 and PS-2, respectively, which are highly homologous. Both PS-1 and PS-2, having 476 and 448 amino acids, respectively, are polytopic membrane proteins and have 10 hydrophobic domains, of which 9 are proposed to span the membrane [22]. PS-2 differs from PS-1 in that it lacks four amino acid residues between amino acids 26 and 29, close to the amino terminus [23]. The N-terminal catalytic site of PS is embedded in a conserved YD motif, whereas the C-terminal active site domain contains the equally conserved GXGD motif.

Nicastrin is a type I single-span membrane protein containing 709 amino acids, with a large extracellular domain, heavily glycosylated and tightly folded on maturation [24]. Aph-1 is apparently a 7-TMD protein (30 k Da) with its amino and carboxyl termini located on the luminal and cytoplasmic sides, respectively [25]. It has been reported that Aph-1 can exist as two splice variants: Aph-1a and Aph-1b [26]. Pen-2 (12 k Da) spans the membrane twice like a hairpin, with both the amino and carboxyl termini being found on the luminal side [27], and it is the smallest component of γ-secretase complex. The overall stoichiometry of γ-secretase complex has been discovered to be likely 1:1:1:1, monomeric containing one copy of each of the four components as understood by various experimental studies [28]. Both the docking site and the active site lay at the interface between the presenilin NTF and CTF [29, 30]. Studies with helical peptide inhibitors suggest that the docking and active sites are relatively close, within the length of three amino acid residues [30]. The extracellular domain of nicastrin also plays a role in recognition of the substrate, binding to the N-terminus [31]. The first known structure of γ-secretase complex is the solution structure of presenilin-1 CTF subunit that is available in PDB, (PDB ID 2KR6) [32]. Further analysis of the structure of γ-secretase complex by cryo-electron microscopy and single-particle image reconstruction at 12 Å resolution revealed several domains on the extracellular side, three low-density cavities, and a surface groove in the transmembrane region of the complex [33].

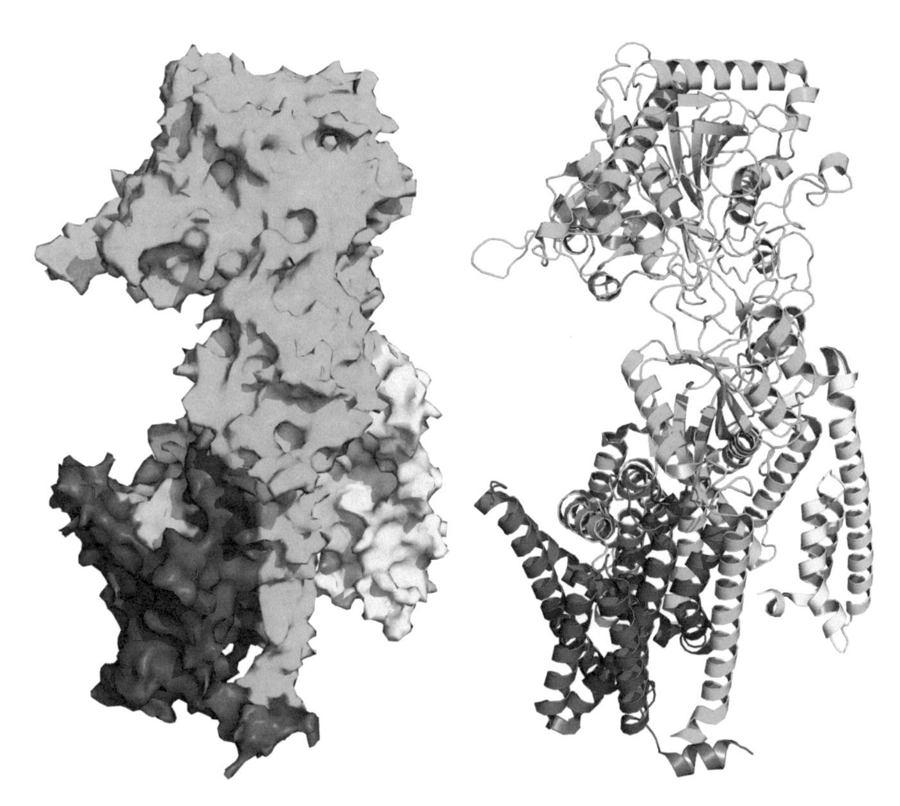

Fig. 2 The gamma-secretase complex (PDB id. 5A63) is shown in surface (*left*) and cartoon (*right*) models. The subunits are colored differently, nicastrin (*green*), presnilin-1 (*cyan*), anterior pharynx-defective 1 (APH-1) (*magenta*), and presenilin enhancer 2 (PEN2) (*yellow*). Figure generated in house using PyMOL

Finally, an atomic structure of human γ-secretase was resolved at 3.4 Å resolution by single-particle cryo-electron microscopy [34]. There are other groups also, successful in identifying gamma-secretase complex by using cryo-electron microscopy [35, 36]. The cryo-EM structure is better than the earlier structures that have good resolution for all four components of γ-secretase. The surface and cartoon model of gamma-secretase complex is shown in Fig. 2.

1.4 γ-Secretase Inhibitor Development: An Overview

γ-Secretase inhibition has been and remains to be an active field in Alzheimer's disease therapy. The first reported compounds, shown to inhibit γ-secretase activity, were peptide aldehyde-type calpain and proteasome inhibitors [36]. Thereafter, many orally bioavailable, brain-penetrating γ-secretase inhibitors (GSIs) have been identified and have been shown to decrease Aβ production in both human and mouse brains [37]. The GSIs can be classified into three subgroups, depending on where they bind to the γ-secretase complex: (i) active site-binding GSIs, (ii) substrate docking site-binding GSIs, and (iii) alternative binding site GSIs.

The design of active site-binding GSIs is inspired by the knowledge that γ-secretase is an aspartyl protease and the availability of active site-binding inhibitors of the HIV protease, another aspartyl protease [38]. These efforts led to identification of first-generation active site-binding GSIs (L-685,458) from the Merck group [39] and other related inhibitors from Wolfe's group [40]. Employing high-throughput screening (HTS), the Bristol-Myers Squibb group identified the peptidomimetic lead with $A\beta_{total}$ inhibition ($IC_{50} = 5000$ nM) in hAPPH4 neuroglioma cells, which contains the hydroxyethylene moiety found in other active site GSIs [41].

The identification of a substrate docking site on the outer surface of GS that is occupied before the active site on the interior of GS has led to the design of a novel class of GSIs, the so-called substrate docking site-based GSIs [42]. Wolfe's group has designed first-generation helical peptides by incorporating the helix-inducing residue α-aminoisobutyric acid (Aib) that mimic the C-99 substrate conformation and inhibit $\varsigma\beta$ production in the low micromolar range in APP-transfected CHO cells [43]. Unfortunately, this series of inhibitors also potently blocks Notch processing by GS, and, like the active site-directed GSIs, no compound from this class of GSIs has entered AD clinical trials.

Discovery and development of the alternative binding site-based GSIs were actively pursued. Because of the absence of a high-resolution structure of GS which precluded a structure-based design approach, the majority of efforts to identify GSIs have utilized HTS to identify viable leads. Using this method, the Lilly group in collaboration with Elan identified the GSI lead with $A\beta_{total}$ $IC_{50} = 900$ nM [44]. The Wyeth group has utilized molecular modeling and HTS screening approaches to identify GSI leads. Using the ROCS (rapid overlay of chemical structures) program that identifies molecules that have a similar three-dimensional shape, the Wyeth group analyzed the Bristol-Myers Squibb arylsulfonamide. The drug discovery efforts from these pharmaceutical companies saw many compounds (BMS 706,163, MK-0752, ELN-006) entering clinical trials for the treatment of Alzheimer's disease [45].

1.5 Overview of CADD in GS Inhibitors Identification

The computer-aided drug design (CADD), an indispensable method, concerns the use of computers in molecular design and in the lead discovery process. Computers have become widely used in drug research for maintaining databases, statistical processing, molecular modeling, theoretical chemical calculations, and so on. The very first computer-aided approach in drug design was developed in the early 1960s, when Corwin Hansch started the quantitative structure–activity relationships (QSAR) discipline [46]. Since the late 1980s, computer-aided drug design (CADD) techniques have found wide application in the pharmaceutical and biotech industries. CADD now plays a critical role in the search for new molecular entities [47].

CADD method can be divided into ligand-based drug design (LBDD) and structure-based drug design (SBDD). The LBDD approach is usually applied when structural information on the target macromolecule is not known. LBDD relies on the hypothesis that compounds with comparable physicochemical properties behave similarly in biological systems. Pharmacophore models as well as QSAR can therefore be developed based on the analysis of known ligands. Ligand-based approaches commonly consider two- or three-dimensional chemistry, shape, electrostatics, and pharmacophoric features to assess similarity. Ligand similarity approaches (2D or 3D) require only a single active molecule, which may come from the literature, patents, or in-house experimental data [48, 49].

CADD methods are used for the identification of new drugs to cure Alzheimer's disease. However, most of the drug design initiatives were toward BACE1. Compared to BACE1, little computational work has been done to identify inhibitors for γ-secretase. The reason for this huge difference is due to the absence of the 3D structure of γ-secretase. The CADD efforts are no more than ligand-based drug design, due to the only availability of small molecule ligands. Also, researchers apply CADD for γ-secretase, using two strategies, viz., (i) to identify/design gamma-secretase inhibitors (GSIs) and (ii) to identify/design gamma-secretase modulators (GSMs). Here a brief summary of the CADD methods used for the identification of GS inhibitors and/or modulator would be discussed as follows.

The 2D-QSAR methods were developed to identify correlation between the GS inhibitory activity and some of the simple descriptors like lipophilicity parameter (QlogP) and steric factor (molar refractivity, SMR) [50]. Molecular field analysis (MFA) and genetic partial least squares (G/PLS) 3D-QSAR methods were used to build predictive models on a series of benzodiazepine analogous as GS inhibitors. In this study, steric and electrostatic features of the GS–ligand interactions were quantified using molecular field analysis (MFA) and GPLS methods [51].

Ligand three-dimensional shape-based search was used to identify GS inhibitors. Rapid overlay of chemical structures (ROCS), a shape similarity search engine, was used to find new GS inhibitors that are similar to sulfonamide compound, a potential GS inhibitor. Chemical synthesis and biological evaluation of the modeled new GS inhibitors showed moderate activity against GS [52].

A general predictive QSAR model was developed using 233 compounds with diverse chemical classes, collected from ChEMBL database. In this 2D-QSAR study, continuous and categorical QSAR models were developed to obtain relevant QSAR descriptors contributing toward GS inhibitory potency, using two approaches for each type, viz., PLS regression, neural network (NN) for continuous and linear discriminant analysis (LDA), and NN for categorical models. It was interesting to observe from the analysis of heterogeneous database compounds that four descriptors, namely, $PEOE_3$ (partial equalization of orbital electronegativities), number of aliphatic rings, relative number of double bonds, and sssN_Cnt (count of atom-type E-state, >N – i.e., the nitrogen atom which is connected to three heavy atoms with single bonds) are important features for GS inhibition. Electronegative substitution on aryl rings, which increases the value of $PEOE_3$ and substitution of acyclic amines with N-substituted cyclic amines that increases the values of sssN_Cnt and number of aliphatic rings, helps in improving GS inhibition [53].

An integrated GS modulator design approach was developed that includes computational virtual screening and chemical synthesis to reveal the structure–activity relationship followed by in vitro pharmacological characterization. In this study, Pharmacophore Alignment Search Tool (PhAST), a virtual screening tool, was used for the rapid ligand-based hit/lead finding and optimization process [54].

A series of bicyclononane compounds were used to develop 2D-QSAR and 3D-QSAR (CoMFA, CoMSIA, GRIND); finally, consensus QSAR was developed for predicting the new GS inhibitors. In this study, QSAR-guided fragment-based bioisosteric replacement method was used to identify potential lead fragments that would have optimal GS inhibitory activity [55]. In this chapter, the abovementioned method will be discussed in detail.

1.6 QSAR-Guided Fragment Hopping Method

A pharmaceutical context, QSAR-guided bioisosteric replacement strategy that promises to deliver novel GSIs is developed in this study. QSAR, a method of quantitatively correlating physicochemical properties of molecules with their biological activity, has metamorphosed into a widely used tool, substantially contributing to the ligand-based drug discovery process. Many valuable extensions to the classical QSAR paradigm have been explored. A drug discovery project, conducted by Merck Sharp and Dohme research laboratories, identified a series of bicyclononane analogs as potent GS inhibitors, which belong to small molecule non-transition state-based sulfonamide class of GS inhibitors [56–58]. Starting with this available information in hand, further exploration was undertaken in pursuit of identifying additional leads. A targeted library of privileged molecules was enumerated, considering shape, chemistry, and electrostatics of established fragment. Intelligent enumeration involves the rational choice of building blocks. Bioisosteric building blocks that are ostensibly known to yield molecular entities, imparting similar biological properties, were considered as ideal replacements for the marked Markush fragment. Thus, the principle of analog design, widely exploited in medicinal chemistry, was used to create a targeted library. The method envisaged herein

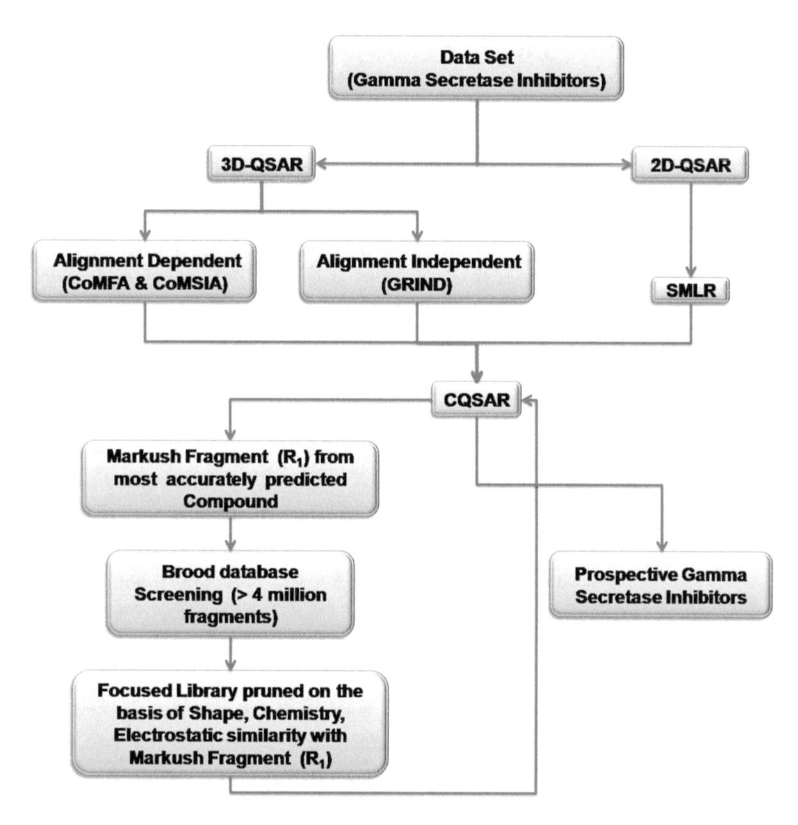

Fig. 3 Computational workflow for the identification of prospective gamma-secretase inhibitors (GSIs) [55] (Reproduced from Molecular Diversity, 2012, with permission from Springer)

would enable lead optimization using a novel approach where QSAR confluences with bioisosteric lead hopping strategy.

Classical QSAR models were developed using the two-dimensional descriptors available in molecular operating environment (MOE) [59] and TSAR [60] and three-dimensional descriptors available in CoMFA [61, 62] and CoMSIA [63, 64] in Sybyl [65] and GRid-INdependent Descriptor (GRIND) [66] in ALMOND [67]. Fitting was carried out using stepwise multiple linear regression (SMLR), partial least squares (PLS) [68], and principal component analysis (PCA) [69] statistical methods. The developed models were subjected to rigorous validation, giving emphasis on model stability and predictivity. Activity forecasting was carried out using a consensus QSAR (CQSAR) model, which was anticipated to provide a better model compared to the arbitrary choice of a single QSAR model [70, 71]. The information gleaned from CQSAR approach was used in an inverse-QSAR (*I*-QSAR) manner to identify ideal Markush fragment, which in turn was employed to screen over four million medicinally relevant fragments. The resultant subset was enumerated into a focused virtual library that was subsequently screened using the developed QSAR models in this study. A key component that needs to be evaluated, particularly when extrapolating QSAR models for lead mining, is to ensure that the predictions come from the domain upon which the model was calibrated. Hence, applicability domain estimation was carried out to ascertain the reliability of the predictions. The computational workflow developed during this study is provided in Fig. 3. The systematic framework presented and implemented in this study describes the process of lead compound optimization with the application of classical QSAR method.

2 Materials and Methods

2.1 Data Set

The data set used in this study comprises a series of 76 gamma-secretase inhibitors; these inhibitors were pooled from literatures reported by the same group. The inhibition potency of the bicyclononane inhibitors, in log $(1/IC_{50})$ values, varies from 5.22 to 10.52 (range 5.30) against gamma-secretase receptor. This is a reasonably large range in inhibition potency concentration from the point of view of constructing statistically significant correlation.

2.2 Molecular Modeling

Molecular modeling is a unifying term given to the application of theoretical and computational chemistry for building, visualizing, and analyzing molecules to understand and/or mimic their behavior. In any molecular modeling and CADD study, the first thing to do is to generate an acceptable model of the molecule in the computer. A simple method of structure generation is the so-called sketch approach. In this method, the 2D structures of the molecules would be drawn and converted into 3D structures, using 2D to 3D conversion programs.

All the molecules were sketched using Sybyl 7.3 [65] in a Linux workstation. Energy minimizations were performed with the steepest descent followed by conjugate gradient method using the MMFF94 [72] force field with a convergence criterion of 0.001 kcal/mol Å. The molecules with obtained reasonable geometry were subjected to rigorous conformational sampling using simulated annealing dynamics. The system was heated at 1000 K for 1 ps and then cooled down to 200 K for 1 ps. The exponential annealing function was used, and seven such cycles were run. Using this method, the conformer with the least energy content was identified and subsequently subjected to further minimization with the same criteria as mentioned above. The final structures were then fully optimized using AM1 semiempirical calculations, implemented in TSAR v3.3 [60]. Care was taken to retain the chirality present in the bicyclononane series of compounds as reported in the literature.

2.3 Data Set Preparation

Data set selection and preparation is very important for obtaining a useful QSAR model. A data set of 76 compounds, with activity range greater than 3 log units and same assay method and experimental conditions, were taken from three literature sources [56–58], reported by the same group. The skewness in the data set was removed by converting IC_{50} (nM) values to pIC_{50} (nM) using a simple logarithmic transformation log $(1/IC_{50})$. Development of predictive QSAR models relies on a multitude of factors, and one such important prerequisite is the rational selection of training and test sets with adequate representation. K-means clustering was performed using SPSS 15.0 [73] for the rational division of training and test sets [74]. Clustering was carried out using more than 100 2D whole molecule descriptors representing topology, structural information, and group counts, calculated using MOE version 2008.10 [59] and TSAR v3.3 [60]. Compounds were manually selected from each cluster to ensure that training set and test set had adequate coverage in terms of activity range and "chemical diversity." Accordingly, 56 compounds served as the training set, and 20 compounds served as the test set. The test set was used for external validation for the SMLR-based 2D-QSAR, CoMFA, and CoMSIA and alignment-independent GRIND-based 3D-QSAR.

2.4 Pharmacophore Modeling and Alignment

Pharmacophore modeling approach is one of the major methods used in drug discovery. The concept of pharmacophore was first introduced in 1909 by Ehrlich [75]. As per the IUPAC definition, a pharmacophore model is "an ensemble of steric and electronic features that is necessary to ensure the optimal supramolecular interactions with a specific biological target and to trigger (or block) its biological response." Typical features include hydrogen bond donors and acceptors, positively and negatively charged groups, hydrophobic regions, and aromatic rings. The use of such features is a natural extension of the concept of bioisosterism, which recognizes that certain functional groups have similar biological, chemical, and physical properties. Traditionally, 3D pharmacophore approaches are used when active molecules are known but where the structure of the target receptor is not available. The process of deriving a 3D pharmacophore is called pharmacophore mapping. Pharmacophore approaches have been used extensively in virtual screening, de novo design, and other applications such as lead optimization and multitarget drug design [48]. Ligand-based pharmacophore modeling has become a key computational strategy for facilitating drug discovery in the absence of a macromolecular target structure. It is usually carried out by extracting common chemical features from 3D structures of a set of known ligands, representative of essential interactions between the ligands and a specific macromolecular target.

In order to generate pharmacophore hypotheses, the conformations were generated for each molecule using Poling algorithm [76], implemented in Catalyst [77] software. To derive a better hypothesis, best conformational analysis option was used to generate conformers. The maximum number of conformer generation was set at 250, with an energy range 20 kcal/mol above the estimated global minimum for each molecule. The HipHop [78] algorithm was used to generate common feature hypotheses, 7 compounds with diverse structures were selected as the training set, and the rest of the 69 compounds were retained as the test set. The structural features of the compounds, selected as the training set, cover the entire data set. The chosen test set provides both structure-wise and activity-wise overall representation of the compounds used in the model building procedure. By using this training set, the pharmacophore model is generated, and the test set is used to validate the developed pharmacophore model. All the molecules were flexibly overlaid to the obtained pharmacophore hypothesis using the best flexible option available within the Catalyst software. Intuition was applied to manually select the possible lowest energy conformer with maximum fit value, which may represent the bioactive conformer. Some of the intuitive criteria for selecting bioactive conformers are correct overlapping of compounds' R-groups to 3D pharmacophoric features and reasonably low-energy conformers. Single-point charges were calculated for all the superimposed compounds using semiempirical AM1 partial charges by using the VAMP electrostatic charge calculation module, present in TSAR.

2.5 2D-QSAR (Stepwise MLR)

Multiple linear regression (MLR) aims to establish a model which quantitatively describes the relationship between data variables and can be used for predicting new compounds. MLR is applied when more than one independent variable are used to derive model. Stepwise regression, a variant of MLR approach that uses a combination of forward and backward MLR, was used as a chemometric method for variable selection and statistical fitting. SMLR identifies an initial model and proceeds by altering the model by adding and removing the explanatory variable in accordance with an F criterion which controls the inclusion or exclusion of explanatory variables until an optimum model is found. SMLR analyses were performed using TSAR with F to include set at 4 and F to exclude set at 3.5.

2.6 3D-QSAR

2.6.1 Alignment-Dependent 3D-QSAR

CoMFA and CoMSIA

The alignment-dependent 3D-QSAR methods need the molecules under investigation to be aligned or superimposed, assuming bioactive conformation, prior to the descriptor calculation. Hence obtaining a bioactive conformation alignment is very important in alignment-dependent 3D-QSAR studies. Comparative molecular field analysis (CoMFA) is one of the most significant developments in the field of alignment-dependent 3D-QSAR. In this method, compounds with common binding mode are superimposed and placed within a regular lattice; the interaction energy between the molecule and a series of probes is calculated. These interaction energies are stored as independent variables and correlated with biological activity to derive an equation. Comparative molecular similarity analysis (CoMSIA) is a similar method like CoMFA, but instead of using arbitrary cutoffs for treating long-range interactions, similarity values were used in CoMSIA, thus reducing the loss of information and results in superior contour maps.

The CoMFA and CoMSIA calculations were performed using Sybyl 7.3. The default settings were used to derive the CoMFA and CoMSIA descriptor fields; a 3D cubic lattice with grid spacing of 2 Å in *x*, *y*, and *z* directions was created to encompass the aligned molecules. To calculate the steric and electrostatic fields, an sp3 carbon probe atom was used with a van der Waals radius of 1.52 Å and a charge of +1.0 to generate steric (Lennard-Jones 6–12 potential) field energies and electrostatic (Coulombic potential) fields with a distance-dependent dielectric at each lattice point. The values of steric and electrostatic energies were truncated at 30 kcal/mol. The CoMFA steric and electrostatic fields thus generated were scaled by the CoMFA-STD method. CoMSIA was performed to calculate the physicochemical properties of each molecule. For the steric, electrostatic, hydrophobicity, hydrogen bond acceptor and donor features, the same probe atoms, grid density, and parameters were used as in the CoMFA model, and the attenuation factor was set to 0.3 for the Gaussian functions. Partial least squares (PLS) regression analysis method was used in CoMFA and CoMSIA model development. Column filtering, for any column of computed energies with a variation less than 2.0 kcal.mol^{-1}, was applied to reduce computation time without negatively affecting the quality of the models.

2.6.2 Alignment-Independent 3D-QSAR

GRIND (ALMOND)

GRIND (GRid-INdependent Descriptors) are obtained from unaligned molecules. These descriptors are insensitive to the position and orientation of the molecular structure in 3D space. These descriptors are easy to interpret by referring back to the compounds under investigation. GRIND were generated, analyzed, and interpreted using the program ALMOND version 3.0. The most important three steps in GRIND calculations are (i) computing a set of molecular interaction fields (MIF), (ii) filtering the MIF to extract the most relevant regions that define the virtual receptor site (VRS), and (iii) encoding the VRS into the GRIND variables. The details about principles behind GRIND calculation can be found elsewhere [66]. In the first step, MIFs were computed using the program GRIND [79]. The probes used to calculate MIF are O (sp2-hybridized oxygen) representing HBA groups, N1 (amide nitrogen) representing HBD groups, DRY representing hydrophobic groups, and TIP [80] (shape descriptor) representing shape of the molecule. A default setting of 0.5 Å was assigned for the grid spacing with the grid extending 5 Å beyond a molecule. The next step is filtering the most relevant MIF, the regions characterized by intense favorable energies of interaction. The relevant regions were extracted using an optimization algorithm, selecting from each MIF a fixed number of nodes optimizing a scoring function. This function includes two optimality criteria: the intensity of the field at a node and the mutual node-node distances between the chosen nodes. The number of nodes to extract and the percentage relative weights of the fields were set as default. The width of the node-node distance range used to discretize the distances was set to 0.8 grid unit. The final step is encoding the geometrical relationships between the extracted relevant regions or virtual receptor site (VRS) into the GRIND in such a way that they are no longer dependent upon their positions in 3D space, thus making the GRIND completely alignment independent.

Multivariate techniques like principal component analysis (PCA) and PLS, present in the ALMOND software, were used to analyze and to build the alignment-independent 3D-QSAR model. PCA was used to analyze the similarity pattern within the compounds, considered for this study. Both PLS and PCA approximate the multivariate data by projecting it onto a lower-dimensional variable space, called latent variables or principal components (PC). PLS regression analysis was used to describe the relation between the experimental pIC_{50} (*Y*) variable and GRIND (*X*) variables. No scaling of *X* variables was applied. Fractional factorial design (FFD) [81]-based variable selection procedure was used to improve model's performance.

2.7 Model Quality Assessment

Complete validation of the generated QSAR model is mandatory in order to rely on the QSAR model. The generated models were assessed for four important qualities, namely, goodness of fit, model stability, predictive ability, and domain applicability. Routine standard metrics like explained variance, CV-LOO, predicted variance, and F statistics were used to judge the quality of the models. One of the foremost methods of model assessment is to calculate squared correlation coefficient or R^2. In order to overcome the problems involved in R^2, cross-validation method is used to assess the models predictability. The leave-one-out cross-validation is performed in which one compound is eliminated from the data set either randomly or in a systematic manner, and then the excluded compound is predicted by the corresponding model. The extrapolative ability of the models is validated by predicting a reserved test set that was not considered during model building process. The extrapolative ability of the model to predict new compounds can be assessed by applicability domain.

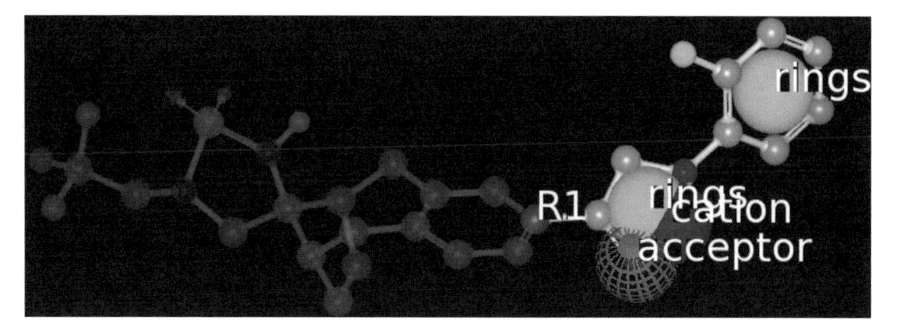

Fig. 4 Query fragment used for bioisosteric replacement and guided virtual screening of Brood fragment database (Reproduced from Molecular Diversity, 2012, with permission from Springer)

2.8 Consensus QSAR Model

Traditionally, single regression QSAR models are used to predict new compounds. Relying solely on an individual QSAR model poses a threat of overemphasizing some physicochemical properties of compounds under investigation, underestimating other properties, and ignoring many key features completely. Hence, the use of consensus QSAR model, derived from different individual models, seems reasonable and provides superior statistical fit and predictive ability as compared to the individual QSAR model [70, 71].

In this study, consensus QSAR model was built from four individual models, i.e., one SMLR, two alignment-dependent 3DQSAR models (CoMFA and CoMSIA), and one alignment-independent 3DQSAR model (GRIND). As evident from the consensus QSAR (CQSAR) results, the compounds with cyclic sulfamides were reliably predicted. Out of these, the highly predicted compound shown in Fig. 4 is having the ideal R-group fragment (R_1 substituent) that would help in achieving a subtle balance between 2D- and 3D-QSAR properties as identified from the overall predicted IC_{50} values, obtained from CQSAR method. This potential R-group (Markush fragment) with mapped 3D features (acceptor, cation, and two rings) (see Fig. 4) is used as the key R-group motif for analog design.

3 Inverse QSAR-Guided Virtual Screening and Bioisosteric Library Design

The objective of inverse-QSAR (I-QSAR) method is to find a relevant molecular descriptor, from which one can construct a molecule with new structure that possesses an improved and/or desired activity or property [82]. One of the bottlenecks in I-QSAR is to find a meaningful descriptor that can be used to construct a new compound. Bioisosteres are chemical substituents or groups with chemical or physical similarities that result in similar biological effects. A particular lead compound may have excellent inhibitory activity for a particular target and also may have undesirable side effects, characteristic that limits its bioavailability. The bioisosteric replacement is an approach used for the rational modification of lead compounds into safer and more clinically efficient agents.

Bioisosteric analogs were identified from a bioisosteric database containing more than 4 million fragments of synthetic tractability, using the Brood [83] software. Medicinally relevant bioisosteres with best overlap of shape, atom type, and attachment geometry to query fragment were considered for library design. A robust in silico screening workflow (Fig. 1), which includes bioisosteric library generation, facilitated the identification of potential leads with predicted pIC_{50} values comparable to that of the most active compound. In a retrospective fashion, the obtained QSAR models were used to predict the activity of the enumerated compounds. Twenty topmost ranked hits were identified at the end of this virtual screening exercise. On the whole, we could obtain some promising molecules with an overall activity (pIC_{50}) ranging from 10.18 to 9.45. The reported pIC_{50} values are based on consensuses QSAR predictions, subjected to rigorous validation with extrapolation limited to the applicability domain. This ensures the highest possible accuracy in forecasting the biological activity of the compounds outside of the training/test set.

The new bioisosteric analogs, gleaned from our robust in silico screening method, were used to map the top-ranked pharmacophore hypothesis for validation. Three topmost ranked hits were precisely overlaid with bioactive pharmacophoric features as shown in Fig. 5, and Catalyst pharmacophore screening of designed bioisosteric library ranked these hits as prospective leads. As an additional measure of confidence, we used similarity metrics to establish the degree of analogy between the designed bioisosteric analogs and its progenitor molecule with regard to shape, chemistry, and electrostatics, altogether termed as electroform. Electroform calculations were performed using ROCS [83] and EON [83]. ROCS is a 3D similarity-based molecule search method. It searches for optimal shape overlay-based comparison and superposition of ligands. Its working methodology is based on the idea that molecules have similar shape, if their volumes overlay well and any volume mismatch is considered as a measure of dissimilarity. Although ROCS is primarily a shape-based method, user-specified definitions of chemistry can be included into the superposition and similarity analysis process, which facilitates the identification of those compounds which are similar both in shape and chemistry [84]. An atom-centered Gaussian shape-based overlay process optimized with respect to chemical features, as defined by Mills and Dean force field [85], was employed to measure the shape and the chemical similarity with the reference compound. The electrostatic potential maps of molecules were compared using EON, which employs partial charges, calculated using MMFF94 [72] force field. The very high shape Tanimoto similarity (T_{Shape}) and electrostatic similarity (ET_{Shape}) values, obtained for the top-ranked three

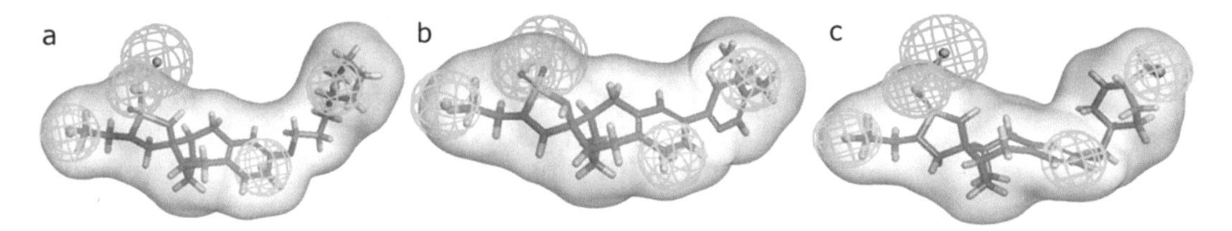

Fig. 5 Three top-ranked hits overlaid on best pharmacophore hypothesis, (**a**) LD1, (**b**) LD2, and (**c**) LD3 (Molecules shown with molecular surface reproduced from Molecular Diversity, 2012, with permission from Springer)

prospective leads, reinforce the fact that they are true electroforms, which are more probable to be active. The color score is the sum of shape similarity and attachment point atom-type similarity with respect to query fragment. Leads obtained with the high color score highlight that the compounds closely represent true bioisosteres, within medicinal chemistry notion.

4 Conclusion

An inverse QSAR-guided fragment hopping method, discussed in this chapter, is the only approach which uses a combination of QSAR methods for the designing of new GS inhibitors. The key strategic advantages of the present study are its ability to explore chemical and structural property space of the designated fragment lead molecule, without risk of patent violation and commercialization. Moreover, consensus QSAR results presented in this work would immensely benefit the interest of medicinal chemists, involved in GS-targeted drug designing in AD therapeutics.

This method only relies on the 2D and 3D properties of bicyclononane series of GS inhibitors. The novel method is derived from a combination of QSAR approaches using a set of 76 non-transition state-based sulfonamide class of GS inhibitors. The present method utilizes some simple 2D descriptors identified through 2D-QSAR-based SMLR model, which reveals the importance of fused cyclic sulfamide in GS inhibition. The 2D-QSAR results were in good agreement with CoMFA and CoMSIA results. The Bicyclononane inhibitor's bioactive conformation is identified through the HipHop common feature hypothesis algorithm that reveals the importance of three hydrophobic and a hydrogen bond acceptor pharmacophoric features responsible in GS inhibition. Reliable CoMFA and CoMSIA statistical results testify that the bioactive conformation predicted by the HipHop method is authentic and these models can be used for GS-targeted virtual screening and drug designing. CoMFA steric and electrostatic and CoMSIA donor, acceptor, and hydrophobic contours results developed during this study put forth a set of 3D features to enhance the GS inhibition. The alignment-independent GRIND-based shape model developed during the course of our study reveals that the shape feature of the inhibitors investigated here plays an important role in GS inhibition. The GRIND-based 3D-QSAR results were in good agreement with alignment-dependent 3D-QSAR (CoMFA and CoMSIA) studies. Multivariate structure–activity correlation results and 2D-QSAR results presented in this study can be exploited for identifying new GS inhibitors.

References

1. Alzheimer (1987) About a peculiar diseases of the cerebral cortex. By Alois Alzheimer, 1907 (Translated by L. Jarvik and H. Greenson). Alzheimer Dis Assoc Disord 1:3–8
2. Selkoe DJ (2001) Alzheimer's disease: genes, proteins, and therapy. Physiol Rev 81:741–766
3. Cummings JL (2004) Alzheimer's disease. N Engl J Med 351:56–67
4. Glenner GG, Wong CW (1984) Alzheimer's disease: initial report of the purification and characterization of a novel cerebrovascular amyloid protein. Biochem Biophys Res Commun 120:885–890
5. Brookmeyer R, Johnson E, Ziegler-Graham K et al (2007) Forecasting the global burden of Alzheimer's disease. Alzheimers Dement 3:186–191
6. Prince M, Bryce R, Albanese E et al (2013) The global prevalence of dementia: a systematic review and meta-analysis. Alzheimers Dement 9:63–75
7. Hardy JA, Higgins GA (1992) Alzheimer's disease: the amyloid cascade hypothesis. Science 256:184–185
8. Sisodia SS, St. George-Hyslop PH (2002) γ -secretase, notch, aβand Alzheimer's disease: where do the presenilins fit in? Nat Rev Neurosci 3:281–289
9. John V, Beck JP, Bienkowski MJ et al (2003) Human beta-secretase (bace) and bace inhibitors. J Med Chem 46:4625–4630
10. Lahiri DK, Ghosh C, Ge YW (2003) A proximal gene promoter region for the β-amyloid precursor protein provides a link between development, apoptosis and Alzheimer's disease. Ann N Y Acad Sci 1010:643–647
11. Walsh DM, Selkoe DJ (2004) Deciphering the molecular basis of memory failure in Alzheimer's disease. Neuron 44:181–193
12. Tolia A, De Strooper B (2009) Structure and function of γ-secretase. Semin Cell Dev Biol 20:211–218
13. Page RM, Gutsmiedl A, Fukumori A et al (2010) β-amyloid precursor protein mutants respond to γ-secretase modulators. J Biol Chem 285:17798–17810
14. Iwatsubo T, Odaka A, Suzuki N et al (1994) Visualization of a beta 42(43) and a beta 40 in senile plaques with end-specific a beta monoclonals: evidence that an initially deposited species is a beta 42(43). Neuron 13:45–53
15. Beel A, Sanders C (2008) Substrate specificity of γ-secretase and other intramembrane proteases. Cell Mol Life Sci 65:1311–1334
16. Gordon WR, Arnett KL, Blacklow SC (2008) The molecular logic of notch signaling–a structural and biochemical perspective. J Cell Sci 121:3109–3119
17. Barten DM, Guss VL, Corsa JA et al (2005) Dynamics of β-amyloid reductions in brain, cerebrospinal fluid, and plasma of β-amyloid precursor protein transgenic mice treated with a γ-secretase inhibitor. J Pharmacol Exp Ther 312:635–643
18. Comery TA, Martone RL, Aschmies S et al (2005) Acute γ-secretase inhibition improves contextual fear conditioning in the Tg2576 mouse model of Alzheimer's disease. J Neurosci 25:8898–8902
19. Wolfe MS, Kopan R (2004) Intramembrane proteolysis: theme and variations. Science 305:1119–1123
20. Edbauer D, Winkler E, Regula JT et al (2003) Reconstitution of gamma-secretase activity. Nat Cell Biol 5:486–488
21. De Strooper B, Saftig P, Craessaerts K et al (1998) Deficiency of presenilin-1 inhibits the normal cleavage of amyloid precursor protein. Nature 391:387–390
22. Laudon H, Hansson EM, Melen K et al (2005) A nine-transmembrane domain topology for presenilin 1. J Biol Chem 280:35352–35360
23. Li H, Wolfe MS, Selkoe DJ (2009) Toward structural elucidation of the γ-secretase complex. Structure 17:326–334

24. Fagan R, Swindells M, Overington J et al (2001) Nicastrin, a presenilin-interacting protein, contains an amino-peptidase/transferring receptor superfamily domain. Trends Biochem Sci 26:213–214
25. Fortnam PC, Crystal AS, Morais VA et al (2004) Membrane topology and nicastrin-enhanced endoproteolysis of APH-1, a component of the γ-secretase complex. J Biol Chem 279:3685–3693
26. Serneels L, Dejaegere T, Craessaerts K et al (2005) Differential contribution of the three aph1 genes to γ-secretase activity in vivo. Proc Natl Acad Sci U S A 102:1719–1724
27. Crystal AS, Morais VA, Pierson TC et al (2003) Membrane topology γ-secretase component pen-2. J Biol Chem 278:20117–20123
28. Kimberly WT, LaVoie MJ, Ostaszewski BL et al (2003) Gamma-secretase is a membrane protein complex comprised of presenilin, nicastrin, aph-1, and pen-2. Proc. Natl Acad Sci U S A 100:6382–6387
29. Wolfe MS, Xia W, Ostaszewski BL et al (1999) Two transmembrane aspartates in presenilin-1 required for presenilin endoproteolysis and gamma-secretase activity. Nature 398:513–517
30. Kornilova AY, Bihel F, Das C et al (2005) The initial substrate binding site of γ-secretase is located on presenilin near the active site. Proc Natl Acad Sci U S A 102:3230–3235
31. Shah S, Lee SF, Tabuchi K et al (2005) Nicastrin functions as a gamma-secretase-substrate receptor. Cell 122:435–447
32. Sobhanifar S, Schneider B, Löhr F et al (2010) Structural investigation of the c-terminal catalytic fragment of presenilin 1. Proc Natl Acad Sci 107:9644–9649
33. De Strooper B, Iwatsubo T, Wolfe MS (2012) Presenilins and γ-secretase: structure, function, and role in Alzheimer disease. Cold Spring Harb Perspect Med 2:a006304
34. Bai XC, Yan C, Yang G et al (2015) An atomic structure of human γ-secretase. Nature 525:212–217
35. Sun L, Zhao L, Yang G et al (2015) Structural basis of human γ-secretase assembly. Proc Natl Acad Sci U S A 112:6003–6008
36. Klafki H, Abramowski D, Swoboda R et al (1996) The carboxyl termini of β-amyloid peptides 1-40 and 1-42 are generated by distinct γ-secretase activities. J Biol Chem 271:28655–28659
37. Bateman RJA (2009) γ-secretase inhibitor decreases amyloid-β production in the central nervous system. Ann Neurol 66:48–54
38. Nguyen J-T, Hamada Y, Kimura T et al (2008) Design of potent aspartic protease inhibitors to treat various diseases. Arch Pharm 341:523–535
39. Shearman MS, Beher D, Clarke EE et al (2000) L-685,458, an aspartyl protease transition state mimic, is a potent inhibitor of amyloid β-protein precursor γ-secretase activity. Biochemistry 39:8698–8704
40. Wolfe MS, Xia W, Moore CL et al (1999) Peptidomimetic probes and molecular modeling suggest that Alzheimer's γ-secretase is an intramembrane-cleaving aspartyl protease. Biochemistry 38:4720–4727
41. Wallace OB, Smith DW, Deshpande MS et al (2003) Inhibitors of aβ production: solid-phase synthesis and sar of r-hydroxycarbonyl derivatives. Bioorg Med Chem Lett 13:1203–1206
42. Wolfe MS (2008) γ-Secretase inhibition and modulation for Alzheimer's disease. Curr Alzheimer Res 5:158–164
43. Das C, Berezovska O, Diehl TS et al (2003) Designed helical peptides inhibit an intramembrane protease. J Am Chem Soc 125:11794–11795
44. Dovey HF et al (2001) Functional gamma-secretase inhibitors reduce beta-amyloid peptide levels in brain. J Neurochem 76:173–181
45. Kreft AF, Martone R, Porte A (2009) Recent advances in the identification of γ-secretase inhibitors to clinically test the aβ oligomer hypothesis of Alzheimer's disease. J Med Chem 52:6169–6188
46. Hansch C, Fujita T (1964) ρ-σ-π Analysis: a method for the correlation of biological activity and chemical structure. J Am Chem Soc 86:1616–1626
47. Klebe G (2006) Virtual ligand screening: strategies, perspectives and limitations. Drug Discov Today 11:580–594
48. Leach AR, Gillet VJ (2003) An introduction to chemoinformatics. Springer, Dordrecht
49. Reddy AS, Pati SP, Kumar PP et al (2007) Virtual screening in drug discovery – a computational perspective. Curr Protein Pept Sci 8:329–351
50. Ravi Keerti A, Ashok Kumar B, Parthasarathy T et al (2005) Qsar studies–potent benzodiazepine gamma-secretase inhibitors. Bioorg Med Chem 13:1873–1878
51. Sammi T, Silakari O, Ravikumar M (2009) Three-dimensional quantitative structure-activity relationship (3d-qsar) studies of various benzodiazepine analogues of gamma-secretase inhibitors. J Mol Model 15:343–348

52. Gundersen E, Fan K, Haas K et al (2005) Molecular-modeling based design, synthesis, and activity of substituted piperidines as gamma-secretase inhibitors. Bioorg Med Chem Lett 15:1891–1894
53. Ajmani S, Janardhan S, Viswanadhan VN (2013) Toward a general predictive qsar model for gamma-secretase inhibitors. Mol Divers 17:421–434
54. Zettl H, Ness J, Hähnke V et al (2012) Discovery of γ-secretase modulators with a novel activity profile by text-based virtual screening. ACS Chem Biol 7:1488–1495
55. Manoharan P, Ghoshal N (2012) Rationalizing lead optimization by consensus 2d- comfa comsia grind (3d) qsar guided fragment hopping in search of γ-secretase inhibitors. Mol Divers 16:563–577
56. Lewis SJ, Smith AL, Neduvelil JG et al (2005) A novel series of potent gamma-secretase inhibitors based on a benzobicyclo[4.2.1] nonane core. Bioorg Med Chem Lett 15:373–378
57. Sparey T, Beher D, Best J et al (2005) Cyclic-sulfamide gamma-secretase inhibitors. Bioorg Med Chem Lett 15:4212–4216
58. Keown LE, Collins I, Cooper LC et al (2009) Novel orally bioavailable gamma-secretase inhibitors with excellent in vivo activity. J Med Chem 52:3441–3444
59. Molecular operating environment (MOE) (2009) Chemical Computing Group, Montreal
60. TSAR, Version 3.3 (2007) Accelrys Inc, San Diego
61. Cramer RD III, Bunce JD (1988) Comparative molecular field analysis (comfa) 1.Effect of shape on binding of steroids to carrier proteins. J Am Chem Soc 110:5959–5967
62. Cramer RD, De Priest SA, Patterson DE et al (1993) In: Kubinyi H (ed) The developing practice of comparative molecular field analysis. in3dqsar in drug design: theory methods and applications. ESCOM, Leiden, pp 443–485
63. Klebe G, Abraham U, Mietzner T (1994) Molecular similarity indices in a comparative analysis (comsia) of drug molecules to correlate and predict their biological activity. J Med Chem 37:4130–4146
64. Klebe G (1998) Comparative molecular similarity indices analysis: comsia. Perspect Drug Discovery Des 12-14:87–104
65. Sybyl, version 7.3 (2009) Tripos International, St. Louis, 63144
66. Pastor M, Cruciani G, McLay I et al (2000) Grid-independent descriptors (grind): a novel class of alignment-independent three-dimensional molecular descriptors. JMed Chem 43:3233–3243
67. Cruciani G, Fontaine F, Pastor M (2004) Almond, 3.3.0. Molecular Discovery Ltd, Perugia
68. Hoskuldsson A (1988) Pls regression methods. J Chemom 2:211–228
69. Carey RN, Wold S, Westgard JO (1975) Principal component analysis: an alternative to "referee" methods in method comparison studies. Anal Chem 47:1824–1829
70. Gramatica P, Giani E, Papa E (2007) Statistical external validation and consensus modeling: a qspr case study for Koc prediction. J Mol Graph Model 25:755–766
71. Ganguly M, Brown N, Schuffenhauer A et al (2006) Introducing the consensus modeling concept in genetic algorithms: application to interpretable discriminant analysis. J Chem Inf Model 46:2110–2124
72. Halgren TA (1996) Merck molecular force field. I. Basis, form, scope, parameterization and performance of mmff94. J Comput Chem 17:490–451
73. Spssversion 15.0 (2009) SPSS Inc, Chicago
74. Vijayan RS, Ghoshal N (2008) Structural basis for ligand recognition at the benzodiazepine binding site of GABAA alpha 3 receptor, and pharmacophore-based virtual screening approach. J Mol Graph Model 27:286–298
75. Ehrlich P (1909) Ueber den jetzigenstand der chemotherapie. Ber Dtsch Chem Ges 42:17–47
76. Smellie A, Teig SL, Towbin P (1995) Poling: promoting conformational variation. J Comput Chem 16:171–187
77. Catalyst, version 4.11 (2005) Accelrys Inc, San Diego
78. Barnum D, Greene J, Smellie A et al (1996) Identification of common functional configurations among molecules. J Chem Inf Comput Sci 36:563–571
79. Goodford PJ (1985) A computational procedure for determining energetically favorable binding sites on biologically important macromolecules. J Med Chem 28:849–857
80. Fontaine F, Pastor M, Sanz F (2004) Incorporating molecular shape into the alignment-free grid-independent descriptors. J Med Chem 47:2805–2815
81. Baroni M, Costantino G, Cruciani G et al (1993) Generating optimal linear pls estimations (golpe): an advanced chemometric tool for handling 3d-qsar problems. Quant Struct Act Relat 12:9–20

82. Cho SJ, Zheng W, Tropsha A (1998) Rational combinatorial library design. 2. Rational design of targeted combinatorial peptide libraries using chemical similarity probe and the inverse qsar approaches. J Chem Inf Comput Sci 38:259–268
83. OpenEye (2006) OpenEye Scientific Software, Santa Fe
84. Nicholls A, McGaughey GB, Sheridan RP et al (2010) Molecular shape and medicinal chemistry: a perspective. J Med Chem 53:3862–3886
85. Mills JE, Dean PM (1996) Three-dimensional hydrogen-bond geometry and probability information from a crystal survey. J Comput Aided Mol Des 6:607–622

Chapter 13

Molecular Modeling of Tau Proline-Directed Protein Kinase (PDPK) Inhibitors

Carlos Navarro-Retamal and Julio Caballero

Abstract

Proline-directed protein kinases (PDPKs) are protein kinases (PKs) that phosphorylate serine or threonine preceding a proline residue (S/TP motif). It is known that these enzymes are responsible for tau hyperphosphorylation, which is related to the formation of amyloid plaques (APs) and neurofibrillary tangles (NFTs) in Alzheimer's disease (AD) patients. In this sense, the inhibition of certain PDPKs is considered a promising strategy to elaborate therapies against AD. The structures of the most important PDPKs are available as high-resolution X-ray crystals; many of them are co-complexed with inhibitors. This information is very valuable because it constitutes a source for creating additional structural information with the aid of molecular modeling methods. Molecular modeling includes methodologies such as docking, molecular dynamics (MD), free energy calculations, etc. that allow accessing to other properties such as the effects of water media, conformational sampling, and binding energy estimations. More importantly, these methods allow to use the available structural information for generating the structures of novel complexes, which can contribute to the rational design of novel inhibitors. Here, we extensively reviewed how molecular modeling methods have contributed to the study of complexes between PDPKs and their inhibitors. We also analyzed the interaction fingerprints (IFPs) generated from the available complexes for each PDPK. IFPs capture the most important interactions between proteins and their ligands which can be used for guidance in docking experiments and in silico screening of novel candidates.

Key words Tau protein kinases, Docking, Molecular dynamics, GSK3, CDK5, MAPK

1 Introduction: Tau Protein Kinases

Alzheimer's disease (AD) is a chronic neurodegenerative pathology that progresses slowly but incessantly in different stages [1]. AD patients could have difficulty in remembering recent events in the early stages, including disorientation, problems with language, etc. In more advanced stages, it could lead to patient withdrawal from family and society, and when bodily functions are impaired, it could lead to death. AD is characterized by the presence of two types of neuropathological hallmarks: amyloid plaques (APs) and neurofibrillary tangles (NFTs) [2–4]. APs are insoluble deposits of the toxic protein peptide β-amyloid accumulated in the spaces between

Kunal Roy (ed.), *Computational Modeling of Drugs Against Alzheimer's Disease*, Neuromethods, vol. 132, DOI 10.1007/978-1-4939-7404-7_13, © Springer Science+Business Media LLC 2018

the brain's nerve cells. NFTs are insoluble paired helical filaments aggregated inside the brain's cells and composed of abnormally hyperphosphorylated tau. In normal conditions, tau assists with formation and stabilization of the microtubule, which helps transport nutrients and other important substances from one part of the nerve cell to another. However, during AD, the hyperphosphorylated tau cannot bind to the microtubule and it collapses.

The molecular mechanisms leading to tau lesions and their link to AD remain unclear. However, there are evidences of the role of phosphorylation in tau aggregation [5]. For example, Fischer et al. [6] found that tau phosphorylation at specific sites such as S262, S293, S324, and S356 induces tau conformational change and decreases tau binding to microtubules. In other report, Sengupta et al. [7] observed that tau phosphorylation at T231, S235, and S262 also contributes to the overall inhibition of tau binding to microtubules by 26%, 9%, and 33%, respectively.

Phosphorylation is the most common tau modification described, and there is ample evidence that the increase in tau phosphorylation reduces its affinity for microtubules. Tau has 85 phosphorylation sites (Fig. 1): 45 serines (53%), 35 threonines (41%), and 5 tyrosines (6%); 28 sites are only phosphorylated in AD brains, 16 are phosphorylated both in AD and in healthy brains, 31 are phosphorylated in physiological conditions, and 10 are tau putative phosphorylation sites without an identified protein kinase (PK) [8].

A balance between various PKs and phosphatases regulates tau phosphorylation in an equilibrium [8]; when this equilibrium is lost, tau aggregates. Each tau phosphorylation site can be subjected to the action of one or more PKs. These PKs are grouped into three classes: proline-directed PKs (PDPKs), non-PDPKs, and tyrosine PKs. Most of the PKs involved in tau phosphorylation are part of the PDPKs; because of this, many efforts have focused on the elucidation of the three-dimensional (3D) structures of these proteins, their catalytic mechanism, and the development of potent inhibitors which can be occupied in anti-AD therapies.

2 Proline-Directed PKs (PDPKs) and Their Inhibitors

PDPKs are PKs targeting serine and threonine residues preceding aproline residue (SP and TP motifs). Tau contains multiple repeats of SP and TP motifs in its entire sequence; most of them are phosphorylated in AD brains (Fig. 1). PDPK group comprises the PKs glycogen synthase kinase-3 (GSK3), cyclin-dependent kinase-5 (CDK5), and mitogen-activated PKs (MAPK) such as p38, extracellular signal-regulated kinases 1 and 2 (ERK1/2), and c-Jun N-terminal kinases (JNK). Alterations in the expression and the activity of these enzymes have been found in the brains of AD patients,

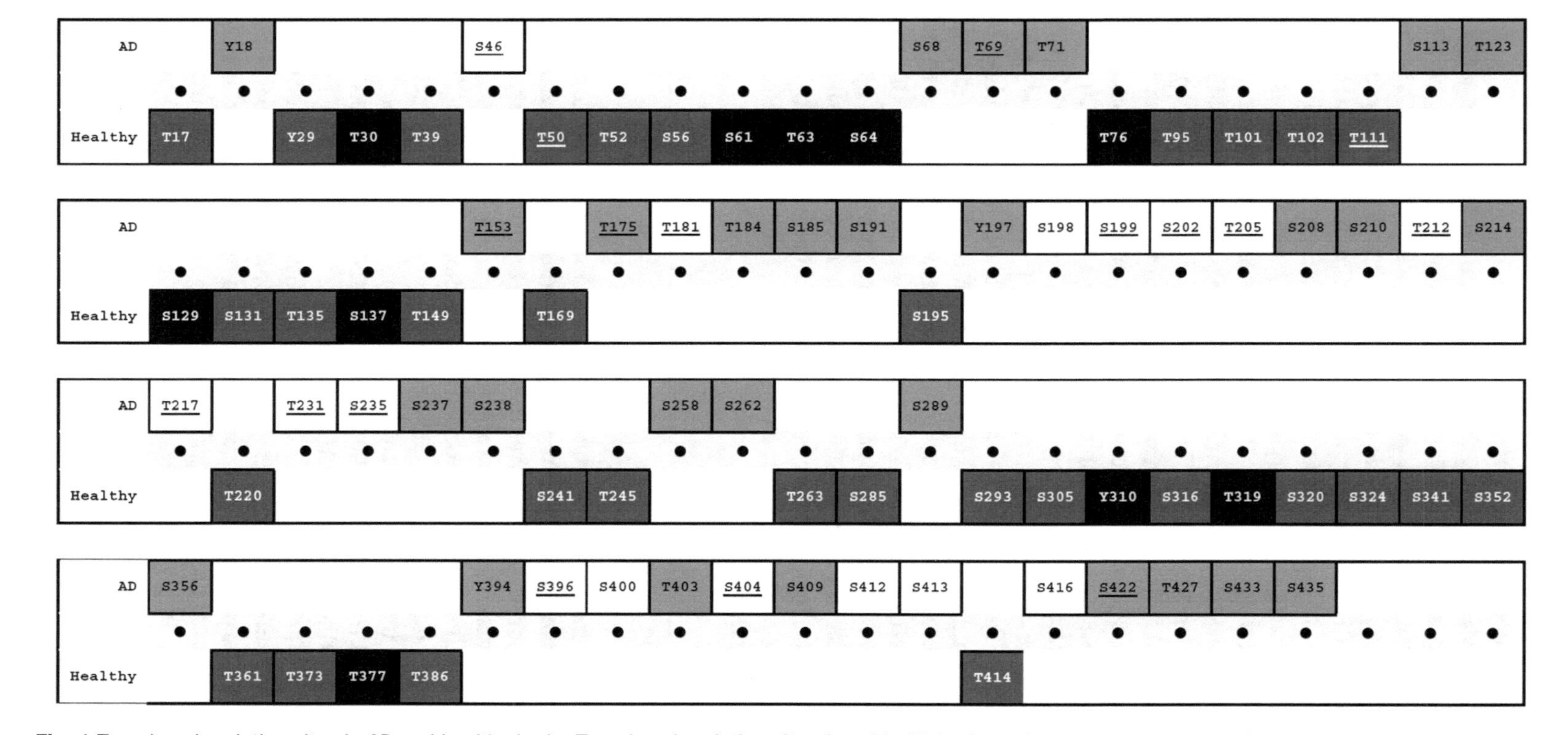

Fig. 1 Tau phosphorylation sites in AD and healthy brain. Tau phosphorylation sites found in AD brains (*black letters* on *gray background*), those found in healthy brain (*white letters* on *dark gray background*), and those present both in healthy and AD brains (*black letters* on *white background*) are represented. Putative phosphorylation sites that have not yet been proven to be phophorylated in vitro or in vivo are in *white letters* on *black background* (information based in Ref. [5]). *Underlined letters* correspond to SP or TP motifs

suggesting that one or several of them could contribute to the hyperphosphorylation of tau protein [9].

Knowledge of the physiological roles and regulation modes of tau phosphorylation is the first step for creating protection strategies against tau aggregation. In order to achieve this objective, in the last decades, tau proline-directed phosphorylatable sites and the different PDPKs that phosphorylate these sites have been identified (Table 1) [10–18]. PDPKs can phosphorylate tau at several sites; in several cases, phosphorylation occurs when tau was pre-phosphorylated by other PK [7]. For instance, when the PDPK GSK3β phosphorylates Thr231, the follow-up phosphorylation by the same enzyme at Ser235 is facilitated, and this process plays a critical role in regulating the binding of tau to microtubules [19].

PDPKs have drawn attention as drug targets in decreasing tau hyperphosphorylation. Inhibition of specific tau PDPKs forms part of the strategy of reversing this process during AD. Efforts in this issue have been directed to obtain 3D structural information of PDPKs. These endeavors have led to a better understanding of PDPK binding sites and the design and synthesis of various organic molecules which rely on complementary shapes and charge distributions to their binding sites. It is hoped that optimization of these interactions guarantees a potent inhibitory activity of the synthesized molecules. The availability of 3D structures of PDPKs helps in these endeavors, but more information could be generated by using molecular modeling methods to understand affinity and selectivity of the ligands.

3 Molecular Modeling Methods

Molecular modeling methods use the 3D crystallographic structures as the starting points. Fortunately, the 3D structures of the most important PDPKs have been already reported; in many cases, these structures are co-complexed with known inhibitors. When PDPK-inhibitor complexes are not available, computational models using docking can be derived [20–22].We will briefly examine the most important molecular modeling methods and their utility as part of the tools for studying protein-inhibitor complexes.

Docking is a method that predicts the orientations of ligands within a protein binding site. Docking algorithms confront two tasks: conformational space search and ranking of potential solutions. The protein surface can be described by mathematical frameworks, based on geometrical shape descriptors, or by construction of a grid. In general, it is possible to get acceptable solutions by considering rigid treatment of the protein surface, although there are strategies to consider protein flexibility [23]. Current docking methods predict right bound associations when reliable atomic coordinates of the target protein and ligands are available [24].

Table 1
Tau SP or TP phosphorylation sites catalyzed by PDPKs

	S46	T50	T69	T111	T153	T181	S199	S202	T205	T212	T217	T231	S235	S396	S404	S422	References
GSK3β	×	×				×	×	×	×	×	×	×					[10–13]
CDK5						×	×	×	×	×	×	×	×	×	×		[11, 14]
p38α		×	×		×	×	×	×	×		×		×	×	×	×	[15]
p38β		×	×	×	×	×	×	×	×	×	×		×	×	×	×	[15]
p38γ		×	×	×	×		×	×	×	×			×	×	×	×	[15]
p38δ		×	×		×			×	×	×			×		×	×	[15]
ERK1/2	×	×			×	×	×	×	×	×	×		×	×	×	×	[16]
JNK						×	×	×	×	×	×			×	×	×	[17, 18]

References show details of the used tau isoforms

However, scoring functions, implemented within docking programs for ranking the solutions, are not sufficiently accurate to evaluate binding energies. In this sense, it is not a good idea to use docking methods for reproducing the difference between experimental biological activities of bioactive compounds [25]. It is common in the use of re-scoring protocols, such as MM/PBSA or MM/GBSA methods, for correctly rank solutions [26, 27]. Considering the research in drug discovery, docking is a key tool for hit finding and optimization during a virtual screening [28]. During this process, databases of commercially available compounds are explored with the aim of generating a subset enriched in active compounds (the hit list). The scoring function of the docking method is commonly used as the filter for such selection.

Ligand-enzyme complexes are available in PDB (reported as X-ray crystals) or they are the results of docking experiments, in both cases as frozen structures. A conformational sampling method is needed to get a more realistic description of the complexes; molecular dynamics (MD) is such method [29]. By moving individual atoms, MD can simulate the natural motions of the complexes and gives temporal trajectories that reflect the available conformational changes of the molecular systems under study. In this sense, this method facilitates the understanding of the dynamics of ligand-enzyme complexes with details on atomistic scales. It is noteworthy that MD is typically applied to the ligand-protein system embedded in an explicit water box containing ions in a selected concentration and with conditions of neutrality; the use of this kind of models is ideal to replicate wet experiments.

MD works with a classic description of interactions. For this, it evaluates the potential energy of the system as a function of point-like atomic coordinates. Motions of every single atom are calculated by considering the interaction forces between all the atoms presented in the system. These forces are described by bonded and nonbonded parameters in force fields, which have been optimized for their application to the study of biomolecular structures. Among the many force fields used for the academic community, we can mention CHARMM [30], OPLS [31], AMBER [32], and Gromos [33], because of their extended use.

MD simulations can be used to derive macroscopic properties if NPT or NVT ensemble is used. Under these conditions, the conformational space of the ligand-protein system could be averaged to get an estimation of the binding affinity of the ligand. In terms of the development of novel drugs, the possibility of having a computational accurate estimator of the binding energy of compounds should be very useful. Estimators designed so far have limitations, but the best options available for free energies calculations are MD-derived methods [34].

Below, we review the use of molecular modeling methods, mainly docking and MD simulations, to investigate PDPK-inhibitor complexes. We will show how 3D structural information of PDPK has been employed to create valuable knowledge useful for medicinal chemists who are attempting to discover potent PDPK inhibitors as anti-AD agents.

4 Modeling of PDPKs and Their Complexes with Inhibitors

Most of the inhibitors designed for PKs (including PDPKs) get bound to the ATP-binding site. Many of them are placed exclusively in the ATP-binding site (type I inhibitors), and others extend to a deeper allosteric hydrophobic binding pocket which is available after a movement of the phenylalanine side chain in the highly conserved DFG motif (type II inhibitors) [35].

The majority of PDPK subdomains contain residues in contact with the ATP-binding site or the allosteric hydrophobic binding pocket. The information of how ligands interact with these sites can be summarized by using interaction fingerprints (IFPs) [36], where IFPs capture interactions between protein and ligand in the form of different chemotypes. In this manner, IFPs leverage the information present in X-ray crystal structures, namely, the interactions between ligand and proteins, for establishing constraints in docking or virtual screening experiments.

The following sections of this review (4.1, 4.2, and 4.3) are focused on the structural characteristics of the complexes between PDPKs and their inhibitors and a brief description of the most recent works that used molecular modeling methods for studying different structural aspects of these systems. IFPs were used for describing the contacts between PDPKs and their inhibitors. IFPs discriminate between contacts with backbone and side-chain functional groups of PDPK residues. In addition, contacts were classified into polar (P), hydrophobic (H), HBs where the residue is acceptor (A), HBs where the residue is donor (D), aromatic (Ar), and electrostatic interactions with charged groups (Ch).

IFPs were applied to the solved crystal structures of PDPK-ligand complexes, specifically for the complexes of GSK3β, CDK5, p38α, ERK2, and JNK1/3. We found that a total of 48 residues (sequence-aligned residues) are involved in the contacts with ligands; these residues and their distribution in subdomains are represented for the abovementioned PDPKs in Figs. 2, 3, 4, 5, and 6.

Firstly, the invariant residues or invariant chemical functions found in all PDPK binding sites were analyzed. All PDPK ATP binding sites contain two hydrophobic walls oriented toward the N- and C-lobes and two polar groups in front of the pocket between these walls. The N-lobe hydrophobic wall is formed by the residues Ile/Val (first circle at subdomain I in Figs. 2, 3, 4, 5, and 6), Val (last circle at subdomain I in Figs. 2, 3, 4, 5, and 6), and

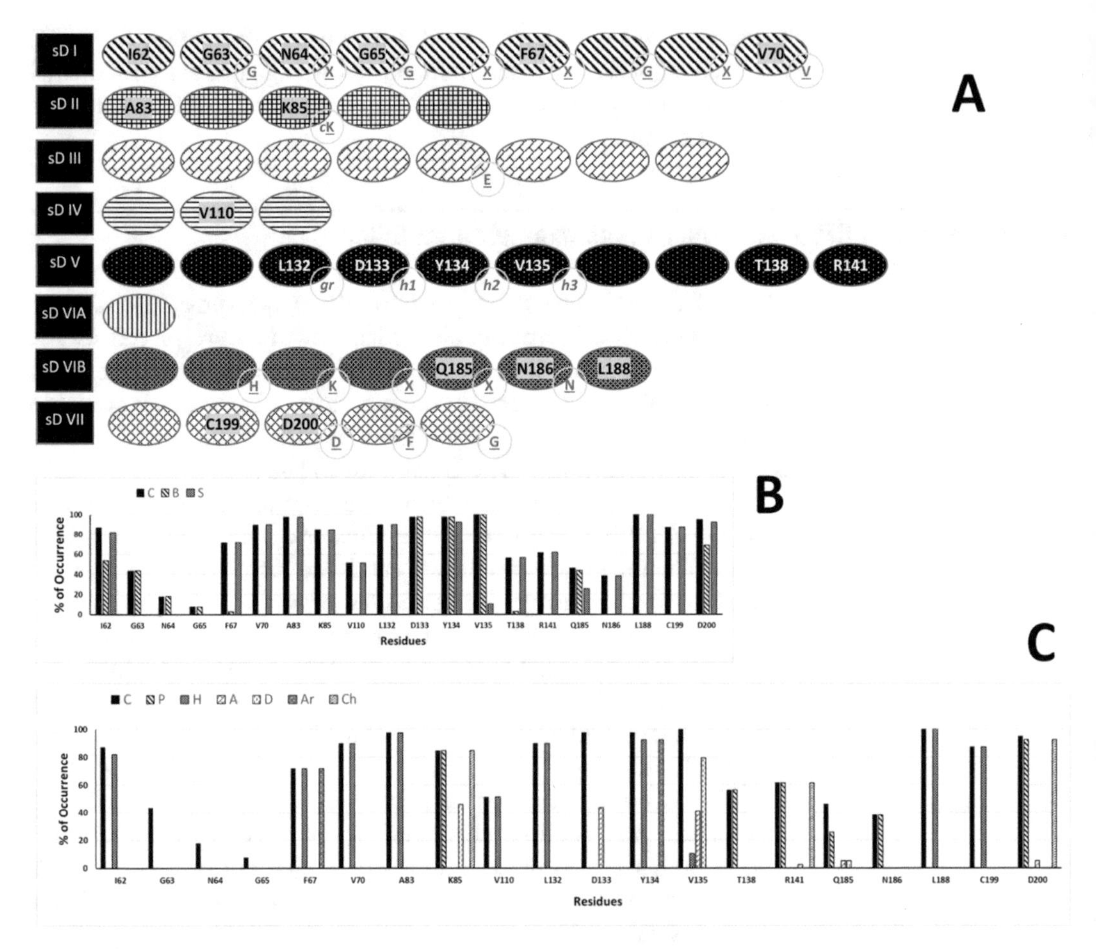

Fig. 2 Occurrence of interaction types at the GSK3β-ligand binding interface considering 39 complexes reported in PDB. (**a**) The GSK3β residues with observed interactions, their position in PK subdomains, consensus motifs, and invariant residues are indicated (H and KXXN are only represented of the HRDLKXXN consensus motif at subdomain VIB). The catalytic lysine is labeled as *c*K, the gatekeeper residue is labeled as *gr*, and hinge residues are labeled as *h1*, *h2*, and *h3*. (**b**) % of occurrence of contacts (*C*), interactions with the backbone of the residue (*B*), and interactions with the side chain of the residue (*S*). (**c**) % of occurrence of chemical interactions: contacts (*C*), polar (*P*), hydrophobic (*H*), HBs where the residue is acceptor (*A*), HBs where the residue is donor (*D*), aromatic (*Ar*), and electrostatic with charged groups (*Ch*)

Ala (first circle at subdomain II in Figs. 2, 3, 4, 5, and 6). The C-lobe hydrophobic wall is formed by the residues Val/Ile (second circle at subdomain IV in Figs. 2, 3, 4, 5, and 6), Leu/Ala/Val (last circle at subdomain VIB in Figs. 2, 3, 4, 5, and 6), and Cys/Leu (second circle at subdomain VII in Figs. 2, 3, 4, 5, and 6). The invariant polar interactions are provided by the backbone NH and CO groups of the hinge region third residue (*h3* in Figs. 2, 3, 4, 5, and 6), which acts as HB donor and/or acceptor, and the side-chain NH_3^+ group of the catalytic lysine (*c*K in the Figs. 2, 3, 4, 5, and 6), which acts as HB donor.

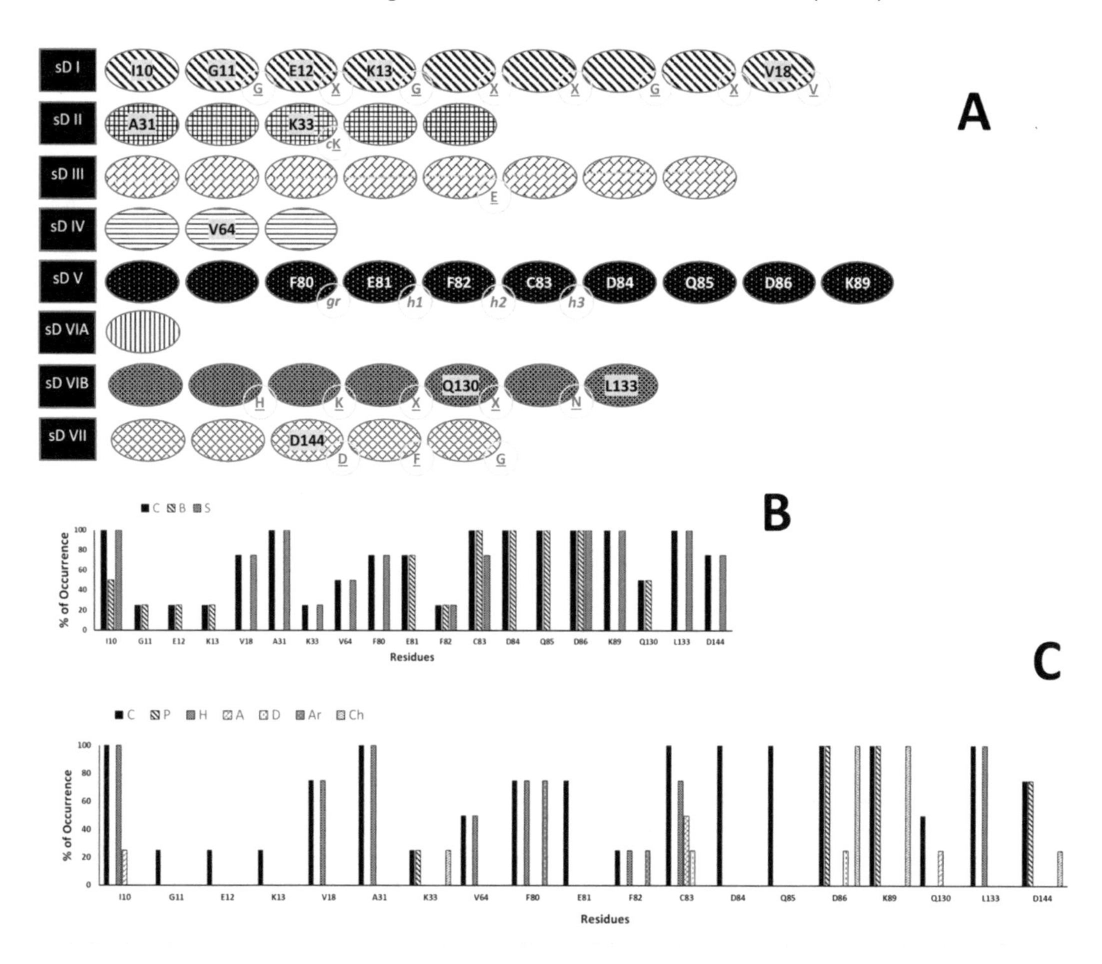

Fig. 3 Occurrence of interaction types at the CDK5-ligand binding interface considering four complexes reported in PDB. (**a**) The CDK5 residues with observed interactions; their position in PK subdomains, consensus motifs, and invariant residues are indicated (H and KXXN are only represented of the HRDLKXXN consensus motif at subdomain VIB). The catalytic lysine is labeled as *c*K, the gatekeeper residue is labeled as *gr*, and hinge residues are labeled as *h1*, *h2*, and *h3*. (**b**) % of occurrence of contacts (*C*), interactions with the backbone of the residue (*B*), and interactions with the side chain of the residue (S). (**c**) % of occurrence of chemical interactions: contacts (*C*), polar (*P*), hydrophobic (*H*), HBs where the residue is acceptor (*A*), HBs where the residue is donor (*D*), aromatic (*Ar*), and electrostatic with charged groups (*Ch*)

IFPs also help in determining the interactions that are unique for each PDPK. These interactions will be mentioned in the following chapters.

4.1 Modeling of GSK3-Inhibitor Complexes

GSK3 isoforms α and β share about 85% sequence homology [37]. Both isoforms are involved in embryo development, neural processes, cell proliferation, immune response, oncogenesis, and apoptosis [38]. GSK3 phosphorylates 42 tau sites; 29 of them are present in AD brains [8]. Recent reports give clear information about the role of GSK3β in AD [39]. The effects of GSK3β overexpression in transgenic mice, cell lines, and animal models were

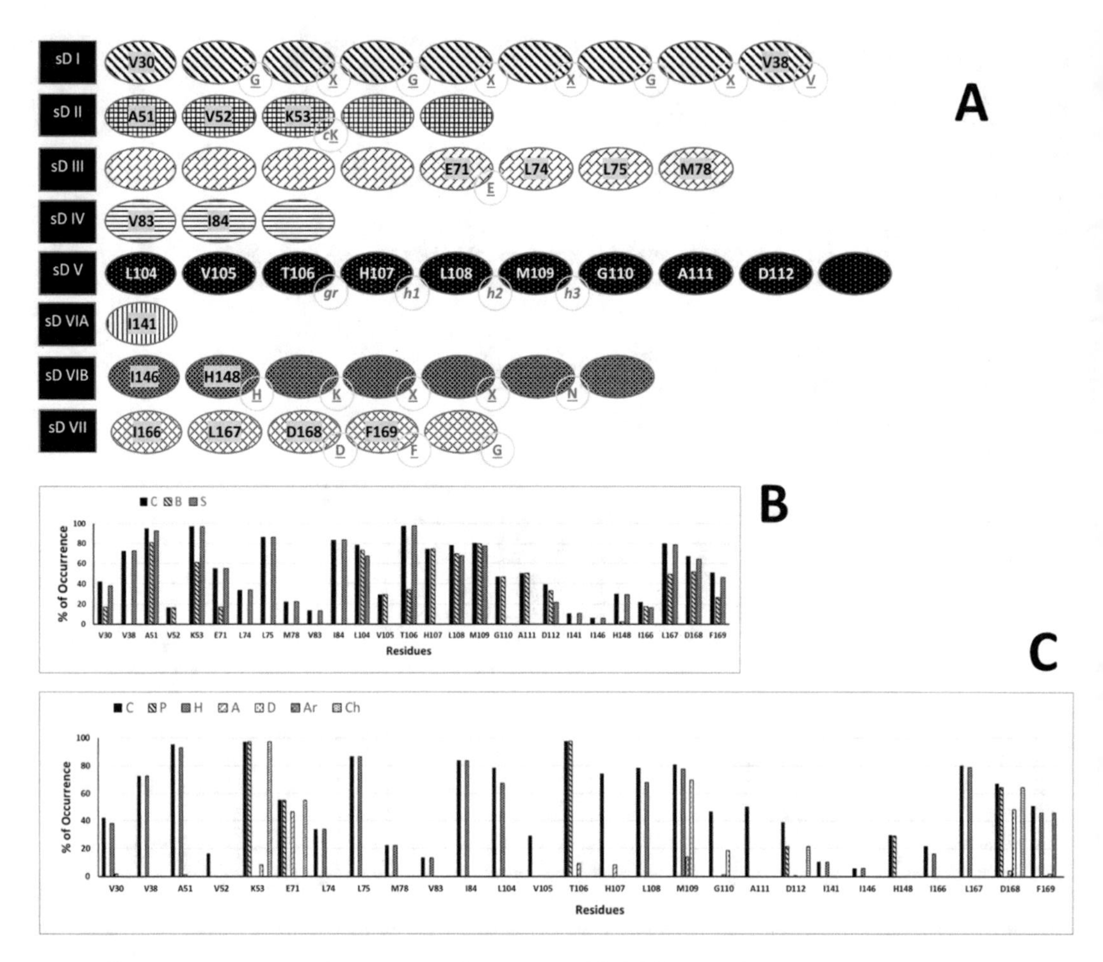

Fig. 4 Occurrence of interaction types at the p38α-ligand binding interface considering 171 complexes reported in PDB. (**a**) The p38α residues with observed interactions; their position in PK subdomains, consensus motifs, and invariant residues are indicated (H and KXXN are only represented of the HRDLKXXN consensus motif at subdomain VIB). The catalytic lysine is labeled as *c*K, the gatekeeper residue is labeled as *gr*, and hinge residues are labeled as *h1*, *h2*, and *h3*. (**b**) % of occurrence of contacts (*C*), interactions with the backbone of the residue (*B*), and interactions with the side chain of the residue (*S*). (**c**) % of occurrence of chemical interactions: contacts (*C*), polar (*P*), hydrophobic (*H*), HBs where the residue is acceptor (*A*), HBs where the residue is donor (*D*), aromatic (*Ar*), and electrostatic with charged groups (*Ch*)

studied, and its involvement in the acceleration of tau pathology was demonstrated [40–42]. In this sense, GSK3β was identified as a promising therapeutic target against AD [43], and rational drug design protocols are commonly employed for developing novel GSK3β inhibitors.

GSK3β inhibitors are commonly designed for competing with ATP. IFPs for GSK3β were performed considering 39 GSK3β-ligand complexes reported in PDB (Fig. 2). The invariant interactions are formed by:

(a) The N-lobe hydrophobic wall: I62, V70, and A83

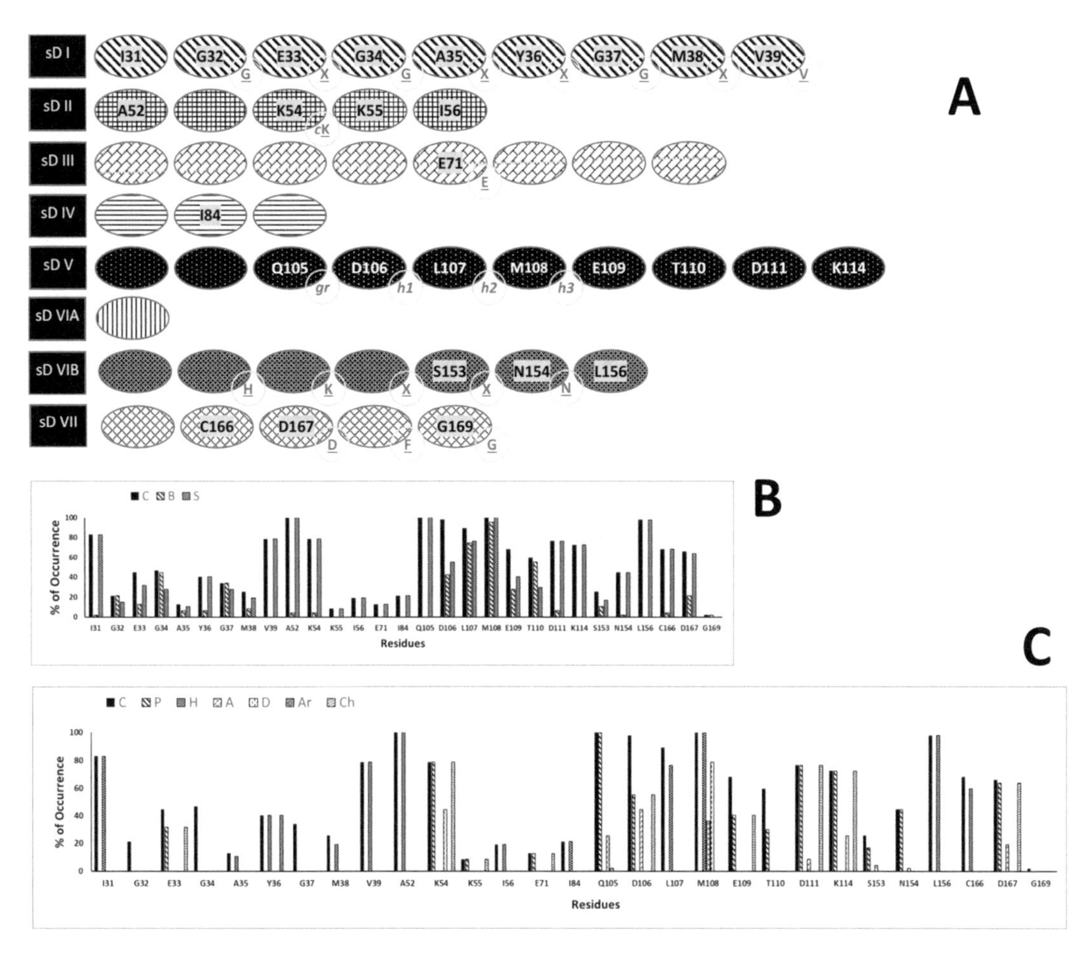

Fig. 5 Occurrence of interaction types at the ERK2-ligand binding interface considering 47 complexes reported in PDB. (**a**) The ERK2 residues with observed interactions; their position in PK subdomains, consensus motifs, and invariant residues are indicated (H and KXXN are only represented of the HRDLKXXN consensus motif at subdomain VIB). The catalytic lysine is labeled as *c*K, the gatekeeper residue is labeled as *gr*, and hinge residues are labeled as *h1*, *h2*, and *h3*. (**b**) % of occurrence of contacts (*C*), interactions with the backbone of the residue (*B*), and interactions with the side chain of the residue (*S*). (**c**) % of occurrence of chemical interactions: contacts (*C*), polar (*P*), hydrophobic (*H*), HBs where the residue is acceptor (*A*), HBs where the residue is donor (*D*), aromatic (*Ar*), and electrostatic with charged groups (*Ch*)

(b) The C-lobe hydrophobic wall: V110, L188, and C199

(c) *h3*: V135

(d) *c*K: K85

The unique GSK3β characteristics can be observed in Fig. 2 and are described below:

(a) Aromatic interactions of the residue F67at the sixth circle of the subdomain I (motif GXGXXGXV) have more than 70% of occurrence in GSK3β. This interaction occurs when Phe is near the N-lobe hydrophobic wall (for instance, in the

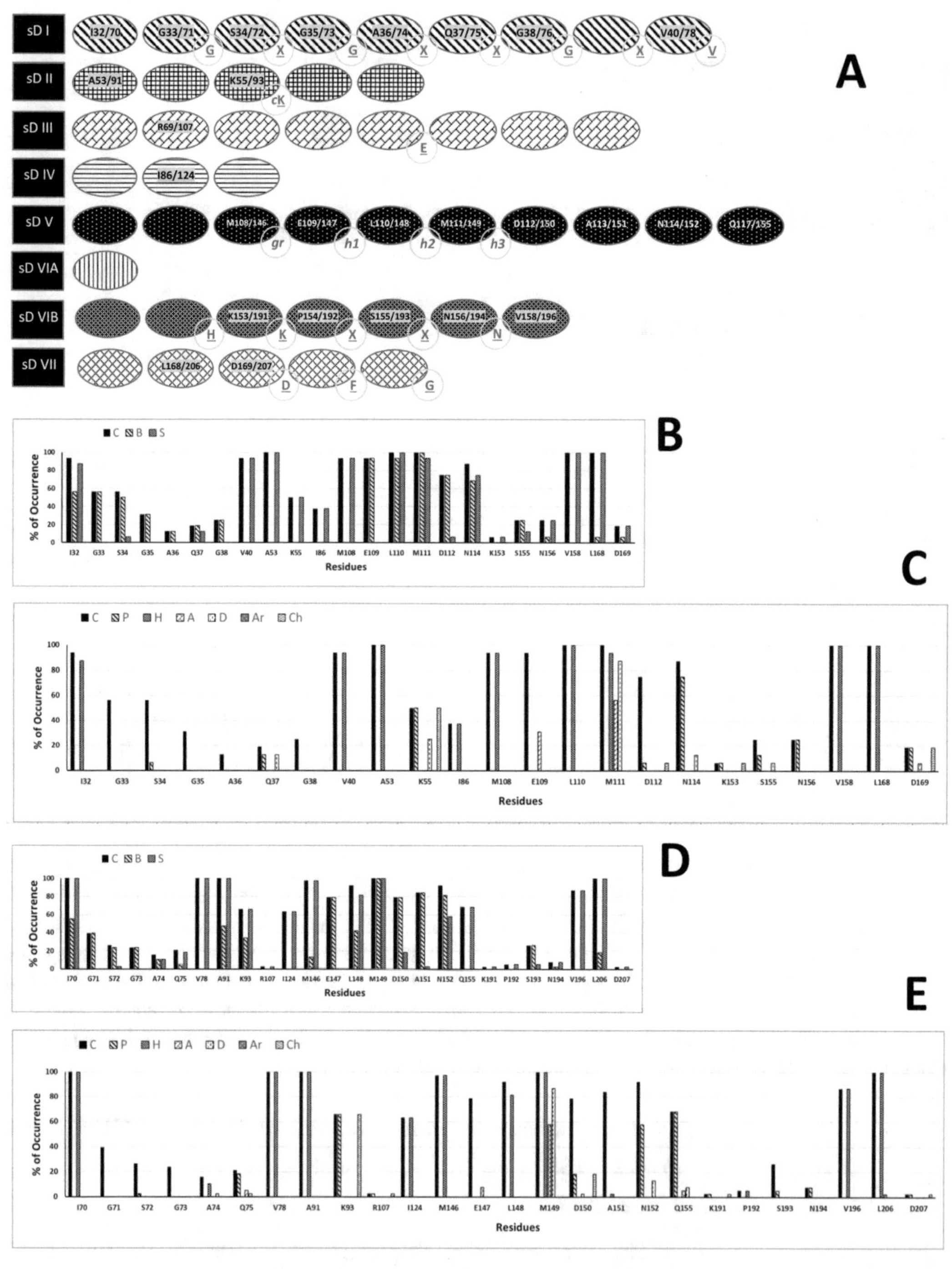

Fig. 6 Occurrence of interaction types at the JNK1/3-ligand binding interface considering 16 and 38 complexes, respectively, reported in PDB. (**a**) The JNK1/3 residues with observed interactions; their position in PK subdomains, consensus motifs, and invariant residues are indicated (H and KXXN are only represented of the HRDLKXXN consensus motif at subdomain VIB). The catalytic lysine is labeled as *c*K, the gatekeeper residue is labeled as *gr*, and hinge residues are labeled as *h1*, *h2*, and *h3*. (**b**) For JNK1: % of occurrence of contacts (*C*), interactions with the backbone of the residue (*B*), and interactions with the side chain of the

structure with PDB code 2JLD). This interaction is observed in ERK (40% in ERK2) when Tyr is present, and it is rarely observed in JNK (<20% in JNK3) when Gln is present.

(b) The gatekeeper residue L132 contributes with hydrophobic interactions in more than 80% of the GSK3β-ligand structures.

(c) The hinge residue *h1* (D133) acts as HB acceptor in more than 40% of the GSK3β-ligand structures.

(d) The hinge residue *h2* (Y134) contributes with aromatic interactions in more than 95% of the GSK3β-ligand structures.

(e) Polar interactions of the residue T138 at the ninth circle of the subdomain V have ~60% of occurrence in GSK3β. This interaction is observed in JNK (~60% in JNK3) when Asn is present and sparsely observed in p38α (~10%) when Glu is present. This polar interaction is also electrostatic in CDK5 and ERK2 because of a presence of a Glu residue.

(f) Electrostatic interactions of the residue R141 at the last circle of the subdomain V have ~60% of occurrence in GSK3β. This interaction is observed in CDK5and ERK2 (100% and 75%, respectively) when Lys is present, and it is replaced by polar interaction in JNK3 (~70%) when Gln is present.

(g) Polar interactions of the residue Q185 at the fifth circle of the subdomain VIB (motif HKXXN) have ~25% of occurrence in GSK3β. This interaction is sparsely observed in ERK and JNK (<20%) when Ser is present.

(h) The HKXXN Asn residue (N186) contributes with polar interactions in have ~40% of the GSK3β-ligand structures.

(i) The DFG Glu residue (D200) contributes with electrostatic interactions in more than 90% of the GSK3β-ligand structures.

The above described chemical features could be considered for the design of novel potent GSK3β ATP-competitive inhibitors.

In the last years, several reports have employed computational molecular modeling methods for studying GSK3βmolecular functions. For instance, Mou et al. studied the role of N-terminal residues Arg4 and Arg6 in the auto-inhibition mechanism of GSK3β [44]. In addition, several authors studied the interactions of GSK3β with substrates [45, 46]. Other studies have been focused to the atomistic description of the interactions with

Fig. 6 (Continued) residue (*S*). (**c**) For JNK1: % of occurrence of chemical interactions: contacts (*C*), polar (*P*), hydrophobic (*H*), HBs where the residue is acceptor (*A*), HBs where the residue is donor (*D*), aromatic (*Ar*), and electrostatic with charged groups (*Ch*). (**b**) For JNK3: % of occurrence of contacts (*C*), interactions with the backbone of the residue (*B*), and interactions with the side chain of the residue (*S*). (**c**) For JNK3: % of occurrence of chemical interactions: contacts (*C*), polar (*P*), hydrophobic (*H*), HBs where the residue is acceptor (*A*), HBs where the residue is donor (*D*), aromatic (*Ar*), and electrostatic with charged groups (*Ch*)

different molecular systems that modulate GSK3β activity. For instance, Lu et al. studied the inhibitory effect of Ca^{2+} [47] and the differences between the group I metal ions in inhibiting GSK3β [48]. Other works studied the binding mode of peptides that interact with GSK3β [49], including the effects of mutations on these peptides [50]. For instance, Lu et al. [51] performed MD simulations and MM/GBSA calculations to explore the structural influence of the double mutations K214A–E215Q of FRATide (a peptide corresponding to residue 188–226 of FRAT1) on the GSK3β-FRATide complex. The authors found that double mutations of the FRATide do not affect the GSK3β activity toward the primed substrates, and their binding free energy calculations indicate that less energy-favorable GSK3β-FRATide complex is observed in the mutant than in the WT complex, which would lead to the nonbinding of the FRATide to GSK3β. The same authors performed MD simulations and MM/GBSA calculations on both the wild type (WT) and the K85 M mutation of the GSK3β-FRATide complex to elucidate the mechanisms concerning kinase inactivation and nonbinding of FRATide to GSK3β [52].

Computational molecular modeling methods are commonly used for describing the interactions between GSK3β and its inhibitors. Docking methods are frequently used for medicinal chemists for describing the interactions of the molecules they synthesize with GSK3β [53]. Recent works that investigate GSK3β-inhibitor interactions are listed below (they are also summarized in Table 2).

In a recent report, Wang et al. studied 192 GSK3β inhibitors (3-aryl-4-(arylhydrazono)-1*H*-pyrazol-5-one analogs and indazoles) using the 3D quantitative structure-activity relationship (QSAR) methodologies like CoMFA and CoMSIA, docking, and MD [54]. The authors reported reliable 3D-QSAR models and used docking and MD simulation for revealing the detailed structures of the GSK3β complexes of the studied compounds. In other recent work, Caballero et al. studied 57 3-amino-1*H*-indazoles as GSK3β inhibitors using docking and 3D-QSAR methods [55]. The authors constructed the QSAR models based on two alignment schemes (binding site constraint and template-based alignments). They reported reliable 3D-QSAR models and described the orientations adopted for these compounds in the binding site. The same authors studied binding modes and constructed robust QSAR models for describing differential activities of 42 oxadiazole derivatives [56] and 77 maleimide derivatives [57].

The work carried out by Sangu and coworkers used stepwise variable selection k-nearest neighbor molecular field QSAR method to study quinazoline derivatives as GSK3β inhibitors [58]. These authors also described the binding orientations of these compounds using docking. Following a very interesting protocol, Mazanetz et al. applied the method active site pressurization (ASP) [59] for describing the dynamic nature of GSK3β [60].

Table 2
Molecular modeling works applied to investigate GSK3β-inhibitors

Authors	Ligands	Molecular modeling protocol	Findings	Reference
Wang et al.	192 3-aryl-4-(arylhydrazono)-1*H*-pyrazol-5-one (AHP) analogs and indazoles (ID)	3D-QSAR (CoMFA and CoMSIA), docking and MD	Good internal and external predictive abilities. $Q^2 = 0.551$; $R^2_{test} = 0.698$ for AHPs and $Q^2 = 0.511$; $R^2_{test} = 0.791$ for IDs. AHPs form HBs with D133 and V135. IDs form HB with V135 and P136	[54]
Caballero et al.	57 3-amino-1*H*-indazoles	Docking and 3D-QSAR (CoMFA CoMSIA)	Good internal and external predictive abilities. $Q^2 = 0.646$ and 0.672 for docking and template alignment, respectively. HBs with D133 and V135	[55]
Caballero et al.	42 oxadiazole derivatives. 37 and 5 compounds in training and test sets, respectively	Docking and two QSAR methods (using 3D vectors and SMILES-based descriptors)	Good internal and external predictive abilities for both QSAR methods. HB with V135	[56]
Caballero et al.	77 maleimide derivatives. 64 and 13 compounds in training and test sets, respectively	Docking and 3D-QSAR (CoMSIA)	Good internal and external predictive abilities for both QSAR methods ($Q^2 = 0.539$). HBs with D133 and V135	[57]
Sangu et al.	85 quinazoline derivatives	K-nearest neighbor molecular field QSAR and docking	Good internal and external predictive abilities $Q^2 = 0.81$; $R^2_{test} = 0.75$. HBs with K183, S203, and Q185	[58]

(continued)

Table 2 (continued)

Authors	Ligands	Molecular modeling protocol	Findings	Reference
Mazanetz et al.	46 and 11 compounds in training and test set, respectively	ASP [59] and 3D-QSAR	The 3D-QSAR model determined from the interactions between a dataset of GSK3β inhibitors, and the ASP model had $Q^2 = 0.96$ and $R^2_{test} = 0.89$. The model was built from a training set of 46 compounds (open triangles) and had a non-cross validated R^2 of 0.98; the test set was composed of 11 compounds (boxes) and had an R^2 of 0.89 with a root mean square error of prediction of 0.55	[60]
Lee et al.	7-(4-hydroxyphenyl)-2-pyridin-4-yl-5 h-thieno[3,2-c]pyridin-4-one analogs	Thermodynamic integration MD simulation	Calculated ΔΔG in reasonably good agreement with the experimental value. In silico design of analogs with significant enhancement in affinity	[61]
Pradeep and Rajanikant	Search in ZINC database	Pharmacophore modeling, screening, docking, and MD	They found ZINC49538628, ZINC17194074, and ZINC36621457 as potent and selective candidates	[62, 63]
Lu et al.	Ten inhibitors contained in X-ray crystal structures	QM/MM, MD, docking	Results elucidate the importance of bridging water molecules at the GSK3β-inhibitor interfaces	[64]
Zhang et al.	The marine sponge constituents hymenialdisine and dibromocantharelline	Docking and MD	Reveal the structural basis for the selectivity of the studied compounds	[65]
Li et al.	Valmerin-19	Docking, MD, and MM/PBSA	Reveal the structural basis for the dual inhibitory activity of the studied compound	[66]

The ASP method allowed to examine the inherent flexibility of the protein and the development of a robust QSAR model. This study highlighted the need to consider GSK3β plasticity when designing models to describe its interactions with ligands.

Other interesting works are cited here: Lee et al. applied thermodynamic integration MD simulation to reproduce relative protein-ligand binding free energy of a pair of analogous GSK3β inhibitors and employed the same protocol to propose analogous inhibitors with a stronger affinity [61]. In other work, Pradeep and Rajanikant [62] performed pharmacophore modeling, screening, and docking to develop a new methodology to identify selective GSK3β inhibitors from ZINC database. They found the best docking scores for compounds ZINC49538628, ZINC17194074, and ZINC36621457, which also demonstrated greater selectivity for GSK3β against CDK1 and CDK2. In an additional effort, the same authors carried out MD simulation studies on the previously identified inhibitors docked to the ATP-binding domains of GSK3β, CDK1, and CDK2 for describing the interaction profiles crucial for the exclusion ability of GSK3β inhibitors against CDKs [63]. The work carried out by Lu and coworkers reported quantum mechanics/molecular mechanics (QM/MM) calculations, MD simulations, and molecular docking studies to investigate the effect of bridging water molecules at the GSK3β-inhibitor interfaces [64].

Computational methods have been also used for studying the selectivity of analogs between GSK3β and CDK5. For instance, Zhang et al. [65] studied the binding and selectivity of the marine sponge constituents hymenialdisine (no selective for GSK3β and CDK5) and dibromocantharelline (selective for GSK3β) using docking methods and MD. They suggest that similar binding modes of hymenialdisine explain the poor selectivity of this compound toward GSK3β and CDK5, and they found that the shape and electrostatic configuration of CDK5 is not adequate for dibromocantharelline binding. More recently, Li et al. [66] studied the structural basis of the compound valmerin-19 as dual inhibitors of GSK3β and CDK5 by using molecular docking, MD, and MM/PBSA calculations. In a very interesting work, Mazanetz et al. tested the ASP method [59] in the PDPKs GSK3β, CDK5, and ERK2 to look for PK-specific patterns of flexibility that could be leveraged for the design of selective inhibitors [67]. The authors used ASP to examine the intrinsic flexibility of the PDPKs ATP-binding pockets, and they observed that the method is able to induce significant conformational changes when compared with X-ray crystal structures. They also provide relevant information, such as the shape of the ATP-binding site and the rigidity of the ATP-binding pocket, which can be exploited in the design of selective inhibitors.

4.2 Modeling of CDK5-Inhibitor Complexes

CDK5 is enzymatically active in presence of the coactivators p39/p35. P35 and p39 are located in the nervous system; therefore, CDK5 is active only in neurons [68]. CDK5 has a key role in the development of the central nervous system and is involved in the regulation of neuronal movements, synaptic functions, and memory consolidation [69]. Upon an increase of calcium concentrations, proteolytic cleaving of p35 and p39 by calpains generates p25 and p29, respectively, which form more stable complexes with CDK5. When CDK5 is associated with p25, CDK5 becomes hyperactive and causes abnormal hyperphosphorylation of various substrates such as tau, resulting in formation of senile plaques, formation of NFTs, and other characteristic symptoms of AD [70].

CDK5/p25 is overexpressed in AD brains [71]. This complex phosphorylates 11 sites of tau, and all of them have been found phosphorylated in AD brains. Under this context, effective inhibition of CDK5/p25 activity could contribute to prevent tau aberrant phosphorylation. Therefore, CDK5 represents a promising therapeutic target for AD.

CDK5 inhibitors are commonly designed for competing with ATP. IFPs for CDK5 were performed considering only four CDK5-ligand complexes reported in PDB (Fig. 3). Lack of data available limits the accuracy of IFP analysis; however, we will consider the information we found to date. The invariant interactions in CDK5 are formed by:

(a) The N-lobe hydrophobic wall: I10, V18, and A31
(b) The C-lobe hydrophobic wall: V64, L133, and A143
(c) *h3*: C83
(d) *c*K: K33

The unique CDK5 characteristics can be observed in Fig. 3 and are described below:

(a) The gatekeeper residue F80 contributes with hydrophobic interactions in 75% of the CDK5-ligand structures.
(b) The hinge residue *h2* (F82) contributes with aromatic interactions in 25% of the CDK5-ligand structures.
(c) Electrostatic interactions of the residue D86 at the ninth circle of the subdomain V have 100% of occurrence in CDK5. This interaction is also observed in ERK2. Analog Glu residue in p38α has polar (not electrostatic) interactions only in ~20% of the structures. Glu is replaced in GSK3β and JNK for Thr and Asn, respectively.
(d) Electrostatic interactions of the residue K89 at the last circle of the subdomain V have 100% of occurrence in CDK5. This interaction is observed in GSK3β and ERK2 (~60% and 75%,

respectively) when Arg or Lys is present, and it is replaced by polar interaction in JNK3 (~70%) when Gln is present.

(e) HB acceptor interactions of the residue Q130 at the fifth circle of the subdomain VIB (motif HKXXN) have 25% of occurrence in CDK5. Polar interactions of the analog residue at this position are sparsely observed in GSK3β, ERK, and JNK (~20%) when Gln or Ser is present.

(f) The DFG Glu residue (D144) contributes with electrostatic interactions in 25% of the CDK5-ligand structures.

The chemical features described for CDK5 are not enough for establishing valid constraints when novel compounds are designed. It is necessary to recall that characteristics that have 25% of occurrence are present in one of only four available CDK5-ligand structures.

Over the past decades, several reports have employed computational molecular modeling methods for studying CDK5 structural aspects and its interactions with p25 and other biomolecules. For instance, Tan et al. [72] used MD simulations and energy decomposition to investigate the deregulation mechanisms of CDK5 inhibitory peptide (CIP) on CDK5 activity. CIP, which is the result of truncation of p25, specifically inhibit CDK5/p25 activity by reducing the phosphorylation of tau. The authors found that truncation of the N- and C-terminals of p25 introduces important conformational changes into a hydrophobic pocket that is crucial for accommodating I153 on the activation loop of CDK5. In addition, they found that such truncations lead to displacement of the activation loop affecting binding of the substrate peptide. More recently, Cardone et al. [73] investigated the molecular basis of the CDK5 inhibitory properties of the peptide p5 by using MD and complementary methods. It is known that this peptide reduces the CDK5/p25 activity without affecting the endogenous CDK5/p35 activity. The authors described the binding site and pharmacophore and discussed options to increase the binding affinity and selectivity in the design of drug-like compounds against AD.

Computational molecular modeling methods are commonly used for describing the interactions between CDK5 and its inhibitors. Docking methods are frequently used for medicinal chemists for describing the interactions of synthesized molecules [74, 75] or natural products [76]. Recent works that investigate CDK5-inhibitor interactions are listed below (they are also summarized in Table 3).

In a recent report, Pitchuanchom et al. [77] developed a template for the active binding site of CDK5 and validated it by redocking with crystallographic ligands but also with 36 CDK5 inhibitors; the results were compared between the calculation of binding energy and the value of biological activity. Their results

Table 3
Molecular modeling works applied to investigate CDK5 inhibitors

Authors	Ligands	Molecular modeling protocol	Findings	Reference
Pitchuanchom et al.	36 ligands	Docking	The constructed template is a good model system for predicting ligand binding orientations and binding affinities	[77]
Chatterjee et al.	Phase database (2.84 million compounds)	Virtual screening	Potent ATP noncompetitive CDK5/p25 inhibitors were identified	[78]
Wang et al.	Five 2-aminothiazoles	Docking, MD simulations, and MM/PBSA	Rank of calculated binding free energies is consistent with experimental result. HBs with C83 favor binding	[79]
Dong et al.	Four roscovitine derivatives	Docking, MD simulation, and MM/PBSA	HBs with C83 and K33 stabilize the inhibitors in binding site. VDW interactions with I10 and L133 have larger contributions to the binding free energy	[80]
Wu et al.	(R)-roscovitine(R)-roscovitine, aloisine-A and indirubin-3′-oxime	MD and MM/PBSA	Strong HB with C83. VDW interactions with I10, V18, and, L133 constitute a substantial component of binding modes	[81]
Patel et al.	Nine compounds	Steered MD	SMD qualitatively discriminate binders from nonbinders, while it failed to properly rank series of inhibitors	[82]
Ul Haq et al.	41 and 7 in training and test set, respectively	Docking and 3D-QSAR	CoMSIA model: $Q^2 = 0.779$	[83]
Zhang et al.	1990 compounds from the NCI diversity database	MM-PBSA, virtual screening, MD simulations	The identified molecules do not share a general HB or VDW interaction network	[84]

indicate that the constructed CDK5 template is a good model system for predicting ligand binding orientations and binding affinities. In other work, Chatterjee et al. [78] employed a protocol comprised of e-pharmacophore models and virtual screening workflow to identify CDK5 inhibitors from a commercial database containing 2.84 million compounds. These authors identified thienoquinolone derivatives as selective and ATP noncompetitive CDK5/p25 inhibitors.

In one such study, Wang et al. [79] performed docking, MD simulations, and MM/PBSA to reveal differences in the binding affinities between five 2-aminothiazole inhibitors and CDK5. A similar protocol was used by Dong et al. [80]. These authors applied docking, MD simulation, and MM/PBSA analysis to reveal the detailed binding mechanism of four roscovitine derivatives with CDK5. Similarly, Wu et al. [81] studied the binding mechanism of (R)-roscovitine, aloisine-A, and indirubin-3′-oxime to CDK5 by using MD simulations and MM/PBSA. The predicted binding affinities obtained by using free energy analysis are consistent with the experimental data.

In a very interesting work, Patel et al. [82] applied steered MD (SMD) simulations to investigate the unbinding mechanism of nine CDK5 inhibitors. The study declared two objectives: (i) to create a correlation between the unbinding force profiles and the inhibition activities of the studied compounds (IC_{50} values) and (ii) to reveal atomistic insights of the unbinding process. The authors found that SMD qualitatively discriminate binders from nonbinders, while it failed to properly rank series of inhibitors, particularly when IC_{50} values were too similar. Additionally, the authors noted that SMD provided useful insights related to transient and dynamical interactions, which complement static description obtained by X-ray crystallography experiments.

Following a combined protocol, Ul Haq et al. [83] applied docking-based 3D-QSAR approaches (CoMFA and CoMSIA) to understand the structural requirements of 4-amino-5-(2-thienyl)-4H-1,2,4-triazole-3-thiol derivatives as CDK5/p25 inhibitors. On the other hand, finally, work carried out by Zhang et al. [84] used a combined strategy of MM/PBSA-based alanine-scanning calculations, virtual screening, MD simulations, and bio-assays to explore novel inhibitors capable of interrupting the interactions between CDK5 and p25, and consequently of inhibiting the PK activity. After this protocol, they identified two compounds without highly specific patterns.

4.3 Modeling of MAPK-Inhibitor Complexes

The members of the MAPK family, p38, ERK1/2, and JNK1/2/3, are implicated in abnormal tau phosphorylation at specific sites found in AD brain tissues [85]. In particular, activated p38, ERK1/2, and JNK1/2/3 expression have been observed in increased levels in AD brains [86, 87]. The use of inhibitors for avoiding MAPK activities represents a promising strategy for avoiding tau phosphorylation, with implications for therapies against AD. The structural information derived from X-ray determinations and the use of molecular modeling methods for supporting the design of novel MAPK inhibitors are exposed below.

4.3.1 p38

Research in the last decades demonstrated the involvement of the p38 MAPK in the pathogenesis of AD. P38 is involved in the inflammatory response associated with AD [88]; in AD brains the p38 active form co-localizes with NFTs [89]. P38 phosphorylates GSK3β [90] and phosphorylates tau at 21 putative tau phosphorylation sites; 15 of them are present in AD patient brains [8].

The crystallographic structures of p38α, p38β, p38γ, and p38δ are solved; however, there are no complexes between p38γ and ligands. On the other hand, only three and two structures of protein-ligand complexes are solved for p38β and p38δ; therefore, the application of IFP analysis to these systems does not give reliable information. The best option for creating IFPs to study this family is p38α, since 171 protein-ligand complexes are solved for this PDPK. IFPs for p38α are reported in Fig. 4. The invariant interactions in p38α are formed by:

(a) The N-lobe hydrophobic wall: V30, V38, and A51
(b) The C-lobe hydrophobic wall: I84, A157, and L167
(c) *h3*: M109
(d) *c*K: K53

The unique p38α characteristics can be observed in Fig. 4 and are described below:

(a) The conserved Glu residue in the αC-helix (E71) acts as HB acceptor and negatively charged group in more than 45% of the p38α-ligand structures.
(b) Hydrophobic interactions of the residues L74 and L75 at the sixth and seventh circles of the subdomain III have ~35% and ~85% of occurrence in p38α, respectively. These residues are in the allosteric hydrophobic binding pocket, which is exposed to ligands in DFG-out structures.
(c) Hydrophobic interactions of the residue L104 at the first circle of the subdomain V have ~65% of occurrence in p38α. This residue is in contact with ligands due to the presence of a small gatekeeper residue. The side chain of small gatekeeper residues does not limit the size of the cavity; instead, the cavity expands toward residues at β-sheet 5 and the αC-helix.
(d) The gatekeeper residue T106 contributes with polar interactions in more than 95% of the p38α-ligand structures and can accept HBs.
(e) The hinge residue *h1* (H107) acts as HB acceptor in <10% of the p38α-ligand structures.
(f) The hinge residue *h2* (L108) contributes with hydrophobic interactions in ~65% of the p38α-ligand structures.

(g) Polar interactions of the residue H148 at the second circle of the subdomain VIB (motif HKXXN) have ~30% of occurrence in p38α. This residue is in the allosteric hydrophobic binding pocket, which is exposed to ligands in DFG-out structures.

(h) The DFG Glu residue (D168) contributes with electrostatic interactions in ~65% of the p38α-ligand structures. It can contribute also with the backbone NH group as HB donor.

(i) The DFG Phe residue (F169) contributes with aromatic interactions in ~45% of the p38α-ligand structures. DFG Phe side chain orients to the ATP site in DFG out structures, with possibilities of establishing interactions with inhibitors.

The above described chemical features could be considered for the design of novel potent p38α ATP-competitive inhibitors.

P38 PKs have been widely studied by using molecular modeling methods, especially p38α isoform. In the last years, several reports have employed computational molecular modeling methods for studying p38 interactions with other proteins. For instance, Yang et al. studied the protein–protein interaction between p38 MAPK and MAPK-activated PK 2 (MK2) by using MD simulation and binding free energy calculation [91]. Computational molecular modeling methods are also used for describing the interactions between p38 and its inhibitors. For example, docking approaches have been used complementarily with experimental structure-activity relationships for describing the interactions in the ATP binding site of novel reported p38 inhibitors [92–96]. Recent works that investigated the bioactive conformations and chemical interactions of organic molecules that inhibit p38are listed below (they are also summarized in Table 4).

In one such study, Choi et al. [97] performed virtual screening with protein-ligand docking to identify novel inhibitors of p38α MAPK. After this protocol, they found compound *1–3* with a moderate inhibitory activity (IC_{50} values ranging from 0.7 to 20 μM). In other report, Willemen et al. performed virtual screening to identify docking-groove-targeted p38 MAPK inhibitors [100]. After this protocol, they found several benzo-oxadiazole compounds with low micromolar inhibitory activity in a p38 MAPK activity assay. In other report, Poon et al. designed and synthesized azastilbene-based compounds with vicinal 4-fluorophenyl/4-pyridine rings as p38α MAPK inhibitors using docking [101]. They performed the design based on the known imidazole-based SB203580 inhibitor ($IC_{50} = 48$ nM), and found two active compounds with IC_{50} values of 111 and 137 nM. In still another recent work, Pinsetta et al. performed virtual screening, MD, molecular interaction fields studies, shape and electrostatic similarities, and toxicity predictions for identifying p38α MAPK inhibitors [102]. They identified 13 compounds that meet the criteria of

Table 4
Molecular modeling works applied to investigate p38-inhibitors

Authors	Ligands	Molecular modeling protocol	Findings	Reference
Choi et al.	Database comprising about 240,000compounds	Docking-based virtual screening	Compounds with a moderate inhibitory activity (IC_{50} values ranging from 0.7 to20 μM) were identified	[97]
Willemen et al.	~10^6 nonredundant molecules obtained from the ZINC database [98]	VSDMIP automated platform [99] (docking, MD, MM/GBSA)	Several benzo-oxadiazole compounds were identified with low micromolar inhibitory activity in a p38 MAPK activity assay	[100]
Poon et al.	Design based on SB203580	Docking	Docking suggests that several different azastilbenes mimic the binding mode of the known imidazole-based SB203580 inhibitor. Two novel inhibitors were found	[101]
Pinsetta et al.	Databases comprising more than 500,000 compounds	Virtual screening, docking, MD	13 compounds meet the criteria of low or no toxicity potential and high predicted activities	[102]
He et al.	21 structural diverse inhibitors extracted from the literature	Pharmacophore model, CoMFA, and docking	Pharmacophore model was found to be highly predictive for identifying p38 inhibitors over drug-like non-inhibitors	[103]
Chang et al.	39 4-benzoyl-5-aminopyrazole derivatives	3D-QSAR, MD, MM/GBSA	Best CoMFA model: $Q^2 = 0.684$; best CoMSIA model: $Q^2 = 0.726$. MM/GBSA free energies were strongly correlated with experimental activities and explained the selective inhibition of p38α and p38β, but not p38γ and p38δ	[104]
Adasme-Carreño et al.	13 congeneric p38 inhibitors	Docking, MD, HB network-based conformational selection, and MM/GBSA	The protocol successfully ranked the studied compounds. The importance of including the target flexibility by means of MD simulation on protein-ligand complexes was evidenced	[27]

(continued)

Table 4 (continued)

Authors	Ligands	Molecular modeling protocol	Findings	Reference
Vinh et al.	46known inhibitors containing a heterocyclic core substituted by pyridine and fluorophenyl rings (structurally related to SB203580) and a set of decoy compounds	Virtual screening, docking conducted on a number of p38α MAPK crystal structures	Identified four crystal structures that distinguish active compounds from drug-like decoy compounds	[105]
Ebadi et al.	GK6	MD	Asp168, Leu167, Met109 and Glu71 had most contribution in binding to ligand	[106]
Lan et al.	59 fused pyrazole derivatives	3D-QSAR and docking	Best CoMFA model: $Q^2 = 0.725$; best CoMSIA model: $Q^2 = 0.609$. HBs: M109, D168	[107]
Astolfi et al.	Initial set of 86 crystal structures of p38α MAPK in complex with type I inhibitors	Docking and virtual screening	The different orientation of these structural elements influence the size and shape of the binding site. The dynamic nature of the active site of p38α MAPK would lead to the loss of much information in single rigid structure docking protocols	[108]

low or no toxicity potential and predicted activities comparable with those obtained for the reference compounds, while maintaining the main interactions observed for the most potent inhibitors.

Work carried out by He and coworkers employed virtual ligand-based 3D pharmacophore model, CoMFA, and docking to identify novel p38α MAPK inhibitors [103]. A more complex protocol was employed by Chang et al., who studied 4-benzoyl-5-aminopyrazole derivatives by using 3D-QSAR (CoMFA and CoMSIA), to highlight the structural requirements of these compounds as p38 MAPK inhibitors [104]. Furthermore, the authors employed MD simulations and MM/GBSA to compare the binding modes and binding free energies of a potent and selective compound interacting with p38α, p38β, p38γ, and p38δ MAPK in detail.

Other interesting works are cited below: Adasme-Carreño et al. combined MD simulations on protein-ligand complexes, HB network-based conformational selection, and MM/GBSA to rank 13 congeneric p38 inhibitors [27]. The approach makes use of protein-ligand complexes obtained from X-ray crystallographic data as well as from docking calculations. Vinh and coworkers performed a virtual screening to develop models for the binding of aryl-substituted heterocycles to p38α MAPK [105]. The authors used conformationally flexible target protein scheme, and they identified four crystal structures (1BL7, 1OZ1, 2EWA, and 3FMK) that effectively distinguish active compounds belonging to a particular structural class, exemplified by SB203580, from drug-like decoy compounds. They observed that p38α MAPK is in DFG-in conformation in these model structures, and Tyr35 side chain, which is directed toward the ligand in the binding pocket, seems to be an important residue for interacting with inhibitors. In other work, Ebadi et al. studied the interactions between a biphenyl amide (GK6) and p38α MAPK by using MD [106]. They used amino acid decomposition analysis to evaluate the contributions of several residues in binding to the studied ligand. In other recent report, Lan et al. performed 3D-QSAR (CoMFA and CoMSIA) and docking studies on 59 fused pyrazoles as p38α MAPK inhibitors [107]. The authors reported QSAR models with good predictive statistics and reported the interactions between the studied compounds and the binding site. More recently, Astolfi et al. [108] analyzed the available crystal structures of p38α MAPK in complex with ATP competitive type I inhibitors, by using cross-docking and virtual screening, giving insights into ATP binding site conformation and its influence on automated molecular docking results. They found that use of target ensembles, rather than single conformations, resulted in a performance improvement in both the ability to reproduce experimental bound conformations and the capability of mining active molecules from compound libraries.

4.3.2 ERK1/2

ERK1 and ERK2 have significant homology (83%); both proteins regulate cell division, growth, proliferation, and apoptosis [109]. ERK1/2 phosphorylates tau at 16 sites; 15 of them are present in AD patient brains [110]. The crystal structures of ERK1 and ERK2 are solved; however, only two structures of protein-ligand complexes are solved for ERK1; therefore, the application of IFP analysis to this system does not give reliable information. The best option for creating IFPs to study this family is ERK2, since 47 protein-ligand complexes are solved for this PDPK. IFPs for ERK2 are reported in Fig. 5. The invariant interactions in ERK2 are formed by:

(a) The N-lobe hydrophobic wall: I31, V39, and A52
(b) The C-lobe hydrophobic wall: I84, L156, and C166

(c) *h3*: M108

(d) *c*K: K54

The unique ERK2characteristics can be observed in Fig. 5 and are described below:

(a) Polar interactions of the residue E33 at the third circle of the subdomain I (motif GXGXXGXV) have ~30% of occurrence in ERK2.

(b) Aromatic interactions of the residue Y36 at the sixth circle of the subdomain I (motif GXGXXGXV) have ~40% of occurrence in ERK2. This interaction occurs when Tyr is near the N-lobe hydrophobic wall (for instance, in the structure with PDB code 4QP8). This interaction is observed in GSK3β (>70%) when Phe is present, and it is rarely observed in JNK (<20% in JNK3) when Gln is present.

(c) The gatekeeper residue Q105 contributes with polar interactions in 100% of the ERK2-ligand structures, and can accept HBs.

(d) The hinge residue *h1* (D106) acts as HB acceptor in >40% of the ERK2-ligand structures.

(e) The hinge residue *h2* (L107) contributes with hydrophobic interactions in ~75% of the ERK2-ligand structures.

(f) Polar interactions of the residues E109 and T110 at the seventh and eighth circles of the subdomain V have ~40% and ~30% of occurrence in ERK2, respectively.

(g) Electrostatic interactions of the residue D111 at the ninth circle of the subdomain V have ~75% of occurrence in ERK2. This interaction is also observed in CDK5. Analog Glu residue in p38α has polar (not electrostatic) interactions only in ~20% of the structures. Glu is replaced in GSK3β and JNK for Thr and Asn, respectively.

(h) Electrostatic interactions of the residue K114 at the last circle of the subdomain V have ~70% of occurrence in ERK2. This interaction is observed in GSK3β and CDK5 (~60% and 100%, respectively) when Arg or Lys are present, and it is replaced by polar interaction in JNK3 (~70%) when Gln is present.

(i) The HKXXN Asn residue (N154) contributes with polar interactions in ~45% of the ERK2-ligand structures.

(j) The DFG Glu residue (D167) contributes with electrostatic interactions in ~65% of the ERK2-ligand structures. It can contribute also with the backbone NH group as HB donor.

The above described chemical features could be considered for the design of novel potent ERK2 ATP-competitive inhibitors.

In the last years, several reports have employed computational molecular modeling methods for studying ERK structural aspects. For instance, Misiura and Kolomeisky [111] studied the mechanism of ERK2 enzymatic catalysis. In contrast to traditional enzymes, ERK2 utilizes additional interactions with substrates outside the active sites during the enzymatic catalysis, these interactions enhance their specificity. The authors propose that the role of the docking interactions is not only to increase the enzymatic specificity but also to optimize the dynamics of the catalytic process. In other interesting application, Barr et al. [112] performed a series of MD simulations of ERK1 and ERK2 in various stages of activation as well as several constitutively active ERK2 mutants to identify the mechanism of ERK autoactivation. The authors demonstrated the importance of domain closure for autoactivation and activity regulation and made predictions of the effect of mutants on the activation process.

Over the last decades, computational molecular modeling methods have been commonly used for describing the interactions between ERK and its inhibitors. There are reports that use docking methods for describing the interactions between molecules synthesized by medicinal chemists and their ERK protein targets [113–115]. Recent works that investigate ERK-inhibitor interactions are listed below (they are also summarized in Table 5).

In one recent work, Kinoshita et al. [116] developed an in silico biased screening using the focused library derived from the ZINC database and the ERK2 structure. This protocol rendered two compounds (Hit-1 and Hit-2 in the manuscript) with inhibitory activities of $IC_{50} < 100$ μM, which revealed a concentration-dependent competition with the peptide-type ERK2 inhibitor denoted as PEP. PEP is bound to the allosteric site of ERK2 by occupation of a negative charge pool and the shallow hydrophobic pocket, without interruption of the ATP binding pocket. The authors reported that Hit-1 and Hit-2 are specifically competitive with PEP; docking experiments revealed that both compounds mainly interacted with the negative charge pool. In other work, Larif et al. [117] developed a 3D-QSAR pharmacophore model on the basis of experimentally known inhibitors, and they used this model for virtual screening of potential inhibitors from several databases. They subject the resulting hit compounds to docking and MD for identifying the binding site regions essential for better potency of the inhibitors. In a very interesting work, Niu et al. [118] used MM/GBSA, SMD, and adaptive biasing force (ABF) simulations to study the complexes between ERK2 and the inhibitors SCH772984, VTX-11e, FR180204, and 5-iTU. The proposed protocol makes it possible to calculate the binding free energies, investigate the molecular mechanism of binding, and identify the contributions of the residues. Following a very complex

Table 5
Molecular modeling works applied to investigate ERK inhibitors

Authors	Ligands	Molecular modeling protocol	Findings	Reference
Kinoshita et al.	ZINC database	Virtual screening, docking	Compounds Hit-1 and Hit-2, which are allosteric inhibitors	[116]
Larif et al.	39 compounds for creating pharmacophore model, ZINC, DrugBank, NCI, Maybridge, and Chembank databases for virtual screening	3D-QSAR pharmacophore model, virtual screening, docking, MD	Model includes three HBs and one hydrophobic site. Compounds have to interact with catalytic site, glycine-rich loop, hinge region, gatekeeper region, and ATP site entrance residues	[117]
Niu et al.	Four inhibitors	MM/GBSA, SMD, and ABF	SCH772984 has different unbinding mechanism compared with the other three inhibitors. SCH772984 needs to overcome two energy barriers: a π-π stacking interaction with Y55 and hydrophobic interaction at the ATP active site. VTX-11e, FR180204, and 5-iTU just need to overcome the hydrophobic interaction at the ATP active site	[118]
Wu et al.	Two inhibitors	Thermodynamic integration MD	ΔΔG values were in good agreement with experimental values	[119]
Rastelli et al.	Hypothemycin	Docking and MD	Model is in agreement with subsequent X-ray determination	[120]
Liu et al.	DrugBank and ZINC databases	Docking and virtual screening	Compound BL-EI001	[121]

protocol, Wu et al. [119] carried out thermodynamic integration MD simulation for a pair of analogous inhibitors binding with ERK. Their calculated ΔΔG values were in good agreement with experimental values.

The work carried out by Rastelli and coworkers [120] reports modeling of the structures of ERK2-hypothemycin reversible and covalent complexes using docking and extended MD simulations. It is known that this inhibitor forms a covalent adduct with a conserved cysteine (C164 in ERK2) in the ATP binding site of their target kinases. Subsequently, the authors determined the 2.5 Å resolution crystal structure of the complex that was in excellent accord with the modeled structure. In a multidisciplinary work, Liu et al. [121] screened computationally and experimentally a series of small-molecule inhibitors targeting ERK. Subsequently, they synthesized some candidates and identified the ERK inhibitor BL-EI001.

4.3.3 JNK1/2/3

JNK1/2/3 are expressed in the brain; recent data have supported their involvement in AD and various neurodegenerative abnormalities [122]. JNK activation is associated with tau-induced neurodegeneration [123] and β-amyloidpathology [124]. High JNK levels are detected in tau aggregates [125]; it is already a well-known fact that JNKs phosphorylate tau at 12 sites and all are present in AD brains [110].

The crystallographic structures of JNK1, JNK2, and JNK3 are solved; however, only two structures of protein-ligand complexes are solved for JNK2; therefore, the application of IFP analysis to this system does not give reliable information. Therefore, IFPs were created for JNK1 and JNK3 to study this family, since 16 and 38 protein-ligand complexes are solved for these PDPKs. IFPs for JNK1 and JNK3 are reported in Fig. 6. The invariant interactions in JNK1/3 are formed by:

(a) The N-lobe hydrophobic wall: I32/70, V40/78, and A53/91
(b) The C-lobe hydrophobic wall: I86/124, V158/196, and L168/206
(c) *h3*: M111/149
(d) *c*K: K55/93

The unique JNK1/3 characteristics can be observed in Fig. 6 and are described below:

(a) The gatekeeper residue M108/146 contributes with hydrophobic interactions in ~95% of the JNK1/3-ligand structures.
(b) The hinge residue *h1* (E109/147) acts as a HB acceptor in ~30% and 10% of the JNK1 and JNK3-ligand structures, respectively.
(c) The hinge residue *h2* (L110/148) contributes with hydrophobic interactions in 100% and 80% of the JNK1 and JNK3-ligand structures, respectively.
(d) Polar interactions of the residue N114/152 at the ninth circle of the subdomain V have ~75% and 60% of occurrence in JNK1 and JNK3-ligand structures, respectively. This interaction is observed in GSK3β (~60%) when Thr is present and sparsely observed in p38α (~10%) when Glu is present. This

polar interaction is also electrostatic in CDK5 and ERK2 because a presence of a Glu residue.

(e) Polar interactions of the residue Q117/155 at the last circle of the subdomain V have <10 and ~70% of occurrence in JNK1 and JNK3-ligand structures, respectively. This polar interaction is also electrostatic in GSK3β, CDK5, and ERK2 (60%, 100%, and 75% of occurrence, respectively) when Arg or Lys is present.

The above described chemical features could be considered for the design of novel potent JNK ATP-competitive inhibitors.

In the last years, computational molecular modeling methods have been commonly used for describing the interactions between JNK and its inhibitors. It is common that medicinal chemists use docking methods during the course of their work to describe the interactions between their synthesized molecules and JNKs [126]. Recent works that investigate JNK-inhibitor interactions are listed below (they are also summarized in Table 6).

In one such study, Dykstra et al. [127] developed molecular docking and MD simulations to examine binding interactions between JNK and both zuonin A enantiomers for identifying differences in their binding modes. Zuonin A is a non-ATP competitive inhibitor of the JNKs; while (−)-zuonin A inhibits the activity of JNK toward c-Jun by 80% when saturating, (+)-zuonin A only inhibits by 15%. Molecular docking and MD simulations provide a better understanding of how these inhibitors interact with JNK and explain differences in the inhibitory properties of the studied enantiomers. In other work, Ngoei et al. [128] performed in silico screens of commercially available chemical databases to identify JNK1-interacting compounds and tested their in vitro JNK inhibitory activity. Using this protocol, they identified the compound, 4′-methyl-N^2-3-pyridinyl-4,5′-bi-1,3-thiazole-2,2′-diamine, which inhibited JNK1 activity toward a range of substrates. In still another recent work, Zhuo et al. [129] employed an integrated in vitro-in silico protocol to rationally design kinase-peptide interaction specificity for JNK over p38 based on crystal structures of both PDPKs. In the procedure, a simulated annealing iteration optimization strategy is used to improve peptide selectivity for JNK. Thereafter, the authors performed MD simulations and binding free energy analysis. Finally, they confirmed the theoretical findings and computational designs by fluorescence polarization assays. Using this protocol the authors successfully obtain three decapeptide ligands that exhibit both potent affinity to JNK and high selectivity for JNK over p38. In a very recent work, Katari et al. [130] performed receptor-based pharmacophore modeling with 18 co-crystal structures of JNK1 in PDB. The authors used the pharmacophore models as 3D query for shape-based similarity screening against more than 1 million small molecules to generate a JNK1

Table 6
Molecular modeling works applied to investigate JNK inhibitors

Authors	Ligands	Molecular modeling protocol	Findings	Reference
Dykstra et al.	Zuonin A enantiomers	Docking and MD	Differences in the inhibitory properties of the studied enantiomers are explained	[127]
Ngoei et al.	Molecules from different databases, JNK1 inhibitors	Virtual screening,	Compound 4′-methyl-N2-3-pyridinyl-4,5′-bi-1,3-thiazole-2,2′-diamine	[128]
Zhuo et al.	Complexes between JNK3 and JIP1 peptide	Simulated annealing, MD, and MM/GBSA	Three decapeptides that exhibit both potent affinity to JNK and high selectivity for JNK over p38	[129]
Katari et al.	More than 1 million molecules from different databases	Pharmacophore modeling, virtual screening, docking, MD	17 leads	[130]
Shaikh et al.	34 and 10 compounds in the training and test sets	3D-QSAR and docking	$Q^2 > 0.6$ using both methods	[131]
Kim et al.	23 and 6 compounds in the training and test sets	3D-QSAR and docking	CoMFA: $Q^2 = 0.795$, CoMSIA: $Q^2 = 0.700$	[132]
Wu et al.	106 compounds	3D-QSAR and docking	CoMFA: $Q^2 = 0.752$, CoMSIA: $Q^2 = 0.789$	[133]
Madhavan et al.	29 and 6 compounds in the training and test sets	3D-QSAR, docking, MD, virtual screening	3D-QSAR results considering docking poses: CoMFA: $Q^2 = 0.605$, CoMSIA: $Q^2 = 0.587$	[134]
Hierold et al.	Quercetagetin	Docking	Discovery of two potent analogs	[135]

ligand library. They identified several leads, and they studied them with quantum-polarized ligand docking, induced fit docking, and MM/GBSA. Finally, the authors described the conformational space of the best candidate forming complex with JNK1 by using 50-ns MD simulations.

The work carried out by Shaikh and coworkers [131] reported 3D-QSAR models (molecular field analysis and receptor surface analysis) for benzothiazole-2-yl acetonitrile derivatives as JNK3 inhibitors. They reported predictive models and performed additional docking of several active compounds inside JNK3 binding site. Similarly, Kim et al. [132] developed 3D-QSAR models (CoMFA and CoMSIA) for describing 1,2-diaryl-1H-benzimidazole derivatives as selective JNK3 inhibitors. They aligned the studied molecules in the JNK3 binding site by using the docking method. In other similar application, Wu et al. [133] applied a combined molecular docking and 3D-QSAR method to understand the structural factors affecting the activities and binding modes of 106 JNK3 inhibitors. Madhavan et al. [134] also used docking and 3D-QSAR for describing anilinopyrimidine derivatives as JNK1 inhibitors. They applied QSAR methods by using different alignment schemes, including receptor-guided alignment that considers docking poses. The authors also used MD simulations to study several systems under more realistic conditions and performed virtual screening analysis against NCI database to propose potentials hits.

Following a multidisciplinary protocol, Hierold et al. [135] used the X-ray structure of JNK1 forming complex with the natural flavonoid quercetagetin for designing novel derivatives. They used docking to predict that 5′-hydroxy- and 5′-hydroxymethyl-quercetagetin inhibit JNK1 more actively than the parent compound. They tested their theoretical predictions and found that both compounds strongly suppress JNK1 activity.

5 Conclusions

This review focuses on the use of molecular modeling methodologies for studying PKs involved in AD, particularly the study of PDPKs. These PKs produce aberrant tau phosphorylation, and this process is related to AD pathology; therefore, their effective inhibition is one of the strategies to elaborate therapies against AD.

Structural information of PDPKs is a source for generating novel drugs that inhibit these proteins. As can be seen from the applications mentioned in previous sections, many researchers have applied computational methodologies to study the interactions between PDPKs and their inhibitors. Molecular modeling methods have evolved over the last decades to produce important tools that may help to investigate proteins with emphasis on their functions, available conformations, interactions with other proteins and small molecules, and identification of druggable sites. Their applications to the study of PDPKs have contributed to the discovery of novel inhibitors which are potential drugs for the treatment of AD. At the same time, these methods have increased our insight into the PDPK active sites.

It is clear that there are many research groups now interested in the design of novel PDPK inhibitors. An important theme is to "learn from available structural information," and the process of this information requires of molecular modeling methods. PDB contains vast information about PDPKs and their interactions with their inhibitors. Molecular modeling methods process this information to give rules that help in the future design of novel inhibitors. Keeping in view the vast amount of information available in PDB, fingerprints reported here are a source of support for predicting correct docking poses of inhibitors. It is necessary to emphasize that a good estimation of ligand orientation is the first and more important step for calculating more complex parameters, such as stability of interactions and binding free energies. For several PDPKs, such as CDK5, ERK1, and JNK2, available structural information of their interactions with inhibitors is not extensive. Thus, construction of reliable fingerprints may require the generation of larger datasets on PDPK-inhibitor complexes. The availability of more reliable fingerprints will contribute to the generation of more reliable docking predictions for these PDPKs.

This work summarizes the studies on the design of PDPK inhibitors, which are supported by molecular modeling methodologies. It is demonstrated the increasing relevance of these methods for supporting research related to these proteins.

References

1. Hickman RA, Faustin A, Wisniewski T (2016) Alzheimer disease and its growing epidemic: risk factors, biomarkers, and the urgent need for therapeutics. Neurol Clin 34:941–953. doi:10.1016/j.ncl.2016.06.009
2. Luo J, Wärmländer SKTS, Gräslund A, Abrahams JP (2016) Cross-interactions between the Alzheimer disease amyloid-β peptide and other amyloid proteins: a further aspect of the amyloid cascade hypothesis. J Biol Chem 291:16485–16493. doi:10.1074/jbc.R116.714576
3. Edwards G, Moreno-Gonzalez I, Soto C (2016) Amyloid-beta and tau pathology following repetitive mild traumatic brain injury. Biochem Biophys Res Commun 483 (4):1137–1142. doi:10.1016/j.bbrc.2016.07.123
4. Bukar Maina M, Al-Hilaly YK, Serpell LC (2016) Nuclear tau and its potential role in Alzheimer's disease. Biomol Ther 6:9. doi:10.3390/biom6010009
5. Martin L, Latypova X, Terro F (2011) Post-translational modifications of tau protein: implications for Alzheimer's disease. Neurochem Int 58:458–471. doi:10.1016/j.neuint.2010.12.023
6. Fischer D, Mukrasch MD, Biernat J, Bibow S, Blackledge M, Griesinger C, Mandelkow E, Zweckstetter M (2009) Conformational changes specific for pseudophosphorylation at serine 262 selectively impair binding of tau to microtubules. Biochemistry 48:10047–10055. doi:10.1021/bi901090m
7. Sengupta A, Kabat J, Novak M, Wu Q, Grundke-Iqbal I, Iqbal K (1998) Phosphorylation of tau at both Thr 231 and Ser 262 is required for maximal inhibition of its binding to microtubules. Arch Biochem Biophys 357:299–309. doi:10.1006/abbi.1998.0813
8. Hanger DP, Anderton BH, Noble W (2009) Tau phosphorylation: the therapeutic challenge for neurodegenerative disease. Trends Mol Med 15:112–119. doi:10.1016/j.molmed.2009.01.003
9. Chung S-H (2009) Aberrant phosphorylation in the pathogenesis of Alzheimer's disease. BMB Rep 42:467–474
10. Wang JZ, Wu Q, Smith A, Grundke-Iqbal I, Iqbal K (1998) Tau is phosphorylated by GSK-3 at several sites found in Alzheimer disease and its biological activity markedly inhibited only after it is prephosphorylated by A-kinase. FEBS Lett 436:28–34

11. Liu F, Iqbal K, Grundke-Iqbal I, Gong C-X (2002) Involvement of aberrant glycosylation in phosphorylation of tau by cdk5 and GSK-3beta. FEBS Lett 530:209–214
12. Johnson GV, Hartigan JA (1999) Tau protein in normal and Alzheimer's disease brain: an update. J Alzheimers Dis 1:329–351
13. Godemann R, Biernat J, Mandelkow E, Mandelkow EM (1999) Phosphorylation of tau protein by recombinant GSK-3beta: pronounced phosphorylation at select Ser/Thr-Pro motifs but no phosphorylation at Ser262 in the repeat domain. FEBS Lett 454:157–164
14. Pei JJ, Grundke-Iqbal I, Iqbal K, Bogdanovic N, Winblad B, Cowburn RF (1998) Accumulation of cyclin-dependent kinase 5 (cdk5) in neurons with early stages of Alzheimer's disease neurofibrillary degeneration. Brain Res 797:267–277
15. Feijoo C, Campbell DG, Jakes R, Goedert M, Cuenda A (2005) Evidence that phosphorylation of the microtubule-associated protein tau by SAPK4/p38δ at Thr50 promotes microtubule assembly. J Cell Sci 118:397–408. doi:10.1242/jcs.01655
16. Wang J-Z, Liu F (2008) Microtubule-associated protein tau in development, degeneration and protection of neurons. Prog Neurobiol 85:148–175. doi:10.1016/j.pneurobio.2008.03.002
17. Pei J-J, Braak E, Braak H, Grundke-Iqbal I, Iqbal K, Winblad B, Cowburn RF (2001) Localization of active forms of C-jun kinase (JNK) and p38 kinase in Alzheimer's disease brains at different stages of neurofibrillary degeneration. J Alzheimers Dis 3:41–48
18. Yoshida H, Hastie CJ, McLauchlan H, Cohen P, Goedert M (2004) Phosphorylation of microtubule-associated protein tau by isoforms of c-Jun N-terminal kinase (JNK). J Neurochem 90:352–358. doi:10.1111/j.1471-4159.2004.02479.x
19. Cho J-H, Johnson GVW (2004) Primed phosphorylation of tau at Thr231 by glycogen synthase kinase 3beta (GSK3beta) plays a critical role in regulating tau's ability to bind and stabilize microtubules. J Neurochem 88:349–358
20. Sabatini S, Manfroni G, Barreca ML, Bauer SM, Gargaro M, Cannalire R, Astolfi A, Brea J, Vacca C, Pirro M, Massari S, Tabarrini O, Loza MI, Fallarino F, Laufer SA, Cecchetti V (2015) The pyrazolobenzothiazine core as a new chemotype of p38 alpha mitogen-activated protein kinase inhibitors. Chem Biol Drug Des 86:531–545. doi:10.1111/cbdd.12516
21. Boulahjar R, Ouach A, Bourg S, Bonnet P, Lozach O, Meijer L, Guguen-Guillouzo C, Le Guevel R, Lazar S, Akssira M, Troin Y, Guillaumet G, Routier S (2015) Advances in tetrahydropyrido[1,2-a]isoindolone (valmer ins) series: potent glycogen synthase kinase 3 and cyclin dependent kinase 5 inhibitors. Eur J Med Chem 101:274–287. doi:10.1016/j.ejmech.2015.06.046
22. De SK, Stebbins JL, Chen L-H, Riel-Mehan M, Machleidt T, Dahl R, Yuan H, Emdadi A, Barile E, Chen V, Murphy R, Pellecchia M (2009) Design, synthesis, and structure-activity relationship of substrate competitive, selective, and in vivo active triazole and thiadiazole inhibitors of the c-Jun N-terminal kinase. J Med Chem 52:1943–1952. doi:10.1021/jm801503n
23. Fischer M, Coleman RG, Fraser JS, Shoichet BK (2014) Incorporation of protein flexibility and conformational energy penalties in docking screens to improve ligand discovery. Nat Chem 6:575–583. doi:10.1038/nchem.1954
24. Halperin I, Ma B, Wolfson H, Nussinov R (2002) Principles of docking: an overview of search algorithms and a guide to scoring functions. Proteins 47:409–443. doi:10.1002/prot.10115
25. Ramírez D, Caballero J (2016) Is it reliable to use common molecular docking methods for comparing the binding affinities of enantiomer pairs for their protein target? Int J Mol Sci 17:525. doi:10.3390/ijms17040525
26. Lyne PD, Lamb ML, Saeh JC (2006) Accurate prediction of the relative potencies of members of a series of kinase inhibitors using molecular docking and MM-GBSA scoring. J Med Chem 49:4805–4808. doi:10.1021/jm060522a
27. Adasme-Carreño F, Muñoz-Gutierrez C, Caballero J, Alzate-Morales J (2014) Performance of the MM/GBSA scoring using a binding site hydrogen bond network-based frame selection: the protein kinase case. Phys Chem Chem Phys 16:14047–14058. doi:10.1039/C4CP01378F
28. Verdonk ML, Berdini V, Hartshorn MJ, Mooij WTM, Murray CW, Taylor RD, Watson P (2004) Virtual screening using protein-ligand docking: avoiding artificial enrichment. J Chem Inf Comput Sci 44:793–806. doi:10.1021/ci034289q
29. Caballero J, Alzate-Morales JH (2012) Molecular dynamics of protein kinase-inhibitor complexes: a valid structural information. Curr Pharm Des 18:2946–2963

30. Brooks BR, Bruccoleri RE, Olafson BD, States DJ, Swaminathan S, Karplus M (1983) CHARMM: a program for macromolecular energy, minimization, and dynamics calculations. J Comput Chem 4:187–217. doi:10.1002/jcc.540040211
31. Jorgensen WL, Maxwell DS, Tirado-Rives J (1996) Development and testing of the OPLS all-atom force field on conformational energetics and properties of organic liquids. J Am Chem Soc 118:11225–11236. doi:10.1021/ja9621760
32. Cornell WD, Cieplak P, Bayly CI, Gould IR, Merz KM, Ferguson DM, Spellmeyer DC, Fox T, Caldwell JW, Kollman PA (1995) A second generation force field for the simulation of proteins, nucleic acids, and organic molecules. J Am Chem Soc 117:5179–5197. doi:10.1021/ja00124a002
33. Schuler LD, Daura X, van Gunsteren WF (2001) An improved GROMOS96 force field for aliphatic hydrocarbons in the condensed phase. J Comput Chem 22:1205–1218. doi:10.1002/jcc.1078
34. Kollman PA, Massova I, Reyes C, Kuhn B, Huo S, Chong L, Lee M, Lee T, Duan Y, Wang W, Donini O, Cieplak P, Srinivasan J, Case DA, Cheatham TE (2000) Calculating structures and free energies of complex molecules: combining molecular mechanics and continuum models. Acc Chem Res 33:889–897. doi:10.1021/ar000033j
35. Zuccotto F, Ardini E, Casale E, Angiolini M (2010) Through the "gatekeeper door": exploiting the active kinase conformation. J Med Chem 53:2681–2694. doi:10.1021/jm901443h
36. Deng Z, Chuaqui C, Singh J (2004) Structural interaction fingerprint (SIFt): a novel method for analyzing three-dimensional protein-ligand binding interactions. J Med Chem 47:337–344. doi:10.1021/jm030331x
37. Shaw PC, Davies AF, Lau KF, Garcia-Barcelo M, Waye MM, Lovestone S, Miller CC, Anderton BH (1998) Isolation and chromosomal mapping of human glycogen synthase kinase-3 alpha and -3 beta encoding genes. Genome 41:720–727
38. Beurel E, Michalek SM, Jope RS (2010) Innate and adaptive immune responses regulated by glycogen synthase kinase-3 (GSK3). Trends Immunol 31:24–31. doi:10.1016/j.it.2009.09.007
39. Hernández F, Gómez de Barreda E, Fuster-Matanzo A, Lucas JJ, Avila J (2010) GSK3: a possible link between beta amyloid peptide and tau protein. Exp Neurol 223:322–325. doi:10.1016/j.expneurol.2009.09.011
40. Lucas JJ, Hernández F, Gómez-Ramos P, Morán MA, Hen R, Avila J (2001) Decreased nuclear beta-catenin, tau hyperphosphorylation and neurodegeneration in GSK-3beta conditional transgenic mice. EMBO J 20:27–39. doi:10.1093/emboj/20.1.27
41. Hetman M, Cavanaugh JE, Kimelman D, Xia Z (2000) Role of glycogen synthase kinase-3beta in neuronal apoptosis induced by trophic withdrawal. J Neurosci 20:2567–2574
42. Hoshi M, Takashima A, Noguchi K, Murayama M, Sato M, Kondo S, Saitoh Y, Ishiguro K, Hoshino T, Imahori K (1996) Regulation of mitochondrial pyruvate dehydrogenase activity by tau protein kinase I/glycogen synthase kinase 3beta in brain. Proc Natl Acad Sci U S A 93:2719–2723
43. Hooper C, Killick R, Lovestone S (2008) The GSK3 hypothesis of Alzheimer's disease. J Neurochem 104:1433–1439. doi:10.1111/j.1471-4159.2007.05194.x
44. Mou L, Li M, Lu S-Y, Li S, Shen Q, Zhang J, Li C, Lu X (2014) Unraveling the role of Arg4 and Arg6 in the auto-inhibition mechanism of GSK3β from molecular dynamics simulation. Chem Biol Drug Des 83:721–730. doi:10.1111/cbdd.12286
45. Ilouz R, Kowalsman N, Eisenstein M, Eldar-Finkelman H (2006) Identification of novel glycogen synthase kinase-3beta substrate-interacting residues suggests a common mechanism for substrate recognition. J Biol Chem 281:30621–30630. doi:10.1074/jbc.M604633200
46. Lu S-Y, Jiang Y-J, Zou J-W, Wu T-X (2011) Molecular modeling and molecular dynamics simulation studies of the GSK3β/ATP/substrate complex: understanding the unique P +4 primed phosphorylation specificity for GSK3β substrates. J Chem Inf Model 51:1025–1036. doi:10.1021/ci100493j
47. Lu S-Y, Huang Z-M, Huang W-K, Liu X-Y, Chen Y-Y, Shi T, Zhang J (2013) How calcium inhibits the magnesium-dependent kinase gsk3β: a molecular simulation study. Proteins 81:740–753. doi:10.1002/prot.24221
48. Lu S-Y, Jiang Y-J, Zou J-W, Wu T-X (2011) Dissection of the difference between the group I metal ions in inhibiting GSK3β: a computational study. Phys Chem Chem Phys 13:7014–7023. doi:10.1039/c0cp02498h
49. Tang X-N, Lo C-W, Chuang Y-C, Chen C-T, Sun Y-C, Hong Y-R, Yang C-N (2011) Prediction of the binding mode between GSK3β

and a peptide derived from GSKIP using molecular dynamics simulation. Biopolymers 95:461–471. doi:10.1002/bip.21603

50. Zhang N, Jiang Y, Zou J, Zhuang S, Jin H, Yu Q (2007) Insights into unbinding mechanisms upon two mutations investigated by molecular dynamics study of GSK3beta-axin complex: role of packing hydrophobic residues. Proteins 67:941–949. doi:10.1002/prot.21359
51. Lu S-Y, Jiang Y-J, Zou J-W, Wu T-X (2012) Effect of double mutations K214/A-E215/Q of FRATide on GSK3β: insights from molecular dynamics simulation and normal mode analysis. Amino Acids 43:267–277. doi:10.1007/s00726-011-1070-4
52. Lu S, Jiang Y, Lv J, Zou J, Wu T (2011) Mechanism of kinase inactivation and nonbinding of FRATide to GSK3β due to K85M mutation: molecular dynamics simulation and normal mode analysis. Biopolymers 95:669–681. doi:10.1002/bip.21629
53. Lee S-C, Shin D, Cho JM, Ro S, Suh Y-G (2012) Structure-activity relationship of the 7-hydroxy benzimidazole analogs as glycogen synthase kinase 3β inhibitor. Bioorg Med Chem Lett 22:1891–1894. doi:10.1016/j.bmcl.2012.01.065
54. Wang F, Liu M, Liu J (2012) In silico prediction of inhibitory effects of pyrazol-5-one and indazole derivatives on GSK3β kinase enzyme. J Mol Struct 1024:94–103. doi:10.1016/j.molstruc.2012.05.018
55. Caballero J, Zilocchi S, Tiznado W, Collina S, Rossi D (2011) Binding studies and quantitative structure-activity relationship of 3-amino-1H-indazoles as inhibitors of GSK3β. Chem Biol Drug Des 78:631–641. doi:10.1111/j.1747-0285.2011.01186.x
56. Quesada-Romero L, Caballero J (2014) Docking and quantitative structure–activity relationship of oxadiazole derivates as inhibitors of GSK3beta. Mol Divers 18:149–159. doi:10.1007/s11030-013-9483-5
57. Quesada-Romero L, Mena-Ulecia K, Tiznado W, Caballero J (2014) Insights into the interactions between maleimide derivates and GSK3β combining molecular docking and QSAR. PLoS One 9:e102212. doi:10.1371/journal.pone.0102212
58. Sangu S, Vema A, Bigala R (2014) 3D-QSAR and molecular docking studies of quinazoline derivatives as glycogen synthase kinase-3β (Gsk-3β) inhibitors. Pharm Lett 6:289–296
59. Withers IM, Mazanetz MP, Wang H, Fischer PM, Laughton CA (2008) Active site pressurization: a new tool for structure-guided drug design and other studies of protein flexibility. J Chem Inf Model 48:1448–1454. doi:10.1021/ci7004725
60. Mazanetz MP, Withers IM, Laughton CA, Fischer PM (2008) Exploiting glycogen synthase kinase 3beta flexibility in molecular recognition. Biochem Soc Trans 36:55–58. doi:10.1042/BST0360055
61. Lee H-C, Hsu W-C, Liu A-L, Hsu C-J, Sun Y-C (2014) Using thermodynamic integration MD simulation to compute relative protein-ligand binding free energy of a GSK3β kinase inhibitor and its analogs. J Mol Graph Model 51:37–49. doi:10.1016/j.jmgm.2014.04.010
62. Pradeep H, Rajanikant GK (2012) A rational approach to selective pharmacophore designing: an innovative strategy for specific recognition of Gsk3β. Mol Divers 16:553–562. doi:10.1007/s11030-012-9387-9
63. Hanumanthappa P, Krishnamurthy RG (2014) A comparative molecular dynamics simulation study to assess the exclusion ability of novel GSK3β inhibitors. Med Chem Res 23:3092–3095. doi:10.1007/s00044-013-0889-5
64. Lu S-Y, Jiang Y-J, Lv J, Zou J-W, Wu T-X (2011) Role of bridging water molecules in GSK3β-inhibitor complexes: insights from QM/MM, MD, and molecular docking studies. J Comput Chem 32:1907–1918. doi:10.1002/jcc.21775
65. Zhang N, Zhong R, Yan H, Jiang Y (2011) Structural features underlying selective inhibition of GSK3β by dibromocantharelline: implications for rational drug design. Chem Biol Drug Des 77:199–205. doi:10.1111/j.1747-0285.2010.01069.x
66. Li X, Wang X, Tian Z, Zhao H, Liang D, Li W, Qiu Y, Lu S (2014) Structural basis of valmerins as dual inhibitors of GSK3β/CDK5. J Mol Model 20:2407. doi:10.1007/s00894-014-2407-1
67. Mazanetz MP, Laughton CA, Fischer PM (2014) Investigation of the flexibility of protein kinases implicated in the pathology of Alzheimer's disease. Molecules 19:9134–9159. doi:10.3390/molecules19079134
68. Paglini G, Cáceres A (2001) The role of the Cdk5–p35 kinase in neuronal development. Eur J Biochem 268:1528–1533. doi:10.1046/j.1432-1327.2001.02023.x
69. Quan H, Wu X, Tian Y, Wang Y, Li C, Li H (2014) Overexpression of CDK5 in neural stem cells facilitates maturation of embryonic neurocytes derived from rats in vitro. Cell

Biochem Biophys 69:445–453. doi:10.1007/s12013-014-9816-8

70. Liu S-L, Wang C, Jiang T, Tan L, Xing A, Yu J-T (2016) The role of Cdk5 in Alzheimer's disease. Mol Neurobiol 53:4328–4342. doi:10.1007/s12035-015-9369-x
71. Camins A, Verdaguer E, Folch J, Canudas AM, Pallàs M (2006) The role of CDK5/P25 formation/inhibition in neurodegeneration. Drug News Perspect 19:453–460. doi:10.1358/dnp.2006.19.8.1043961
72. Tan VBC, Zhang B, Lim KM, Tay TE (2010) Explaining the inhibition of cyclin-dependent kinase 5 by peptides derived from p25 with molecular dynamics simulations and MM-PBSA. J Mol Model 16:1–8. doi:10.1007/s00894-009-0514-1
73. Cardone A, Brady M, Sriram R, Pant HC, Hassan SA (2016) Computational study of the inhibitory mechanism of the kinase CDK5 hyperactivity by peptide p5 and derivation of a pharmacophore. J Comput Aided Mol Des 30:513–521. doi:10.1007/s10822-016-9922-3
74. Demange L, Abdellah FN, Lozach O, Ferandin Y, Gresh N, Meijer L, Galons H (2013) Potent inhibitors of CDK5 derived from roscovitine: synthesis, biological evaluation and molecular modelling. Bioorg Med Chem Lett 23:125–131. doi:10.1016/j.bmcl.2012.10.141
75. Dehbi O, Tikad A, Bourg S, Bonnet P, Lozach O, Meijer L, Aadil M, Akssira M, Guillaumet G, Routier S (2014) Synthesis and optimization of an original V-shaped collection of 4-7-disubstituted pyrido[3,2-d]pyrimidines as CDK5 and DYRK1A inhibitors. Eur J Med Chem 80:352–363. doi:10.1016/j.ejmech.2014.04.055
76. Shrestha S, Natarajan S, Park J-H, Lee D-Y, Cho J-G, Kim G-S, Jeon Y-J, Yeon S-W, Yang D-C, Baek N-I (2013) Potential neuroprotective flavonoid-based inhibitors of CDK5/p25 from Rhus parviflora. Bioorg Med Chem Lett 23:5150–5154. doi:10.1016/j.bmcl.2013.07.020
77. Pitchuanchom S, Boonyarat C, Forli S, Olson AJ, Yenjai C (2012) Cyclin-dependent kinases 5 template: useful for virtual screening. Comput Biol Med 42:106–111. doi:10.1016/j.compbiomed.2011.10.014
78. Chatterjee A, Cutler SJ, Doerksen RJ, Khan IA, Williamson JS (2014) Discovery of thienoquinolone derivatives as selective and ATP non-competitive CDK5/p25 inhibitors by structure-based virtual screening. Bioorg Med Chem 22:6409–6421. doi:10.1016/j.bmc.2014.09.043
79. Wang W, Cao X, Zhu X, Gu Y (2013) Molecular dynamic simulations give insight into the mechanism of binding between 2-aminothiazole inhibitors and CDK5. J Mol Model 19:2635–2645. doi:10.1007/s00894-013-1815-y
80. Dong K, Wang X, Yang X, Zhu X (2016) Binding mechanism of CDK5 with roscovitine derivatives based on molecular dynamics simulations and MM/PBSA methods. J Mol Graph Model 68:57–67. doi:10.1016/j.jmgm.2016.06.007
81. Wu Q, Kang H, Tian C, Huang Q, Zhu R (2013) Binding mechanism of inhibitors to CDK5/p25 complex: free energy calculation and ranking aggregation analysis. Mol Inform 32:251–260. doi:10.1002/minf.201200139
82. Patel JS, Berteotti A, Ronsisvalle S, Rocchia W, Cavalli A (2014) Steered molecular dynamics simulations for studying protein-ligand interaction in cyclin-dependent kinase 5. J Chem Inf Model 54:470–480. doi:10.1021/ci4003574
83. Ul Haq Z, Uddin R, Wai LK, Wadood A, Lajis NH (2011) Docking and 3D-QSAR modeling of cyclin-dependent kinase 5/p25 inhibitors. J Mol Model 17:1149–1161. doi:10.1007/s00894-010-0817-2
84. Zhang B, Corbel C, Guéritte F, Couturier C, Bach S, Tan VBC (2011) An in silico approach for the discovery of CDK5/p25 interaction inhibitors. Biotechnol J 6:871–881. doi:10.1002/biot.201100139
85. Atzori C, Ghetti B, Piva R, Srinivasan AN, Zolo P, Delisle MB, Mirra SS, Migheli A (2001) Activation of the JNK/p38 pathway occurs in diseases characterized by tau protein pathology and is related to tau phosphorylation but not to apoptosis. J Neuropathol Exp Neurol 60:1190–1197
86. Pei J-J, Braak H, An W-L, Winblad B, Cowburn RF, Iqbal K, Grundke-Iqbal I (2002) Up-regulation of mitogen-activated protein kinases ERK1/2 and MEK1/2 is associated with the progression of neurofibrillary degeneration in Alzheimer's disease. Brain Res Mol Brain Res 109:45–55
87. Zhu X, Castellani RJ, Takeda A, Nunomura A, Atwood CS, Perry G, Smith MA (2001) Differential activation of neuronal ERK, JNK/SAPK and p38 in Alzheimer disease: the "two hit" hypothesis. Mech Ageing Dev 123:39–46
88. Munoz L, Ammit AJ (2010) Targeting p38 MAPK pathway for the treatment of Alzheimer's disease. Neuropharmacology 58:561–568. doi:10.1016/j.neuropharm.2009.11.010

89. Zhu X, Rottkamp CA, Boux H, Takeda A, Perry G, Smith MA (2000) Activation of p38 kinase links tau phosphorylation, oxidative stress, and cell cycle-related events in Alzheimer disease. J Neuropathol Exp Neurol 59:880–888
90. Thornton TM, Pedraza-Alva G, Deng B, Wood CD, Aronshtam A, Clements JL, Sabio G, Davis RJ, Matthews DE, Doble B, Rincon M (2008) Phosphorylation by p38 MAPK as an alternative pathway for GSK3beta inactivation. Science 320:667–670. doi:10.1126/science.1156037
91. Yang Y, Liu H, Yao X (2012) Understanding the molecular basis of MK2-p38α signaling complex assembly: insights into protein-protein interaction by molecular dynamics and free energy studies. Mol BioSyst 8:2106–2118. doi:10.1039/c2mb25042j
92. Ma Z, Pan Y, Huang W, Yang Y, Wang Z, Li Q, Zhao Y, Zhang X, Shen Z (2014) Synthesis and biological evaluation of the pirfenidone derivatives as antifibrotic agents. Bioorg Med Chem Lett 24:220–223. doi:10.1016/j.bmcl.2013.11.038
93. Heo J, Shin H, Lee J, Kim T, Inn K-S, Kim N-J (2015) Synthesis and biological evaluation of N-cyclopropylbenzamide-benzophenone hybrids as novel and selective p38 mitogen activated protein kinase (MAPK) inhibitors. Bioorg Med Chem Lett 25:3694–3698. doi:10.1016/j.bmcl.2015.06.036
94. de Oliveira LR, Romeiro NC, de Lima CKF, Louback da Silva L, de Miranda ALP, Nascimento PGBD, Cunha FQ, Barreiro EJ, Lima LM (2012) Docking, synthesis and pharmacological activity of novel urea-derivatives designed as p38 MAPK inhibitors. Eur J Med Chem 54:264–271. doi:10.1016/j.ejmech.2012.05.006
95. Khoshneviszadeh M, Ghahremani MH, Foroumadi A, Miri R, Firuzi O, Madadkar-Sobhani A, Edraki N, Parsa M, Shafiee A (2013) Design, synthesis and biological evaluation of novel anti-cytokine 1,2,4-triazine derivatives. Bioorg Med Chem 21:6708–6717. doi:10.1016/j.bmc.2013.08.009
96. Vinh NB, Devine SM, Munoz L, Ryan RM, Wang BH, Krum H, Chalmers DK, Simpson JS, Scammells PJ (2015) Design, synthesis, and biological evaluation of tetra-substituted thiophenes as inhibitors of p38α MAPK. ChemistryOpen 4:56–64. doi:10.1002/open.201402076
97. Choi H, Park HJ, Shin JC, Ko HS, Lee JK, Lee S, Park H, Hong S (2012) Structure-based virtual screening approach to the discovery of p38 MAP kinase inhibitors. Bioorg Med Chem Lett 22:2195–2199. doi:10.1016/j.bmcl.2012.01.104
98. Irwin JJ, Shoichet BK (2005) ZINC–a free database of commercially available compounds for virtual screening. J Chem Inf Model 45:177–182. doi:10.1021/ci049714+
99. Cabrera ÁC, Gil-Redondo R, Perona A, Gago F, Morreale A (2011) VSDMIP 1.5: an automated structure- and ligand-based virtual screening platform with a PyMOL graphical user interface. J Comput Aided Mol Des 25:813–824. doi:10.1007/s10822-011-9465-6
100. Willemen HLDM, Campos PM, Lucas E, Morreale A, Gil-Redondo R, Agut J, González FV, Ramos P, Heijnen C, Mayor F, Kavelaars A, Murga C (2014) A novel p38 MAPK docking-groove-targeted compound is a potent inhibitor of inflammatory hyperalgesia. Biochem J 459:427–439. doi:10.1042/BJ20130172
101. Poon J-F, Alao JP, Sunnerhagen P, Dinér P (2013) Azastilbenes: a cut-off to p38 MAPK inhibitors. Org Biomol Chem 11:4526–4536. doi:10.1039/c3ob27449g
102. Pinsetta FR, Taft CA, de Paula da Silva CHT (2014) Structure- and ligand-based drug design of novel p38-alpha MAPK inhibitors in the fight against the Alzheimer's disease. J Biomol Struct Dyn 32:1047–1063. doi:10.1080/07391102.2013.803441
103. He L, Dai R, Zhang XR, Gao SY, He YY, Wang LB, Gao X, Yang LQ (2015) Ligand-based 3D pharmacophore design, virtual screening and molecular docking for novel p38 MAPK inhibitors. Med Chem Res 24:797–809. doi:10.1007/s00044-014-1158-y
104. Chang H-W, Chung F-S, Yang C-N (2013) Molecular modeling of p38α mitogen-activated protein kinase inhibitors through 3D-QSAR and molecular dynamics simulations. J Chem Inf Model 53:1775–1786. doi:10.1021/ci4000085
105. Vinh NB, Simpson JS, Scammells PJ, Chalmers DK (2012) Virtual screening using a conformationally flexible target protein: models for ligand binding to p38α MAPK. J Comput Aided Mol Des 26:409–423. doi:10.1007/s10822-012-9569-7
106. Ebadi SA, Razzaghi-Asl N, Khoshneviszadeh M, Miri R (2015) Detailed atomistic molecular modeling of a potent type II p38α inhibitor. Struct Chem 26:1125–1137. doi:10.1007/s11224-015-0568-x

107. Lan P, Huang Z-J, Sun J-R, Chen W-M (2010) 3D-QSAR and molecular docking studies on fused pyrazoles as p38α mitogen-activated protein kinase inhibitors. Int J Mol Sci 11:3357–3374. doi:10.3390/ijms11093357
108. Astolfi A, Iraci N, Sabatini S, Barreca ML, Cecchetti V (2015) p38α MAPK and type I inhibitors: binding site analysis and use of target ensembles in virtual screening. Molecules 20:15842–15861. doi:10.3390/molecules200915842
109. Medina MG, Ledesma MD, Domínguez JE, Medina M, Zafra D, Alameda F, Dotti CG, Navarro P (2005) Tissue plasminogen activator mediates amyloid-induced neurotoxicity via Erk1/2 activation. EMBO J 24:1706–1716. doi:10.1038/sj.emboj.7600650
110. Sergeant N, Bretteville A, Hamdane M, Caillet-Boudin M-L, Grognet P, Bombois S, Blum D, Delacourte A, Pasquier F, Vanmechelen E, Schraen-Maschke S, Buée L (2008) Biochemistry of tau in Alzheimer's disease and related neurological disorders. Expert Rev Proteomics 5:207–224. doi:10.1586/14789450.5.2.207
111. Misiura MM, Kolomeisky AB (2016) Theoretical investigation of the mechanisms of ERK2 enzymatic catalysis. J Phys Chem B 120:10508–10514. doi:10.1021/acs.jpcb.6b08435
112. Barr D, Oashi T, Burkhard K, Lucius S, Samadani R, Zhang J, Shapiro P, MacKerell AD, van der Vaart A (2011) Importance of domain closure for the autoactivation of ERK2. Biochemistry 50:8038–8048. doi:10.1021/bi200503a
113. Blake JF, Gaudino JJ, De Meese J, Mohr P, Chicarelli M, Tian H, Garrey R, Thomas A, Siedem CS, Welch MB, Kolakowski G, Kaus R, Burkard M, Martinson M, Chen H, Dean B, Dudley DA, Gould SE, Pacheco P, Shahidi-Latham S, Wang W, West K, Yin J, Moffat J, Schwarz JB (2014) Discovery of 5,6,7,8-tetrahydropyrido[3,4-d]pyrimidine inhibitors of Erk2. Bioorg Med Chem Lett 24:2635–2639. doi:10.1016/j.bmcl.2014.04.068
114. Choi W-K, El-Gamal MI, Choi HS, Baek D, Oh C-H (2011) New diarylureas and diarylamides containing 1,3,4-triarylpyrazole scaffold: synthesis, antiproliferative evaluation against melanoma cell lines, ERK kinase inhibition, and molecular docking studies. Eur J Med Chem 46:5754–5762. doi:10.1016/j.ejmech.2011.08.013
115. Jin F, Gao D, Wu Q, Liu F, Chen Y, Tan C, Jiang Y (2013) Exploration of N-(2-aminoethyl)piperidine-4-carboxamide as a potential scaffold for development of VEGFR-2, ERK-2 and Abl-1 multikinase inhibitor. Bioorg Med Chem 21:5694–5706. doi:10.1016/j.bmc.2013.07.026
116. Kinoshita T, Sugiyama H, Mori Y, Takahashi N, Tomonaga A (2016) Identification of allosteric ERK2 inhibitors through in silico biased screening and competitive binding assay. Bioorg Med Chem Lett 26:955–958. doi:10.1016/j.bmcl.2015.12.056
117. Larif S, Ben Salem C, Hmouda H, Bouraoui K (2014) In silico screening and study of novel ERK2 inhibitors using 3D QSAR, docking and molecular dynamics. J Mol Graph Model 53:1–12. doi:10.1016/j.jmgm.2014.07.001
118. Niu Y, Pan D, Yang Y, Liu H, Yao X (2016) Revealing the molecular mechanism of different residence times of ERK2 inhibitors via binding free energy calculation and unbinding pathway analysis. Chemom Intell Lab Syst 158:91–101. doi:10.1016/j.chemolab.2016.08.002
119. Wu K-W, Chen P-C, Wang J, Sun Y-C (2012) Computation of relative binding free energy for an inhibitor and its analogs binding with Erk kinase using thermodynamic integration MD simulation. J Comput Aided Mol Des 26:1159–1169. doi:10.1007/s10822-012-9606-6
120. Rastelli G, Rosenfeld R, Reid R, Santi DV (2008) Molecular modeling and crystal structure of ERK2-hypothemycin complexes. J Struct Biol 164:18–23. doi:10.1016/j.jsb.2008.05.002
121. Liu B, Fu L, Zhang C, Zhang L, Zhang Y, Ouyang L, He G, Huang J (2015) Computational design, chemical synthesis, and biological evaluation of a novel ERK inhibitor (BL-EI001) with apoptosis-inducing mechanisms in breast cancer. Oncotarget 6:6762–6775. doi:10.18632/oncotarget.3105
122. Mehan S, Meena H, Sharma D, Sankhla R (2011) JNK: a stress-activated protein kinase therapeutic strategies and involvement in Alzheimer's and various neurodegenerative abnormalities. J Mol Neurosci 43:376–390. doi:10.1007/s12031-010-9454-6
123. Dias-Santagata D, Fulga TA, Duttaroy A, Feany MB (2007) Oxidative stress mediates tau-induced neurodegeneration in Drosophila. J Clin Invest 117:236–245. doi:10.1172/JCI28769

124. Colombo A, Repici M, Pesaresi M, Santambrogio S, Forloni G, Borsello T (2007) The TAT-JNK inhibitor peptide interferes with beta amyloid protein stability. Cell Death Differ 14:1845–1848. doi:10.1038/sj.cdd.4402202
125. Ferrer I, Barrachina M, Puig B (2002) Anti-tau phospho-specific Ser262 antibody recognizes a variety of abnormal hyperphosphorylated tau deposits in tauopathies including Pick bodies and argyrophilic grains. Acta Neuropathol 104:658–664. doi:10.1007/s00401-002-0600-2
126. Christopher JA, Atkinson FL, Bax BD, Brown MJB, Champigny AC, Chuang TT, Jones EJ, Mosley JE, Musgrave JR (2009) 1-Aryl-3,4-dihydroisoquinoline inhibitors of JNK3. Bioorg Med Chem Lett 19:2230–2234. doi:10.1016/j.bmcl.2009.02.098
127. Dykstra DW, Dalby KN, Ren P (2013) Elucidating binding modes of zuonin A enantiomers to JNK1 via in silico methods. J Mol Graph Model 45:38–44. doi:10.1016/j.jmgm.2013.08.008
128. Ngoei KRW, Ng DCH, Gooley PR, Fairlie DP, Stoermer MJ, Bogoyevitch MA (2013) Identification and characterization of bi-thiazole-2,2′-diamines as kinase inhibitory scaffolds. Biochim Biophys Acta 1834:1077–1088. doi:10.1016/j.bbapap.2013.02.001
129. Zhuo Z-H, Sun Y-Z, Jin P-N, Li F-Y, Zhang Y-L, Wang H-L (2016) Selective targeting of MAPK family kinases JNK over p38 by rationally designed peptides as potential therapeutics for neurological disorders and epilepsy. Mol BioSyst 12:2532–2540. doi:10.1039/C6MB00297H
130. Katari SK, Natarajan P, Swargam S, Kanipakam H, Pasala C, Umamaheswari A (2016) Inhibitor design against JNK1 through e-pharmacophore modeling docking and molecular dynamics simulations. J Recept Signal Transduct Res 36:558–571. doi:10.3109/10799893.2016.1141955
131. Shaikh AR, Ismael M, Del Carpio CA, Tsuboi H, Koyama M, Endou A, Kubo M, Broclawik E, Miyamoto A (2006) Three-dimensional quantitative structure-activity relationship (3 D-QSAR) and docking studies on (benzothiazole-2-yl) acetonitrile derivatives as c-Jun N-terminal kinase-3 (JNK3) inhibitors. Bioorg Med Chem Lett 16:5917–5925. doi:10.1016/j.bmcl.2006.06.039
132. Kim M-H, Ryu J-S, Hah J-M (2013) 3D-QSAR studies of 1,2-diaryl-1H-benzimidazole derivatives as JNK3 inhibitors with protective effects in neuronal cells. Bioorg Med Chem Lett 23:1639–1642. doi:10.1016/j.bmcl.2013.01.082
133. Wu X-X, Dai D-S, Zhu X, Li X-F, Yuan J, Wu X-F, Miao M-S, Zeng H-H, Zhao C-L (2014) Molecular modeling studies of JNK3 inhibitors using QSAR and docking. Med Chem Res 23:2456–2475. doi:10.1007/s00044-013-0782-2
134. Madhavan T, Chung JY, Kothandan G, Gadhe CG, Cho SJ (2012) 3D-QSAR studies of JNK1 inhibitors utilizing various alignment methods. Chem Biol Drug Des 79:53–67. doi:10.1111/j.1747-0285.2011.01168.x
135. Hierold J, Baek S, Rieger R, Lim T-G, Zakpur S, Arciniega M, Lee KW, Huber R, Tietze LF (2015) Design, synthesis, and biological evaluation of quercetagetin analogues as JNK1 inhibitors. Chemistry 21:16887–16894. doi:10.1002/chem.201502475

Chapter 14

Computational Modelling of Kinase Inhibitors as Anti-Alzheimer Agents

Mange Ram Yadav, Mahesh A. Barmade, Rupesh V. Chikhale, and Prashant R. Murumkar

Abstract

Alzheimer's disease is one of the leading causes of deaths in both developed as well as developing countries. It is characterized mainly by the deposition of extracellular β-amyloid plaques and intracellular neurofibrillary tangles. To date various hypotheses regarding the aetiology of AD have been reported in the literature, but the exact cause of AD still remains unknown. Hyperphosphorylation of tau protein is considered as one of the most accepted reasons by the scientists and researchers throughout the world. Currently, four FDA-approved drugs (rivastigmine, galantamine, donepezil, memantine) are available in the market, and the fifth one (tacrine) is withdrawn due its hepatotoxicity concerns. Protein kinases are well-known targets in cancer, and several reports are available regarding their critical role in the hyperphosphorylation of tau protein in AD. Although not even a single molecule is available in the market for the treatment of AD as a kinase inhibitor, a number of studies are in the pipeline targeting the kinases. In this chapter efforts have been made to discuss the computational studies carried out till date on various protein kinases in search of potential anti-Alzheimer's agents.

Key words Alzheimer's disease, Kinase inhibitors, Molecular modelling studies, Protein kinases, Tau phosphorylation

1 Introduction

Alzheimer's disease (AD) is a chronic, irreversible and the most fatal form of senile dementia characterized by extravagant amount of extracellular β-amyloid plaques and intracellular neurofibrillary tangles [1]. According to the latest WHO report, there are 47.5 millions of people worldwide suffering from dementia, and on an average 7.7 million, new cases are reported every year. Individually AD contributed to 60–70% of overall dementia cases, and it is one of the leading and most fatal forms of dementia among various neurodegenerative diseases [2]. Latest key finding in the World Alzheimer Report 2016 states that poor access to healthcare system to the dementia patients in high- to low-income countries has

Kunal Roy (ed.), *Computational Modeling of Drugs Against Alzheimer's Disease*, Neuromethods, vol. 132, DOI 10.1007/978-1-4939-7404-7_14, © Springer Science+Business Media LLC 2018

muddled the eradication of dementia which has become a social and economic burden throughout the world [3, 4].

AD is broadly classified according to the age of onset of the disease and whether it has developed spontaneously or acquired hereditarily as (1) early-onset AD and (2) late-onset AD. Half of the early-onset AD is accounted for by three genes, namely, *β-protein precursor*, *presenilin* 1 and *presenilin* 2, and the late-onset AD is gauged by *apolipoprotein* E-4 genetic risk factor [5]. Although the exact cause of AD is not fully known till date, a number of hypotheses regarding the origin of AD have been well reviewed in the past, and these are (a) amyloid cascade hypothesis, (b) tau hypothesis, (c) cholinergic hypothesis, (d) dendritic hypothesis, (e) mitochondrial cascade hypothesis, (f) metabolic hypothesis and (g) other hypotheses that implicate oxidative stress and neuroinflammation [6]. Among these hypotheses, amyloid cascade hypothesis was vastly evidenced genetically and pathologically [7]. Tau hypothesis is one of the most studied hypotheses among the researchers till date, and it is associated with abnormal hyperphosphorylation of tau protein which leads to the formation of paired helical filaments and neurofibrillary tangles [8]. As far as the cholinergic hypothesis is concerned, four (**1**–**4**) out of the five FDA-approved drugs (**1**–**5**) for the treatment of AD are developed on the basis of this hypothesis, e.g. tacrine (**1**), donepezil (**2**), rivastigmine (**3**) and galantamine (**4**), and the fifth one memantine (**5**) is an NMDA receptor antagonist (Fig. 1). Tacrine (**1**) was withdrawn from the market due to its hepatotoxicity [9]. The currently available medicine box (**2**–**5**) for the treatment of AD contains mainly acetylcholinesterase inhibitors (**2**–**4**) and NMDA antagonist (**5**). These drugs improve only the cognitive functions, and the treatment of patients having noncognitive neuropsychiatric symptoms still remains a question mark [10].

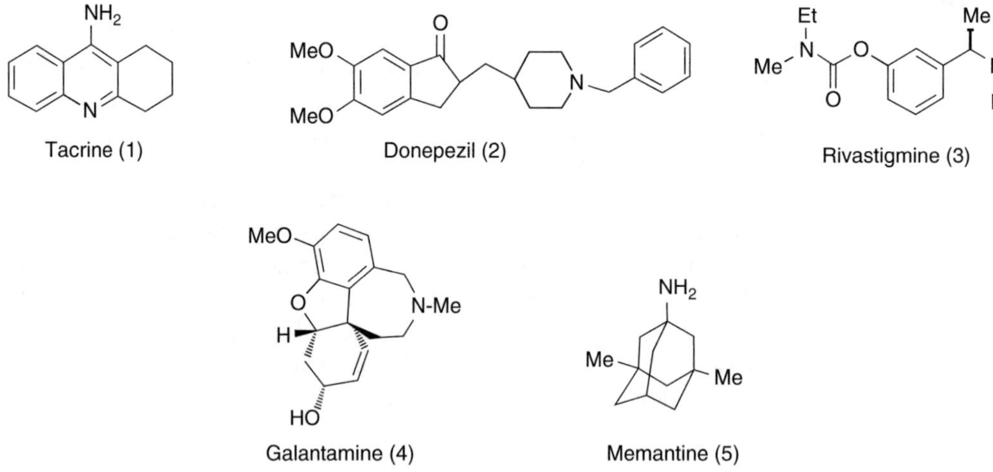

Tacrine (1) Donepezil (2) Rivastigmine (3)

Galantamine (4) Memantine (5)

Fig. 1 FDA-approved drugs (**1**–**5**) for the treatment of Alzheimer's disease

It is more than a decade that FDA approved a drug (last in 2003) for the treatment of AD, and day-after-day treatment of AD is becoming a challenging task for the neurologists, psychiatrists, geriatricians and family doctors. Hence, there is an urgent need for the development of newer treatment strategies as well as therapeutics to tackle this life-threatening disease [11]. In spite of the fact that there are no strict guidelines for the treatment of AD and the existing guidelines vary country-wise, the conventional therapeutic regimen recommends the use of six classes of drugs in AD, namely, acetylcholinesterase inhibitors (AchEI), *N*-methyl-*D*-aspartate (NMDA) receptor antagonists, monoamine oxidase inhibitors (MAOI), antioxidants, metal chelators and anti-inflammatory agents [12].

Currently, scientists and researchers worldwide are focusing their efforts mainly on the development of novel lead molecules having a tendency to specifically bind to the proteins, receptors and different signalling pathways related to AD [13]. Among these, protein kinases are regarded as the growing drug targets mainly in the peripheral tissues. Exhaustive reports are available regarding their role in AD, and they have been well reviewed in the literature [14–16]. Although imatinib (Gleevec) and nilotinib (Tasigna) are well-known marketed anticancer drugs from the class of protein kinase inhibitors, but till date not even a single drug molecule has emerged from this class for the treatment of AD. This situation has motivated scientists and researchers to work on this target for the development of small-molecule kinase inhibitors for AD therapeutics [17–20]. These protein kinases are associated with phosphorylation of tau protein, and they are regarded as tau kinases. They are further classified in to the proline-directed kinases (PDKs), e.g. glycogen synthase kinase (GSK) and cyclin-dependent protein kinase (CDK), dual-specificity tyrosine phosphorylation-regulated kinase (DYRK), p38α mitogen-activated protein kinase (MAPK) and non-PDKs such as casein kinase, calcium calmodulin-activated protein kinase II (CaM kinase II) and protein kinase A (PKA) [21]. Approximately 20 protein kinases are known to phosphorylate the tau protein in vitro and in the cells [22–25].

1.1 Role of Protein (Tau) Kinases in the Phosphorylation of Tau Proteins

Tau proteins play an essential role in the stabilization of microtubules in the axon. These water-soluble proteins are abundantly present in the neurons. Tau proteins are endowed with large phosphorylation sites, i.e. 80 Ser/Thr sites, which are phosphorylated by various protein kinases [26, 27]. Phosphorylation of tau is an essential process for the binding of tau to the microtubules and stabilization of the microtubules. This process of phosphorylation would be termed as a sum of functions of the protein kinases and phosphatases, and any imbalance between the functions of these two enzymes causes hyperphosphorylation of tau. Hyperphosphorylation of tau results into the formation of insoluble paired helical filaments and consequently formation of neurofibrillary tangles, loss of binding capacity of tau to the microtubules and finally neuron death. Both the classes of kinases such as PDKs and NPDKs are responsible for causing the phosphorylation of tau proteins [28, 29].

1.2 Proline-Directed Kinases (PDKs)

1.2.1 Glycogen Synthase Kinase (GSK)

Glycogen is a polysaccharide of glucose that serves as an energy reservoir in animals and certain fungi. Glycogen synthase converts glucose to glycogen, a process also known as glycogenesis. Glycosyltransferase catalyses the polymerization reaction leading to the formation of $(1,4\text{-}\alpha\text{-}D\text{-glucosyl})_{n+1}$ via α-(1 $\rightarrow$ 4) glycosidic bond. Glycogen synthase is regulated by phosphorylation-dephosphorylation mechanism which is catalysed by several protein kinases. Three glycogen synthase kinases are known (GSK-1/2/3), but only GSK-3 is implicated in AD. Cohen et al. reported glycogen synthase kinase (GSK-3) from rabbit skeletal muscles in 1980 [30]. GSK-3 is highly specific for glycogen synthesis; its activity remains unaffected by cAMP, cGMP, EGTA, calcium, calmodulin or any other inhibitor of protein. The roles and functions of GSK-3 have been extensively detailed in the past [31–34]. It exists in two isoforms, as GSK-3α and GSK-3β that share almost 98% homology in their catalytic domain and are ubiquitously expressed in all cells and tissues, and have similar properties (Fig. 2). These are serine/threonine kinases, highly active in cells and are affected by extracellular stimuli. Phosphorylation of Ser21 in GSK-3α and Ser9 in GSK-3β located in the N-terminal domain inhibits the activity of GSK-3 [32–34].

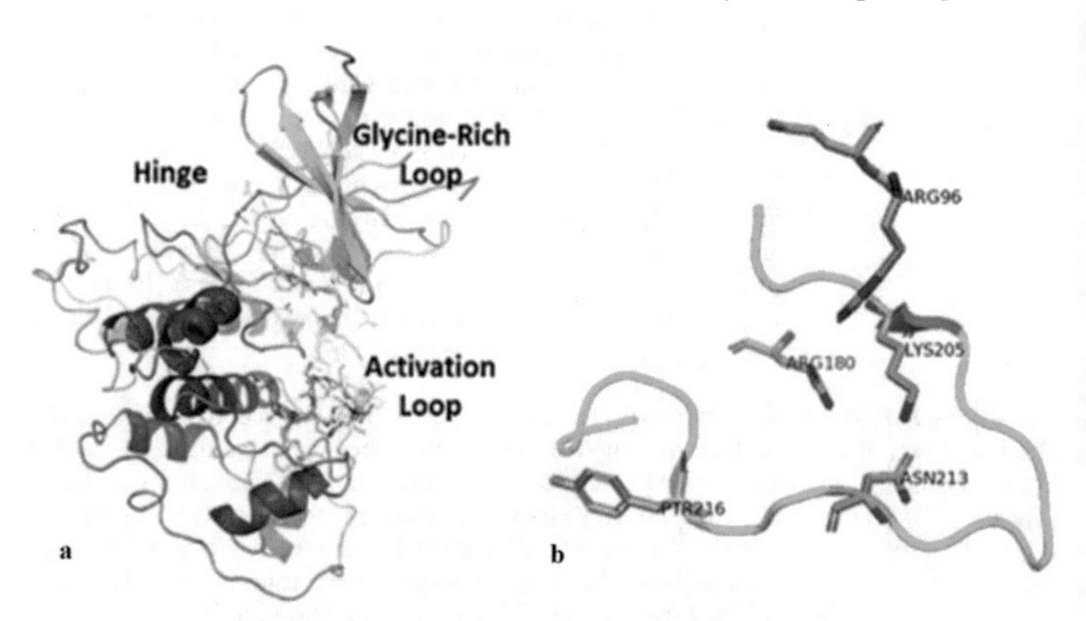

Fig. 2 (**a**) Structure of GSK-3β, three distinct regions are shown as activation loop, hinge region and glycine-rich loop. (**b**) Activation loop with backbone is coloured *green*, and the protein residues are coloured *light orange* [31] (Used with permission from Ombrato et al. J Chem Info Model 2015, 55, 2540–2551 [copyright© 2015, American Chemical Society])

1.2.2 Cyclin-Dependent Protein Kinase (CDK)

Cyclins are a large family of proteins consisting of approximately 30 members having mass range from 35 to 60 kDa. Cyclin-dependent protein kinases belong to the class of serine/threonine kinases that need cyclin domain for the activity [35]. CDKs play a crucial role in the cell division and transcription process. CDKs are further subclassified into CDK1, CDK2, CDK3, CDK4, CDK5, CDK7, CDK8, CDK9, CDK11 and CDK20 [36]. Among the subfamilies of CDKs, CDK-5 has already proved its crucial role in phosphorylation of tau protein, leading to AD [24, 37]. CDK5 having a mass of 33 kDa phosphorylates mainly serine/threonine residues. Activation of CDK5 is dependent on cofactors p35 or p39, and these cofactors are converted into p25 or p29, respectively, by the proteolytic action of calpain in the presence of calcium. The conversion of p35 to p25 or p39 to p29 is subjected to the protracted activation and mislocalization of CDK5, resulting into the hyperphosphorylation of tau protein [38]. Phosphorylation of tau by CDK5 can be transpiring at Ser202, Thr205, Thr212, Thr217, Ser235, Ser396 and Ser404, the pathologically relevant residues [39]. Crystal structures of CDK5/p25 co-crystallized with three inhibitors such as *R*-roscovitine, indirubin-2′-oxime and Aloisine have been well documented in the literature [40–44].

1.2.3 Dual-Specificity Tyrosine Phosphorylation-Regulated Kinases (DYRKs)

Dual-specificity tyrosine phosphorylation-regulated kinases (DYRK) are the members of CMGC group that play an important role in the proliferation, survival and development of cells. These kinases are not dependent on any cofactor and need no activation motifs [45]. DYRKs are again subclassified into two classes depending upon the presence of C/N-terminal motif. The first class consists of DYRK1A and 1B kinases having PEST region at C-terminal, and the N-terminal has nuclear localization signal. Second class consisting of DYRK 2, 3 and 4 kinases, supposed to lack both C/N-terminal motifs, are predominantly cytosolic [46]. DYRK1A is expressed in human foetal and adult brains, and overexpression of DYRK1A is associated with mental retardation syndromes such as Down syndrome and also AD [47]. As the name indicates 'dual-specificity', DYRK1A has the ability to phosphorylate dual substrates, i.e. phosphorylation of tau protein at various sites of serine/threonine as well as autophosphorylation at tyrosine321 (for self-activation) residue in the catalytic domain of the active site [48]. X-ray crystal structure (PDB Id: 2VX3) of DYRK1A revealed the presence of a disulphide bridge between the two thiol groups, and it also showed that arginines R325 and R328 were interacting with the phosphorylated Y321 residue [49].

1.2.4 p38α Mitogen-Activated Protein Kinase (p38α MAPK)

Mitogen-activated protein kinases (MAPKs) are a class of serine/threonine kinases endowed with the functioning of integration and processing of extracellular stimuli through a series of phosphorylation cascades [50]. A set of three tired MAPKs is accessible in its simplest form, i.e. a MAPK (ERKs, c-Jun N-terminal kinases (JNKs) and p38α) and two upstream components (a MAPK kinase and a MAPK kinase kinase) which are responsible for the activation of MAPK through the process of phosphorylation. Activated MAPKs have the ability to phosphorylate a number of substrates and are responsible for stimulus-specific response [51]. MAPKs have several isoforms with diverse functions; one of the isoforms p38α plays a key role in CNS disorders and has emerged as a key target in the management of CNS diseases including AD [52]. p38α MAPK pathway is activated in the neurons and in the glial cells of brain, and its overexpression leads to phosphorylation of tau protein as well as synaptic dysfunction [53, 54].

1.3 Non-proline-Directed Kinases (NPDKs)

1.3.1 Casein Kinases (CK)

Casein kinase, member of a family of monomeric serine/threonine protein kinases, is involved in diverse cellular cascades such as cell division, DNA repair, nuclear localization and regulation of membrane transport [55, 56]. Casein kinases are further subdivided into casein kinase 1(CK-1) and casein kinase 2 (CK-2), and both of them have high homology in the active site of the catalytic domain [57]. Casein kinases are divergent from other protein kinases due to the presence of S-I-N sequence instead of A-P-E in kinase domain VIII [58]. Casein kinase family members contain nearly 290 residues in the N-terminal, and these residues are coupled to the C-terminal residues whose size ranges from 40 to 180 amino acids [59]. One of the interesting things about the casein kinases is that their phosphorylation process is ATP dependent, and they can phosphorylate only those substrates which have already been prephosphorylated [60]. Casein kinases have similar substrate specificity in vitro, and S/Tp-X-X-<u>S/T</u> is referred to as one of the phosphorylation sites (S/Tp denotes the phospho-serine or phosphothreonine; X can be any amino acid, and the underlined residue S/T represents the target site) [61].

1.3.2 Calcium Calmodulin-Activated Protein Kinase II (CaM-PK II)

CaM-PK II is a member of calcium-/calmodulin-stimulated protein kinases family which is characterized by its broad substrate specificity, and it is abundantly present in hippocampus [62]. CaM-PK II is a holoenzyme having two subunits, α and β with masses of 50 and 60 kDa, respectively, and both the subunits have 85% of identical primary amino acid sequence [63]. Activation of CaM-PK II is catalysed by calcium/calmodulin (Ca^{2+}/CaM) and in some instances also by NMDA receptors at synapse [64]. ATP-dependent calcium/calmodulin also initiates autophosphorylation of CaM-PK II, particularly at Thr-286 residue, and causes conversion of Ca^{2+}/CaM-dependent state to Ca^{2+}/CaM-independent one, which is essential for the creation of spatial memory [65, 66]. Significant evidences are available regarding the disruption of this Ca^{2+} homeostasis and dysregulated CaM-PK II in hippocampus in AD patients, and these are the main contributors for the formation of amyloid-β and hyperphosphorylation of tau protein in AD [67].

1.3.3 Protein Kinase A (PKA)

Protein kinase A (PKA) is also known as cyclic AMP (cAMP)-dependent protein kinase because the activation of PKA depends on the cellular levels of cAMP [68]. PKA is a tetrameric holoenzyme, and in the absence of cAMP, it consists of two catalytic (C) subunits and two regulatory (R) subunits. [69]. Activation of PKA by cAMP dissociates this holoenzyme and triggers the translocation of fraction of the C subunit to the nucleus [70]. Generally, two cAMP molecules are required for the activation of single PKA regulatory subunit [71]. cAMP-PKA phosphorylates several serine or threonine residues of tau protein and diminishes the binding affinity of tau protein to the tubulin [72].

Fyn kinase, from protein-tyrosine kinase (Src) family, is also proved to have active role in the phosphorylation of tyrosine18 of tau protein [73, 74], and a number of reports are available claiming increased levels of Fyn kinase in AD [75].

Keeping in mind the unmet need of novel lead molecules for the treatment of AD, researchers are working on one or the other target. Exhaustive literature is available now concerning protein kinases for the design and development of novel anti-Alzheimer's agents. Following sections cover various computational studies reported worldwide for the development of novel small molecules as kinase inhibitors for AD. These studies have been classified on the basis of various kinases used as potential therapeutic targets for the treatment of the AD.

2 Computational Studies on Glycogen Synthase Kinase (GSK) and Its Inhibitors as Potential Anti-Alzheimer's Agents

Meijer et al. reported a semisynthetic derivative of indirubin, 6-bromoindirubin-3′-oxime (6-BIO, **6**) with remarkable selectivity and inhibition capability of GSK-3 [76]. This derivative displayed promising activity in the in vitro studies ($IC_{50} = 0.005\ \mu M$) with inhibition of both GSK-3α and GSK-3β, and it also displayed promising inhibitory activity in the in vivo Xenopus models. Co-crystal of BIO with GSK-3β was grown and analysed (Fig. 3) wherein the ligand

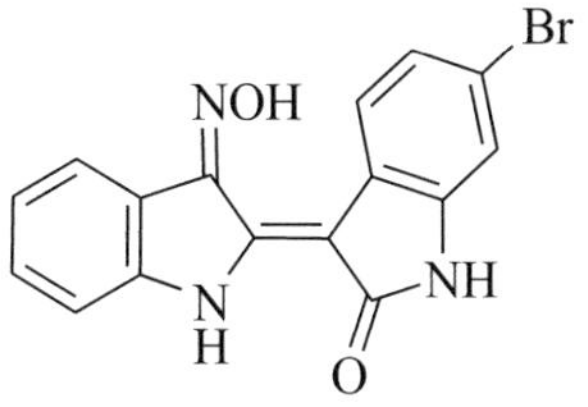

Br
NOH
N
H
NH
O

6

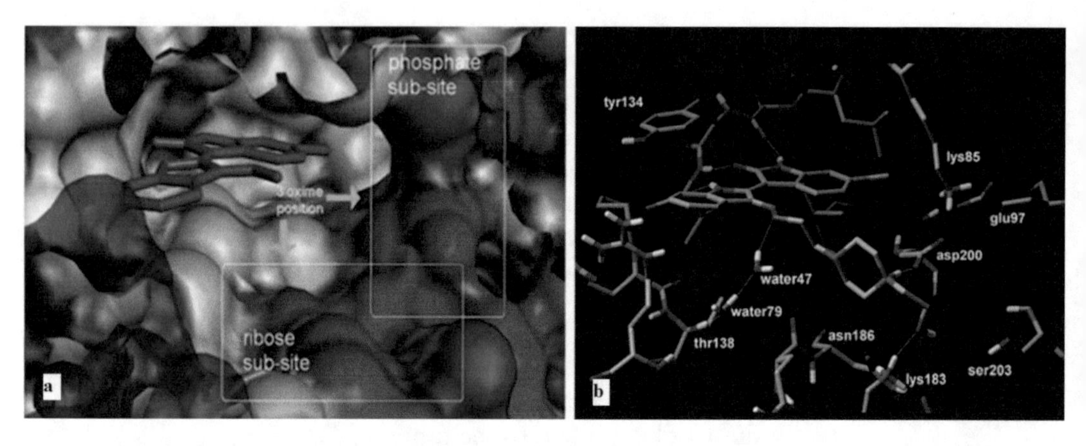

Fig. 3 (**a**) GSK-3β phosphate-binding subsite and ribose subsite, (**b**) co-crystal structure of GSK-3β/6BIO [77] (Used with permission from Vougogiannopoulou et al. J Med Chem 2008, 51, 6421–6431 [copyright© 2008, American Chemical Society])

occupied the activation loop of GSK-3, and it was found to form two hydrogen bonds with Val135 and one hydrogen bond with Asp133. The two hydrogen bonds with Val135 were formed between pyrrole-NH of compound (**6**) and oxygen of valine with a bond length of 2.93 Å and between lactam-oxygen of compound (**6**) and nitrogen of valine with a bond length of 2.42 Å. The interaction between Asp133 of GSK-3β and compound (**6**) was formed by lactam-NH and oxygen of Asp133 with a bond length of 2.90 Å. This complex also showed several van der Waals interactions between Ala83, Pro136, Ile62, Arg141, Val70, Cys199, Leu132, Tyr134, Thre138 and Leu188, making it clear that this molecule had good interaction with the enzyme [76].

Vougogiannopoulou et al. successfully improved the solubility profile of indirubins based on GSK-3/indirubins co-crystal structures [77]. They designed novel analogs of 6-bromoindirubin-3′-oxime (**6**) with an extended amino acid chain at 3′ position which led to increase in hydrophilicity, enhanced selectivity and potent inhibitory activity. Initially, the authors mapped the crystal structure of GSK-3/indirubins and determined the binding mode of **6** in the bound state. It was found that two new regions, the phosphate subsite and the ribose subsite, are interesting portions of GSK-3 where the ATP molecule binds (Fig. 3). These sites can provide polar environment that would help in accommodating favourable electrostatic or hydrogen bond interactions with the potential inhibitors. On the basis of this hypothesis, the authors designed several analogs of **6**. Substitutions were possible only at the 3′-oxime position so as to retain inherent structural properties of **6**. Accordingly, syntheses were carried out to obtain about 26 new derivatives. These were evaluated for their GSK-3 inhibitory activity, and it was found that compounds (2′Z-3′E)-6-

bromoindirubin-3′-O-[2-(4-methylpiperazin-1-yl)ethyl]oxime (**7**) and (2′Z-3′E)-6-bromoindirubin-3′-(O-(2-[4-(2-hydroxyethyl) piperazin-1-yl]ethyl)oxime) (**8**) showed potent GSK-3β inhibitory activity with improved physicochemical properties. Molecular docking studies revealed that the piperazine substituents of compounds (**7**, **8**) interacted with Asp200 and Lys183 located at the phosphate subsite by forming hydrogen bonds (Fig. 3a, b). These observations justified the improvement seen in inhibitory potential of these analogs. Compound (**6**) was further studied in detail by Chaluveelaveedu et al. wherein the docking and ADME studies proved it to be a suitable candidate as GSK-3β inhibitor [78].

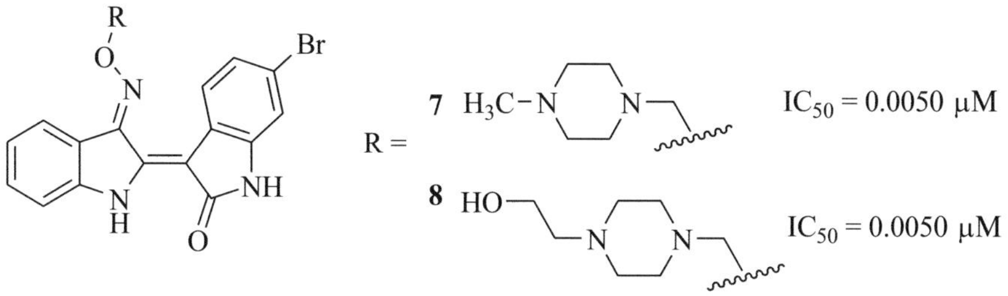

7 and **8**

Voigt et al. reported a series of novel 1-aza-9-oxafluorenes as selective GSK-3β inhibitors for the treatment of Alzheimer's disease [79]. The designed compounds had two main structural features, 4-phenyl and 6-hydroxy functions. These features were considered essential for kinase inhibitory activity. One of the compounds from this series 6-hydroxy-*N*,*N*-dimethyl-4-phenylbenzofuro[2,3-*b*] pyridine-3-carboxamide (**9**) showed the highest GSK-3β selectivity ($K_i = 1.5\ \mu M$) from this set of compounds. Docking studies carried out for the compound (**9**) showed hydrogen bonding between the 6-hydroxy function of the scaffold with the NH group of Cys199 of GSK-3β protein backbone (Fig. 4).

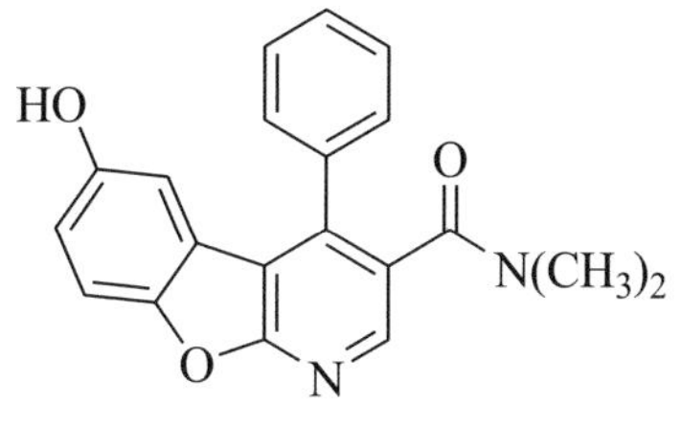

9

Chioua et al. reported a series of 3,6-diamino-1*H*-pyrazolo [3,4-*b*]pyridine derivatives with potent GSK-3β inhibitory activity for their use in the treatment of AD [80]. One of the derivatives from this series, 3,6-diamino-4-phenyl-1*H*-pyrazolo[3,4-*b*]

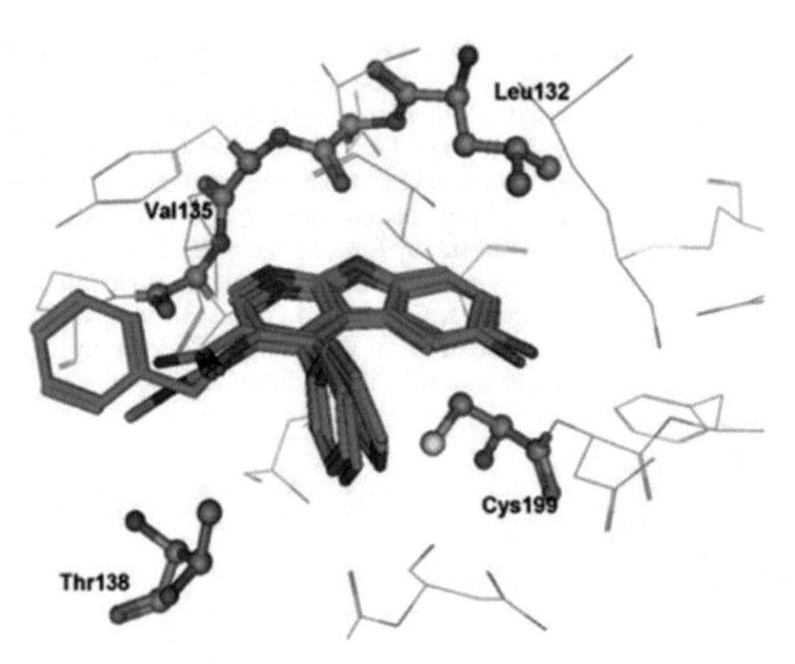

Fig. 4 Docking pose of compound (**9**) within GSK-3β [79] (Used with permission from Voigt et al. ChemMedChem 2008, 3, 120–126 [Copyright © 2007, John Wiley and Sons])

pyridine-5-carbonitrile (**10**), was found to be highly promising (GSK-3 $IC_{50} = 1.5$ μM) and selective inhibitor of GSK-3.

H_2N CN N N H N NH_2

10

The authors have carried out molecular docking studies on this series of compounds and found that in compound (**10**), the 6-amino group on the pyridine ring formed a hydrogen bond with Asp133 of GSK-3 facilitating formation of bonds between the Val135 and NH of pyrazole ring and the nitrogen of the pyridine ring (Fig. 5a). Compound (**10**) is positioned close to the hinge region that allows the 4-phenyl group (Fig. 5b) to get accommodated near the lipophilic residues of GSK-3.

Saithoh et al., in a series of studies, explored oxadiazole moiety for its selective GSK-3 inhibitory activity [81–83]. They started with a high-throughput screening experiment of some proprietary

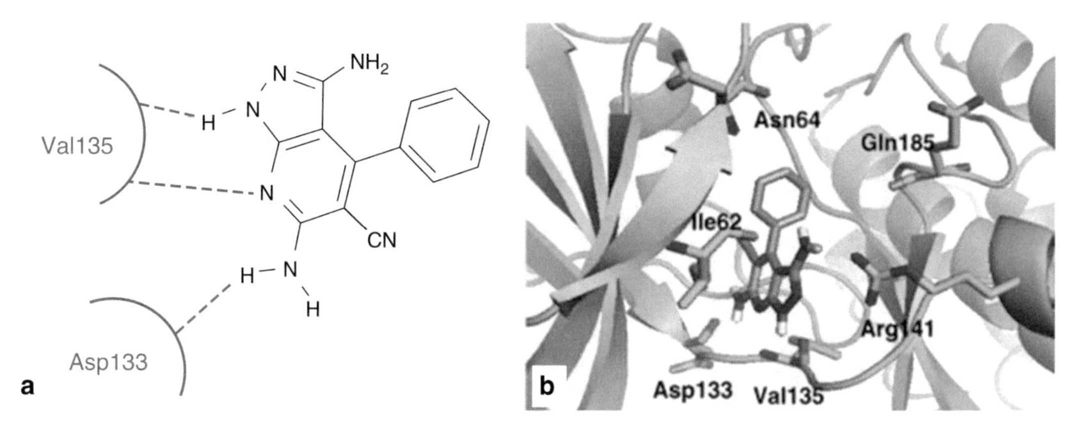

Fig. 5 (**a**) Interaction of compound (**10**) with Val135 and Asp133 residues. (**b**) Position of compound (**10**) near the hinge region of GSK-3β [80] (Used with permission from Chioua et al. Bioorg Med Chem Lett, 2009, 19, 4566–4569 [Copyright© 2009 Elsevier Ltd.])

compounds database and found 2-(3-fluoro-4-methoxybenzylthio)-5-(benzo[*d*][1,3]dioxol-5-yl)-1,3,4-oxadiazole (**11**) as a lead molecule ($IC_{50} = 65$ nM). The crystal structure of this molecule shows interaction with the hydrogen of NH and carbonyl oxygen of Val135 in GSK-3. Other substituents of this molecule were also found to interact with Lys85, Glu97 and Asp200 fragments (Fig. 6).

11

On the basis of these observations, several 1,3,4-oxadizaole derivatives were designed and synthesized [81]. 3-((5-(3-(4-Methoxyphenyl)-3*H*-benzo[*d*]imidazol-5-yl)-1,3,4-oxadiazol-2-ylthio) methyl)benzonitrile (**12**) was found to be the most potent inhibitor with a high degree of selectivity for GSK-3β ($IC_{50} = 2.3$ nM). This compound was further co-crystallized with GSK-3β, but it was not fully characterized (Fig. 7). However, it was found that the

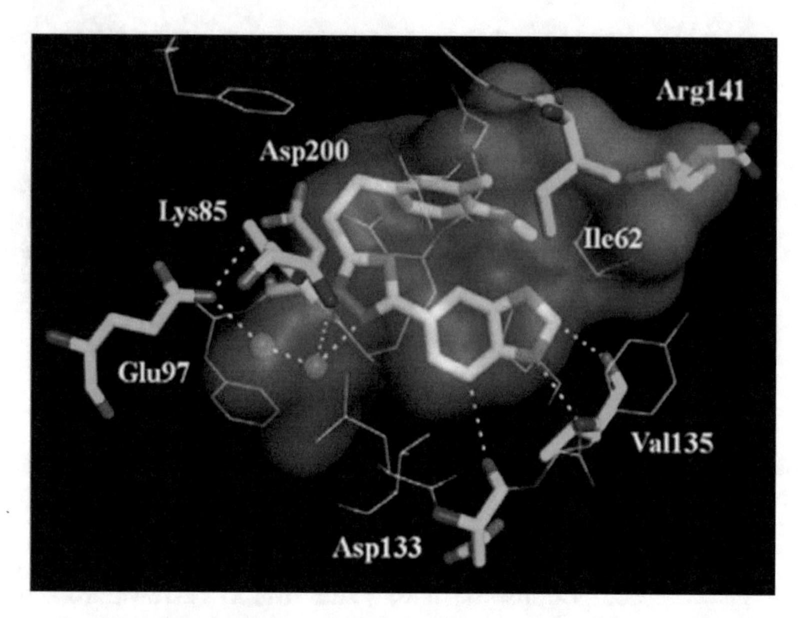

Fig. 6 Co-crystal of compound **(11)** with GSK-3β (PDB Id: 3F7Z) [81] (Used with permission from Saithoh et al. Bioorg Med Chem 2009, 17, 2017–2029 [Copyright© 2009 Elsevier Ltd.])

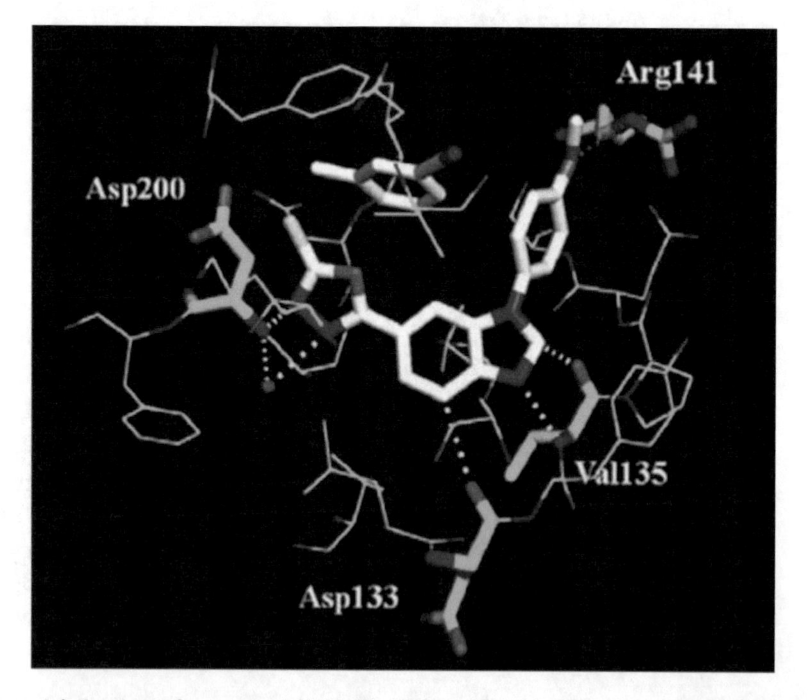

Fig. 7 X-ray crystal structure of compound **(12)** with GSK-3β (PDB Id: 3F88) [81] (Used with permission from Saithoh et al. Bioorg Med Chem 2009, 17, 2017–2029 [Copyright© 2009 Elsevier Ltd.])

benzimidazole core of the molecule interacted with the hinge region and the oxadiazole ring formed hydrogen bond with Asp200. 4-Methoxy group on the N-phenyl

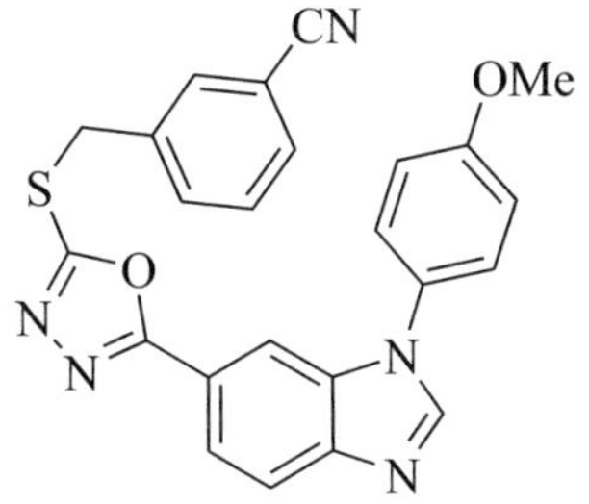

CN
OMe
S
O
N
N
N
N

12

substituent of benzimidzole offered interactions with Arg141 of GSK-3β. Compound (**12**) was found to be highly lipophilic and hence showed poor metabolic stability and bioavailability. The authors explored alternatives of methoxy group interacting with Arg141 to improve the solubility profile of this compound.

In optimization attempts [83], the 1,3,4-oxadizole and the fused heterocyclic ring system of the molecule along with the phenyl ring were conserved. The designed compounds 2-methyl-5-(3-(4-(methylsulfinyl)phenyl)-1-benzofuran-5-yl)-1,3,4-oxadiazole (**13**) and 2-(3-(4-(ethylsulfinyl)phenyl)-1-benzofuran-5-yl)-5-methyl-1,3,4-oxadiazole (**14**) were found to be highly potent GSK-3β inhibitors with an IC_{50} value of 35 nM each. These molecules also showed an improved solubility, metabolic stability and bioavailability and favourable BBB penetration profile. The authors have also studied co-crystal of **13**/GSK-3β (Fig. 8) and observed that the sulfoxide group on phenyl ring of compound (**13**) interacted with Arg141 and the C2 hydrogen atom of benzofuran ring formed hydrogen bond with Val135. The hydrogen atom of C7 also formed hydrogen bond with the carbonyl of Asp133. Thus, the authors successfully optimized the lead molecule (**11**) to obtain compounds (**13** and **14**) with good pharmacological and pharmacokinetic properties using X-ray crystallography.

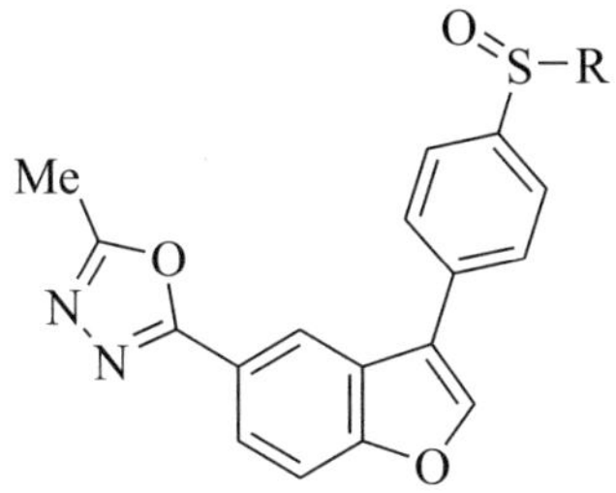

O
S−R
Me
O
N
N
O

13 and **14**

R= **13:** methyl; **14**: ethyl

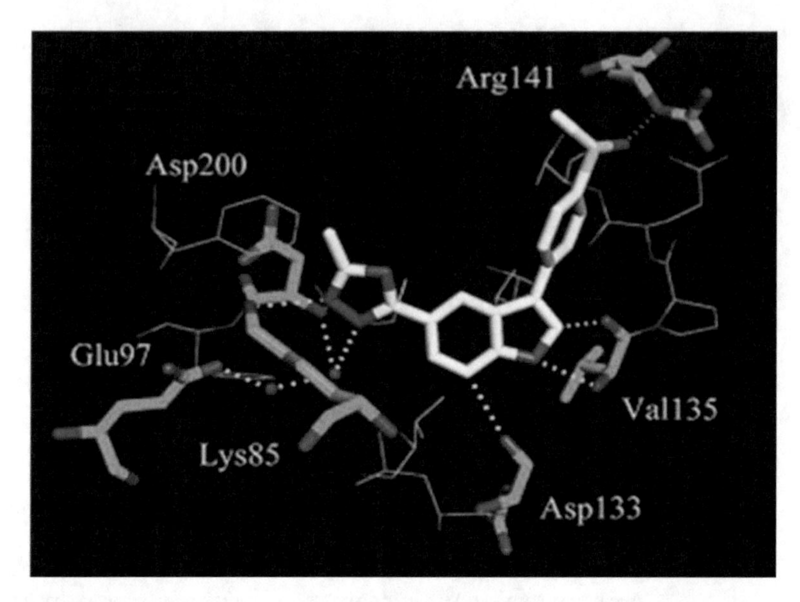

Fig. 8 X-ray co-crystal structure of compound **(13)** and GSK-3β (PDB Id: 3GB2) [82] (Used with permission from Saithoh et al. J Med Chem 2009, 52, 6270–6286 [Copyright © 2009, American Chemical Society])

The authors further studied the pharmacological activities of compound (**13**) on GSK-3 in detail, as it inhibited tau protein phosphorylation in the in vitro and in vivo models [83]. Compound (**13**) significantly decreased hippocampal tau protein phosphorylation at the GSK-3 sites in a transgenic mouse model. This compound significantly improved the performance of transgenic AD mice in the behavioural studies assessed with the help of Y-maze test and object recognition tests. These results adequately prove the potential of compound (**13**) as a drug candidate for AD treatment [82].

Perez et al. reported a series of thienylhalomethylketones as the first irreversible inhibitors of GSK-3β. These compounds were found to bind covalently with the cysteine residue present in the ATP-binding site [84]. 2-Chloro-1-(4,5-dibromothiophen-2-yl) ethanone (**15**) was found to possess significant GSK-3β inhibitory activity (IC_{50} = 1 μM) with high selectivity. Molecular docking studies of this series of compounds suggested that compound (**15**) was bound in the ATP-binding site of the GSK-3β. It fits into the hydrophobic cavity of the receptor which is delineated by hydrophobic residues. The irreversible competitive nature of this compound is supposed to be due to the formation of bond between the

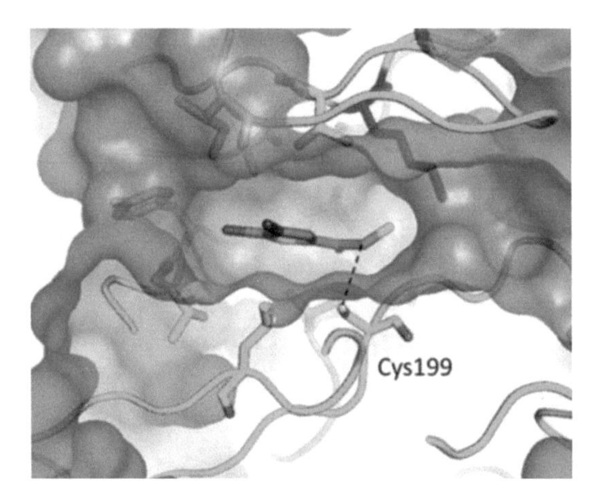

Fig. 9 Molecular docking pose of compound (**15**) in the ATP-binding site of GSK-3β [84] (Used with permission from Perez et al. Bioorg Med Chem 2009, 17, 6914–6925 [Copyright© 2009 Elsevier Ltd.])

carbon atom bearing the chloro-group and the sulphur atom of Cys199 (Fig. 9), whereby the chloro-group remains in proximity to the amino group of Lys85.

15

Monte et al. reported the designing of several selective GSK-3 inhibitors on the basis of a highly selective known GSK-3 inhibitor AR-A014418 (AstraZeneca, **16**) [85]. On the basis of the structure of this molecule and molecular docking studies, the authors designed and synthesized several benzylureas possessing GSK-3β inhibitory activity. Among the reported series, compound 1-(6-(1*H*-tetrazol-5-yl)benzo[*d*]thiazol-2-yl)-3-(4-methoxybenzyl) urea (**17**) was found to be the most potent with 69% and 18% GSK-3β inhibition at 10 and 1 μM concentrations, respectively. When compound (**17**) was subjected to molecular docking studies with GSK-3β

16

17

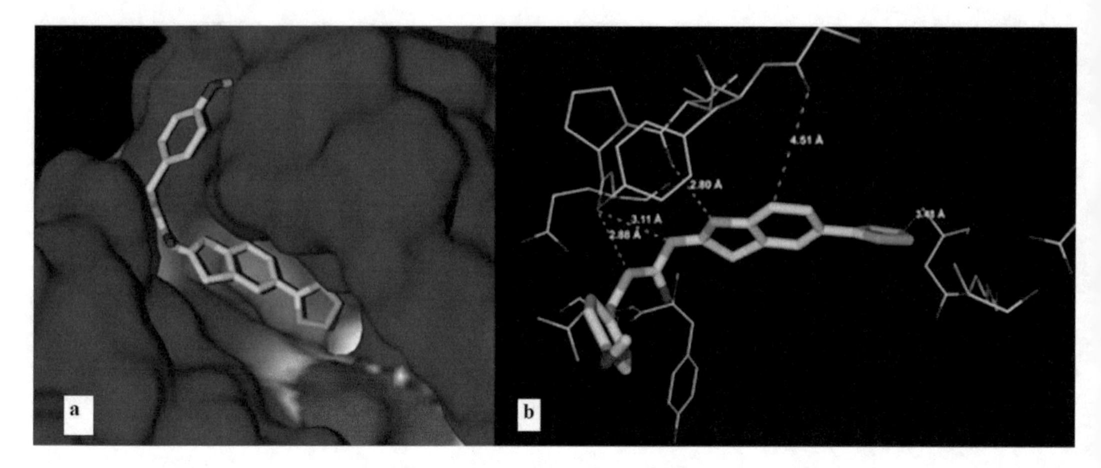

Fig. 10 (**a**) Binding mode of compound (**17**), (**b**) Molecular interactions between compound (**17**) and GSK-3β [85] (Used with permission from Monte et al. Bioorg Med Chem Lett 2011, 21, 5610–5615 [Copyright© 2011 Elsevier Ltd.])

(on PDB structure 1Q5K), it was found to bind to the ATP-binding pocket in a fashion similar to AR-A014418 (**16**). Two hydrogen bonds were observed between compound (**17**) and Glu137 of GSK-3β. Other interactions were observed between carbonyl of Asp133 and a hydrogen of phenyl group and tetrazole moiety and a polar pocket consisting of Lys85, Glu97 and Asp200, which could be contributing to the potency of this compound (Fig. 10). This group of researchers further reported several novel scorpion shaped GSK-3 inhibitors with noteworthy GSK-3α selectivity [86]. Two compounds from this series, 4′-((5-(benzo[*d*][1,3]dioxol-5-yl)-1,3,4-oxadiazol-2-ylthio)-methyl)biphenyl-4-carbonitrile (**18**) and 4′-((5-(2,3-dihydrobenzo[*b*][1,4]dioxin-6-yl)-1,3,4-oxadiazol-2-ylthio)methyl)-4-fluorobiphe- nyl-2-carbonitrile (**19**) were found to be highly potent and selective inhibitors. Compound (**18**) displayed IC_{50} values of 0.006 and 0.316 μM for GSK-3α and β, respectively, whereas compound (**19**) offered the IC_{50} values of 0.002 and 0.185 μM for GSK-3α and β, respectively.

18 **19**

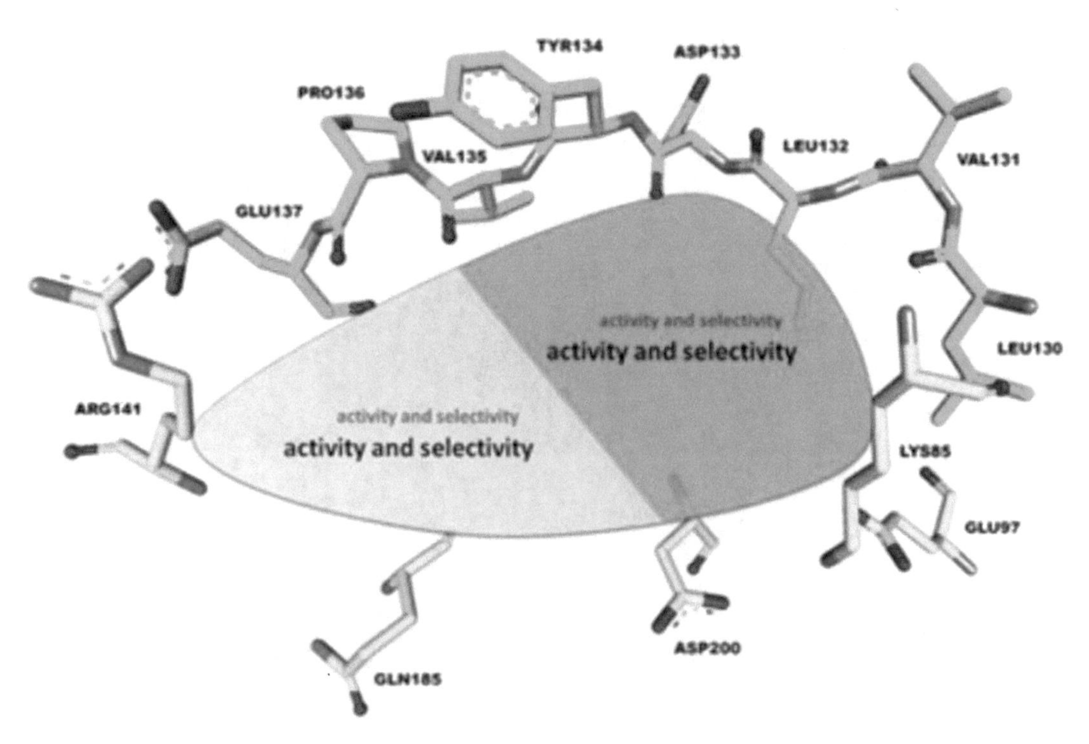

Fig. 11 Schematic overview of the GSK-3 ATP-binding pocket. The green marked area reveals the interaction site of the heterocycle and the oxadiazole ring. The area authors wanted to occupy by the their structures is marked light blue [87] (Used with permission from Monte et al. Eur J Med Chem 2013, 61, 26–40 [Copyright© 2013 Elsevier Masson SAS])

The authors targeted the ATP-binding region of GSK-3 (Fig. 11) as per the developed hypothesis [87]. Based on prior knowledge, they designed a series of oxadiazoles as potent and selective GSK-3 inhibitors. Two of the most potent compounds, 4′-((5-(benzo[*d*][1,3]dioxol-5-yl)-1,3,4-oxadiazol-2-ylthio) methyl)-4-methoxybiphenyl-2-carbonitrile (**20**) showed an IC_{50} value of 2 nM for GSK-3α and 17 nM for GSK-3β, whereas compound *N*-(4-(5-((2-cyanobiphenyl-4-yl)methylthio)-1,3,4-oxadiazol-2-yl)pyridin-2-yl) acetamide (**21**) showed an IC_{50} of 35 nM for GSK-3α.

20 **21**

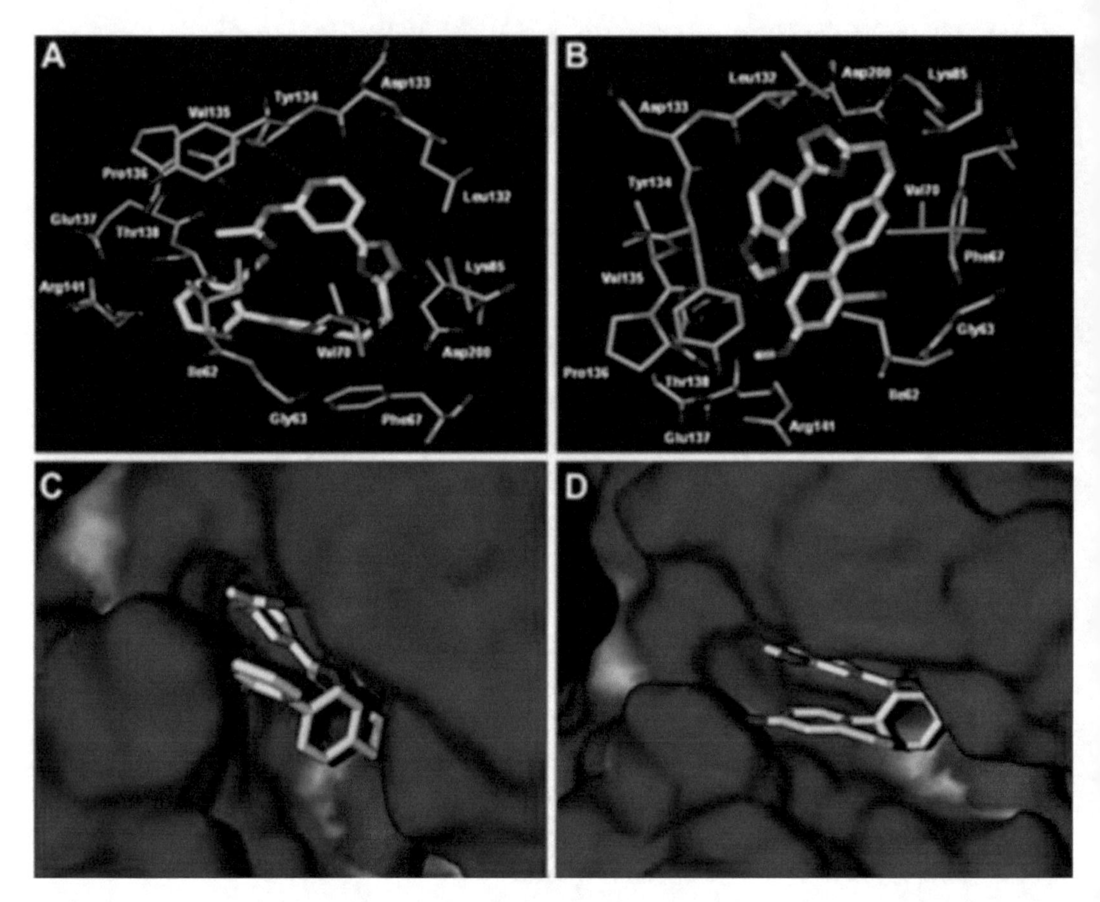

Fig. 12 Docking of compounds (**20** and **21**) into PDB structure 3F88 of GSK-3b; hydrogen bond interactions of compounds **21** (**a**) and **22** (**b**) with the amino acids of the ATP-binding pocket; surface illustration of the ATP-binding pocket with compound **21** (**c**) and **20** (**d**). The inhibitors are shown in *yellow* [87] (Used with permission from Monte et al. Eur J Med Chem 2013, 61, 26–40 [Copyright© 2013 Elsevier Masson SAS])

Compounds (**20** and **21**) were docked into 3F88 structure of GSK-3β, and this study disclosed that these compounds occupied the ATP-binding sites of the receptor (Fig. 12). The oxadiazole ring interacts with the polar pocket of the receptor consisting of Lys85 and Asp200. In case of compound (**20**), the amide functional group binds with the hinge region of the ATP-binding region; biphenyl system makes interactions with Ile62, Gly63, Phe67 and Val70, the glycine-rich loop of GSK-3. Figure 12b also shows that methoxy group of compound (**20**) is located near the salt bridge formed by residues Glu137 and Arg141, contributing to the activity of this compound. Due to the lack of co-crystals of GSK-3α, docking studies were carried out on GSK-3β co-crystals. As both the isoforms of GSK-3 have 98% structural identity, these observations on GSK-3β may be applicable to GSK-3α as well.

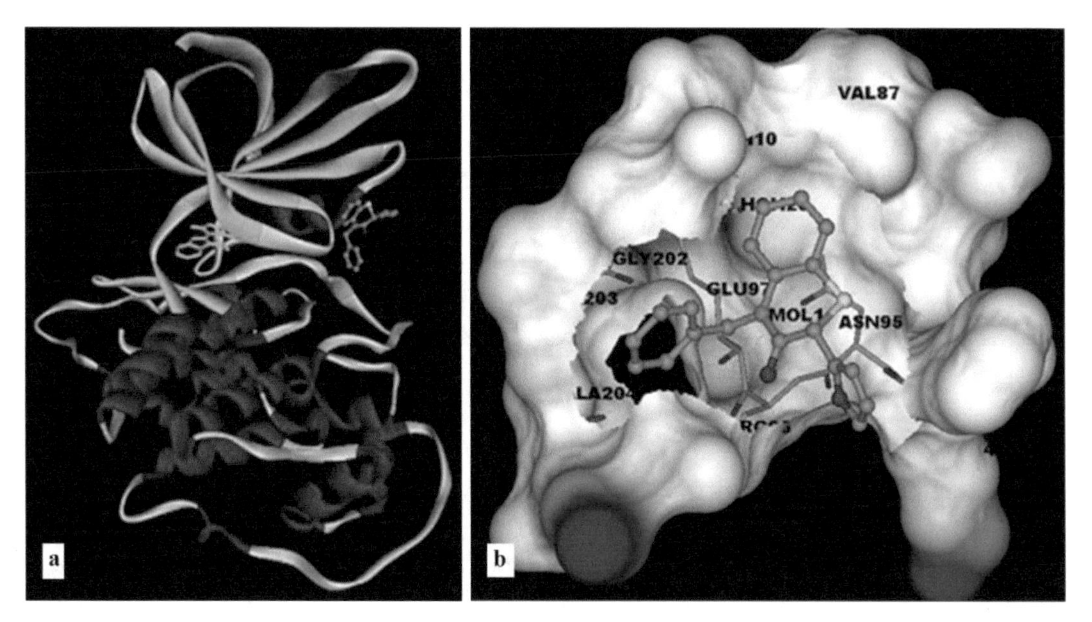

Fig. 13 (**a**) Molecular docking of compound (**22**), (**b**) binding of compound (**22**) in the ATP non-competitive site [88] (Used with permission from Zhang et al. Bioorg Med Chem Lett 2012, 22, 7232–7236 [Copyright© 2012 Elsevier Ltd.])

Zhang et al. identified a novel scaffold of benzothiazepinones as non-ATP-competitive GSK-3β inhibitors through virtual screening [88]. To accomplish this work, the authors screened the Maybridge database using AutoDock programme, and from this screening benzothiazepinones were found to be the most efficient compounds. These benzothiazepinones were synthesized and screened for their GSK-3β inhibitory activity. It was found that compound (**22**) showed significant activity and selectivity to GSK-3β with an IC_{50} value of 77 μM. Molecular docking studies were performed using AutoDock software on this compound with compound (**6**) already attached to the receptor. The authors did not remove this ATP-competitive inhibitor to study the action and binding site of compound (**22**) (Fig. 13). The docking simulation showed that compound (**22**) docked with energy of −7.82 kcal mol^{-1} in a cavity consisting of Phe67, Phe93, Arg96 and Lys205.

However, in this report the authors did not validate the docked complex on the basis of molecular dynamics simulation for complex stability.

22

Zhang et al. further investigated and reported a compound, 2,3-dihydro-2-benzyl-5-(2-nitrobenzyl)-1,5-benzothiazepin-4 (5*H*)-one (**23**) as a non-ATP-competitive inhibitor which effectively inhibited GSK-3β at 23.0 μM with high selectivity in comparison to other 13 protein kinases [89]. Molecular docking studies on compound (**23**) show that it binds to non-ATP-binding pocket of GSK-3β (Fig. 14). Its affinity is attributed to the hydrogen bonds

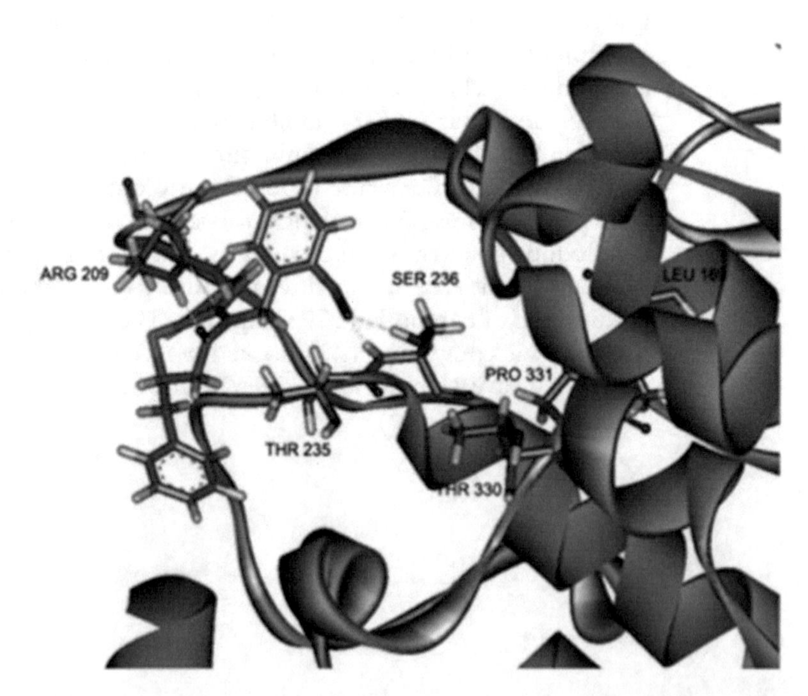

Fig. 14 Suggested binding mode for compound (**23**) in GSK-3b (PDB ID: 1PYX) [89] (Used with permission from Zhang et al. Eur J Med Chem 2013, 61, 95–103 [Copyright© 2013 Elsevier Masson SAS])

formed by carbonyl and nitro groups of **23** with Arg209 and Ser236, thus making it a selective and reversible inhibitor of GSK-3β.

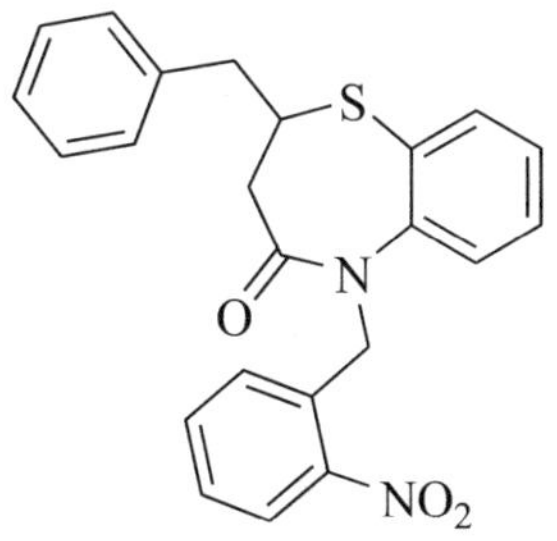

23

Berg et al. reported the synthesis of some novel pyrazines as highly selective GSK-3β inhibitors for AD [90]. One of the compounds from this series 3-amino-6-(4-(4-methylpiperazin-1-ylsulfonyl)phenyl)-*N*-(pyridin-3-yl)pyrazine-2-carboxamide (**24**) offered an IC_{50} value of 4.9 nM for GSK-3β. It had an IC_{50} value of 540 nM for CDK2 which made it 317-fold more selective for GSK-3β. Compound (**24**) was further studied on the basis of X-ray crystallography to determine its binding mode. Co-crystal structure of compound (**24**) showed that it was binding within the ATP-binding site and formed interactions with the active site residues. The nitrogen of pyridine forms bond with Lys85, and the amine group on pyrazine and the adjoining nitrogen show interactions with Asp133 and Val135, respectively (Fig. 15).

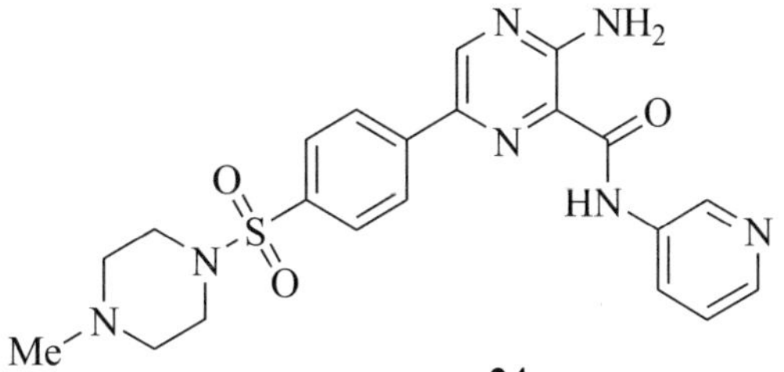

24

In an attempt to develop selective GSK-3β inhibitors, Uehara et al. carried out an in silico screening of their in-house chemical library [91]. Hits obtained from this screening were further subjected to cell-free enzyme inhibition assay resulting into a hit compound, 2-methyl-6-(pyridin-4-yl)pyrimidin-4(3*H*)-one (**25**). The authors performed molecular docking of this compound with GSK-3β and found that it interacted with residues of the ATP-binding site like Asp200, Lys85 and Tyr134. On this basis, several new compounds were designed and tested for their inhibitory activity on GSK-3β. One of the compounds from this series, 3-methyl-2-((2-oxo-2-phenylethyl)amino)-6-(pyridin-4-yl)pyrimidin-4(3*H*)-

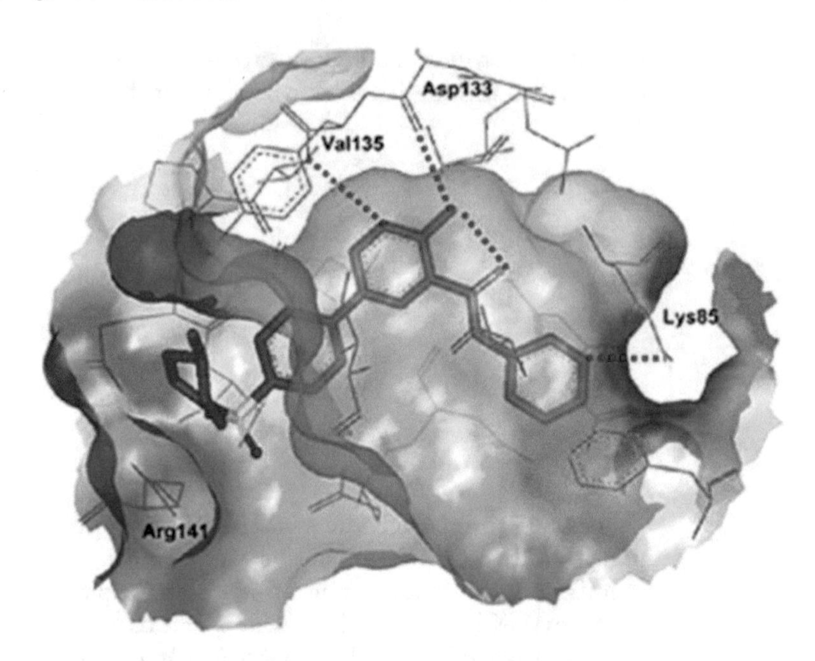

Fig. 15 Top view of X-ray crystal structure of compound (**24**) in the GSK3β ATPsite [90] (Used with permission from Berg et al. J Med Chem 2012, 55, 9107–9119 [Copyright© 2012, American Chemical Society])

one (**26**) was found to be highly active and selective GSK-3β inhibitor with an IC_{50} value of 8.9 nM. This compound showed time-bound decrease in tau protein phosphorylation in rats at a dose of 100 mg/kg, ip and good brain permeability. Molecular docking studies of compound (**26**) revealed that it was bound to the ATP-binding site through hydrogen bond formation with Lys85 and Asp200, similar to the hit compound (**25**) (Fig. 16). The newly introduced phenacyl group on the second position of pyrimidine ring was disposed near to Arg141, which was considered important for the GSK-3β inhibitory activity.

Optimisation

25
Hit compound

26

La Pietra et al. reported a set of 1-phenylpyrazolo[3,4-*e*]pyrrolo[3,4-*g*]indolizine-4,6(1*H*,5*H*)-diones as new GSK-3β inhibitors [92]. Compound 1-(2′,4′-dichlorophenyl)-3-methylpyrazolo

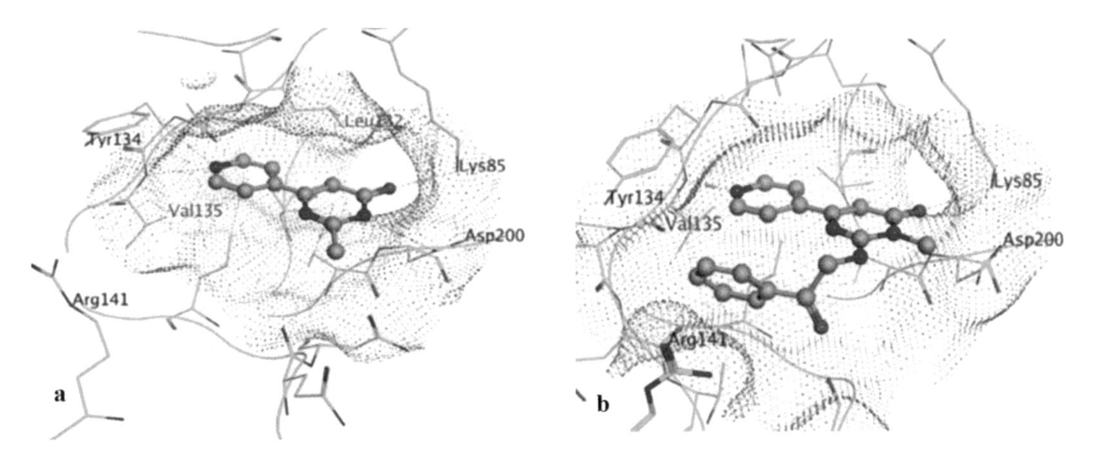

Fig. 16 (**a**) Docking study of compound (**25**) with GSK-3β. (**b**) Docking of compound (**26**) with GSK-3β [91] (Used with permission from Uehara et al. Bioorg Med Chem Lett 2013, 23, 6928–6932 [Copyright© 2013 Elsevier Ltd.])

[3,4-*e*]pyrrolo[3,4-*g*]indolizine-4,6(1*H*,5*H*)-dione (**27**) was found to be a highly potent and selective inhibitor with an IC_{50} value of 0.24 μM that inhibited tau protein phosphorylation in cell-based assay. This compound was highly selective towards GSK-3 in a panel of 17 protein kinases and displayed>ten fold higher selectivity against CDK2. It diffuses passively across the blood-brain barrier as per the calculated physicochemical properties and Volsurf predictions. Molecular docking studies performed using Glide 5.5. Programme on Maestro 9.0.211 revealed that methyl group at C7 position was inserted in the N-lobe pocket, where it formed hydrophobic interactions with Leu132, Met101, Val70 and Lys85 (Fig. 17). The two electron-withdrawing chloro-groups near the C-lobe residues boost the hydrophobic contacts.

27

Ye et al. reported synthesis of 3-benzisoxazolyl-4-indolylmaleimides as potent and selective GSK-3β inhibitors [93]. One of the most potent compounds 3-(1-(3-(1*H*-imidazol-1-yl)propyl)-1*H*-indol-3-yl)-4-(benzo[*d*]isoxazol-3-yl)-1*H*-pyrrole-2,5-dione (**28**) (IC_{50} = 0.73 nM) was found to be highly selective for GSK-3β

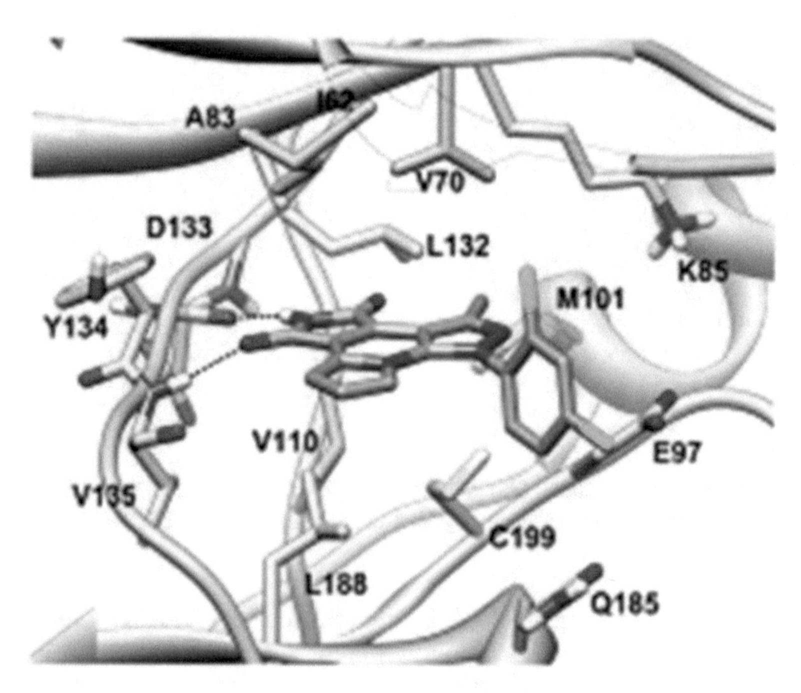

Fig. 17 Putative binding pose of compound (**27**) in the active site of GSK-3β shown as a cornflower stick and tan cartoon, respectively. Interacting residues are shown as sticks and are coloured by atom type, whereas the H-bond is represented by a *black dashed line* [92] (Used with permission from La Pietra et al. J Med Chem 2013, 56, 10066–10078 [Copyright© 2013, American Chemical Society])

over other kinases such as CDK2. It reduced the Aβ-induced tau hyperphosphorylation by inhibiting GSK-3β in a cell-based functional assay. Molecular docking was carried out on C-Docker programme of Discovery Studio 2.1 with PDB Id: 1Q3D for GSK-3β. This compound binds to the ATP-binding site where NH and carbonyl groups of maleimide ring form two hydrogen bonds with Asp133 and Val135 of GSK-3β. The 3-position nitrogen atom of the imidazole ring forms another hydrogen bond with Lys-183. Compound (**28**) showed dock energy of −20.821 Kcl/mol which was the best in the described series of compounds (Fig. 18). This compound (**28**) was further studied in detail by

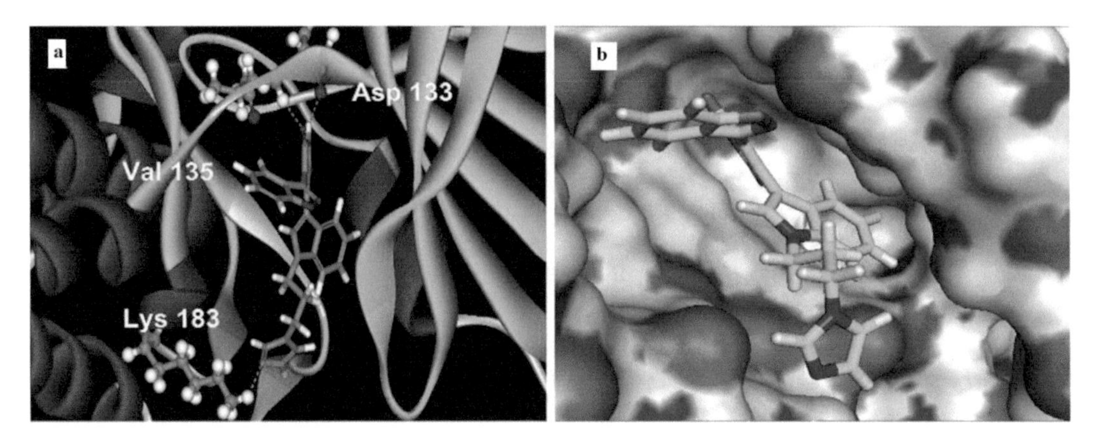

Fig. 18 Docking of compound (**28**) into GSK-3β crystal structure. (**a**) Ribbon show of compound (**28**) bound to GSK-3β. (**b**) Surface show of compound (**28**) docked into GSK-3β [93])

Pang et al. and found it to be a novel neuroprotective for ischemic stroke as it increased Nrf2 and antioxidative signalling [94].

28

Sivaprakasam et al. reported several acylaminopyridines on the basis of structure-guided design of pyrrolopyridinone core [95]. Several of these compounds showed GSK-3β inhibition at sub-nanomolar concentrations. These compounds were developed on the basis of molecular docking studies of publically available data bases. The hits so obtained with acylaminopyridines and hydrophobic core like the pyrrolopyridinones were optimized to get several scaffolds. Compound 2-(2-(cyclopropanecarboxamido)pyridin-4-yl)-4-methoxythiazole-5-carboxamide (**29**) showed an IC_{50} value of 1.1 nM against GSK-3β, which was one of the best potencies. The co-crystal structure of this compound shows hydrogen bond

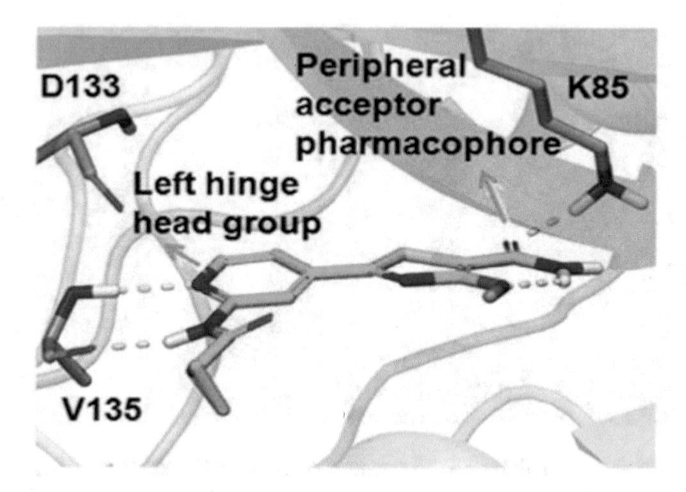

Fig. 19 X-ray co-crystal structure of compound (**29**) in the kinase domain of GSK-3b. For clarity the non-polar hydrogen atoms of the ligands are not shown. Only the ATP-binding site where the ligands engage in critical hydrogen bonding is shown. The ligand is rendered as a stick model in *green* carbons, and hydrogen bonds are shown as *yellow dotted lines*. The methoxy group in **29** is hand modelled. Figures are made using pymol [95] (Used with permission from Sivaprakasam et al. Bioorg Med Chem Lett 2015, 25, 1856–1863 [Copyright © 2015, Elsevier Ltd.])

between nitrogen of pyridine ring and Val135, while the oxygen of carboxamide forms another hydrogen bond with Lys85 of the ATP-binding site of the GSK-3β (Fig. 19).

29

Yue et al. reported benzoisoindole-1,3-diones as selective GSK-3β inhibitors on the basis of structure-based design that activated Wnt/β-catenin pathway [96]. The authors improved the binding interactions of an already known GSK-3 inhibitor (**30**) with an IC_{50} value of 1.01 μM by structural modifications, as evident from the docking studies (Fig. 20).

The authors developed several molecules on the basis of this hypothesis and evaluated these compounds using Kinase-Glo Luminescent Kinase Assay. One of these compounds, 7,8-dimethoxy-5-((1-(methylsulfonyl)piperidin-4-ylamino)methyl)-1*H*-benzo[*e*]isoindole-1,3-(2*H*)-dione (**31**, R = Me) was found to be highly active at sub-micromolar concentration (IC_{50} value of 0.34 μM), without showing any significant inhibition of CDK2 (3.4% inhibition at 100 μM). Molecular docking studies revealed

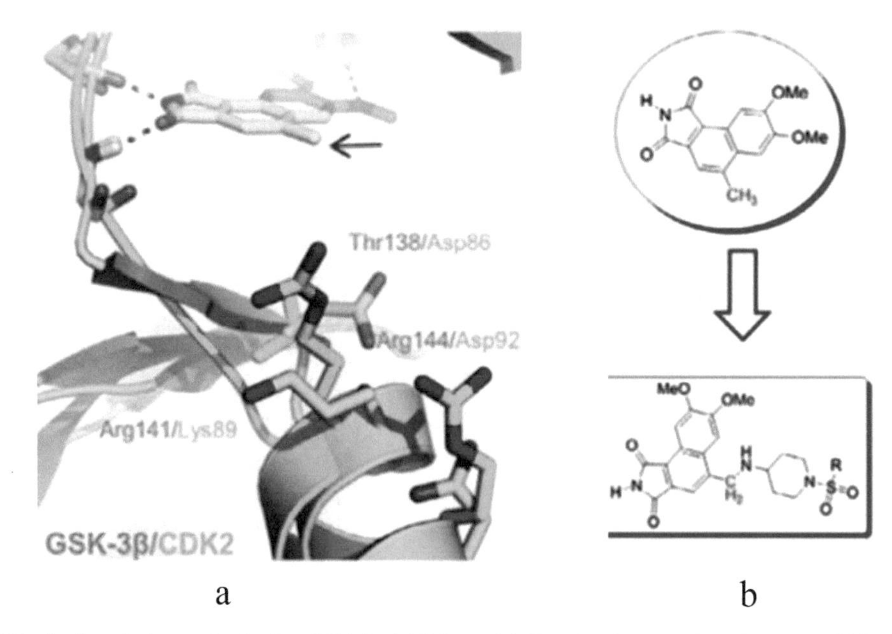

Fig. 20 Designing strategy of selective GSK-3β inhibitors. (**a**) Structural comparison between GSK-3b (*green*) and CDK2 (*cyan*), different residues at the helix D were shown in stick representation. (**b**) Chemical modifications of the parent compound (**30**) to selective GSK-3β inhibitors (**31**) [96] (Used with permission from Yue et al. Bioorg Chem 2015, 61, 21–27 [Copyright © 2015, Elsevier Inc.])

hydrogen bonding between oxygen of sulfonyl moiety and Arg141 and Arg144 of the GSK-3β (Fig. 21). The authors proved the use of structure-based design by reducing the IC_{50} of the reference molecule (**30**) ($IC_{50} = 1.01$ μM) to the designed final derivative (**31**) ($IC_{50} = 0.34$ μM).

OMe OMe O HN O Me **30**

OMe OMe O HN O H N N–S(=O)(=O)–R **31**

A novel class of 1*H*-indazole-3-carboxamides was discovered on the basis of structure-based discovery by Ombrato et al. as GSK-3 inhibitors [31]. They carried out virtual screening of about 8000 compounds from the Angelini library out of which eight compounds were selected for molecular docking studies, in vitro activity and X-ray crystallographic studies. One of the compounds *N*-((1-(4-hydroxyphenethyl)piperidin-4-yl)methyl)-5-methoxy-1*H*-indazole-3-carboxamide (**32**) offered the best IC_{50} value of 0.35 μM. Molecular docking studies were performed on Glide module of Schrödinger molecular simulation package, which showed that

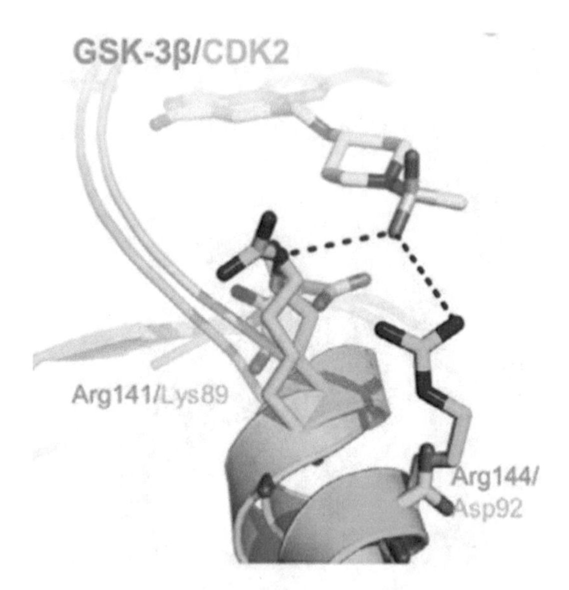

Fig. 21 Docked structure of (**31**) with GSK-3b (*green*), CDK2 (*cyan*) is also shown for analysis [96] (Used with permission from Yue et al. Bioorg Chem 2015, 61, 21–27 [Copyright© 2015, Elsevier Inc.])

methoxy group at position 5 of the indazole ring was directed towards the ribose-binding site and partially filled it but did not form a bond with Lys85. The first nitrogen of indazole ring forms hydrogen bond with Asp133, nitrogen of the amide linkage forms a hydrogen bond with Val135 residue and the hydroxy group of terminal phenyl ring forms a hydrogen bond with Gln185 (Fig. 22). These compounds bind to the ATP-binding site of GSK-3β and show promising inhibitory activity in micromolar range.

32

Di Martino et al. reported discovery of several curcumin analogs as GSK-3β and BACE-1 dual inhibitors [96a]. Curcumin exhibits the potential to inhibit both the targets at higher concentrations (BACE1: IC_{50} = 343 μM; GSK-3β: IC_{50} = 17.95 μM). The authors designed and tested several analogs of curcumin, one of the derivatives from the series (1E,4Z,6E)-1-(4-(benzyloxy)phenyl)-5-hydroxy-7-(4-hydroxy-3-methoxyphenyl)-hepta-1,4,6-trien-3-one (**33**) exhibited potent dual activity (BACE1: IC_{50} = 0.97 μM; GSK-3β: IC_{50} = 0.90 μM). When curcumin was

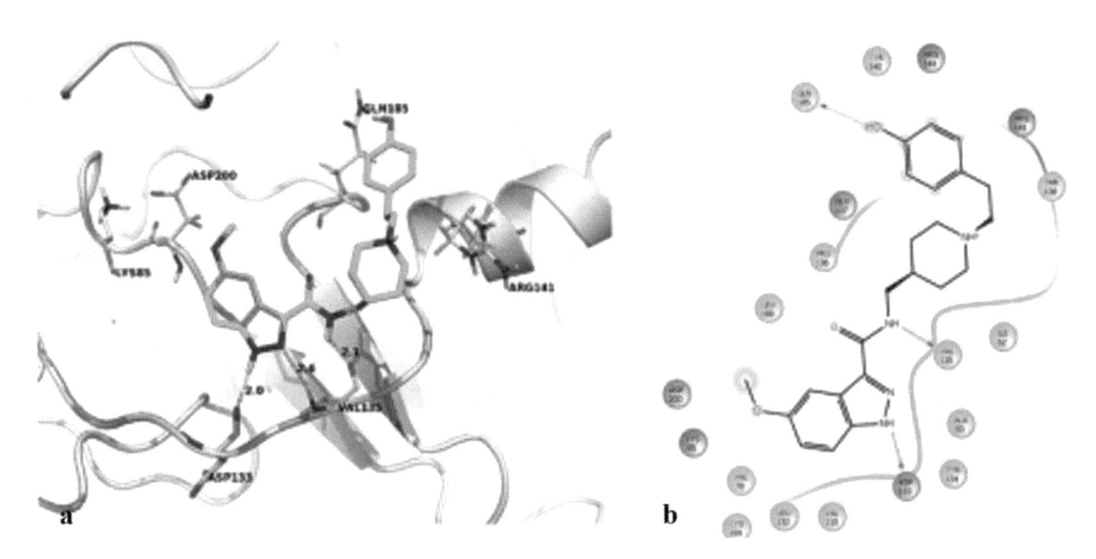

Fig. 22 (**a**) Predicted binding modes of six hit compounds into PDB entry 1Q3D and (**b**) receptor-ligand (**32**) interaction representation [31] (Used with permission from Ombrato et al. J Chem Info Model 2015, 55, 2540–2551 [Copyright© 2015, American Chemical Society])

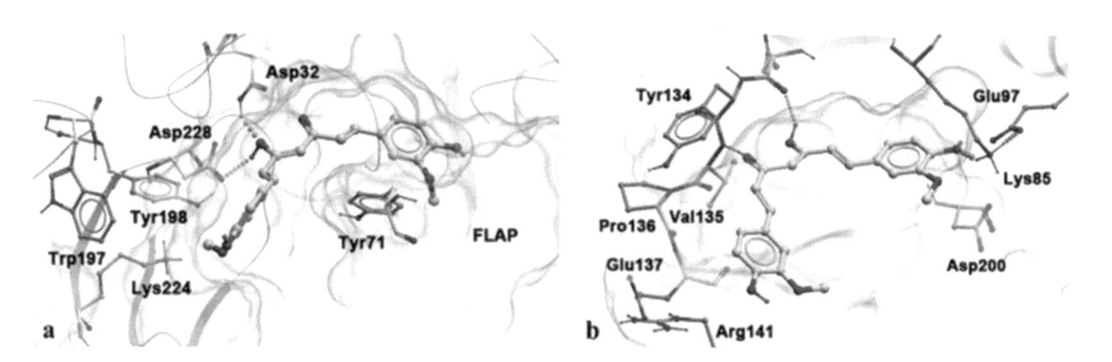

Fig. 23 Curcumin docked into the catalytic region of (**a**) BACE-1 and (**b**) GSK-3β [31] (Used with permission from Di Martino et al. J Med Chem 2016, 59, 531–544 [Copyright© 2016, American Chemical Society])

docked with BACE-1 and GSK-3β, the hydroxy function on carbon 12 of curcumin showed hydrogen bonding with Asp32 and Asp228 residues. Docking of curcumin with GSK-3β showed hydrogen bond interactions between hydroxy and methoxy functions on 22 and 23 carbons with Glu97 and Asp200 (Fig. 23). Molecular docking studies of compound (**33**) showed that BACE-1 formed hydrogen bonds between hydroxy function on carbon 12 with Asp32 and Asp228. In case of GSK-3β, hydrogen bonds were observed between hydroxyl function of phenyl ring with Asp200 (Fig. 24). It indicates that the interactions between BACE1 and curcumin

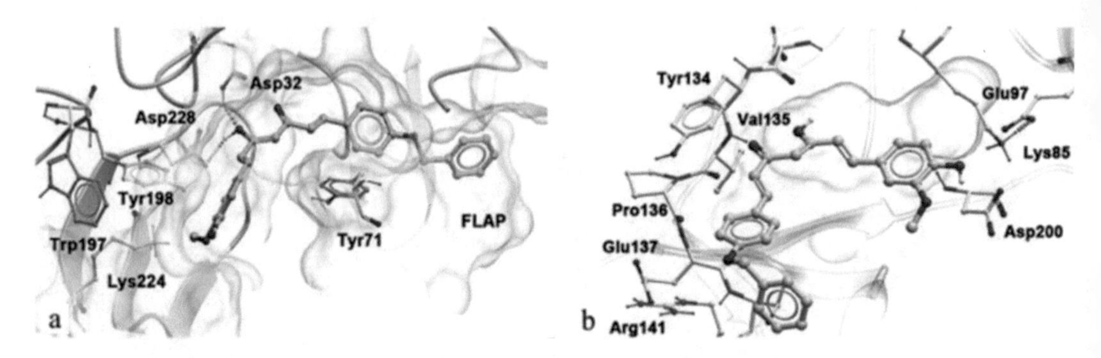

Fig. 24 The predicted bound conformation of compound (**33**) at (**a**) the binding site of BACE-1 and at (**b**) the binding site of GSK-3β [*97* supplementary material] (Used with permission from Di Martino et al. J Med Chem 2016, 59, 531–544 [Copyright© 2016, American Chemical Society])

analogs (e.g. **33**) are the same as observed in curcumin but it differs in case of GSK-3β.

33

Lin et al. performed virtual screening of a database consisting of 1.1 million compounds from zinc and in-house library [97]. A few compounds were found to possess the ability to reduce tau protein aggregation via GSK-3β inhibition. Compound (5E, 8E, 11E, 14E)-*N*-(4-hydroxyphenyl)icosa-5,8,11,14-tetraenamide (**34**), when studied in Tet-On ΔK280 and tau_{RD}-DsRed SH-SY5Y cells, enhanced HSPB1 and GRP78 resulting into neurite outgrowth. Thus, compound (**34**) inhibits GSK-3β activity and reduces tau protein aggregation. Molecular docking studies of **34** revealed that hydroxy function of phenyl ring formed a hydrogen bond with Asp133 of the ATP-binding hinge region of GSK-3β (Fig. 25). This inhibitor significantly varies in structure from the earlier reported compounds due to its long non-polar tail and polar head.

34

Liang et al. reported isolation and identification of a 6-C-glycosylflavone, isoorientin (**35**) as selective inhibitor of GSK-3β in in vitro model with IC_{50} value of 185 μM [98]. Enzyme kinetics

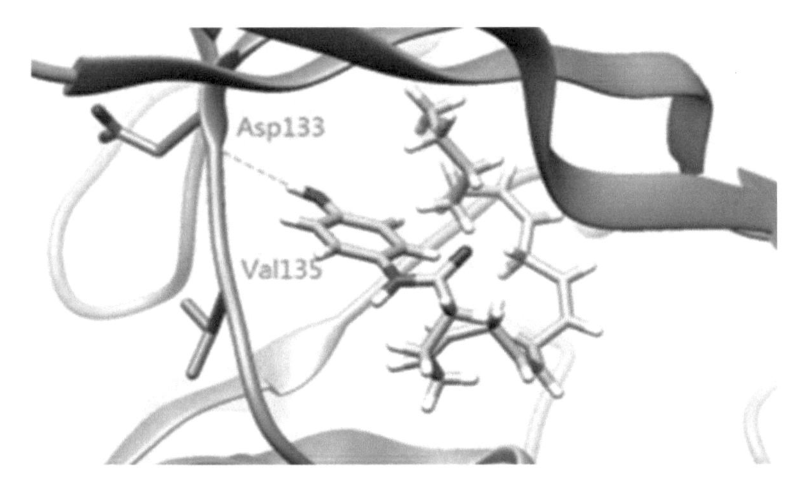

Fig. 25 Docked conformation of compound (**34**). Hydrogen bonds are shown in dashed line [97] (Used with permission from Lin et al. Eur J Pharm Sci 2016, 89, 11–19 [Copyright© 2016 Elsevier B.V.])

assay demonstrated that compound (**35**) specifically inhibited GSK-3β via ATP non-competitive mechanism. It attenuated GSK-3β mediated tau protein phosphorylation and was neuroprotective against β-amyloid-induced tau hyperphosphorylation and neurotoxicity in SH-SY5Y cells. Molecular docking studies suggested that this compound was binding in the substrate-binding site of GSK-3β which was an allosteric cavity present immediately near the ATP-binding site. The C-glycogen of **35** makes a polar interaction with Leu88, Gln89, Asp90, Lys94 and Arg96 forming hydrogen bonds. The 7-hydroxyl oxygen atom on A-ring of the flavone core interacted with Asn95 by forming a hydrogen bond. The 3′-hydroxy oxygen atom on the phenyl of B-ring potentially stabilizes these binding interactions. The Mg^{2+}cofactor causes weakening of binding affinity of ATP with GSK-3β. Interactions of compound (**35**) with key residues of GSK-3β are explained for its activity and selectivity (Fig. 26).

35

Arfeen et al. reported fragment-based design, synthesis and evaluation of iminothiazolidin-4-one derivatives as selective GSK-3β inhibitors [99]. Compound 5-(4-methoxy-3-nitrobenzylidene)-2-(benzylimino)thiazolidin-4-one (**36**) was found to be a highly potent and selective inhibitor with an IC_{50} value of

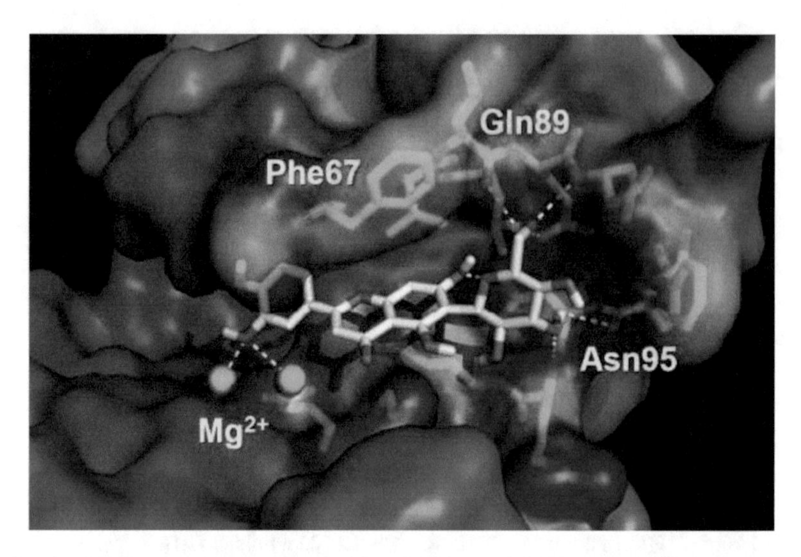

Fig. 26 Isoorientin (**35**) is a substrate competitive inhibitor of GSK3β. Predicted docking structures of isoorientin [98] (Used with permission from Liang et al. ACS Chem Neurosci 2016, 7, 912–923 [Copyright© 2016, American Chemical Society])

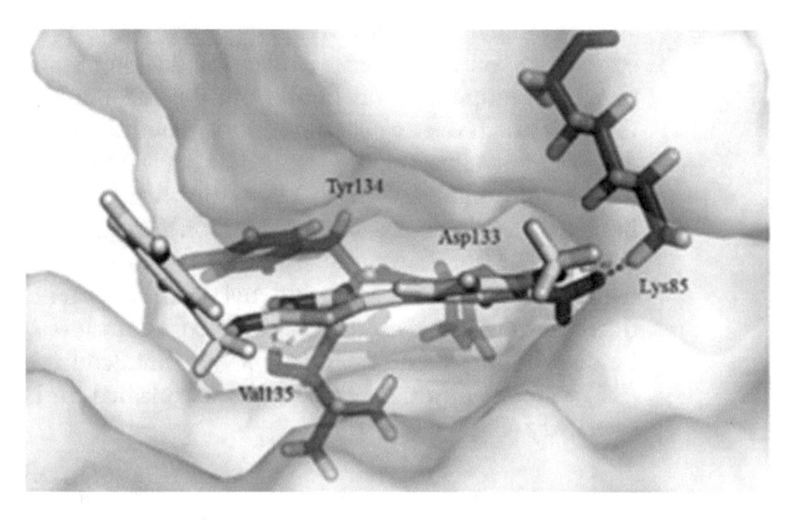

Fig. 27 Docked pose of compound (**36**) with GSK-3β [99] (Used with permission from Arfeen et al. Eur J Med Chem 2016, 121, 727–736 [Copyright© 2016 Elsevier Masson SAS])

2.1 nM. This compound was subjected to molecular docking studies, where it was found to form hydrogen bond with Val135 with a Glide dock score of −8.196 and Emodel energy of −69.30 making it a compound with the highest affinity towards GSK-3β (Fig. 27). The authors further performed molecular dynamics studies on the enzyme-ligand complex using Amber 11 programme with generalized Amber force field 'leaprc.ff03' for 24 ns. The MM-

GBSA energy decompositions per residue were analysed showing a difference of 4.10 kcal/mol in the free binding energy for **36** in the solvent phase, which favoured GSK-3β selectivity over CDK-2.

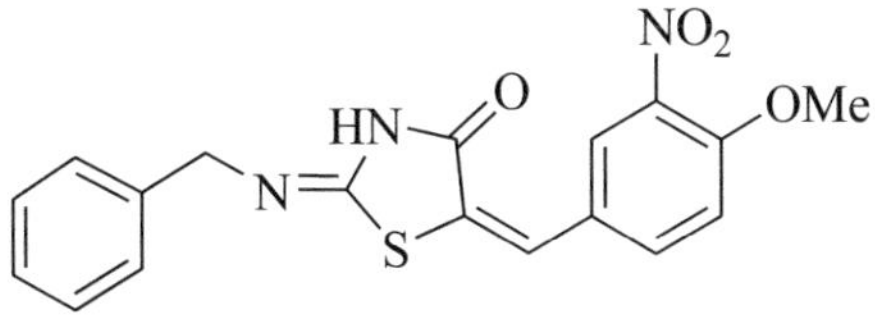

36

Fukunaga et al. reported several 2-(2-phenylmorpholin-4-yl) pyrimidin-4(3*H*)-ones as a novel class of GSK-3β inhibitors with high selectivity and oral activity [100]. Compound 2-(2-(4-fluorophenyl)morpholino)-3-methyl-6-(pyrimidin-4-yl)pyrimidin-4 (3*H*)-one (**37**) was found to be the most potent compound with GSK-3β inhibition having an IC_{50} value of 12 nM. This compound was tested for the kinase selectivity and was found to be 1000-fold more selective over 50 different kinases except for CK_1 (IC_{50} = 1.5 μM). Dose-response study on the compound (**37**) showed a significant decrease in tau phosphorylation after 2 h of administration (dose 30 mg/kg), and the compound also effectively inhibited in vivo tau protein phosphorylation on oral administration. Molecular docking studies support these observations for this series of derivatives. One of the compounds 3-methyl-2-(2-phenylmorpholino)-6-(pyridin-4-yl)pyrimidin-4(3*H*)-one (**38**) was docked into GSK-3β (PDB Id: 1Q3W). This scaffold displayed two hydrogen bonds with the enzyme, one between pyridine moiety and the N-H-bond of Val135 and the other between carbonyl oxygen of pyrimidinone and the side chain of Lys85. The phenyl ring forms a π-bond interaction with Arg141 in the hydrophobic site of GSK-3β (Fig. 28).

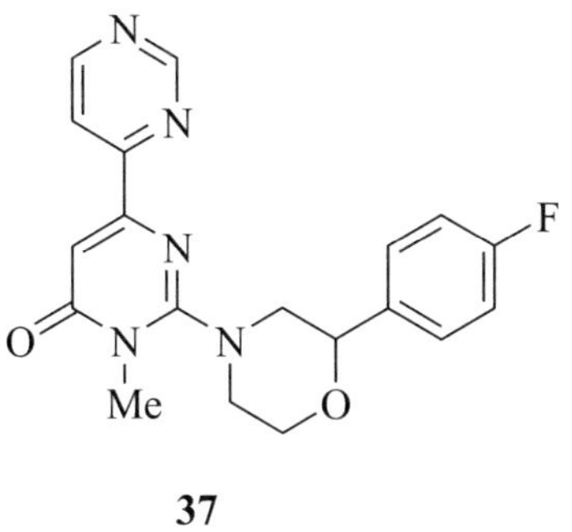

37 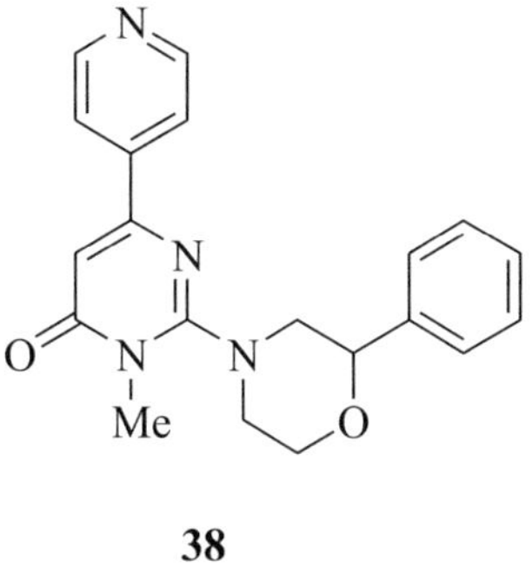 **38**

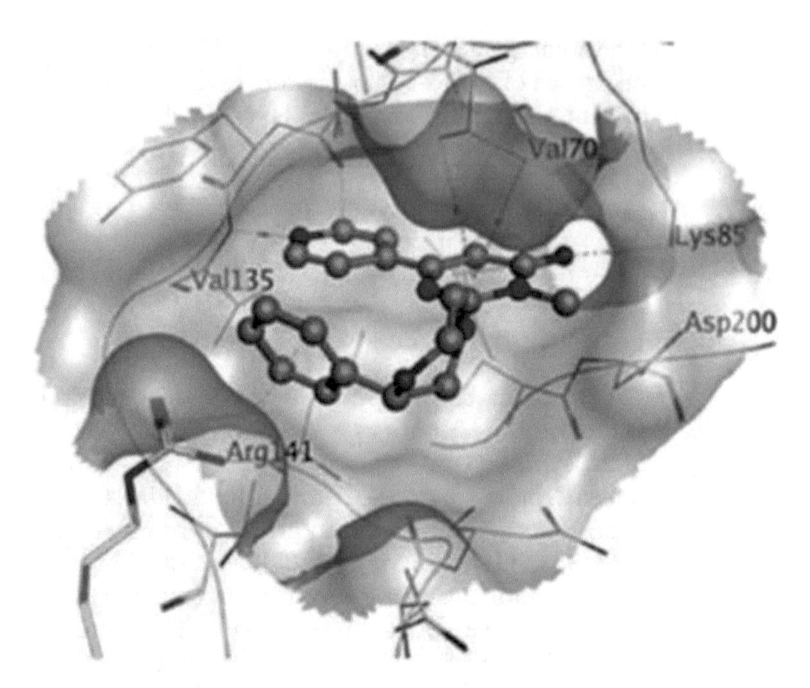

Fig. 28 Structures of compounds (**37** and **38**) along with predicted binding mode of **38** with GSK-3b by docking calculation [100] (Used with permission from Fukunaga et al. Bioorg Med Chem Lett 2013, 23, 6933–6937 [Copyright © 2013 Elsevier Ltd.])

3 Computational Studies on Cyclin-Dependent Kinase (CDK) and Its Inhibitors as Potential Anti-Alzheimer's Agents

Application of high-throughput screening protocol with CDK5/p25 complex has resulted into the discovery of *N*-(5-isopropylthiazol-2-yl)isobutyramide analog (**39**). Compound (**39**) proved to be an equipotent inhibitor of CDK5 and CDK2/cyclin E complex with IC_{50} values of 321 and 318 nM, respectively. Modifications of the obtained hit (**39**) at amide side chain and 5-isopropyl group led to a number of moderate to potent compounds. Molecular docking studies of the obtained analogs within the ATP-binding site of CDK5/p25 complex revealed that these compounds have the ability to form hydrogen bonding between aminothiazole NH and heterocyclic N with the carbonyl and NH of Cys83, respectively. The main challenge was to achieve selectivity over CDK2 due to the similarity of their active site residues, as only two out of 29 ATP-binding pocket residues were different, i.e. Leu83 and His84 in CDK2 versus Cys83 and Asp84 in CDK5 [101].

Me Me N S N H O Me Me

39

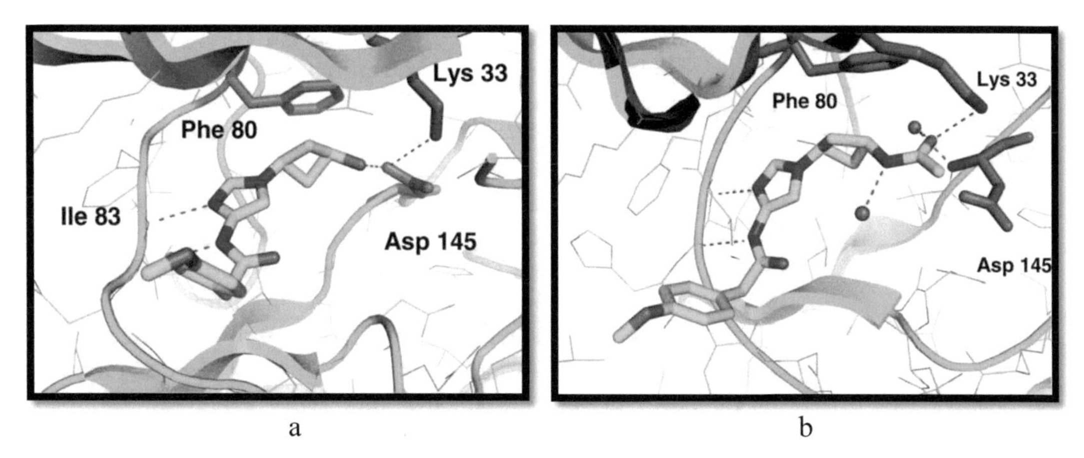

Fig. 29 Binding interactions of compounds (**40**) (**a**) and (**41**) (**b**) within the ATP-binding site of CDK2. Hydrogen bonds are indicated by dashed lines [102] (Used with permission from Helal et al. Bioorg Med Chem Lett 2009, 19, 5703–5707 [Copyright© 2009 Elsevier Ltd.])

In continuation of their efforts to screen various scaffolds as cyclin-dependent kinase inhibitors for the treatment of Alzheimer's disease, Helal et al. have reported a series of 4-aminoimidazole analogs. Based upon the results obtained from previous studies [101], the authors replaced 2-aminothiazole scaffold of compound (**39**) with 4-aminoimidazole to overcome metabolism-mediated toxicity, and efforts were also made to maintain the key hydrogen bonds crucial for inhibitory activity. Preliminary modelling studies suggested that polar groups in the inhibitor molecule tended to have interactions with the polar amino acids within the ATP-binding site, which could be useful for potent inhibitory activity. On the basis of this assumption, the authors have synthesized some compounds possessing polar groups, i.e. *cis* analogs (**40** and **41**) which showed selectivity ratio (CDK2 IC_{50}/CDK5 IC_{50}) of 0.7 and 7, respectively. Docking studies of compounds (**40** and **41**) as depicted in Fig. 29 showed key interactions with Lys33 and Asp145 revealing that these polar interactions were essential for binding of the inhibitors to the ATP-binding sites of both CDK2 and CDK5 [102].

Modification were carried out to overcome the metabolically mediated toxicity

39

40 and **41**

R= **40**: -OH; **41**: -NHCOMe

Rzasa et al. carried out screening of a library of in-house compounds for the design of novel CDK5 inhibitors. On the basis of obtained hits, the authors have disclosed a series of acyclic ureas (**42–44**). Preliminary biological screening (on **42** and **43**)

and on the basis of crystallographic data (of compound **44**), the authors disclosed 3,4-dihydro-1*H*-quinazolin-2-one scaffold containing derivatives as effective inhibitors of CDK5. To analyse the structure-activity relationship, different substitution pattern on AD rings were carried out. Modification on B-ring gave compound (**45**) as the most potent compound with an IC_{50} value of 79 nM. Docking studies divulged that increase in the activity (**45** vs **42**; Fig. 30) was mainly attributed to van der Waals interactions (with Gln85, Lys89, Asp86 and Leu134) of A-ring (in **45**) and better positioning of the phenyl ring (of **45**) through puckering of the fused B ring. C-Ring modification in compound (**45**) maintaining the thiazole analog resulted into compound (**46**, $IC_{50} = 77$ nM) with equipotent activity to compound (**45**). Both the compounds (**45** and **46**) were found to have maintained good interactions with Lys33-Asp144 salt bridge. Modifications using different substituents (R_1 to R_3) in A-ring were also carried out, and compound (**47**) with fluoro-substituent at R_3 position showed the best activity with an IC_{50} value of 16 nM. Docking studies of compound (**47**) showed that the R_3 substituent exhibited van der Waals interactions with Gln85, Lys89 and Asp86 residues. Finally, modifications in the D-ring were made by changing the position of nitrogen atom in

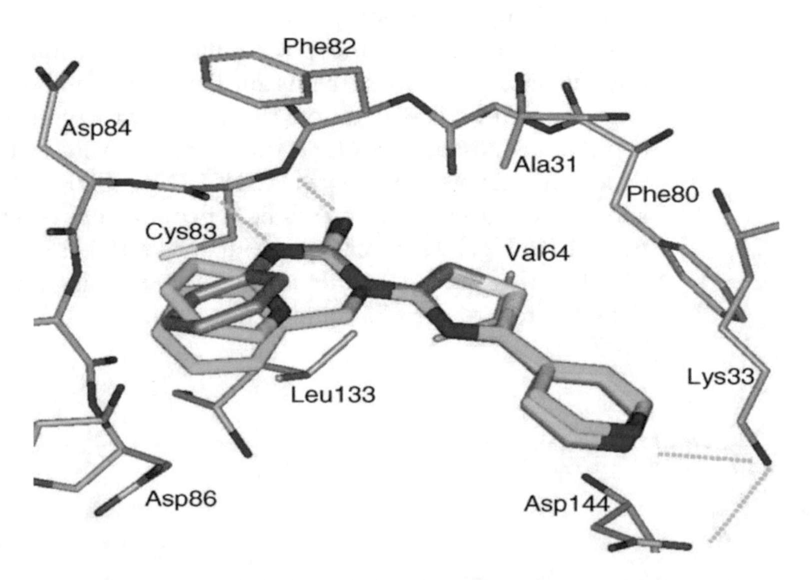

Fig. 30 Docking pose of compounds (**42**, **43**, and **45**) in the active site of CDK5. Nitrogen, oxygen and sulphur atoms are shown in *blue*, *red* and *yellow* colour, respectively. Carbon atoms of the active site residues are represented in *brown* colour. Carbon atoms of compounds (**42**, **43**, and **45)** are shown in *orange*, *cyan* and *green*, respectively. Ligand-protein H-bonds are shown in *green* [103] (Used with permission from Rzasa et al. Bioorg Med Chem 2007, 15, 6574–6595 [Copyright© 2007 Elsevier Ltd.])

the D-ring, and the importance of hydrogen bonding between D-ring of the inhibitor and Asp144 salt bridge was clearly demonstrated by the docking studies [103].

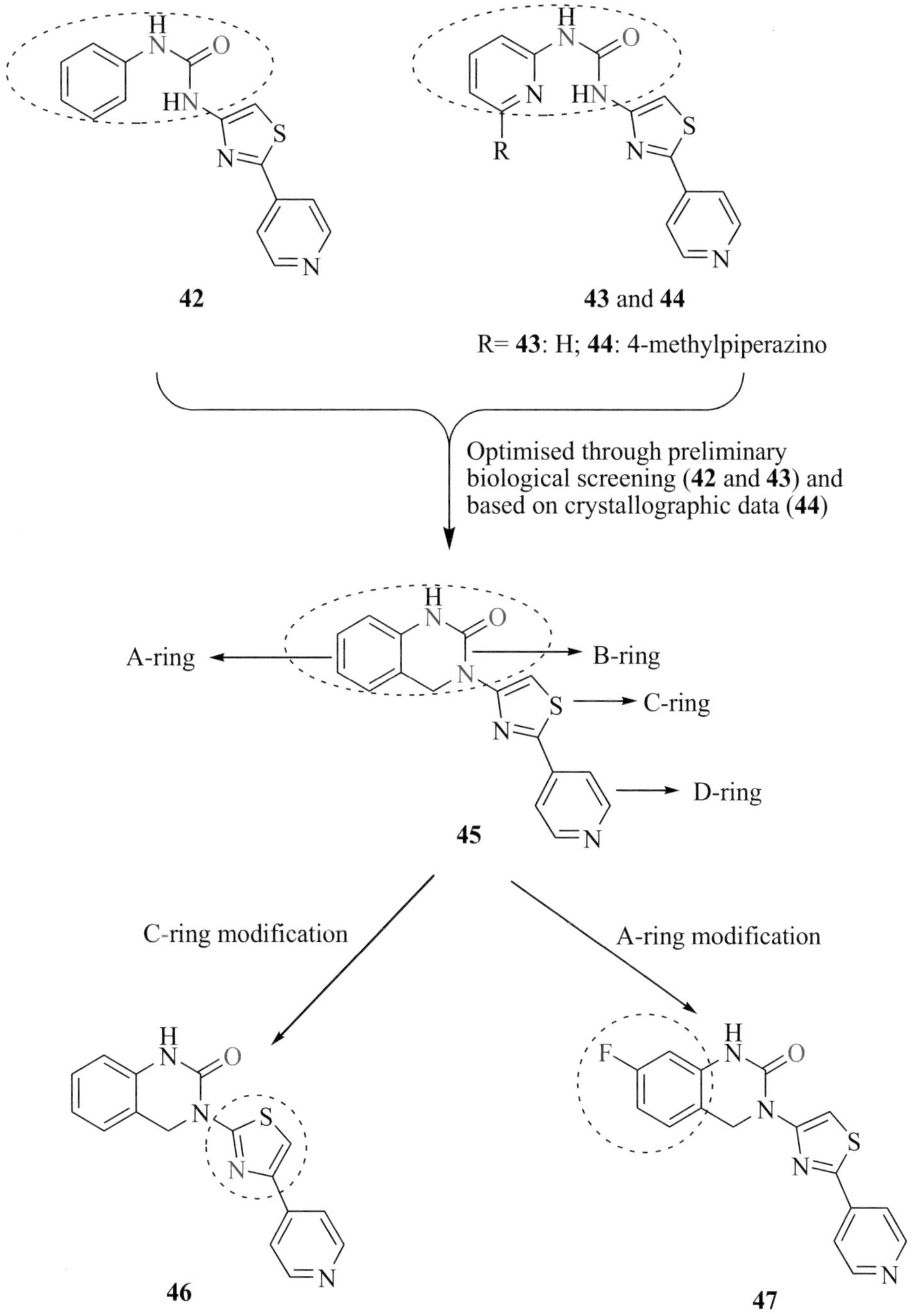

42

43 and **44**

R= **43**: H; **44**: 4-methylpiperazino

Optimised through preliminary biological screening (**42** and **43**) and based on crystallographic data (**44**)

A-ring

B-ring

C-ring

D-ring

45

C-ring modification

A-ring modification

46

47

Laha et al. reported a series of 2,4-diaminothiazoles as CDK/p25 kinase inhibitors on the basis of the previously reported compound (**48**) which was obtained through colorimetric enzyme-

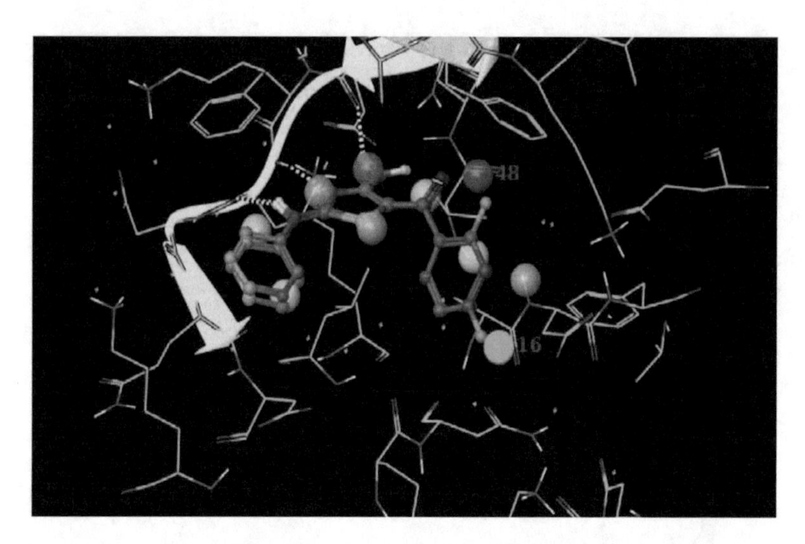

Fig. 31 Docked pose of compound (**49**) within the active site of CDK/p25. Hydration sites were represented using spheres [104] (Used with permission from Laha et al. Bioorg Med Chem Lett 2011, 21, 2098–2101 [Copyright© 2011, Elsevier Ltd.])

linked immunosorbent assay (ELISA)-based high-throughput screening protocol for CDK5. Among the reported series, compound (**49**) proved to be the most potent one with an IC_{50} value of 30 nM. SAR study of the analogs revealed that 2-substituted phenyl analogs, e.g. **49**, were more important for biological activity rather than the 4-substituted phenyl analogs. In order to support their observation, the authors carried out analysis of protein hydration sites using WaterMap. This study in vicinity of the ATP-binding pocket disclosed site 16 as stable one with ΔG of -3.2 kcal/mol and site 48 with unstable ΔG of 6.2 kcal/mol as shown in Fig. 31 for compound (**49**). Intrinsic oxidative stability of the compound (**49**) towards CYP450 3A4 and 2D6 was also predicted using computational models, and 6-position of the 3-pyridyl group was identified as the site of oxidation [104].

48 —Optimised through HTS of CDK5 and protein hydration sites→ **49**

Efforts were made by Jain et al. for the structural modifications of (*R*)-roscovitine (**50**) scaffold to design and develop CDK5 inhibitors with increased potency and increased selectivity towards the CDK2 enzyme. These efforts led to the development of 1-isopropyl-4-aminobenzyl-6-ether linked benzimidazole derivatives with significant inhibition of CDK5/p25 enzyme. Among the

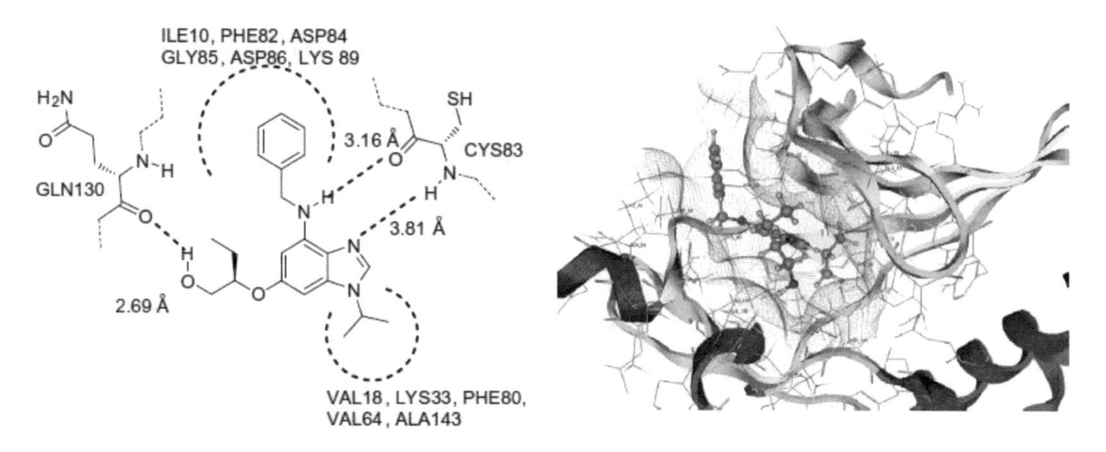

Fig. 32 Docked pose of compound (**51**) within the binding pocket of CDK5 enzyme [105] (Used with permission from Jain et al. Bioorg Med Chem 2011, 19, 359–373 [Copyright© 2011, Elsevier Ltd.])

series, compound (**51**) proved to be the most potent one with an IC_{50} value of 13 μM. Molecular modelling data suggested that replacement of purine scaffold and exocyclic 2-NH substituent at C-2 of (*R*)-roscovitine with benzimidazole and ether linkage, respectively, was found to be helping H-bonding along with additional hydrophobic interactions with Ile10 and Leu33 residues (Fig. 32). As per the authors' claim, increase in selectivity towards CDK2 enzyme could be because of substituting the exocyclic NH at C-2 position of (*R*)-roscovitine with ether linkage which could be responsible to cause H-bond interaction with Glu31 residue of CDK2 [105].

Increase selectivity towards CDK2

Help to retain H-bonding and add hydrophobic interaction

(*R*)-Roscovitine (50) **51**

Malmström et al. applied a high-throughput screening protocol to identify novel CDK5 kinase inhibitors. This study led to the discovery of compound (**52**) as a moderate inhibitor of CDK5 kinase with an IC_{50} value of 551 nM. The obtained hit (**52**) was co-crystallized with CDK5, and the structure was solved with a resolution of 1.90 Å. The X-ray structure revealed that the ligand lacked direct binding with the backbone residues Glu81 and Cys83 within the ATP-binding site of the kinase. Interestingly, a water molecule tends to form three bridging interactions in between

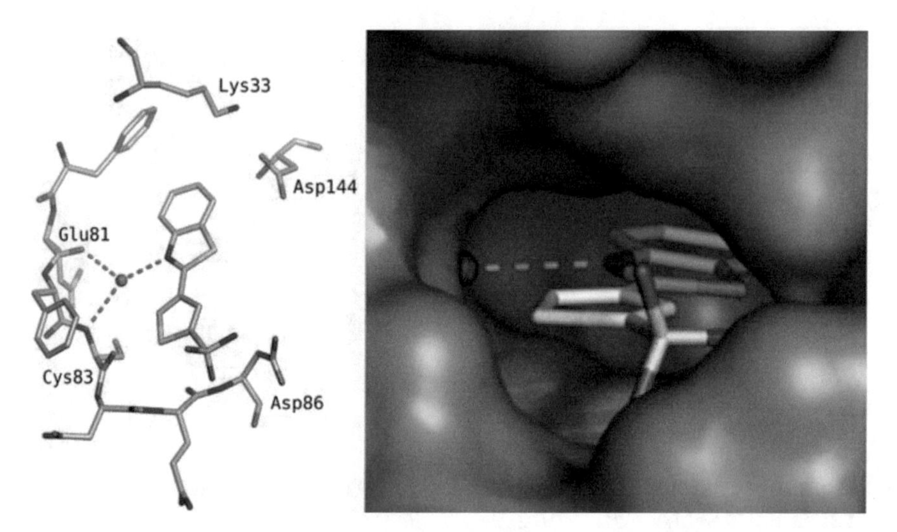

Fig. 33 Structure of compound (**52**) and its X-ray structure within the ATP-binding site of CDK5 [106] (Used with permission from Malmstrom et al. Bioorg Med Chem Lett 2012, 22, 5919–5923 [Copyright© 2012, Elsevier Ltd.])

ligand and the hinge backbone, i.e. with Glu81, Cys83 and nitrogen of the benzothiazole ring as depicted in Fig. 33. The sulfonamide group of compound (**52**) forms hydrogen bonding with Asp86 and Ile10, and substituting the sulfonamide group proved to be detrimental for the inhibitory activity. On the other hand, benzothiazole ring tends to form a bond with gatekeeper amino acid Phe80 through van der Waals interactions. Less bulky substituents are more favourable on benzothiazole ring as larger substituents might clash with the gatekeeper residue Phe80 [106].

O NH$_2$ N S O S S

52

Purine scaffold is one of the most widely used scaffolds among the heterocyclic ring systems in medicinal chemistry. One of the promising CDK inhibitor (*R*)-roscovitine (**50**) developed by Cyclacel Pharmaceuticals has a purine scaffold. On the basis of this scaffold, Demange et al. reported a series of 2,6,9-trisubstituted purine derivatives as potential CDK5 inhibitors. Among the reported series, sixth position-substituted compound (**53**) emerged as the most prominent CDK5 inhibitor with an IC_{50} value of 17 nM. Docking studies of compound (**53**) demonstrated that H-bonding between Lys89 and nitrogen of pyridine ring exhibited key interactions for biological activity (Fig. 34). Presence of a hydrophobic pocket was also confirmed through docking

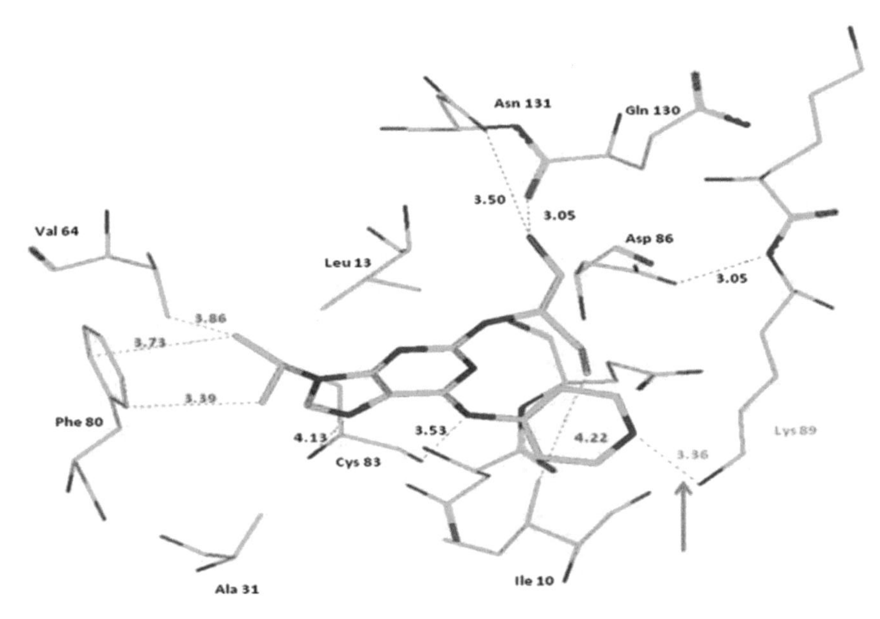

Fig. 34 Docked pose of compound (**53**) within the ATP-binding pocket of CDK5. One of the key interactions between aromatic nitrogen and Lys89 was shown by *black dotted line* (indicated by *orange arrow*) [107] (Used with permission from Demange et al. Bioorg Med Chem Lett 2013, 23, 125–131 [Copyright© 2013, Elsevier Ltd.])

studies as exemplified by interactions with Ala30, Phe80 and Val64. As per the authors' claim, the results obtained from the study were consistent with structural homology of ATP-binding site of the previously reported kinase inhibitors [107].

Modification at 6-position

(*R*)-Roscovitine (50) **53**

Sanphui et al. reported cyclin-dependent kinase 4 inhibitors containing two scaffolds, i.e. naphthalene and phenanthrene. Among these two series, compounds containing the phenanthrene scaffold proved to be superior in inhibitory activity than the naphthalene-based derivatives. Compound (**54**), a phenanthrene-based small-molecule inhibitor, showed CDK4 kinase inhibitory potential with neuroprotective effect at 5 μM dose. Docking studies

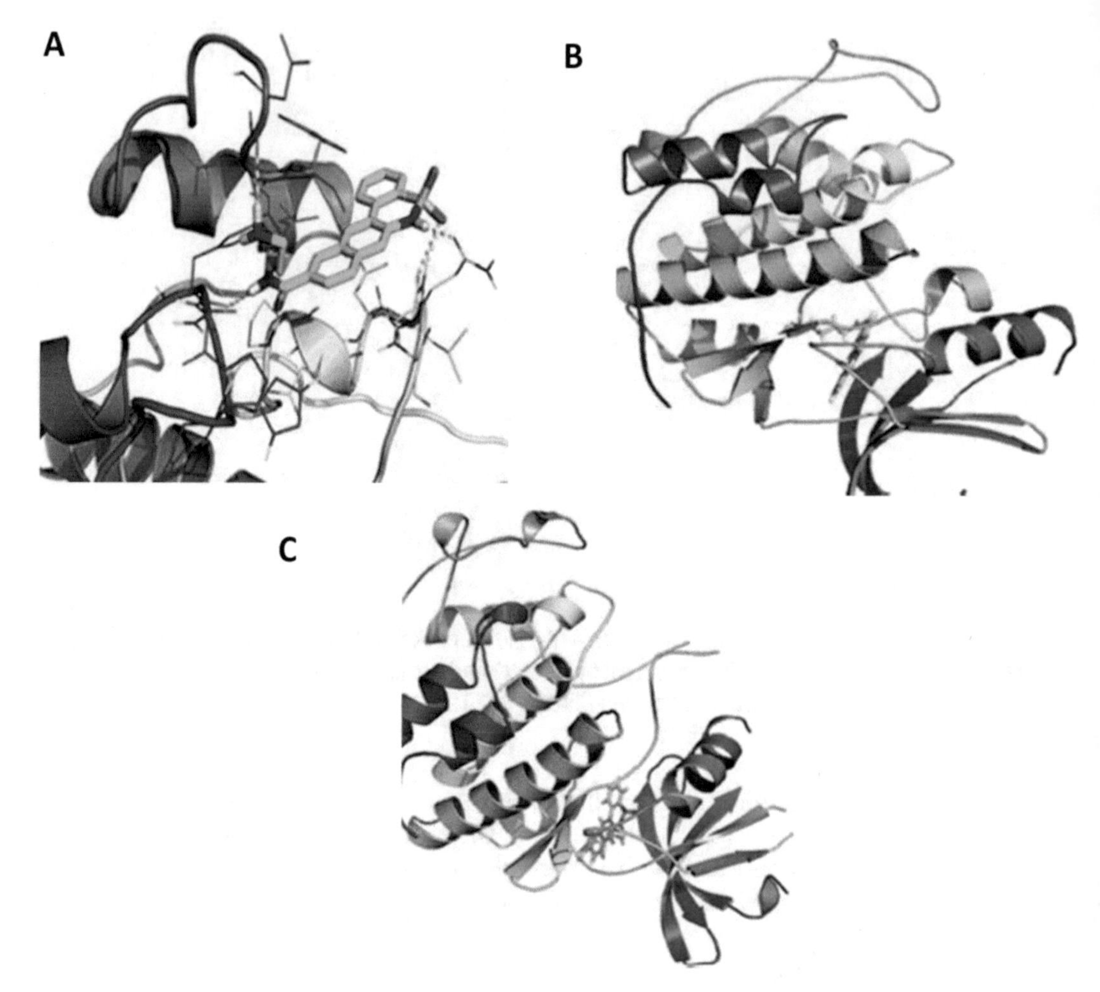

Fig. 35 Docked pose of compound (**54**) within the active site of (**a**) CDK4, (**b**) CDK2 and (**c**) CDK5 [108]

carried out using AutoDock indicated that compound (**54**) formed five hydrogen bonds within the active site of CDK4 enzyme, and Arg62, Glu43 and Phe69 residues were found to be responsible for the protein-ligand interactions as depicted in Fig. 35a. Strong hydrogen bonding of compound (**54**) with the key amino acid residues (Arg62, Glu43 and Phe69) also encouraged the authors to study the changes in accessible surface area and binding-free energies of the compound, and the calculated values were found to be 73.12% and -30 kJ mol^{-1}, respectively. Compound (**54**) was also docked within the active site of both the enzymes CDK2/CDK5.

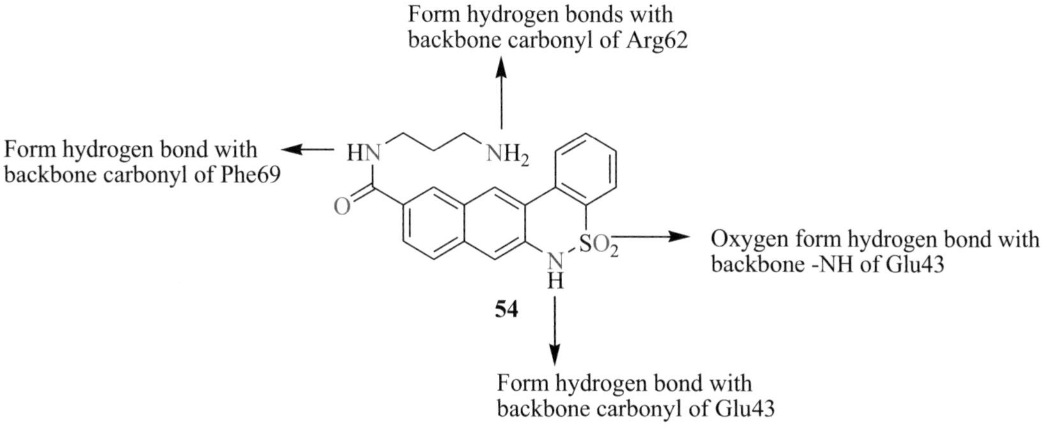

Form hydrogen bonds with backbone carbonyl of Arg62

Form hydrogen bond with backbone carbonyl of Phe69

Oxygen form hydrogen bond with backbone -NH of Glu43

54

Form hydrogen bond with backbone carbonyl of Glu43

Binding modes were observed to be different from those obtained from CDK4 (Fig. 35a–c). The results clearly indicated the specificity of compound (**54**) for CDK4 instead of CDK2/CDK5 enzymes [108].

4 Computational Studies on p38α Mitogen-Activated Protein Kinase (MAPK) and Its Inhibitors as Potential Anti-Alzheimer's Agents

Scaffold repurposing, a pharmacoinformatics approach, was applied that disclosed compound (**55**) as a novel p38α MAPK inhibitor (Ki value of 101 nM) by Roy et al. Crystallographic structure elucidation concluded that the nitrogen atom of pyridine ring of **55** has hydrogen bond interactions with the amide of Leu108-Met109 backbone residues (Fig. 36a, b). Naphthyl ring of **55** acted as a greater space-filling constituent and played a key role in the specificity of compound (**55**) towards the p38α MAPK target (Fig. 36c) [109].

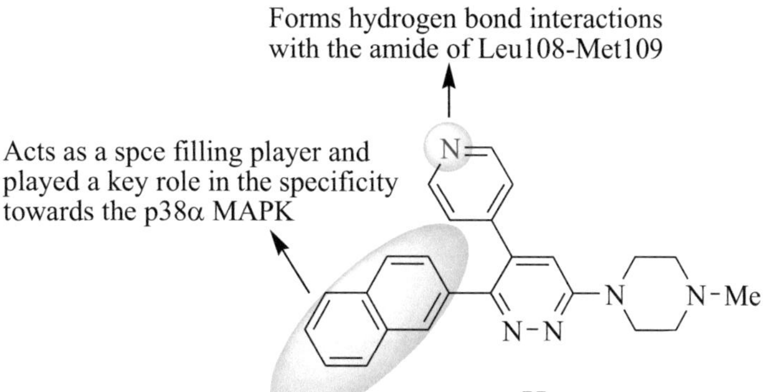

Forms hydrogen bond interactions with the amide of Leu108-Met109

Acts as a spce filling player and played a key role in the specificity towards the p38α MAPK

55

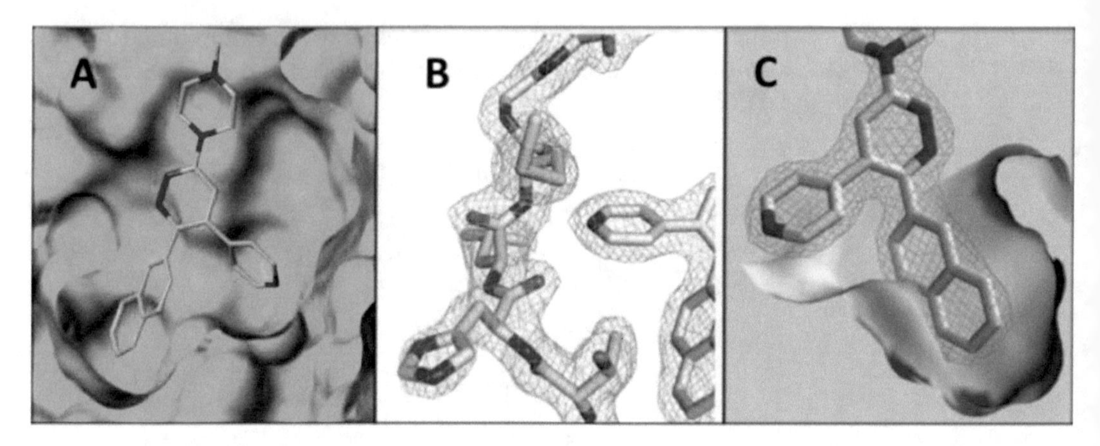

Fig. 36 (**a**) Docked conformation of **55** within the active site of p38α MAPK. (**b**) Close-up view of interaction between the nitrogen of pyridine of **55** with the Met109 residue of hinge region. (**c**) View of space-filling contributor naphthyl group of **55** [109]

Markandeyan et al. applied a virtual screening approach on various phytochemicals present in *Morinda citrifolia* plant and screened them for their inhibitory activity against p38α MAPK enzyme. Out of a total 32 phytoconstituents, 22 were selected as they satisfied the conditions of Lipinski's rule of five. Crystal structure of p38α MAPK (PDB ID 4F9Y) was used, and five cavities were detected in it with the help of Molegro Virtual Docker (MCD). Out of these five cavities, cavity 1 with MW181 and cavity 2 with GG5 as reference ligands were selected for docking purposes. Among the docked compounds, isoprincepin (**56**, −141.14 kcal/mol) was found to have the best docking score in cavity 1 compared to the reference ligand MW181 (−111.29 kcal/mol). In this study, isoprincepin (**56**) showed hydrogen bond interaction with amide of Met109 residue (Fig. 37) and also caused glycine flip phenomenon. In cavity 2, which is bound with GG5 as the reference ligand, isoprincepin (**56**) and balanophonin (**57**) showed the best docking scores of −177.52 and −135.27 kcal/mol, respectively, compared to that of GG5 (−120.31 kcal/mol) [110].

Form hydrogen bond interaction with amide of Met109

Isoprincepin (56)

Balanophonin (57)

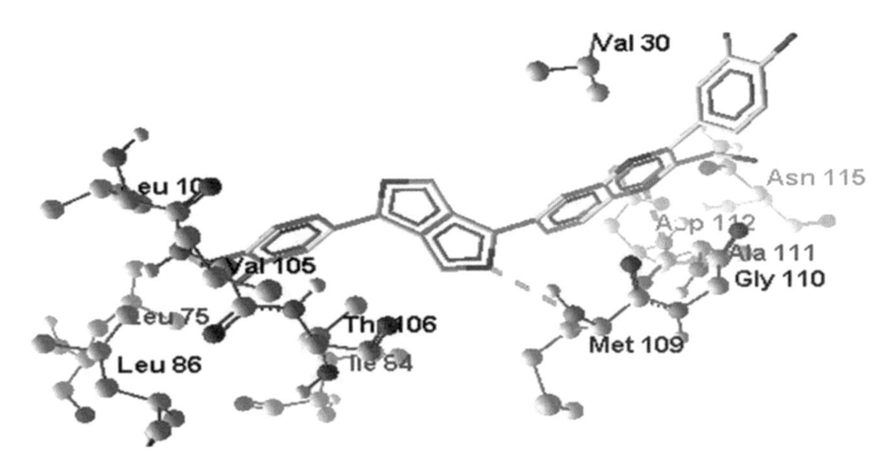

Fig. 37 Docked pose of isoprincepin (**56**) within the active site of p38α MAPK enzyme. *Dotted line* represents the hydrogen bond with amide of Met109 residue [110]

5 Computational Studies on Dual-Specificity Tyrosine-Regulated Kinases (DYRKs) and Their Inhibitors as Potential Anti-Alzheimer's Agents

Gourdain et al. reported a series of 3,5-diaryl-1*H*-pyrrolo[2,3-*b*] pyridines (DANDY) analogs having inhibitory activity against DYRK1A kinase. Among the reported DANDY analogs, compounds with hydroxyl substituent on the aryl moieties proved to be the most active as exemplified by compound (**58**, IC_{50} value of 3 nM), and the methoxy substituent proved detrimental for the inhibitory activity. Putative binding sites for the compound (**58**) were determined through molecular modelling studies within the ATP-binding site of DYRK1A. The 7-azaindole scaffold showed H-bond interactions with the carbonyl of Glu239 (1.68 Å) and NH of Leu241 (1.86 Å) residues. Affinity of compound (**58**) was mainly due to the combined result of interactions between C3-*para* phenol with Lys188 and a 3-centred bond network due to the interactions between 3,4-diol at C-5 with Asp247 residue (Fig. 38) [111].

OH → Played key role for high affinity of compound **58**

Essential for potential inhibitory activity against DYRK1A kinase ← HO, HO

→ Showed H-bond interactions with Glu239 and Leu241 residues

N, N, H

58

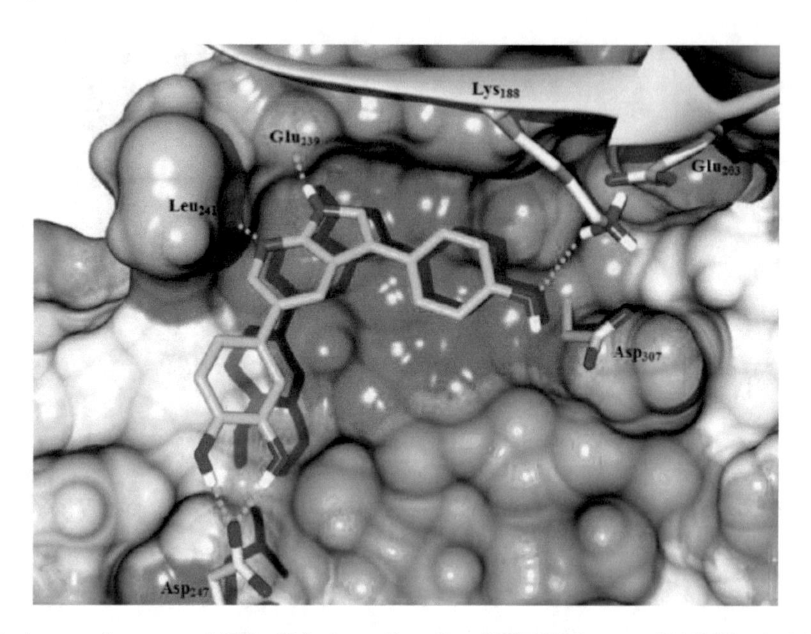

Fig. 38 Docked pose of compound (**58**) within the active site of DYRK1A enzyme [111] (Used with permission from Gourdain et al. J Med Chem 2013, 56, 9569–9585 [Copyright© 2013, American Chemical Society])

Myrianthopoulos et al. strategically converted a non-selective bis-indole indirubin scaffold into a selective DYRK2 inhibitor (**59**; IC_{50} value of 0.13 μM) through the combined substituents, bromo and acidic functionality at 7- and 5′- positions, respectively. Putative binding sites observed through docking studies revealed that indirubin scaffold of **59** adopted inverse pose within the DYRK2 enzyme, and it also showed nonstandard kinase-binding mode due to the steric clash between the bulky group (bromo) at seventh position of **59** and the hinge region. Enhanced inhibitory activity of **59** was achieved through the acidic functionality at 5′-position which mainly displaced unstable water molecule and formed salt bridge between 5′-carboxylate and Lys178. Compound (**59**) also showed two hydrogen bonds between (i) carbonyl of lactam ring (**59**) and NH of backbone residue Leu231 and (ii) NH of lactam core and carbonyl group of backbone Leu231 residue. As per the authors' claim, the reported selectivity profile was obtained for ATP-competitive inhibitors and not for the ATP-mimetic inhibitors through a variety of drug design approaches [112].

Responsible for enhanced inhibitory activity

NOH

HOOC

Br

Caused steric clash with hinge region residue and responsible for nonstandard binding pose of **59**

N
H

NH

O

Showed H-bonds with backbone Leu231 residue

59

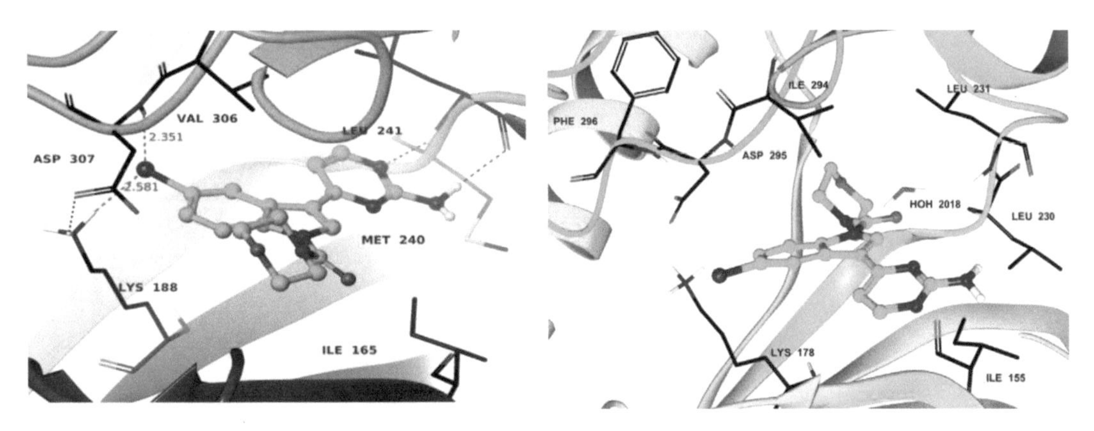

Fig. 39 Docked pose of compound (**60**) within the active site of (**a**) DYRK1A, (**b**) DYRK2 [113] (Used with permission from Yadav et al. Bioorg Med Chem Lett 2015, 25, 2948–2952 [Copyright© 2015, American Chemical Society])

Yadav et al. reported a series of N1-substituted and C-ring modified meridianin analogs as prominent inhibitors of DYRK1A enzyme and also studied selectivity of these compounds towards the DYRKs, i.e. DYRK2 and DYRK3. Among the reported series, morpholinoyl-linked compound (**60**) proved to be the most potent one with IC_{50} values of 0.5, 1.4 and 2.2 μM, respectively, against DYRK1A, DYRK2 and DYRK3 kinases. Molecular modelling studies were performed on compound (**60**) to elucidate the key binding interactions with DYRK1A and DYRK2 (Fig. 39) responsible for the potential inhibitory activity. This study suggested that 2-aminopyrimidine is a key component which interacted with the ATP-binding site of the DYRKs. Bulky groups, i.e. the bromo substituent (as in **60**), interacted with Asp307 and Lys188 residues of the side chain through hydrophobic and ionic interactions. Further hydrophobic interactions were displayed by Val306 and DFG motif (Asp307-Phe308-Gly309), and these hydrophobic interactions could be the possible cause for DYRK1A inhibition by **60**. Lower activity of compound (**60**) against DYRK2 versus DYRK1A was due to the loss of interaction of 2-aminopyrimidine ring with the hinge region residues Leu230 and Leu231. Overall, this study also concluded that compound (**60**) possessed three- and fourfold selectivity for DYRK1A in comparison to DYRK2 and DYRK3 [113].

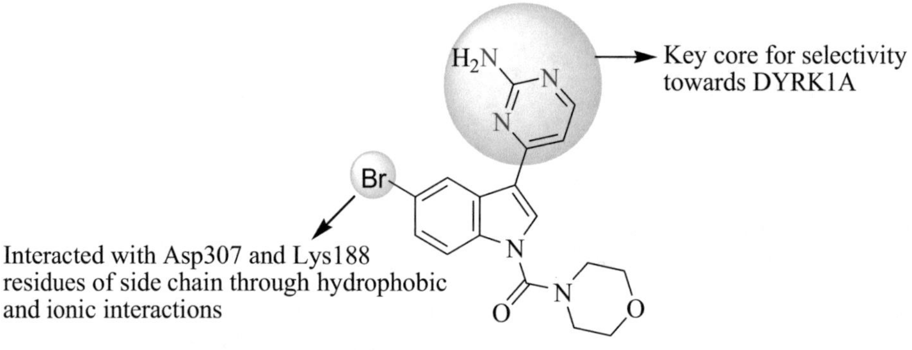

Key core for selectivity towards DYRK1A

Interacted with Asp307 and Lys188 residues of side chain through hydrophobic and ionic interactions

60

6 Computational Modelling Studies on Miscellaneous Kinases and Their Inhibitors as Anti-Alzheimer's Agents

6.1 Fyn Kinase Inhibitors

Tintori et al. have successfully applied structure-based drug design approach for the screening of commercially available and in-house compounds library for their inhibitory activity against Fyn kinase. Two hits (**61** and **62**, Ki 2 μM each) from commercial library and one hit (**63**) from in-house library were obtained having ki inhibition value of 0.9 μM. Hit-to-lead optimization approach and slight modification on compound (**63**) resulted into two more potent compounds (**64** and **65**) with Ki values of 70 and 95 nM, respectively. Molecular docking studies of the compounds (**64** and **65**) showed that these compounds formed van der Waals contacts within the hydrophobic region of Fyn kinase due to the presence of Cl (in **64**) and CH_3 (in **65**) substituents at the *para* position of the C3 phenyl ring. A weak hydrogen bond interaction between the nitrogen (N5) of the inhibitors (**64** and **65**) and the backbone NH of Met345 residue was observed. A stable hydrogen boding has also been observed between the carbonyl group of the backbone residue Glu343 with exocyclic amine (N4), and this stable hydrogen bond weakens the intermolecular interactions between the side chain of the backbone containing Thr342 and Glu343. This effect was seen in both of the compounds (**64** and **65**). The most active compounds (**64** and **65**) were stabilized through the (i) salt bridge between Lys299 and the side chain Asp408 and (ii) hydrogen bonding between Glu314 and the backbone NH of Phe409 [114].

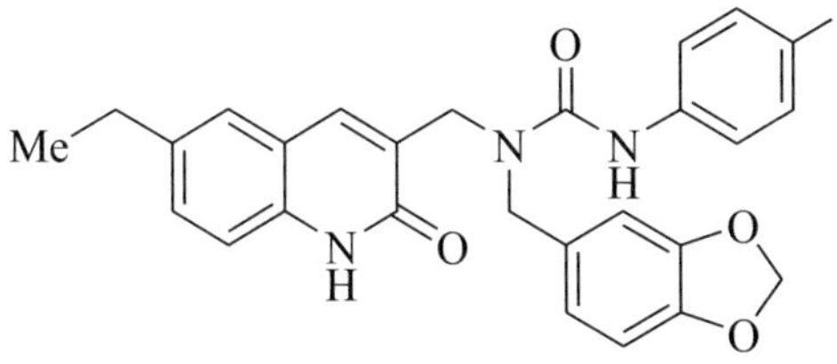

61

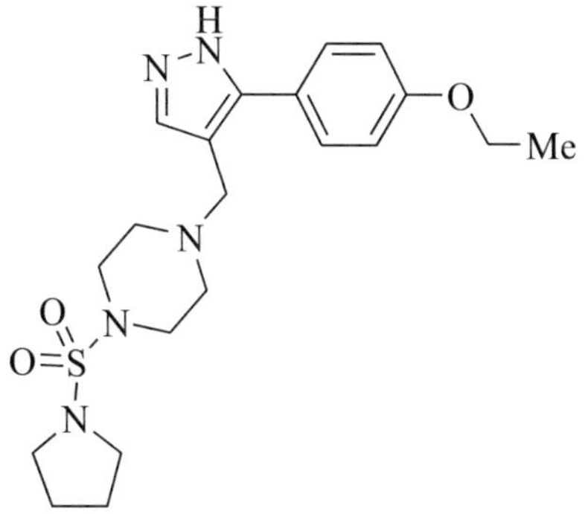

62

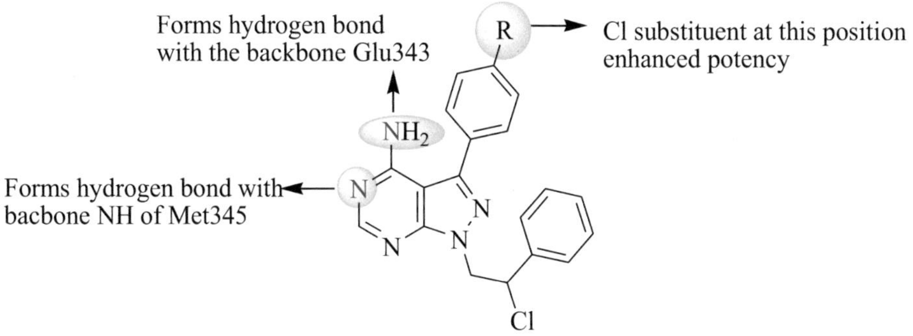

Forms hydrogen bond with the backbone Glu343

Cl substituent at this position enhanced potency

Forms hydrogen bond with bacbone NH of Met345

63-65

R= **63**: H; **64**: Cl; **65**: CH_3

6.2 Dual Kinase Inhibitors

A series of compounds based on unsubstituted 6-phenyl[5*H*]pyrrolo[2,3-*b*]pyrazine scaffold (Aloisine) was reported by Mettey et al. having inhibitory activities against different kinases. Aloisine B (**66**), the most prominent dual inhibitor of CDK/GSK-3 kinases, was docked into the active site of CDK2 kinase. The docking study demonstrated that the two nitrogen atoms of the phenylpyrrolo [2,3-*b*]pyrazine scaffold formed two hydrogen bond interactions with the backbone hydrogen and oxygen atoms of Leu83 residue, and another nitrogen atom interacted with the Lys33 residue (Fig. 40). The structure-activity relationship study concluded that

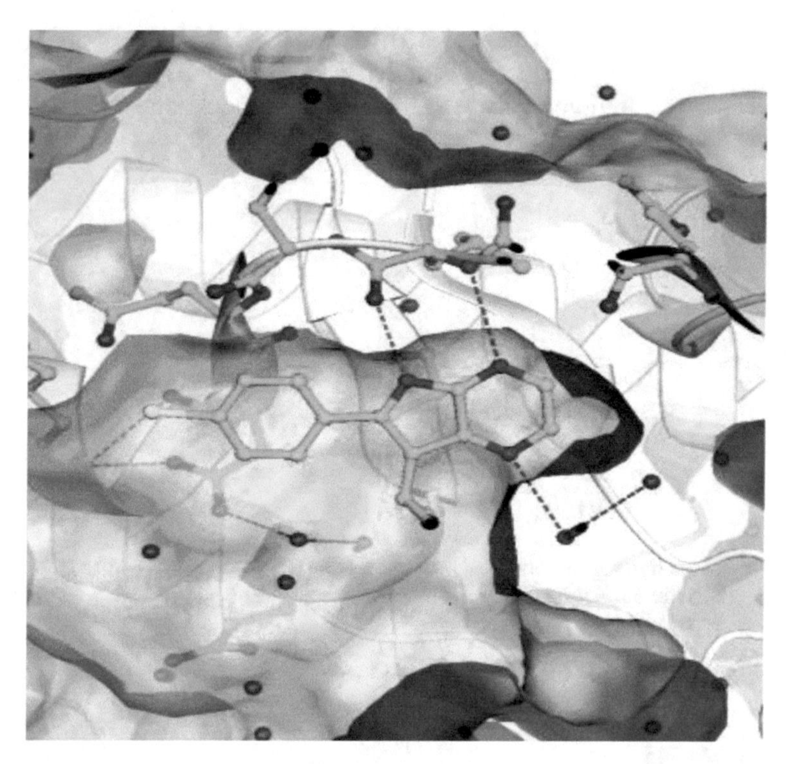

Fig. 40 Hydrogen bond interactions of Aloisine B (**66**) with the residues of active site of CDK2 [43] (Used with permission from Mettey et al. J Med Chem 2003, 46, 222–236 [Copyright© 2003, American Chemical Society])

replacement of any of the nitrogen atoms with carbon would result into diminished inhibitory activity of Aloisine B (**66**) [43].

Aloisine B (66)

Polychronopoulos et al. reported a series of naturally occurring indirubin derivatives as potent and selective dual inhibitors of GSK-3/CDK kinases. Various 6-substituted and 5,6-disubstituted derivatives of indirubin were synthesized. Among the reported series, compound (**6**) with 6-bromo and 3′-oxime functions proved to be the most potent compound with IC_{50} values of 0.320, 0.083 and 0.005 μM against CDK1/cyclin B, CDK5/p25 and GSK-3α/β kinases, respectively. To gain insights into the binding interactions of the most potent compound (**6**) in the binding sites of both of the kinases CDK5/p25 and GSK-3α/β, docking studies were performed. The oxime group played an important role in the stabilization of the inhibitor within the ATP-binding site of the kinases

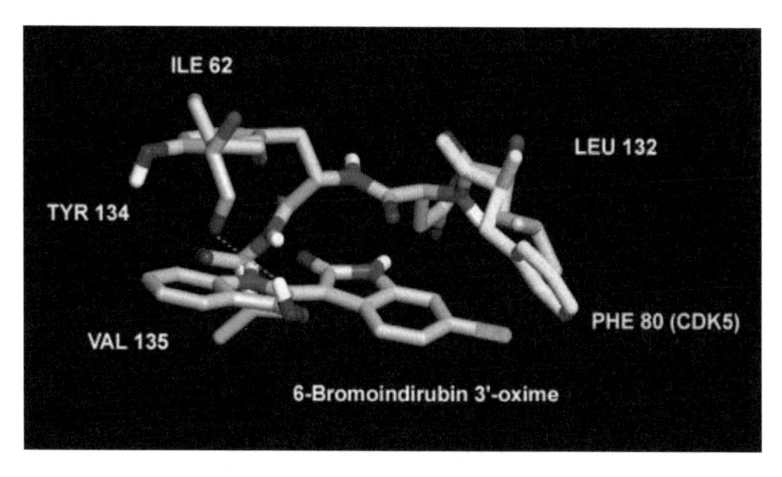

Fig. 41 Docked pose of compound (**6**) within the ATP-binding pocket of GSK-3α/β [115] (Used with permission from Polychronopoulos et al. J Med Chem 2004, 47, 935–946 [Copyright© 2004, American Chemical Society])

through electrostatic interaction and hydrogen bonding. These studies suggested that the substituent at sixth position of indirubin was responsible for the selectivity towards the kinases, and the substituent at 3′-position was essential for the binding affinity. As far as the selectivity of the compound (**6**) was concerned, it inhibited GSK-3α/β kinases more selectively over CDK5 due to steric hindrance between Phe80 residue of CDK5 and the bromine atom of compound (**6**) (Fig. 41) [115].

Essential for the stabilization of inhibitor inside the ATP binding site of kinases

Br → Responsible for steric clash with Phe80 residue of CDK5

NOH

NH

N
H

O

6

Oumata et al. reported a series of 2,6,9-trisubstituted purine analogs on the basis of the structure of (*R*)-roscovitine which mainly differed with a different substituent at C6 position. The synthesized compounds were screened for their potential inhibitory activity on various kinases like CDK1/cyclin B, CDK5/p25, GSK-3α/β and CK1δ/ε. Among the reported analogs, compound (**67**) proved to be the most potent one with IC_{50} values of 0.22, 0.08 and 0.014 μM, against the respective CDK1, CDK5 and CK1 kinases. Structural rationale for the obtained biological data of compound (**67**) against these kinases has been elucidated using

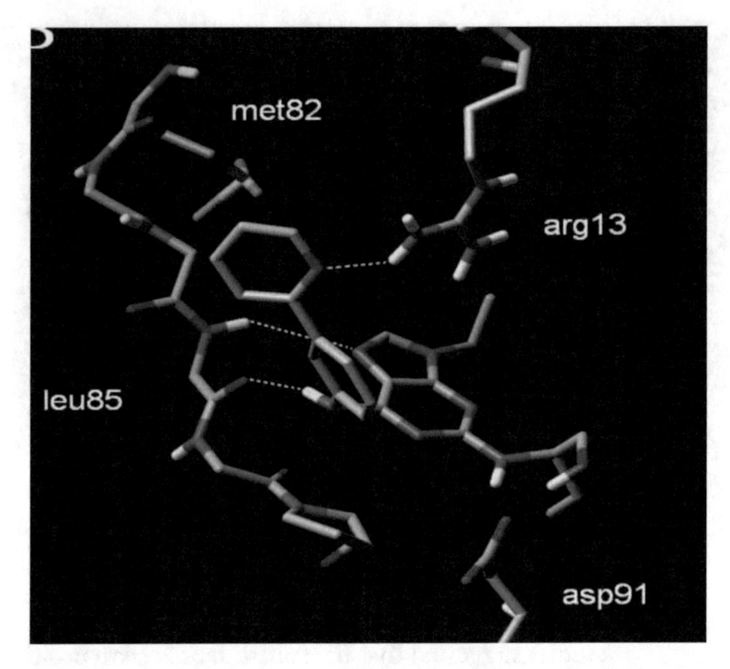

Fig. 42 Key interaction of compound (**67**) within the ATP-binding site of CK1 [116] (Used with permission from Oumata et al. J Med Chem 2008, 51, 5229–5242 [Copyright© 2008, American Chemical Society])

molecular docking studies. Docked pose of compound (**67**) within the ATP-binding pocket of CK1 kinase is shown in Fig. 42. Compound (**67**) showed two distinct binding modes with adenine type of interaction between the ligand and the receptor. Pyridine nitrogen of the inhibitor (**67**) showed hydrogen bond interaction with guanidine moiety of the active site, whereas the phenyl ring formed stable pi-pi interactions through parallel orientation with guanidinium ion of Arg. 3-Pyridinylphenylamino substituent was responsible for hydrophobic interactions with Pro87 and also for pi-pi stacking with the Phe95 residue. The overall study concluded that the 3-pyridinyl substituent was essential for inhibitory activity for CK1. As per the authors' claim, dual CDK-CK1 specificity character of analog (**67**) is due to the enhanced hydrogen bonding potential of N6 carrying two aromatic rings [116].

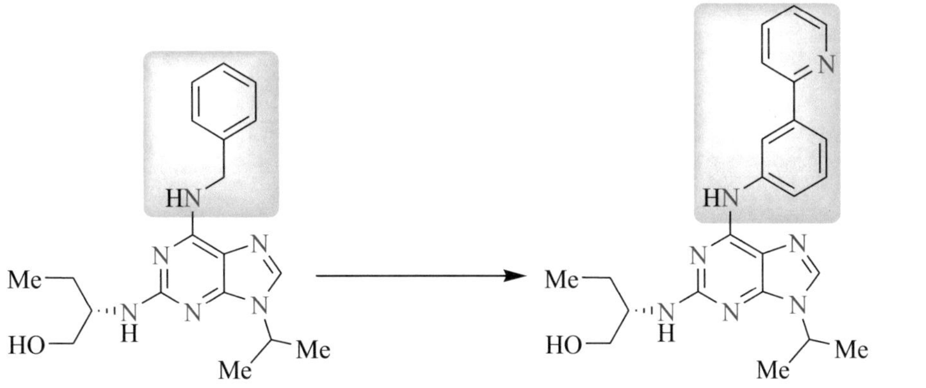

(*R*)-Roscovitine (50)

CDK5: IC_{50} = 0.200 μM
CK1: IC_{50} = 2.300 μM

67

CDK5: IC_{50} = 0.080 μM
CK1: IC_{50} = 0.014 μM

In continuation of their research on the development of novel lead compounds as dual inhibitors of CDK1/GSK-3, Loge et al. reported a series of 9-oxothiazolo[5,4-*f*]quinazoline-2-carbonitrile analogs. Among the reported series, an imidate analog (**68**) proved to be a prominent dual inhibitor of CDK1 and GSK-3*β* with IC_{50} values of 0.15 and 0.13 μM, respectively. Binding pose of compound (**68**) within the GSK-3*β* showed that N-6 positioned nitrogen of the quinazoline ring formed hydrogen bonding with the backbone NH of Val135 residue in the hinge region. Molecular interactions of compound (**68**) were studied with the help of molecular electrostatic potential (MEP), and the study concluded that analogs with more electronegative potentials might be better suited for binding inside the active pocket of CDK1 as exemplified by **68** (IC_{50} value of 0.15 μM), compared to the electropositive amidine analog (**69**, IC_{50} value of 9 μM). The results of MEP study also concluded that the charge distribution might affect the binding of the inhibitors inside the ATP-binding pocket of CDK1 rather than the GSK-3*β* [117].

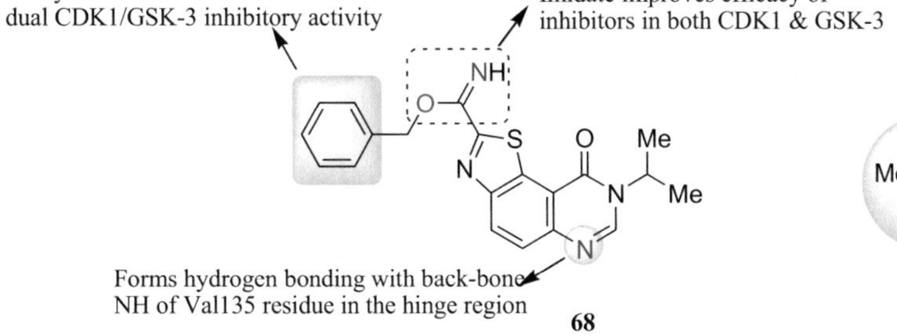

Bulky substituent most favoured for dual CDK1/GSK-3 inhibitory activity

Imidate improves efficacy of inhibitors in both CDK1 & GSK-3

Forms hydrogen bonding with back-bone NH of Val135 residue in the hinge region

68

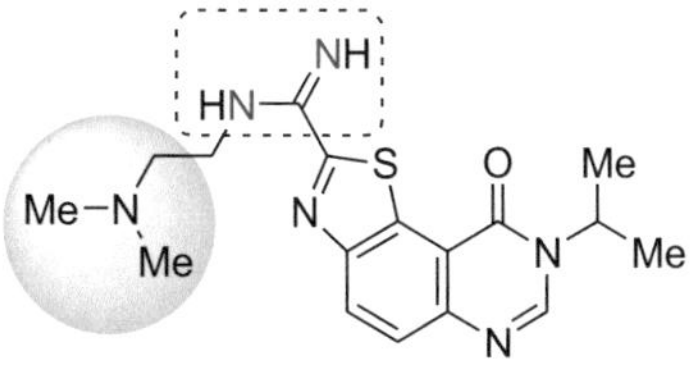

69

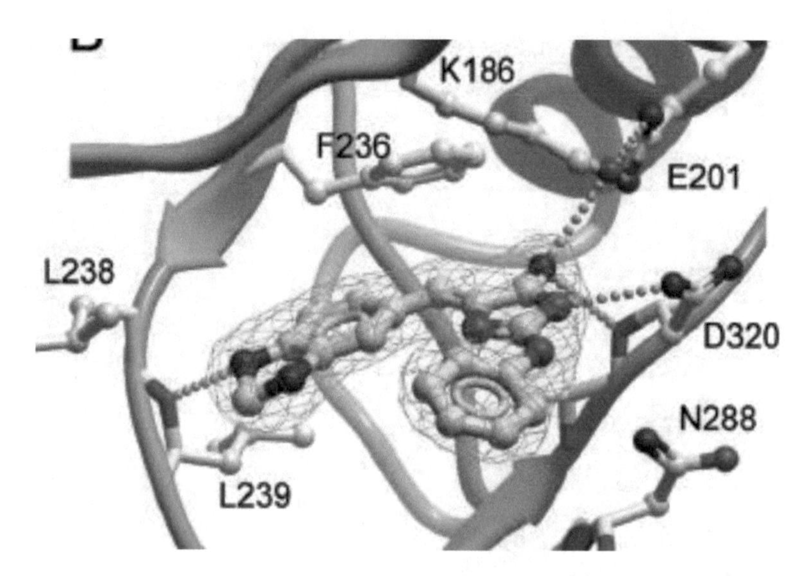

Fig. 43 Docked pose of leucettamine L41 (**71**) within the ATP-binding site of CLK3 [118] (Used with permission from Debdab et al. J Med Chem 2011, 54, 4172–4186 [Copyright© 2011, American Chemical Society])

Debdab et al. reported a series of leucettine derivatives on the basis of the structure of naturally occurring compound leucettamine B (**70**) obtained from sponge *Leucetta microraphis* whose significant biological activity was not reported till 2011. In this study the authors initially reported the results of preliminary biological screening of leucettamine B (**70**) with an important finding that this compound possessed inhibitory activity over protein kinases, which could be used for the treatment of Alzheimer's disease. Among the disclosed derivatives, leucettamine L41 (**71**) proved its affinity for DYRK1A, CLK1 and CLK3 kinases with IC_{50} values of 0.040, 0.015 and 4.5 μM, respectively. To understand the binding mode of leucettamine L41 (**71**), docking studies were performed using the protein structure of CLK3 kinase which revealed that 2-aminoimidazoline scaffold formed hydrogen bonding with lysine residue (E201), and the motifs, i.e. αD, αF and αG, anchored the leucettamine L41 (**71**) in the ATP-binding site (Fig. 43). The benzodioxole ring of leucettamine L41 (**71**) interacted through hydrogen bonding with the amide of hinge region residue L239 offering inhibitory activity against the kinases. Docking of leucettamine L41 (**71**) within the ATP-binding of DYRK1A showed that NH of leucettamine L41 (**71**) interacted with the C = O of the lactam of kinase hinge region, and this must be responsible to improve its affinity for DYRK1A and also for CLK1 [118].

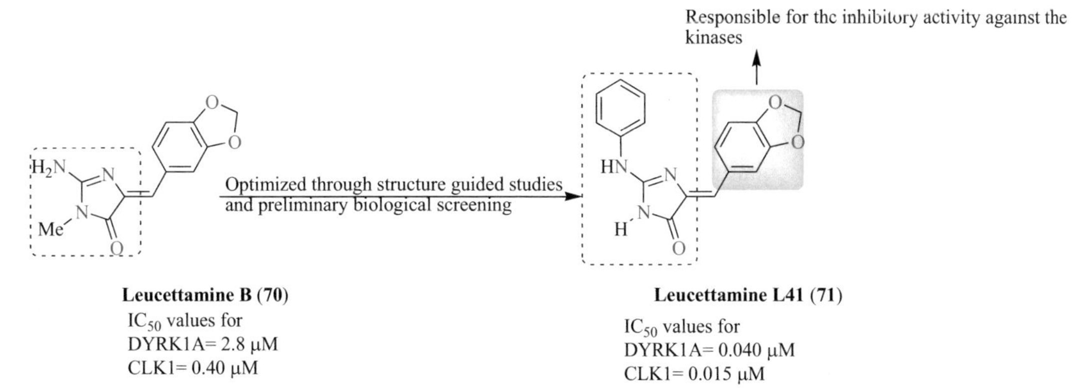

Through structure-guided studies and on the basis of preliminary biological screening of various leucettamine B (**70**) derivatives, leucettamine L41 (**71**) has been reported in the past as dual-specific inhibitor of DYRK and CLK kinases. The co-crystal structure of leucettamine L41 (**71**) with CLK3 has been reported previously [118]. In their report [119], with an aim to gain better understanding of kinase-inhibitor interaction, the authors docked leucettamine L41 (**71**) within the ATP-binding pocket of various kinases like DYRK1A, DYRK2 and CLK3. Molecular docking studies revealed that leucettamine L41 (**71**) formed two direct polar bonds with both DYRK1A and DYRK2 kinases (Fig. 44a, b). One hydrogen bond was formed with Leu241 and another one with Lys188 residues. This study also disclosed that ATP-binding

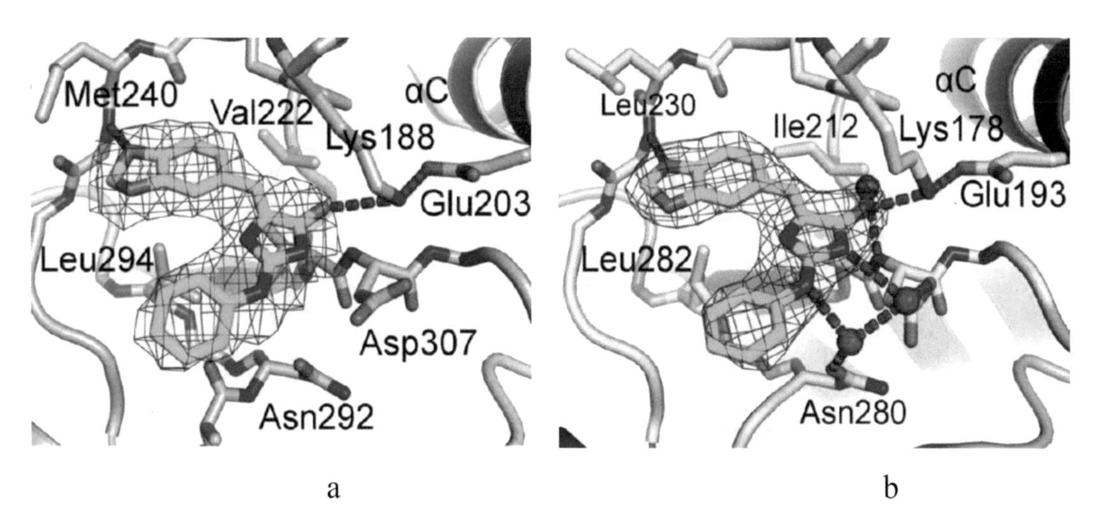

Fig. 44 Docked pose of leucettamine L41 (**71**) within the ATP-binding pocket of (**a**) DYRK1A and (**b**) DYRK2. *Dotted lines* represent the key polar interactions [119] (Used with permission from Tahtouh et al. J Med Chem 2012, 55, 9312–9330 [Copyright© 2012, American Chemical Society])

sites of both the kinases, i.e. DYRK1A and DYRK2, were similar except for three different residues (Val222 vs isoleucine, Met240 vs leucine and Val306 vs Ile294 in DYRK1A vs DYRK2 kinases, respectively) [119].

A series of 1-aza-9-oxafluorene derivatives have been reported by Tell et al. through a systematic structure-directed substitution pattern (on 1-aza-9-oxafluorene scaffold) to get potent kinase inhibitors for the treatment of Alzheimer's disease. 3-Benzyloxy-substituted compound (**72**) has emerged as a potent inhibitor of both CDK1/B and CDK5/p25 kinases with *K*i values of 0.09 and 2.1 μM, respectively. Docking studies were carried out to elucidate the binding modes of compound (**72**) with both the kinases, i.e. CDK1/B and CDK5/p25 (Fig. 45a, b). The results of the docking studies demonstrated that the nitrogen atom of the pyridine ring tended to form hydrogen bonds with the NH amide of Leu83 (CDK1) and of Cys83 (CDK5) residues within the hinge regions of both the kinases. 3-Benzyloxy substituent of **72** was oriented inside the hydrophobic pocket close to the gatekeeper residue Phe80 of both the kinases (CDK1 and CDK5). Hydroxyl group of the compound (**72**) forms hydrogen bond with the C = O of Ile10 in both CDK1 and CDK5 kinases [120].

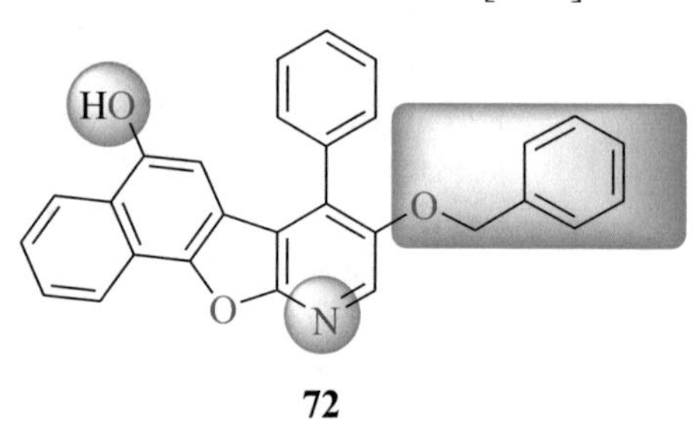

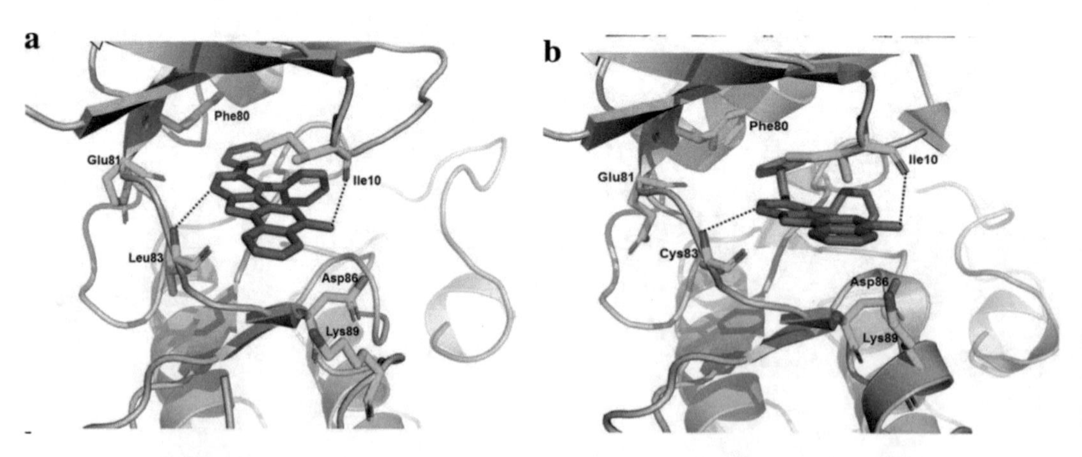

Fig. 45 Key interactions of compound (**72**) within the ATP-binding site of both the kinases CDK1 and b) CDK5 [120] (Used with permission from Tell et al. MedChemComm 2012, 3, 1413–1418 [Copyright © 2012, Royal Society of Chemistry])

Waiker et al. disclosed a novel series of 4-anilinoquinazoline derivatives and evaluated them against five protein kinases, namely, CDK5/p25, CK1δ/ε, GSK-3α/β, DYRK1A and CLK1. These protein kinases act as druggable targets for the treatment of Alzheimer's disease. Among the reported series, compounds (**73** and **74**) showed promising activity against CLK1 with IC_{50} values of 1.5 and 7.6 μM, respectively. Compound (**73**) inhibited GSK-3α/β kinases with an IC_{50} value of 3 μM, while no inhibition of GSK-3α/β kinase was reported by compound (**74**). Molecular docking studies of compounds (**73** and **74**) were performed to elucidate the binding pattern of these compounds within the ATP-binding sites of the two kinases (CLK1 and GSK-3α/β). Docked poses of compounds (**73** and **74**) within the active site of CLK1 kinase as depicted in Fig. 46a, b showed that nitrogen (N1) of quinazoline ring of both the compounds (**73** and **74**) facilitated hydrogen bonding with Leu241 residue. This H-bonding was considered to be one of the important interactions for inhibitory activity. Compound (**73**) also tends to form crucial hydrogen bonds with Leu241, Glu292 and Asn293 residues within the active site of CLK1 enzyme. Docking studies of compound (**73**) within the ATP-binding site of GSK-3α/β kinase revealed that amino group between the two aryl rings tended to form important hydrogen bond with Val135 (Fig. 47), which was considered as a key interaction for the inhibition of GSK-3α/β kinase [121].

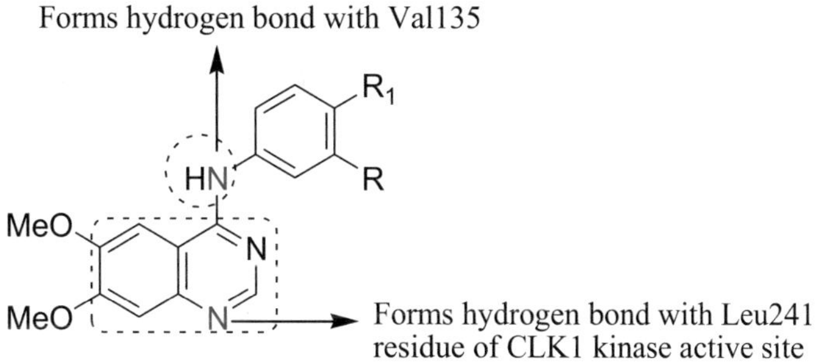

Forms hydrogen bond with Val135

R_1 R HN MeO MeO N N

Forms hydrogen bond with Leu241 residue of CLK1 kinase active site

73 and **74**

73: R=R1= OMe;
74: R= Cl, R_1= F

Based on their previous report [122] of *N*-arylbenzothieno[3,2-*d*]pyrimidin-4-amines and their pyrido- and pyrazino-analogs as multitargeted kinase inhibitors, Loidreau et al. disclosed a new

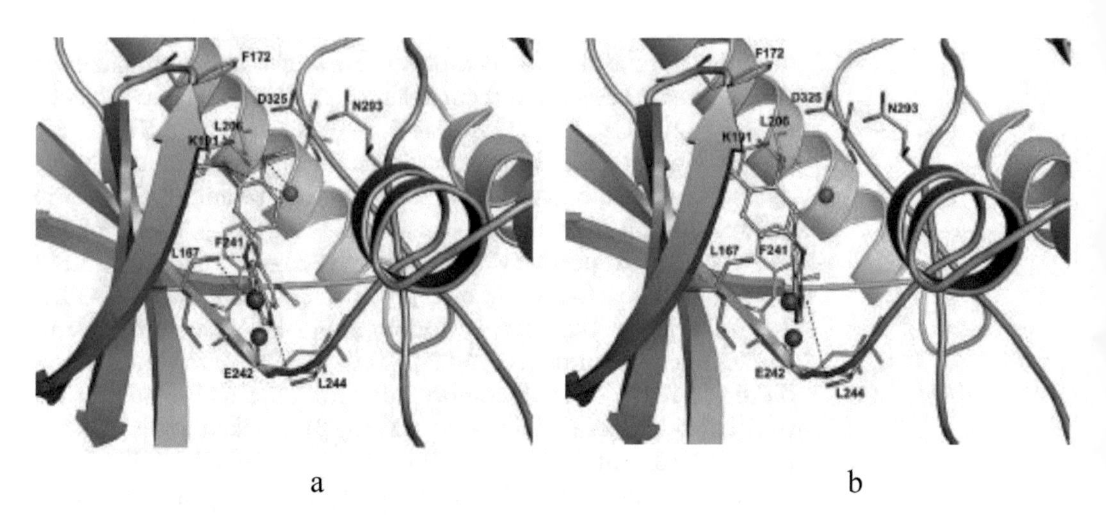

Fig. 46 Docked poses of compounds (**73**) (a) and (**74**) (b) within the active site of CLK1 [121] (Used with permission from Waiker et al. Bioorg Med Chem 2014, 22, 1909–1915 Copyright © 2014 Elsevier Ltd.])

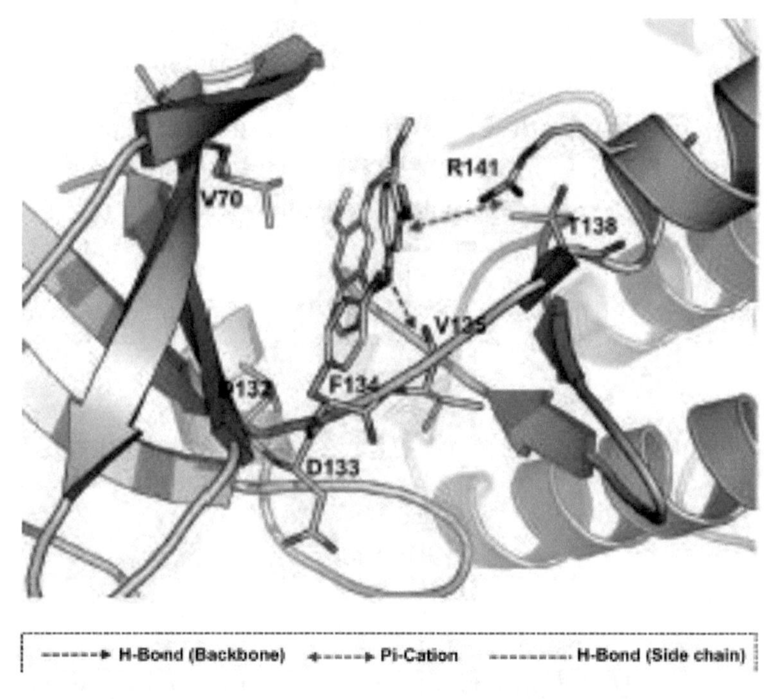

Fig. 47 Docked pose of compound (**73**) within the ATP-binding site of GSK-3α/β [121] (Used with permission from Waiker et al. Bioorg Med Chem 2014, 22, 1909–1915. [Copyright © 2014 Elsevier Ltd.])

series of C8-substituted pyrido[3′,2′:4,5]thieno[3,2-*d*]pyrimidines. In the reported series, compound (**75**) emerged as the most potent compound against both CK1δ/ε and CLK1 kinases with IC_{50} values of 0.22 and 0.088 μM, respectively. 7-Substituted isomer (**76**) of the most potent compound (**75**) showed somewhat low activity against both the kinases, i.e. CK1δ/ε and CLK1 with IC_{50} values of 0.22 and 0.11 μM, respectively. Binding modes of compound (**75**) within the active sites of both the enzymes, i.e. CK1δ/ε and CLK1, have been studied with the help of molecular modelling technique. The docking study revealed that 2,4-dichlorophenyl group of **75** was inclined towards the phosphate-binding loop (P-loop), which is considered to be one of the most flexible regions of the kinases. As far as interaction of this group within the CLK1 and CK1δ/ε kinases is concerned, pi-pi interaction with Phe172 within CLK1 was observed. Whereas in CK1δ/ε kinase the corresponding amino acid Phe20 was far away from the active site, it might not had any impact on the binding affinity. Docking studies of compound (**76**) revealed that the substituent on seventh position led to steric clash with the P-loop of CK1 kinase which resulted into complete loss of its activity, and against CLK1 it showed only partially sparing activity. Overall, this study concluded that affinity against the kinases depended upon the type of heteroaryl substituents [123].

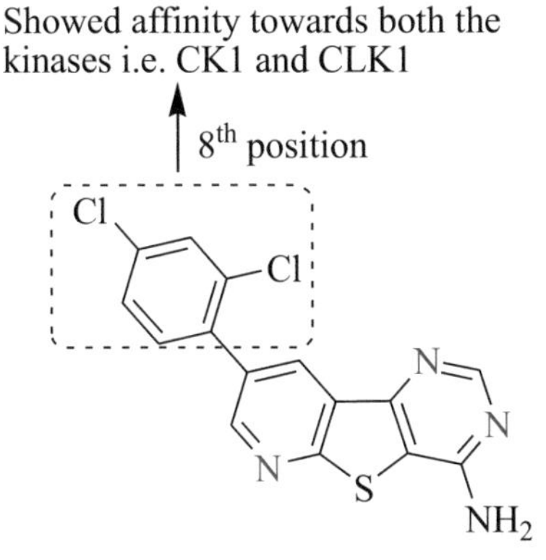

Showed affinity towards both the kinases i.e. CK1 and CLK1

8th position

Causes steric clash and complete loss of activity gainst CK1 kinase

7th position

75 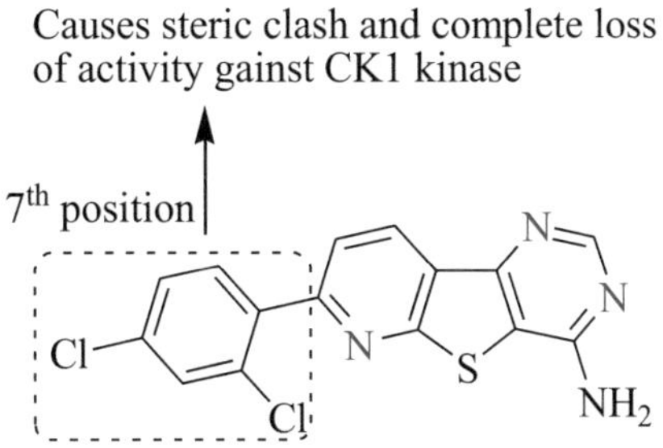 **76**

A novel series of imidazo[1,2-*α*]pyridine analogs obtained through the reaction between 2-aminopyridine and 2-bromoacetophenone have been reported by Lawson et al. having inhibitory activity against DYRK1A and CLK1 kinases. To gain structural insights for the designed prototypes, compounds were synthesized with modifications in ring A or B. Among the disclosed series, compound (77) was identified as the most potent compound with IC_{50} values of 0.7 and 2.6 μM, respectively, against CLK1 and DYRK1A kinases. Molecular modelling studies of compound (77) within CLK1 indicated that imidazo[1,2-*α*]pyridine core tended to form two hydrogen bonds with the backbone residues, i.e. with carbonyl of Glu242 and NH of Leu244. Bromo substituent of compound (77) acts as the hydrogen bond acceptor group present in most of the kinase inhibitors by forming a 2.7 Å distance from the backbone carbonyl of Glu242, and as per the authors' claim, this could be one of the possible reasons for its good binding affinity. Hydroxybenzamide core was also found to be responsible for the formation of two hydrogen bonds, i.e. one between NH_2 of the amide and the carbonyl of the side chain Asn293 and the other between the phenol and the backbone carbonyl of Glu169 residue. Putative binding sites of DYRK1A with compound (77) were similar to that of CLK1 except that the Phe170 in DYRK1A protruded outside the loop, whereas it pointed inside in the CLK1 [124].

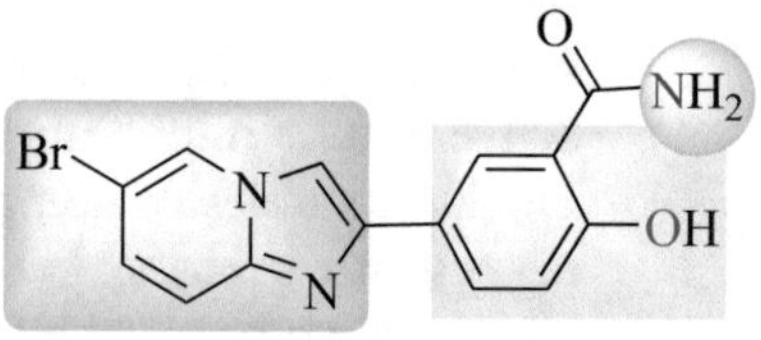

77

7 Selectivity of Kinase Inhibitors: Computational Strategies and Related Issues for the Development of Selective Kinase Inhibitors

Aberrant protein kinases are considered as progenitors for a number of diseases like inflammation, diabetes, cardiac disorders, Alzheimer's and for causing cancer in particular [well documented, *125*]. Kinases have emerged as the major therapeutic targets for the discovery of new drugs for these diseases in the past one decade [125]. Some 518 protein kinases are reportedly identified at present in the mammalian genome [126]. The first report on the crystal structure of protein kinase domain in 1991 opened up new avenues for the development of kinase inhibitors as potential drugs [127]. After the discovery of imatinib, the first small-molecule kinase inhibitor, researchers got more curious to map the kinase-inhibitor interactions [128]. As these kinases are linked to the development of a large variety of diseases, target selectivity is an important issue to design a drug molecule for that target. Absolute selectivity for a particular kinase may not be an essential feature for efficacy of an inhibitor for therapeutic purposes if the inhibitor shows higher selectivity for a limited number of clinically relevant targets than targeting a single kinase with the aim to achieve greater therapeutic efficacy [129, 130]. In today's paradigm, majority of the selective kinase inhibitors available in the market have evolved serendipitously, and optimization of strategies for the development of selective kinase inhibitors remains an exigent task [128].

Availability of co-crystallized three-dimensional structures of kinases with diverse inhibitors has helped in identifying the binding sites in the kinases. Type I inhibitors are ATP-mimetic compounds that bind to the active ATP-binding site (DFG in state), while Type II inhibitors bind to the ATP-binding site (inactive, DFG-out state) as well as to an allosteric pocket adjacent to the ATP-binding site. Type III inhibitors are allosteric inhibitors that bind outside the ATP site in such a way that leads to displacement of ATP molecule from its binding site in the kinase. Type IV inhibitors bind to some surface pockets and target the kinases outside the catalytic domain interfering with binding of an interacting kinase regulator protein. ATP-binding active catalytic domain is well defined and considerably more rigid than the inactive state of the kinase. The rigid nature of the active state can be exploited to design inhibitors with excellent shape complementarity. Additionally, some kinases have unique active features such as some specific amino acid(s) in key positions that can be targeted for designing irreversible kinase inhibitors. One such key residue in the ATP-binding pocket is the so-called gatekeeper amino acid located close to the hinge region [128]. Kinase inactive states are structurally highly diverse and dynamic suggesting that inhibitors (Type II) targeting such states may have a better chance of being selective. But, the DFG-out conformation alone and its dynamic adjustable surface per se offer no apparent advantage for the development of specific inhibitors [128]. The phosphate-binding loop (P-loop) is a flexible structural component. In the inhibitor-kinase complex, the P-loop may fold into the ATP site, forming aromatic stacking interactions. The folded P-loop conformation has been strongly correlated with favourable inhibitor selectivity [131]. Stringent requirements for inhibitor binding to the kinase may make this binding mode useful for the development of selective inhibitors for a particular kinase. Type III and Type IV inhibitors are usually very selective because their targeted binding sites are usually unique to a particular kinase [132]. The main challenge in developing Type III inhibitors is that either the allosteric site is not present in the kinase structure or it is not known that targeting a certain binding cavity will result in kinase inhibition.

Research on employing computational studies for resolving the selectivity aspect of the inhibitors for different kinases has remained slow due to two major factors. Kinase-inhibitor selectivity is generally not reported in literature for majority of the tested compounds because absolute inhibitor specificity towards an individual kinase is not necessary to achieve a therapeutic effect, and the kinase research has been principally focused on a small subset of the kinome. Secondly, the traditional kinase-inhibitor screening process, in which the activity (inhibition of phosphorylation activity or binding affinity) of a compound (inhibitor) is determined against the entire kinome, is a low-throughput process. However, the application of computational methods have shown an upward trend in recent years due to the availability of results of high-throughput studies of large compound data sets against large subsets of human kinome. These studies could identify target specificities of a large set of kinase inhibitors for specific kinases, which were used for computational learning techniques. Additionally, a large and ever-increasing number of 3D structures of whole kinases or kinase domain are available now (some as co-crystallized structures with inhibitors) in Protein Data Bank, which provide a rich source of information about kinase-binding sites [133].

Computational approaches beyond conventional QSAR approaches, providing multidimensional structure-activity relationship for hundreds of targets at the same time, have been employed. One type of approach uses numerical descriptors of physicochemical, structural and/or geometrical properties of both the ligands and the targets for machine-learning methods for classification (binding/non-binding) or regression on the measured inhibition descriptors. Niijima et al. and Cao et al. proposed a similar kinase/inhibitor deconvolution approach in which the whole kinase sequences on the ATP-binding pockets were deconstructed into residues and compounds into chemical fragments or in topological day light fingerprints using kinase SARfari and Metz data sets [134, 135]. Using GVK Biosciences kinase-inhibitor database, Yabuuchi et al. developed a method called CGBVS (chemical genomics-based virtual screening) in which compounds were represented by a large set of substructure descriptors and physicochemical properties, and the protein descriptors were computed from the protein sequence dipeptide composition using a string kernel [136]. Schurer and Muskal employed the Eidogen-Sertanty KKB Q4 2009 release involving more than 4,30,000 kinase-compound pairs for filtering, standardization and clustering procedures for classifying the compounds into actives and inactives using extended connectivity fingerprints [137].

Another type of approach has utilized 3D structures of the kinases to obtain more accurate representation of kinase-binding sites. Caffrey et al. utilized a subset of kinase-inhibitor pairs extracted from the Fabian and Karaman dataset. An algorithm was developed to predict specificity determinants for the design of drugs with a desired specificity [138]. X-React^{KIN}, a machine-learning method, was developed for assessment of cross reactivity of the inhibitors for various kinases. Similarity between the kinases was computed by different metrics using structure-, sequence- and ligand-binding profiles [139]. Using similarity between pairs of kinases, a network of kinase-binding sites was constructed which recapitulated well a network based on the similarity between the inhibition profiles in the Karaman dataset. Integration of the binding site similarity network with the inhibition profile network led to inferences of off-target interactions [140]. Using Karaman dataset for in silico docking procedures, cluster-specific residues acting as interaction hot spots were identified and converted into a series of descriptors used for RF training. The RF was then used for the prediction of novel kinase-inhibitor relationships [141].

Data from different resources can be integrated in order to provide a unified view on kinome inhibition. The whole kinase-inhibitor data can be represented as a network where binding can be treated as a binary on-off relation or weighed by the affinity or strength of the inhibitory effect. This type of network aids in the identification and rationalization of secondary effects of the inhibitors. KIDFamMap [142] and K-Map [143] are two such networks reported in public domain. Tang et al. integrated the Metz, Davis and Anastassiadis datasets with ChEMBL and STITCH datasets into a network resulting in a kinase-inhibition bioactivity map comprising of 467 kinases and more than 50,000 compounds [144].

One of the foremost criteria for designing of selective kinase inhibitors is to identify the off-targets which affect the selectivity of the inhibitor for a particular kinase [145]. Reverse docking is a potential computational tool to identify the off-targets [146]. Strategies based on virtual screening along with X-ray co-crystal structure-guided design have been applied for the development of selective Aurora kinase A inhibitors [147]. Reverse pharmacophore mapping approach has been used for selecting compounds acting as tyrosine kinase inhibitors [148, 149].

8 Expert View

AD is one of the most catastrophic and prime causes of deaths in the old age people. It affects patient's ability to judge, recognize and perform everyday activities along with the impairment of process of thought and language. WHO anticipated that AD will affect 75.6 million people by 2030, and this numerical value will multiply thrice by 2050. Although extensive research on AD is under way, uncovering the exact causes of AD is anxiously awaited.

Various targets for the treatment of AD are reported in the literature, but inhibition of some aberrant protein kinases is regarded as an important therapeutic strategy due to the active participation of these kinases in the hyperphosphorylation of tau protein. Computational techniques have always proved informative to the researchers as they provide a peeping window into the molecular interactions taking place inside the target site with the ligand molecules. The present chapter offers a cascade of information about the computational studies performed on a variety of organic molecules acting as kinase inhibitors having the potential to be developed as future drugs for the treatment of AD.

GSK-3β is the most highly explored target among the protein kinases involved in the aetiology of AD. It has been seen that fused heterocyclics such as indoles, benzazoles, indazoles and simple heterocycles like oxazoles, pyrazines and pyrimidines have offered highly active compounds for the inhibition of GSK-3β kinase. Molecular docking studies revealed that most of the test compounds bind to the ATP-binding site of GSK-3β and thus acted as ATP-competitive inhibitors. Compound (**6**) showed higher selectivity towards GSK-3β kinase due to the existence of van der Waals interactions between the bromo group of the inhibitor (**6**) and Leu132 residue in the active site of GSK-3β kinase. The same molecule (**6**) has also been tested for its selectivity towards the CDKs, but it showed poor selectivity revealing the fact that the presence of phenylalanine residue in the ATP-binding cavity of CDKs hinders the binding of bromo group into the active site of CDKs. As GSK has two isoforms GSK-α and β, selectivity of inhibitors for these isoforms is dependent upon a combination of different substituents present in the inhibitor molecules. Two most active compounds (**28** and **29**) having IC_{50} values of 0.73 and 1.1 nM, respectively, revealed that Asp133, Val135 and Lys183 are the key active site residues essential for both potency and selectivity of the inhibitors for the GSKs. Other potent compounds (**6–8**) also showed the same interactions with these residues.

As far as the inhibitors of cyclin-dependent kinases are concerned, aminothiazole, aminoimidazole, benzimidazole and diaminothiazole containing derivatives and (*R*)-roscovitine have proved their potential. Through virtual screening and on the basis of preliminary biological data, one (**47**) of the most potent compounds of CDK5 has been discovered with an IC_{50} value of 16 nM. Docking studies of compound (**47**) revealed that Gln85, Lys89 and Asp86 residues within the ATP-binding site of CDK5 were essential for activity. A significant number of reports are available about (*R*)-roscovitine, and its derivatives having potential activity against CDK5, e.g. compound (**53**), have shown an IC_{50} value of 17 nM. Some reports are also available regarding reversal of the selectivity of compounds from CDK5 to CDK2 through the replacement of exocyclic NH in (*R*)-roscovitine moiety with ether linkage as seen in compound (**51**) showing the importance of exocyclic NH in (*R*)-roscovitine for selectivity towards CDK5.

Scaffold repurposing and crystallographic structure elucidation studies of compound (**55**, Ki value of 101 nM) proved it a potent inhibitor of p38α MAPK. The naphthyl ring of compound (**55**) played an important role in selectively inhibiting p38α MAPK as it formed H-bonds with Leu108-Met109 backbone residues.

In dual-specificity tyrosine-regulated kinases (DYRKs), DANDY analog (**58**) was found to be a potent inhibitor with an IC_{50} value of 3 nM, and the affinity of this analog for the target was mainly due to combined interactions with Lys188 and Asp247 residues. Other scaffolds that have provided significant inhibitors of DYRKs are meridianin analog (**60**) and bis-indole indirubin (**59**).

Fyn kinase is the least-explored class of protein kinases involved in the aetiology of AD, and this target could prove to be useful to strengthen the ongoing research on kinase inhibitors as exemplified by two compounds (**64** and **65**). These compounds (**64** and **65**) offered Ki values of 70 and 95 nM, respectively, for this kinase.

Some publications have also come out in the form of dual inhibitors of protein kinases, and a significant number of such analogs have been reported with potential use in AD. In dual kinase inhibitors, indirubin analog (**6**) potentially inhibited GSK-3α/β, CDK1 and CDK5 with IC_{50} values of 0.005, 0.32 and 0.083 μM, respectively, and it showed high selectivity for GSK-3α/β. Another dual inhibitor of kinases, leucettamine L41 (**71**) showed a significant inhibition of DYRK1A and CLK1 kinases possessing the respective IC_{50} values of 0.040 and 0.015 μM.

Due to the involvement of abnormally working kinases in various diseased states in our body, selectivity of the inhibitors for particular kinase or set of kinases is an important issue. With the deposition of large number of 3D structures of kinases in the Protein Data Bank including some co-crystallized structures with a large number of ligands, computational studies have been reported to resolve the selectivity issue. Multidimensional structure-activity relationships for hundreds of targets simultaneously have been reported. Various machine-learning methods have been developed for off-target interactions. The kinase-inhibitor data, represented as a network, has been used for determining on-off relation of the ligands with the target kinases.

Overall, the literature covered in this chapter divulges important findings involving computational modelling of organic compounds as kinase inhibitors which could afford vital leads for discovering new molecules as novel AD therapeutics to be used for the management of AD. With the advancements in computational techniques, it is hoped that selectivity issues for AD-targeted kinases would be resolved favourably to offer selective kinase inhibitors for the treatment of AD in very near future. With a better understanding of the aetiology of AD and involvement of computational technique, the day is not far off when we have a kinase inhibitor as a drug to treat AD.

Acknowledgement

The authors thank Department of Science and Technology, Government of India, for bestowing DST-FIST status to the Department.

References

1. Chinag K, Koo EH (2014) Emerging therapeutics for Alzheimer's disease. Annu Rev Pharmacol Toxicol 84:381–405
2. Face sheet on Dementia (2016) http://www.who.int/mediacentre/factsheets/fs362/en/. Accessed on 14 Jan 2017
3. The world Alzheimer report 2015: the global impact of dementia. https://www.alz.co.uk/research/world-report-2015. Accessed on 14 Jan 2017
4. The world Alzheimer report 2016: improving healthcare for people living with dementia: coverage, quality and costs now and in the future. https://www.alz.co.uk/research/world-report-2016. Accessed on 14 Jan 2017
5. Blacker D, Tanzi RE (1998) The genetics of Alzheimer disease. Arch Neurol 55:294–296
6. Folch J et al (2016) Current research therapeutic strategies for Alzheimer disease treatment. Neural Plast 2016:Article id 8501693
7. Selkoe DJ, Schenk D (2003) Alzheimer's disease: molecular understanding predicts amyloid-based therapeutics. Annu Rev Pharmacol Toxicol 43:545–584
8. Iqbal K, Gong C, Liu F (2014) Microtubule-associated protein tau as a therapeutic target in Alzheimer's disease. Expert Opin Ther Targets 18:1–12
9. Shidore M et al (2016) Benzylpiperidine-linked diarylthiazoles as potential anti-Alzheimer's agents: synthesis and biological evaluation. J Med Chem 59:5823–5846
10. Citron M (2010) Alzheimer's disease: strategies for disease modification. Nat Rev 9:387–398
11. Scarpini E, Schelten S, Feldman H (2003) Treatment of Alzheimer's disease: current status and new perspectives. Lancet Neurol 2:539–547
12. Mitra A, Dey B (2013) Therapeutic interventions in Alzheimer's disease. In: Uday K (ed) Neurodegenerative disease chapter 12, Intech, Rijeka, Croatia, pp 291–317
13. Mizuno S et al (2012) Alzpathway: a comprehensive map of signalling pathways of Alzheimer's disease. BMC Syst Biol 6:article 52
14. Chico LK, Eldik LJ, Watterson DM (2009) Targeting protein kinases in central nervous system disorders. Nat Rev Drug Discov 8:892–909
15. Pietri M et al (2013) PDK1 decreases TACE-mediated α-secretase activity and promotes disease progression in prion and Alzheimer's disease. Nat Med 19:1124–1131
16. Ma T et al (2013) Suppression of eIF2α kinases alleviates Alzheimer's disease-related plasticity and memory deficits. Nat Neurosci 16:1209–1305
17. Borders AS, de Almeida L, Van Eldik LJ, Watterson DM (2008) The p38α mitogen-activated protein kinase as a central nervous system drug discovery target. BMC Neurosci 9:S12
18. Martinez A, Perez DI (2008) GSK-3 inhibitors: a ray of hope for the treatment of Alzheimer's disease? J Alzheimers Dis 15:181–191
19. Mueller BK, Mack H, Teusch N (2005) Rho kinase, a promising drug target for neurological disorders. Nat Rev Drug Discov 4:387–398
20. Schumacher AM, Velentza AV, Watterson DM (2002) Death associated protein kinase as a potential therapeutic target. Exp Opin Ther Targets 6:497–506
21. Iqbal K et al (2005) Tau pathology in Alzheimer's disease and other tauopathies. Biochim Biophys Acta 1739:198–210
22. Hanger DP et al (2007) Novel phosphorylation sites in tau from Alzheimer brain support a role for casein kinase 1 in disease pathogenesis. J Biol Chem 282:23645–23654
23. Mandelkow EM et al (2004) MARK/PAR1 kinase is a regulator of microtubule-dependent transport in axons. J Cell Biol 167:99–110
24. Cruz JC, Tsai LH (2004) CDK5 deregulation in the pathogenesis of Alzheimer's disease. Trends Mol Med 10:452–458
25. Anderton BH et al (2001) Sites of phosphorylation in tau and factors affecting their regulation. Biochem Soc Symp 67:73–80
26. West S., Bhugra P. (2015) Emerging drug targets for Aβ and tau in Alzheimer's disease: a systematic review. Br J Clin Pharmacol 80:221–234
27. Shefet-Carasso L, Benhar I (2015) Antibody-targeted drugs and drug resistance—challenges and solutions. Drug Resist Updat 18:36–46
28. Boutajangout A, Sigurdsson EM, Krishnamurthy PK (2011) Tau as therapeutic target for Alzheimer's disease. Curr Alzheimer Res 8:666–667
29. Jayapalan S, Subramanian D, Natarajan J (2016) Computational identification and analysis of neurodegenerative disease associated protein kinases in hominid genomes. Genes Diseases 3:228–237

30. Embi N, Rylatt DB, Cohen P (1980) Glycogen synthase kinase-3 from rabbit skeletal muscle. Eur J Biochem 107:519–527
31. Ombrato R et al (2015) Structure-based discovery of 1 h-indazole-3-carboxamides as a novel structural class of human GSK-3 inhibitors. J Chem Inf Model 55:2540–2551
32. Grimes CA, Jope RS (2001) The multifaceted roles of glycogen synthase kinase 3β in cellular signaling. Prog Neurobiol 65:391–426
33. Woodgett JR (2001) Judging a protein by more than its name: GSK-3. Sci STKE 100:8–12
34. Frame S, Cohen P (2001) GSK3 takes centre stage more than 20 years after its discovery. Biochem J 359:1–16
35. Lim S, Kaldis P (2013) CDKs, cyclins and CKIs: roles beyond cell cycle regulation. Development 140:3079–3093
36. Malumbres M (2014) Cyclin-dependent kinases. Genome Biol 15:122
37. Crus JC et al (2006) p25/cyclin-dependent kinases 5 induces production and intraneuronal accumulation of amyloid β in vivo. J Neurosci 26:10536–10541
38. Paglini G, Caceres A (2001) The role of the CDK5–p35 kinase in neuronal development. Eur J Biochem 268:1528–1533
39. Flaherty DB et al (2000) Phosphorylation of human tau protein by microtubule-associated kinases: GSK3beta and CDK5 are key participants. J Neurosci Res 62:463–472
40. Azevedo WF et al (1997) Inhibition of cyclin-dependent kinases by purine analogues: crystal structure of human CDK2 complexed with roscovitine. Eur J Biochem 243:518–526
41. Meijer L, Kim SH (1997) Chemical inhibitors of cyclin-dependent kinases. Methods Enzymol 283:113–128
42. Meijer L, Raymond E (2003) Roscovitine and other purines as kinase inhibitors. From starfish oocytes to clinical trials. Acc Chem Res 36:417–425
43. Mettey Y et al (2003) Aloisines, a new family of CDK/GSK-3 inhibitors. SAR study, crystal structure in complex with CDK2, enzyme selectivity, and cellular effects. J Med Chem 46:222–236
44. Hoessel R et al (1999) Indirubin, the active constituent of a Chinese antileukaemia medicine, inhibits cyclin-dependent kinases. Nat Cell Biol 1:60–67
45. Ferrer I et al (2005) Constitutive DYRK1A is abnormally expressed in Alzheimer's disease, down syndrome, pick disease, and related transgenic models. Neurobiol Dis 20:392–400
46. Varjosalo M et al (2008) Application of active and kinase-deficient kinome collection for identification of kinases regulating hedgehog signalling. Cell 133:537–548
47. Aranda S et al (2011) DYRK family of protein kinases: evolutionary relationships, biochemical properties, and functional roles. FASEB J 25:449–462
48. Panzer P et al (2001) Identification of the autophosphorylation sites and characterization of their effects in the protein kinase DYRK1A. Biochem J 359:497–505
49. Becker W, Sippl W (2011) Activation, regulation, and inhibition of DYRK1A. FEBS J 278:246–256
50. Kyriakis JM, Avruch J (2001) Mammalian mitogen-activated protein kinase signal transduction pathways activated by stress and inflammation. Physiol Rev 81:808–869
51. Shaul YD, Seger R (2006) The MEK/ERK cascade: from signalling specificity to diverse functions. Biochim Biophys Acta 1773:1213–1226
52. Cuenda A, Rousseau S (2007) p38 MAP-kinases pathway regulation, function and role in human diseases. Biochim Biophys Acta 1773:1358–1375
53. Sheng JG et al (2001) Interleukin-1 promotion of MAPK-p38 overexpression in experimental animals and in Alzheimer's disease: potential significance for tau protein phosphorylation. Neurochem Int 39:341–348
54. Sun A et al (2003) p38 MAP kinase is activated at early stages in Alzheimer's disease brain. Exp Neurol 183:394–405
55. Gross SD, Anderson RA (1998) Casein kinase I: spatial organization and positioning of a multifunctional protein kinase family. Cell Signal 10:699–711
56. Knippschild U et al (2005) The casein kinase 1 family: participation in multiple cellular processes in eukaryotes. Cell Signal 17:675–689
57. Schittek B, Sinnberg T (2014) Biological functions of casein kinase 1 isoforms and putative roles in tumorigenesis. Mol Cancer 13:231
58. Hanks SK, hunter T. (1995) Protein kinases 6. The eukaryotic protein kinase superfamily: kinase (catalytic) domain structure and classification. FASEB J 9:576–596

59. Flajolet M et al (2007) Regulation of Alzheimer's disease amyloid-β formation by casein kinase I. PNAS 104:4159–4164
60. Knippschild U et al (2005) The role of casein kinase 1 (ck1) family in different signalling pathways linked to cancer development. Onkologie 28:508–514
61. Flotow H, Roach PJ (1989) Synergistic phosphorylation of rabbit muscle glycogen synthase by cyclic AMP-dependent protein kinase and casein kinase I. Implications for hormonal regulation of glycogen synthase. J Biol Chem 264:9126–9128
62. O'Day DH, Eshak K, Myre MA (2015) Calmodulin binding proteins and Alzheimer's disease. J Alzheimers Dis 46:553–569
63. Chin D, Means AR (2000) Calmodulin: a prototypical calcium sensor. Trends Cell Biol 10:322–328
64. Lee SJ et al (2009) Activation of CaMKI in single dendritic spines during long-term potentiation. Nature 458:299–304
65. Terry RD et al (1991) Physical basis of cognitive alterations in Alzheimer's disease: synapse loss is the major correlate of cognitive impairment. Ann Neurol 30:572–580
66. Lucchesi W, Mizuno K, Giese KP (2011) Novel insights into CaMKII function and regulation during memory formation. Brain Res Bull 85:2–8
67. Ghosh A, Giese KP (2015) Calcium/calmodulin-dependent kinase II and Alzheimer's disease. Mol Brain 8:78
68. Hallows KR et al (2009) AMP-activated protein kinase inhibits alkaline pH- and PKA-induced vacuolar H+−ATPase accumulation in epididymal clear cells. Am J Physiol Cell Physiol 296:C672–C681
69. Liang Z et al (2007) Down-regulation of cAMP-dependent protein kinase by over-activated calpain in Alzheimer disease brain. J Neurochem 103:2462–2470
70. Shi J et al (2011) Cyclic AMP-dependent protein kinase regulates the alternative splicing of tau exon 10. J Biol Chem 286:14639–14648
71. Amini E et al (2015) Paradoxical role of PKA inhibitor on amyloidβ-induced memory deficit. Physiol Behav 149:76–85
72. Litersky JM et al (1996) Tau protein is phosphorylated by cyclic AMP-dependent kinase and calcium/calmodulin-dependent protein kinase II within its microtubule-binding domains at Ser-262 and Ser-356. Biochem J 316:655–660
73. Resh MD (1998) Fyn, a Src family tyrosine kinase. Int J Biochem Cell Biol 30:1159–1162
74. Lee G et al (2004) Phosphorylation of tau by Fyn: implications for Alzheimer's disease. J Neurosci 24:2304–2312
75. Nygaard HB, Dyck CH, Strittmatter SM (2014) Fyn kinase inhibition as a novel therapy for Alzheimer's disease. Alzheimers Res Ther 6:8
76. Meijer L et al (2003) GSK-3-selective inhibitors derived from Tyrian purple indirubins. Chem Biol 10:1255–1266
77. Vougogiannopoulou K et al (2008) Soluble 3′, 6-substituted indirubins with enhanced selectivity toward glycogen synthase kinase-3 alter circadian period. J Med Chem 51:6421–6431
78. Nisha CM et al (2016) Docking and ADMET prediction of few GSK-3 inhibitors divulges 6-bromoindirubin-3-oxime as a potential inhibitor. J Mol Graph Model 65:100–107
79. Voigt B et al (2008) Probing novel 1-Aza-9-oxafluorenes as selective GSK-3β inhibitors. ChemMedChem 3:120–126
80. Chioua M et al (2009) Synthesis and biological evaluation of 3, 6-diamino-1*H*-pyrazolo [3, 4-*b*] pyridine derivatives as protein kinase inhibitors. Bioorg Med Chem Lett 19:4566–4569
81. Saitoh M et al (2009) Design, synthesis and structure–activity relationships of 1, 3, 4-oxadiazole derivatives as novel inhibitors of glycogen synthase kinase-3β. Bioorg Chem 17:2017–2029
82. Saitoh M et al (2009) 2-{3-[4-(Alkylsulfinyl) phenyl]-1-benzofuran-5-yl}-5-methyl-1, 3, 4-oxadiazole derivatives as novel inhibitors of glycogen synthase kinase-3β with good brain permeability. J Med Chem 52:6270–6286
83. Onishi T et al (2011) A novel glycogen synthase kinase-3 inhibitor 2-methyl-5-(3-{4-[(S)-methylsulfinyl] phenyl}-1-benzofuran-5-yl)-1, 3, 4-oxadiazole decreases tau phosphorylation and ameliorates cognitive deficits in a transgenic model of Alzheimer's disease. J Neurochem 119:1330–1340
84. Perez DI et al (2009) Thienylhalomethylketones: irreversible glycogen synthase kinase 3 inhibitors as useful pharmacological tools. Bioorg Med Chem 17:6914–6925
85. Monte FL et al (2011) Synthesis and biological evaluation of glycogen synthase

kinase 3 (GSK-3) inhibitors: an fast and atom efficient access to 1-aryl-3-benzylureas. Bioorg Med Chem Lett 21:5610–5615
86. Monte FL et al (2012) Identification of glycogen synthase kinase-3 inhibitors with a selective sting for glycogen synthase kinase-3α. J Med Chem 55:4407–4424
87. Monte FL et al (2013) Structure-based optimization of oxadiazole-based GSK-3 inhibitors. Eur J Med Chem 61:26–40
88. Zhang P et al (2012) Identification of novel scaffold of benzothiazepinones as non-ATP competitive glycogen synthase kinase-3β inhibitors through virtual screening. Bioorg Med Chem Lett 22:7232–7236
89. Zhang P et al (2013) Design, synthesis and biological evaluation of benzothiazepinones (BTZs) as novel non-ATP competitive inhibitors of glycogen synthase kinase-3β (GSK-3β). Eur J Med Chem 61:95–103
90. Berg S et al (2012) Discovery of novel potent and highly selective glycogen synthase kinase-3β (GSK3β) inhibitors for Alzheimer's disease: design, synthesis, and characterization of pyrazines. J Med Chem 55:9107–9119
91. Uehara F et al (2013) 6-(4-Pyridyl) pyrimidin-4 (3*H*)-ones as CNS penetrant glycogen synthase kinase-3β inhibitors. Bioorg Med Chem Lett 23:6928–6932
92. La Pietra G et al (2013) Design, synthesis, and biological evaluation of 1-phenylpyrazolo [3, 4-*e*] pyrrolo [3, 4-*g*] indolizine-4, 6 (1 *H*, 5 *H*)-diones as new glycogen synthase kinase-3β inhibitors. J Med Chem 56:10066–10078
93. Ye Q et al (2013) Synthesis and biological evaluation of 3-benzisoxazolyl-4-indolylmaleimides as potent, selective inhibitors of glycogen synthase kinase-3β. Molecules 18:5498–5516
94. Pang T et al (2016) A novel GSK-3β inhibitor YQ138 prevents neuronal injury induced by glutamate and brain ischemia through activation of the Nrf2 signaling pathway. Acta Pharm Sin 37:741–752
95. Sivaprakasam P et al (2015) Discovery of new acylaminopyridines as GSK-3 inhibitors by a structure guided in-depth exploration of chemical space around a pyrrolopyridinone core. Bioorg Med Chem Lett 25:1856–1863
96. Yue H et al (2015) Structure-based design of benzo [*e*] isoindole-1, 3-dione derivatives as selective GSK-3β inhibitors to activate Wnt/β-catenin pathway. Bioorg Chem 61:21–27
96a. Martino D et al (2016) Versatility of curcumin scafold; discovery of potent and balanced dual BACE-1 andGSK-3β inhibitors. J Med Chem 39:531–544
97. Lin CH et al (2016) Identifying GSK-3β kinase inhibitors of Alzheimer's disease: virtual screening, enzyme, and cell assays. Eur J Pharm Sci 89:11–19
98. Liang Z et al (2016) C-glycosylflavones alleviate tau phosphorylation and amyloid neurotoxicity through GSK3β inhibition. ACS Chem Nerosci 7:912–923
99. Arfeen M et al (2016) Design, synthesis and biological evaluation of 5-benzylidene-2-iminothiazolidin-4-ones as selective GSK-3β inhibitors. Eur J Med Chem 121:727–736
100. Fukunaga K et al (2013) 2-(2-Phenylmorpholin-4-yl) pyrimidin-4 (3H)-ones; a new class of potent, selective and orally active glycogen synthase kinase-3β inhibitors. Bioorg Med Chem Lett 23:6933–6937
101. Helal CJ et al (2004) Discovery and SAR of 2-aminothiazole inhibitors of cyclin-dependent kinase 5/p25 as a potential treatment for Alzheimer's disease. Bioorg Med Chem Lett 14:5521–5525
102. Helal CJ et al (2009) Potent and cellularly active 4-aminoimidazole inhibitors of cyclin-dependent kinase 5/p25 for the treatment of Alzheimer's disease. Bioorg Med Chem Lett 19:5703–5707
103. Rzasa RM et al (2007) Structure–activity relationships of 3,4-dihydro-1*H*-quinazolin-2-one derivatives as potential CDK5 inhibitors. Bioorg Med Chem 15:6574–6595
104. Laha JK et al (2011) Structure-activity relationship study of 2,4-diaminothiazoles as CDK5/p25 kinase inhibitors. Bioorg Med Chem Lett 21:2098–2101
105. Jain P et al (2011) Design, synthesis and testing of an 6-O linked series of benzimidazole based inhibitors of CDK5/p25. Bioorg Med Chem 19:359–373
106. Malmstrom J et al (2012) Synthesis and structure-activity relationship of 4-(1,3-benzothiazol-2-yl)-thiophene-2-sulfonamides as cyclin-dependent kinase 5 (CDK5)/p25 inhibitors. Bioorg Med Chem Lett 22:5919–5923
107. Demange L et al (2013) Potent inhibitors of CDK5 derived from roscovitine: synthesis, biological evaluation and molecular modelling. Bioorg Med Chem Lett 23:125–131
108. Sanphui P et al (2013) Efficacy of cyclin dependent kinase 4 inhibitors as potent neuroprotective agents against insults relevant to Alzheimer's disease. PLoS One 8:e78842
109. Roy SM et al (2015) Targeting human central nervous system protein kinases: an isoform selective p38αMAPK inhibitor that attenuates disease progression in Alzheimer's disease

mouse models. ACS Chem Nerosci 6:666–680

110. Markandeyan D et al (2015) Virtual screening of phytochemicals of *Morinda citrifolia* anti-inflammatory and anti-Alzheimer agents using MOLEGRO virtual docker on p38α mitogen-activated protein kinase enzyme. Asian J Pharm Clin Res 8:141–145
111. Gourdain S et al (2013) Development of DANDYs, new 3,5-diaryl-7-azaindoles demonstrating potent DYRK1A kinase inhibitory activity. J Med Chem 56:9569–9585
112. Myrianthopoulos V et al (2013) Novel inverse binding mode of indirubin derivatives yields improved selectivity for DYRK kinases. ACS Med Chem Lett 4:22–26
113. Yadav RR et al (2015) Meridianin derivatives as potent DYRK1A inhibitors and neuroprotective agents. Bioorg Med Chem Lett 25:2948–2952
114. Tintori C et al (2015) Studies on the ATP binding site of Fyn kinase for the identification of new inhibitors and their evaluation as potential agents against tauopathies and tumors. J Med Chem 58:4590–4609
115. Polychronopoulos P et al (2004) Structural basis for the synthesis of indirubins as potent and selective inhibitors of glycogen synthase kinase-3 and cyclin-dependent kinases. J Med Chem 47:935–946
116. Oumata N et al (2008) Roscovitine-derived, dual-specificity inhibitors of cyclin-dependent kinases and casein kinases 1. J Med Chem 51:5229–5242
117. Loge C et al (2008) Novel 9-oxo-thiazolo [5,4-f]quinazoline-2-carbonitrile derivatives as dual cyclin-dependent kinase 1 (CDK1)/ glycogen synthase kinase-3 (GSK-3) inhibitors: synthesis, biological evaluation and molecular modeling studies. Eur J Med Chem 43:1469–1477
118. Debdab M et al (2011) Leucettines, a class of potent inhibitors of cdc2-like kinases and dual specificity, tyrosine phosphorylation regulated kinases derived from the marine sponge leucettamine b: modulation of alternative pre-rna splicing. J Med Chem 54:4172–4186
119. Tahtouh T et al (2012) Selectivity, cocrystal structures, and neuroprotective properties of leucettines, a family of protein kinase inhibitors derived from the marine sponge alkaloid leucettamine B. J Med Chem 55:9312–9330
120. Tell V et al (2012) Novel aspects in structure–activity relationships of profiled 1-aza-9- oxafluorenes as inhibitors of Alzheimer's disease-relevant kinases CDK1, CDK5 and GSK3β. Med Chem Commun 3:1413–1418
121. Waiker DK et al (2014) Synthesis, biological evaluation and molecular modelling studies of 4-anilinoquinazoline derivatives as protein kinase inhibitors. Bioorg Med Chem 22:1909–1915
122. Loidreau Y et al (2012) First synthesis of 4-aminopyrido[2′,3′:4,5]furo[3,2-*d*]pyrimidines. Tetrahedron 53:944–947
123. Loidreau Y et al (2015) Synthesis and molecular modelling studies of 8-arylpyrido [30,20:4,5]thieno[3,2-*d*]pyrimidin-4-amines as multitarget Ser/Thr kinases inhibitors. Eur J Med Chem 92:124–134
124. Lawson M et al (2016) Synthesis, biological evaluation and molecular modeling studies of imidazo[1,2-α]pyridines derivatives as protein kinase inhibitors. Eur J Med Chem 123:105–114
125. Cohen P (2002) Protein kinases- the major drug targets of the twenty-first century? Nat Rev Drug Discov 1:309–315
126. Ma H, Deacon S, Horiuchi K (2008) The challenge of selecting protein kinase assays for lead discovery optimization. Expert Opin Drug Discovery 3:607–621
127. Knighton DR et al (1991) Structure of a peptide inhibitor bound to the catalytic subunit of cyclic adenosine monophosphate-dependent protein kinase. Science 253:414–420
128. Müller S et al (2015) The ins and outs of selective kinase inhibitor development. Nat Chem Biol 11:818–821
129. Morphy R (2010) Selectively nonselective kinase inhibition: striking the right balance. J Med Chem 53:1413–1437
130. Mencher K, Wang G (2005) Promiscuous drugs compared to selective drugs (promiscuity can be a virtue). BMC Clin Pharmacol 5:3
131. Guimarães CR et al (2011) Understanding the impact of the P-loop conformation on kinase selectivity. J Chem Inf Model 51:1199–1204
132. Davis MI et al (2011) Comprehensive analysis of kinase inhibitor selectivity. Nat Biotechnol 29:1046–1051
133. Berman HM et al (2013) How community has shaped the Protein Data Bank. Structure 21:1485–1491
134. Niijima S, Shiraishi A, Okuno Y (2012) Dissecting kinase profiling data to predict activity and understand cross-reactivity of kinase inhibitors. J Chem Inf Model 52:901–912
135. Cao DS et al (2013) Large-scale prediction of human kinase/inhibitor interactions using protein sequences and molecular topological structures. Anal Chim Acta 792:10–18

136. Yabuuchi H et al (2011) Analysis of multiple compound-protein interactions reveals novel bioactive molecules. Mol Syst Biol 7:472
137. Schürer SC, Muskal SM (2013) Kinome-wide activity modeling from diverse public high-quality data sets. J Chem Inf Model 53:27–38
138. Caffrey DR, Lunney EA, Moshinsky DJ (2008) Prediction of specificity-determining residues for small-molecule kinase inhibitors. BMC Bioinformatics 9:491
139. Brylinski M, Skolnick J (2010) Cross-reactivity virtual profiling of the human kinome by X-react(KIN): a chemical systems biology approach. Mol Pharm 7:2324–2333
140. Huang D et al (2010) Kinase selectivity potential for inhibitors targeting the ATP binding site: a network analysis. Bioinformatics 26:198–204
141. Anderson PC et al (2012) Identification of binding specificity-determining features in protein families. J Med Chem 55:1926–1939
142. Chiu YY et al (2013) KIDFamMap: a database of kinase/inhibitor-disease family maps for kinase inhibitor selectivity and binding mechanisms. Nucleic Acids Res 41: D430–D440
143. Kim J et al (2013) K-map: connecting kinases with therapeutics for drug repurposing and development. Hum Genomics 7:20
144. Tang J et al (2014) Making sense of large-scale kinase inhibitor bioactivity data sets: a comparative and integrative analysis. J Chem Inf Model 54:735–743
145. Manning G et al (2002) The protein kinase complement of the human genome. Science 298:1912–1934
146. Fan S et al (2012) Clarifying off-target effects for torcetrapib using network pharmacology and reverse docking approach. BMC Syst Biol 6:152
147. Coumar MS et al (2009) Structure-based drug design of novel Aurora kinase A inhibitors: structural basis for potency and specificity. J Med Chem 52:1050–1062
148. Chen Z et al (2011) Acenaphtho[1,2-*b*]pyrrole-based selective fibroblast growth factor receptors 1 (FGFR1) inhibitors: design, synthesis, and biological activity. J Med Chem 54:3732–3745
149. Prada-Gracia D et al (2016) Application of computational methods for anticancer drug discovery, design, and optimization. Bol Med Hosp Infant Mex 73:411–423

Chapter 15

Computational Modeling of Drugs for Alzheimer's Disease: Design of Serotonin 5-HT_6 Antagonists

Ádám A. Kelemen, Stefan Mordalski, Andrzej J. Bojarski, and György M. Keserű

Abstract

The 5-hydroxytryptamine receptor 6 (5-HT_6R) represents one of the most avowed targets for alleviating cognitive, learning, and memory deficits related to Alzheimer's disease (AD). Ligand- and structure-based computational modeling methods serve as main tools at the initial stages of drug discovery projects to underlie and to understand small molecule targeting of the receptor. Here, we describe the currently known 5-HT_6R antagonists in clinical trials and at discovery stages. We analyze existing ligand-based information and disposable pharmacophore models, quantitative structure-activity relationship methods, usable crystal structure templates, homology models, and molecular docking approaches. Our goal is to provide the reader with guidelines on how to utilize the existing knowledge and ligand- and structure-based methods for the design of new 5-HT_6R antagonists and to highlight advantages and limitations of corresponding approaches and computational modeling tools in the field of 5-HT_6R drug design.

Key words 5-Hydroxytryptamine receptor 6, Alzheimer's disease, Pharmacophore modeling, Quantitative structure-activity relationship, Bioisosteres, Homology modeling, Molecular dynamics, Site-directed mutagenesis, Molecular docking

1 Introduction

1.1 Physiological and Therapeutic Potential

The 5-hydroxytryptamine receptor 6 (5-HT_6R) is a member of the Class A G-protein-coupled receptors (aminergic family) [1]. The serotoninergic subfamily consists of seven receptor types (5-HT_{1-7}), among which 5-HT_6R is the only one without known isoforms; however, the C267T polymorphism can be observed in some studies associated with depression and schizophrenia [2]. Unlike other serotonin receptors, subtype 6 is almost exclusive for the central nervous system (CNS) [3], with infinitesimal occurrence in peripheral organs. 5-HT_6Rs are expressed rather on dendrites, cell bodies, and postsynaptic sites of GABAergic, cholinergic, and glutamatergic neurons than on astrocytes. Their expression is also well characterized for different brain areas, including the striatum,

Kunal Roy (ed.), *Computational Modeling of Drugs Against Alzheimer's Disease*, Neuromethods, vol. 132, DOI 10.1007/978-1-4939-7404-7_15, © Springer Science+Business Media LLC 2018

caudate nucleus, nucleus accumbens > hippocampus, amygdala > prefrontal cortex, thalamus, subthalamic nucleus, and substantia nigra [4].

The localization and neurochemistry of 5-HT$_6$R receptors support different central nervous system indications, such as cognitive, learning, and memory disorders related to Alzheimer's disease (AD) [5], Parkinson's disease [6], and schizophrenia [7]. Other CNS indications such as analgesia [8], obesity [9], anti-drug abuse [10], sleep-wake regulation [11], self-transcendence [12], executive cognition [13], and major depressive disorder [14] have also been suggested.

Blockade of the 5-HT$_6$R mediates the cholinergic, glutamatergic, and GABAergic systems by increasing acetylcholine and glutamate release. The initial discovery of 5-HT$_6$R was related to the finding that antipsychotics and tricyclic antidepressants display a high affinity for this receptor [15]. In the case of AD patients, a significant reduction in 5-HT$_6$R density in cortical areas has been described, but it showed no clear correlation with the cognitive status [16]. In contrast, while 5-HT$_6$R blockade induces acetylcholine release, the expression of 5-HT$_6$R is reduced in the nervous system for disinhibition of this process [17].

Experimental approaches were evaluated to study the effects of 5-HT$_6$R blockade on cognition in animal models. In vivo experiments that were aimed to alleviate cognitive deficits (provoked, e.g., by applying scopolamine) induced memory loss rather than increasing their function in healthy animals [5]. Interestingly, there was an unclear concordance between 5-HT$_6$R agonists and antagonists regarding the cognitive enhancement of memory and learning [17, 18]. Some results have shown that agonists and antagonists can synergistically induce memory improvements, although other studies demonstrated a blockade of agonistic effects by antagonists. A possible reason for this synergism could be an auto-amplifying mechanism, in which the antagonist increases acetylcholine release by stimulating cholinergic transmission. By introducing mAChR inhibitors, it has been confirmed that the cholinergic system mediates the effect of 5-HT$_6$R on memory. Another explanation deals with the different signaling pathways induced by agonists and antagonists, such as $G\alpha_s$, $G\alpha_q$, $G\alpha_{i/o}$, and Ca^{2+} signaling [16]. The 5-HT$_6$ receptors co-localize with GABA neurons in approximately 20% of all nerve cells, and thus, the modulation of cholinergic and glutamatergic systems is possible via the disinhibition of GABA neurons. Despite this synergistic effect, most drug discovery efforts have been focused on 5-HT$_6$R antagonists.

Although 5-HT$_6$R is considered to be a therapeutically relevant drug target, there are a number of challenges in tackling the discovery of a promising 5-HT$_6$ antagonist chemotype. The main goals and challenges include penetration through the blood-brain barrier (BBB), selectivity toward 5-HT subtypes and other targets, a low number of drug metabolites, rapid desensitization of the receptor, signaling interference, and unclear relation between the signaling pathways.

1.2 5-HT$_6$R Antagonists in Clinical Trials

The growing number of AD patients worldwide (approximately 35 M according to 2014 statistics) [19] represents the clear and unmet need for new therapeutic agents. An analysis of clinicaltrials.gov [20] by Cummings and colleagues concluded that for neuroprotective agents, 25 drug candidates were under investigation for AD in phase 1 (2), 2 (19), and 3 (4) clinical stages (Table 1). The following four drug candidates are currently under clinical investigation: SUVN-502 (phase 2, Suven Life Sciences), PF-0521377/SAM-760 (phase 2, Pfizer), Lu-AE58054/idalopirdine/SGS-518 (phase 3, Lundbeck), and RVT-101/SB-742457 (phase 3; GSK, Axovant). In vivo preclinical studies have revealed that SUVN-502 [21] is a potent, selective, orally bioavailable, efficacious, and safe 5-HT$_6$R antagonist. Clinical investigations are targeting co-administration with donepezil and memantine for the treatment of moderate AD. Phase 2a is currently in progress with 537 patients across 57 sites in the USA, progressing until 2017 [22]. PF-0521377 [23] has undergone successful phase 1 trials from 2009 to 2014 in 230 healthy volunteers followed by scopolamine-induced cognitive impairment studies and multiple-ascending dose studies on safety and pharmacokinetics. These promising results led to phase 2 clinical studies. Efficacy studies were initiated in 342 AD patients to evaluate 30 mg doses in addition to stable donepezil dosing, but the trials were terminated in 2015 due to a lack of efficacy [24]. After terminating the phase 2 trials with Lu-AE-58054 [25] in schizophrenia-related cognitive deficits, the compound was proposed for indications in AD dementia as a symptomatic adjunct therapy to cholinesterase inhibitor treatment. Co-administration with donepezil was studied in multiple courses of phase 3 trials from 2014 to 2016. Recently, Lundbeck announced that idalopirdine did not reach the primary endpoint due to weak efficacy. One of the most promising 5-HT$_6$R antagonists to date, intepirdine/RVT-101/SB-742457 [26], has been reported to enhance cognition in AD deficits in vivo [27]. GSK has demonstrated success in safety, tolerability, and pharmacokinetics during an initial phase 1 course followed by a phase 2 trial examining the dose-dependent benefits of reversing AD dementia in 371 patients [28]. The compound was acquired by Axovant in 2014 and was renamed RVT-101. Phase 3 trials that are currently underway indicate promising results in patients with mild-to-moderate AD.

1.3 Discovery Stage of 5-HT$_6$R Ligands

Here, the latest release (Release 22) of the ChEMBL database [63] was evaluated for 5-HT$_6$R bioactivity data, providing 7269 data points on 4472 compounds. To date, ChEMBL contains 1586

Table 1
Summary of prominent 5-HT$_6$R antagonists (compounds in shaded boxes have not yet reached the clinic)

Structure	Name/synonyms	Developer	5-HT$_6$ K_i (nM)	Clinical phase	Scaffold group	Year	Pharmacophore	References
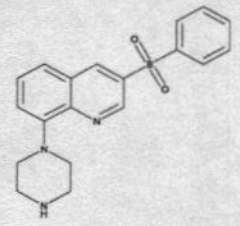	SB-742457, RVT-101, intepirdine	GSK, Axovant	0.234	Phase 3 2017	Aryl-sulfone, quinoline	2008	PM1-PI	[29]
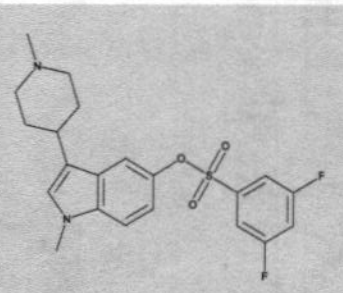	SGS-518, Ly 483518	Lundbeck, Lilly	–	Phase 2	Aryl-sulfonate	2007	PM2-PI-HS	[30]
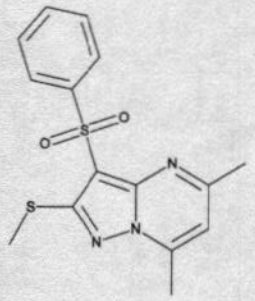	AVN-211	Avineuro	2.1	Phase 2 active	Aryl-sulfone, pyrazolopyrimidine	2011	PM4-noPI-HS	[31]
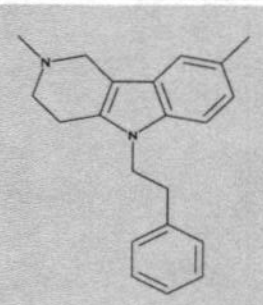	AVN-101	Avineuro	2.0	Phase 2 active	Pyridoindole	2016	Dimebon-like	[32]

Undisclosed	SYN-120, landipirdine	Roche, Biotie Therapies	0.2	Phase 2	–	–	–	[33]
Undisclosed	SAM760, PF-05212377, WYE-103760	Pfizer	–	Phase 2	–	–	–	[34]
Undisclosed	ABT-354	AbbVie	–	Phase 1	–	–	–	[35]
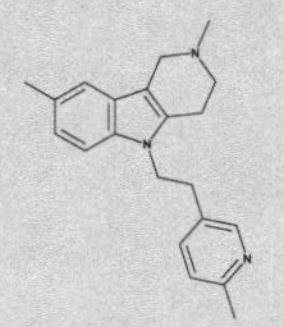	Dimebon (latrepirdine)	Pfizer	26.0	Phase 3 failed	Tryptamine	1983	Dimebon-like	[36]
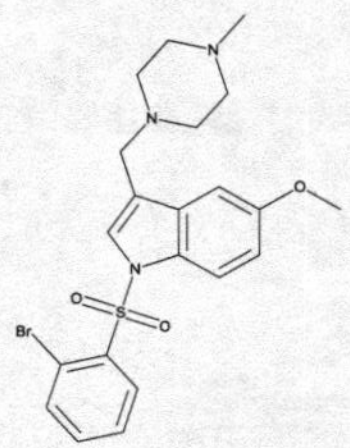	SUVN-502	Suven Life Sciences	1.71	Phase 3 failed	Aryl-sulfonamide (ring-N), tryptamine	2011	PM2-PI-HS	[21]
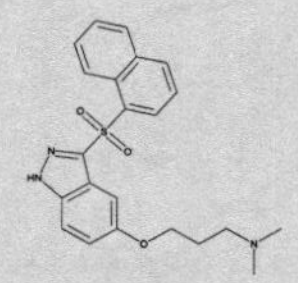	PF-05212365, SAM-531, cerlapirdine	Pfizer	1.3	Phase 3 failed	Aryl-sulfone, benzoxazole	2011	PM2-PI-HS	[23]

(continued)

Table 1 (continued)

Structure	Name/synonyms	Developer	5-HT_6 K_i (nM)	Clinical phase	Scaffold group	Year	Pharmacophore	References
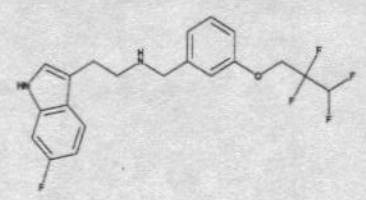	Idalopirdine, Lu_AE_58054,	Lundbeck Pharma	0.83	Phase 3 failed	Tryptamine	2010	PM2-PI-HS (without HBA)	[25, 37]
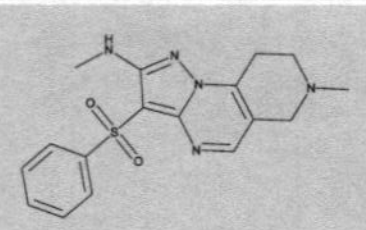	AVN-322	Avineuro	0.39	Phase 2 failed	Aryl-sulfone, pyrazolopyri dopyrimidine	2011	PM3-PI-HS-iHBD	[38]
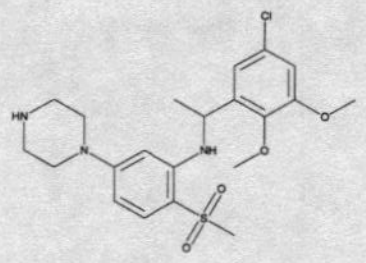	PRX-07034	Epix Pharma	4.0	Phase 2 ter minated	Mesylate	2008	PM2-PI-HS	[39]
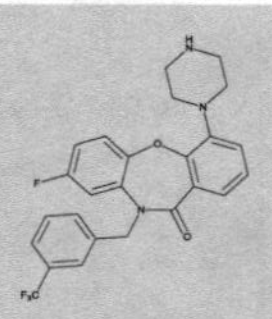	SYN-114	Synosia Therapeutics, Biotie Therapies	0.3	Phase 1 failed	Dibenzoxazepine	2014	PM2-PI-HS	[40, 41]

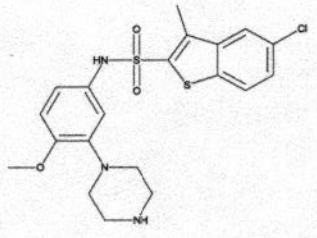	SB-271046	GSK	1.26	Phase 1 failed	Aryl-sulfonamide, benzothiophene	1999	PM2-PI-HS	[42, 43]
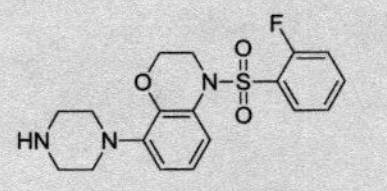	R-1485	Roche	1.26	Phase 1 failed	Aryl-sulfonamide, benzoxazine	2007	PM2-PI-HS	[44]
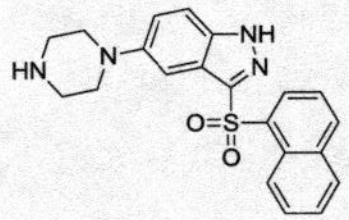	SAM-315	Pfizer	1.1	Preclinical	Aryl-sulfone, alternative PI position	-	PM2-PI-HS	[45]
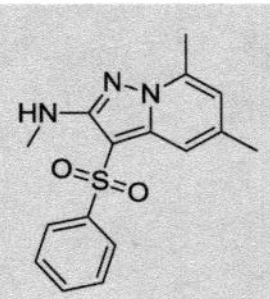	AVN-216	Avineuro	9.59	Preclinical	Pyrazolopyridine	2014	PM5-noPI-HS-iHBD	[46]
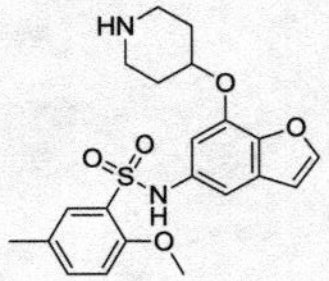	BVT-74316	Biovitrum	1.2	Preclinical	Aryl-sulfonamide, benzofuran	2006	PM2-PI-HS	[47]

(continued)

Table 1 (continued)

Structure	Name/synonyms	Developer	5-HT_6 K_i (nM)	Clinical phase	Scaffold group	Year	Pharmacophore	References
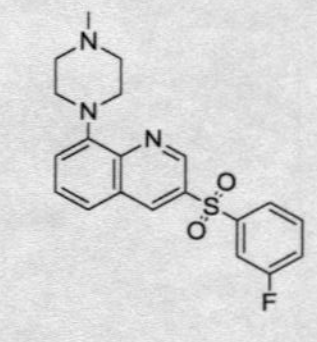	^{11}C-LuAE60157	Lundbeck Pharma	0.2	Preclinical	Quinoline	2010	PM2-PI-HS	[48]
Undisclosed	^{11}C-GSK215083	GSK	0.34	Preclinical	Quinoline	2012	-	[49]
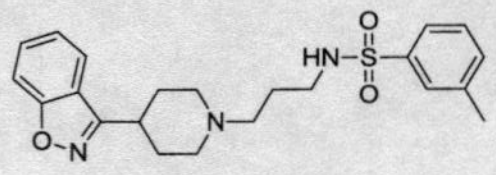	ADN-1184	Adamed	16.0	Preclinical	Long-chain benzoxazole piperidine	2014	PM2-PI-HS	[50]
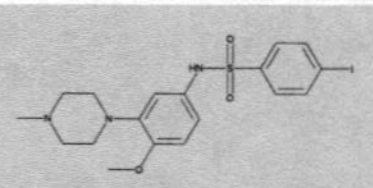	SB-258585	GSK	2.51	Preclinical	Aryl-sulfone	-	PM2-PI-HS	[51]
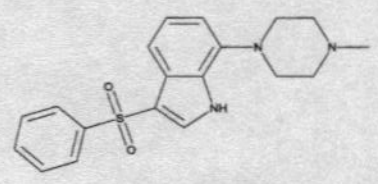	Ro 4368554	Roche	0.5	Preclinical	Aryl-sulfone, alternative PI position	2005	PM2-PI-HS	[52]

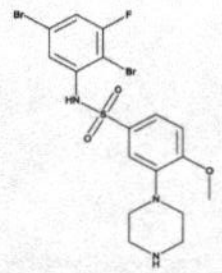	SB-357134	GSK	3.16	Preclinical	Aryl-sulfonamide, reversed sulfonamide	2002	PM2-PI-HS	[53]
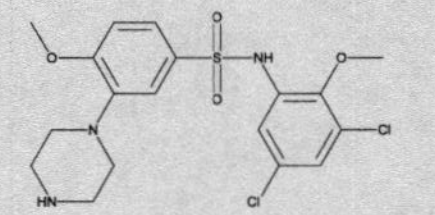	SB-399885	GSK	0.7	Preclinical	Aryl-sulfonamide, reversed sulfonamide	2006	PM2-PI-HS	[54]
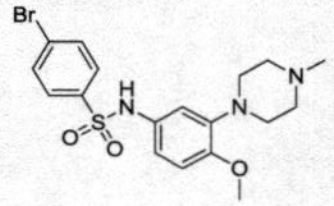	SB-214111	GSK	5.0	Preclinical	Aryl-sulfonamide	1999	PM2-PI-HS	[43]
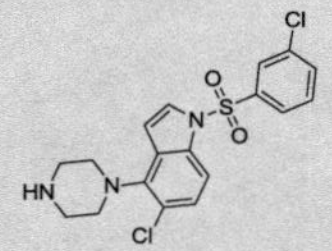	SB-699929	GSK	0.3	Preclinical	Aryl-sulfonamide (ring-N), alternative PI position	–	PM2-PI-HS	[55]
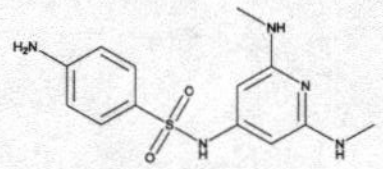	Ro 63-0563	Roche	12.0	Preclinical	Aryl-sulfonamide	1998	PM2-PI-HS	[56]

(continued)

Table 1 (continued)

Structure	Name/synonyms	Developer	5-HT_6 K_i (nM)	Clinical phase	Scaffold group	Year	Pharmacophore	References
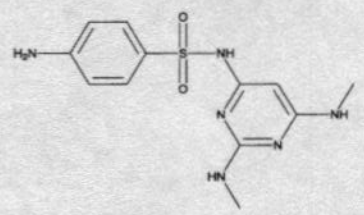	Ro 04-6790	Roche	50.12	Preclinical	Aryl-sulfonamide	1998	PM2-PI-HS	[56]
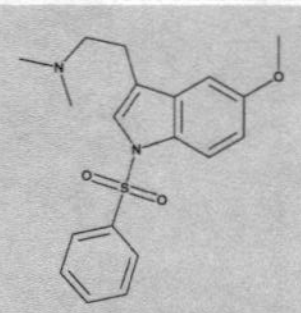	MS-245	Merck	2.3	Preclinical	Aryl-sulfonamide (ring-N), 5-HT	2000	PM2-PI-HS	[57, 58]
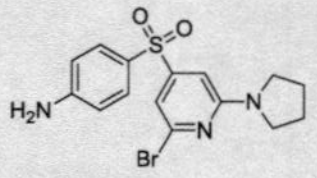	Ro-66-0074	Roche	1.0	Preclinical	Aryl-sulfone	2003	PM2-PI-HS	[59]
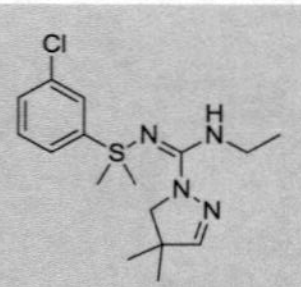	CMP X	Abbott	25.0	Preclinical	Sulfonyl-guanidine	2011	PM4-noPI-HS	[60]

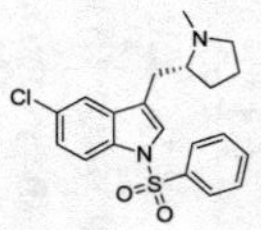	ALX-1161	Allelix-NPS Pharmaceuticals	1.4	Preclinical	Aryl-sulfonamide, tryptamine-like	1999	PM2-PI-HS	[61]
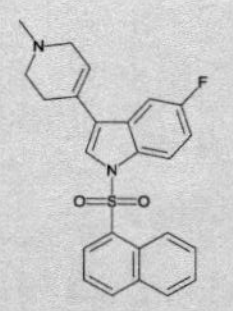	ALX-1175	Allelix-NPS Pharmaceuticals	1.0	Preclinical	Aryl-sulfonamide, tryptamine-like	2000	PM2-PI-HS	[62]
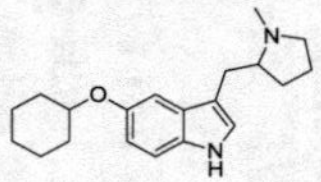	ALX-0440	Allelix-NPS Pharmaceuticals	0.93	Preclinical	Serotonin-like, non-sulfonyl	1999	PM2-PI-HS	[61]

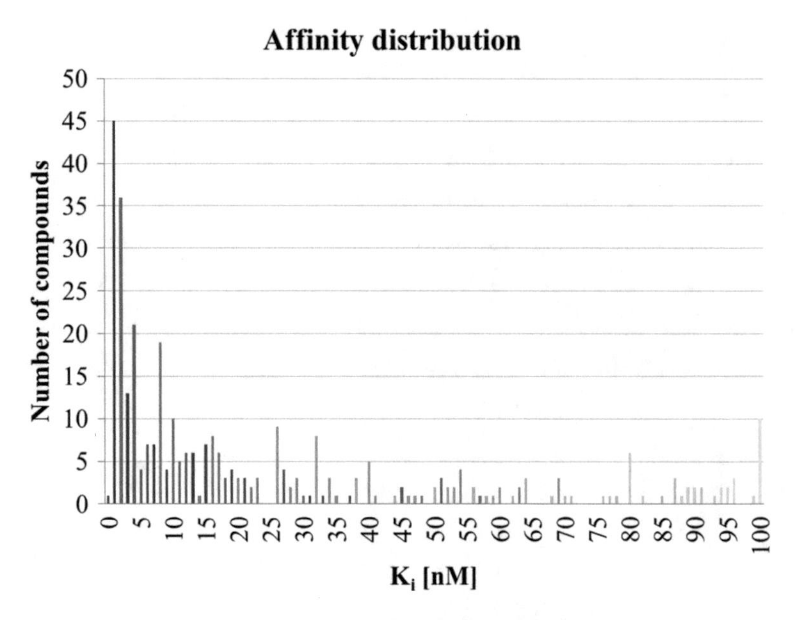

Fig. 1 Distribution of 5-HT$_6$R antagonist activities in the ChEMBL database

binding affinity and functional data entries (K_i, IC_{50}) on 5-HT$_6$R. Among these 650 compounds, 345 unique antagonists and 105 agonists showed 100 nM or better activity toward 5-HT$_6$R. Figure 1 shows the distribution of binding activities across the 5-HT$_6$R antagonists (mean = 24.902 ± 29.643 nM, median = 11.7 nM).

The occurrence of published active antagonists shows that the number of introduced compounds in the ChEMBL database peaked between the years 2010 and 2012 (Fig. 2), which is consistent with the recently published analysis of the emergence of new groups of ligands for different 5-HT GPCRs [64].

2 Materials

2.1 Main Scaffolds and Pharmacophores for 5-HT$_6$R Antagonists

Clustering of the antagonist structures available in ChEMBL using RDKit Hierarchical Clustering (Knime Extension pack, fingerprint-based, Tanimoto distance, complete linkage [65], Fig. 3) revealed that the aryl (phenyl, naphthyl, and heteroaryl) moiety is a hallmark of the 5-HT$_6$R antagonists. Furthermore, the protonable basic amino moiety clearly does not manifest as an indispensable feature for those ligands. Calculation of basic pK_a values for the strongest basic centers of all 345 compounds of the ChEMBL 5-HT$_6$R antagonist set revealed that 14% of the compounds did not contain positively ionizable groups (7% contained secondary amines that acted as intramolecular hydrogen-bond donors). This observation also infers that 5-HT$_6$R selectivity over other aminergic receptors may be accessed by abandoning the basic nitrogen. Another

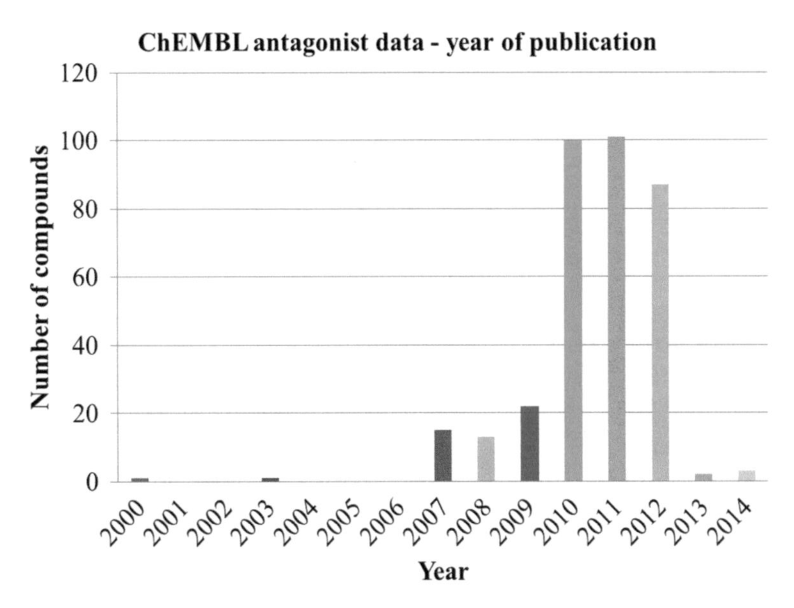

Fig. 2 Distribution of publication years for 5-HT$_6$R antagonists in the ChEMBL database

characteristic feature of selective 5-HT$_6$R antagonists (and agonists) is the presence of sulfonyl or sulfonamide groups located between two aromatic systems. Of the total ChEMBL 5-HT$_6$R antagonist set, 97% contain a type of SO_2 moiety. The remaining 3% consists of compounds that nonetheless satisfy the pharmacophore coverage.

Cluster 1 represents the scaffold of arylsulfonylindazoles with 44 members, which contains either piperidinyl or methylpiperazinyl moieties in the fifth position. Another unique scaffold of the 5-HT$_6$R antagonists is represented by Clusters 2 and 5, in which the aryl-sulfonyl portion is connected through the ring nitrogen of the indole-like scaffolds. Clusters 3 and 4 consist of bridgehead nitrogen heterocycles. The thienotriazolo-pyrimidines are the result of 3D pharmacophore model-based screening (Catalyst/HipHop Module 3D query) of the ChemDiv GPCR focused library developed by Kim and colleagues [30], resulting in single-digit nanomolar hits (more are described in Sect. 3.1.1).

2.2 Materials: Structure-Based Design

2.2.1 Crystal Structure Templates for Homology Modeling of 5-HT$_6$R

Rapid progress in the crystallization of GPCRs, as initiated in 2008, has resulted in an increasing availability of structural templates for homology modeling. To date, 154 crystal structures are available for 36 different GPCR targets [66]. However, the resolution, activation state, and auxiliary proteins fused to the crystallized GPCR render the selection of appropriate templates for homology modeling, a process that must be conducted meticulously.

For homology modeling of the 5-HT$_6$R, many crystallized targets with suitable identity and similarity are available for prediction of the spatial structure. Following the overall sequence

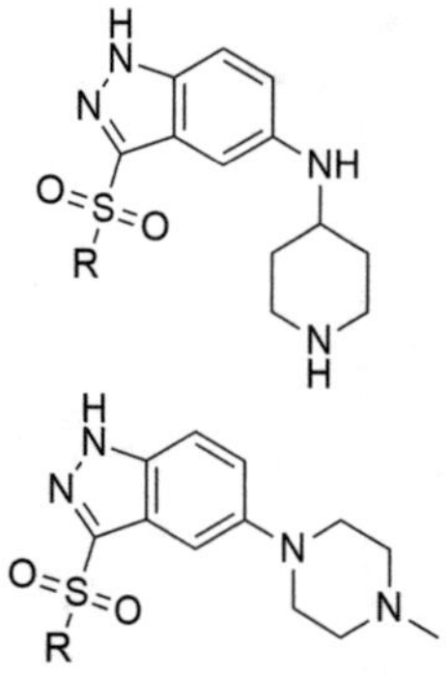

Cluster 1 (44 members)
R = naphthyl w. different conn.

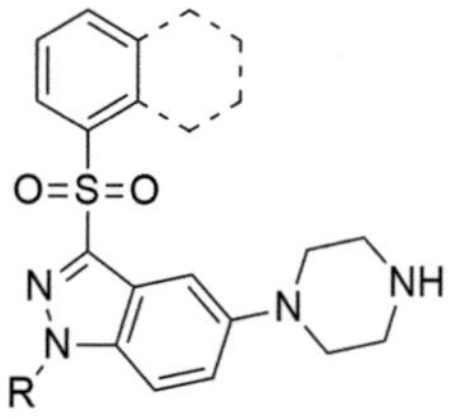

Cluster 1b (14 members)
R = n-alkyl/i-alkyl, benzyl

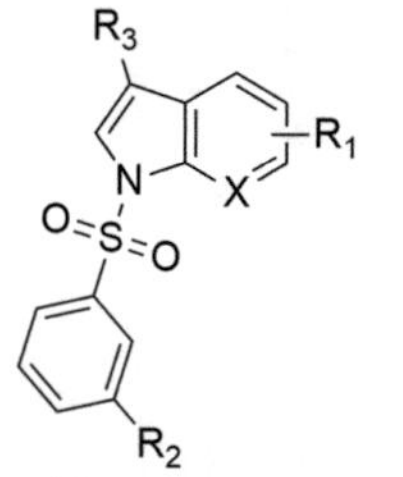

Cluster 2 (56 members)
R1 = halogen, methoxy, cond. azepino
R2 = H, methyl, methylpiperidinylamino
R3 = H, methyl, chloro, piperidinyl
X = C, N

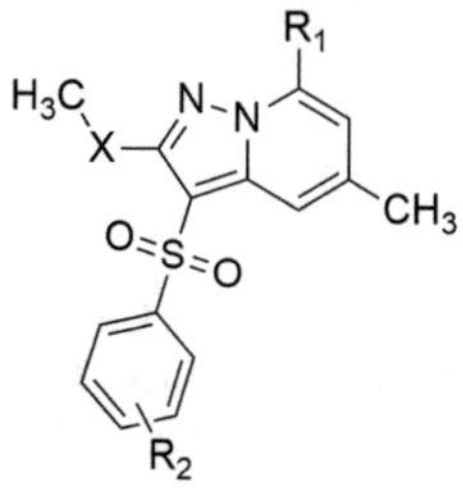

Cluster 3 (39 members)
R1 = alkyl, alkylamino
R2 = H, halogen
X = NH, S

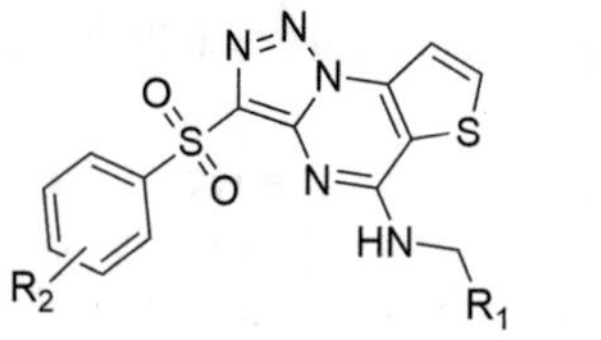

Cluster 4 (13 members)
R1 = i-alkyl, phenyl, heterocyclic
R2 = H, halogen, methyl

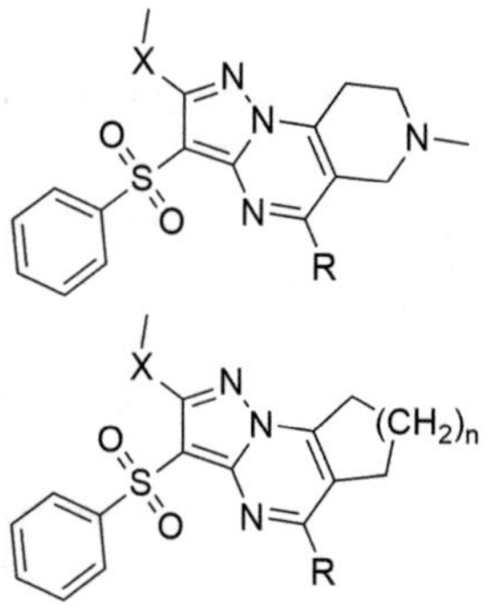

Cluster 4b (16 members)
R1 = H, methyl
X = S, NH
n = 1,2

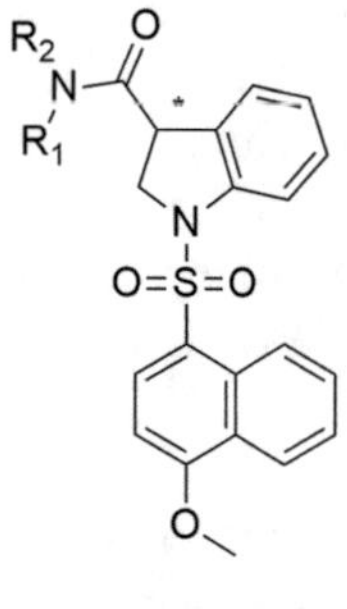

Cluster 5 (12 members)
R1 = H, n-alkyl
R2 = phenyl-piperidine

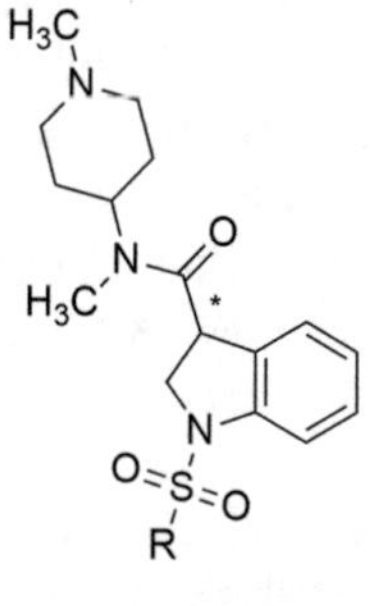

Cluster 5b (11 members)
R = phenyl, naphthyl, heteroaryl

Fig. 3 Main clusters of 5-HT_6R antagonists in ChEMBL

identity/similarity ranking between the receptors, the following four crystallized targets are suitable for homology modeling: β_1-AR, β_2-AR, 5-$HT_{1B}R$, and 5-$HT_{2B}R$ (Table 2). Again, the

Table 2
A comparison of the best available crystal structure templates of 5-HT$_6$R

	Full sequence		Transmembrane				
Target	Sequence identity to 5-HT$_6$R	Sequence similarity to 5-HT$_6$R	Sequence identity to 5-HT$_6$R	Sequence similarity to 5-HT$_6$R	# of crystals available	Highest resolution [Å]	Activation states
β_1-AR	23	35	36	56	18	2.1	Inactive, partially active[a]
β_2-AR	20	33	32	53	18	2.4	Inactive, partially active, active
5-HT$_{1B}$R	19	33	31	51	2	2.7	Partially active
5-HT$_{2B}$R	19	29	31	49	2	2.7	Partially active

[a]In this case, the receptor has the agonist-bound conformation, but it does not exhibit all of the hallmarks of the active state structure.

differences between these receptor targets result in alternative template selection for the construction of models with a distinct purpose.

2.2.2 Structural Features

The β_1-AR structure 4BVN, crystallized with the inverse agonist cyanopindolol, has the best resolution (2.1 Å) among the considered structures, and 2Y00 has the best resolution (2.5 Å) among the agonist-bound structures representing a partially active conformation. All the other β_1-AR structures were obtained at a lower resolution: there were three structures with a resolution lower than 2.5 Å, and the majority of the available 18 structures were within the range of 2.7–3.7 Å. β_2-AR contains only two high-resolution structures, 2RH1 and 5D5A (2.4 and 2.5 Å, respectively), both of which were crystallized with the inverse agonist carazolol. The resolutions of the remaining 16 structures encompass the range of 2.8–4.0 Å. The structures of the serotonin receptors, 5-HT$_{1B}$ and 5-HT$_{2B}$, were solved at 2.7 and 2.8 Å resolutions, respectively, and both were crystallized with ergotamine. Recently, a new 5-HT$_{2B}$R structure bound to lysergic acid diethylamide (LSD) was presented at 2.9 Å resolution [67].

High-resolution structures might support the detection of water molecules and ions within the binding site, providing insights into the process of receptor activation but also indicating bridged interactions with the active site residues. Apart from numerous water molecules within the binding site, the 2.1 Å resolution of the crystal structure of β_1-AR shows Na^+ ion interacting with TM2 (2x50 – GPCRdb format [66]), TM3 (3x39), TM6 (6x48), and TM7 (7x46), acting as the negative allosteric modulator and stabilizing the inactive conformation of the receptor [68]. Water molecules were also found in the crystal structures of β_2-AR, but they are less numerous than in β_1-AR. Furthermore, apart from occupying the cleft around 2x50 (e.g., 2RH1), the water molecules do not interact with the ligands (such interactions were observed for the 4AMJ crystal, wherein water was proxying the interaction between (S)-carvedilol and extracellular loop 2 (ECL2) of β_1-AR). Serotonin 1B and 2B receptors were both crystallized with ergotamine, a biased agonist that activates the G-protein (5-HT_{1B}R) or β-arrestin (5-HT_{2B}R) signaling pathway. The latest, LSD-bound structure of 5-HT_{2B}R, revealed possible structural features of functional selectivity associated to agonist-residence time. The resolution of those crystals allowed for the observation of crystalline waters interacting with ECL2 of the 5-HT_{1B}R [69] and the water particle bridging the ionic lock between the basic nitrogen of ergotamine an D^{3x32} and also between $D^{3x32}$and $Y^{7.42x342}$ of 5-HT_{2B}R, indicating an inactive conformation of the receptor.

2.2.3 Structural Completeness

The crystallization of GPCRs is challenging and often requires sequence modifications. The most common modification is the truncation of the extramembrane N- and C-terminal parts of the protein. Those fragments, which are fairly flexible and long (by up to nearly 100 residues for β_1-AR and 5-HT_{2B}R), are barely crystallizable but may interact with neighboring proteins, affecting the formation of stable crystals. For identical reasons, all of the crystals analyzed herein lack ICL3, which contains 20 residues and 60 residues in β_2-AR and 5-HT_{2B}R-long loop, respectively.

However, the auxiliary proteins (T4 lysozyme or BRIL – B562 cytochrome with thermostabilizing RIL mutations for the positions 98, 102, and 106, respectively, or nanobodies such as NB80) are fused to the target to enable the crystallization process by increasing the hydrophilic surface of the construct [70]. Here, β_2-AR is fused to T4 lysozyme, typically replacing ICL3 of the receptor (e.g., in the 2RH1 structure) and 5-HT crystals with BRIL. β_1-AR crystals do not contain any auxiliary proteins, as the receptor is stabilized by the point mutations (8–11, depending on the crystal). Other crystallized receptors contain significantly fewer (3–4) mutations.

2.2.4 Activation States

Since the potential 5-HT_6R templates were crystallized with ligands having different pharmacological functions, they were provided in different activation states (Table 2). β_1-AR was crystallized with a variety of ligands, but the structures with agonists were obtained via mutations rather than stabilization of the activated conformation with an auxiliary protein. Again, β_1-AR had the crystals with the best resolution among the considered set of templates, with antagonist-bound 4BVN of 2.1 Å. β_2-AR is the most studied receptor in structural terms, with several models of different activation states available. The crystals came with a variety of ligands, ranging from an antagonist (3NYA) to inverse agonists (e.g., 2RH1) and covalently bound agonist (3PDS) to agonists (e.g., 4LDE or 3SN6) and ligand-free (2R4R and 2R4S) structures. In addition, for the agonist-bound structures, the active conformation of the receptor was stabilized by either a G-protein (3SN6) or an engineered nanobody (e.g., 4LDE), mimicking the signaling protein. The richness of the different states and crystals explains why β_2-AR is the reference receptor for comparing/assessing the activation states and serves as the best template for comparisons of active and inactive models. The main disadvantage of β_2-AR as a template is its relatively low resolution of the crystal structures: only six crystals were below 3 Å, including five with inverse agonists (2RH1, 5D5A, 3D4S, 3NY9, and 3NY8) and one with an agonist (covalently bound to the receptor 4LDE). For 5-HT_{1B}R, the ligand, ergotamine, acts as an agonist; thus, the available structures are assumed to be partially active. In this case, the receptor has the agonist-bound conformation, but it does not exhibit all of the hallmarks of the active state structure (e.g., no rotation of TM6 due to BRIL fusion, an ionic lock between 3×50 and 7×42 [69]). However, a trigger P-I-F motif (positions 5×50, 3×40, and 6×44) has the spatial orientation of the activated β_2-AR. 5-HT_{2B}R was crystallized with the same ligand. However, for this receptor, ergotamine has a different pharmacological function; it is a β-arrestin-biased agonist, so the conformation of the receptor and key structural motifs in particular resembles the inactive state conformation of the reference β_2-AR. On the extracellular side of the receptor, the most significant difference between the 1B and 2B subtypes is in the conformation of TM5. 5-HT_{2B}R contains a kink at the end of this helix, which is additionally stabilized by the water molecule between residues 5×33, 5×36, and 5×37 that are missing in the 5-HT_{1B} structures.

2.2.5 Sequence and Structural Similarity

Structure-based sequence alignment performed using gpcrdb.org [66] revealed mismatching helical lengths between the 5-HT_6R and the selected crystal templates (Fig. 4).

The largest discrepancy can be observed for TM1; this segment of 5-$HT_{1B/2B}$R is significantly shorter than that in β-adrenergic receptors and the projected helix of 5-HT_6R by five and nine residues, respectively. It is, however, not observed in all crystal structures of β-AR, and positions $1 \times 24 - 1 \times 29$ are not within

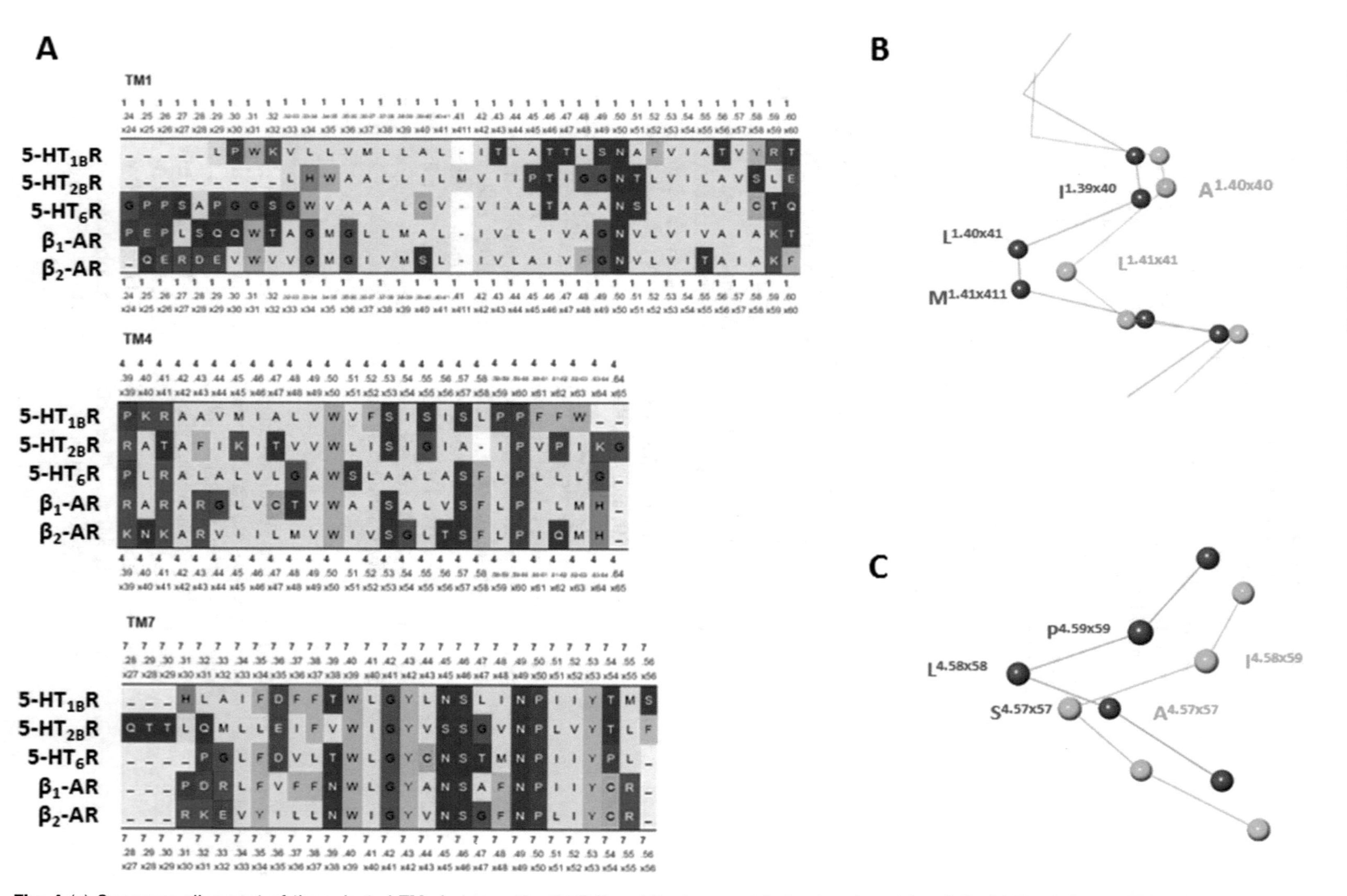

Fig. 4 (**a**) Sequence alignment of the selected TMs between the $5\text{-}HT_6R$ and the top crystal structure templates [66]. (**b**) The bulge on TM1 of $5\text{-}HT_{2B}R$ (*red*) compared with the crystal structure of $5\text{-}HT_{1B}R$ (*green*). (**c**) Constriction in TM4 of $5\text{-}HT_{2B}R$ (*red*) compared with the crystal structure of 5-HT

the membrane plane and thus are more likely part of the N-terminal domain.

The second most mismatched segment is TM7, which for 5-$HT_{2B}R$ is longer by three residues than that of 5-HT_6R, while the other crystal templates are longer by two (5-$HT_{1B}R$) or only one amino acid (β-AR, Fig. 4a). This difference, however, is located in the intracellular part of the receptor and would not affect the conformation of the ligand binding site in homology modeling. Other helices do not exceed two residue deviations from the respective target segment.

2.2.6 Binding Site

The binding sites of the templates are highly similar, occupying the volume between TM3, TM5, TM6, and TM7. Additionally, the depth of the bound ligands is nearly identical in all of the analyzed structures (Fig. 5) that can be rationalized by the similar sizes of the crystallized ligands and the high sequence similarity of the binding sites.

The 5-HT receptors were crystallized with bulky ligands, and so their binding sites expanded toward the exterior of the cell. Second, an even more important difference between serotonin and adrenergic crystals is the extracellular loop 2, which forms three turns of the α-helix in β-ARs and thus hardly interacts with the crystallized ligands. In contrast, ECL2 in serotonin receptors

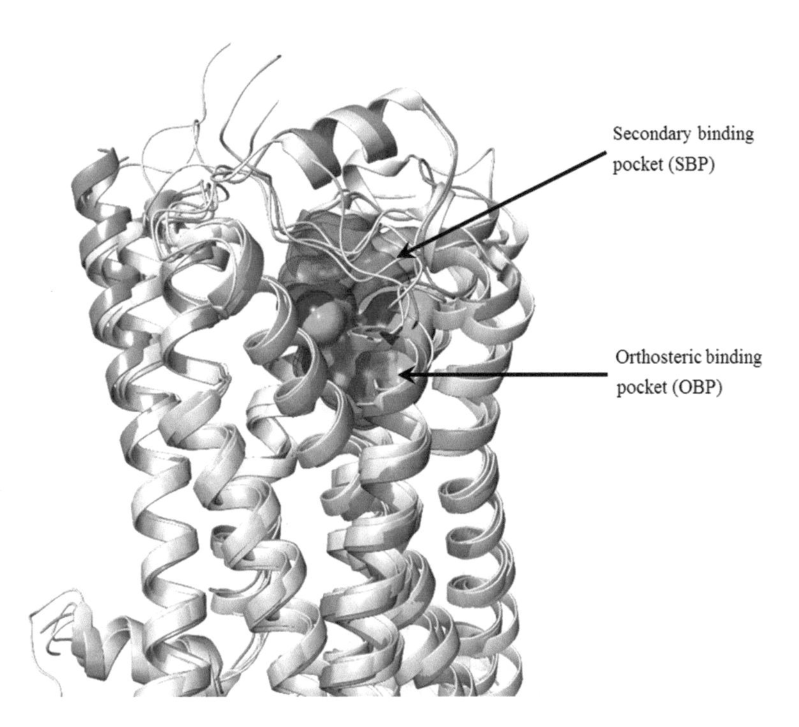

Fig. 5 Comparison of the depths and sizes of the binding sites of the crystal structure templates. The ligand of 5-$HT_{1B/2B}$ (4IAR and 4IB4, respectively) receptors is indicated in the red surface, β_1-AR (4BVN) is shown in *blue*, and β_2-AR (2RH1) is shown in *green*

does not form any regular secondary structure but lies in close proximity to the ligand, forming a number of interactions and thus stabilizing the conformation of the receptor.

The main difference between the active sites of the adrenoceptors and serotonin receptors, as observed in the crystal structures, is the secondary binding pocket (SBP), which is an extension of the orthosteric binding pocket (OBP) that accommodates the aromatic moieties of the ligands. For β-ARs, these secondary contacts have been observed for TM2, TM3, and TM7 and ECL1 and ECL2 (e.g., β_1-AR 4AMI), whereas the corresponding SBP for 5-HT_{2B}R occupies the region of TM5, TM6, and TM7 and ECL2 [71]. The secondary pocket was not observed for 5-HT_{1B}R, and thus, it is postulated that the interactions with SBP result in biased agonism or subtype selectivity.

2.2.7 Overview and Future Prospects of the 5-HT_6R Structural Models

The top crystal structure templates (by sequence similarity/identity) express a high level of structural similarity with nearly identical placement of the ligand binding sites, packing the helices and fairly similar ranges of the transmembrane helices. Three of four of the templates were successfully utilized in computational studies and computer-aided drug discovery experiments (β_2-AR [72], 5-HT_{2B}R [73], and 5-HT_{1B}R [74]). The differences between the structural templates are subtle and imply the purpose of the modeling. Adrenoreceptor templates were provided in different activation states (especially β_2-AR) and could be the basis for comparison of the active and inactive conformations of 5-HT_6R. Serotonin receptors, despite a lower resolution of the crystals, offer better matching of the loops (the particularly important ECL2 for ligand binding) and a greater volume of the binding site, which may be an advantage when studying the antagonist binding mode. Awareness of the structural features and anomalies in crystals (the extra bulge in TM1 and constriction in TM4 in the 5-HT_{2B}R structures) is, however, essential for proper modeling of the target structure. Moreover, for antagonist-focused studies, further processing of the serotonin receptor-based models is required, as these structures are in active-like conformations.

3 Methods

3.1 Ligand-Based Methods

3.1.1 5-HT_6R Antagonist Pharmacophores

Pharmacophore modeling, the canonical ligand-based technique, has been used by a number of research groups to describe the 5-HT_6R antagonists (Table 3). The scope and precision of the pharmacophore models are defined by the training set of compounds and can vary from large-scale virtual screening studies aimed at identifying new chemotypes (broad and diverse training set) to the design of specific compounds with improved 5-HT_6R affinity (class or group of compounds used to build the model).

Table 3
Pharmacophore models in the 5-HT$_6$R antagonist design

References	Year	Methodology	Training set	2D/ 3D	Approach	Results
Holenz et al. [45]	2005	Hypothetical model	7	2D	Hypothetical framework model	N-aminoalkylindoles, 3-aminoalkylindoles, 3-piperidinylindoles, and 3-glyoxamidoindoles
López-Rodríguez et al. [75]	2005	Catalyst/ HypoGen	45	3D	Activity-based alignment	Pharmacophore fragment toolkit
Tasler et al. [76]	2007	4Scan pharmacophore alignment	SB-271046, SB-357134, and Ro 04-6790	2D	Virtual screening of 3.3 M library	4-methyl piperazinyl aniline derivatives
Kim et al. [30]	2008	Catalyst/HipHop, Cerius/CART recursive partitioning	281	3D	Virtual screening of 16,560 ChemDiv GPCR focused library	Phenylsulfonyl thieno[2,3-e][1,2,3]triazolo[1,5-a]pyrimidine derivatives
Liu et al. [77]	2009	Scaffold-hopping	7	2D	–	N1-Aryl-sulfonylindazoles
Tanía de la Fuente et al. [78] crossref: [75]	2010	Catalyst/ HypoGen	45	3D	Activity-based alignment	Piperazinyl-benzimidazoles
Devegowda et al. [79]	2013	Catalyst/ HypoGen	45	3D	Activity-based alignment	Pyrazolo-pyridin-7-one
Hayat et al. [80] crossref: [75]	2013	Catalyst/ HypoGen	45	3D	Activity-based alignment	N-((1-(phenylsulfonyl)-1H-indol-3-yl) methylene) piperazin-1-amine
Ivachtchenko et al. [46]	2014	Hypothetical models	–	–	Classification of different pharmacophore types	–
Flachner et al. [81]	2015	Schrödinger/ PHASE	46	3D	Pharmacophore model based on clustered training set	Chemotype novelty comparison of similarity and pharmacophore-based approaches

Pharmacophore models can drive rational medicinal chemistry, and they also represent 2D or 3D queries for the identification of new compounds retaining specific groups and structural features that are characteristic of the reference ligands (summarized in Table 3). A hypothetical framework model of 5-HT$_6$R antagonists was proposed by Holenz and colleagues [45] in 2005 defining a reference for all pharmacophore and QSAR models in the upcoming years. The main 5-HT$_6$R pharmacophore framework was defined as having two hydrophobic rings or ring systems connected by a hydrogen-bond acceptor portion consisting mostly of a sulfonyl linker. One of the hydrophobic sites may also include a positively ionizable (typically basic amine) group that is able to bind to the TM3 aspartate present in the majority of aminergic GPCR receptors.

In 2008, Kim and colleagues built a 3D common feature pharmacophore model by the HipHop algorithm of Catalyst for highly potent antagonists that were evaluated by recursive partitioning [30]. The authors indicated the limitations of both Catalyst algorithms, highlighting that HypoGen requires a wide range of activities, while HipHop focuses rather on a small ligand set with high activity values, resulting in high ratio of false positives. To overcome the drawbacks of the HipHop model, a recursive partitioning model [82] (CART algorithm/Cerius software) was applied in parallel using known antagonists and decoys by calculating topological property descriptors. The development of the model was not only a follow-up to the study reported by López-Rodríguez and colleagues [75] but also integrated the clustering of antagonists reported by Holenz in 2005 [45]. An important comparison was presented in this study, highlighting the improved performance of the HipHop protocol against the HypoGen algorithm. A reestablished modeling run [80] created by Hayat and colleagues in 2013 renewed the HypoGen/Catalyst hypothesis initially reported by López-Rodríguez et al. in 2005.

An interesting protocol was presented recently by Dobi and colleagues, in which simple 2D fingerprint-based similarity search hits resulting from a preliminary run were iteratively used in combination with known 5-HT$_6$R antagonist templates for the development of pharmacophore hypotheses [81]. An important addition to the combination of similarity search hits and known reference compounds was to search for a common pharmacophore hypothesis by Schrödinger PHASE [83] using the main clusters derived from the training set. The development of pharmacophore models may be systematically diversified by clustering the training set. An additional important aspect of this study was the significant improvement of novelty and diversity of screening hits provided by the clustered pharmacophore-based search in comparison to that performed with the non-clustered pharmacophore model and to the stand-alone similarity search approach.

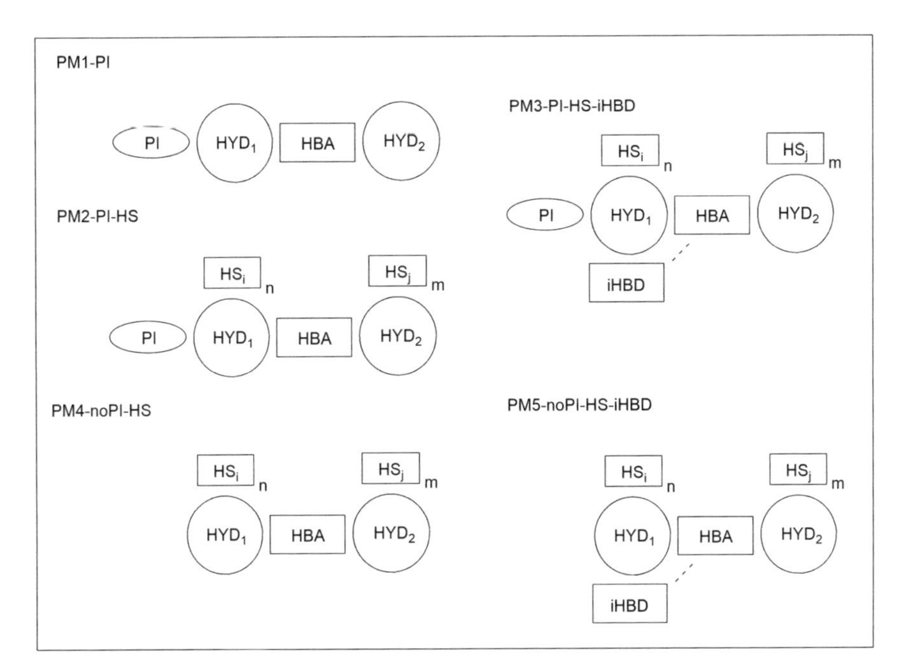

Fig. 6 Schematic representation of alternative 5-HT$_6$R pharmacophore models presented by Ivachtchenko et al. [46]: PM1-PI, pharmacophore model type-1 with positive ionizable group, examples SB-742457 and RVT-101 [26, 28]; PM2-PI-HS, pharmacophore model type 2 containing a number of substituents attached to the hydrophobic rings/scaffolds, example [48]; PM3-PI-HS-iHBD, pharmacophore model type 3 containing an intramolecular hydrogen-donor group, examples AVN561 [84] and AVN-322 [85]; PM4-noPI-HS, pharmacophore model type 4 lack of PI, example [86]; PM5-noPI-HS-iHBD, pharmacophore model type 5 with stabilization of the binding conformation by intramolecular HBD, example [87]; *PI* positive ionizable, *$HYD_{1,2}$* hydrophobic group, *HBA* hydrogen-bond acceptor group, *$HS_{i,j}$* substituents of hydrophobic groups, *iHBD* group forming intramolecular hydrogen bonds with HB

A significant addition to the existing knowledge of 5-HT$_6$R pharmacophore modeling was published by Ivachtchenko in an exhaustive review presented in 2014 [46]. The appraisal of the already existing pharmacophore models showed that the biological activity of new 5-HT$_6$R ligands published from 2005 cannot be explained completely by the canonical pharmacophore patterns. Figure 6 shows five different arrangements of pharmacophores encompassing the current chemical space of the 5-HT$_6$R ligands.

It is important to note that the PM4-noPI-HS pharmacophore (Fig. 6) represents an alternative trend in the discovery and development of the 5-HT$_6$R ligands lacking a basic nitrogen moiety (PI). Evidence shows that ligands without PIs may lose their affinities toward other aminergic targets, and thus selectivity toward 5-HT$_6$R may be tuned through retaining the remaining pharmacophore features. Another advantage of PI removal is a decreased hERG affinity that is typically associated with basic centers.

3.1.2 2D and 3D QSAR Methods

The quantitative structure-activity relationship (QSAR) is a regression (or classification-based) model used for analyses of structural and biological activity data as well as for predictions, providing hints about the synthesis of new potentially active compounds. It is a ligand-based methodology that was founded on the assumption that compounds with similar 2D and/or 3D properties have similar bioactivities. The typical QSAR experiment comprises the following four steps: selection of the training data and descriptors, selection of the variables, and construction and validation of the model. Indeed, the selection of compounds for training is crucial for the outcome and confidence of the results. For example, the selection of a diverse set of known antagonists leads to a more general model [88, 89], thus allowing broader screenings. A different approach is to gather a scaffold-focused template library [90, 91] for training that results in more reliable ascertained relationships between molecular descriptors and affinity.

Ligand-based information is also analyzed at atomic and molecular levels, with steric and electrostatic properties (0D, 1D, and 2D, respectively) [88, 91–93] calculated by CoMFA (comparative molecular field analysis) and CoMSIA (comparative molecular similarity index analysis) descriptors or the consideration of 3D geometry [89, 90, 94, 95] also including pharmacophore features in the model generation.

The number of descriptors infers a number of possible variable combinations, leading to distinct SAR models, and thus selection of the best model is necessary. Selection of the best models is evaluated by regression analysis (MLR, multiple linear regression analysis; ANN, artificial neuronal network; or GA, genetic algorithm, among others) of the descriptor-affinity relationship using significance evaluators of the variables and by analyzing the collinearity of the descriptors [96].

The existing antecedents of 5-HT_6-related ligand-based models (Table 4) may also serve as good starting points and benchmarks for research because they lead to very important observations for the 5-HT_6R antagonists.

One of the most interesting examples of QSAR approaches in 5-HT_6R ligand analysis concerns a very prevalent scaffold of N1-arylsulfonylindoles on a set of 20 similar compounds (CoMFA and CoMSIA), suggesting a negatively charged region near the 5th position of the indole system, a positively charged area near the amine side chain, and a negatively charged region over the phenyl ring of the aryl-sulfonyl group. Furthermore, the CoMSIA contours showed a favored hydrophobic feature in the aryl-sulfonyl region, an unfavored hydrophobic region around the 5th position of the core, and more hydrophobic substituents on the aryl ring of the aryl-sulfonyl region.

More general conclusions were obtained using a hologram-type QSAR study of a set of 30 highly diverse specific and

Table 4
QSAR methods in the field of 5-HT_6R antagonists

Author name	Year	Methodology	Training set size	Descriptors	Type of descriptors	Focused chemical groups
Doddareddy [90]	2004	QSAR	20	CoMFA/ CoMSIA 3D	Steric, electrostatic, hydrophobic, donor/ acceptor	N1-arylsulfonylindoles
Doddareddy [88]	2004	HQSAR	48 specific vs 30 nonspecific	Molecular holograms	Atom types, bond types, connectivity, hydrogens, and chirality	N1-arylsulfonamides
Sharma [91]	2010	MLR QSAR	73	148	0D, 1D, 2D descriptors	Indolyl and piperidinylsulfonamides
Goodarzi [94]	2010	MLR ANN-GA QSAR	52	1497	0D, 1D, 2D, 3D, multidimensional	1-(azacyclyl)-3-arylsulfonyl-1H-pyrrolo[2,3-b]pyridines
Hao [89]	2011	DRAGON, CoMFA/CoMSIA, 3D QSAR, docking, MD	223	2 CoMFA, 5 CoMSIA, 658 2D	3D steric, electrostatic, HYD, HBA/D	General
Choudhary [92]	2014	CP MLR QSAR	30	128	0D, 1D, 2D descriptors	Epiminocyclohepta[b]indoles
Choudhary [93]	2015	DRAGON, CP MLR QSAR	43	471	0D, 1D, 2D descriptors	1-arylsulfonyltryptamines
Velingkar [95]	2016	3D QSAR (PHASE)	46	24 QikProp	3D-pharmacophore, atom-based QSAR	Piperidinylaminomethyl aryl sulfonamides

nonspecific 5-HT_6R antagonists [88]. The sulfonamide structural element was highlighted as a general contributor to affinity, whereas the basic nitrogen was only confirmed in the case of the nonspecific dataset, showing the relevance of nonbasic 5-HT_6R antagonists. Moreover, it was ascertained by aligning SB-357134 on the specific model such that the secondary nitrogen of the piperazine ring was identified as a negative contributor to affinity. Multiple halogen substitutions in the aryl-sulfonyl ring were also observed to be beneficial.

QSAR was also successfully combined with the results of docking and molecular dynamics (MD) experiments [89], expanding beyond the typical ligand-based approach.

3.1.3 Bioisosteres and Privileged Structures

A recently proposed methodology [97] of exchanging certain structural parts based on pharmacophore features may result in new bioisosteric compounds with improved physicochemical properties and may produce new scaffolds. The chemical space of serotonin receptor ligands was covered by classifying known active ligands as original compounds and their replacements based on the Pipeline Pilot bioisosteric compilation. The analysis resulted in a number of cross- and self-bioisosteres, which either occur in multiple different 5-HT-focused ligand subsets or are only present in a single set (self-bioisosteres). The study provided insights regarding the frequently occurring 5-HT_6R ligand bioisosteric replacements such as the 2-substituted pyridine ring that replaces the phenyl ring, nitrile group substituted with any halogen atom, and introduction of the sulfonamide group by exchanging the carbonyl moiety.

An exhaustive bioisosteric query-based virtual screening combined with pharmacophore mapping [98] resulted in nonsulfonamide PI-containing 5-(4-methylpiperazin-1-yl)-2-nitroaniline derivatives with nanomolar inhibitory activities toward 5-HT_6 and multiple-fold selectivity toward 5-HT_{1A}, 5-HT_7, and D_2Rs.

Another case of combining bioisosteric queries with structure-based virtual screening was reported by Staroń et al. [99] to successfully identify 5-HT_6/D_2dual-acting ligands. This study underlined the opportunities associated with common bioisosteres over different aminergic target classes for the optimization of fine-tuned polypharmacology.

3.2 Structure-Based Methods

3.2.1 Homology Modeling

The concept of homology (comparative) modeling is built on structural biology paradigms suggesting that the structure of a protein is determined by its primary sequence [100] and that structural elements are more conserved than the sequence within a given family of proteins [101]. In the case of GPCRs, these paradigms do not imply a linear relationship between sequential similarity and the quality of the structure, as observed for virtual screening applications [101, 102]; however, as a starting point for receptor structure prediction experiments, its effectiveness has been demonstrated [103,104].

The first two steps of a homology modeling protocol are the template(s) selection and the sequence alignment (target to the template(s)), closely related and in many cases, especially for GPCRs, can be swapped because template selection is primarily determined by the sequence alignment. For G-protein-coupled receptors, the general topology of the TM domain is well defined, and the main challenge associated with template selection is to reflect the structural nuances of the helices of the target, such as the bulges and constrictions, which affect the orientation of the key residues for ligand binding. Intra- and extracellular loops connecting the transmembrane helices are the most challenging part of GPCR modeling due to the great variability of the sequences and secondary structural elements as well as to the paucity of crystal templates to effectively probe this structural space. In addition, loop regions are often missing in the available crystal structures. For example, ICL3 is probably the most variable structural segment of the GPCRs, ranging from 4 to 230 amino acids in length, and in many cases, it is replaced by fusion proteins or nanobodies, enabling crystallization. As a result, it has only been crystallized a few times (e.g., rhodopsin or $A_{2A}R$).

In the third step, inferring the coordinates from template(s) to the target, the backbone of the template is used as an anchor for assembling the structure. Therefore, packing of the helices in the model reflects one of the templates. The side chains of the modeled residues may undergo large conformational changes, depending on the algorithm used. Typically, the initial coordinates are subjected to force field-based optimization. For instance, in Modeller [105], which is one of the most commonly used programs for homology modeling, side chain conformations are converged using the spatial restraints derived from the template, and the structure is minimized in the CHARMM force field. However, a new approach in side chain modeling utilizes the fragments of the crystal structures that are directly mapped to the side chain conformations. This knowledge-based approach provided the best model of 5-$HT_{1B}R$ in the community-wide homology modeling contest [106]. An alternative approach is the use of multiple templates for distinct regions of the target protein, as the sequence similarity/identity for, e.g., a single transmembrane helix is higher than that for the overall sequence. This rationale has been utilized in the SSFE database [107] and GPCRM web service [108].

In the case for modeling of the inactive state of 5-HT_6R, the available serotonin receptor crystals are in an active-like conformation but have important water molecules crystallized and offer more effective bases for loop modeling than β-adrenoceptors. In such a case, it is necessary to optimize the model of the investigated receptor to reach the proper conformation of the binding site, for which there are indeed a number of supportive techniques.

In the case of GPCRs, MD simulations and simulated annealing (SA) methods require the construction of a complex model system, including the ligand, the receptor, the membrane, and the solvent. This method has been successfully applied for investigating the ligand-binding mechanism for β_2-AR [109] or observations of the molecular switches in GPCRs [110]. The main disadvantage of this approach is the long computational time required, which can be overcome through the use of GPUs.

Instead of simulating the whole system, it may be sufficient to perform only local optimization of the binding site to allow its adaptation to the bound ligand. Here, only the structure of the ligand and a few surrounding residues undergo the minimization process, being significantly faster than MD or SA. There are a number of available algorithms in popular molecular modeling software packages, e.g., Induced Fit Docking protocol in Maestro [111], LMOD and LLMOD in MacroModel [112], and AMBER [113].

3.2.2 Site-Directed Mutagenesis Data

Point mutations (i) indicate residues of interest and allow the identification of potential hotspots for ligand-receptor interactions and (ii) can and should be used for the evaluation of the homology models as the interacting residues indicate the orientation of the given amino acid.

The 5-HT$_6$R has been widely studied in terms of mutagenesis [18, 77, 113, 114, 115], mapping the binding site of the receptor located between TM3, TM5, and TM6 (Fig. 7) [66]. Positions 3×28, 3×32, and 6×55 have been identified as crucial for interactions with ligands and, indeed, contain the complementary pharmacophore features in the majority of the identified ligand scaffolds (see Materials).

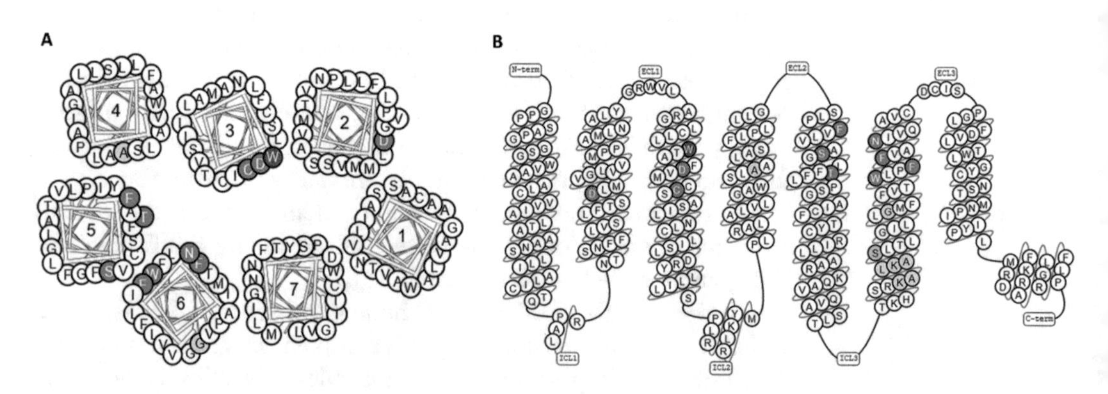

Fig. 7 Mapping of the point mutations of 5-HT$_6$R retrieved from the GPCRdb [66]. (**a**) Visualization of the receptor from the membrane surface, (**b**) *side view*. Residue colors correspond to the change in binding: increased binding/potency, >fivefold; reduced binding/potency, >fivefold, >tenfold; no/low effect (< fivefold)

3.2.3 Docking

There are a number of programs to perform docking experiments, both commercial (like Glide [116] or Gold [117]) and freely available for academics (DOCK [118]). The software differs in performance [118, 119], but the assessment of the best docking program is not as easy as it may seem. The results of benchmarks are affected by the test sets used [120] and the quality of the protein, among other characteristics [121]; thus, it is strongly advised to perform retrospective virtual screening prior to the actual prospective experiments. Sets of known binders [122], as well as artificially generated decoys [122, 123], offer the possibility to test the combination of the current homology model, the docking software, and the type of ligands used for screening. However, the benchmarking studies [124] showed that docking alone does not provide sufficient VS performance, and therefore post-processing by docking analysis techniques has become essential. Interaction fingerprints (IFPs) [125–130] have become a tool of choice for such tasks, offering an enhancement of the screening performance [125, 130] and becoming the final step in the virtual screening protocol.

4 Tips/Notes

4.1 Chemical Structures of Clinical Candidates

Seven drug candidates are currently undergoing clinical investigation for indications in Alzheimer's disease (bolded entries in Table 5) based on a mechanism of action as 5-HT$_6$R inhibitors. Only the chemical structures of five candidates have been avowed (SGS-518, SYN-114, idalopirdine, AVN-211, and AVN-101). The other seven examples have already failed at certain clinical stages (entries marked with "x" in Table 5).

The majority of clinical candidates contain the characteristic sulfonyl linker structural motif represented as sulfonyls, sulfonylamides, and sulfonates. Two examples of ligands lacking the sulfonyl linker failed clinical trials, while SYN-114 remains a promising lead. AVN-211 is an important example of a nonbasic 5-HT$_6$R antagonist, whereas its close analog AVN-216 contains a basic amino group, although this type of aliphatic amino group serves as an intramolecular H-bond donor. Emerging from the scaffold of serotonin are a large number of 5-HT$_6$R agonists and antagonists that are derivatives of the indole ring system and that also support SGS-518 and AVN-101 as prominent clinical candidates under phase 2 studies. However, a significant subset of antagonists possesses different N-, S-, and O-heterocyclic ring systems as bioisosteres and pharmacophore mimics of the indole structure. For example, the clinical candidate AVN-211 (phase 2) contains a pyrazolopyrimidine core, whereas intepirdine (phase 3) is built on a quinoline scaffold. Another set of antagonists contains two simple aryl (mostly phenyl) rings connected by sulfonyl linkage categorized as "carbocyclics," as shown in Table 5.

Table 5
Structural classification of clinical, preclinical, and investigational 5-HT$_6$R antagonists

Sulfonyl containing			Non-sulfonyl containing	Nonbasic	Undisclosed structure	Scaffold		
Sulfone	Sulfonamide	Sulfonate	SYN-114(1x), latrepirdine (3x), idalopirdine (3x), **AVN-101 (2)**	**AVN-211 (2)**, AVN-216*	**ABT-354(1), SYN-120 (2), SAM-760 (2)**	Indole	N/S/O-heterocyclic	Carbocyclic, others
AVN-322(2x), **AVN-211(2),** PRX-07034 (2x), **intepirdine (3),** cerlapirdine (3x), AVN-216*, 11C-LuAE60152*, Ro-4368554*, Ro-66-0074*	SB-271046(1x), SUVN-502(3x), ADN-1184*, SB258585*, SB-357134*, SB-399885*, SB-214111*, SB-699929*, ALX-1161*, ALX-1175*, MS-245*	**SGS-518 (2)**				**SGS-518(2),** latrepirdine(3x), SUVN-502(3x), idalopirdine3x), Ro-4368554*, SB-699929*, MS-245, ALX-1161, ALX-1175, ALX-0440, **AVN-101(2)**	SB_271046, AVN-322, **AVN-211 (2), intepirdine (3),** SAM-531, SAM315, AVN-216, BVT-74316, ^{11}C-LuAE60157, AVN-101, AND-1184, Ro-630563, Ro-046790, Ro-660074, R-1485, SYN-114(1x)	PRX-07034, SB-258585, SB-357134, SB-399885, SB-214111

Bold entries denote drug candidates under running clinical studies, numbers in brackets current clinical phase, and "x" in brackets failed clinical trials. Compound names marked with asterisk stand for non-industrial, investigational compounds.

4.2 Mechanism of Action of Antagonists Versus Agonists

Serotonin is the endogenous agonist of all 5-HT subtypes, affecting (in case of 5-HT_1R and $5\text{-HT}_{4\text{-}7}R$) either negative or positive coupling to the adenylate cyclase second messenger system through the induction of an active conformation of the receptors [132]. Despite the four pharmacophore features of serotonin, the fragment-like size prevents its subtype specificity, even though the binding sites may be distinct in different 5-HT receptor subtypes.

The clustering of ChEMBL 5-HT_6R agonists (Fig. 8) raises the possibility for the comparison with the antagonist scaffold collection, offering insights into the main differences between functionally opposite ligand families and opening a means to the rational design of ligands with a specific function. Seemingly, the indole scaffold is overrepresented in agonists (e.g., tryptamine derivatives), and the fifth position of the indole is commonly substituted (5-hydroxytryptamine analogy). A positive ionizable group is prevalent in almost every agonist. Antagonists are more likely to incorporate bulky tricyclic hydrophobic ring systems, while agonists tend to have 5 + 6 ring systems. The 5-pyrrolo[2,3-b]pyridine scaffold is

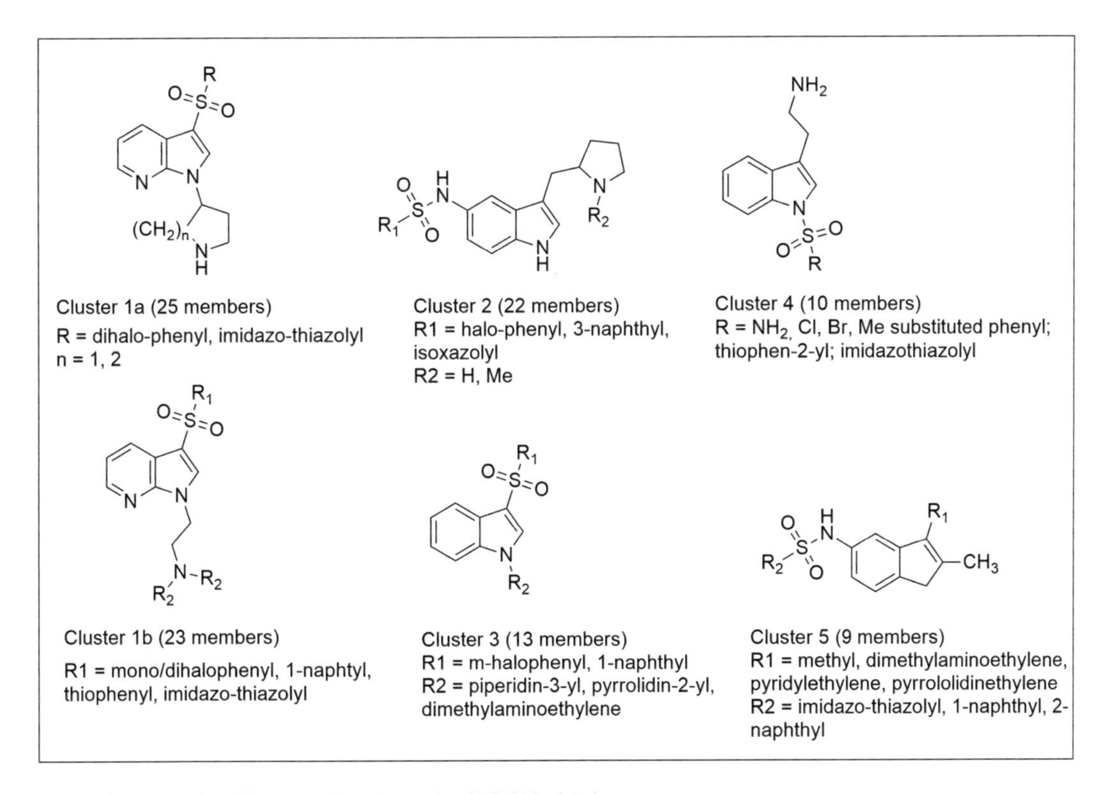

Fig. 8 Clusters of 5-HT_6R agonists from the ChEMBL database

oppositely substituted in the case of the Cluster 1 agonist and the Cluster 2 antagonist.

As discussed above, the indoles are well represented in agonist scaffolds. The agonistic effect might be related to the indole NH that forms an H-bond with the Thr^{5x46}side chain, but N^1-sulfonylindole agonists may bind in a different orientation.

Studying the activation process of aminergic receptors [133], such as the beta-2-adrenergic receptor and, above all, the 5-HT_6 receptor, it has been proposed that agonist binding induces the outward movement of TM6, which is followed by binding of the G-protein. Thus, a conformational change is induced in the NPxxY motif (TM7), which positions the Tyr^{7x53}side chain close to Tyr^{5x58}. There is also an important amino acid pair (Ile^{3x40} – Trp^{6x48}) that plays a role in receptor activation, located in the orthosteric binding pocket, which is part of the connector region and contains a so-called gatekeeper tryptophan [73].

These structural rearrangements of the GPCRs also affect the binding site and imply distinct binding modes for ligands of different functions. Moreover, they emphasize the importance of building homology models with an appropriate state for structure-based VS. Different sizes of agonists and antagonists, as well as subtle differences in their generalized structure (pharmacophore model), also affect the screening procedure. It is again advisable to use retrospective studies to elucidate the binding mode for ligands with a specific function and to evaluate this mode based on experimental findings.

4.3 Evaluation of Ligand- and Structure-Based Methods During the Discovery of 5-HT_6R Antagonists

Generic pharmacophores should be greatly extended for future 5-HT_6R antagonist discovery projects. In addition, attempts to design specific pharmacophore may still serve as starting points as exemplified by Tasler and colleagues, who are using alignment-based screening with a pharmacophore template derived from an agonist, E-6837, followed by a secondary pharmacophore screen using three different antagonists as templates. In another study, Liu and colleagues [69] systematically changed the core scaffold of molecules possessing the main pharmacophore properties. Although the general 2D and 3D pharmacophore models may suggest novel chemotypes, the models overlap significantly. This overlap also expresses the robustness and the confidence of the recognized HYD-SULFONYL-HYD-PI pattern of 5-HT_6R antagonists, which might prevent the design of completely new structural patterns of 5-HT_6R antagonists. Currently, the most comprehensive survey of 5-HT_6R antagonist pharmacophores includes five conceptually diversified pharmacophore patterns. The examples presented in this chapter have shown that diverse and innovative molecular architectures may lead to high-affinity 5-HT_6R antagonists.

The current field of 5-HT_6R antagonist-related pharmacophore models demonstrates the applicability of the existing hypothetical frameworks serving as starting points, and the activity-based pharmacophore alignment models using known antagonists listed in Table 4 also provide viable templates for high-throughput ligand-based virtual screening [30, 75, 76, 79, 80]. The design of new scaffolds and chemotypes for 5-HT_6R antagonists may be elaborated either by running pharmacophore-based screening of existing models or implementing the Catalyst/HypoGen or HipHop module or Schrödinger's PHASE for the design of new models that integrate new antagonists in the training set.

Eight CoMFA, CoMSIA, and QSAR methods and their combinations revealed their usefulness in focusing on generic frameworks of N-aryl-sulfonyl systems (summarized in Table 5). These methodologies [88–90, 93] can be adopted for our own research and are useful for establishing robust structure-activity relationship models based on new expanding sets of known 5-HT_6R antagonists. However, specific scaffold-focused approaches [91, 93–95] have demonstrated their forward-looking potential for use in cases of new chemotypes.

Progress in the crystallization of GPCRs and the wave of new structures supports the structure-based virtual screening of 5-HT_6R, even though the structure of this receptor has not yet been solved. GPCR Dock competition [104, 106] shows the improvement in model resolution over the course of 3 years. The availability of more than 150 crystal structures of GPCRs has improved the quality of sequence alignments and enabled the development of knowledge-based methods for homology modeling. In fact, each new crystal structure improves those methods because it provides better statistical sampling of the conformational space of residues at different positions. However, the improving resolution of the new crystal structures provides information about key cofactors (water molecules, ions) that affect the process of ligand binding.

Improvement of the structure prediction methods infers an increase in the reliability of structure-based methods for virtual screening. Indeed, in the past few years, a number of successful applications of these methods can be noted (Table 6) as supporting the hit discovery projects. It is, however, remarkable, that the structure-based methods are hardly used as the only means of filtering compounds. In most cases, they are used to evaluate the libraries designed by chemists [71, 133], generated combinatorically [135] or as a step of the mixed ligand- and structure-based protocol [77, 135]. A purely structure-based approach to the analysis of the 5-HT_6R antagonists was used to elucidate the molecular mechanism of antagonist binding [72].

Indeed, it is a common approach to use structure-based methods in combination with ligand-oriented algorithms. The

Table 6
Applications of structure-based modeling techniques to the discovery of 5-HT$_6$R antagonists

Author et al.	Year	Method	Template (PDB code)	Approach
Staroń	2015	Multiple homology models, docking	β$_2$-AR (2RH1)	Prospective virtual screening, IFP analysis
Vass	2015	Induced Fit Docking, MD, docking	5-HT$_{2B}$R (4IB4)	Retrospective virtual screening
Kołaczkowski	2015	Multiple homology models, Induced Fit Docking, MD	β$_2$-AR (2RH1)	Prospective virtual screening
Kołaczkowski	2014	Multiple homology models, Induced Fit Docking, MD	β$_2$-AR (2RH1)	Prospective virtual screening
Więckowska	2014	Multiple homology models, Induced Fit Docking, MD, MM-GBSA	5-HT$_{1B}$R (4IAR)	Prospective virtual screening, binding energy estimation for selected compounds
la Fuente	2010	Single homology model, site-directed mutagenesis, docking	β$_2$-AR (2RH1)	Docking supplementing and guiding the in vitro experiments

underlying reason is the computational complexity of running the docking or more sophisticated methods for huge screening libraries containing millions of compounds, although such efforts have been achieved for 5-HT$_{1B}$R [136].

4.4 Reliability of Homology Model Validations

The quality of the homology models can be assessed in a number of ways. The first line of evaluation is an energy-based scoring function, such as the Discrete Optimized Protein Energy (DOPE) score [137], which gives a general overview of the packing of the residues, indicating troublesome regions in the structure when calculated residue by residue. Modeling software packages also offer tools to assess the quality of the protein, allowing the calculation of Ramachandran plots, displaying steric clashes within the structure, and disallowing torsion angles. Visual inspection of the model supported by this type of report is also advised.

Although the energetic perspective is very important, especially when the model is intended for further optimization with MD-based techniques, a careful evaluation of the binding site is a key for virtual screening purposes. Site-directed mutagenesis data are excellent sources of information about key interactions with ligands, and mapping point mutations onto the model shows misaligned residues and potential pitfalls due to the disallowed orientation of important side chains.

The final step is the retrospective validation of the model. Such enrichment studies are mandatory for any virtual screening experiment. Discriminating known binders (true positives) from the set of decoy structures (false positives) reveals the screening performance of the model. Analysis of the binding modes obtained in such experiments can again be cross-checked with the mutagenesis data and determine whether the model returns the proper conformation of the ligand in the binding site.

A general advice for homology modeling is to develop several models and, after the validation process, select a small set of production models. This strategy was a key to success in GPCR Dock 2011 [104] and is the consequence of the naturally occurring receptor flexibility [138], as one universal spatial orientation of the protein may not be optimal for accommodating all types of ligands. Here, methods capitalizing on the ensembles of models [139] or binding pockets [140] can be useful.

4.5 Binding Mode Analysis

As has been previously stressed, an analysis of the binding mode is extremely important for structure-based virtual screening. The elucidation of a reliable binding mode for known binders provides a reference for the evaluation of the candidate ligands found via virtual screening. However, it can provide hints for further mutagenesis studies, and knowledge about the volume, shape, and properties of the binding site might facilitate the design of new compounds or serve as inspiration for synthetic chemists.

Again, docking scores alone are not sufficient to properly classify the screened compounds. The docking results are usually post-processed and always subjected to visual inspection. As has been confirmed in a number of studies [98, 125, 140], structural information inferred from crystal structures along with descriptors encoding ligands with the surrounding residues (and their interactions) provide a means for successful hits in discovery projects.

The binding modes of 5-HT$_6$R described in the literature (Fig. 9) overlap with the consensus binding site of GPCRs [131, 141, 142]. The common core orthosteric binding cleft comprised amino acids D^{3x32}, C^{3x36}, $S^{5.43x44}$, $T^{5.46x461}$, W^{6x48}, F^{6x51}, F^{6x52}, N^{6x55}, $D^{7.36x35}$, and $Y^{7.43x42}$ as well as the most frequently interacting residue positions found in crystal structures [132] and apart from N^{6x55} corresponding to the OBP of the serotonin receptors [71]. The focal points of the binding site are the following residues: D^{3x32}, forming a charge-assisted hydrogen bond with the ligand; $Y^{7.43x42}$, interacting with the ligand or with D^{3x32} (either directly or via a proxy water molecule [143]); and the hydrophobic cluster on TM6 (W^{6x48}, F^{6x51}, and F^{6x52}). The remaining interacting residues are mainly dependent on the studied ligands and form a hydrophobic subpocket corresponding to the SBP [71] with residues A^{4x56}, L^{4x59}, P^{4x60}, L^{4x63}, $V^{5.42x41}$, W^{6x48}, F^{6x51}, F^{6x52}, N^{6x55}, and L182 of the ECL2 or extend to the TM2 and TM7 involving residues $N^{2.64x63}$, $A^{2.65x64}$, $D^{7.36x35}$, and $W^{7.40x39}$ [71, 133, 134], with few contacts with ECL2 R181 and L182, similar to the SBP of

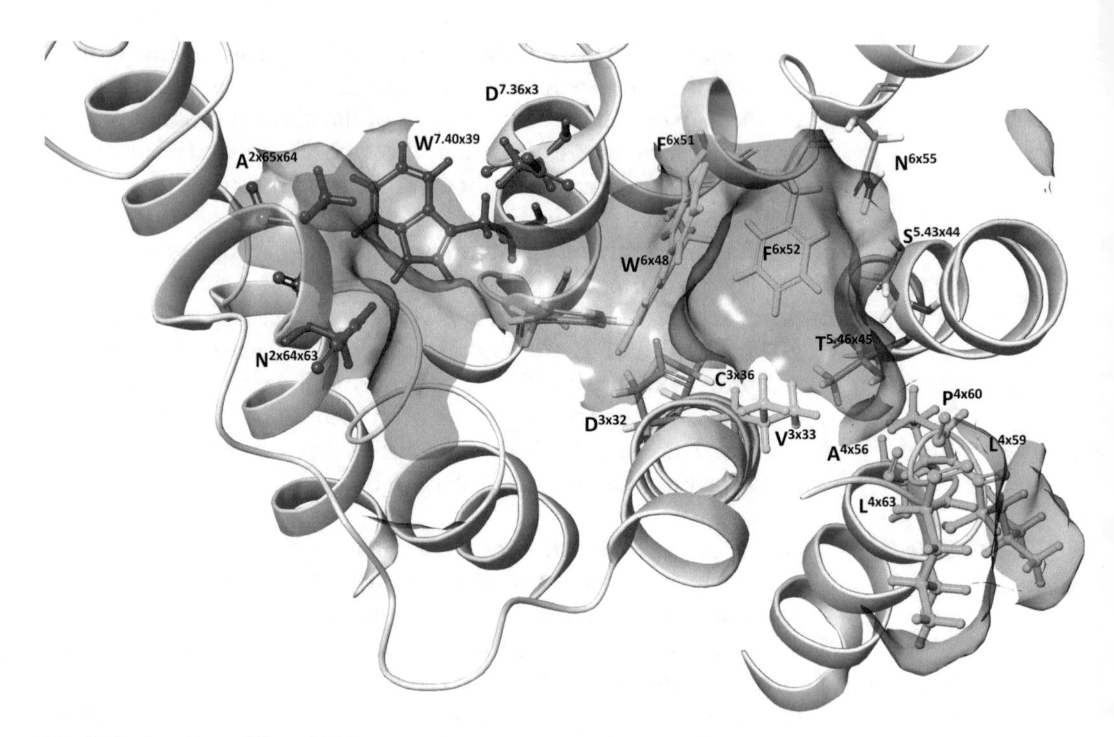

Fig. 9 Binding sites of the 5-HT$_6$R antagonists as extracted from the literature. Residues in *stick* comprise the common binding site from all of the analyzed data; residues in green form the hydrophobic subpockets: *lime green* subpocket I and *green* subpocket II [73]. Amino acids in *red* correspond to the binding site described by Kołaczkowski et al. [72]

adrenoceptors. Here, $D^{7.36x35}$ interacts with the tacrine moiety of the investigated compounds.

Similarly, the evolved pharmacophore and QSAR models assume that the 5-HT$_6$R binding pocket may contain two separate lipophilic cavities. These subpockets can be occupied by hydrophobic groups connected by the characteristic sulfonyl or sulfonylamide linker moiety (or a similar linker group containing hydrogen-bond acceptors similarly to the –SO_2 oxygens). Although the sulfonyl linkage has been confirmed to be a technique for specific 5-HT$_6$R ligands with lower affinity for other 5-HT subtypes and other aminergic off-targets, it is important to mention that this phenomenon is not obligatory. The primary lipophilic subpocket is neighbored by the D^{3x32}side chain, which allows the formation of H-bonding interactions with positive ionizable (classically basic amines) groups of 5-HT$_6$R ligands. However, it is remarkable that the positive ionizable feature is also not imperious. These observations extend the known chemical space of 5-HT$_6$R antagonists and open routes for the design of structurally novel, potent compounds.

4.6 Perspectives of 5-HT$_6$R Antagonists in Drug Discovery

Based on the ligand-based methods presented in the field of 5-HT$_6$R antagonism, a remarkable quantity of data and knowledge has been published in the period from 2004 to 2016. The still growing number of 5-HT$_6$R ligands facilitates the development of new pharmacophores and bioisosteres as well as QSAR models.

As stressed in Sect. 4.5, new structural data are driving the progress of structure-based drug discovery. Better receptor models, new templates for homology modeling, and new crystallized chemotypes expand our knowledge about GPCRs and improve our understanding of the molecular recognition process, facilitating the design of more effective ligands. Of course, the crystal structure(s) of the 5-HT$_6$ receptor is the most desired breakthrough for structure-based VS methods. However, as indicated by the example of β_2-AR, new structures might tend to provoke more questions than they answer.

Although 5-HT$_6$R antagonists (SUVN-502 (Suven Pharma) [21] and intepirdine (GSK/Axovant) [26], both in phase 3) are, in fact, being evaluated in AD clinical trials and progressing further, there remains a significant need for new 5-HT$_6$R leads and drug candidates with the potential to improve cognitive deficits.

Acknowledgments

The authors participate in the European Cooperation in Science and Technology (COST) Action CM1207 – GPCR-Ligand Interactions, Structures, and Transmembrane Signalling: a European Research Network (GLISTEN). This work was supported by the National Brain Research Program KTIA-NAP-13-1-2013-0001. The study was partially supported by the grant OPUS 2014/13/B/NZ7/02210 from the Polish National Science Centre.

References

1. Southan C et al (2016) The IUPHAR/BPS guide to PHARMACOLOGY in 2016: towards curated quantitative interactions between 1300 protein targets and 6000 ligands. Nucleic Acids Res 44:D1054–D1068
2. Dubertret C, Hanoun N, Adès J, Hamon M, Gorwood P (2004) Family-based association study of the serotonin-6 receptor gene (C267T polymorphism) in schizophrenia. Am J Med Genet B Neuropsychiatr Genet 126B:10–15
3. Sleight AJ, Boess FG, Bös M, Bourson A (1998) The putative 5-HT6 receptor: localization and function. Ann N Y Acad Sci 861:91–96
4. Marazziti D et al (2013) Serotonin receptors of type 6 (5-HT6): from neuroscience to clinical pharmacology. Curr Med Chem 20:371–377
5. Upton N, Chuang TT, Hunter AJ, Virley DJ (2008) 5-HT6 receptor antagonists as novel cognitive enhancing agents for Alzheimer's disease. Neurotherapeutics 5:458–469
6. Messina D et al (2002) Association of the 5-HT6 receptor gene polymorphism C267T with Parkinson's disease. Neurology 58:828–829
7. Meltzer HY, Li Z, Kaneda Y, Ichikawa J (2003) Serotonin receptors : their key role in drugs to treat schizophrenia. Prog Neuro-Psychopharmacol Biol Psychiatry 27:1159–1172
8. Kendall I et al (2011) E-6801, a 5-HT6 receptor agonist, improves recognition

memory by combined modulation of cholinergic and glutamatergic neurotransmission in the rat. Psychopharmacology 213:413–430

9. Dudek M, Marcinkowska M, Bucki A, Olczyk A, Kołaczkowski M (2015) Idalopirdine – a small molecule antagonist of 5-HT6 with therapeutic potential against obesity. Metab Brain Dis 30:1487–1494
10. Frantz K, Hansson K, Stouffer D, Parsons L (2002) 5-HT6 receptor antagonism potentiates the behavioral and neurochemical effects of amphetamine but not cocaine. Neuropharmacology 42:170–180
11. Morairty SR, Hedley L, Flores J, Martin R, Kilduff TS (2008) Selective 5HT2A and 5HT6 receptor antagonists promote sleep in rats. Sleep 31:34–44
12. Ham B-J et al (2004) Serotonergic genes and personality traits in the Korean population. Neurosci Lett 354:2–5
13. Lane H-Y et al (2008) Prefrontal executive function and D1, D3, 5-HT2A and 5-HT6 receptor gene variations in healthy adults. J Psychiatry Neurosci 33:47–53
14. Lee S-H et al (2005) Association between the 5-HT6 receptor C267T polymorphism and response to antidepressant treatment in major depressive disorder. Psychiatry Clin Neurosci 59:140–145
15. Frederick JA, Meador-Woodruff JH (1999) Effects of clozapine and haloperidol on 5-HT6 receptor mRNA levels in rat brain. Schizophr Res 38:7–12
16. Yun H-M, Rhim H (2011) The serotonin-6 receptor as a novel therapeutic target. Exp Neurobiol 20:159–168
17. Mitchell ES, Neumaier JF (2005) 5-HT6 receptors: a novel target for cognitive enhancement. Pharmacol Ther 108:320–333
18. Hirst WD et al (2003) Differences in the central nervous system distribution and pharmacology of the mouse 5-hydroxytryptamine-6 receptor compared with rat and human receptors investigated by radioligand binding, site-directed mutagenesis, and molecular modeling. Mol Pharmacol 64:1295–1308
19. Cummings JL, Morstorf T, Zhong K (2014) Alzheimer's disease drug-development pipeline: few candidates, frequent failures. Alzheimers Res Ther 6:37
20. U.S. National Library of Medicine. Home – ClinicalTrials.gov. Available at: https://clinicaltrials.gov/. Accessed 19 Dec 2016
21. Nirogi R et al (2011) SUVN-502: a potent and selective 5-HT6 antagonist, potential drug for the treatment of Alzheimer's disease. Alzheimers Dement 7:S659
22. Suven Life Sciences Limited. SUVN-502 with donepezil and memantine for the treatment of moderate Alzheimer's disease- phase 2a Study – full text view – ClinicalTrials.gov. NCT02580305 Available at: https://clinicaltrials.gov/ct2/show/NCT02580305. Accessed 19 Dec 2016
23. Codony X, Vela JM, Ramírez MJ (2011) 5-HT6 receptor and cognition. Curr Opin Pharmacol 11:94–100
24. Jake King. Pfizer Halts drug trial, investors flee axovant. Available at: http://www.forbes.com/sites/jakeking/2016/02/02/pfizer-halts-drug-trial-investors-flee-axovant/'1546ae6f71f9. Accessed 19 Dec 2016
25. Arnt J et al (2010) Lu AE58054, a 5-HT6 antagonist, reverses cognitive impairment induced by subchronic phencyclidine in a novel object recognition test in rats. Int J Neuropsychopharmacol 13:1021–1033
26. Maher-Edwards G et al (2015) Two randomized controlled trials of SB742457 in mild-to-moderate Alzheimer's disease. Alzheimer's Dement Transl Res Clin Interv 1:23–36
27. de Bruin NM et al (2013) A novel highly selective 5-HT6 receptor antagonist attenuates ethanol and nicotine seeking but does not affect inhibitory response control in Wistar rats. Behav Brain Res 236:157–165
28. Maher-Edwards G et al (2011) SB-742457 and donepezil in Alzheimer disease: a randomized, placebo-controlled study. Int J Geriatr Psychiatry 26:536–544
29. ABOUT INTEPIRDINE(RVT-101) – Axovant sciences – A Dementia Solutions Company. Available at: http://axovant.com/pipeline/intepirdine-rvt-101/. Accessed 5 Jan 2017
30. Kim H-J et al (2008) New serotonin 5-HT(6) ligands from common feature pharmacophore hypotheses. J Chem Inf Model 48:197–206
31. Ivachtchenko AV, Lavrovsky Y, Ivanenkov YA (2016) AVN-211, novel and highly selective 5-HT $_6$ receptor small molecule antagonist, for the treatment of Alzheimer's disease. Mol Pharm 13:945–963
32. Ivachtchenko AV, Lavrovsky Y, Okun I (2016) AVN-101: a multi-target drug candidate for the treatment of CNS disorders. J Alzheimers Dis 53:583–620
33. SYN120 a Dual 5-HT6/5-HT2A antagonist proof of concept study to evaluate its safety, tolerability and efficacy in Parkinson's disease dementia (SYNAPSE) – full text view – ClinicalTrials.gov. Available at: https://

clinicaltrials.gov/ct2/show/NCT02258152. Accessed 5 Jan 2017
34. The Pfizer SAM-760 study | Join dementia research news. Available at: http://news.joindementiaresearch.nihr.ac.uk/sam-760-study/. Accessed 5 Jan 2017
35. Safety, tolerability, and pharmacokinetics of ABT-354 in subjects with mild-to-moderate Alzheimer's disease on stable doses of acetylcholinesterase inhibitors – full text view – ClinicalTrials.gov. Available at: https://clinicaltrials.gov/ct2/show/NCT01908010. Accessed 5 Jan 2017
36. Schaffhauser H et al (2009) Dimebolin is a 5-HT6 antagonist with acute cognition enhancing activities. Biochem Pharmacol 78:1035–1042
37. Herrik KF et al (2016) The 5-HT6 receptor antagonist idalopirdine potentiates the effects of acetylcholinesterase inhibition on neuronal network oscillations and extracellular acetylcholine levels in the rat dorsal hippocampus. Neuropharmacology 107:351–363
38. Ivachtchenko AV, Ivanenkov YA, Veselov MS (2016) Preclinical evaluation of AVN-322, novel and highly selective 5-HT6 receptor antagonist, for the treatment of Alzheimer's disease. Curr Alzheimer Res 14:268–294
39. Mohler EG et al (2012) The effects of PRX-07034, a novel 5-HT6 antagonist, on cognitive flexibility and working memory in rats. Psychopharmacology 220:687–696
40. SYN120 – Biotie therapies. Available at: http://www.biotie.com/product-portfolio/syn120.aspx?sc_lang=en. Accessed 5 Jan 2017
41. Seong CM, Park WK, Park CM, Kong JY, Park NS (2008) Discovery of 3-aryl-3-methyl-1H-quinoline-2,4-diones as a new class of selective 5-HT6 receptor antagonists. Bioorg Med Chem Lett 18:738–743
42. Routledge C et al (2000) Characterization of SB-271046: a potent, selective and orally active 5-HT $_6$ receptor antagonist. Br J Pharmacol 130:1606–1612
43. Bromidge SM et al (1999) 5-Chloro-N-(4-methoxy-3-piperazin-1-yl- phenyl)-3-methyl-2-benzothiophenesulfon- amide(SB-271046): a potent, selective, and orally bioavailable 5-HT6 receptor antagonist. J Med Chem 42:202–205
44. Zhao S-H et al (2007) 3,4-Dihydro-2H-benzo[1,4]oxazine derivatives as 5-HT6 receptor antagonists. Bioorg Med Chem Lett 17:3504–3507
45. Holenz J et al (2004) Medicinal chemistry driven approaches toward novel and selective serotonin 5-HT6 receptor ligands. J Med Chem 48:1781–1795
46. Ivachtchenko AV et al (2014) Sulfonyl-containing modulators of serotonin 5-HT6-receptors and their pharmacophore models. Russ Chem Rev 83:439–473
47. Caldirola P et al (2006) Benzofuranyl derivatives as 5-HT6-receptor inhibitors. WO2006134150 A1
48. Haupt A et al (2010) N-phenyl-(piperazinyl or homopiperazinyl)-benzenesulfonamide or benzenesulfonyl-phenyl-(piperazine or homopiperazine) compounds suitable for treating disorders that respond to modulation of the serotonin 5-ht6 receptor. WO2010125134 A1
49. Parker CA et al (2012) Radiosynthesis and characterization of 11C-GSK215083 as a PET radioligand for the 5-HT6 receptor. J Nucl Med 53:295–303
50. Kołaczkowski M, Mierzejewski P, Bieńkowski P, Wesołowska A, Newman-TancrediA (2014) ADN-1184 a monoaminergic ligand with 5-HT6/7 receptor antagonist activity: pharmacological profile and potential therapeutic utility. Br J Pharmacol 171:973–984
51. Gravius A et al (2011) Effects of 5-HT6 antagonists, Ro-4368554 and SB-258585, in tests used for the detection of cognitive enhancement and antipsychotic-like activity. Behav Pharmacol 22:122–135
52. Monti JM, Jantos H (2011) Effects of the 5-HT6 receptor antagonists SB-399885 and RO-4368554 and of the 5-HT2A receptor antagonist EMD 281014 on sleep and wakefulness in the rat during both phases of the light–dark cycle. Behav Brain Res 216:381–388
53. Stean TO et al (2002) Pharmacological profile of SB-357134: a potent, selective, brain penetrant, and orally active 5-HT6 receptor antagonist. Pharmacol Biochem Behav 71:645–654
54. Hirst WD et al (2006) SB-399885 is a potent, selective 5-HT6 receptor antagonist with cognitive enhancing properties in aged rat water maze and novel object recognition models. Eur J Pharmacol 553:109–119
55. Ahmed M et al (2005) Bicyclic heteroarylpiperazines as selective brain penetrant 5-HT6 receptor antagonists. Bioorg Med Chem Lett 15:4867–4871

56. Sleight AJ et al (1998) Characterization of Ro 04-6790 and Ro 63-0563: potent and selective antagonists at human and rat 5-HT_6 receptors. Br J Pharmacol 124:556–562
57. Tsai Y et al (2000) N1-(Benzenesulfonyl) tryptamines as novel 5-HT6 antagonists. Bioorg MedChem Lett 10:2295–2299
58. Glennon RA et al (2000)2-substituted-tryptamines: agents with selectivity for 5-HT6 serotonin receptors. J Med Chem 43:1011–1018
59. Riemer C et al (2003) Influence of the 5-HT6 receptor on acetylcholine release in the cortex: pharmacological characterization of 4-(2-Bromo-6-pyrrolidin-1-ylpyridine-4-sulfonyl) phenylamine, a potent and selective 5-HT6 receptor antagonist. J Med Chem 46:1273–1276
60. de Bruin et al (2011) Two novel 5-HT6-receptor antagonists ameliorate scopolamine-induced memory deficits in the object recognition and object location tasks in Wistar rats. Neurobiol Learn Mem 96:392–402
61. Slassi A et al (1999) 3-(2-pyrrolidinylmethyl)-indole compounds having 5-HT6 affinity. WO1999047516A1
62. Fludzinski P et al (1987) Indazoles as indole bioisosteres: synthesis and evaluation of the tropanyl ester and amide of indazole-3-carboxylate as antagonists at the serotonin 5HT3 receptor. J Med Chem 30:1535–1537
63. Bento AP et al (2014) The ChEMBL bioactivity database: an update. Nucleic Acids Res 42:D1083–D1090
64. Leśniak D, Jastrzębski S, Podlewska S, Czarnecki WM, Bojarski AJ (2017) Quo vadis G protein-coupled receptor ligands? A tool for analysis of the emergence of new groups of compounds over time. Bioorg Med Chem Lett 27:626–631
65. RDKit: Open-source cheminformatics. http: \crwww.rdkit.org
66. Isberg V et al (2016) GPCRdb: an information system for G protein-coupled receptors. Nucleic Acids Res 44:D356–D364
67. Wacker D et al (2017) Crystal structure of an LSD-bound human serotonin receptor. Cell 168:377–389.e12
68. Miller-Gallacher JL et al (2014) The 2.1 Å resolution structure of cyanopindolol-bound β1-adrenoceptor identifies an intramembrane Na+ ion that stabilises the ligand-free receptor. PLoS One 9:e92727
69. Wang C et al (2013) Structural basis for molecular recognition at serotonin receptors. Science 340:610–614
70. Thorsen TS, Matt R, Weis WI, Kobilka BK (2014) Modified T4 lysozyme fusion proteins facilitate G protein-coupled receptor crystallogenesis. Structure 22:1657–1664
71. Michino M et al (2014) What can crystal structures of Aminergic receptors tell us about designing subtype-selective ligands? Pharmacol Rev 67:198–213
72. Kołaczkowski M et al (2014) Novel arylsulfonamide derivatives with 5-HT6/5-HT7 receptor antagonism targeting behavioral and psychological symptoms of dementia. J Med Chem 57:4543–4557
73. Vass M et al (2015) Dynamics and structural determinants of ligand recognition of the 5-HT6 receptor. J Comput Aided Mol Des 29:1137–1149
74. Benhamú B, Martín-Fontecha M, Vázquez-Villa H, Pardo L, López-Rodríguez ML (2014) Serotonin 5-HT receptor antagonists for the treatment of cognitive deficiency in Alzheimer's disease. J Med Chem 57:7160–7181
75. López-Rodríguez ML et al (2005) A three-dimensional pharmacophore model for 5-hydroxytryptamine6 (5-HT6) receptor antagonists. J Med Chem 48:4216–4219
76. Tasler (2007) Discovery of 5-HT6 receptor ligands based on virtual HTS. Bioorg Med Chem Lett 17:6224–6229
77. Liu KG et al (2009) 1-Sulfonylindazoles as potent and selective 5-HT6 ligands. Bioorg Med Chem Lett 19:2413–2415
78. de la Fuente T et al (2010) Benzimidazole derivatives as new serotonin 5-HT6 receptor antagonists. Molecular mechanisms of receptor inactivation. J Med Chem 53:1357–1369
79. Devegowda VN et al (2013) Synthesis and the 5-HT6 receptor antagonistic effect of 3-arylsulfonylamino-5,6-dihydro-6-substituted pyrazolo[3,4]pyridinones for neuropathic pain treatment. Bioorg Med Chem Lett 23:4696–4700
80. Hayat F et al (2013) Design and synthesis of novel series of 5-HT6 receptor ligands having indole, a central aromatic core and 1-amino-4 methyl piperazine as a positive ionizable group. Bioorg Med Chem 21:5573–5582
81. Dobi K et al (2015) Combination of pharmacophore matching, 2D similarity search, and *in vitro* biological assays in the selection of potential 5-HT antagonists from large commercial repositories. Chem Biol Drug Des 86:864–880
82. Rusinko AI et al (1999) Analysis of a large structure/biological activity data set using

recursive Partitioning. J Med Chem 39:1017–1026
83. Dixon SL, Smondyrev AM, Rao SN (2006) PHASE: a novel approach to pharmacophore modeling and 3D database searching. Chem Biol Drug Des 67:370–372
84. Ivachtchenko A et al (2013) Synthesis of substituted diphenyl sulfones and their structure–activity relationship with the antagonism of 5-HT6 receptors. Bioorg Med Chem 21:4614–4627
85. Ivachtchenko AV et al (2012) Antagonists of 5-HT6 receptors. Substituted 3-(phenylsulfonyl)pyrazolo[1,5-a]pyrido[3,4-e]pyrimidines and 3-(phenylsulfonyl)pyrazolo[1,5-a]pyrido [4,3-d]pyrimidines—synthesis and 'structure–activity' relationship. Bioorg Med Chem Lett 22:4273–4280
86. Ivachtchenko AV et al (2010) Synthesis of cycloalkane-annelated 3-phenylsulfonyl-pyrazolo[1,5-a]pyrimidines and their evaluation as 5-HT6 receptor antagonists. Bioorg Med Chem Lett 20:2133–2136
87. Ivachtchenko AV et al (2013) 5-HT6 receptor antagonists. V. Structure – activity relationship of (4-phenylsulfonyloxazol-5-yl) amines. Pharm Chem J 46:639–646
88. Doddareddy MR et al (2004) Hologram quantitative structure activity relationship studies on 5-HT6 antagonists. Bioorg Med Chem 12:3815–3824
89. Hao M, Li Y, Li H, Zhang S (2011) Investigation of the structure requirement for 5-HT_6 binding affinity of arylsulfonyl derivatives: a computational study. Int J Mol Sci 12:5011–5030
90. Doddareddy MR, Cho YS, Koh HY, Pae AN (2004) CoMFA and CoMSIA 3D QSAR analysis on N1-arylsulfonylindole compounds as 5-HT6 antagonists. Bioorg Med Chem 12:3977–3985
91. Sharma BK, Singh P, Sarbhai K, Prabhakar YS (2010) A quantitative structure-activity relationship study on serotonin 5-HT_6 receptor ligands: indolyl and piperidinyl sulphonamides. SAR QSAR Environ Res 21:369–388
92. Choudhary M, Sharma BK (2014) QSAR rationales for the 5-HT6 antagonistic activity of Epiminocyclohepta [b] indoles. Der Pharma Chem 6:321–330
93. Choudhary M, Deshpande S, Sharma BK (2015) CP-MLR directed QSAR rationales for the 1-aryl Sulfonyl Tryptamines as 5-HT 6 receptor ligands. Br J Pharm Res 8:1–17
94. Goodarzi M, Freitas MP, Ghasemi N (2010) QSAR studies of bioactivities of 1-(azacyclyl)-3-arylsulfonyl-1H-pyrrolo[2,3-b]pyridines as 5-HT6 receptor ligands using physicochemical descriptors and MLR and ANN-modeling. Eur J Med Chem 45:3911–3915
95. Velingkar V et al (2016)3D-QSAR,Synthesis and evaluation of novel piperidinylaminomethyl aryl sulfonamides with memory enhancing activity. JSM Chem 4:1–11
96. Todeschini R, Consonni V, Wiley Inter-Science (Online service) (2008) Handbook of molecular descriptors. Methods and Principles in Medicinal Chemistry Wiley-VCH, New York
97. Warszycki D, Mordalski S, Staroń J, Bojarski AJ (2015) Bioisosteric matrices for ligands of serotonin receptors. ChemMedChem 10:601–605
98. Staroń J et al (2015) Rational design of 5-HT 6R ligands using a bioisosteric strategy: synthesis, biological evaluation and molecular modelling. RSC Adv 5:25806–25815
99. Staroń J et al (2016) Halogen bonding enhances activity in a series of dual 5-HT 6/ D2 ligands designed in a hybrid bioisostere generation/virtual screening protocol. RSC Adv 6:54918–54925
100. Liu T, Tang GW, Capriotti E (2011) Comparative modeling: the state of the art and protein drug target structure prediction. Comb Chem High Throughput Screen 14:532–547
101. Chothia C, Lesk AM (1986) The relation between the divergence of sequence and structure in proteins. EMBO J 5:823–826
102. Tang H, Wang XS, Hsieh J-H, Tropsha A (2012) Do crystal structures obviate the need for theoretical models of GPCRs for structure-based virtual screening? Proteins 80:1503–1521
103. Rataj K, Witek J, Mordalski S, Kosciolek T, Bojarski AJ (2014) Impact of template choice on homology model efficiency in virtual screening. J Chem Inf Model 54:1661–1668
104. Kufareva I et al (2011) Status of GPCR modeling and docking as reflected by community-wide GPCR dock 2010 assessment. Structure (London) 19:1108–1126
105. Fiser A, Sali A (2003) Modeller: generation and refinement of homology-based protein structure models. Methods Enzymol 374:461–491
106. Kufareva I, Katritch V, Stevens RC, Abagyan R (2014) Advances in GPCR modeling evaluated by the GPCR dock 2013 assessment: meeting new challenges. Structure (London) 22:1120–1139

107. Worth CL, Kreuchwig A, Kleinau G, Krause G (2011) GPCR-SSFE: a comprehensive database of G-protein-coupled receptor template predictions and homology models. BMC Bioinformatics 12:185
108. Latek D, Pasznik P, Carlomagno T, Filipek S (2013) Towards improved quality of GPCR models by usage of multiple templates and profile-profile comparison. PLoS One 8: e56742
109. Dror RO et al (2011) Pathway and mechanism of drug binding to G-protein-coupled receptors. Proc Natl Acad Sci USA 108:13118–13123
110. Trzaskowski B et al (2012) Action of molecular switches in GPCRs–theoretical and experimental studies. Curr Med Chem 19:1090–1109
111. Sherman W, Beard HS, Farid R (2006) Use of an induced fit receptor structure in virtual screening. Chem Biol Drug Des 67:83–84
112. Schrödinger Release 2016-4: MacroModel, Schrödinger, LLC, New York (2016)
113. Case DA, Babin V, Berryman JT, Betz RM, Cai Q, Cerutti DS, Cheatham TE III, Darden TA, Duke RE, Gohlke H, Goetz AW, Gusarov S, Homeyer N, Janowski P, Kaus J, Kolossváry I, Kovalenko A, Lee TS, LeGrand S, Luchko T, Luo R, Madej B, Merz KM, Paesani F, Roe DR, Roitberg A, Sagui C, Salomon-Ferrer R, Seabra G, Simmerling CL, Smith W, Swails J, Walker RC, Wang J, Wolf RM, Wu X, Kollman PA (2014) AMBER 14. University of California, San Francisco
114. Zhang J et al (2006) Effects of mutations at conserved TM II residues on ligand binding and activation of mouse 5-HT6 receptor. Eur J Pharmacol 534:77–82
115. Harris RN et al (2010) Highly potent, nonbasic 5-HT6 ligands. Site mutagenesis evidence for a second binding mode at 5-HT6 for antagonism. Bioorg Med Chem Lett 20:3436–3440
116. Halgren TA et al (2004) Glide: a new approach for rapid, accurate docking and scoring. 2. Enrichment factors in database screening. J Med Chem 47:1750–1759
117. Jones G, Willett P, Glen RC, Leach AR, Taylor R (1997) Development and validation of a genetic algorithm for flexible docking. J Mol Biol 267:727–748
118. Allen WJ et al (2015) DOCK 6: impact of new features and current docking performance. J Comput Chem 36:1132–1156
119. Plewczynski D, Łaźniewski M, Augustyniak R, Ginalski K (2011) Can we trust docking results? Evaluation of seven commonly used programs on PDBbind database. J Comput Chem 32:742–755
120. Lagarde N et al (2015) Benchmarking data sets for the evaluation of virtual ligand screening methods: review and perspectives. J Chem Inf Model 55:1297–1397
121. Chaput L,Martinez-Sanz J, Saettel N, Mouawad L (2016) Benchmark of four popular virtual screening programs: construction of the active/decoy dataset remains a major determinant of measured performance. J Cheminform 8:56
122. Gaulton A et al (2012) ChEMBL: a large-scale bioactivity database for drug discovery. Nucleic Acids Res 40:D1100–D1107
123. Mysinger MM, Carchia M, Irwin JJ, Shoichet BK (2012) Directory of useful decoys, enhanced (DUD-E): better ligands and decoys for better benchmarking. J Med Chem 55:6582–6594
124. Huang N, Shoichet BK, Irwin JJ (2006) Benchmarking sets for molecular docking. J Med Chem 49:6789–6801
125. Deng Z, Chuaqui C, Singh J (2004) Structural interaction fingerprint (SIFt): a novel method for analyzing three-dimensional protein-ligand binding interactions. J Med Chem 47:337–344
126. Vass M et al (2016) Molecular interaction fingerprint approaches for GPCR drug discovery. Curr Opin Pharmacol 30:59–68
127. Da C, Kireev D (2014) Structural protein-ligand interaction fingerprints (SPLIF) for structure-based virtual screening: method and benchmark study. J Chem Inf Model 54:2555–2561
128. Clark AM, Labute P (2007) 2D depiction of protein-ligand complexes. J Chem Inf Model 47:1933–1944
129. Desaphy J, Raimbaud E, Ducrot P, Rognan D (2013) Encoding protein-ligand interaction patterns in fingerprints and graphs. J Chem Inf Model 53:623–637
130. Marcou G, Rognan D (2007) Optimizing fragment and scaffold docking by use of molecular interaction fingerprints. J Chem Inf Model 47:195–207
131. Smusz S et al (2015) Multi-step protocol for automatic evaluation of docking results based on machine learning methods – a case study of serotonin receptors 5-HT6 and 5 HT7. J Chem Inf Model 55:823–832
132. Venkatakrishnan AJ et al (2013) Molecular signatures of G-protein-coupled receptors. Nature 494:185–194
133. Dror RO et al (2011) Activation mechanism of the β2-adrenergic receptor. Proc Natl Acad Sci U S A 108:18684–18689

134. Kołaczkowski M et al (2015) Novel 5-HT6 receptor antagonists/D2 receptor partial agonists targeting behavioral and psychological symptoms of dementia. Eur J Med Chem 92:221 235
135. Więckowska A et al (2016) Novel multi-target-directed ligands for Alzheimer's disease: combining cholinesterase inhibitors and 5-HT6 receptor antagonists. Design, synthesis and biological evaluation. Eur J Med Chem 124:63–81
136. Rodríguez D, Brea J, Loza MI, Carlsson J (2014) Structure-based discovery of selective serotonin 5-HT(1B) receptor ligands. Structure 22:1140–1151
137. Fan H et al (2011) Statistical potential for modeling and ranking of protein-ligand interactions. J Chem Inf Model 51:3078–3092
138. Nygaard R et al (2013) The dynamic process of β2-adrenergic receptor activation. Cell 152:532–542
139. Mordalski S, Witek J, Smusz S, Rataj K, Bojarski AJ (2015) Multiple conformational states in retrospective virtual screening – homology models vs. crystal structures: beta-2 adrenergic receptor case study. J Cheminform 7:13
140. Rueda M, Totrov M, Abagyan R (2012) ALiBERO: evolving a team of complementary pocket conformations rather than a single leader. J Chem Inf Model 52:2705–2714
141. Kooistra AJ et al (2016) Function-specific virtual screening for GPCR ligands using a combined scoring method. Sci Rep 6:28288
142. Gloriam DE, Foord SM, Blaney FE, Garland SL (2009) Definition of the G protein-coupled receptor transmembrane bundle binding pocket and calculation of receptor similarities for drug design. J Med Chem 52:4429–4442
143. Wacker D et al (2013) Structural features for functional selectivity at serotonin receptors. Science 340:615–619

Chapter 16

Computational Modeling of Diagnostic Imaging Agents for Alzheimer's Disease: Molecular Imaging Agents for the In Vivo Detection of Amyloid Plaques in Alzheimer's Disease

Dionysia Papagiannopoulou and Dimitra Hadjipavlou-Litina

Abstract

In the presented review, the authors provide the basic background about the radionuclide imaging techniques and the molecular imaging agents for the in vivo detection of amyloid plaques in Alzheimer's disease. Furthermore, the authors review SAR, 2D-QSAR, and 3D-QSAR as well as other computational modeling studies performed on the diagnostic imaging agents for Alzheimer's disease. The information provided includes chemical scaffold and discussion for the role of the implicated physicochemical and will be useful in the development of new PET and SPECT imaging agents with clinical applicability to detect Aβ plaques.

Key words Modeling, QSAR, Alzheimer's disease, Diagnostic imaging agents

1 Radionuclide Imaging Techniques

The radiotracer imaging techniques use radiopharmaceuticals that emit gamma radiation from within the body. After systemic administration, the radiopharmaceuticals accumulate to the target tissue or organ, driven by their chemical structure and their biological properties. The external measurement of the radiation that is emitted from the target organ gives information about the related pathophysiology and provides noninvasive diagnosis of the disease. For example, lower or excessive radiotracer uptake from the target organ signifies malfunction.

There are two radionuclide imaging techniques that are used in nuclear medicine. The first one uses radiotracers that emit a single energy gamma photon which is detected by an external gamma camera. The cameras rotate around the patient and collect the data from different angles and then reconstitute them in a 3D image on the computer (single photon emission computed tomography). The second radionuclide imaging technique is the positron

Kunal Roy (ed.), *Computational Modeling of Drugs Against Alzheimer's Disease*, Neuromethods, vol. 132, DOI 10.1007/978-1-4939-7404-7_16, © Springer Science+Business Media LLC 2018

emission tomography (PET) which uses radiotracers that emit positrons. Positrons (β^+) interact with electrons and generate two annihilation gamma rays simultaneously with opposite directions. These coincidence gamma rays are detected by gamma cameras located opposite to each other. The data collected by this technique give images of high resolution. Nowadays, the SPECT and PET cameras are combined with computed tomography (CT) or magnetic resonance imaging (MRI) which enables anatomically accurate fused images.

The power of radionuclide imaging lies in its high efficiency to detect metabolic and functional abnormalities at the nanomolar or subnanomolar level, which is unparalleled by other imaging modalities. The radiotracers are administered in very low concentrations approximately in the nanomolar range in the radiopharmaceutical preparations, which cause limited side effects and exhibit no pharmacological effects at the target site. Other imaging techniques such as MR, X-ray, or CT, although they provide superior anatomical information, are not as efficient in providing functional information.

Traditionally, brain imaging has been performed by ^{99m}Tc radiopharmaceuticals (Ceretec or Neurolite) with SPECT cameras and later by ^{18}F–FDG with PET [1]. Nowadays, new emerging radiopharmaceuticals for brain imaging are specialized for particular diseases, and they are mostly radiolabeled with short-lived PET radionuclides, ^{11}C and ^{18}F [2]. Alternatively, the SPECT radionuclide ^{123}I is also used.

2 Labeling with Imaging Radionuclides

Radiolabeling with imaging radionuclides requires specialized personnel and radiation licensed laboratories. Radiolabeling methods are optimized to achieve high radiochemical yields in short reaction times and under mild conditions [3]. In the case of short-lived PET isotopes, automated modules are used. Carbon-11 is a cyclotron-produced radionuclide that emits positrons. It has a very short half-life of $t_{1/2} = 20$ min, and therefore it can only be used in-house in cyclotron-equipped hospitals. The most common method of radiolabeling with ^{11}C is by [^{11}C]-methylation. It is produced as [^{11}C]-CO_2 and by automated syntheses is converted to [^{11}C]-CH_3I as the methylating agent [3].

The most widely used PET tracer for brain imaging is ^{18}F, which is also a positron emitter with a half-life of $t_{1/2} = 109.8$ min. The half-life of this radionuclide is long enough to allow transport to small distances and its use is more widespread. There are various methods of ^{18}F radiolabeling. When produced from enriched [^{18}O]-H_2O in the form of ^{18}F–fluoride ion, it is introduced via aliphatic nucleophilic substitution on an activated pharmacophore. Activation of the pharmacophore requires the introduction of a suitable leaving group (such as sulfonate esters, e.g., triflate-CF_3SO_3-) at the position of radiofluoride substitution. Alternatively, aromatic nucleophilic substitutions are performed by using nitro- (-NO_2) or trimethylammonium- (-$^{+}NMe_3$) as leaving groups [3].

The alternative radionuclide for SPECT imaging is ^{123}I which is more long-lived compared to the PET ones, with a half-life of $t_{1/2} = 13.2$ h and gamma energy of 159 keV. Its half-life allows shipment from the producer to the hospital. Radioiodine is produced from a cyclotron in the form of ^{123}I-iodide and is introduced on aromatic rings via electrophilic substitution. The protocol involves the use of an oxidizing agent to convert iodide (I^-) to the electrophile I^+ [4].

The radiotracers cannot be analyzed by standard spectroscopic methods for structural characterization, because of their low concentration. Therefore, prior to producing the radiotracer, the nonradioactive analogue must be prepared and characterized. The radiolabeling mixture is then analyzed chromatographically (by radiometric detection) and compared to the chromatographic behavior of the nonradioactive analogue, which should be identical. Furthermore, if the radiochemical purity of the product is lower than 95%, purification must be performed. The imaging agents must be chemically and biologically stable, in order for the radiotracer to reach the target site intact. Therefore, the position where the radionuclide is introduced is crucial for the stability of the product. Additionally, the imaging agent must be designed in such a way that its physicochemical characteristics allow distribution in the target area. In order for brain imaging agents, the lipophilicity and the ability of the compounds to penetrate the blood–brain barrier (BBB) is critical. Finally, the biological activity of the radiolabeled molecules must be established by in vitro and in vivo experiments to ensure binding of the radiotracer to the desired biological target.

3 Imaging Agents for Alzheimer's Disease

Alzheimer's disease (AD) is a neurodegenerative disease that leads to memory loss affecting primarily the older population. AD is characterized histologically by the deposition of neuritic plaques composed of beta-amyloid ($A\beta$) aggregates and neurofibrillary tangles composed of tau proteins [5]. Noninvasive detection techniques for AD include magnetic resonance imaging and nuclear medicine imaging with radiopharmaceuticals. Nuclear medicine imaging of the regional cerebral blood flow (rCBF) can be performed by brain perfusion radiopharmaceuticals, like ^{99m}Tc-HMPAO (Ceretec) or ^{99m}Tc-ECD (Neurolite) with the use of SPECT, where in AD patients decreased regional radiotracer uptake is observed in the affected areas [6]. Alternatively, the regional cerebral metabolic rate for glucose ($rCMR_{glc}$) is measured by ^{18}F-FDG (fluorodeoxyglucose) with PET [5, 7]. Glucose hypometabolism is observed in the areas where amyloid plaques are deposited and can be correlated with the stage of the disease and therefore is considered to be a reliable method for AD analysis, although it is not specific in detecting the characteristic element of AD, which is the amyloid deposits.

The predominant hypothesis for the development of AD is the "amyloid hypothesis" by which the formation of *Aβ* plaques precedes the other events that ultimately lead to synaptic dysfunction and neurodegeneration. This hypothesis has spurred the development of imaging agents for the detection of *Aβ* plaques [8]. The agents developed are either thioflavin or stilbene analogues and were radiolabeled with SPECT or PET radionuclides such as ^{11}C, ^{18}F, and ^{123}I [9–12]. The first agent developed that was widely used in humans in the clinic was the thioflavin T analogue [^{11}C]-PIB, known as Pittsburgh Compound B (Fig. 1). With [^{11}C]-PIB, it was made possible to detect noninvasively the *Aβ* levels of healthy and non-healthy individuals [13]. High amyloid levels can be detected in humans with diagnosed AD; however, lower *Aβ* levels can also be detected in patients with mild cognitive impairment (MCI) as well as in some healthy individuals with no symptoms of cognitive impairment. Because of its short half-life, [^{11}C]-PIB was recently replaced by similar imaging agents labeled with the longer half-life radionuclide ^{18}F that is more suitable for clinical use and which received Food and Drug Administration (FDA) and European Medicines Agency (EMA) approval between 2011 and 2014. These PET imaging probes include [^{18}F]-florbetaben (NeuraCeq) [14], [^{18}F]-florbetapir (AMYViD) [15], and [^{18}F]-flutemetamol (VIZAMYL) [16] (Fig. 1) and are used for the noninvasive detection of *Aβ* deposits in individuals with cognitive impairment who are being evaluated for AD and other causes of cognitive impairment.

[^{11}C]-PIB

[^{18}F]-flutemetamol

[^{18}F]-florbetaben

[^{18}F]-florbetapir

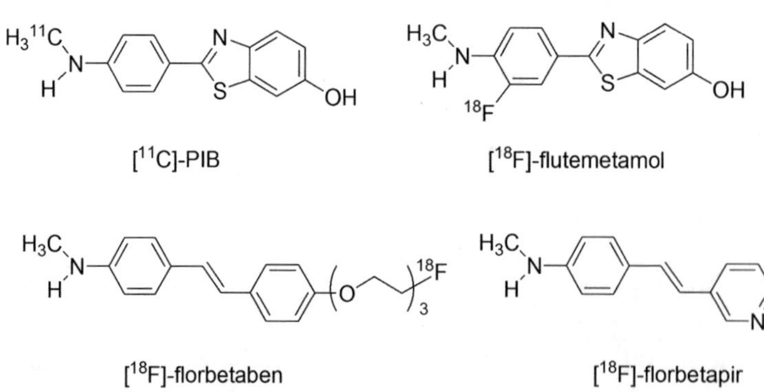

Fig. 1 PET radiotracers for imaging beta-amyloid

4 Technetium Complexes as Potential SPECT Agents for *Aβ* Plaques

Among the radionuclides used for PET and SPECT imaging, ^{99m}Tc ($t_{1/2}$ = 6.01 h, 141 keVγ) has been the most widely used in diagnostic nuclear medicine because it has ideal nuclear properties such as optimal detection for SPECT imaging, suitable half-life for biological localization, and availability at low cost from $^{99}Mo/^{99m}Tc$ generators. All these advantages have spurred further interest in the development of ^{99m}Tc-labeled *Aβ* imaging probes. Labeling with a radiometal such as ^{99m}Tc requires the conjugation of a suitable chelator on the pharmacophore moiety. As technetium is a metal with rich coordination chemistry, there is a wide selection of chelators that lead to neutral complexes able to cross the BBB. The structural characterization of the ^{99m}Tc complexes is commonly performed with the nonradioactive rhenium analogues, as rhenium and technetium are considered to be congener metals.

In these efforts, various ^{99m}Tc-labeled tracers were developed based on Congo red, curcumin, aurone, flavone, and chalcone derivatives [17] as well as a series of ^{99m}Tc-labeled benzothiazole [18], benzoxazole [19], and benzofuran [20] derivatives as thioflavin T analogues (Fig. 2). The derivatives of chalcones, aurones, and flavones were prepared by conjugation of a bisaminethiol (N_2S_2-type) chelator on the flavonoid where after radiolabeling the neutral oxotechnetium (V) complex was formed. These ^{99m}Tc-labeled tracers exhibited high binding to the *Aβ* plaques and the ability to cross the BBB. Similarly, benzofuran, benzoxazole, and benzothiazole derivatives were conjugated to N_2S_2-type chelators as well as in some cases to triaminethiol (SN_3-type) or thiol diamine phenol (SN_2O-type) chelators, all leading to neutral oxotechnetium (V) complexes upon complexation. These complexes exhibited the ability to bind to the *Aβ* plaques that was dependent on the position of the chelator as well as the ability to cross the BBB.

Although the properties of the ^{99m}Tc tracers show promise, no clinical studies in humans have been performed at this point.

5 Structure–Activity Relationship, 2D-Quantitative Structure–Activity Relationships (2D-QSAR), 3D-Quantitative Structure–Activity Relationships (3D-QSAR), and Modeling and Docking Studies

Thioflavin T and Congo red are both dyes used in the histopathological detection of *Aβ* plaques of AD, postmortem. Modifications based on these structures, led to the development of specific radiotracers with high binding affinity for the beta-amyloid as well as optimal biodistribution properties for in vivo imaging. Specifically, in order to achieve a good contrast (high target/nontarget ratio),

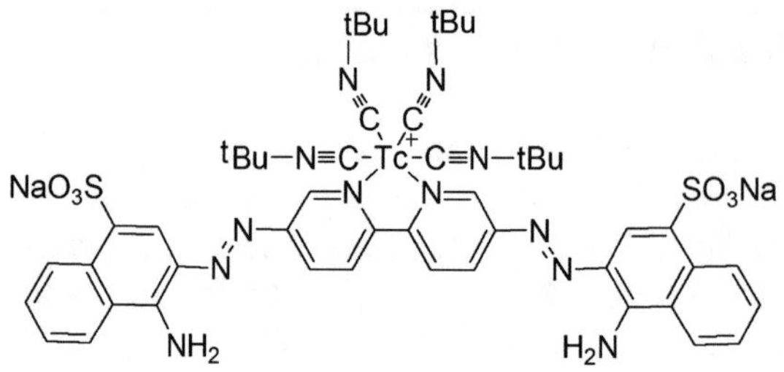

^{99m}Tc(I)-Congo Red derivative

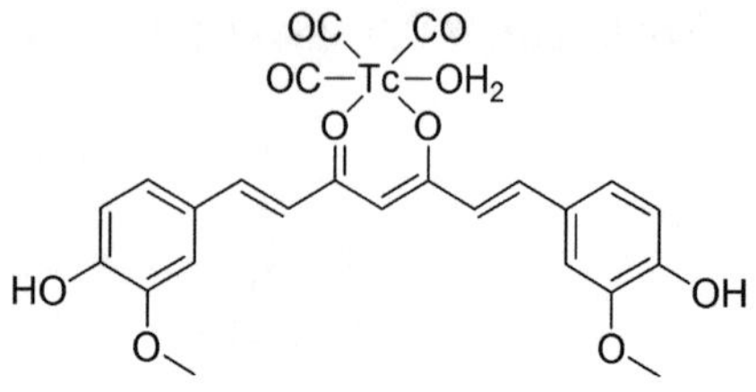

^{99m}Tc(I)-Curcumin derivative

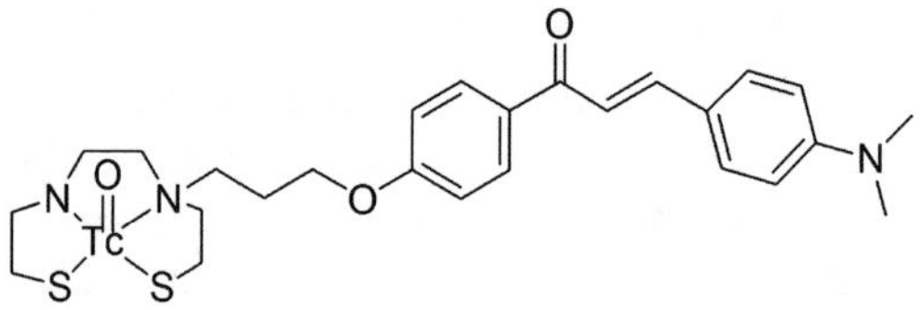

^{99m}TcO(V)-N_2S_2-chalcone

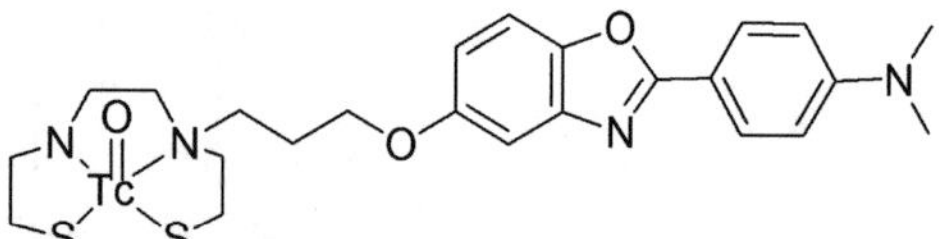

^{99m}TcO(V)-N_2S_2-benzoxazole

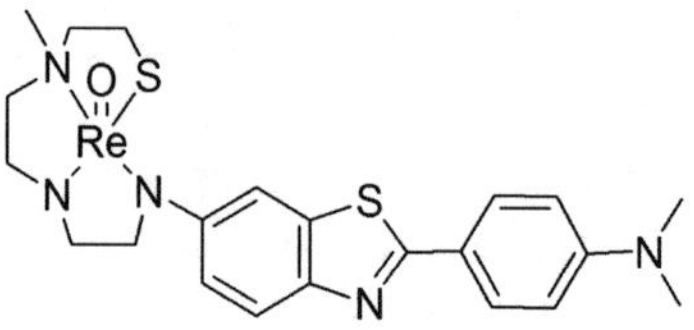

ReO(V)-N_3S-benzothiazole

Fig. 2 Technetium and rhenium complexes for *Aβ* imaging

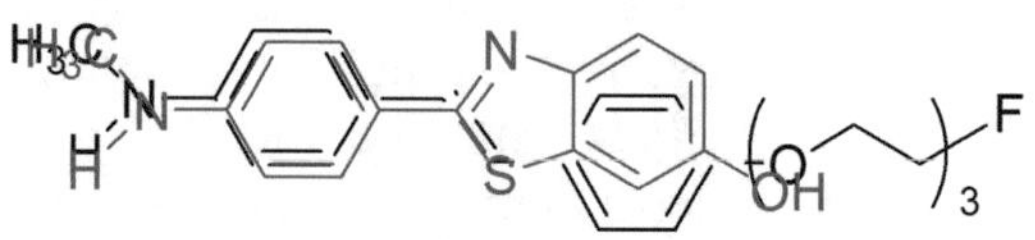

Fig. 3 Benzothiazole and stilbene structural analogy

the radiotracer should exhibit the following properties: (a) fast entry in the brain, (b) fast washout of the unbound radiotracer, and (c) low nonspecific binding. In addition to the above properties, the radiotracers should maintain the ability to be inserted within the beta sheets of the *Aβ* plaques. These prerequisites were met by small, planar, and polyaromatic systems, based on 2-arylbenzothiazoles as well as on stilbenes. In addition, the presence of two electron-donating moieties, a methyl-amino group on one end and a hydroxyl group on the other end of the polyaromatic system, was important for improving the *Aβ* binding [21, 22]. Both, the benzothiazole and the stilbene analogues have similar dimensions and are able to bind at the same *Aβ* sites, as shown in

Fig. 3. In addition, further modification of the stilbene analogues with a chain of three ethylene glycol groups improved the lipophilicity and the washout of the unbound radioactivity from the brain.

In the course of these efforts, many compounds have been screened as *Aβ* plaque-specific imaging agents in industrial laboratories, and a number of active compounds with novel structures are undergoing clinical trials. However, the 2D-/3D-QSAR results for this type of active compounds are very limited. Furthermore, crystal structures of ligand–protein complexes are still not available.

To detect *Aβ* plaques with PET and/or with SPECT, it is necessary to develop probes with a high affinity for *Aβ* plaques. Many ligands with high affinity have been developed and found to bind to mutually exclusive sites on the *Aβ* plaques [23–25]. However, little information is known about the interaction between the ligands and *Aβ* plaques so far with the exception of comparative molecular field analysis (*CoMFA*) and comparative molecular similarity indices analysis (*CoMSIA*) study on serotonin transporter (SERT) ligands [26, 27].

The following 2D-/3D-QSAR examples will highlight the more significant implicated physicochemical properties.

It is known that a successful *Aβ* plaque-specific imaging agent must have good brain permeability and excellent *Aβ* selectivity. Also the initial brain uptake is one of the main factors required for an *Aβ* imaging agent. Liu et al. wanted to delineate [28] experimentally why some AD imaging agents of 2-(4'-aminophenyl)benzothiazole (BTA) (Fig. 4) derivatives have higher whereas others have lower initial brain uptake. A set of 13 compounds as *Aβ* imaging agents with initial brain uptake (Bu) values at 2-min post-injection were selected for quantitative structure–activity relationship (QSAR) study. The initial structures with minimum energy were constructed using the PM3 method available in MOPAC 7.0, and the minimum RMS gradient was 0.0001. Molecule volume (Vm) and dipole moment (Dp) were calculated after geometry optimization using B3LYP method by G98 W software. A multiple linear regression QSAR analysis was performed. Three QSAR models were obtained:

$$\begin{aligned}\text{LogBu} = {} & 1.270 + 0.05147\text{logVm} + 0.04497\text{logP} \\ & - 1.651\text{logD}\end{aligned} \tag{1}$$

$$(n = 13, r = 0.937, \text{SD} = 0.1222, F = 21.619, \text{Sig.} = 0.000)$$

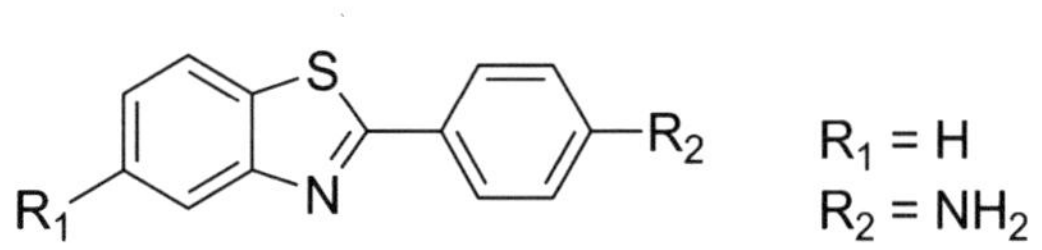

R_1 = H
R_2 = NH_2

Fig. 4 2-(4′-aminophenyl)benzothiazole (BTA)

$$\text{LogBu} = 1.374 + 0.04686\text{logVm} - 1.644\text{logDp} \tag{2}$$

$$(n = 13, r = 0.933, \text{SD} = 0.1194, F = 33.704, \text{Sig.} = 0.000)$$

$$\text{LogBu} = 1.403 - 1.554\text{logDp} \tag{3}$$

$$(n = 13, r = 0.922, \text{SD} = 0.1227, F = 62.333, \text{Sig.} = 0.000)$$

where Bu is the brain uptake at 2-min postinjection, Dp is the molecular dipole moment, n is the number of compounds used in the fit, r is the correlation coefficient, SD is the standard error of the estimate, F is the value of F-statistic test, and Sig. is the statistical significance level.

Equations 1, 2, and 3 indicated that logBu was correlated linearly with logVm, logP, and logDp. It seems that parameter Dp is very significant for initial brain uptake. Lower Dp values are correlated to higher initial brain uptake (Bu).

Recently, Liu et al. [29] synthesized and screened as radiotracers for SPECT imaging of Aβ plaques a class of ^{125}I-labeled flexible benzyloxybenzene derivatives without highly rigid planarity. To quantitatively illustrate the structure–activity correlation of benzyloxybenzene derivatives, 3D-QSAR modeling was performed. The structures of 21 compounds and imidazole-pyrazole derivative IMPY (Fig. 5) were geometry-optimized at the B3LYP/6-31G and 3-21G level in the water phase. 3D-QSAR studies successfully yielded two statistically reliable models, including comparative molecular field analysis (CoMFA: r^2, 0.947; q^2, 0.606) and comparative molecular similarity indices analysis (CoMSIA: r^2, 0.988; q^2, 0.777), which were able to predict binding abilities of novel derivatives accurately. The deviations of the calculated pKi values (ligand-binding affinities in a negative logarithmic scale, $-\log Ki$) from the corresponding actual data in both models were all smaller than 0.5 log unit. The CoMFA study shows that bulky substituents are desirable at the para-position of the phenyl ring. All the ortho-substituted ligands exhibited significant encroachment. More positively charged substituents at the ortho- and para-position may enhance the binding abilities, whereas more negatively charged groups are favorable at the same positions. Hydrophobic para substitution improves binding affinities. Hydrogen bond acceptors seem to be unfavorable.

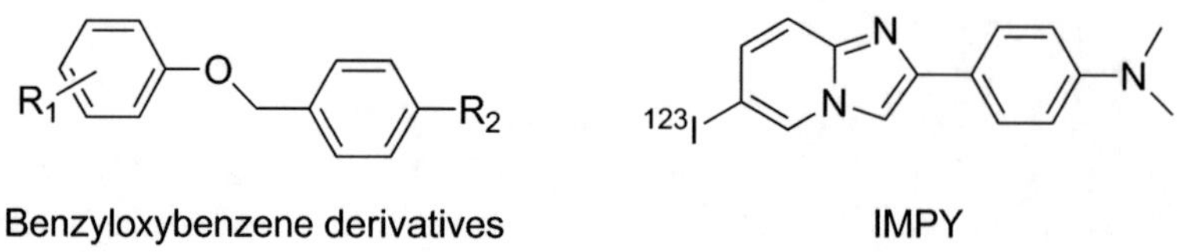

R1 O R2 ^{123}I N N N

Benzyloxybenzene derivatives IMPY

Fig. 5 Benzyloxybenzene derivatives and imidazole-pyrazole derivative IMPY

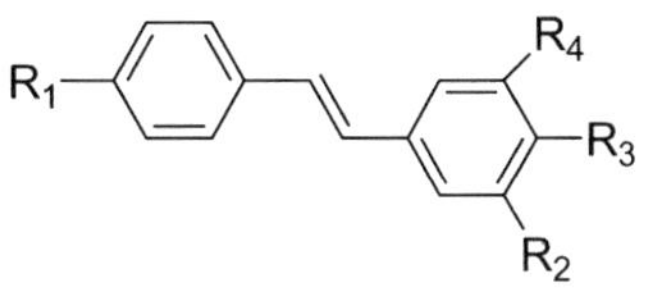

Fig. 6 TSB based analogs

Molecular docking studies were also carried out. A previous publication revealed that the hydrophobic cleft formed between Val18 and Phe20 was the most probable and feasible binding site [30]. Computational docking results revealed that all the benzyloxybenzene analogues, similar to IMPY, inserted into the hydrophobic Val18_Phe20 channel with long molecular axes oriented longitudinally to the fiber axis. It seems that benzyloxybenzene ligands shared the same binding site with IMPY. All the flexible para substituted benzyloxybenzene derivatives tended to be locked into a near-flat conformation. The near-flat characteristic of these ligands favored their impaction into the hydrophobic channel with low-binding energies, which go in parallel to their high binding affinities. On the contrary, the ortho-substituted analogues are harder to be accommodated in the shallow binding channel, resulting in high binding energies and low-binding affinities to the binding pocket. These results are in good agreement with the experimental data and 3D-QSAR model.

Liu et al. [31] utilized the above results and models to design and synthesize new dual imaging agents. The predicted Ki values are in good accordance with the biologically determined ones.

Kung et al. [22] developed some trans-stilbene (TSB) derivatives (Fig. 6). These are simple and relatively small molecules, which are hopeful as probes for in vivo evaluation of $A\beta$ plaques. Xiangji Chen [32a] carried out a QSAR analysis and molecular docking to investigate their interaction mode. Twenty-two TSB-based analogues were optimized using the DFT method. The local minimum-energy and the lowest-energy conformations of each molecule were obtained, followed by an energy minimization. The electronic descriptors were obtained from the results of density functional calculations. Molecular volume and partition coefficient (cLogP) were calculated using the QSAR module in HyperChem [32b]. The dependent variable was defined as the inverse log of the Ki value. The relationship between the *pKi* and a series of molecular descriptors was quantitatively studied by the means of partial least squares (PLS) analysis.

Four descriptors highly correlated with binding activity (pKi): HOMO energy (E_{HOMO}) and charge sum of *trans*-stilbene segment (Charge) are electronic descriptors, molecular volume (Volume) is shape-dependent descriptor, and partition coefficient (cLogP) is a thermodynamic descriptor. The predicted binding affinities (pKi) with the QSAR model are found to be in good agreement with the experimental data within a statically tolerable error range correlation coefficient of $r^2 = 0.857$.

$$\begin{aligned} pKi = & -6.198(\pm 1.535) + 20.693\ (\pm 3.885)\mathrm{E_{HOMO}} \\ & + 0.502(\pm 0.105)\mathrm{cLogP} \\ & - 0.00463((\pm 0.001)\ \mathrm{Volume} \\ & + 2.474((\pm 0.365)\mathrm{Charge} \end{aligned} \quad (4)$$

$$N = 22, S = 0.221, r^2 = 0.857, F = 25.396$$

S is the standard error, r^2 is the correlation coefficient, and F is the testing factor of the reliability. The QSAR model (4) points to the fact that large molecular volume would lower the binding affinity. Moreover, hydrophobicity of the ligand plays an important role in the biological activity, as it is coincident with literature [21], in which structure–activity relationship studies showed a strong correlation between the lipophilicity of the iodinated BTA compounds and the binding affinity. The strong correlation between HOMO energy, charge density, and biological activities (*pKi*) suggest that there might be π–π stacking interaction and electron transfer between ligand and some phenyl groups in the side chain.

Analysis of results from AutoDock [21b] revealed that all ligands entered the hydrophobic channel with their long axis parallel to the long axis of protein. Thus, the ligand could be extended along the two benzene rings without decreasing the binding affinity. The predicted binding-free energies (*DGs*) of the ligands to the *Aβ* protein and the *pKi* were correlated via a linear regression analysis and gave a rather good correlation coefficient $r^2 = 0.881$, which demonstrates that the binding site and binding model of the TSB-based ligands are reasonable.

$$\mathrm{DG} = 2.644 - 1.391\mathrm{pKi}$$

$$N = 8, S = 0.399, r^2 = 0.881 \quad (5)$$

Introduction of some sterically bulky groups in the middle of the molecule would decrease their activities dramatically because it would hinder the ligand molecule from entering the hydrophobic channel.

Chong and co-researchers [33a] constructed a 3D-QSAR model with several PET ligands such as ThioT analogues and stilbene derivatives, which could be applied to predict binding affinity of the structurally related compounds against *Aβ* plaques. (Fig. 7).The compounds were divided into two groups: 63 compounds as a training set and the other 10 compounds as a test set. The training set was used to build 3D-QSAR models with CoMFA and CoMSIA methods, while the test set was used to validate the 3D-QSAR model. All compounds were constructed by the sketch module in SYBYL base [33b] and assigned with Merck molecular force field 94 s (MMFF94s) [33c] charges. In this study, the styrene

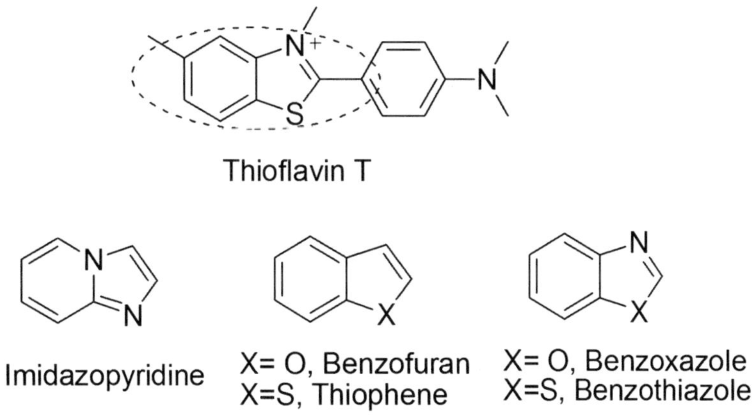

Thioflavin T

Imidazopyridine

X= O, Benzofuran
X=S, Thiophene

X= O, Benzoxazole
X=S, Benzothiazole

Fig. 7 Thio T derivatives

moiety [PhC = X (X = C or N)] commonly found in the ThioT and stilbene derivatives was used as the substructure for alignment. CoMFA was set at standard values, with a sp^3 carbon atom with one positive charge used to probe steric and electrostatic fields. The basic principle of CoMSIA is the same as that of CoMFA, but CoMSIA includes some additional descriptors such as hydrophobicity, hydrogen bond donor, and hydrogen bond acceptor.

Using steric, electrostatic, hydrophobic, and hydrogen bond donor and acceptor properties as descriptors, CoMSIA analysis was performed with a cross-validated q^2 of 0.654 for four components and a conventional r^2 of 0.900.

The CoMFA model with 63 molecules in the training set was consequently a clearly statistically significant model showing a cross-validated q^2-value of 0.631 at 5 components and a good correlation coefficient $r^2 = 0.926$. The corresponding field distributions of these five descriptor variables were 20.4, 5.1, 25.9, 14.9, and 33.7%, respectively, which indicates that steric rather than electrostatic and H-bond acceptor rather than H-bond donor contribute more to the final CoMSIA model, and thus, these fields play crucial roles in determining the binding affinity of the PET ligands to the target protein. The usefulness of a 3D-QSAR model was shown by predicting the activity of the compounds included in the test set. The binding affinities of the test set molecules were predicted reasonably well (residuals from 0.04 to 0.81 for CoMFA model and 0.04–0.54 for $CoMSIA_{all}$ model). The predictive performances of CoMFA and $CoMSIA_{all}$ model *(all = steric + electrostatic + hydrogen bond donor + hydrogen bond acceptor + hydrophobic)* on the test were $r^2_{pred} = 0.74$ and $r^2_{pred} = 0.93$, respectively, which indicated that the CoMSIA model was more reliable and able to predict biological activity of new derivatives more accurately. The coplanarity of the PET ligands would be the key for high binding affinity to the target protein. Thus, substitutions of the phenyl ring

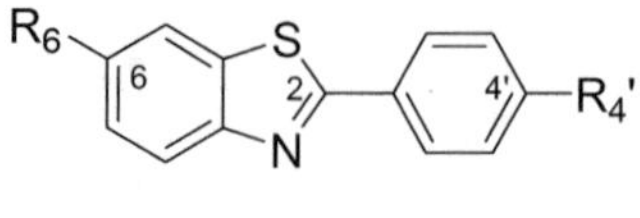

Fig. 8 Common scaffold of the used 2-phenylbenzothiazoles (PBT) derivatives

that would influence coplanarity would result in distortion of the molecule and would not be beneficial to the binding affinity. On the other hand, significant preferences for positive electrostatic interaction around the sulfur atom of the benzothiazolyl system could be found and positively charged groups or electron-donating substituents may increase the binding affinity of ThioT derivatives to the target protein by taking advantage of the electrostatic nature of the environment at this position. Hydrophobic interaction above and beneath the benzothiazole ring is detrimental for binding affinity. The only site available for beneficial substitution with hydrophobic group is expected to be around the position corresponding to the phenolic OH group. Monoalkyl substitution on the amine group (aniline) is preferable for high binding affinity.

In 2015, Ambure and Roy [34] performed a 2D-QSAR and a Group QSAR (G-QSAR) [35] analyses on a congeneric series of 44 2-phenylbenzothiazoles (PBT) derivatives (Fig. 8) [11, 18, 36] collected from the literature in order: (i) to design improved structures and (ii) to predict binding affinity of structurally related new imaging agents. These imaging agents are labeled with different radionuclides, e.g., 17 PET 2-phenylbenzothiazoles are radiolabeled with ^{11}C and 27 SPECT 2-phenylbenzothiazoles are radiolabeled with Re, a congener of ^{99m}Tc. The researchers have exploited different chemometric techniques such as stepwise-multiple linear regression (stepwise-MLR), partial least squares (PLS), genetic function approximation (GFA) linear, GFA spline, and genetic PLS (G/PLS) using Cerius 2 [37] and VLifeMDS software [35]. For 2D-QSAR, the dataset was rationally divided into 34 trainings and 10 test set compounds using the "Dataset Division GUI" developed by Roy's group, based on the Kennard--Stone algorithm [38]. In the G-QSAR methodology, every dataset molecule is considered as a set of fragments, and the fragmentation scheme is either template based or user defined. It can be seen that there are two substitution sites at positions 2 and 6 of the benzothiazole in the congeneric series used in the G-QSAR study. The common scaffold was defined as well as the substitution site(s) and overall 308 descriptors were calculated for both the substitution sites using VLifeMDS version 3 software [35]. After that 387 2D/3D descriptors are calculated for each fragment of the molecule using Cerius 2 version 4.10 [37], PaDEL-Descriptor version 2.11 [38] and Dragon 6 software [39]. For model validation, several requisite statistical metrics that include internal, external, overall validation metrics and the Y-randomization test metric were used.

The statistically significant 2D-QSAR and G-QSAR models derived using the GFA linear technique are given below.

$$pKi = 6.9035(\pm 0.3548) - 0.32805(\pm 0.05713)\text{Atype\_Unknown} + 0.31909(\pm 0.09073)$$

$$\text{S\_aasC} - 0.4404(\pm 0.17307)\text{B05[N} - \text{N]} \quad (6)$$

$$\text{N}_{\text{Training}} = 37, \text{N}_{\text{Test}} = 10, R^2 = 0.812, Q^2 = 0.764, \text{R}^2{}_{\text{pred}} = 0.784$$

where $N_{Training}$ and N_{Test} are the number of compounds in training and test sets, respectively, whereas the computed validation parameters were determination coefficient (R^2), leave-one-out (LOO) cross-validated correlation, coefficient (Q^2), and $R^2{}_{pred}$ (external validation).

QSAR (6) corresponds to the best 2D model, while the selected descriptors imply the requisite structural features for improving the binding affinity.

The descriptor "Atype_Unknown" indicates the presence of atoms that could not be classified according to the authors as any of the defined AlogP98atom [37] types, and it is associated with rhenium element, which is complexed with SN_3, SN_2O, and N_2S_2 chelators. The negative sign points at a decrease of the binding affinity against *Aβ* plaques. However, it should be noticed that the presence of metal is essential since it acts as the radionuclide to perform SPECT imaging, and furthermore the chelators SN_3, SN_2O, and N_2S_2 help in building neutral and lipophilic complexes, able to cross the BBB. "S _ aasC" belongs to the class of estate indices and indicates the sum of the index values for carbon with two aromatic bonds and one single bond encoding information about the topological environment of an atom and its electronic interaction with all other atoms in the molecule [38].The positive contribution is well correlated with the presence or absence of functional groups at 6th position of the benzothiazole ring and at the 4′ position of phenyl ring (Fig. 9). Bulky substituents with less electronegative atoms at these positions improve the binding affinity.

"B05[N-N]" indicates the presence of two nitrogen atoms (N...N) at a topological distance of 5 [38]. It seems that the presence of N-N (one nitrogen from the PBT pharmacophore and other nitrogen from chelator) at a topological distance of 5 in the compounds shows less binding affinity (6–7 *pKi*), whereas compounds without N...N at a topological distance of 5 show more binding affinity (>7 *pKi*). The researchers observed that the presence of SN_2O and N_2S_2 chelators (with the absence of N-N topological distance 5) has shown good binding affinity compared

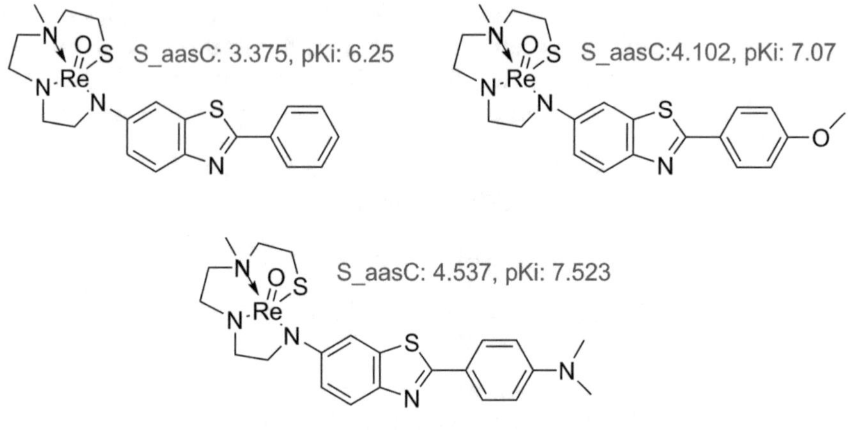

S_aasC: 3.375, pKi: 6.25

S_aasC:4.102, pKi: 7.07

S_aasC: 4.537, pKi: 7.523

Fig. 9 The effect of phenyl substituents on the values of "S _ aasC" and pKi [34]

to SN_3, SN_2O, and N_2S_2 chelator (with the presence of N-N topological distance 5). This finding supports the fact that binding affinity extensively depends on the selection of chelators and the way in which the chelators were integrated with PBT pharmacophore. Also from the point of chemical structure this condition is applicable when the chelator is fused with the phenyl ring and not applicable when it is attached with a single covalent bond at 4′ position or at 6th position.

Equation 7 corresponds to the best G-QSAR model that comprises four descriptors

$$pKi = 7.0731\ (\pm 0.216) + 0.622 \pm 6(\pm 0.13269)\mathrm{R6} - \mathrm{MMFF\_21} - 0.1348(\pm 0.01619)$$

$$\mathrm{R2} - \mathrm{chiV0} + 1.222\ (\pm 0.1431)\mathrm{R2\_k3alpha} + 1.74 \pm (0.3573)\mathrm{R6\_XKMostHydrophilic} \quad (7)$$

$$\mathrm{N_{Training}} = 34; R^2 = 0.862, \mathrm{SEE} = 0.282, Q^2 = 0.817, r^2_{\mathrm{m(training)}} = 0.749, \Delta r^2_{\mathrm{m(training)}} = 0.077, \mathrm{N_{Test}} = 10, R^2_{\mathrm{pred}}\ 0.765, r^2_{\mathrm{m(test)}} = 0.636, \Delta r^2_{\mathrm{m(test)}} = 0.173, r^2_{\mathrm{m(overall)}} = 0.733, \Delta r^2_{\mathrm{m(overall)}} = 0.134.$$

The validation parameters computed were determination coefficient (R^2); adjusted R^2 (R_a^2); standard error of estimate (SEE); leave-one-out (LOO) cross-validated correlation coefficient (Q^2); both scaled and unscaled versions of novel r_m^2 metrics, such as average r_m^2(training) and $\Delta r_m^2{}_{(\mathrm{training})}$ for internal validation, $r_m{}_{(\mathrm{test})}$ and $\Delta r_m^2{}_{(\mathrm{test})}$ for external validation, and $r_m^2{}_{(\mathrm{Overall})}$ and $\Delta r_m^2{}_{(\mathrm{Overall})}$ for overall validation; and R_{pred}^2 (external validation). The descriptor R6-MMFF_21 which implies atom types according to Merck molecular force field (MMFF) at R6 position [40] contributes positively to the binding affinity. Thus, the presence of

–OCH_3 group at R6 position (Fig. 9) leads to better binding against *Aβ* plaques. This observation is in parallel to the contribution of the S_aasC descriptor in the 2D-QSAR model. The descriptor R2-chiV0 implies atomic valence connectivity index of order zero [38] and decreases the binding affinity. The impact of the descriptor R2-chiV0 is found to be corroborated with the descriptor Atype_Unknown contributed in the 2D-QSAR model. The descriptor R2-k3 alpha signifies third alpha modified kappa shape index at the substitution at the second position [38]. It contributes positively to the binding affinity and is well correlated with R4′ on the phenyl ring (Fig. 8). Thus, a bulky substituent like dimethyl substituted amine at 4' position increases the binding. The descriptor R6-XKMostHydrophilic implies the most hydrophilic value on the van der Waals surface determined using the Kellogg logP method [41]. For each fragment, the descriptor value is calculated by generating van der Waals surface of the fragment (R6 position), and a probe (sp^2 carbon) atom is placed at each point on van der Waals surface, and then hydrophilic interactions are calculated. It enhances binding affinity and suggests that relatively less electronegative groups like Br (with maximum descriptor value) at R6 position favor the binding affinity against *Aβ* plaques. The most active compound within the dataset presents a Br at R6 position.

The -$N(CH_3)_2$ group at R4' was found to be significant for the binding since it was present in the most active compounds in the dataset. Thus, the researchers suggested the design of some molecules containing this group and predicted their binding activities using models 6 and 7. Attempts to replace substituents in R6 with bulkier groups' halogens/alkyl led to an enhancement of the in silico predicted activity according to both the models.

6 Conclusion

This chapter offers basic information on the radionuclide imaging techniques and agents applied for Alzheimer's disease detection as well as the essential structural features in PET and SPECT imaging agents for proper binding to the *Aβ* plaques.

The QSAR results revealed that some important molecular characteristics should significantly affect ligands' binding affinities. In general, from the presented SAR, 2D-QSAR, and 3D-QSAR models and computational studies, it can be concluded that:

- Hydrophobic interactions are detrimental for binding affinity.
- Monoalkyl substitution on the amine group (aniline) is preferable for high binding affinity.
- Bulkier groups improve the binding affinity for both the PET and SPECT imaging agents.

- The binding affinities of the SPECT imaging agents are influenced by the chelators and the way in which the chelators are integrated.

The developed models are useful tools to predict the $A\beta$ binding properties of nonmetallic PET agents as well as of metallated derivatives for SPECT imaging.

References

1. Silverman DHS (2004) Brain ^{18}F-FDG PET in the diagnosis of neurodegenerative dementias: comparison with perfusion SPECT and with clinical evaluations lacking nuclear imaging. J Nucl Med 45(4):594–607
2. Johnson K, Minoshima S, Bohnen NI et al (2013) Appropriate use criteria for amyloid PET: a report of the amyloid imaging task force, the Society of Nuclear Medicine and Molecular Imaging, and the Alzheimer's association. Alzheimers Dement 54(3):476–491
3. Elsinga P (2002) Radiopharmaceutical chemistry for positron emission tomography. Methods 27(3):208–217
4. Adam MJ, Wilbur DS (2005) Radiohalogens for imaging and therapy. Chem Soc Rev 34:153–163
5. Nordberg A (2004) Reviews PET imaging of amyloid in Alzheimer 's disease. Lancet Neurol 3:519–527
6. Donnemiller E, Heilmann J, Wenning GK et al (1997) Brain perfusion scintigraphy with $_{99m}$Tc-HMPAO or ^{99m}Tc-ECD and ^{123}I-beta-CIT single-photon emission tomography in dementia of the Alzheimer-type and diffuse Lewy body disease. Eur J Nucl Med 24 (3):320–325
7. Chételat G, Desgranges B, de la Sayette V, Viader F, Eustache F, Baron JC (2003) Mild cognitive impairment: can FDG-PET predict who is to rapidly convert to Alzheimer's disease? Neurology 60:1374–1377
8. Moghbel MC, Saboury B, Basu S et al (2012) Amyloid-β imaging with PET in Alzheimer's disease: is it feasible with current radiotracers and technologies? Eur J Nucl Med Mol Imaging 39(2):202–208
9. Kung H (2003) Iodinated tracers for imaging amyloid plaques in the brain. Mol Imaging Biol 5(6):418–426
10. Ono M, Wilson A, Nobrega J et al (2003) ^{11}C-labeled stilbene derivatives as Aβ-aggregate-specific PET imaging agents for Alzheimer's disease. Nucl Med Biol 30(6):565–557
11. Mathis C, Wang Y, Holt DP et al (2003) Synthesis and evaluation of ^{11}C-labeled 6-substituted 2-Arylbenzothiazoles as amyloid imaging agents. J Med Chem 46:2740–2754
12. Choi SR, Golding G, Zhuang Z et al (2009) Preclinical properties of ^{18}F-AV-45: a PET agent for Abeta plaques in the brain. J Nucl Med 50(11):1887–1894
13. Klunk WE, Engler H, Nordberg A et al (2004) Imaging brain amyloid in Alzheimer 's disease with Pittsburgh compound-B. Ann Neurol 55:306–319
14. Villemagne VL, Ong K, Mulligan RS et al (2011) Amyloid imaging with ^{18}F-florbetaben in Alzheimer disease and other dementias. J Nucl Med 52(8):1210–1217
15. Landau SM, Breault C, Joshi AD et al (2013) Amyloid-β imaging with Pittsburgh compound B and florbetapir: comparing radiotracers and quantification methods. J Nucl Med 54 (1):70–77
16. Nelissen N, Van Laere K, Thurfjell L et al (2009) Phase 1 study of the Pittsburgh compound B derivative ^{18}F-flutemetamol in healthy volunteers and patients with probable Alzheimer disease. J Nucl Med 50(8):1251–1259
17. Ono M, Saji H (2015) Recent advances in molecular imaging probes for β-amyloid plaques. Med Chem Commun 6:391–402
18. Lin KS, Debnath ML, Mathis C, Klunk WE (2009) Synthesis and β-amyloid binding properties of rhenium 2-phenylbenzothiazoles. Bioorg Med Chem Lett 19(8):2258–2262
19. Wang X, Cui M, Yu P et al (2012) Synthesis and biological evaluation of novel technetium-99m labeled phenylbenzoxazole derivatives as potential imaging probes for β-amyloid plaques in brain. Bioorg Med Chem Lett 22 (13):4327–4331
20. Cheng Y, Ono M, Kimura H, Ueda M, Saji H (2012) Technetium-99m labeled pyridyl benzofuran derivatives as single photon emission computed tomography imaging probes for β-amyloid plaques in Alzheimer's brains. J Med Chem 55:2279–2286

21. (a) Mathis C, Wang Y, Klunk WE (2004) Imaging β-amyloid plaques and neurofibrillary tangles in the aging human brain. Curr Pharm Des 10:1469–1492; (b) Autodock Vina, version 1.1.1; The Scripps Research Institute: San Diego, 2010
22. Kung HF, Choi SR, Qu W et al (2010) ^{18}F stilbenes and Styrylpyridines for PET imaging of Aβ plaques in Alzheimer's disease: a Miniperspective. J Med Sci 53(3):933–941
23. Agdepp ED, Kepe V, Liu J et al (2003) 2-Dialkylamino-6-acylmalonitrile-substituted naphthalenes (DDNP analogs): novel diagnostic and therapeutic tools in Alzheimer's disease. Mol Imaging Biol 5:404–417
24. Lockhart A, Ye L, Judd DB et al (2005) Evidence for the presence of three distinct binding sites for the thioflavin T class of Alzheimer's disease PET imaging agents on β-amyloid peptide fibrils. J Biol Chem 280:7677–7684
25. Ye L, Morgenstern JL, Gee AD et al (2005) Delineation of positron emission tomography imaging agent binding sites on beta amyloid peptide fibrils. J Biol Chem 280:23599–23604
26. Wellsow J, Machulla H-J, Kovar K-A (2002) 3D-QSAR of serotonin transporter ligands: CoMFA and CoMSIA studies. Quant Struct Act Relat 21:577–589
27. Wellsow J, Kovar KA (2002) Molecular modeling of potential new and selective PET radiotracers for the serotonin transporter. Positron emission tomography. J Pharm Pharm Sci 5:245–247
28. Wang W, Zhang J, Liu B (2005) QSAR study of ^{125}I-labeled 2-(4-aminophenyl)benzothiazole derivatives as imaging agents for β-amyloid in the brain with Alzheimer's disease. JRNC 266:107–111
29. Yang Y, Cui M, Zhang X et al (2014) Radioiodinatedbenzyloxybenzene derivatives: a class of flexible ligands target to β-amyloid plaques in Alzheimer's brains. J Med Chem 57:6030–6042
30. Cook NP, Ozbil M, Katsampes C et al (2013) Unraveling the photoluminescence response of light-switching ruthenium(II) complexes bound to amyloid-β. J Am Chem Soc 135:10810–10816
31. Yang Y, Zhang X, Cui M et al (2015) Preliminary characterization and In Vivo studies of structurally identical ^{18}F- and ^{125}I-Labeled-Benzyloxybenzenes for PET/SPECT imaging of β-amyloid plaques. Sci Rep 5:12084. doi:10.1038/srep12084
32. (a) Chen X (2006) QSAR and primary docking studies of trans -stilbene (TSB) seriesof imaging agents for β-amyloid plaques. J MolStruct: THEOCHEM 763:83–89; (b) HyperChem 7.0, Hypercube Inc., Gainesville, 2002
33. (a) Kim MK, Choo IH, Lee HS et al (2007) 3D–QSAR of PET Agents for Imaging β -Amyloid in Alzheimer's Disease. Bull Korean ChemSoc 28:1231–1234; (b) SYBYL 7.2 Tripos Associates, St. Louis; (c) http:rrccl.osc.edurccardatarMMFF94
34. Ambure P, Roy K (2015) Exploring structural requirements of imaging agents against Aβ plaques in Alzheimer's disease: a QSAR approach. Comb Chem High Throughput Screen 18:411–419
35. Vlife MDS (2008) software package, version 3.0, supplied by Vlifescience technologies Pvt. Ltd, Pune
36. Pan J, Mason NS, Debnath ML (2013) Design, synthesis and structure-activity relationship of rhenium 2-arylbenzothiazoles as beta-amyloid plaque binding agents. Bioorg med Chem Lett 23:1720–1726; Kennard RW, Stone LA (1969) Computer aided design of experiments. Technometrics 11:137–148
37. Cerius2 Version 410 (2005) Accelrys, Inc: San Diego
38. Yap CW (2011) PaDEL-descriptor: an open source software tocalculate molecular descriptors and fingerprints. J Comput Chem 32:1466–1474
39. Todeschini R, Consonni V, Mauri A et al (2004) DRAGON Software for the calculation of molecular descriptors. Web Version 3
40. http://padel.nus.edu.sg/software/padeldescriptor
41. Kellogg EG, Abraham DJ (2000) Hydrophobicity: is LogP o/w more than the sum of its parts? Eur J Med Chem 35:651–661

Part IV

Special Topics

Chapter 17

Computational Approaches for Therapeutic Application of Natural Products in Alzheimer's Disease

Manika Awasthi, Swati Singh, Sameeksha Tiwari, Veda P. Pandey, and Upendra N. Dwivedi

Abstract

Alzheimer's disease (AD) is characterized by deterioration of cognitive functions and behavioral changes eventually leading to cell death and dementia. Till date, several efforts have been made to analyze the causes, symptoms, and cure of AD and also to identify biochemical changes and pathogenesis of the brain affected with AD. Several attempts have been made to treat AD by combined drug therapy directed against its various targets. However, the side effects due to long-term administration of these drugs have motivated attempts for development of a new generation of therapeutics based on natural products. Drug discovery based on plant-derived natural products has a long and successful history. To further enhance the identification of novel drugs from natural sources, natural product research has been increasingly combined with computer-aided drug design approaches. This chapter reviews the recent advances in the application of chemoinformatic methods to quantify the chemical diversity and structural complexity of natural products and analyze their arrangement in chemical space with respect to the treatment of AD. The computational approaches involved in the whole drug discovery pipeline from target identification and mechanism of action to identification of novel leads and drug candidates have been depicted and discussed, with the aim to provide a general view of computational tools and databases available. The advancement in virtual screening has also been discussed to systematically identify bioactive compounds from natural product databases, and the progress in target identification methods has also been presented to discover molecular targets of compounds from natural origin.

Key words Alzheimer's disease, Natural products, Computer-aided drug design, Target identification and validation, Lead identification and optimization

1 Introduction

1.1 Overview of Alzheimer's Disease

Alzheimer's disease (AD) is a progressive neurodegenerative disorder characterized by deterioration of memory and cognitive functions, leading to the most common form of dementia worldwide and affecting a large number of demented individuals across all nations [1]. It is characterized by progressive and irreversible decline of memory and other cognitive functions including language, judgment, and reasoning along with progressive loss of

Kunal Roy (ed.), *Computational Modeling of Drugs Against Alzheimer's Disease*, Neuromethods, vol. 132, DOI 10.1007/978-1-4939-7404-7_17, © Springer Science+Business Media LLC 2018

physical functioning and associated neuropsychiatric symptoms, which become severe enough to impede social or occupational functioning. AD has been considered as the fifth leading cause of death for the older population and sixth leading cause of death in the United States. In addition to being a leading cause of death, it is also a leading cause of disability and morbidity [2]. The estimates of dementia prevalence are higher in Asia and Latin America while lower in India and sub-Saharan Africa. Epidemiological studies have revealed that AD is the most common cause of dementia in India, affecting 2.7% of the population [3]. It is a matter of great concern that 1000 new cases of AD are reported daily throughout the United States [4]. As a consequence of the steady growth of the aging population, the prevalence of this disease is expected to further increase in future in both developed and developing countries.

Despite decades of research and many significant advances, the neuropathology of AD is still not completely understood; however, amyloid plaques and neurofibrillary tangles are considered as the two primary pathological hallmarks of the disease [5]. Histopathological studies of the AD brain revealed prominent ultrastructural changes triggered by two classical lesions, the senile plaques, mainly composed of amyloid beta (Aβ) peptides, and the neurofibrillary tangles, composed of hyperphosphorylated tau proteins. Amyloid plaques are insoluble, dense cores of 5–10 nm fibrils containing aggregates of 42 amino acid-long ςβ peptides as found in AD patients. On the other hand, neurofibrillary tangles contain aggregates of hyperphosphorylated tau, a microtubule-associated protein, which causes the degeneration of the microtubule network required for neuronal survival [6]. The neuronal loss caused by these two factors leads to global neurotransmitter deficiencies, specifically in norepinephrine and acetylcholine [7]. Thus, current models for AD suggest a prodromal period of amyloid accumulation, followed by a progression of tau pathology, inflammation, and neurodegeneration that tracks cognitive decline. Oxidative damage to proteins, lipids, and DNA in the brains of AD patients also accompanies the widespread inflammation [8].

Preclinical AD, mild cognitive impairment (MCI), and dementia have been defined as the three main stages of pathological progression of AD. Preclinical AD can be diagnosed by the presence of a biomarker such as an imaging marker like an amyloid scan or volumetric hippocampal analysis or presence of a genetic risk factor like the ApoE4 gene or tau/amyloid levels in cerebrospinal fluid (CSF). Symptoms of preclinical AD range from those who are completely asymptomatic but are at risk of developing AD to those with subtle cognitive symptoms not severe enough for an MCI diagnosis. On the other hand, MCI is diagnosed once cognitive impairments become noticeable on cognitive testing but are not severe enough to significantly interfere with daily routine activities and can be classified as prodromal AD if a biomarker is present, whereas dementia diagnosis requires significant functional impairment to be present in addition to cognitive impairment. Accordingly, the strategies applied for the treatment of dementia will vary depending upon these three stages of AD [5].

Although several compounds have been proposed to decrease neurodegeneration in AD models, there is still no permanent cure for AD to reverse the damage caused to brain cells. Currently, only five drugs are available for the management of AD patients which are approved by the US Food and Drug Administration (FDA) for its symptomatic treatments that may help lessen or stabilize cognitive symptoms such as memory loss and confusion, for a limited time, by affecting certain chemicals involved in carrying messages among the brain's nerve cells. Four of these drugs, namely, donepezil (Aricept®), galantamine (Razadyne®), rivastigmine (Exelon®), and tacrine (Cognex®), are acetylcholinesterase (AChE) inhibitors, while one, namely, memantine (Namenda®), acts as an antagonist for N-methyl-D-aspartic acid (NMDA) receptors that improves memory by restoration of homeostasis in the glutamatergic system [9]. These drugs are prescribed to treat symptoms related to memory, thinking, language, judgement, and ability to perform simple tasks as well as to treat moderate to severe AD. However, these drugs have some adverse side effects that commonly include nausea, vomiting, loss of appetite, headache, constipation, and dizziness [10]. Tacrine has been reported to be discontinued by the US government agencies in relation to its side effects.

1.2 Natural Products as Promising Drug Candidates Against AD

The side effects and toxicity exhibited by synthetic drugs and various other therapeutic strategies have led to an increased demand of plant-derived (herbal) medicines which are either approved or in different stages of clinical trials for a number of diseases in the last few decades. Although synthetic chemistry dominates the present scenario of drug discovery and manufacturing, still the contribution of plant-derived compounds for treatment and prevention of various diseases could not be ignored. Plant products being natural are often considered as safe, though some side effects have also been reported for some plant products. Such adverse effects of plant-derived natural products may include allergic reactions, toxicity, and plant pharmaceutical interactions with synthetic/semisynthetic drugs [11]. In spite of these side effects of some of the natural products, these plant-derived natural products have been considered to be less toxic and relatively safer substitutes to synthetic drugs against a number of harmful diseases, including AD [12]. Moreover, a number of plant-derived natural compounds have been shown to provide better patient safety and tolerance and increase in the chances of acceptance by the patient. Hence, both the basic researchers and pharmaceutical companies are intensively working for identification of plant-derived natural products against deadly diseases.

Medicinal plants have also been reported as important source of potential therapeutic compounds against AD. The effect of crude plant extracts as well as purified compounds, having neuroprotective effects, in AD has been reported [13]. Medicinal plants have always been the important medical reservoirs, with considerable number of modern FDA-approved drugs having been derived from natural sources. In the field of Alzheimer's, several data on the possibility of natural product leads have already been supported by the experimental outcomes as the majority of the compounds examined till date for treatment of AD are primarily from plants [14]. Still till date, the greatest success has been from plant-based AChE inhibitors, which have provided two of the five currently approved drugs for the treatment of AD. Hence, multiple factors are likely the driving force for increased interest in natural products and their use in AD and other neurodegenerative diseases.

There are so many raw plant extracts or herbal formulations that find enormous uses as natural remedies in the treatment of AD and other neurodegenerative diseases. Most of the traditional natural medical systems including Indian, Chinese, Native American, and European have had various memory enhancers [15]. Vegetables such as pumpkin and carrot and spices like ginger, sesame, and sunflower seeds are very useful for enhancing the function of the brain along with blueberries, grapes, and pomegranate juice which has recently been proven to have beneficial effect in AD [16, 17]. Food supplements of vitamin B6 and B12, folic acid, vitamin E, vitamin C, and coenzyme Q10 also have been found to exhibit beneficial effect in AD patients [18].

Numerous literatures are available on the pharmacological and clinical studies of natural products used in the treatment of AD and related diseases, including their medicinal efficacy, safety, and other significant properties. In recent years, a number of review articles have also been published detailing on the naturally occurring products of varying scaffolds that show efficient potential against AD and other neurodegenerative disorders [5, 19–23]. More than 200 compounds of natural origin are reported so far to exhibit neuroprotective activities among which few representative chemical agents are presented in Table 1 to get a sight of the major advances in this direction. Antibodies which are being used against AD and undergoing clinical trials are also included in Table 1. Furthermore, a list of medicinal herbs used for the treatment of AD, along with their effects, is presented in Table 2. In addition to the natural products which can be directly used as drug entities for medicinal application, the structure of these active substances can serve as templates for the design and synthesis of a wide range of potential therapeutic compounds for treating diseases. Despite a number of new approaches for drug discovery involving synthetic

Table 1
Natural product-based drugs and antibodies against AD undergoing various stages of clinical trials along with their mode of action

Name of drug	Plant source along with family / antibodies	Stages of clinical trial	Mode of action	Reference
Physostigmine	*Physostigma venensosum Balf.* (Leguminosae)	Phase IIIb	Potent and reversible inhibitor of AChE and improves cognitive functions in vivo	[24]
Galantamine	*Lycoris radiata Herb.*, *Galanthus nivalis L.*, *Narcissus* spp. (Amaryllidaceae)	Phase IIIb	Reversible inhibition of AChE and allosteric potentiation of nicotinic acetylcholine receptors	[25]
Huperzine A	*Huperzia serrata Trevis.* (Lycopodiaceae)	Phase II	Inhibits the enzyme AChE reversibly and selectively	[26]
ZT-1	Prodrug of huperzine	Phase II	Blocking the action of AChE and restores adequate levels of acetylcholine	[27]
Curcumin	*Curcuma longa L.* (Zingiberaceae)	Phase II	Inhibits Aβ aggregation and Aβ-induced inflammation as well as the activities of β-secretase and AChE	[28]
Resveratrol	*Vitis vinifera L.* (Vitaceae)	Phase III	Promotes the decomposition and clearance of intracellular Aβ aggregates	[29]
Bryostatin-1	*Bugula neritina L.* (Bugulidae)	Phase II	Enhances α-secretase activation in human fibroblast cells, reduces Aβ42 levels, and reduces mortality of transgenic AD mice	[30]
Rifampicin	*Amycolatopsis rifamycinica* (Pseudonocardiaceae)	Phase IIIa	Inhibits Aβ aggregation in vitro	[31]
Scyllo-cyclohexanehexol (ELND005, AZD-103)	*Cornus florida L.* (Cornaceae) and *Cocos nucifera L.* (Arecaceae)	Phase II	Breaks down neurotoxic fibrils, allowing amyloid peptides to clear the body rather than form amyloid plaques	[32]
Bapineuzumab (AN-1792)	Humanized monoclonal antibody	Phase III	Antibody to the Aβ plaques	[33]
Ponezumab (PF-04360365)	Humanized monoclonal antibody	Phase II	Antibody to the Aβ plaques	[34]
Solanezumab	Monoclonal antibody	Phase III	Neutralizes Aβ plaques	[35]

(continued)

Table 1 (continued)

Name of drug	Plant source along with family / antibodies	Stages of clinical trial	Mode of action	Reference
Crenezumab	Monoclonal antibody	Phase II	Antibody to the Aβ plaques	[36]
Gantenerumab	Humanized monoclonal antibody	Phase III	High capacity to bind and remove Aβ plaques in the brain	[37]
BAN2401	Humanized immunoglobulin-1 monoclonal antibody	Phase II	Binds selectively to large, soluble Aβ oligomers and neutralizes their damage to brain cells	[38]
Cyclosporin A	*Tolypocladium inflatum (fungus)*	Phase IIa	Most powerful neuroprotectant in stroke and traumatic brain injury	[39]

combinatorial chemistry, it would not be possible to ignore the importance of natural products as many of the scaffolds for the designing of synthetic drugs are based on natural products.

In spite of the huge literature supporting the immense role of natural products in drug discovery, little is known about the structure-activity or structure-property relationships of these natural products. The comparison of the structures of rationally designed and naturally occurring anti-AD agents showed that both kinds of compounds differ in their scaffolds, the latter showing a seamless framework, while the former is composed of two or more isolated parts, linked by spacers of different lengths, each part aiming at a particular target. Thus, the natural strategy for designing of anti-AD agents needs to be explored [58]. The therapeutic potential agents from natural sources are categorized according to their therapeutic targets for AD such as cholinesterase inhibitors, nicotinic acetylcholine receptor agonists, secretase inhibitors, anti-aggregation and clearance promoters, kinase modulators, and antioxidants [59].

It is noteworthy that the structures of the above-mentioned natural agents mostly consist of phenolic groups. It is well known that phenolic hydroxyls are the most potent groups to neutralize reactive oxygen species (ROS) through donating hydrogen atoms and also effective to chelate transition metal ions. However, these natural phenolics go beyond scavenging ROS or chelating transition metals as they can inhibit various enzyme activities and prevent protein aggregation. The major underlying reason for this may be that phenolic hydroxyls act as both hydrogen bond acceptors and hydrogen bond donors simultaneously, which facilitates their binding with protein targets. This explanation is supported by a meta-

Table 2
A list of medicinal herbs used for the treatment of AD along with their family names and beneficial effects on patients

Name of medicinal plants	Family	Beneficial effects on AD patients	Reference
Curcuma longa L.	Zingiberaceae	Anti-inflammatory and antioxidant activity	[40]
Bacopa monniera Wettst.	Scrophulariaceae	Improve cognitive functions	[41]
Centella asiatica L.	Umbelliferae	Memory improvement	[42]
Ginkgo biloba L.	Ginkgoaceae	Antioxidant and protection against Aβ-induced oxidative damage	[43]
Salvia officinalis	Lamiaceae	Protect the brain from oxidative damage	[44]
Rosmarinus officinalis	Lamiaceae	Anti-inflammatory and antioxidant activity	[45]
Matricaria recutita	Asteraceae	Stimulate the brain, dispel weariness, calm the nerves, counteract insomnia, aid in digestion, break up mucus in the throat and lungs, relieve anxiety, and aid the immune system	[46]
Melissa officinalis L.	Lamiaceae	Sharpens memory, improves cognitive decline, inhibits AChE	[47]
Glycyrrhiza glabra	Fabaceae	Protective effect against apoptotic neuronal cell death induced by Aβ fragments	[48]
Galanthus nivalis L.	Amaryllidaceae	Long-acting and specific inhibitor of AChE, potentiates cholinergic nicotinic neurotransmission by allosterically modulating the nicotinic acetylcholine receptors	[49]
Huperzia serrata	Lycopodiaceae	Reversible, potent, and selective AChE inhibitor; improves memory, concentration, and learning capacity; reduces the abnormally high radical activity in the brains of elderly animals as well as in the blood of AD patients	[50]
Commiphora whighitti	Burseraceae	Potential cognitive enhancer for improvement of memory in scopolamine-induced memory deficits	[51]
Lepidium Meyenii Walp	Brassicaceae	Improves experimental memory impairment due to its antioxidant and AChE inhibitory activities	[52]
Panax Ginseng	Araliaceae	Memory-enhancing action for the learning impairment induced by scopolamine, enhances the psychomotor and cognitive performance, improves the brain cholinergic function, and repairs the damaged neuronal networks	[53]

(continued)

Table 2 (continued)

Name of medicinal plants	Family	Beneficial effects on AD patients	Reference
Acorus calamus L.	Araceae	Memory-enhancing property for memory impairment, learning performance, and behavior modification, inhibits AChE, anti-inflammatory, antioxidant, antispasmodic, cardiovascular hypolipidemic, immunosuppressive, cytoprotective, antidiarrheal, antimicrobial, and anthelmintic activities	[51]
Angelica archan gelica L.	Umbelliferae	Increases blood flow in the brain and inhibits AChE in vitro	[54]
Tinospora cordifolia	Menispermaceae	Memory-enhancing property for learning and memory in normal and memory-deficit animals	[51]
Magnolia officinalis	Magnoliaceae	Treatment of neurosis, anxiety, stroke, and dementia, inhibits memory impairment induced by scopolamine through the inhibition of AChE activity in vitro and in vivo	[51]
Collinsonia canadensis	Lamiaceae	Prevents the breakdown of acetylcholine	[55]
Bertholletia excelsa	Lecythidaceae	Enhances the concentration of acetylcholine in AD patients	[56]
Urtica dioica L.	Clusiaceae	Treats allergy symptoms, reduces inflammation, enhances the levels of estrogen, which can be beneficial in short-term memory	[55, 56]
Withania somnifera	Solanaceae	Antioxidant function which is accomplished by increasing the activities of superoxide dismutase, catalase, and glutathione peroxidise, rejuvenates the cells and boosts energy significantly and inhibits AChE	[57]

analysis on phenol-protein binding patterns which revealed that more than 70% phenolic hydroxyls form intermolecular hydrogen bonds with targeted proteins [60].

Another feature of the naturally occurring anti-AD agents is that their structures contain more than one conjugated phenolic rings and most of the conjugated systems are still flexible. Thus, these molecules reach a good balance between rigidity and flexibility, which must benefit their binding with various targets. Finally, since p-stacking plays an important role in protein amyloid formation, the aromaticity of phenolic ring is favorable to prohibit amyloid fibril formation [61]. Furthermore, the blood-brain barrier (BBB), the semipermeable protective shield surrounding the brain to restrict substances in the blood from entering the central nervous system (CNS), can present developmental challenges for drugs in AD. Crossing the BBB requires a drug with low molecular weight and lipid solubility. Few permeability studies have been conducted on the natural products discussed above, and curcumin was found to be most effective at crossing the BBB [62].

1.3 Computational Approaches for Design and Development of Potent Drugs Against AD

Development of cheap and efficient drugs has always been a desirable feature of drug discovery. Research-based pharmaceutical companies on an average spend about 20% of their sales on research and development (R&D). Despite these enormous expenditures, there has been a steady decline in the number of drugs introduced each year into human therapy. This decline can be attributed to increased demand on safety for drugs as the average number of clinical trials per new drug application (NDA) increased drastically along with the prolonged duration of the drug development process. A benchmark report estimates that the cost of bringing a drug to market has more than doubled in the past 10 years. A new report published by the Tufts Center for the Study of Drug Development (CSDD) estimates that the total cost to develop and gain marketing approval for a new drug is about US $ 2.6 billion and it can take 10–12 years for that new drug to get through the Food and Drug Administration's (FDA) approval process and hit the market (Fig. 1). Moreover, once the drug has made it to market, there is often post-approval research and tests to evaluate dosing strength and a host of other factors. About 75% of this cost (US $ 1.95 billion) is attributable to failure during development as 90% of all drug development candidates fail to make it to market [63, 64]. The conventional drug discovery process, based on random trial and error, possessed low success rate despite high cost and time investment.

Thus, methods that enhance the drug discovery process and reduce failure rates are highly desirable. In this direction, tools of

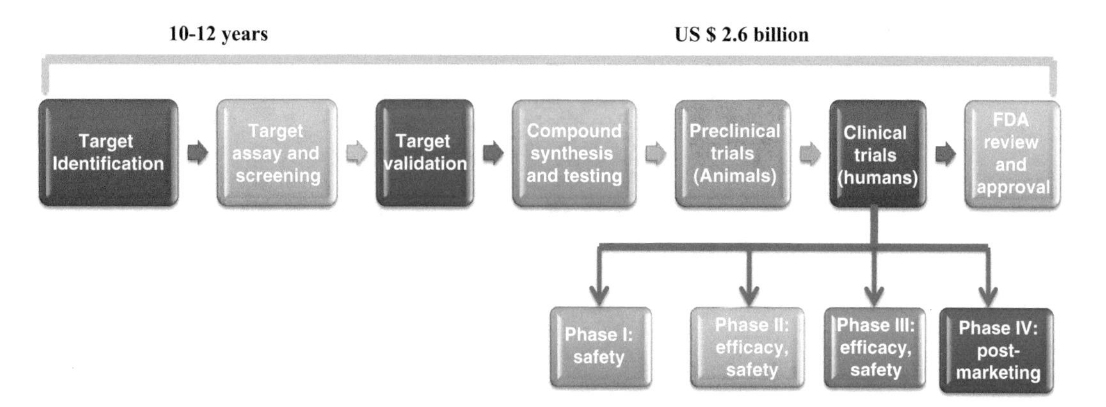

Fig. 1 Conventional drug discovery and development process (from target identification to market release) along with average cost and time span involved

bioinformatics have opened up a new area of research called as rational drug designing or computational drug designing. This has led to the shifting of the focus of the pharmaceutical industry from trial and error process of drug discovery to a rational, structure-based drug designing. Such a successful and reliable drug designing process would reduce the time and cost of developing useful therapeutic molecules by increasing the success rate along with less animal sacrifice. It provides highly accurate prediction of drug efficacy narrowing down to compounds with high novelty and high affinity [65]. Current trends in drug discovery focus on disease mechanisms and their understanding, followed by target identification and lead compound discovery. In the present era of "omics," a system of personalized medicine that is based on molecular states and changes has become important in drug discovery. The molecular characterization of disease along with environmental influences and the gut microbiome is necessary to build such a system [66]. In this context, the present chapter summarizes the computational tools used for reliable target identification and validation in cooperation with drug discovery methods which will pave the way to more efficient computer-aided drug design (CADD). Furthermore, new network-based computational models and systems biology integrated with omics databases and combinational regimens of drug development with special reference to plant-derived natural products against AD have been presented. Also, computational approaches commonly used in CADD (Fig. 2), namely, target identification and validation and lead identification and optimization, have been presented.

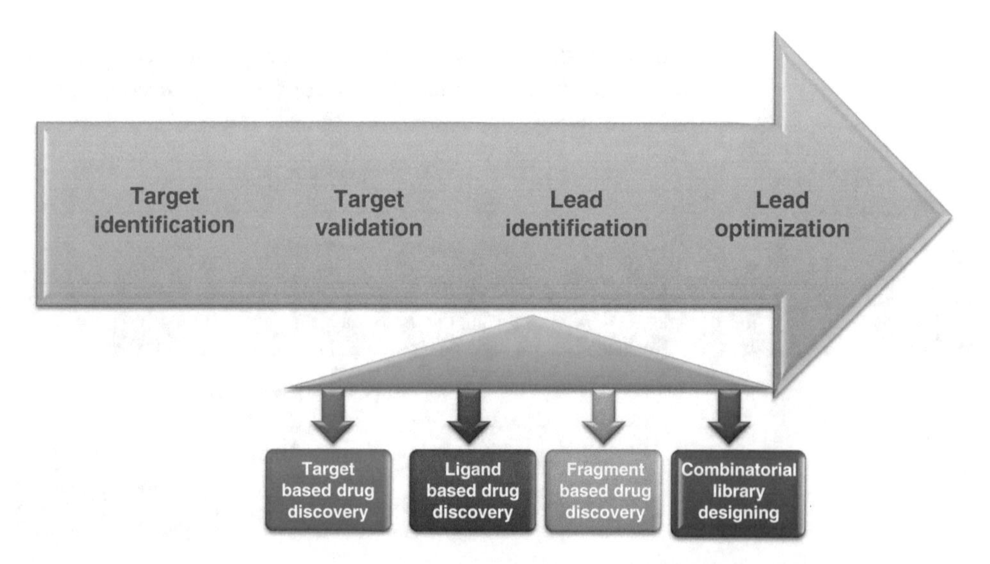

Fig. 2 Computer-aided drug design process which reduces the cost and time on an average by 30% as compared to the conventional process (Fig. 1)

2 Computer-Aided Drug Design (CADD)

Steps involved in computer-aided drug design may be divided into four major categories, namely, target identification, target validation, lead identification, and lead optimization each of which is described under following sections.

2.1 Target Identification

The identification of new, clinically relevant, molecular targets is of most importance to the discovery of innovative drugs. It has been estimated that up to ten genes contribute to multifactorial diseases. Typically these disease genes are linked to other five to ten gene products in physiological circuits which are also suitable for pharmaceutical intervention. If these numbers are multiplied with the number of diseases that pose a major medical problem in the industrial world, then there are ~5000 to 10,000 potential drug targets [67]. Current therapy is based upon less than 500 molecular targets, 45% of which are G-protein-coupled receptors, 28% are enzymes, 11% are hormones and factors, 5% ion channels, and 2% nuclear receptors [68]. Besides classical methods of cellular and molecular biology, new techniques of target identification are becoming increasingly important which include genomics, proteomics, and bioinformatics. Bioinformatic methods are used to transform the raw sequence into meaningful information and to compare whole genomes. The bioinformatic approach helps in in silico identification of novel drug targets by systematically searching for paralogs (related proteins within an organism) of known drug targets. To facilitate gaining detailed knowledge of disease mechanisms or phenotypes, information technologies are greatly needed. The study of disease mechanisms or phenotypes has turned from investigating a particular gene or protein into the analysis of entire sets of biomolecules. The advent of omics technologies further complicates storing, visualizing, and analyzing large biological data for which information technologies provide the means toward extensive data storage and interpretation [69].

Another important field in target identification is the semantic field, which could give insights to associations between heterogeneous information of diseases and drug targets. Recently, these network-based computational approaches have gained popularity by suggesting novel therapeutic targets and interpreting disease mechanisms [70]. However, little effort has been devoted to investigating associations among drugs, diseases, and genes in an integrative manner which could result in the interpretation of novel research hypotheses important for translational medicine research and personalized medicine. The identification of new biological targets has also been benefited from the genomic approach which provide blueprint of all the proteins by sequencing of the human genome. It has also become feasible to compare the entire genome of pathogenic and nonpathogenic strains of a microbe and identify genes/proteins associated with pathogenicity. Using gene expression microarrays and gene chip technologies, a single device can be used to evaluate and compare the expression of up to 20,000 genes of healthy and diseased individuals at once [71]. It is evident that the complexity of biological systems is present at the protein level and that genomics alone will not be sufficient to understand these systems since at the protein level only disease becomes manifest at which most (91%) drugs act. Therefore, the analysis of proteins (including protein-protein, protein-nucleic acid, and protein-ligand interactions) will be most important for target discovery.

Target identification also involves network-based drug discovery, where different levels of information in drug-protein and protein-disease networks are integrated. This approach involves collaboration of databases and genomics, proteomics, metabolomics, transcriptomics, microbiome, and pharmacogenomics, which are highly dependent on the development of relevant computational and systems biology tools for data interpretation [72]. The approaches relating pharmacological and genomic characteristics can be used to develop computational frameworks for drug target identification. The integration of large-scale structural genomics and disease association studies, to generate three-dimensional human interactome, is a recent network-based application that resulted in the identification of candidate genes for unknown disease-to-gene associations with proposed molecular mechanisms [73].

2.2 Target Validation

Target validation is a time-consuming and expensive process that must ensure that the identified target possesses relevance to a desirable molecular process, biological pathway, or pathogenicity. This can be obtained with knockout or knock-in animal models, ribozymes, small molecule inhibitors, neutralizing antibodies, and antisense nucleic acid constructs [74]. Knowledge of the three-dimensional structure of proteins facilitates the prediction of the residues involved in binding and interaction of the target with that of the ligand and therefore gives precision of the interaction while demonstrating the druggability of the ligand. The target validation efficiency can be greatly improved by using combination of strict data filtering and statistics as high-throughput screening provides insight into the cellular responses in disease models [75]. Comparison of the network of interest to other random networks generated using random shuffling of the graph can be performed in network validation [76]. Genome-wide approaches and functional screens can be used to validate gene function or gene regulatory networks [77, 78]. In the presence of the electronic medical records and clinical trial data, the interindividual variability upon drug administration can be tracked and analyzed. In addition to the molecular and clinical data, free text data presented in literature through various databases are also useful in drug discovery via extensive

Table 3
Various computational approaches and databases commonly used in CADD

Computational approaches	Databases
Ligand-based interaction fingerprint (LIFt)	Human Metabolome Database (HMDB)
Protein-ligand interaction fingerprint (PLIF)	Potential Drug Target Database (PDTD)
Shape-based screening	Kyoto Encyclopedia of Genes and Genomes (KEGG)
Bidimensional pharmacophoric fingerprint	Gene Ontology Consortium
Artificial neural network (ANN)	Reactome Pathway Database
RNA interference	PANTHER-Gene List Analysis
Homology modeling	BioCarta Pathway Database
Molecular docking	Ingenuity Pathway Analysis
Quantitative structure activity relationship (QSAR)	ChEMBL-European Bioinformatics Institute
Pharmacophore modeling	The PubChem Project
Molecular dynamic simulation	RCSB Protein Data Bank (PDB)
Fingerprint-based similarity searching	Repositioning, Recombination and Discovery of Drugs (R2D2) System-DrugMap Central
Polypharmacology modeling	The Comparative Toxicogenomics Database (CTD)
Proteochemometric modeling (PCM)	The Toxin and Toxin Target Database (T3DB)
Molecular mechanics/ Poisson-Boltzmann surface area (MM-PBSA)	The Human Protein Atlas
Free energy perturbation (FEP)	Therapeutic Target Database (TTD)
Linear interaction energy (LIE)	Semantic MEDLINE Database
Pharmacokinetic analysis (ADMET and drug likeness)	ONCOMINE: A Cancer Microarray Database and Integrated Data
Pharmacodynamic analysis	Biological General Repository for Interaction Datasets (BioGRID)

data mining processes. Chemoinformatic tools possess incredible potential for advancing in silico drug design and discovery as they provide the integration of information in several levels to enhance the reliability of the results. A list of computational approaches and databases involved in target identification and validation has been given in Table 3 [79].

2.3 Lead Identification

After identification of the desired target, the in silico designing of drug can be initiated. The lead identification method depends on the availability of the target or the ligand. In the past decade, CADD has offered several tools for the identification of

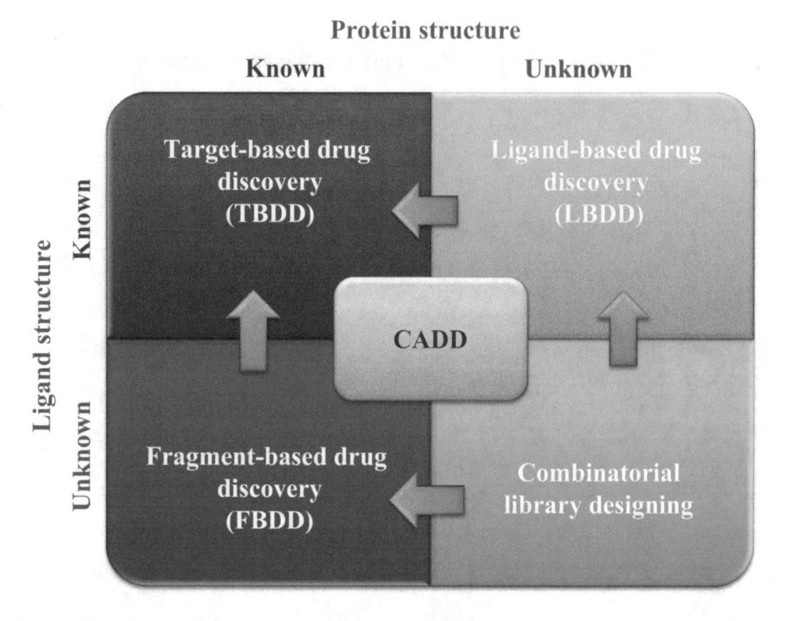

Fig. 3 Phases of computational lead identification (ligand-based, target-based, and fragment-based drug design and combinatorial library designing) depending upon the availability of the three-dimensional (3D) structure of the target and corresponding ligand

compounds, reducing the risk of discontinuation of lead compounds at a later stage. Success rates of the hits obtained from high-throughput screening (HTS) are often very low, and many of the identified compounds are later rejected due to their physicochemical properties [80]. CADD plays a significant role in high success rates of hit compound identification as well as in the arrangement of HTS-derived compounds according to their priority. There are generally four phases of computational lead identification, depending upon the availability of the three-dimensional (3D) structure of the target and its ligand, namely, ligand-based, target-based, and fragment-based drug design and combinatorial library designing (Fig. 3).

In the absence of the information on the structure of the target, computational methods for new molecule identification are based on information derived from known active or inactive compounds. Thus, the ligand-based methods depend on the information of molecular properties of known ligands without taking into account explicitly the interactions of the ligands with their target protein [81]. Thus, ligand-based methodologies can only be applied when known ligands exist. Target-based approaches, instead, are based on the direct calculation of protein-ligand interactions and can be applied only when the structure of the target protein is available experimentally or generated through computational modeling [82]. The combination of ligand and target-based molecular modeling methods has become a useful approach in virtual screening through sequential, parallel, or hybrid approaches [83]. These methods can also be used in combination with network-based approaches toward drug discovery. In addition to this, other combinatorial tools can be implemented as well for the aim of multi-targeted drug design.

2.3.1 Ligand-Based Drug Design (LBDD)

Ligand-based CADD methods are based on the information on binding affinities, chemical structure, and physicochemical properties of small molecules interacting with the target in order to identify novel potent compounds. LBDD methods are considered more useful than target-based methods in some cases. A ligand-based approach involves the selection of new compounds based on chemical similarity of known active compounds using fingerprint methods which allow the representation of a compound in a unique way so that they can be effectively compared to other molecules. These methods depend on the chemical information of the known active compounds providing a highly useful approach in the search of more potent ligands [64].

In addition to similarity search, another significant approach of LBDD is the quantitative structure-activity relationship (QSAR) , where a statistical model is used to describe a correlation between structures of a set of compounds and their activities [84]. The general QSAR methodology involves collection of a set of active and inactive molecules against a particular target and generation of their descriptors describing their structural and physicochemical properties. The model is further used to correlate these structural features with the experimental activity resulting in a predictive tool for novel molecules. QSAR algorithms are continuously developing from the implementation of several 2D and 3D descriptors to multiple levels of increasing complexity [85]. Thus, the QSAR along with the target-based approaches provides an efficient tool for virtual screening of a large number of dataset and has been widely used for identifying novel chemical compounds. The evolution of QSAR techniques has widened the scope of rationalized drug design and the search for the mode of action of the drugs. Furthermore, they can also be used in design of virtual compound libraries, computational optimization of compounds, and combinatorial library designing exhibiting improved absorption, distribution, metabolism, and excretion properties.

2.3.2 Target-Based Drug Design (TBDD)

Three-dimensional structures of proteins and DNA are being used from the past three decades in the process of drug design. The largest depository of biomolecule structure information, the Protein Data Bank (PDB), contains structures determined mostly by X-ray crystallography and NMR techniques. The number of structures deposited in PDB in the year 1998 was 2058 which has been increasing by ~7.5% each year, resulting in a total of 125,463 structures in 2016 [86]. The structure-based drug design process for the past years has been abundantly benefited from this enormous source of structural information in academic as well as the pharmaceutical industry.

In the cases where the structure of a protein target for drug design is not yet solved, the predictive tools for building comparative models provide great help. Comparative modeling using structural template of a homologous structure is used to predict the structure of a protein [87]. The most common computational technique used for this purpose is homology modeling which develops a protein model after identification of a structural template protein having similar sequence, alignment of their sequences, and using coordinates of the aligned regions. The structure is checked for missing atoms, and finally model building and refinement is done. The most commonly used freely available software for homology modeling are MODELER [88] and SWISS-MODEL [89].

Docking methods are used to predict the different orientations of a small molecule into the binding site of a protein when bound to each other to form a stable complex [90]. Depending on the method, either the ligand or the protein can be considered as flexible or rigid during the docking process. The most commonly used method usually takes ligand flexibility under consideration, while the protein is held rigid which is prepared by application of molecular dynamic force fields. Various software packages available for docking include Gold [91], Autodock [92], AutoDock Vina [93], DOCK [94], etc. The results of docking generate a list of scores which is an evaluation of the binding affinity of the complex as determined by scoring functions based on molecular mechanics, empirical knowledge, or consensus functions. In consensus scoring, the binding affinities of compounds are predicted for a particular target by using a number of scoring algorithms [95].

Another method used for determining the binding capability of a small molecule to protein binding site is pharmacophore modeling. A pharmacophore is defined as the three-dimensional spatial arrangement of steric, electronic, and hydrophobic features of a molecule to ensure the optimal molecular interactions with the corresponding biological target structure. The three-dimensional mapping of the interactions of an active compound within its target protein can represent its geometrical and chemical properties which include hydrogen bond acceptors and donors, partial charge, basic and acidic groups, aliphatic and aromatic hydrophobic moieties, and steric as well as electrostatic properties [96]. It has been mostly employed in virtual screening process to identify potential ligands based on these interactions. Several software packages such as LigandScout [97] and The Pocket [98] have been developed to perform pharmacophore modeling in protein-ligand complex for mapping of interactions between target and ligand. Docking and pharmacophore modeling have been widely used in virtual screening to identify novel compounds against various drug targets. Pharmacophore modeling has also been useful in high-throughput virtual screening for identification of novel active compounds [99].

Recently, a field gaining popularity in the area of computational drug discovery is proteochemometric and polypharmacology modeling. Proteochemometric modeling is a combination of both ligand and target information in a single model used to predict desirable output. Ligand information of the system also involves the information of its biological effect, and merging this data with the target information into the frame of a machine learning model provides the prediction of the most suitable pharmacological treatment for a given genotype which ligand-only and protein-only approaches alone are not able to perform [100].

2.3.3 Fragment-Based Drug Design (FBDD)

Fragment-based drug design (FBDD) process is employed for finding lead compounds when the computational hit identification methods produce compounds of low affinity. FBDD involves virtual modeling and optimization of a lead structure within the active site of its target, and thus, it can be performed only when the 3D structure of the corresponding target is available [101]. In view of the fact that commercial databases of compounds are limited with the representation of only a small amount of the available chemical space [102], the FBDD is used to increase the list of hit compounds and produce more potent lead compounds. The two commonly used approaches involved in FBDD for the optimization of the hits are fragment placing and fragment linking [103]. Fragment placing involves addition of functional groups to the active fragment scaffold in order to optimize interactions with the binding site of the target. Fragment linking is a less commonly used method where the fragments that bind in adjacent sites of a target protein are linked to generate high-affinity leads. The computational tools available to perform fragment placing and linking are LUDI [104], CLIX [105], CAVEAT [106], LeapFrog [107], etc.

2.3.4 Combinatorial Library Designing

The first combinatorial libraries of diversified peptides were discovered by Mario Geysen in early 1980s [108]. The process of combinatorial library designing aims to reduce the number of molecules to be synthesized, without decreasing the structural and chemical diversity. It can be used for rapid identification of potential leads as smaller number of molecules needs to be tested. The process of design is carried out by combining molecular building blocks to create a very large number of different molecules using a scaffold and reference sets of compounds which together combine by placing different structures on attachment positions [109]. In comparison to the traditional synthetic chemistry, the combinatorial chemistry method allows rapid synthesis of a very large number of molecules at lower cost. Combinatorial chemistry libraries consist of one or more structures with a few number of R-group positions where alternative groups are placed. Combinatorial library methods, first applied to peptides and oligonucleotides, have now been extended to include proteins, synthetic oligomers, small molecules, and oligosaccharides. The basic steps involved in combinatorial library designing consist of preparation of the library, screening of the library components, and determination of the chemical structures of active compounds [110]; however, the steps involved in library preparation vary according to the desired type of library.

2.3.5 Molecular Dynamic Simulation

Biological activity of a molecule corresponds to its time-dependent interactions with other molecules or its environment. These microscopic interactions occur at the interfaces such as protein-protein, protein-nucleic acid, and protein-ligand and can be calculated by molecular dynamic (MD) simulation which is a computer simulation of physical movements of atoms and molecules. In MD simulation, the atoms and molecules are allowed to interact for a given period of time, in the physiological environment created computationally, and the results are analyzed in the form of trajectories [111]. MD simulations can calculate a trajectory of conformations of a given system as a function of time based on molecular mechanics and force field. The chemical bonds and atomic angles are determined using virtual springs and dihedral angles using a sinusoidal function. Non-covalent interactions such as van der Waals interactions and the Lennard-Jones and Coulomb's potential are used to evaluate hydrophobic and electrostatic interactions. The results of MD simulations combined with experimental data provide a significant tool for drug discovery process. The free energy calculations for prediction of binding affinities can also be performed using MD simulations with the help of methods such as molecular mechanics Poisson-Boltzmann surface area (MM/PBSA) [112], linear interaction energy (LIE) [113], and free energy perturbation methods (FEP) [114].

In addition to this, MD simulation has a very useful application in understanding of the process of Aβ oligomer formation for the prevention of plaque formation and their disintegration. In view of the different conformational states and complex self-associations of Aβ monomers leading to generation of non-covalently linked metastable oligomers, the experimental exploration of the process of Aβ monomer folding and oligomerization becomes difficult. Furthermore, the quantitation and fractionation of non-covalently linked oligomers becomes difficult due to SDS-induced dissociation [115]. Thus, the computational determination of Aβ structure for understanding the process of Aβ monomer folding and development of therapeutic compounds to prevent Aβ oligomerization can be done using MD simulation. All the computational approaches and databases involved in CADD have been listed in Table 3 [79].

2.4 Lead Optimization

Leads, once identified, are characterized with respect to their physiochemical properties, pharmacodynamic properties such as efficacy and potency, pharmacokinetic properties, as well as toxicological aspects. If the potential therapeutic compounds fulfill certain criteria possessed by the given properties, they can be considered as leads. The criteria for pharmacodynamic properties can originate from their efficacy, potency, and selectivity, for physiochemical properties from water solubility and chemical stability, and for pharmacokinetic properties from metabolic stability, drug likeness, and toxicological aspects including chemical optimization potential, i.e., ease of chemical synthesis and patentability [116]. Thus, maintenance of this process requires simultaneous optimization of all the given parameters experimentally and is thus a time-consuming and expensive process. The modification of a lead compound can be performed with the help of molecular modeling, quantitative structure-activity relationship (QSAR), and structure-based drug designing techniques. Along with the characterization of the compound based on its potency and selectivity, in vitro assays and in silico analyses for the prediction of their pharmacokinetic and pharmacodynamic properties must also be performed. The identified compounds possessing desirable in vitro properties are then further characterized using in vivo models.

2.4.1 ADMET Analysis

Long-term success of drug depends on its efficient interaction with the proposed target along with its favorable therapeutic effect without showing any undesirable toxicity. The drugs fail both at early and late pipeline stages generally due to undesired pharmacokinetic and toxicity properties suggesting absorption, distribution, metabolism, excretion, and toxicity (ADMET) properties as crucial part of drug discovery process. Many FDA-approved drugs, including those for AD such as tacrine, have been discontinued at a later stage due to their undesirable pharmacokinetic properties and side effects [117]. In this direction, computational tools are very useful in screening a large number of potential therapeutic molecules for their long-term success based on their pharmacokinetic properties. ADMET analysis involves several pharmacokinetic descriptors to study the movement of drug, from the time of administration to the point of elimination from the body. The in silico ADMET descriptors used to describe these properties include aqueous solubility, BBB penetration, CYP2D6 binding, human intestinal absorption (HIA), plasma protein binding (PPB), and hepatotoxicity [118]. In view of the fact that the treatment of neurological diseases like AD requires CNS permeability, BBB is considered as the major drawback for the long-term administration of CNS affecting drugs and requires highest priority while designing of drug candidates against AD.

2.4.2 Drug Likeness

Drug-like properties of the identified leads can be analyzed on the basis of Lipinski's rule of five [119] which is characterized by four properties, namely, molecular weight, partition coefficient, number of hydrogen bond acceptors, and number of hydrogen bond donors. The potential therapeutic compounds or leads, for their long-term application as drug without failure, must exhibit number of hydrogen bond donors ≤ 5, number of hydrogen bond acceptors ≤ 10, molecular weight < 500, and partition coefficient (log *P* value) ≤ 5. The compounds which successfully pass the Lipinski's rule of five are easily absorbed by the body and are more likely to be membrane permeable suggesting their use as potential drug candidates [120]. However, it is noteworthy that there are few exceptions to this rule. Thus, orally active therapeutic classes of molecules such as antibiotics, antifungals, vitamins, and cardiac glycosides violate Lipinski's rule of five because they have structural features that allow them to act as substrates for naturally occurring transporters [120]. Some common examples of such category of molecules being used as drugs include cyclosporine A, a large cyclic peptide that is being used as an immunosuppressant. Despite its molecular mass of 1202 Da, a partition coefficient of 14.4, and a polar surface area of 279 $Å^2$, approximately 30% of every oral dose of cyclosporine A makes its way through the intestinal wall into our bloodstream [39]. Similarly, taxol, a naturally obtained anticancer drug, also violates Lipinski's rule of five.

3 Application of Computational Approaches for the Design and Development of Potent Drugs Based on Natural Products Against AD

As discussed in Sect. 1.2, the plant-derived natural compounds have been proven to be a major source of therapeutics with relatively lower toxicity. In this direction, in silico approaches have provided a great opportunity toward identification of novel potential lead compounds, saving the cost and time of performing clinical trials and increasing the success rate of new drugs. Thus, a number of plant-derived secondary metabolites have been screened using various computational approaches for designing potential therapeutic compounds against neurodegeneration. Docking studies of 13 plant-derived alkaloids with AChE, a well-known target for AD, revealed pleiocarpine as the most promising and potential anticholinergic compound, while its amino derivative exhibited about sixfold higher potential than pleiocarpine against AChE. Moreover, when compared to the commonly used drugs for AChE, namely, tacrine and rivastigmine, pleiocarpine and its amino derivative were found to be better inhibitors, suggesting development of these molecules as potential therapeutics against AD in the future [121]. Similarly, 100 terpenoids, selected from the relevant databases as well as literature, were analyzed for their inhibitory potential against acetylcholinesterase and Aβ monomeric peptide. Due to the failure of drugs owing to undesired pharmacokinetic and toxicity properties, all the hundred terpenoids were subjected to screening based on ADMET properties and Lipinski's rule of five for their successful therapeutic application. Finally, 25 screened terpenoids were analyzed and compared with commonly used anti-cholinergic drug, namely, galantamine, and known inhibitor of Aβ, namely, curcumin, for their relative potential against AD through in silico molecular docking and MD simulation approaches. Based on docking and MD simulation analyses, nimbolide was found to be most potent and stable inhibitor for both AChE and Aβ. Thus, further clinical and experimental studies using nimbolide may provide a potential lead compound targeting AChE and Aβ for the treatment of AD [122].

In addition, a series of 30 resveratrol derivatives proved as Aβ aggregation inhibitors were subjected to a comparative molecular field analysis (CoMFA) and comparative molecular similarity indices analysis (CoMSIA)-based three-dimensional quantitative structure-activity relationship (3D-QSAR) studies. The two CoMFA models and thirty one CoMSIA models, corresponding to all possible field combinations were generated using a training set of twenty compounds and test set of ten compounds. The obtained models showed very good predictability as the predicted activity values of all the compounds were very close to the observed activity values. Contour maps obtained from these models identified several key features, which explain the wide range of activities among selected resveratrol derivatives. These studies provided a better understanding of the resveratrol structural elements that are essential for Aβ binding and guide the future structural modifications for the design of resveratrol-related Aβ disruptors with enhanced activity (unpublished data) [123].

Another computational analysis of AD drug targets, acetylcholinesterase, and beta-secretase enzymes led to designing of multi-target-directed potential candidates using molecular docking. In addition, the 3D-QSAR model for 4,3-hydroxy ethylamine derivatives of beta-secretase inhibitors developed using CoMFA and CoMSIA approaches provided insight into the effects of the substitution pattern of the drugs that are related to the biological activity of anti-Alzheimer compounds [124]. In an additional study while performing molecular modeling, docking, and ADMET analysis of a novel hybrid drug for the treatment of AD, it was observed that the molecular hybrids of tacrine with donepezil would be a useful proposal for future treatment of AD [125]. These kinds of molecular modifications based on the molecular hybridization have led to the elaboration of libraries of new potentially more active therapeutic derivatives by optimization of a scaffold.

In addition to this, QXP, a powerful, rapid computer algorithm for structure-based drug design, has also been used to optimize inhibitors of human beta-secretase, which is an important therapeutic target for treating AD by diminishing amyloid deposits from the brain [126]. A homology model of M1 acetylcholine receptor, having therapeutic potential for treating dementia, including AD and cognitive impairment, was used for virtual screening of a corporate compound collection, which revealed approximately 1000 putative hits using docking into a previously known allosteric binding site. The optimization of this target for improving potency and selectivity led to development of a series of novel 1-(N-substituted piperidin-4-yl) benzimidazolones, which resulted in identification of potent, CNS penetrant, and orally active compounds against the receptor [127]. In a QSAR study combined with synthetic optimization for inhibitors of autotaxin, an autocrine motility factor linked to AD, the analogs of the lead compound showed two- to three-fold higher affinity for autotaxin than the lead [128]. Furthermore, AD-related H3 antagonists were also used to generate a pharmacophore model with four features including a distal positive charge, an electron-rich position, a central aromatic ring, and either a second basic amine or another aromatic [129].

4 Preclinical and Clinical Development of Drugs

Preclinical studies involve experimental (in vitro) tests and trials on animal populations. Various doses of the potential lead are administered to the cell line or animal in order to obtain preliminary efficacy and pharmacokinetic data. Once the lead successfully completes preclinical studies, it goes through the four major phases of clinical trials during which the safety and efficacy of the lead on healthy as well as diseased humans are determined. The trial results of the acceptable lead compounds are documented, and a complete description of the manufacturing procedures, formulation details, etc. is included which is further reviewed by FDA [130].

The phases of clinical trials start with phase I followed by phase II, phase III, and phase IV. Phase I trials commonly focus on safety, bioavailability, and tolerability of the lead compound where the drug is administered to a small number (tens) of healthy volunteers. Phase II trials are focused on the efficacy of the lead in large number of patients (several hundred) suffering from the disease which the drug is supposed to treat. Phase II trials are performed globally and provide information on efficacy and safety of the lead compound in large population. Phase II trials are sometimes divided into phase IIa (pilot trials) and IIb (well-controlled trials). Similar to the phase II, phase III clinical trials are also intended to determine lead efficacy involving several thousand patients and evaluate their effect in comparison with the drugs currently used for the treatment of that particular disease. Phase III trials are divided into IIIa, where the trials are conducted after efficacy of the medicine is demonstrated but prior to regulatory submission of NDA, and IIIb, where the trials are conducted after regulatory submission of NDA but prior to the medicine's approval and launch. Phase IV trials are commonly referred as post-marketing trials which are performed after the medicine has been approved by FDA. These trials provide information about long-term risk and benefit of the drug. These trials involve many thousands of patients and may continue for many years [131].

5 Conclusions and Future Directions

In view of the increasing number of AD patients globally, this is a major concern for society, especially the aging population. The limited success of the currently available therapies in AD can be attributed to the drawbacks in understanding of the pathogenesis of the disease, development of target specific drugs, and subsequent failure of drugs in clinical trials due to their side effects. Thus, from a long time, natural products have been a major component in drug discovery providing safer lead compounds approved for clinical use and inspiring the synthesis of chemical libraries. In view of the huge literature supporting the immense role of natural products in drug discovery, these compounds need to be explored to provide new lead compounds as well as drug targets that are strongly connected with the pathogenesis of AD.

The high expectations of drug discovery and development from natural sources have been due to its increasing number of molecules with new scaffolds and interesting biological activities. These molecules do not only target one pathophysiological aspect of AD but also ameliorate several AD-related pathologies, suggesting that increased consumption of these compounds might lead to a safe strategy to delay the onset of AD. Thus, the continuous investigation of the potential of these compounds is necessary as they are promising to yield a possible remedy for this disease. In spite of the large number of natural products and related compounds being in clinical and preclinical trials against AD, a large portion of the flora remains unexplored. Therefore, new approaches are needed to identify the plant that produces these secondary metabolites and also to identify gene clusters responsible for biosynthesis of these compounds.

Chemoinformatic methods, which characterize the chemical space of the available natural products databases by comparing their molecular properties including structural complexity and diversity with approved drugs, have been helpful to further understand the biological activity of these compounds and to guide the chemical synthesis of natural products derivatives with enhanced activity. Virtual screening is being used for the systematic search of bioactive natural products with a molecular target of interest, while target identification approaches have been employed to identify potential molecular targets for natural products. Natural molecules can also be optimized by chemical derivatization and synthesis of analogs for better pharmacokinetic properties and efficacy. In view of the fact that natural product-based drug discovery and development represent a complex process demanding integration of several interdisciplinary approaches, computational methods are playing a major role in the enhancement of these modified drug development. In addition, the recent scientific developments and technologic advances have indicated that natural products are among the most important sources of novel drugs in the future against various diseases including AD. Thus, the combination of experimental natural products research with computational approaches has proved to be very effective in the drug discovery. Regulatory agencies as well as pharmaceutical industries are actively involved in the development of new computational tools that will improve effectiveness and efficiency of drug discovery and development process, reducing the time and cost of developing useful therapeutic molecules and increasing the success rate along with less animal sacrifice. It is expected that the power of computer-aided drug design will keep on growing as the technology continues to evolve.

References

1. Ballard C, Gauthier S, Corbett A et al (2011) Alzheimer's disease. Lancet 377:1019–1031
2. Alzheimer's Association, Alzheimer's disease facts and figures, Alzheimers Dement. 10 (2016) http://www.alz.org/downloads/Facts_Figures_2016.pdf
3. Kalaria RN, Maestre GE, Arizaga R et al (2008) Alzheimer's disease and vascular dementia in developing countries: prevalence, management, and risk factors. Lancet Neurol 7:812–826
4. Hebert LE, Scherr PA, Bennett DR, Evans DA (2004) Alzheimer disease in the US population: prevalence estimates using the 2000 census. Arch Neurol 61:802–803
5. Awasthi M, Singh S, Pandey VP, Dwivedi UN (2016) Alzheimer's disease: an overview of amyloid beta dependent pathogenesis and its therapeutic implications along with in silico approaches emphasizing the role of natural products. J Neurosurg Sci 361:256–271
6. Sandberg LM, Luheshi S, Sollvander T et al (2010) Stabilization of neurotoxic Alzheimer amyloid-beta oligomers by protein engineering. Proc Natl Acad Sci U S A 107:15595–15600
7. Hardy J, Duff K, Hardy K et al (1998) Genetic dissection of Alzheimer's disease and related dementias: amyloid and its relationship to tau. Nat Neurosci 1:355–358
8. Jack CR Jr, Knopman DS, Jagust WJ et al (2010) Hypothetical model of dynamic biomarkers of the Alzheimer's pathological cascade. Lancet Neurol 9:119–128
9. Smith MA, Perry G, Richey PL et al (1996) Oxidative damage in Alzheimer's. Nature 382:120–121
10. Schneider LS, Mangialasche F, Andreasen N et al (2014) Clinical trials and late-stage drug development for Alzheimer's disease: an appraisal from 1984 to 2014. J Intern Med 275:251–283
11. Loveman E, Green C, Kirby J et al (2006) The clinical and cost-effectiveness of donepezil, rivastigmine, galantamine and memantine for Alzheimer's disease. Health Technol Assess Rep 10:iii–iv, ix–xi, 1–160
12. Harvey AL (2008) Natural products in drug discovery. Drug Discov Today 13:894–901
13. Howes M-JR, Perry NSL, Houghton PJ (2003) Plants with traditional uses and activities, relevant to the management of

Alzheimer's disease and other cognitive disorders. Phytother Res 17:1–18

14. Shu Y-Z (1998) Recent natural products based drug development: a pharmaceutical industry perspective. J Nat Prod 61:1053–1071
15. Howes M-JR, Houghton PJ (2003) Plants used in Chinese and Indian traditional medicine for improvement of memory and cognitive function. Pharmacol Biochem Behav 75:513–527
16. Quid PM (1999) A review of nutrients and botanicals in the integrative management of cognitive dysfunction. Altern Med Rev 4:144–161
17. Perry EK, Pickering AT, Wang WW (1999) Medicinal plants and Alzheimer's disease: from ethnobotany to phytotherapy. J Pharm Pharmacol 51:527–534
18. Quinn R, Isaac M, Tabet N (2008) Vitamin E for Alzheimer's disease and mild cognitive impairment. Cochrane Database Syst Rev 3: CD002854
19. Williams P, Sorribasa A, Howes M-JR (2011) Natural products as a source of Alzheimer's drug leads. Nat Prod Rep 28:48–77
20. McCaleb R (1990) Nature's medicine for memory loss. HerbalGram 23:15
21. Viegas C Jr, Bolzani VS, Barreiro EJ, Fraga CAM (2005) New anti-Alzheimer drugs from biodiversity: the role of the natural acetylcholinesterase inhibitors. Mini Rev Med Chem 5:915–926
22. Houghton PJ, Ren Y, Howes M (2006) Acetylcholinesterase inhibitors from plants and fungi. Nat Prod Rep 23:181–199
23. Wang J, Zhao W, Teplow DB, Ho L (2008) Grape-derived polyphenolics prevent Aβ oligomerization and attenuate cognitive deterioration in a mouse model of Alzheimer's disease. J Neurosci 28:6388–6392
24. Burger A (2003) In: Abraham DJ (ed) Burger's medicinal chemistry and drugs discovery, 6th edn. Wiley, New York, pp 847–900
25. Furukawa S, Yang L, Sameshima H (2014) Galantamine, an acetylcholinesterase inhibitor, reduces brain damage induced by hypoxia-ischemia in newborn rats. Int J Dev Neurosci 37:52–57
26. Qian ZM, Ke Y (2014) Huperzine A: is it an effective disease-modifying drug for Alzheimer's disease? Front Aging Neurosci 6:216
27. Jia JY, Zhao QH, Liu Y, Gui YZ, Liu GY, Zhu DY (2013) Phase I study on the pharmacokinetics and tolerance of ZT-1, a prodrug of huperzine A, for the treatment of Alzheimer's disease. Acta Pharmacol Sin 34:976–982
28. Yang F, Lim GP, Begum AN, Ubeda OJ, Simmons MR, Ambegaokar SS, Chen PP, Kayed R, Glabe CG, Frautschy SA, Cole GM (2005) Curcumin inhibits formation of amyloid β-oligomers and fibrils, binds plaques and reduces amyloid *in vivo*. J Biol Chem 280:5892–5901
29. Marambaud P, Zhao H, Davies P (2005) Resveratrol promotes clearance of Alzheimer's disease amyloid β peptides. J Biol Chem 280:37377–37382
30. Sun MK, Alkon DL (2005) Dual effects of bryostatin-1 on spatial memory and depression. Eur J Pharmacol 512:43–45
31. Tomiyoma T, Kaneko H, Kataoka K, Asano S, Endo N (1997) Rifampicin inhibit the toxicity of pre aggregated amyloid peptides by binding to peptide fibrils and preventing amyloid cell interaction. Biochem J 322:859–865
32. Schenk D, Barbour R, Dunn W, Gordon G, Grajeda H, Guido T (1999) Immunization with amyloid-β attenuates Alzheimer disease like pathology in the PDAPP mouse. Nature 400:173–177
33. Rinne JO, Brooks DJ, Rossor MN, Fox NC, Bullock R, Klunk WE (2010) 11c-PiB PET assessment of change in fibrillar amyloid β load in patients with Alzheimer's disease treated with bapineuzumab: a phase 2 double blind placebo controlled ascending dose study. Lancet Neurol 9:363–372
34. Burstein AH, Zhao Q, Ross J, Styren S, Landen JW, Ma WW, McCush F, Alvey C, Kupiec JW, Bednar MM (2013) Safety and pharmacology of ponezumab (PF-04360365) after a single 10-minute intravenous infusion in subjects with mild to moderate Alzheimer disease. Clin Neuropharmacol 36:8–13
35. https://www.clinicaltrials.gov.ct2/NCT01900665
36. Boxer A (2013) Phase II clinical trials of crenezumab for mild to moderate Alzheimer's disease. Sandler Neurosciences Center. Published by UCSF Memory and Aging Center, California
37. Ostrowitzki S, Deptula D, Thurfjell L, Barkhof F, Bohrmann B, Brooks DJ, Klunk WE, Ashford E, Yoo K, Xu ZX, Loetscher H, Santarelli L (2012) Mechanism of amyloid removal in patients with Alzheimer disease treated with gantenerumab. Arch Neurol 69:198–207
38. https://www.clinicaltrials.gov/show/NCT01767311
39. Babitha K, Vazhayil R, Shanmuga SVM (2014) Natural products and its derived

drugs for the treatment of neurodegenerative disorders: Alzheimer's disease-a review. Br Biomed Bull 2:359–370

40. http://www.naturalremediescenter.com
41. Goswami S, Saoji A, Kumar N, Thawani V, Tiwari M, Thawani M (2011) Effect of *Bacopa monnieri* on cognitive functions in Alzheimer's disease patients. Int J Collab Res Intern Med Public Health 3:285–293
42. Singhal AK, Naithani V, Bangar OP (2012) Medicinal plants with a potential to treat Alzheimer and associated symptoms. Int J Nutr Pharmacol Neurol Dis 2:84–91
43. http://www.herbal-supplement-resource.com
44. http://www.suite101.com
45. Duke JA (2007) The Garden Pharmacy: Rosemary, the herb of remembrance for Alzheimer's disease. Altern Complement Ther 13:287–290
46. http://www.associatedcontent.com
47. Ji YA, Suna K, Sung EJ, Tae YH (2010) Effect of licorice (*Glycyrrhiza uralensis* fisch.) on amyloid-β-induced neurotoxicity in PC12 cells. Food Sci Biotechnol 19:1391–1395
48. Bilge S, Ilkay O (2005) Discovery of drug candidates from some Turkish plants and conservation of biodiversity. Pure Appl Chem 77:53–64
49. lkay O, Gürdal O, Bilge S (2010) An update on plant-originated treatment for Alzheimer's disease. Ethnomed: Source Complem Therap 12: 245–265
50. http://www.holistictherapypractice.com
51. Lannert H, Hoyer S (1998) Intracerebroventricular administration of streptozotocin causes long-term diminutions in learning and memory abilities and in cerebral energy metabolism in adult rats. Behav Neurosci 112:1199–1208
52. Fu LM, Li JT (2009) A systematic review of single Chinese herbs for Alzheimer's disease treatment. Evid Based Complement Alternat Med 2011:640284
53. Park CH, Kim SH, Choi W, Lee YJ, Kim JS, Kang SS (1996) Novel anticholinesterase and antiamnesic activities of dehydroevodiamine, a constituent of *Evodia ruraecarpa*. Planta Med 62:405–409
54. http://herbal-ayurveda-remedy.com
55. http://www.livestrong.com
56. Keyvan D, Damien DHJ, Heikki V, Raimo H (2007) Plants as potential sources for drug development against Alzheimer's disease. Int J Biomed Pharm Sci 1:83–104
57. Kumar S, Christopher JS, Edward JO (2011) Kinetics of acetylcholinesterase inhibition by an aqueous extract of *Withania somnifera* roots. Int J Pharm Sci Res 2:1188–1192
58. Ji H, Zhang H (2008) Multipotent natural agents to combat Alzheimer's disease. Functional spectrum and structural features. Acta Pharmacol Sin 29:143–151
59. Dev K, Maurya R (2017) Marine-derivedanti-Alzheimer's agents of promise. In: Brahmachari G (ed) Neuroprotective natural products: clinical aspects and mode of action, 1st edn. Wiley-VCHVerlag GmbH & Co. KGaA, Weinheim, Germany, pp 153–184
60. Brahmachari G (2011) Natural products in the drug discovery programmes in Alzheimer's: impacts and prospects. Asia Pac Biotech News 15:14–34
61. Apetz N, Munch G, Govindaraghavan S, Gyengesi E (2014) Natural compounds and plant extracts as therapeutics against chronic inflammation in Alzheimer's disease–a translational perspective. CNS Neurol Disord Drug Targets 13:1175–1191
62. Calcul L, Zhang B, Jinwal UK, Dickey CA, Baker BJ (2012) Natural products as a rich source of tau-targeting drugs for Alzheimer's disease. Future Med Chem 4:1751–1761
63. https://www.scientificamerican.com/article/cost-to-develop-new-pharmaceutical-drug-now-exceeds-2-5b/
64. Stumpfe D, Ripphausen P, Bajorath J (2012) Virtual compound screening in drug discovery. Future Med Chem 4:593–602
65. Zhang S (2011) Computer-aided drug discovery and development. Methods Mol Biol 716:23–38
66. Talele TT, Khedkar SA, Rigby AC (2010) Successful applications of computer aided drug discovery: moving drugs from concept to the clinic. Curr Top Med Chem 10:127–141
67. Zhao S, Li S (2010) Network-based relating pharmacological and genomic spaces for drug target identification. PLoS One 5:e11764. doi:10.1371/ journal.pone.0011764
68. Barabasi A-L, Gulbahce N, Loscalzo J (2011) Networkmedicine: a network-based approach to human disease. Nat Rev Genet 12:56–68
69. Billur EH, Gursoy A, Nussinov R, Keskin O (2014) Network-based strategies can help mono-and poly-pharmacology drug discovery: a systems biology view. Curr Pharm Des 20:1201–1207
70. Wang X, Wei X, Thijssen B et al (2012) Three-dimensional reconstruction of protein

networks provides insight into human genetic disease. Nat Biotechnol 30:159–164

71. Buchan NS, Rajpal DK, Webster Y et al (2011) The role of translational bioinformatics in drug discovery. Drug Discov Today 16:426–434
72. Keiser MJ, Setola V, Irwin JJ et al (2009) Predicting new molecular targets for known drugs. Nature 462:175–181
73. Wang L, Xie X-Q (2014) Computational target fishing: what should chemogenomics researchers expect for the future of in silico drug design and discovery? Future Med Chem 6:247–249
74. Rognan D (2010) Structure-based approaches to target fishing and ligand profiling. Mol Inf 29:176–187
75. Chen B, Butte AJ (2015) Leveraging big data to transform target selection and drug discovery. Clin Pharmacol Ther 99:285–297
76. Shannon P, Markiel A, Ozier O et al (2003) Cytoscape: a software environment for integrated models of biomolecular interaction networks. Genome Res 13:2498–2504
77. Wessel J et al (2015) Low-frequency and rare exome chip variants associate with fasting glucose and type 2 diabetes susceptibility. NatCommun 6. https://doi.org/10.1038/ncomms6897
78. Barrangou R, Birmingham A, Wiemann S et al (2015) Advances in CRISPR-Cas9 genome engineering: lessons learned from RNA interference. Nucleic Acids Res 43:3407–3419
79. Katsila T, Spyroulias GA, Patrinos GP et al (2016) Computational approaches in target identification and drug discovery. Comput Struct Biotechnol J 14:177–184
80. Shekhar C (2008) In silico pharmacology: computer-aided methods could transform drug development. Chem Biol 15:413–414
81. S-s O-Y, Lu J-y, Kong X-q, Liang Z-j et al (2012) Computational drug discovery. Acta Pharmacol Sin 33:1131–1140
82. Sliwoski G, Kothiwale S, Meiler J, Lowe EW (2014) Computational methods in drug discovery. Pharmacol Rev 66:334–395
83. Zhang Q, Muegge I (2006) Scaffold hopping through virtual screening using 2d and 3d similarity descriptors: ranking, voting, and consensus scoring. J Med Chem 49:1536–1548
84. Zhang S (2011) Computer-Aided Drug Discovery and Development. In: Satyanarayanajois SD (ed) Drug design and discovery. Humana Press, New York, pp 23–38
85. Gasparini F, Bilbe G, Gomez-Mancilla B et al (2008) mGluR5 antagonists: discovery, characterization and drug development. Curr Opin Drug Discov Devel 11:655–665
86. Wu B, Chien EYT, Mol CD et al (2010) Structures of the CXCR4 chemokine GPCR with small-molecule and cyclic peptide antagonists. Science 330:1066–1071
87. Tounge BA, Rajamani R, Baxter EW (2006) Linear interaction energy models for β-secretase (BACE) inhibitors: role of van der Waals, electrostatic, and continuum-solvation terms. J Mol Graph Model 24:475–484
88. Eswar N, Webb B, Marti-RenomMA et al (2006) Comparative protein structure modeling using Modeller. Curr Protoc Protein Sci. Current Protocols in Bioinformatics. 15(5.6):5.6.1–5.6.30.
89. Schwede T, Kopp J, Guex N, Peitsch MC (2003) SWISS-MODEL: an automated protein homology-modeling server. Nucleic Acids Res 31:3381–3385
90. Yuriev E, Agostino M, Ramsland PA (2011) Challenges and advances in computational docking: 2009 in review. J Mol Recognit 24:149–164
91. Verdonk ML, Cole JC, Hartshorn MJ (2003) Improved protein– ligand docking using GOLD. Proteins Struct Funct Bioinf 52:609–623
92. Morris GM, Huey R, Lindstrom W et al (2009) AutoDock4 andAutoDockTools4: automated docking with selective receptor flexibility. J Comput Chem 30:2785–2791
93. Trott O, Olson AJ (2010) AutoDock Vina: improving the speed and accuracy of docking with a new scoring function, efficient optimization, and multithreading. J Comput Chem 31:455–461
94. Moustakas D, Lang PT, Pegg S et al (2006) Development and validation of a modular, extensible docking program: DOCK 5. J Comput Aided Mol Des 20:601–619
95. Houston DR, Walkinshaw MD (2013) Consensus docking: improving the reliability of docking in a virtual screening context. J Chem Inf Model 53:384–390
96. Meslamani J, Li J, Sutter J et al (2012) Protein–ligand-based pharmacophores: generation and utility assessment in computational ligand profiling. J Chem Inf Model 52:943–955
97. Wolber G, Langer T (2005) LigandScout: 3-D pharmacophores derived from

proteinbound ligands and their use as virtual screening filters. J Chem Inf Model 45:160–169
98. Chen J, Lai L (2006) Pocket v.2: further developments on receptor-based pharmacophore modeling. J Chem Inf Model 46:2684–2691
99. Reymond J-L, Awale M (2012) Exploring chemical space for drug discovery using the chemical universe database. ACS Chem Neurosci 3:649–657
100. Cortes-Ciriano I, Ain QU, Subramanian V et al (2015) Polypharmacology modelling using proteochemometrics (PCM): recent methodological developments, applications to target families, and future prospects. Med Chem Commun 6:24–50
101. Kumar A, Voet A, Zhang KYJ (2012) Fragment based drug design: from experimental to computational approaches. Curr Med Chem 19:5128–5147
102. Cheng Y, Judd TC, Bartberger MD et al (2011) From fragment screening to in vivo efficacy: optimization of a series of 2-aminoquinolines as potent inhibitors of beta-site amyloid precursor protein cleaving enzyme 1 (BACE1). J Med Chem 54:5836–5857
103. Hughes SJ, Millan DS, Kilty IC et al (2011) Fragment based discovery of a novel and selective PI3 kinase inhibitor. Bioorg Med Chem Lett 21:6586–6590
104. Erlanson DA (2012) Introduction to fragment-based drug discovery. Top Curr Chem 317:1–32
105. Lawrence MC, Davis PC (1992) CLIX: a search algorithm for finding novel ligands capable of binding proteins of known three-dimensional structure. Proteins 12:31–41
106. Erlanson DA, McDowell RS, O'Brien T (2004) Fragment-based drug discovery. J Med Chem 47:3463–3482
107. Computational Chemistry: A Tool for Feeding R&D Pipelines (2007) Wiley handbook of current and emerging drug therapies. Wiley, Hoboken, pp 5–8
108. Dolle RE (2004) Comprehensive survey of combinatorial library synthesis: 2003. J Comb Chem 5:623–679
109. Xu H, Agrafiotis DK (2002) Retrospect and Prospect of virtual screening in drug discovery. Curr Top Med Chem 2:1305–1320
110. Lengauer T, Lemmen C, Rarey M et al (2004) Novel technologies for virtual screening. Drug Discov Today 9:27–34
111. Liu W, Schmidt B, Voss G, Müller-Wittig W (2008) Accelerating molecular dynamics simulations using graphics processing units with CUDA. Comput Phys Commun 179:634–641
112. Hou T, Wang J, Li Y, Wang W (2010) Assessing the performance of theMM/PBSA andMM/GBSA methods.The accuracy of binding free energy calculations based on molecular dynamics simulations. J Chem Inf Model 51:69–82
113. Hansson T, Marelius J, Åqvist J (1998) Ligand binding affinity prediction by linear interaction energy methods. J Comput Aided Mol Des 12:27–35
114. Singh UC, Brown FK, Bash PA, Kollman PA (1987) An approach to the application of free energy perturbation methods using molecular dynamics: applications to the transformations of CH3OH – N CH3CH3, H3O+–N NH4+, glycine – N alanine, and alaninephenylalanine in aqueous solution and to H3O+(H2O)3– N NH4+(H2O)3 in the gas phase. J Am Chem Soc 109:1607–1614
115. Miyamoto S, Kollman PA (1993) Absolute and relative binding free energy calculations of the interaction of biotin and its analogs with streptavidin using molecular dynamics/free energy perturbation approaches. Proteins Struct Funct Genet 16:226–245
116. Eddershaw PJ, Beresford AP, Bayliss MK (2000) ADME/PK as part of a rational approach to drug discovery. Drug Discov Today 5:409–414
117. Kaitin KI (2008) Obstacles and opportunities in new drug development. Clin Pharmacol Ther 83:210–212
118. Roberts SA (2003) Drug metabolism and pharmacokinetics in drug discovery. Curr Opin Drug Discov Devel 6:66–80
119. Bleicher KH, Bohm HJ, Muller K (2003) Alanine AI. Hit and lead generation: beyond high-throughput screening. Nat Rev Drug Discov 2:369–378
120. Lipinski CA, Lombardo F, Dominy BW, Feeney PJ (2012) Experimental and computational approaches to estimate solubility and permeability in drug discovery and development settings. Adv Drug Deliv Rev 64:4–17

121. Naaz H, Singh S, Pandey VP, Singh P, Dwivedi UN (2013) Anti-cholinergic alkaloids as potential therapeutic agents for Alzheimer's disease: an *in silico* approach. Indian J Biochem Biophys 50:120–125
122. Awasthi M, Upadhyay AK, Singh S, Pandey VP, Dwivedi UN (2017) Terpenoids as promising therapeutic molecules against Alzheimer's disease: amyloid beta and acetylcholinesterase directed pharmacokinetic and molecular docking analyses. Mol Simul. https://doi.org/10.1080/08927022.2017.1334880
123. Awasthi M, Singh S, Pandey VP, Dwivedi UN (2017) CoMFA and CoMSIA based designing of resveratrol derivatives asamyloid-beta-aggregation inhibitors against Alzheimer's disease (communicated)
124. Gupta S, Pandey A, Tyagi A, Mohan GA (2010) Computational analysis of Alzheimer's disease drug targets. Curr Res Inf Pharm Sci 11:1–10
125. da Silva CH, Campo VL, Carvalho I, Taft CA (2006) Molecular modeling, docking and ADMET studies applied to the design of a novel hybrid for treatment of Alzheimer's disease. J Mol Graph Model 25:169–175
126. Malamas MS, Erdei J, Gunawan I, Barnes K, Johnson M, Hui Y, Turner J, Hu Y, Wagner E, Fan K (2009) Aminoimidazoles as potent and selective human beta-secretase (BACE1) inhibitors. J Med Chem 52:6314–6323
127. Budzik B, Garzya V, Walker G, Woolley-Roberts M, Pardoe J, Lucas A, Tehan B, Rivero RA, Langmead CJ (2010) Novel N-substituted benzimidazolones as potent, selective, CNS-penetrant, and orally active M (1) mAChR agonists. ACS Med Chem Lett 1:244–248
128. Umemura K, Yamashita N, Yu X, Arima K, Asada T, Makifuchi T, Murayama S, Saito Y, Kanamaru K, Goto Y (2006) Autotaxin expression is enhanced in frontal cortex of Alzheimer-type dementia patients. Neurosci Lett 400:97–100
129. Witkin JM, Nelson DL (2004) Selective histamine H3 receptor antagonists for treatment of cognitive deficiencies and other disorders of the central nervous system. Pharmacol Ther 103:1–20
130. DeMets D, Friedman L, Furberg C (2010) Fundamentals of clinical trials, 4th edn. Springer, Berlin. ISBN 978-1-4419-1585-6
131. Poole AV, Peterson AM (2005) Pharmacotherapeutics for advanced practice: a practical approach. Lippincott Williams & Wilkins, Philadelphia. ISBN 0-7817-5784-3

Chapter 18

In Silico Studies Applied to Natural Products with Potential Activity Against Alzheimer's Disease

Luciana Scotti and Marcus T. Scotti

Abstract

The neurodegenerative disease, named Alzheimer's disease (AD) after its discoverer, is today considered the most common form of dementia. AD represents 60–70% of dementia cases in patients of 65 years of age or older. It leads to complete dementia and death. AD's causes are unknown, yet it evidences mutations in several genes, factors such as the amyloid precursor protein (APP), presenilins (PS1, PS2), apolipoprotein E, etc. In silico methods or CADD (computer-aided drug design) studies are increasingly being used in both industry and universities. They involve an understanding of the molecular interactions from both qualitative and quantitative points of view. These methods generate and manipulate three-dimensional (3D) molecular structures; calculate descriptors and the independent molecular properties, followed by model constructions; and employ other tools that encompass computational drug research. Analysis of the molecular structure of a given system allows relevant information to be extracted and to predict the potential of bioactive compounds.

We found mainly research on the inhibitory activity of flavonoids, alkaloids, and xanthones in key enzymes of the biochemical changes that occur in AD: acetylcholinesterase (AChE; EC 3.1.1.7), butyrylcholinesterase (BChE; EC 3.1.1.8), and monoamine oxidase (MAO; EC 1.4.3.4). Recent research has used QSAR and docking for selection of multifunctional compounds that are drugs on multitargets. The multitarget QSAR model can simultaneously predict activity or classify compounds as actives or inactives against different targets, such as the proteins (amyloid-A4 protein (ABPP), glycogen synthase kinase-3 alpha, glycogen synthase kinase-3 beta (GSK-3β), monoamine oxidase B (MAO-B), and presenilin-1 (PSN-1)).

This chapter will discuss several in silico studies reported considering the complexity of this neurodegenerative disease, the possible multifactorial origin, and the pharmacological potential of natural products.

Key words Alzheimer's disease, In silico, secondary metabolites, Flavonoids, Alkaloids, Terpene

1 Introduction

Natural products are chemical compounds resulting from evolution processes in plants, marine organisms, and fungi. Many such chemicals have been produced and optimized under organism/predator selective coevolution forces. Such natural compounds have been utilized by humans since ancient times for treatments and cures of many diseases. The use of these compounds has been the single

Kunal Roy (ed.), *Computational Modeling of Drugs Against Alzheimer's Disease*, Neuromethods, vol. 132, DOI 10.1007/978-1-4939-7404-7_18, © Springer Science+Business Media LLC 2018

most successful strategy in the discovery of novel medicines, and many medical breakthroughs are based on such natural products. Half of the top 20 best-selling drugs are natural chemical compounds, and their total sales amounted to US $ 16 billion dollars. These numbers suggest that natural chemicals are pre-optimized to be potentially bioactive and therefore to possess "drug-like properties" [1–3].

In silico methods or studies in CADD (computer-aided drug design) are increasingly being used, both in industry and in universities. They consist of representations and manipulations of three-dimensional (3D) molecular structures, calculation of descriptors and molecular properties of the dependent structures, and model construction with other tools that encompass the computational involvement in drug research. These studies require understanding the molecular interaction from qualitative and quantitative points of view. Analysis of the molecular structure of a given system allows relevant information to be extracted, with possible prediction of the potential of bioactive compounds [4].

Theoretical studies have aided in the process of drug discovery. Technological advances in the areas of structural characterization, computational science, and molecular biology have contributed to the speed and productive planning of new molecules. Many chemoinformatics studies exist which show that a large fraction of natural products are either (hopefully) "drug-like" or, at least, "lead-like" with structural and physicochemical properties that render them useful as potential drugs or leads [4].

Alzheimer's disease was discovered in 1907 by the German neuropathologist, Alois Alzheimer. One of his patients presented memory loss, behavioral disorders, and cognitive deficits. On biopsy, extensive brain damage was observed, with loss of neurons and synapses [5–8], mainly in the hippocampus and neocortex association regions, senile plaques, and neuron tangles, called neurofibrillary tangles and spindles [5, 6]. Hippocampal atrophy is the starting point of Alzheimer's disease pathogenesis. It is an irreversible, progressive neurodegenerative disorder of the central nervous system and is characterized by gradual loss of cognitive function, progressive memory loss, disorientation, language impairment, abnormal behavior, and personality changes occasioned by selective neuronal death and abnormal formation of neurotic and core plaques in the cerebral cortex [5–8].

One of the main biochemical changes observed in AD is the decline of the cholinergic system, as shown in Fig. 1. There is a decrease in the activity of acetylcholinesterase and acetylcholine transferase, along with decreased levels of acetylcholine. Acetylcholine (ACh) is synthesized from choline, with acetyl CoA as a substrate in a reaction catalyzed by the enzyme choline acetyltransferase (ChAT). This reaction is limited by the concentration of choline, which owes roughly half of its concentration to

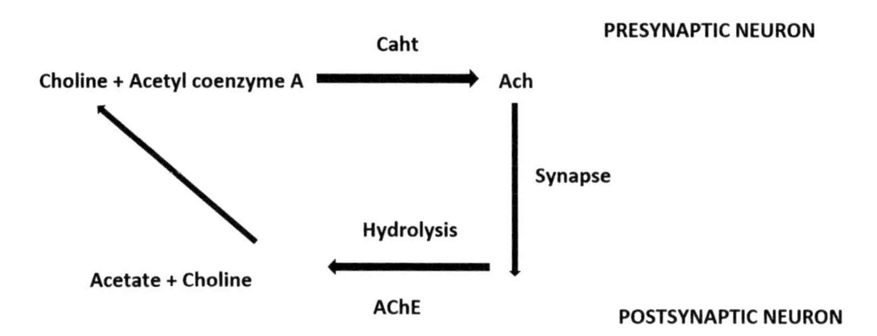

Fig. 1 Synthesis and reuptake of acetylcholine

hydrolysis by choline acetylcholinesterase (AChE). Secondly, ACh is released by fusion of presynaptic neuron membranes with synaptic vesicles storing ACh, where ACh reacts with the receptors of postsynaptic neurons to cause electrochemical signaling. Subsequently, ACh is hydrolyzed by AChE, producing acetate and choline [9].

Sufficient acetylcholine (ACh) is required for proper brain functioning, but due to decreased ACh in AD patients, they suffer from progressive decline in cognitive functioning (thinking, remembering, and reasoning) and behavioral abilities. The symptoms start usually at age of 65 years with short-term memory loss and then long-term memory loss and eventually lead to death due to multiple organ failures within approximately 8 years of onset. There is no effective treatment available till date, but inhibition of ACh breakdown by blocking the AChE has been proved to be helpful in slowing down the disease progression.

Alzheimer's disease (AD) is the most common cause of dementia in elderly people, and it is characterized by progressive neurodegenerative alterations, with 35.6 million people living with dementia worldwide in 2010, a projection of 65.7 million by 2030, and one million new cases per year until 2050 [11–14].

Amyloid in senile plaques is the product of cleavage by α-, β-, and γ-secretases, responsible for generating one particular β-amyloid peptide, the $\alpha\beta_{42}$—that is 42 amino acids in length and has pathogenetic importance. It can form insoluble toxic fibrils and accumulates in the senile plaques isolated from the brains of patients with Alzheimer's disease [15].Treatment strategies against AD are γ-, β- secretase inhibitors, Aβ vaccination, Cu-Zn chelators, cholesterol lowering drugs, statins, and nonsteroidal anti-inflammatory drugs (NSAIDs) [16]. Also neurotrophins (NTs) can be used in the treatment of AD because neuronal death is prevented by endogenous neurotrophic factors (NFs) [17].

Khan et al. [18] reported a molecular docking study for 4-acetoxy-plakinamine B (4APB), a natural marine product with inhibitory effects against AChE [9, 10]. In scientific literature, many reviews have reported studies using natural products as potential drug candidates for treatment of Alzheimer's disease [18, 19]. The most significant natural metabolites studied against AD will be reviewed in the following sections below.

2 Flavonoids

The beneficial activities of the polyphenols against AD are reported in several scientific studies [20, 21]. Their most obvious and easy to understand action is to decrease the cell damage caused by free radicals, as also observed in AD [22, 23]. Rodacka et al. [24] observed the neuroprotective effects of the polyphenols resveratrol and tiron through inactivation of glyceraldehyde-3-phosphate dehydrogenase (GAPDH) induced by superoxide anion radical. Both ligands were studied through docking with the enzyme and showed that tiron and resveratrol radicals are more efficient in GAPDH inactivation than the superoxide anion alone; and tiron was the greater antioxidant.

Lakey-Beitia et al. [25] based on examples of other scientific studies using water extracts reported that extracts from leaves of *Caesalpinia crista* prevented Aβ aggregation from monomers and disintegrated preformed Aβ fibril [26], *Centella asiatica* prevented synuclein aggregation [27], *Ginkgo biloba* extract inhibited the formation of oligomers [28], and extracts prepared from the medicinal herb *Paeonia suffruticosa* inhibited Aβ fibril formation and also destabilized preformed amyloid fibrils [28] and developed a hypotheses on how polyphenols may modulate APP processing, prevent fibrils, and Aβ aggregation.

Mora-Pale et al. [29] reported that polyphenolic glycosides such as naringenin, rutin, and apigenin reduce β-amyloid aggregation.

A polyphenolic xanthone derived from *Garcinia mangostana* Linn, α-Mangostin, was studied by Wang et al. 2012 [30]. This compound is utilized in several studies [31, 32] and reported to have many activities such as antimalarial [33], membrane-protective [34], antimicrobial [35], antiviral [36], etc. Through molecular docking with residues of Aβ structures (PDB ID: 1BA4), the xanthone showed a binding energy of 68.76 kcal/mol, a phenolic hydroxyl group of α-Mangostin-formed hydrogen bonds with Lys16 and Asp23, and interacted with Glu22 of Aβ. The authors observed neuroprotective effects similar to resveratrol [37, 38] and curcumin [39, 40].

In AD, tau proteins are the major constituents of intraneuronal and glial fibrillar lesions, causing dementia and increasing the neurodegenerative process. Madeswaran et al. [41] studied the tau protein kinase I inhibitory activity of 6 flavonoids through molecular docking. The investigated compounds were acacatechin, catechin, galangin, scopoletin, silbinin, and memantine. Parameters like molecular formula, molecular weight, aromatic carbons, rotatable bonds, and amount of torsions were calculated and evaluated with Lipinski's rule. Silbinin and galangin showed the lowest energies, and these could be used in other studies for AD treatment.

Cyclin-dependent kinase 5 (CDK5) belongs to the serine/threonine cyclin-dependent kinase (CDK) family, which regulates neuronal activity, increases GABAergic neurotransmission, and impairs synaptic plasticity, and its inhibitors prevent the pathological phosphorylation of tau and neurofibrillary pathology [42–44]. Thus, CDK5 inhibition is considered as a new target for drugs against AD.

In a study by Shrestha et al. [45], six flavonoid compounds, sulfuretin, aureusidin, ureusidin-6-O-b-D-glucopyranoside, hovetrichoside C, quercetin-3-O-b-D-galactopyranoside, and cupressuflavone, isolated from of the indigenous medicinal shrub *Rhus parviflora* (Anacardiaceae) were revealed to be CDK inhibitors through non-radiometric in vitro CDK5/p25 inhibitory activity. The compounds were submitted to molecular docking. Three-dimensional structures of CDK5/p25 were obtained from the RCSB protein data bank, and derivative flavonoid compound structures were selected from the Pubchem compound database of the National Center for Biotechnology Information. All of the flavonoids showed good interactions with the receptor through hydrogen bonding with active-site residues at CYS83 and GLN130. However, aureusidin and auroras showed better energy value results.

Acetylcholinesterase inhibitors were the first category of drugs used to treat Alzheimer's disease. Some of the most used AChE inhibitors, like physostigmine and tacrine, had revealed specific problems during clinical studies such as problem with oral administration route and hepatotoxic liability. Studies focused on finding a new type of acetylcholinesterase inhibitors then discovered donepezil hydrochloride, which inaugurated a new class of AChE inhibitors with longer and more selective action and manageable adverse effects. In the work of Sugimoto and coauthors, CADD studies were applied including QSAR-3D (quantitative structure-activity relationship—in three dimensions), comparative molecular field analysis (CoMFA), and docking to the search for new inhibitors. QSAR modeling on the benzamine set suggests that the intrinsic activity of cis isomers of the compound is higher than for the trans isomers, since there is a negative correlation between IC_{50} and DE(C-T).

Currently, researchers use secondary metabolites: flavonoids, alkaloids, and xanthones as AChE inhibitors. An example is the work of Yang and co-workers, who synthesized a set of aporphine alkaloid analogs. The anti-acetylcholinesterase activity of the synthesized compounds was measured using Ellman's method (modified), as described in the work published in 2012. The investigative study, using molecular docking, found two highly active compounds (nuciferine derivatives): 1, 2-dihydroxyaporphine and dehydronuciferine which interacted as ligands with the AChE site [46, 47].

Changes in the cholinergic system are a major concern occurring in Alzheimer's disease. The biggest difference between the two cholinesterases is the substrate: AChE is found in the blood and neural synapses which hydrolyzes acetylcholine, while butyrylcholinesterase (BChE) is found in the liver which hydrolyzes butyrylcholine [48–50]. Acetylcholine can be degraded by two types of cholinesterases, and inhibition of both enzymes by a dual inhibitor should result in higher brain ACh levels and provide greater clinical efficacy. Research on AChE inhibitors includes drug candidates like tetrahydroaminoacridine (tacrine), physostigmine, velnacrine, rivastigmine, and donepezil [51, 52].

Li et al. [53] studied a set of 20 compounds (13a–u; structures not shown here) tacrine-flavonoid hybrids, searching for multifunctional compounds, with acetylcholinesterase inhibition, β-amyloid-reduction, and metal-chelating properties.

Amyloid plaques in brains are one of the major pathological hallmarks of AD [54]. The major protein component of the plaques is the amyloid β-peptide (Aβ), which is a peptide with 39–43 amino acids. Aβ deposition generates ROS, which are involved in Alzheimer's inflammatory and neurodegenerative pathology [55, 56]. A main component of amyloid plaques is the amyloid-β peptide (Aβ) generated by proteolytic cleavage of amyloid precursor protein (APP), by β-and γ-secretases. Many authors attribute cognitive loss to this accumulation of amyloid plaque, a characteristic of AD, but there is a theory that plaque formation is a consequence of metabolic decline in the brain of Alzheimer's patients [57]. Recent studies indicated that AChE could promote amyloid fibril formation by interaction through the PAS of AChE, giving stability to the AChE-Aβ complex, which are more toxic than single Aβ peptides alone.

Li et al. [53] studied the set compounds (13a–u) against both cholinesterases using a spectrophotometric method, with tacrine and galanthamine as reference compounds. Aβ (1–42) self-induced aggregation inhibition, kinetic ChE inhibition, and metal-chelating effects were studied, using curcumin as the reference. From these assays, compound 13 s was the most potent inhibitor of AChE; the authors then submitted it to molecular docking with *Tc*AChE id PDB: 2ckm. The flavonoid ring interacted with the indole ring of Trp 279 and Tyr70, through the π-π interactions with distances of

4.24 and 5.05 Å. Despite 13 s being a strong inhibitor, 13 k showed better multifunctional properties.

Research in multifunctional agents against AD is reported in many studies, several using natural products [58–60]. Lou et al. [61] synthetized and evaluated seventeen 4-dimethylamine flavonoid derivatives against these cholinesterases, Aβ aggregation, and reactive oxygen species. The authors investigated the ligands with two enzymes: AChE (code ID: 1ACJ) and human BChE (code ID: 1POI), from the Protein Data Bank. Also in this study, π-π interactions in the residue Trp 279 were observed. The results showed inhibitory potentials for both AChE and BChE, at levels similar to or better than those of rivastigmine.

Goyal et al. [62] studied 24 flavonoids through 3D quantitative structure-activity relationship (QSAR) and pharmacophore-based virtual screening to elucidate the structural features of flavonoid derivatives that contribute to the inhibitory activity of AChE. A QSAR-3D model was developed, and binding modes of action with the enzyme were then investigated using detailed docking studies. The best pharmacophore hypothesis had one hydrogen-bond acceptor, two hydrophobic regions, and two aromatic rings. From this study, four lead compounds were identified; one of these was found highly apt to inhibit the dual AChE-binding site.

QSAR models were built by Chakraborty et al. [63] to investigate the anti-amyloidogenic activity of polyphenols. Eighteen compounds were submitted to CODESSA for quantum chemical and thermodynamic calculations to generate constitutional, topological, geometrical, electrostatics, quantum chemical, and thermodynamic descriptors. Then a database of 200 phytochemicals was calculated using the best QSAR model. This approach is a way to screen for new leads.

The flavonoid Morin is reported for its anti-aggregation effect on Aß monomers and dimers [64] and for its inhibitory capacity on the enzyme acetylcholinesterase (AChE) [65]. Through molecular dynamic simulations, the hydrophobicity, aromaticity, and hydrogen-bonding capacity of morin were observed that might affect binding energies. Morin affects the tertiary and quaternary structure of Aβ. Like quercetin, myricetin, galangin, and fisetin, morin's AChE inhibition was investigated with molecular docking. Afterwards, Remya and collaborators [66] developed the structure-activity relationships, as shown in Fig. 2.

Silibinin, a flavonoid extracted from *Silybum marianum*, was studied to determine its potential as a dual inhibitor of acetylcholinesterase (AChE) and αβ peptide aggregation for AD treatment. The authors used molecular docking and molecular dynamics simulations to examine the affinity of silibinin with Aβ and AChE *in silico* [67]. Figure 3 shows the flavonoid binds to the active sites of AChE.

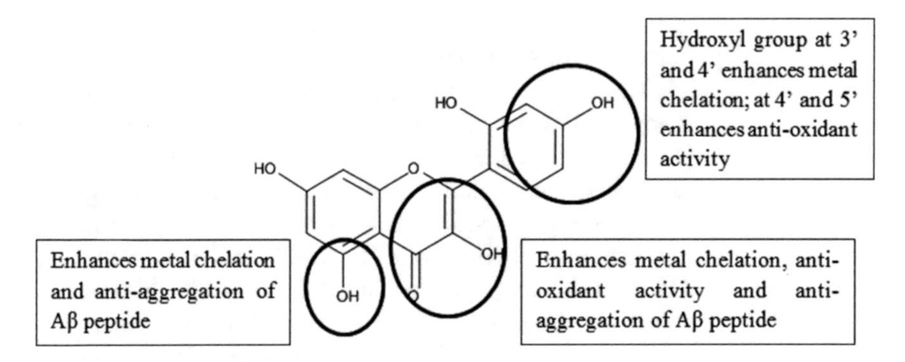

Fig. 2 Structure-activity relationships developed by Remya et al. [66]

3 Alkaloids

Alkaloids are nitrogen-containing heterocyclics typically isolated from several plant families. Most alkaloids are toxic (alkaloids include morphine, strychnine, quinine, ephedrine, and nicotine), but studies have reported biological actives [68], such as anti-HV [69], antimalarial [70], fungal infections treatment [71], and. in AD, inhibition of acetylcholinesterase [72–75].

In a review discussion by Konrath et al. [76], steroidal/triterpenoidal, quinolizidine, isoquinoline, and indole classes of Buxaceae, Amaryllidaceae, and Lycopodiaceae are considered important alkaloid inhibitors of AChE and BChE. Their structural diversity and physicochemical properties are reported, highlighting both structure-activity relationships (SARs) and docking studies. The authors show that one of the AChE-binding sites occurs which shows interaction of a positively charged nitrogen.

Another enzyme related to AD is retinoblastoma-associated protein (RbAp48), which (when deficient) provokes memory loss. Huang et al. [77] reported docking studies using bittersweet alkaloid II, eicosandioic acid, and Perivine (Fig. 4) from traditional Chinese medicine.

Makowska and co-workers [78] studied the affinity of trigonelline to the Aβ(1–42) peptide through hydrophilic interaction liquid chromatography and molecular docking. The authors observed that trigonelline interacted with the His6, His13, His14, and Tyr10 residues present in the Aβ(1–42) protein.

It was reported that the alkaloids 3-epimacronine and lycoramine present in *Zephyranthes carinata* herb show AChE and BChE inhibition and are interesting for potential treatment of AD. In a docking study, Cortes and collaborators [79] observed that 3-epimacronine derivatives were selective ligands for AChE. Analytical studies, using gas chromatography-mass spectrometry (GC-MS) and an assay measuring inhibition of AChE activity, indicated that the alkaloids produced by other species of the same botanical

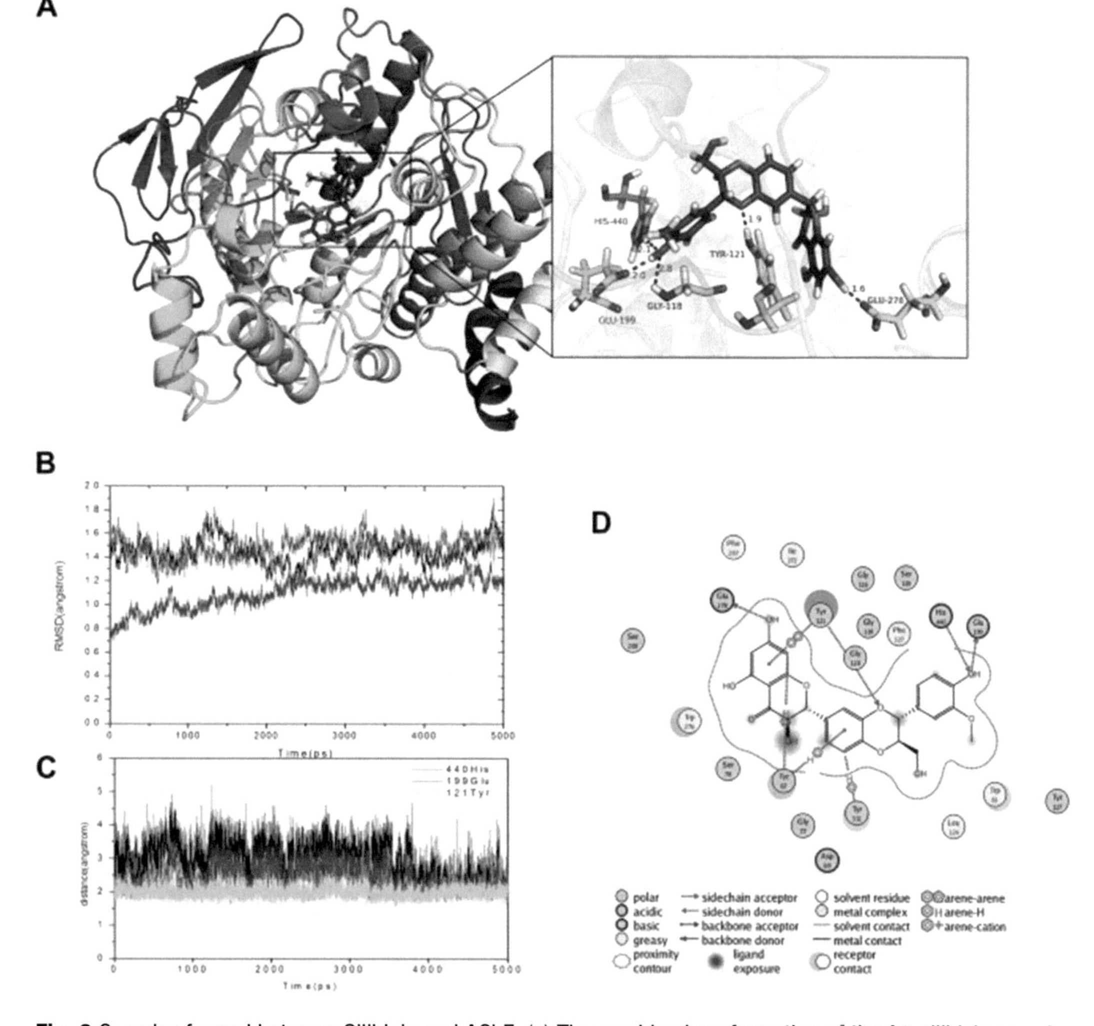

Fig. 3 Complex formed between Silibinin and AChE. (**a**) The combined conformation of the Aβ-silibinin complex shows that silibinin binds to the active sites of AChE. (**b**) The root mean square distance of the AChE-silibinin complex. The red, blue, and black are three repeating trajectories for the MD simulations. (**c**) The key hydrogen-bond distances between AChE and silibinin with time dependence. (**d**) The interactions between silibinin and AChE. The π-H and π-π interactions between silibinin and residues within the active sites of AChE are identified. Abbreviations: *Ab* amyloid beta, *AChE* acetylcholinesterase, *CAS* catalytic active site

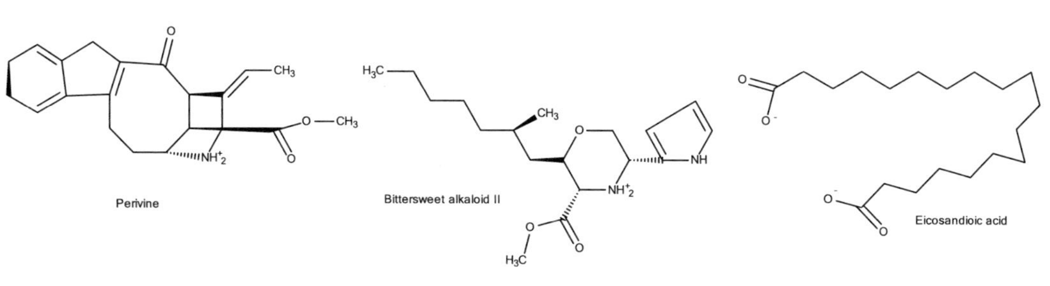

Perivine

Bittersweet alkaloid II

Eicosandioic acid

Fig. 4 Ligands tested with RbAp48 [77]

family (Amaryllidaceae family) also exhibit AChE and BChE inhibitory activities [80–83].

Holarrhena antidysenterica of the Apocynaceae family is widely used in traditional medicine, and its alkaloids show several activities, such as anticancer [84], antibacterial [85], antimalarial [86], and anti-gastrointestinal disorder [87]. Yang and co-workers [88] studied five alkaloids isolated from *H. antidysenterica:* conessine, isoconessimine, conessimin, conarrhimin, and conimin. The AChE inhibitory activity was tested in vitro, and using the results, the authors developed an SAR model. The compounds were submitted to docking with the enzyme. The inhibition of AChE increased with the number of methyl groups, and all ligands had interactions in the hydrophobic pocket of the protein.

The alkaloid physostigmine is an acetylcholinesterase inhibitor [89–93], but this natural product presents short half-life, variable bioavailability, and low clinical efficacy. Physostigmine analogs were studied by Ul-Haq et al. [94]. The compounds were selected from the literature observing their values of AChE inhibitor activity IC_{50} (9.77–20892.96 nM); the values of IC_{50} were converted to pIC_{50} ($-logIC_{50}$). The authors used thirty-two alkaloids for the training set and eight for the test set, and they were also divided into two groups: first being phenyl and alkyl chains and the second being ionizable nitrogen of morpholine moiety. The compound structures were energy minimized and then used to generated 3D-QSAR (PLS regression analysis), CoMSIA, and CoMFA models. Through these tools, coordinates of atoms, charges, physicochemical properties (steric, electrostatic, hydrophobic, and hydrogen-bond donor and acceptor), and steric and electrostatic potentials of the physostigmine analogs were evaluated. The models showed both high predictive powers and structure-activity correlations which facilitates and encourages the use of these compounds against AD.

In the study of Amat-ur-Rasool and Ahmed [94], heterodimers were designed as combinations of natural alkaloids berberastine and berberine with tacrine and pyrimidine rings that are active components of top synthetic leads. These heterodimers can be potential second-generation AD drugs in future that can be used as inhibitors of the human enzyme AChE. Figure 5 shows visual results of docking of best designed AD drug candidate, berberastine-5C-pyrimidine, into modeled hAChE.

4 Terpenoids

These compounds are formed from two main precursors, isopentenyl pyrophosphate (IPP) and dimethylallyl pyrophosphate (DMAPP), which are subclassified by their structure: monoterpenes (C10), sesquiterpenes (C15), diterpenes (C20), triterpenes

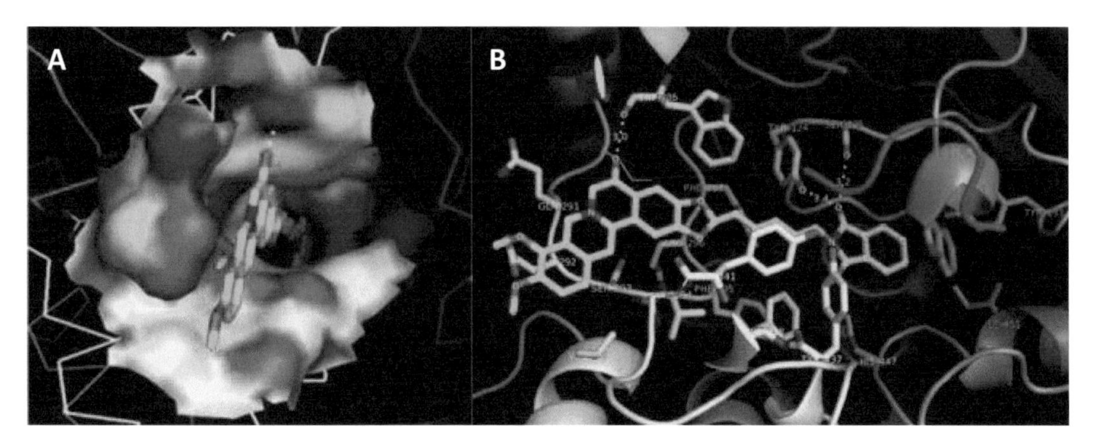

Fig. 5 (**a**) Ligand attachment conformation along the active-site groove shown by solid surface ligand site. (**b**) Ligand attached with different residues of catalytic site, represented by sticks, and colored according to the element type, and element name is labeled. Active-site residues are represented by sticks and enzyme by cartoons [94]

(C30), tetraterpenes (C40), and polyterpenes. Despite the diversity of classes, plenty of compounds and several beneficial effects are attributed to these natural products [95–98], including against AD [99–102]; little is needed in rational planning.

Bidon-Chanal and co-workers [103] reported isolation of the furano-sesquiterpene palinurin, a marine natural product. The authors observed its inhibitory effect on glycogen synthase kinase 3b (GSK-3b). The studied enzyme is a potential target to aid against Alzheimer's disease [104], neurodegenerative disorders [105], myocardial ischemia [106], and cancer [107]. Through molecular docking, the authors identified two pockets in the receptor, one at the C-terminal and one in the N-terminal lobe which is a potential allosteric binding site. The inhibitor activity of the compound occurs on site binding.

The first report of terpenoids as effective inhibitors of BACE1 was with two asperterpenes isolated from the culture broth of *Aspergillus terreus.* The compounds were investigated (Fig. 6) through molecular docking, with eight significant enzymes involved in AD processes (PSH, GlpG, KMO, amyloid fibril, APOE4, acetylcholinesterase, NMDA receptor, and BACE1) [108].

Targets with lower calculated binding energies are considered, and BACE1 showed higher binding affinity for the ligands than the other targets. Figure 7 shows the complexes formed with the asperterpenes.

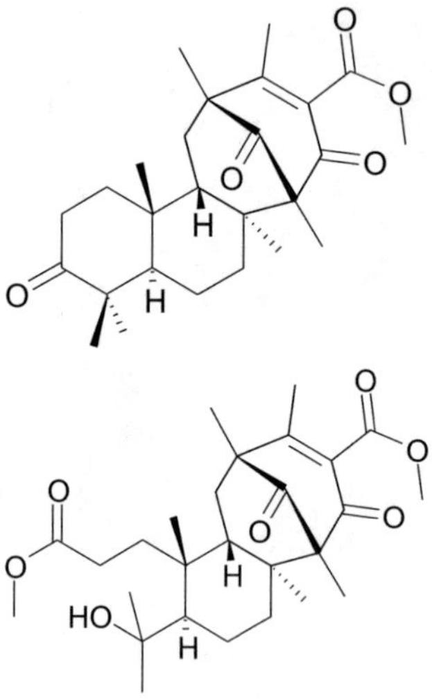

A

B

Fig. 6 (**A**, **B**) Asperterpenes

5 Conclusion

Alzheimer's affects about 80% of individuals aged 65 years or more; dementia only occurs in a small percentage of individuals at this age, yet the prevalence of dementia in Alzheimer increases to 25% in individuals aged 80 years.

In this chapter we reviewed studies of CADD using secondary metabolites in the search for new medicines [109] for the treatment of AD.

We emphasize the key compromises involved in these diseases and, consequently, the main focus of study:

- Decline in the cholinergic system, there is a decrease in the activity of the enzymes acetylcholinesterase and acetylcholine transferase, with decreased levels of acetylcholine.
- An accumulation of amyloid-beta (β) peptides, causing dementia.
- The oxidative damage caused by reactive oxygen species (ROS), such as superoxide anion, hydrogen peroxide radicals, and maneuvering.
- Unbalance of catecholamines.

The studies of CADD mostly use molecular docking. Generally, flavonoids, alkaloids, and xanthones are investigated as potential inhibitors of enzymes COX, AChE, BChE, MAO-A, and MAO-B, trying to reduce the imbalance of neurotransmitters and decrease inflammation. One can also find alkaloids complexed with β (1–42) peptides, inhibiting the aggregation.

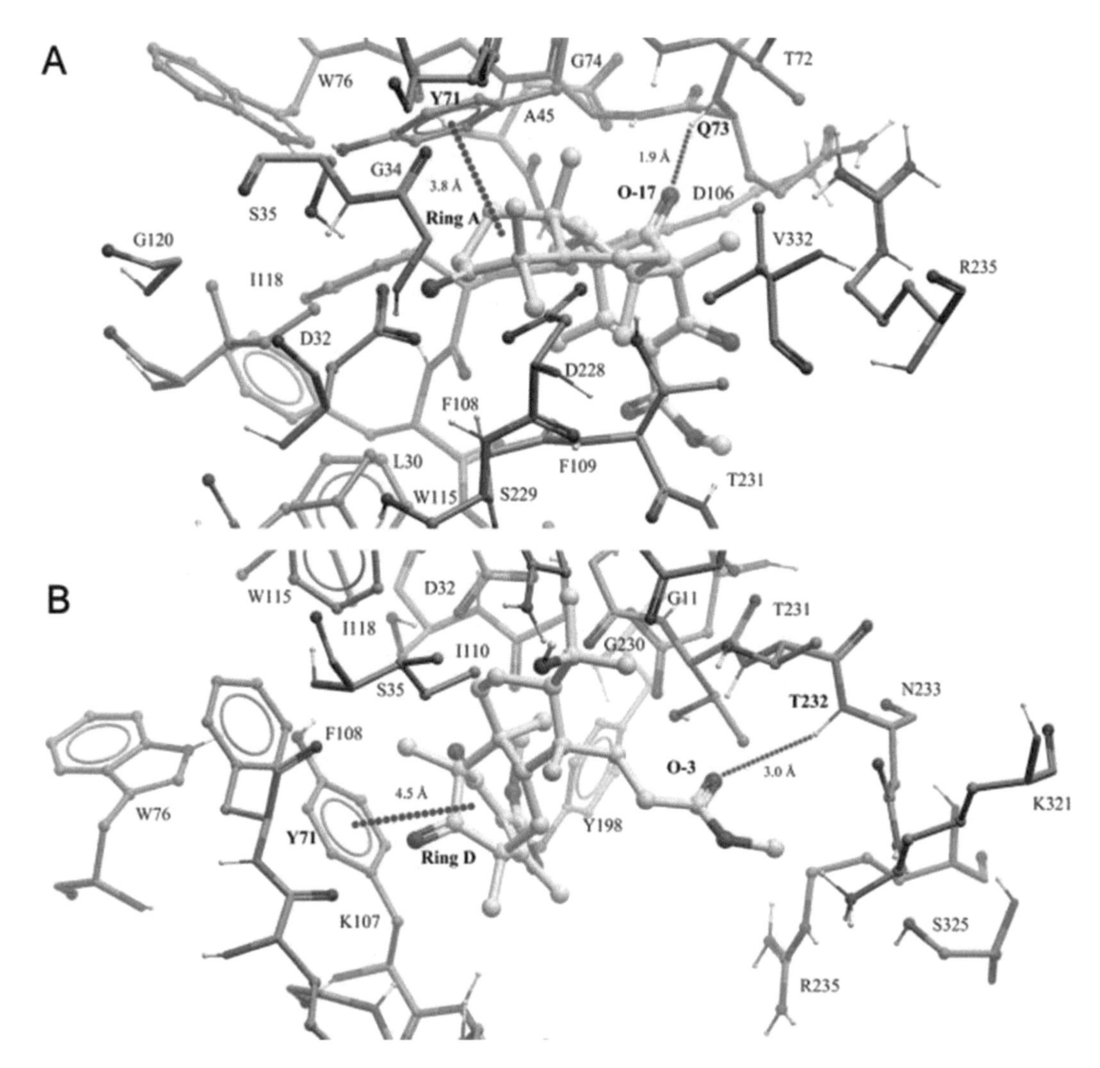

Fig. 7 The binding modes of compounds 1 (**a**) and 2 (**b**) modeled in silico with BACE1 were predicted in the previous experimental assays (red dashed lines represent hydrogen bonds, and blue dashed lines represent Π-Πinteractions) [103]

Acknowledgments

The authors wish to acknowledge the ConselhoNacional de Desenvolvimento Científico e Tecnológico (CNPq).

References

1. Clement JA (2014) Recent progress in medicinal natural products drug discovery. Curr Top Med Chem 14:2758
2. Wang BC, Deng J, Gao YM, Zhu LC, He R, Xu YQ (2011) The screening toolbox of bioactive substances from natural products: a review. Fitoterapia 82:1141–1151
3. Kennedy DA, Hart J, Seely D (2009) Cost effectiveness of natural health products: a systematic review of randomized clinical trials.

Evid-based Complement Altern Med 6:297–304

4. Leitao A, Montanari CA, Donnici CL (2000) The use of chemometric methods on combinatorial chemistry. Quim Nova 23:178–184
5. Sabuncu MR, Desikan RS, Sepulcre J, BTT Y, Liu H, Schhmansky NJ, Buckner RL, Sperling RA, Fischl B (2011) The dynamics of cortical and hippocampal atrophy in Alzheimer disease. Arch Neurol 68:1040–1048
6. Byun CJ, Seo J, Jo SA, Park YJ, Klug M, Rehli M, Park M-H, Jo I (2012) DNA methylation of the 50-untranslated region at +298 and +351 represses BACE1 expression in mouse BV-2 microglial cells. BiochemBioph Res Co 417:387–392
7. Hamdan AC (2008) Avaliação neuropsicológica na doença de Alzheimer e no comprometimento cognitivo leve. Psicol Argum 26:183–192
8. Dhikav V, Anand K (2011) Potential predictors of hippocampal atrophy in Alzheimer's disease. Drugs Aging 28:1–11
9. Pohanka M (2015) Biosensors containing acetylcholinesterase and butyrylcholinesterase as recognition tools for detection of various compounds. Chem Pap 69(1):4–16
10. Ferrer I (2012) Defining Alzheimer as a common age-related neurodegenerative process not inevitably leading to dementia. Prog-Neurobiol 97:38–51
11. Mecocci P, Polidori MC (2012) Antioxidant clinical trials in mild cognitive impairment and Alzheimer's disease. BiochimBiophActa 1822:631–638
12. Reitz C, Brayne C, Mayeux R (2011) Epidemiology of Alzheimer disease. Nat Rev Neurosci 7:137–152
13. Kumar A, Singh A, Ekavali (2015) A review on Alzheimer's disease pathophysiology and its management: an update. Pharmacol Rep 67(2):195–203
14. Gold BT (2015) Lifelong bilingualism and neural reserve against Alzheimer's disease: a review of findings and potential mechanisms. Behav Brain Res 281:9–15
15. Nussbaum RL, Ellis CE (2003) Alzheimer's disease and Parkinson's disease. N Engl J Med 348:1356–1364
16. Akagi M, Matsui N, Akae H, Hirashima N, Fukuishi N, Fukuyama Y, Akagi R (2015) Nonpeptide neurotrophic agents useful in the treatment of neurodegenerative diseases such as Alzheimer's disease. J Pharmacol Sci 127(2):155–163
17. Sarvaiya J, Agrawal YK (2015) Chitosan as a suitable nanocarrier material for anti-Alzheimer drug delivery. Int J BiolMacromol 72:454–465
18. Catala-Lopez F, Tabares-Seisdedos R (2015) Alzheimer's disease and cancer: the need of putting research into context with previous published systematic reviews. J Cancer Res Clin Oncol 141(3):569–570
19. Choi DY, Choi H (2015) Natural products from marine organisms with neuroprotective activity in the experimental models of Alzheimer's disease, Parkinson's disease and ischemic brain stroke: their molecular targets and action mechanisms. Arch Pharm Res 38 (2):139–170
20. Khan I, Samad A, Khan AZ, Habtemariam S, Badshah A, Abdullah SM, Ullah N, Khan A, Zia-Ul-Haq M (2013) Molecular interactions of 4-acetoxy-plakinamine B with peripheral anionic and other catalytic subsites of the aromatic gorge of acetylcholinesterase: computational and structural insights. Pharm Biol 51 (6):722–727
21. Richard T, Papastamoulis Y, Waffo-Teguo P, Monti JP (2013) 3D NMR structure of a complex between the amyloid beta peptide (1-40) and the polyphenol epsilon-viniferin glucoside: implications in Alzheimer's disease. Biochim Biophys Acta-Gen Subj 1830 (11):5068–5074
22. Crichton GE, Bryan J, Murphy KJ (2013) Dietary antioxidants, cognitive function and dementia - a systematic review. Plant Food Hum Nutr 68(3):279–292
23. Devore EE, Grodstein F, van Rooij FJA, Hofman A, Stampfer MJ, Witteman JCM, Breteler MMB (2010) Dietary antioxidants and long-term Risk of dementia. Arch Neurol 67 (7):819–825
24. Rodacka A, Strumillo J, Serafin E, Puchala M (2014) Effect of resveratrol and Tiron on the inactivation of glyceraldehyde-3-phosphate dehydrogenase induced by superoxide anion radical. Curr Med Chem 21(8):1061–1069
25. Lakey-Beitia J, Berrocal R, Rao KS, Durant AA (2015) Polyphenols as therapeutic molecules in Alzheimer's disease through modulating amyloid pathways. Mol Neurobiol 51 (2):466–479
26. Ramesh BN, Indi SS, Rao KSJ (2010) Anti-amyloidogenic property of leaf aqueous extract of Caesalpinia Crista. Neurosci Lett 475(2):110–114
27. Berrocal R, Vasudevaraju P, Indi SS, Rao K, Rao KS (2014) In vitro evidence that an aqueous extract of Centella Asiatica modulates alpha-Synuclein aggregation dynamics. J Alzheimers Dis 39(2):457–465

28. Perez-Jimenez J, Neveu V, Vos F, Scalbert A (2010) Systematic analysis of the content of 502 polyphenols in 452 foods and beverages: an application of the phenol-explorer database. J Agric Food Chem 58(8):4959–4969
29. Mora-Pale M, Sanchez-Rodriguez SP, Linhardt RJ, Dordick JS, Koffas MAG (2013) Metabolic engineering and in vitro biosynthesis of phytochemicals and non-natural analogues. Plant Sci 210:10–24
30. Wang Y, Xia Z, Xu JR, Wang YX, Hou LN, Qiu Y, Chen HZ (2012) α-Mangostin, a polyphenolic xanthone derivative from mangosteen, attenuates β-amyloid oligomers-induced neurotoxicity by inhibiting amyloid aggregation. Neuropharmacology 62(2):871–881
31. Morelli CF, Biagiotti M, Pappalardo VM, Rabuffetti M, Speranz G (2015) Chemistry of alpha-mangostin. Studies on the semisynthesis of minor xanthones from Garcinia Mangostana. Nat Prod Res 29(8):750–755
32. Ibrahim MY, Hashim NM, Mohan S, Abdulla MA, Abdelwahab SI, Arbab IA, Yahayu M, Ali LZ, Ishag OE (2015) α-Mangostin from Cratoxylum arborescens: an in vitro and in vivo toxicological evaluation. Arab J Chem 8 (1):129–137
33. Chaijaroenkul W, Na-Bangchang K (2015) The in vitro antimalarial interaction of 9-hydroxycalabaxanthone and alpha-mangostin with mefloquine/artesunate. Acta Parasitol 60(1):105–111
34. Buravlev EV, Shevchenko OG, Kutchin AV (2015) Synthesis and membrane-protective activity of novel derivatives of alpha-mangostin at the C-4 position. Bioorg Med Chem Lett 25(4):826–829
35. Koh JJ, Lin SM, Aung TT, Lim F, Zou HX, Bai Y, Li JG, Lin HF, Pang LM, Koh WL, Salleh SM, Lakshminarayanan R, Zhou L, Qiu SX, Pervushin K, Verma C, Tan DTH, Cao DR, Liu SP, Beuerman RW (2015) Amino acid modified Xanthone derivatives: novel, highly promising membrane-active antimicrobials for multidrug-resistant gram-positive bacterial infections. J Med Chem 58 (2):739–752
36. Choi M, Kim YM, Lee S, Chin YW, Lee C (2014) Mangosteen xanthones suppress hepatitis C virus genome replication. Virus Genes 49(2):208–222
37. Ahmad MZ, Ahmad J, Amin S, Rahman M, Anwar M, Mallick N, Ahmad FJ, Rahman Z, Kamal MA, Akhter S (2014) Role of Nanomedicines in delivery of anti-acetylcholinesterase compounds to the brain in Alzheimer's disease. CNS Neurol Disord-Drug Targets 13(8):1315–1324
38. Moretti G, Mosc L (2013) Mixture useful for treatment of diseases e.g. inflammatory, diabetic, cardiac, neurodegenerative, atherosclerotic pathologies, viral, metabolic, cardiovascular diseases comprises resveratrol; and carboxymethylglucan. EP2674155-A1; IT1412486-B, EP2674155-A1 18 Dec 2013 A61K-031/05 201404
39. Kumar P, Choonara YE, Modi G, Naidoo D, Pillay V (2014) Cur(Que)min: a neuroactive permutation of Curcumin and Quercetin for treating spinal cord injury. Med Hypotheses 82(4):437–441
40. Shen BJ, Truong J, Helliwell R, Govindaraghavan S, Sucher NJ (2013) An in vitro study of neuroprotective properties of traditional Chinese herbal medicines thought to promote healthy ageing and longevity. BMC Complement Altern Med 13:8
41. Madeswaran A, Umamaheswari M, Asokkumar K, Sivashanmugam T, Subhadradevi V, Jagannath P (2013) Computational drug discovery of potential TAU protein kinase I inhibitors using in silico docking studies. J Pharmacol 8(2):131–135
42. Odemuyiwa SO, Ilarraza R, Davoine F, Logan MR, Shayeganpour A, Wu YQ, Majaesic C, Adamko DJ, Moqbel R, Lacy P (2015) Cyclin-dependent kinase 5 regulates degranulation in human eosinophils. Immunology 144(4):641–648
43. Rouget R, Sharma G, LeBlanc AC (2015) Cyclin-dependent kinase 5 phosphorylation of familial prion protein mutants exacerbates conversion into amyloid structure. J Biol Chem 290(9):5759–5771
44. Rudenko A, Seo J, Hu J, Su SC, de Anda FC, Durak O, Ericsson M, Carlen M, Tsai LH (2015) Loss of Cyclin-dependent kinase 5 from Parvalbumin interneurons leads to Hyperinhibition, decreased anxiety, and memory impairment. J Neurosci 35 (6):2372–2383
45. Shrestha S, Natarajan S, Park JH, Lee DY, Cho JG, Kim GS, Jeon YJ, Yeon SW, Yang DC, Baek NI (2013) Potential neuroprotective flavonoid-based inhibitors of CDK5/p25 from Rhus Parviflora. Bioorg Med Chem Lett 23(18):5150–5154
46. Yang Z, Song Z, Xue W, Sheng J, Shu Z, Shi Y, Lang J, Yao X (2014) Synthesis and structure–activity relationship of nuciferine derivatives as potential acetylcholinesterase inhibitors. Med Chem Res 23:3178–3186
47. Yang ZD, Zhang X, Du J, Ma ZJ, Guo F, Li S, Yao XJ (2012) An aporphine alkaloid from Nelumbo Nucifera as an acetylcholinesterase inhibitor and the primary investigation for

structure-activity correlations. Nat Prod Res 26(5):387–392
48. Chatonnet A, Lockridge O (1989) Comparison of butyrylcholinesterase and acetylcholinesterase. Biochem J 260:625–634
49. Gao D, Zhan C-G (2005) Modeling effects of oxyanion hole on the Ester hydrolysis catalyzed by human Cholinesterases. J PhysChem 109(B):23070–23076
50. Bishara D, Harwood D (2014) Safe prescribing of physical health medication in patients with dementia. Int J Geriatr Psychiatry 29 (12):1230–1241
51. Singh DB, Gupta MK, Kesharwani RK, Sagar M, Dwivedi S, Misra K (2014) Molecular drug targets and therapies for Alzheimer's disease. Transl Neurosci 5(3):203–217
52. Wong KY, Mercader AG, Saavedra LM, Honarparvar B, Romanelli GP, Duchowicz PR (2014) QSAR analysis on tacrine-related acetylcholinesterase inhibitors. J Biomed Sci 21:8
53. Li SY, Wang XB, Xie SS, Jiang N, Wang KDG, Yao HQ, Sun HB, Kong LY (2013) Multifunctional tacrine flavonoid hybrids with cholinergic, beta-amyloid-reducing, and metal chelating properties for the treatment of Alzheimer's disease. Eur J Med Chem 69:632–646
54. Qu J, Zhou Q, Du Y, Zhang W, Bai M, Zhang Z, Xi Y, Li ZY, Miao JT (2014) Rutin protects against cognitive deficits and brain damage in rats with chronic cerebral hypoperfusion. Br J Pharmacol 171(15):3702–3715
55. Grosso C, Valentao P, Ferreres F, Andrade PB (2013) The use of flavonoids in central nervous system disorders. Curr Med Chem 20 (37):4694–4719
56. Henry MS, Passmore AP, Todd S, McGuinness B, Craig D, Johnston JA (2013) The development of effective biomarkers for Alzheimer's disease: a review. Int J Geriatr Psychopharmacol 28:331–340
57. Holmes C (2013) Review: systemic inflammation and Alzheimer's disease. Neuropath ApplNeuro 39:51–68
58. Hyde C, Peters J, Bond M et al (2013) Evolution of the evidence on the effectiveness and cost-effectiveness of acetylcholinesterase inhibitors and memantine for Alzheimer's disease: systematic review and economic model. Age Ageing 42:14–20
59. Anand P, Singh B (2013) A review on cholinesterase inhibitors for Alzheimer's disease. Arch Pharm Res 36:375–399
60. Stuble RG, Ala T, Patrylo PR, Brewer GJ, Yan X-X (2010) Is brain amyloid production a cause or a result of dementia of the Alzheimer type? J Alzheimers Dis 22:393–399
61. Luo W, Su YB, Hong C, Tian RG, Su LP, Wang YQ, Li Y, Yue JJ, Wang CJ (2013) Design, synthesis and evaluation of novel 4-dimethylamine flavonoid derivatives as potential multi-functional anti-Alzheimer agents. Bioorg Med Chem 21(23):7275–7282
62. Goyal M, Grover S, Dhanjal JK, Goyal S, Tyagi C, Grover A (2014) Molecular modelling studies on flavonoid derivatives as dual site inhibitors of human acetyl cholinesterase using 3D-QSAR, pharmacophore and high throughput screening approaches. Med Chem Res 23(4):2122–2132
63. Chakraborty S, Basu S (2014) Insight into the anti-amyloidogenic activity of polyphenols and its application in virtual screening of phytochemical database. Med Chem Res 23 (12):5141–5148
64. Lemkul JA, Bevan DR (2012) Morin inhibits the early stages of amyloid beta-peptide aggregation by altering tertiary and quaternary interactions to produce off-pathway structures. Biochemist 51(30):5990–6009
65. Lemkul JA, Bevan DR (2010) Destabilizing Alzheimer's a beta(42) Protofibrils with Morin: mechanistic insights from molecular dynamics simulations. Biochemist 49 (18):3935–3946
66. Remya C, Dileep KV, Tintu I, Variyar EJ, Sadasivan C (2012) Design of potent inhibitors of acetylcholinesterase using morin as the starting compound. Front Life Sci 6 (3–4):107–117
67. Duan S, Guan X, Lin R et al (2015) Silibinin inhibits acetylcholinesterase activity and amyloid β peptide aggregation: a dual-target drug for the treatment of Alzheimer's disease. Neurobiol Aging 36:1792–1807
68. Boratynski PJ (2015) Dimeric cinchona alkaloids. Mol Divers 19(2):385–422
69. Kong DG, Zhao Y, Li GH, Chen BJ, Wang XN, Zhou HL, Lou HX, Ren DM, Shen T (2015) The genus Litsea in traditional Chinese medicine: an ethnomedical, phytochemical and pharmacological review. J Ethnopharmacol 164:256–264
70. Vale VV, Vilhena TC, Trindade RCS, Ferreira MRC, Percario S, Soares LF, Pereira WLA, Brandao GC, Oliveira AB, Dolabela MF, De Vasconcelos F (2015) Anti-malarial activity and toxicity assessment of Himatanthus articulatus, a plant used to treat malaria in the Braz Amazon. Malar J 14:132. doi:10.1186/s12936-015-0643-1

71. Larsson S, Ronsted N (2014) Reviewing Colchicaceae alkaloids - perspectives of evolution on medicinal chemistry. Curr Top Med Chem 14(2):274–289
72. Jiang WW, Su J, Wu XD, He J, Peng LY, Cheng X, Zhao QS (2015) Geissoschizine methyl ether N- oxide, a new alkaloid with antiacetylcholinesterase activity from Uncaria rhynchophylla. Nat Prod Res 29(9):842–847
73. Liu W, Shi XY, Yang YD, Cheng XM, Liu Q, Han H, Yang BH, He CY, Wang YL, Jiang B, Wang ZT, Wang CH (2015) In vitro and in vivo metabolism and inhibitory activities of Vasicine, a potent acetylcholinesterase and Butyrylcholinesterase inhibitor. PLoS One 10 (4):35
74. Liew SY, Khaw KY, Murugaiyah V, Looi CY, Wong YL, Mustafa MR, Litaudon M, Awang K (2015) Natural indole butyrylcholinesterase inhibitors from Nauclea officinalis. Phytomedicine 22(1):45–48
75. Shaikh S, Zainab T, Shakil S, Rizvi SMD (2015) A neuroinformatics study to compare inhibition efficiency of three natural ligands (Fawcettimine, Cernuine and Lycodine) against human brain acetylcholinesterase. Netw-Comput Neural Syst 26(1):25–34
76. Konrath EL, Passos CD, Klein LC, Henriques AT (2013) Alkaloids as a source of potential anticholinesterase inhibitors for the treatment of Alzheimer's disease. J Pharm Pharmacol 65 (12):1701–1725
77. Huang HJ, Lee CC, Chen CYC (2014) Lead discovery for Alzheimer's disease related target protein RbAp48 from traditional Chinese medicine. Biomed Res Int 2014:764946
78. Makowska J, Szczesny D, Lichucka A, Gieldon A, Chmurzynski L, Kaliszan R (2014) Preliminary studies on trigonelline as potential anti-Alzheimer disease agent: determination by hydrophilic interaction liquid chromatography and modeling of interactions with beta-amyloid. J Chromatogr B 968:101–104
79. Cortes N, Alvarez R, Osorio EH, Alzate F, Berkov S, Osorio E (2015) Alkaloid metabolite profiles by GC/MS and acetylcholinesterase inhibitory activities with binding-mode predictions of five Amaryllidaceae plants. J Pharm Biomed Anal 102:222–228
80. Cahlikova L, Valterova I, Macakova K, Opletal L (2011) Analysis of Amaryllidaceae alkaloids from Zephyranthes Grandiflora by GC/MS and their cholinesterase activity. Revista Brasileira De Farmacognosia-Brazilian J Pharmacog 21(4):575–580
81. Kulhankova A, Cahlikova L, Novak Z, Macakova K, Kunes J, Opletal L (2013) Alkaloids from Zephyranthes Robusta baker and their acetylcholinesterase- and Butyrylcholinesterase-inhibitory activity. Chem Biodivers 10 (6):1120–1127
82. Larsen MM, Adsersen A, Davis AP, Lledo MD, Jager AK, Ronsted N (2010) Using a phylogenetic approach to selection of target plants in drug discovery of acetylcholinesterase inhibiting alkaloids in Amaryllidaceae tribe Galantheae. Biochem Syst Ecol 38 (5):1026–1034
83. Jahn S, Seiwert B, Kretzing S, Abraham G, Regenthal R, Karst U (2012) Metabolic studies of the amaryllidaceous alkaloids galantamine and lycorine based on electrochemical simulation in addition to in vivo and in vitro models. Anal ChimActa 756:60–72
84. Sharma V, Hussain S, Bakshi M, Bhat N, Saxena AK (2014) In vitro cytotoxic activity of leaves extracts of Holarrhena Antidysenterica against some human cancer cell lines. Indian J Biochem Biophys 51(1):46–51
85. Chusri S, Na-Phatthalung P, Siriyong T, Paosen S, Voravuthikunchai SP (2014) Holarrhena Antidysenterica as a resistance modifying agent against Acinetobacter Baumannii: its effects on bacterial outer membrane permeability and efflux pumps. Microbiol Res 169(5–6):417–424
86. Dua VK, Verma G, Singh B, Rajan A, Bagai U, Agarwal DD, Gupta NC, Kumar S, Rastogi A (2013) Anti-malarial property of steroidal alkaloid conessine isolated from the bark of Holarrhena Antidysenterica. Malar J 12:6
87. Kadir MF, Bin Sayeed MS, Mia MMK (2013) Ethnopharmacological survey of medicinal plants used by traditional healers in Bangladesh for gastrointestinal disorders. J Ethnopharmacol 147(1):148–156
88. Yang ZD, Duan DZ, Xue WW, Yao XJ, Li S (2012) Steroidal alkaloids from Holarrhena Antidysenterica as acetylcholinesterase inhibitors and the investigation for structure-activity relationships. Life Sci 90 (23–24):929–933
89. Plaschke K, Muller AK, Kopitz J (2014) Surgery-induced changes in rat IL-1 beta and acetylcholine metabolism: role of physostigmine. Clin Exp Pharmacol Physiol 41 (9):663–670
90. Killi UK, Wsol V, Soukup O, Kuca K, Winder M, Tobin G (2014) In vitro functional interactions of acetylcholine esterase inhibitors and muscarinic receptor antagonists in the urinary

bladder of the rat. Clin Exp Pharmacol Physiol 41(2):139–146

91. Yokota SI, Nakamura K, Ando M, Kamei H, Hakuno F, Takahashi SI, Shibata S (2014) Acetylcholinesterase (AChE) inhibition aggravates fasting-induced triglyceride accumulation in the mouse liver. FEBS Open Bio 4:905–914
92. Shaikh S, Verma A, Siddiqui S, Ahmad SS, Rizvi SMD, Shakil S, Biswas D, Singh D, Siddiqui MH, Shakil S, Tabrez S, Kamal MA (2014) Current acetylcholinesterase-inhibitors: a Neuroinformatics perspective. CNS Neurol Disord-Drug Targets 13 (3):391–401
93. Ul-Haq Z, Mahmood U, Jehangir B (2009) Ligand-based 3D-QSAR studies of Physostigmine analogues as acetylcholinesterase inhibitors. ChemBiol Drug Des 74(6):571–581
94. Amat-ur-Rasool H, Ahmed M (2015) Designing second generation anti-Alzheimer compounds as inhibitors of human acetylcholinesterase: computational screening of synthetic molecules and dietary phytochemicals. PLoS One 10(9):e0136509
95. Scotti L, Scotti MT, Ishiki H, Junior F, dos Santos PF, Tavares JF, da Silva MS (2014) Prediction of anticancer activity of Diterpenes isolated from the Paraiban Flora through a PLS model and molecular surfaces. Nat Prod Commun 9(5):609–612
96. Scotti L, Tavares JF, da Silva MS, Falcao EV, de Morais e Silva L, da Silva Soares GC, Scotti MT (2012) Chemotaxonomy of three genera of the Annonaceae Family using self-organizing maps and C-13 Nmr data of Diterpenes. Quim Nova 35(11):2146–2152
97. Scotti MT, Fernandes MB, Ferreira MJP, Emerenciano VP (2007) Quantitative structure-activity relationship of sesquiterpene lactones with cytotoxic activity. Bioorg Med Chem 15(8):2927–2934
98. Scotti MT, Emerenciano V, Ferreira MJP, Scotti L, Stefani R, da Silva MS, Mendonca FJB (2012) Self-organizing maps of molecular descriptors for Sesquiterpene lactones and their application to the chemotaxonomy of the Asteraceae Family. Molecules 17 (4):4684–4702
99. Galipoglu M, Erdal MS, Gungor S (2015) Biopolymer-based transdermal films of donepezil as an alternative delivery approach in Alzheimer's disease treatment. AAPS PharmSciTech 16(2):284–292
100. Guo QQ, Ma XJ, Wei SG, Qiu DY, Wilson IW, Wu P, Tang Q, Liu LJ, Dong SK, Zu W (2014) De novo transcriptome sequencing and digital gene expression analysis predict biosynthetic pathway of rhynchophylline and isorhynchophylline from Uncaria rhynchophylla, a non-model plant with potent anti-alzheimer's properties. BMC Genomics 15:16
101. Xie HY, Wang JR, Yau LF, Liu Y, Liu L, Han QB, Zhao ZZ, Jiang ZH (2014) Quantitative analysis of the flavonoid glycosides and Terpene Trilactones in the extract of Ginkgo Biloba and evaluation of their inhibitory activity towards fibril formation of beta-amyloid peptide. Molecules 19(4):4466–4478
102. Yin Y, Zhang Y, Li P, Ma H, Zhang X, Pan G, Wu D, Shen B (2014) Pharmaceutical composition useful for treating Alzheimer's syndrome, comprises Lycium barbarum polysaccharide-III, medlar polysaccharide-IV, resveratrol, terpene-3-beta-alcohol, Astragalus polysaccharide A2 and motherwort saponin A. CN103751207-A, CN103751207-A 30 Apr 2014 A61K-031/715 201448
103. Bidon-Chanal A, Fuertes A, Alonso D, Perez DI, Martinez A, Luque FJ, Medina M (2013) Evidence for a new binding mode to GSK-3: allosteric regulation by the marine compound palinurin. Eur J Med Chem 60:479–489
104. Sivaprakasam P, Han X, Civiello RL, Jacutin-Porte S, Kish K, Pokross M, Lewis HA, Ahmed N, Szapiel N, Newitt JA, Baldwin ET, Xiao H, Krause CM, Park H, Nophsker M, Lippy JS, Burton CR, Langley DR, Macor JE, Dubowchik GM (2015) Discovery of new acylaminopyridines as GSK-3 inhibitors by a structure guided in-depth exploration of chemical space around a pyrrolopyridinone core. Bioorg Med Chem Lett 25 (9):1856–1863
105. Ye Q, Mao WL, Zhou YB, Xu L, Li Q, Gao YX, Wang J, Li CH, Xu YZ, Xu Y, Liao H, Zhang LY, Gao JR, Li J, Pang T (2015) Synthesis and biological evaluation of 3-(1,2,4 triazolo 4,3-a pyridin-3-yl)-4-(indol-3-yl)-maleimides as potent, selective GSK-3 beta inhibitors and neuroprotective agents. Bioorg Med Chem 23(5):1179–1188
106. Kalakech H, Hibert P, Prunier-Mirebeau D, Tamareille S, Letournel F, Macchi L, Pinet F, Furber A, Prunier F (2014) RISK and SAFE signaling pathway involvement in Apolipoprotein A-I-induced Cardioprotection. PLoS One 9(9):7
107. Feng H, Yu Z, Tian Y, Lee YY, Li MS, Go MYY, Cheung YS, Lai PBS, Chan AML, To KF, Chan HLY, Sung JJY, Cheng ASL (2015) A CCRK-EZH2 epigenetic circuitry drives hepatocarcinogenesis and associates with

tumor recurrence and poor survival of patients. J Hepatol 62(5):1100–1111

108. Qi C, Bao J, Zhu H, Xue Y et al (2016) Asperterpenes a and B, two unprecedented meroterpenoids from Aspergillus terreus with BACE1 inhibitory activities. Chem Sci 7:6563–6572

109. Scotti L, Ishiki H, Mendonca FJB, Santos PF, Tavares JF, Silva MS, Scotti MT (2014) Theoretical research into anticancer activity of Diterpenes isolated from the Paraiban Flora. Nat Prod Commun 9(7):911 914

Chapter 19

Computational Modeling of Multi-target-Directed Inhibitors Against Alzheimer's Disease

Akhil Kumar and Ashok Sharma

Abstract

Alzheimer's disease is a neurodegenerative disorder mostly occurring in the elderly. The socioeconomic impact and death rate due to AD are alarming. Thus, one target-one ligand hypothesis which is successful in drug discovery may not be suitable for multifactorial diseases like AD. Recently, researchers have successfully identified dual- or multi-target inhibitors which halt multiple disease-causing pathways and improve the disease conditions. Computational methods such as virtual screening, docking, QSAR, molecular dynamics, etc., are helpful tools to design and identify new MTDL entities. We have described various computational methods to screen and identify top hits and molecular dynamics to ensure the affinity in terms of binding free energy of the receptor ligand complex to design multi-target-directed ligands for Alzheimer's disease.

Key words Alzheimer's disease (AD), Neuritic plaques (SNP), Neurofibrillary tangles (NFT), β-amyloid protein (Aβ), Glycogen synthase kinase-3β (GSK-3β), β-secretase (BACE-1), Multi-target-directed ligands (MTDLs), Structural-based virtual screening (SBVS), Ligand-based virtual screening (LBVS), Quantitative structure-activity relationship (QSAR), Absorption distribution metabolism, Excretion and toxicology (ADME/T), Computer-aided drug design (CADD), Blood-brain barrier (BBB), Molecular dynamics (MD), Root-mean-square deviation (RMSD), Radius of gyration (Rg), Root-mean-square fluctuations (RMSF), Solvent accessible surface area (SASA)

1 Introduction

1.1 Alzheimer's Disease: Introduction, Prevalence, and Biological Pathways

The clinical and neuropathological characteristics of the Alzheimer's disease (AD) were first presented by Alois Alzheimer on 3 November 1906 [1]. The most common symptoms of AD are loss of memory, language, and problem-solving ability, and it further affects a person's ability to perform day-to-day activities [2]. The data showed that AD is the sixth leading cause of death in the United States. It was estimated that since 2000, deaths from heart disease have decreased by 14%, while deaths from AD have increased by 89% [3]. From the economical point of view, it was estimated that in 2017, Alzheimer's and other dementias cost about $259 billion, and by 2050, these costs could rise as high as

Kunal Roy (ed.), *Computational Modeling of Drugs Against Alzheimer's Disease*, Neuromethods, vol. 132, DOI 10.1007/978-1-4939-7404-7_19, © Springer Science+Business Media LLC 2018

$ 1.1 trillion [3]. Such an alarming finding indicates the severity of AD and a need for a solution to prevent Alzheimer's and other related dementias.

Although the research till now has uncovered a great deal about Alzheimer's diseases, many issues related to the disease are still unclear. Biological changes that trigger AD, why it progresses more quickly in some people, and how the disease can be prevented or stopped completely are some of the major concerns [3]. The advancement of AD etiology and pathogenesis has made a significant progress. The major pathologies of AD are the accumulation of the abnormal peptide fragments that include senile neuritic plaques (SNPs) and neurofibrillary tangles (NFTs) [4]. SNPs are insoluble aggregates of β-amyloid peptide (Aβ) deposited outside the neurons, and NFTs are paired helical filaments of hyperphosphorylated tau protein inside the neurons [4]. These changes eventually lead to the death of neural cell [4]. Recent neurodegenerative disease research suggests that AD occurs due to multiple factors such as genetic, environmental, endogenous factors, etc. Moreover, now AD is considered as a multifactorial disease, i.e., more than one pathway are responsible for the disease condition. In AD patients' brain, one can observe protein aggregation, oxidative stress due to free radical formation, metal dyshomeostasis, mitochondrial dysfunction, and tau hyperphosphorylation occurring at the same time [5]. Various hypotheses, viz., amyloid [6], cholinergic [7, 8], glutamatergic [9], oxidative stress [10], metal [11], and inflammatory [12], have been forwarded to explain the underlying causes of disease progression, but none of them alone is sufficient to explain the root cause of AD. Among all the hypotheses, cholinergic hypothesis is the oldest. According to this hypothesis, decrease in acetylcholine concentration leads to periodic forgetfulness in AD patients [7]. Most of the AD drugs are based on this hypothesis, which inhibit the acetylcholinesterase, e.g., donepezil (Aricept, Eisai/Pfizer), galantamine (Razadyne, Johnson & Johnson), and rivastigmine (Exelon, Novartis) [6]. The effects of these drugs are limited as they improve only the symptoms and not the main causes of the disease. Now it is clear that cholinergic dysfunction may not cause cognitive impairment directly but is involved indirectly in disease progression [7].

After the cholinergic hypothesis, a popular amyloid hypothesis was proposed. According to the amyloid hypothesis, Aβ deposition and aggregation are the key players in initial disease conditions [6, 13]. Aβ is generated by the cleavage of β-secretase enzyme [14], and it is a short peptide with variable length. Aβ mostly exists in two forms, Aβ-40 and Aβ-42, which are 40 and 42 amino acid residues long [15]. In healthy individuals, 90% of the total Aβ is Aβ-40, whereas in AD, the ratio of Aβ42:Aβ40 is elevated, leading to increased amyloid aggregation and toxicity [16]. Aβ is neurotoxic, and it induces stress and inflammatory response which ultimately lead to brain cell death [17]. This deposition and aggregation trigger disease conditions and toxicity over the time. Anti-Aβ therapy majorly involves the small molecules that selectively bind to Aβ, inhibit Aβ aggregate, clear the already generated Aβ, reduce in Aβ generation, and destabilize the preformed Aβ fibrils [14, 18, 19]. Various molecules have been earlier reported to inhibit or destabilize preformed fibril [20–24].

Another hallmark in AD patient's brain is NFTs. NFT generation is directly linked to glycogen synthase kinase-3β (GSK-3β) also called as tau phosphorylating kinase. This hypothesis proposes that tau protein is hyperphosphorylated by the GSK-3β [25]. After hyperphosphorylated tau begins to pair with each other, it ultimately forms neurofibrillary tangles inside nerve cells. This results into microtubule disintegration, destroying the structure of the cell's cytoskeleton and other abnormalities leading to brain cell death [26, 27]. Apart from cholinergic, amyloid and tau hypothesis, neurovascular hypothesis, poor functioning of blood-brain barrier [28], oxidative stress [10, 29, 30], metal homeostasis [11, 31, 32], and environmental risk [33] are other hypotheses explaining the AD pathologies. The cholinergic hypothesis and popular amyloid hypothesis-related drug targets are classical drug targets, whereas other drug targets such as β-, γ-, and α-secretases, sirtuin-2 (SIRT2), caspases, and glycogen synthase kinase-3 targets are related to nonclassical drug targets.

1.2 Therapeutics of AD

Although AD is known from last 100 years, only four FDA-approved cholinesterase inhibitors are reported so far. Currently, various therapeutic strategies are suggested and used for the treatment of AD. Few important therapies involve modulating neurotransmission, calcium homeostasis, intracellular signaling cascade, reduction of oxidative stress, and mitochondrial-targeted and anti-inflammatory-based therapy. Amyloid-based strategies and tau-based therapy are important and well studied. They involve secretase enzyme modulation and amyloid transport and clearance, preventing amyloid aggregation and immunotherapy-based molecule against amyloid. Tau-based therapy majorly focuses on tau phosphorylation inhibition and microtubule stabilization, inhibiting tau oligomerization, anti-tau aggregating molecules, and immunotherapy. Various other therapeutic strategies are discussed in "Therapeutics of Alzheimer's disease: Past, present, and future" [34]. Apart for this hypothesis, a new concept known as multi-target-directed ligands (MTDL) has been reported for AD [35]. MTDLs were designed specifically to act on multiple therapeutic targets for the treatment of multifactorial diseases like AD and to provide symptoms as well as disease-modifying-based benefits.

1.3 The Rise of Multi-target Drugs

Most of the drug failures occur since they target a single disease-causing pathway with the assumption of "one drug-one target" hypothesis. This hypothesis might not always affect complex systems, even if a drug completely inhibited or changed the behavior of the target. Thus, a more suitable approach has recently emerged to fight terminal and multifactorial diseases like cancer and AD which is known as a polypharmacological approach [36]. Polypharmacology covers both multiple drugs that act independently on different targets and a single molecule binding to multiple targets of disease-causing pathways [36, 37]. Primarily it involves a combination of two to three drugs given simultaneously for different targets. This approach had various advantages like (1) simple and immediate action; (2) effective in various diseases like cancer, HIV infection, etc.; (3) drug synergies; and (4) lower doses of each compound can be used, resulting in better therapeutic selectivity in preventing drug resistance. However, it also has significant limitations and challenges such as ensuring pharmacokinetic and dose compatibility as well as drug-drug interactions and off-target side effects for multiple molecules [36]. Another major issue was making a combination of one, two, or more drugs in a single pill [36]. MTDLs on the other hand are successful for cancer and depression and have convinced that modulating multiple pathways can be helpful in treating complex disorders like AD [38]. In the case of AD, a major drawback is that it is an age-related disease occurring mostly in older people. Older people are mostly on medication for other diseases; thus the chances of drug-drug interaction increase. The mean number of drugs taken daily in multiple drug approaches has been estimated at 4.0 in men and 4.7 in women [38, 39]. Thus, the polypharmacology concept of multiple drugs shifted toward the multi-target approach. Moreover, single multi-target molecule is simpler than the development of new combination therapies [36].

In medicinal chemistry research, various chemical moieties are known to act as multi-target molecules. Evans and colleagues described the benzodiazepine moiety as MTDL agent, which binds efficiently at different proteins [40]. Many drugs were rediscovered as multi-target drugs which lead to the concept of polypharmacology from single molecules [41]. However, unplanned interactions with other targets may cause side effect [42]. Regardless of the theoretical promises delivered by MTDLs, the logical and practical implementation of designing multi-target drugs with predefined biological properties can be difficult. The major challenges are identification of biologically relevant validated combinations of drug targets which are related to disease condition, hit identification, and optimization stages for multiple parameters. Another major difficulty is in lead optimization; one has to balance multiple activities while retaining drug-like properties and also controlling other off-target side effects. Although MTDLs represent a new field, it has grown remarkably over the last decade. Every year new agents are reaching the clinical trial and market specifically for cancer [43] and depression [44].

1.4 Single-Target Versus Multi-target Approach in AD

During the past two decades, drug discovery research followed the lock and key model which mainly focused on the selective targeting, i.e., single target for particular disease. It involved identification of single selective target responsible for the disease pathway and one ligand which interacts or binds with the target to modulate its activity. This approach has been widely applied and successful in the past and will be used in the future also. Various successful stories were reported for the computer-aided drug designing against AD [45]. The scope of single target-one molecule concept is very simple to treat for the complex diseases like AD, cancer, HIV, etc. In multifactorial diseases like AD, drugs hitting a single target may be inadequate because multiple pathways are involved in the disease [46]. The accumulating evidences suggested that one molecule hitting multiple targets provides better treatment and effective strategy [47]. The MTDL approach has already been proven successful in the treatment of similarly complex diseases such as cancer, HIV, and hypertension [46]. One of the studies on neurodegenerative diseases suggested the MTDLs are better for neurodegenerative disorder, and these are valuable tools for hitting the multiple targets implicated in AD. Various earlier compounds under clinical trials such as curcumin and other polyphenols appear to be useful in AD not only because of their dual function as anti-inflammatory and antioxidant agents, but they can inhibit other AD drug targets which support the MTDL strategy against AD. Thus, polypharmacology approach is suitable for AD drug discovery [48–50]. In polypharmacology, major disadvantages include off-target side effects which can lead to adverse effects [50, 51]. This off-target binding of known drug is also related to polypharmacology, and the approach to identify off target is called as drug repurposing [52, 53]. Computational methods are now validated and play a major role in drug repurposing such as network pharmacology, machine learning techniques, and chemogenomics [54, 55]. One of the state-of-the-art studies described the computational method to identify multi-target ligands when structural and sequence homologies are not present [56].

1.5 Hurdles in the CNS (Central Nervous System) Drug Discovery

CNS drug discovery is very important and challenging as most of the CNS drugs tend to fail late in clinical development phase. Only 8% of CNS drugs ever make it to clinical trials, i.e., roughly 1/2 the rate of drugs in other fields [57, 58]. In one of the study, it was estimated that probability of success for a new CNS molecule is low about 2.85% [59], expensive, and time-consuming. Some of the major drawbacks include insufficient knowledge of nervous system disorders and the underlying biological mechanism of the disease.

The average rate of successful translation from animal models to clinical trials is less due to poor pharmacokinetics and pharmacology properties, absorption, distribution, metabolism, excretion, and toxicity (ADME/T) of ligand, blood-brain barrier (BBB) permeability, selectivity of potent ligands, etc.

Computer-aided drug design (CADD) is a widely used method by the industry and academics in the drug discovery and designing projects [60]. Some of the challenges can be overcome by CADD. The process of drug discovery, design, and development is complex and costly. By applying CADD, one may reduce the time and cost at the primary step and also predict the toxicity and optimal activity for the ligand, which reduces the chance of clinical failure.

2 Materials Required for Drug Designing Against AD Using Computational Methods

2.1 Target or Receptor Identification and Their 3D Structures

In the beginning of the drug discovery R&D process, it is mandatory to know whether a chosen biological target is druggable or not. Druggability can be defined as the likelihood of being able to modulate a target with a small-molecule drug [61]. It was reported that only 10% of the human genome represents druggable targets in which only half is being relevant to diseases [61]. Druggable targets and druggability of the particular novel identified target can also be computationally predicted [61–64]. In the case of AD, various druggable targets and their inhibitors have been identified [65]. Major AD drug targets are broadly classified under two categories: (1) drug targets for symptomatic-based therapies (classical drug targets) and (2) drug targets for disease-modifying-based therapies (nonclassical drug targets). Some important and most studied targets against AD are listed below (Table 1).

2.2 Large Diverse Compound Libraries for Screening Against Therapeutic Drug Targets

Lead structures are small chemical molecules that are able to bind with the chosen therapeutic drug targets and produce desired biological effects. Small-molecule drugs maintain their action by binding to specific therapeutic targets to modulate its action in a disease-modifying manner. The good binding depends on the shape, polarity, and chemical requirement between the ligand and receptor. The primary success of small molecule as lead hits depends on the size matches between most biologically relevant binding sites and ligands. One of the studies suggested that the total number of possible small molecules varies from 10^8 to 10^{200} depending upon the criteria [66]. Thus, chemical space is huge, and screening of all "chemical space" is practically impossible [58]. Millions of compounds are currently available for screening in public or private databases. The screening of all available databases is time-consuming, expensive, and biologically irrelevant, because for a particular target, only a small subregion is biologically active or a little chemical space has drug-like properties [67]. Thus,

Table 1
List of important therapeutic strategies and related biological targets

Symptomatic-based drug target	Disease-modifying-based drug target	
Acetylcholinesterase inhibition	Reduction in Aβ generation	Alpha secretase activator
NMDA receptor modulation		BACE-1 inhibition
Nicotinic acetylcholine receptor activation		Gamma secretase inhibition and modulation
Gamma-aminobutyric acid receptor blockade	Degradation of preformed Aβ	Neprilysin enhancement Insulin-degrading enzyme activator
Serotonin receptor activation and blockade	Aβ clearance	Immunization (vaccination and passive) Receptor-mediated removal of Aβ Prevent entry from periphery
Histamine H3 receptor blockade	Reduction in Aβ-induced toxicity	Anti-aggregation agent binds with Aβ Inhibition of oligomerization
Phosphodiesterase inhibition	Tau	Inhibition of tau aggregation Block tau hyperphosphorylation Microtubule stabilization

computational methods add advantage over traditional methods as they are fast and cost-effective. Apart from synthetic derivatives and analogues, natural compounds also have vast diversity and have new scaffolds which are more complex in nature. A large number of natural molecules have been reported for various drug targets including AD [68]. Coumarins, flavonoids, stilbenes, naturally occurring alkaloids, and other natural products are also important sources of AChE inhibitors [69]. Various shikimate-derived, polyketides, terpenoids, and alkaloid natural molecules are reported as secretase inhibitors [70]. Some of the natural molecules promoting anti-aggregation and clearances were tested into various clinical stages. *Scyllo*-cyclohexanehexol [71], colostrinin [72], and polyketides such as tetracycline and minocycline are reported for antiamylogenic properties [70, 73]. Natural molecules have an extra advantage over synthetic derivatives and produce positive effects in AD patients as a large number of natural molecules have antioxidant or anti-inflammatory or both properties and they bind effectively with one of the AD drug targets. Thus, a MTDL approach that has antioxidant property in combination with ability to hit other targets might be a more rational approach for dementia treatment [70]. Several earlier reports suggested the use of natural moiety against AD. Till date, plant-based AChE discovery programs showed great success, as two of the four currently approved drugs for the treatment of AD are plant derived. Natural molecules

Table 2
Major synthetic and natural libraries for screening against various drug targets

	Name	Database size	Characteristics
1	The CAS registry	125 million	Unique organic and inorganic chemical substances
2	PubChem [74]	Compounds 82.6 million entries	Characterized chemical compounds
		Substances 198 million entries	Contains mixtures, extracts, complexes, and uncharacterized substances
		Bioassay results 1.1 million	High-throughput screening programs data with bioactivity
3	ChemSpider [75]	58 million structures	Providing fast text and structure search
4	ZINC*12* [76] ZINC*15* [77]	35 million compounds	Ready-to-dock format and pursuable compounds
		100 million compounds	Different categories and with many filter criteria
5	Traditional Chinese Medicine Database [78]	37,170 compound entries	Natural molecule based on traditional Chinese plants for virtual screening
6	Super natural II [79]	325,508 natural compounds	Natural product information regarding mechanism of action with respect to structurally similar drugs and their target proteins
7	AfroDb [80]	954 compound entries	Compounds from African medicinal plants
8	TIPdb [81]	Approx. 9500 natural molecules	Database of anticancer, antiplatelet, and antituberculosis phytochemicals from indigenous plants in Taiwan
9	UNDP [82]	19,7201	The largest noncommercial and freely available database for natural products of 3D structure from Chinese traditional medicinal herbs
10	NuBBE [83]	640	Natural product database from the biodiversity of Brazil
11	KNApSAcK-3D [84]	–	3D Structure database of plant metabolites
12	3DMET [85]	–	3D Structure database of natural metabolites

may be also suitable as MTDL candidates because they have extra advantage. Thus, screening of natural molecules against nonclassical or disease-modifying AD drug target may result to a better therapeutic lead. Some of the most cited ligand libraries for the screening against various drug targets have been listed below (Table 2).

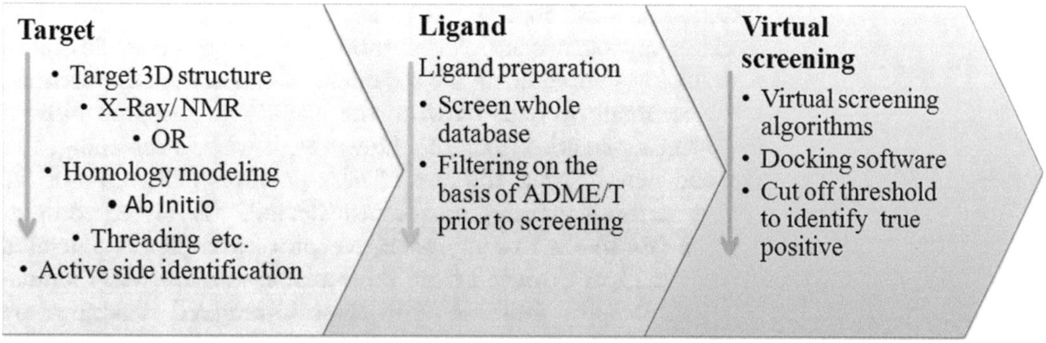

Target
- Target 3D structure
  - X-Ray/NMR
    - OR
- Homology modeling
  - Ab Initio
  - Threading etc.
- Active side identification

Ligand
Ligand preparation
- Screen whole database
- Filtering on the basis of ADME/T prior to screening

Virtual screening
- Virtual screening algorithms
- Docking software
- Cut off threshold to identify true positive

Fig. 1 Major steps in drug discovery project from target identification to virtual screening

2.3 Virtual Screening and Docking Software: Algorithms and Scoring Functions

2.3.1 Virtual Screening

Due to money and resource limitations for performing biological tests of a large number of commercially available compounds, a fast and reliable approach is required. Thus, computer-based method called virtual screening (VS) has been developed. The VS approach is fast enough to screen large databases containing millions of compounds with reasonable accuracy. VS is basically used to find out chemical compounds from large databases, which are predicted to bind well with a given receptor or target structure with reduced off-target effects. In the past, such receptor-based virtual screening faced several fundamental challenges [86]. However, during recent years, it proved to be very promising. In several cases, the hit rates (ligands discovered per molecules tested) are high than that with high-throughput screening. New benchmark methods for structure-based virtual screening have been developed, and standard docking screening process has been discussed [87, 88]. A recent review discusses benchmarking data sets for the evaluation of virtual ligand screening methods [89]. The basic difference is that in docking calculations, we look for maximum accuracy as compared to virtual screening in which we prefer fast calculations with limited accuracy. The virtual screening algorithms must be extremely fast because a large number of molecules are being screened. It is necessary to use techniques taking very small time in analyzing each molecule. Various steps are required for virtual screening (Fig. 1). The first step of VS is to identify the biological therapeutic drug targets. The molecular structure can be obtained from the protein database or can be generated using computational methods such as homology modeling. Further, we identify the active site or binding site of the drug target, and with the help of docking software, virtual screening is performed.

Structure-based virtual screening (SBVS) and ligand-based virtual screening (LBVS) are two main methods of virtual screening. An unexplored aspect of SBVS is the flexibility of target protein structure [90]. To deal with flexibility, all the conformational structures of the target protein are clustered, and all molecules are screened against representative structures of each cluster. Soft docking approach is another method to deal with target flexibility. Soft docking implements the softening of van der Waals potentials allowing small overlaps between the ligand and receptor without large steric penalties [91, 92]. However, in virtual screening, this method may increase the rate of false positives [93]. One of the recent methods focusing on receptor flexibility is relaxed complex scheme (RCS). In this approach, receptor ensemble is generated from structures extracted from molecular dynamics (MD) simulations of receptor molecule, and these ensembled structures are screened against ligand libraries [94]. The next step is to screen chemical libraries in databases against particular drug target.

2.3.2 Molecular Docking

Molecular docking simulations may be used for protein interaction, protein-ligand interaction, DNA and RNA protein interaction, ligand pose prediction into the active site, intermolecular interactions with the target that stabilizes the ligand-receptor complex, etc. [95]. Molecular docking is mostly used to predict the conformation of ligand into the active site of the receptor. In the drug discovery process, the receptor is usually a biological macromolecule like protein, nucleic acid, etc., and the ligand is a small molecule. Furthermore, molecular docking algorithms predict binding energy between protein-ligand complexes and rank the generated conformations of the ligand.

The main components of docking are ligand conformation generation. This conformational change can be further divided into two aspects, ligand sampling and protein flexibility. Another component in docking is scoring that is prediction of the binding for individual ligand/conformations into the active site with a physical or empirical energy function [96]. Protein flexibility is crucial in molecular docking for better accuracy in pose prediction. The problem of protein flexibility can be fixed by several methods. Ligand binding commonly induces protein conformational changes that assist the binding. Large size and many degrees of freedom of proteins and ligand are the major issues in molecular docking [96]. To deal with ligand and protein flexibility, major methods are soft docking, side-chain flexibility, molecular relaxation, and protein ensemble docking [97–99].

Soft docking is the simplest and relatively old technique which accounts for only small conformational changes and considers protein flexibility. The soft docking methods are computationally efficient and easy. The molecular relaxation method uses rigid-body docking and is used to place the ligand into the binding site and then relaxing the protein backbone and side-chain atoms nearby. Then, the formed complexes are relaxed or minimized by molecular dynamics simulations [84, 88]. However, the relaxation method is computationally expensive and time-consuming because it involves side-chain as well as backbone flexibility, and inaccurate scoring function may lead to artifacts in the relaxed protein conformations [96]. Docking of multiple protein structures or ensemble docking (ED) utilizes an ensemble of protein structures representing the different possible conformations of the targets. In addition to experimental methods such as NMR and X-ray, crystal structures of proteins with different ligands and ensembles of protein conformations can also be generated by molecular dynamics simulations and normal mode analysis with other methods. Selection of the ideal crystal structure of a receptor target is the most crucial first step in the computational drug discovery process. However, 3D X-ray-derived crystal structures provide a single conformation that does not provide any information about protein's active or inactive conformations, and often conformation of the target is influenced by the crystallization conditions. If the crystal structure of the target is bound with the ligand within the active site, then the ligand may induce a significant change in the active site. Three-dimensional crystal structures are required for SBDD. But in some cases, 3D structure provides misleading information. Thus, multiple starting structures are preferable because a clear picture of protein flexibility can be understood and multiple representative structures of target may be used for docking. ED comprises docking a single ligand library against multiple rigid receptor conformations, but it takes a longer time [101].

2.3.3 Scoring Functions and Search Algorithms

A good docking software must rapidly and accurately evaluate the protein-ligand complex interaction energy by generating a large number of ligand conformations and calculating energy for each conformation. Thus, to evaluate the protein-ligand interactions, different scoring functions are applied to rank each conformation. In general, scoring function evaluates each ligand conformation on the basis of protein-ligand interaction energies and gives the best optimal conformation in the active site of the protein. Different types of scoring function are reported, viz., (1) force field scoring functions, (2) empirical scoring functions, and (3) knowledge-based scoring functions. Protein-ligand docking is computationally difficult because a ligand can adopt different conformations into the active site. If the ligand has a large number of rotatable bonds and the protein active site residues are flexible, then the number of possibilities grows exponentially. Docking searching algorithms try to find out the most stable state (global minimum) [102, 103]. For the search of optimal conformation in docking, two different approaches are followed: (1) a full search space and (2) a gradual guided progression through solution space. The second method searches only part of the solution space in a partially random and partially guided manner, e.g., simulated annealing, molecular dynamics (MD), and evolutionary algorithms. Various available software use different search algorithms (Table 3). Different docking tools with their strengths and weaknesses useful for drug discovery project have been listed (Table 4). However, high-rank

Table 3
Software and their search algorithms

Software	Search algorithm
AutoDock4.2 [104]	Genetic algorithm
DOCK6 [105]	Incremental construction
ZDOCK [106]	Fast shape matching
MS-DOCK [107]	Fast shape matching
MCDOCK [108]	Monte Carlo simulations
ICM [109]	Monte Carlo simulations
GOLD [110]	Genetic algorithm
Surflex [111]	Incremental construction
FLEXX-PHARM [112]	Incremental construction
EUDOC [113]	Fast shape matching
FLOG [114]	Incremental construction

Table 4
Docking tools with their strengths and weaknesses useful for drug discovery project

Name	Strengths	Weaknesses
AutoDock [104]	Small flexible and hydrophobic molecules Fast	Large number of rotatable bond
DOCK [105]	Suitable for small binding sites and small hydrophobic molecules	Flexible ligands Highly polar ligands
GOLD [115]	Suitable for small binding sites and small hydrophobic molecules	Ranking very polar ligands Ranking ligands in large cavities
Surflex [111]	Suitable for large and small opened cavities Deal with large rotatable bond (flexible ligand)	Low speed for large ligands
FLEXX [112]	Suitable for small binding sites and small hydrophobic molecules	Very flexible ligands
GLIDE [116]	Deal with flexible ligands and small hydrophobic ligands	Ranking very polar ligand low speed
SLIDE [117]	Deal with side-chain flexibility	Sensitivity to ligand input coordinates
IFRED [118]	Large binding sites Flexible and small hydrophobic ligands Fast	Small, polar, buried ligands

conformations may have false positives, and thus predicted top conformations need to be cross-checked by other parameters. Various search algorithms and scoring functions in docking experiment have been discussed by several researchers [90, 91, 103].

2.4 Ligand-Based Virtual Screening via Pharmacophore Models

Pharmacophore can be defined as "the ensemble of steric and electronic features that is necessary to ensure the optimal supramolecular interaction with a specific biological target structure and to trigger or to block its biological response" [119]. In general, pharmacophore is a 3D arrangement of functional groups of a ligand that interacts with the target and produces desirable biological response. The pharmacophore also provides information about noncovalent interaction and interatomic distance between these functional groups. Pharmacophore is a ligand-based method. Pharmacophore model can be built by overlaid multiple active molecules in such a manner that maximum number of functional groups overlap [120]. Pharmacophore superpositioning and pattern matching, history, and development have been widely discussed [120–124]. In brief, the common features used to define pharmacophore maps are positive and negative charge, aliphatic and aromatic hydrophobic moieties, and hydrogen bond acceptors and donors. Some common pharmacophore software packages and their algorithms are listed (Table 5).

2.4.1 Quantitative Structure-Activity Relationship (QSAR) Models

QSARs are mathematical regression or classification models used to predict the activity of a new chemical. In QSAR modeling for drug discovery, physicochemical or theoretical molecular descriptors of ligands are used to correlate the biological activity of the same ligands. That model is further used to predict activity of new chemical entities. The general workflow of QSAR includes (i) selection of the data set and calculation of descriptors, (ii) variable selection, (iii) model construction, and (iv) model validation. QSAR models can be categorized into three types: (1) fragment based (group contribution), (2) 3D-QSAR (like comparative molecular field analysis), and (3) chemical descriptor based. Fragment-based QSAR includes various molecular fragments of known inhibitors in relation to the variation in biological activity. 3D-QSAR models are built on the 3D properties of the ligands such as the steric group and the electrostatic fields which are correlated by means of partial least squares regression. The third approach is different from other methods: in descriptor-based method, the descriptors are computed for the system as a whole rather than from the properties of individual fragments. Compared to the 3D-QSAR approach, the descriptors are computed from scalar quantities rather than from 3D fields. Various software packages are available for QSAR model generation, e.g., R, libSVM, Orange, RapidMiner, Weka, KNIME, and Tanagra. These packages are mostly based on various algorithms such as random forest,

Table 5
Pharmacophore software packages and unique features

No.	Name	Features
1	Phase [125]	Tree-based partitioning algorithm is used for alignment and feature extraction, considers the volume of heavy atom overlap, and also handles molecular flexibility
2	MOE	Computational scaffold replacement techniques
3	Catalyst [126]	Alignment and feature extraction are identified by common chemical features arranged in 3D space
4	LigandScout [127]	Virtual screening based on 3D chemical features of pharmacophore models
5	DISCO [128]	Alignments are based on the spatial orientation of common features among all active compounds

k-nearest neighbors algorithm, support vector machine, artificial neural network, decision tree algorithm, etc. QSAR models were prepared for various AD drug targets like Aβ, BACE-1 AChE, BChE, GSK-3β, M1 receptor, and $5HT_6R$ being discussed in great detail [129]. In one study, QSAR method was used to explore the structural requirements of imaging agents against Aβ plaques [130]. In another study, QSAR study of piperidine scaffolds suggested the structural requirements of AChE inhibition [131]. As AD drug discovery is shifting toward MTDL approach, various in silico strategies have been discussed to build multi-QSAR model [132].

2.5 Molecular Dynamics Simulation Software for Protein-Ligand Complex Stability

In MD, we consider the flexibility of the receptor as well as ligand molecules in the presence of a solvent, and the charges are neutralized by adding counter ions. In MD of protein-ligand docked complex, ligand binding may cause subtle changes in the side-chain rotation or may lead to large deviation in the backbone structure of the protein [133]. Various conformational snapshots of different poses during the production runs are saved in MD trajectories. All conformational changes in the protein and ligand can be assessed using the molecular dynamics trajectories. These MD trajectories are usually generated in nanosecond scale (usually 10–100 ns), containing multiple snapshots used for conformational investigations and binding free-energy calculations. In MD simulations, all the forces acting on each atom of a biomolecular system are calculated using Newtonian physics. The energy contribution for every atom includes bonded interactions and nonbonded interactions. The nonbonded interactions include electrostatic interactions calculated using Coulomb's law and van der Waals interactions using Lennard Jones Potentials (L-J potential). The bonded interaction incorporates bond distances, bond angles,

Table 6
List of molecular dynamics simulation software packages

S. No	Packages	Link	Supported force fields
1	CHARMM	www.charmm.org	CHARMM (E/I, AA/UA), Amber
2	Amber	amber.scripps.edu	Amber (E/I, AA)
3	GROMOS	www.igc.ethz.ch/GROMOS	Gromos (E/vacuum, UA)
4	Gromacs	www.gromacs.org	Amber, Gromos, OPLS (all E)
5	NAMD	www.ks.uiuc.edu/Research/namd	CHARMM, Amber, Gromos
6	Desmond	https://www.deshawresearch.com/resources_whatdesmonddoes.html	CHARMM, Amber, OPLS

E explicit solvent, *I* implicit solvent, *AA* all atom, *UA* united atom (apolar H omitted)

dihedral angles, and improper angles, which are calculated using classical physical equations. The chemical bonds and bond angles are represented as virtual springs, and dihedral angles are modeled as sinusoidal functions. All energy terms are optimized to fit the quantum-mechanical calculations to reproduce the real behavior of atomic motions of molecules. Some of the simulation packages with supported force fields are listed below (Table 6).

3 Methods

Computational tools play an important role in drug discovery process and are extensively and successfully used in AD drug discovery process. CADD is applied to fulfill three purposes: (1) screening of large databases against therapeutic targets and to predict the activity of top hits to reduce the experiment cost and time; (2) ADME/T predictions, where most of the drug candidate fails under clinical trials (in AD drug discovery, blood-brain barrier permeability is also a major hurdle, and BBB permeability prediction is a very important aspect); and (3) fragment-based drug discovery in which two moieties can join together to create novel molecules with better properties and affinity. Some important computational methods used to design anti-Alzheimer leads are discussed below.

3.1 Ligand Preparations for Virtual Screenings

Computational virtual screening methods are very fast and effective, and they yield nearly 35% hit rate as compared to 0.021% with HTS [99, 100]. Ligand preparation is a critical and important step for the drug discovery processes. Virtual screening performed to sort potential actives from inactive molecules on the basis of score is called enrichment. Due to computational cost and large screening of millions of compounds, we consider the protein as a rigid body and ligand as a flexible entity. However, if we add some receptor flexibility or active site residue side-chain mobility, results may improve further. This may be done through explicit sampling or docking to an ensemble of receptor conformations [134–138]. The preparation process involves adding hydrogens, bond orders, and formal charges to the starting protein and ligand molecules. Further ligand geometry and energy minimization may be done after the generation of three-dimensional (3D) coordinates for the ligands. Ionization/tautomeric state generation is an important step in structure-based virtual screening (SBVS). One of the state-of-the-art studies suggested the impact of ligand preparation on virtual screening against β-secretase (BACE-1), a potential drug target in AD drug discovery [139]. Several researchers have discussed ligand preparation and virtual screening strategies [93, 140–142]. The primary goal of SBVS is to identify as many as possible diverse hits and not the most potent molecules. SBVS has some limitations as well [134], viz., the role of solvent is not considered in virtual screening, while the water molecule has a prominent role in hydrogen bond interaction. The criteria for selecting the compounds after VS are ambiguous. Most of the fundamental errors may occur at the first step of VS when converting molecular structure to another format because various software have different protocols which are non-standardized. Another serious concern in VS is that alternation in atomic coordinates, chirality, hybridization, and partial charges of the data sets are rarely used in virtual screening. Good coverage of conformational space is also important in VS. Similarly, we have to remove high-energy conformations, which will otherwise be present in the positive hits. Flexibility of ligands is considered in VS, but target flexibility is probably the most unexplored aspect of VS. It should be considered while doing SBVS.

3.1.1 Filtering Criteria Before VS Data Set Preparation

VS of huge compound libraries such as PubChem, ZINC, and ChemSpider is time-consuming and computationally demanding. Therefore, we need to reduce the search space which is often less for drug-like compounds and biologically relevant for particular drug targets. Screening on the basis of ADME/T is a suitable filter to reduce the search space. This is done so that we can only screen the molecules which have drug-like properties. But applying ADME/T earlier in the screening process, one may not get the new chemical scaffold which is more potent.

3.2 Protein Preparation for Virtual Screening

Protein structure preparation is a crucial step for effective virtual screening. The crucial role of protonation states at Asp dyad (Asp32 and Asp228) to identify BACE-1 inhibitors in virtual screening against AD has been dismissed [143]. Various authors have also discussed various steps in protein preparation in the drug discovery process [101, 118, 140]. Few important steps are (1) selecting the 3D structure of protein for docking and further identifying the relevant active site (3D structure of protein derived from X-ray method is preferable with good resolution); (2) checking for missing connectivity information, bond orders, format errors, atoms, and partial charges/protonation states of important active site residues; (3) adding hydrogen atoms and their optimization; (4) creating disulfide bonds and bonds to metals; (5) removing all water molecules; (6) applying a restrained protein energy minimization step; (8) receptor flexibility should be taken into consideration by ensemble docking; and (9) final quality check needs to be performed.

3.3 Binding Site Identification

Binding site identification is the next step after ligand and protein preparation. It is an additional prerequisite for performing SBVS, whereas in ligand-based drug designing, knowledge about the structure of receptor is not essential. If the binding site is not known, then for identification of the binding pocket, largest concave pocket having a hydrophobic and hydrogen bond donor/acceptors site is considered. Various tools are available for active site prediction, viz., FTMAP [144] and fpocket [145]. MDpocket is another tool for molecular dynamics (MDs) trajectories [146]. Similarly, Q-SiteFinder is one of the most cited binding site prediction server. Another approach is mixed, where the detection of ligand binding sites is performed with a chemical probe as in the static approach and, in addition, the putative binding site is evaluated in terms of flexibility, e.g., FTFlex [147].

3.4 Compound Selection After VS (Post-processing)

In virtual screening process, some ligands get very high scores due to factors like shape complementarily, intra-ligand steric clashes, large number of OH group (easily get high score because of large number of hydrogen bond), twisted amides, and imperfect hydrogen-bonding network. These false-positive ligands must be removed. Thus, visual inspection of docking poses is normally needed before the final in vitro or in vivo testing. Various reports have been made to increase the efficiency and the quality of compound selection after virtual screening [148–151].

3.5 Robust Molecular Docking for Selecting Top Hits for Activity Prediction and MD

Docking is broadly classified into three categories on the basis of receptor flexibility: (a) rigid docking, where docking of a rigid receptor molecule is performed with rigid ligand molecule; (b) flexible docking, where receptor protein molecule behaves as a rigid entity, while the ligand molecule has complete flexibility; and (c) full flexible docking, where the ligands have torsion angle flexibility, while the selected amino acid residues of the protein active side chains also remain flexible. Identified top hits can be further docked (blind, partial flexible, or full flexible docking) within the active site with robust parameters as compared to virtual screening. For correct pose prediction for ligand, one should generate a large number of conformations, high number of iterations, and correct docking algorithm. Docking alone is not sufficient, and MD simulation is also required after robust docking. Docking provides static conformation or binding conformation but is unable to provide atomic binding mechanism and interaction analysis. Docking and MD have been extensively used in amyloid fibril studies. Destabilization mechanism of amyloid fibril by morin was studied using MD techniques [152]. Virtual screening and docking studies were used to identify the binding of Congo red toward GNNQQNY protofibril and to identify new aggregation inhibitors [153]. Docking combined with MD was used to study the angiotensin-converting enzyme with inhibitor lisinopril and Aβ peptide [154], protonation states of catalytic Asp dyad in BACE-1-acyl guanidine-based inhibitor complex [155], and the selective inhibition of BACE-1 by Asp32 and Asp228 [156]. Several reports are available on the use of computational methods applied to design new peptide inhibitors for Aβ [157].

3.6 Building of Pharmacophore and QSAR Models to Screen Database or Activity Prediction of Top Hits

The pharmacophore and QSAR models are based on the ligand-based methods. The early concept of structure-based pharmacophore model was called as "hot spot," i.e., hot spot is defined as the shape and the properties that a ligand has to complement in order to bind to the active site [158]. Ligand-based drug designing is mostly preferred if the 3D structure of the target is unknown. Pharmacophore model is useful in terms of providing information about functional group responsible for the biological activity and a common feature in set of active ligands against a particular target. According to the pharmacophore definition, in general the model provides information of steric electronic, hydrogen bond donor acceptor and hydrophobic regions, and molecular interaction in the active site of the target. Pharmacophore does not represent a real molecule or specific functional group and is a purely abstract concept [159]. For good pharmacophore model generation, one should consider 3D structures of compound and conformation generation, molecular alignments, ligand flexibility, alignment techniques, scoring and optimization, comparison between various generated hypotheses, and validation [160]. After model building, authors must validate the model with the internal test set and external test set to validate the model accuracy and predictive power. The main problem in pharmacophore-based approaches is the need to take into account possible adverse steric interactions between inactive compounds in a data set and the target protein counterpart [161]. Pharmacophore validation and limitation have been discussed in detail [162]. Pharmacophore modeling and QSAR have been successfully applied for BACE-1 inhibitors [163, 164].

3.7 Molecular Dynamics Simulation of Protein-Ligand Complex to Gain Molecular Insight into Molecular Interaction

MD is now providing a better understanding of biological phenomena and biological interactions and is also useful in sampling configuration space, determining equilibrium averages, including structural and motional changes, and the thermodynamics of the system [165]. Molecular dynamics simulations of proteins can in principle provide the ultimate details of conformational changes and other atomic details [166]. Molecular dynamics simulation provides information regarding dynamic and atomic insights to the complicated biological systems. Recently, MD studies have revealed the protonation states of catalytic dyad in BACE-1-acyl guanidine-based inhibitor complex to improve the inhibitor design for BACE-1 and to understand the role of protonation states of catalytic Asp32 and Asp228 [155].

3.7.1 MD Analysis of the Drug Target Receptor (Protein) and Ligand Complex

MD simulation generates the conformation trajectory, i.e., multiple snapshots of the system at specific time interval. Generated MD trajectory of protein-ligand complex is utilized for the stability of the system and other analyses. For the stability and compactness of the protein-ligand complex as well as the protein, we calculate root-mean-square deviation (RMSD) and radius of gyration (Rg), respectively. Stable Rg value from MD trajectories presents that unique stable spatial structure of protein is retained without any structural unfolding of protein [99, 133167]. RMSD shows the stability of the protein and/or protein-ligand complex stability in the simulation run. If protein-ligand complex is stable, then we further analyze the amino acid fluctuations through root-mean-square fluctuations (RMSF). RMSF is the standard deviation of the atomic positions in the MD trajectory fitting to a reference structure [134]. RMSF provides the information about the movement of flexible region-like loop and various inserts which help in substrate binding or ligand binding. MD studies of BACE-1 revealed the movement of various inserts and flap movement which is crucial for drug designing [164, 168]. Similarly, in one of the MD studies for another AD target GSK-3β, it was revealed that solvent accessible surface area (SASA) included hydrophobic and hydrophilic accessible surface area of the protein molecule. This provides the total solvent accessible surface area and extent to which the amino acids interact with the solvent. After RMSD and RG analysis, hydrogen bond analysis between protein-ligand complex to measure the interaction between protein and ligand and important amino acid residues is performed. In general, hydrogen bond suggests the strength of protein-ligand binding. In silico studies of BACE-1 with inhibitors have suggested that the hydrogen bonds between flap residue of Tyr71, BACE-1, and ligand provide the stability to flap movement [130]. Hydrogen bond is very important in drug designing as many potent BACE-1 inhibitors form hydrogen bond with the Asp dyad of BACE-1 active site and also with the flap residue [134]. Hydrogen bond interactions between BACE-1 and inhibitors restrict the movement of flap, or in the presence of inhibitor, mostly the flap adopts a closed conformation and covers the active site which restricts the entry of substrate into the active site, and further the protein adopts inactive conformation. One can also measure the distance between important amino acid residues or two atoms throughout MD. Measure of the distance between two groups provides useful information such as distance between protein and ligand, extent of flap movement in HIV protease [169] and 10s loop and flap region flexibility of BACE-1 [164, 168, 170, 171], and distance between amino acid residues (Asp23-Lys28) which form a salt bridge in amyloid fibril which provides the stability to fibril. MD studies of compounds such as polyphenolpolyphenol [172], flavonoid derivative [21, 22], and morin [152] with amyloid protofibril showed that ligand disrupts the salt bridge which leads to an increase in distance between two neighboring chains in protofibril and dissociates the preformed fibril. Computer simulation techniques were used to study amyloid fibril nucleation and Aβ oligomer/drug interactions [173], structural and fluctuation difference between two ends of Aβ amyloid fibril [174], role of the regions 23–28 in Aβ fibril formation [175], and stability of amyloid protofibrils [176]. Recently, researchers reported the tetramer formation and membrane interactions of amyloid β peptide with MD [177]. For protein-ligand complex affinity, we perform the binding free-energy calculation to calculate the various interaction energies between protein-ligand complexes. The higher the binding free-energy is, the more stable is the protein-ligand complex. For binding energy calculations, various robust and computationally intensive methods are available, viz., free-energy perturbation (FEP), thermodynamic integration (TI), linear interaction energy (LIE) method, and molecular mechanics/Poisson-Boltzmann surface area (MM/PBSA) method.

3.7.2 Free-Energy Calculations

Free-energy estimation is an important tool for lead identification and optimization and study of protein-ligand complex. The tool g_mmpbsa calculates components of binding energy using MM-PBSA method. It takes snapshots from a MD trajectory. The MM-PBSA method has been shown to calculate accurate free energies at a moderate computational cost. It has an advantage that single simulation run is sufficient to determine all energy values. It does not include the entropic component of the free energy, which may produce errors in flexible systems. One of the GSK-3β studies used binding free-energy analysis of 7-azaindole GSK-3 β complex and suggested that calculation of binding free-energy change (ΔG) is effective for the lead-optimization process [178]. The docking, molecular dynamics, and MM/PBSA [179] method have been used to identify the anti-aggregating mechanism and binding affinity of flavonoid derivative 2-(4′ benzyloxyphenyl)-3-hydroxy-chromen-4-one on amyloid fibril [21]. Several methods have been discussed in detail for binding free-energy calculation [180] and application in ligand binding and protein flexibility [181].

3.8 In Silico Approaches to Design Multi-target Inhibitors

Traditional drug discovery approaches mainly focused on the selective agents for a specific target. This approach is now generally recognized as too simplistic for multifactorial diseases like AD, cancer, etc., as AD is caused or triggered by dysregulation of multiple pathways. Computational methods have been successfully applied to screen, design, and develop leads against particular targets of AD [45]. Molecular docking is used to identify leads, and molecular dynamics simulation is used to understand amyloid-destabilizing mechanism [21–23]. Virtual screening, pharmacophore mapping, and dynamics have also been successfully applied to identify natural inhibitors from InterBioScreen database for BACE-1 and metallothionein-III [164, 182]. These computational approaches were combined to identify the multi-target-directed inhibitors against AD diseases. These methods can be classified into combinatorial and fragment-based approaches. The combinatorial method applies simple virtual screening of the database for each target, and then it identifies the top hits which interact with the dual or multiple targets. In the case of multifactorial diseases like AD, single-target-directed therapeutic agents are less effective due to several reasons such as drug resistance due to network resistance [183], redundancy [184], cross talk [185], compensatory and neutralizing action [186], anti-target and counter-target activities [187], and on-target and off-target toxicities [188]. Thus, MTDLs were explored to deal with such diseases. MTDLs were previously explored successfully to achieve better results in cancer [149, 155]. Due to complex and multifactorial nature of AD, the same methods have been suggested to be applied to find out improved drugs for various proven therapeutic drug targets [35, 189, 190]. Till now various MTDLs have been identified against AD [191, 192]. In silico methods have been widely used to design single-target inhibitors in drug discovery process [21–23, 164], and other computational methods [193–195] combining with the docking, molecular dynamics, and binding free energy are providing very promising solution for identifying new leads with multiple activities and binding mechanism. Mostly MTDLs are designed by adding known fragments to each other which are already reported to inhibit the target, and this comes under fragment-based methods. Three in silico approaches have been

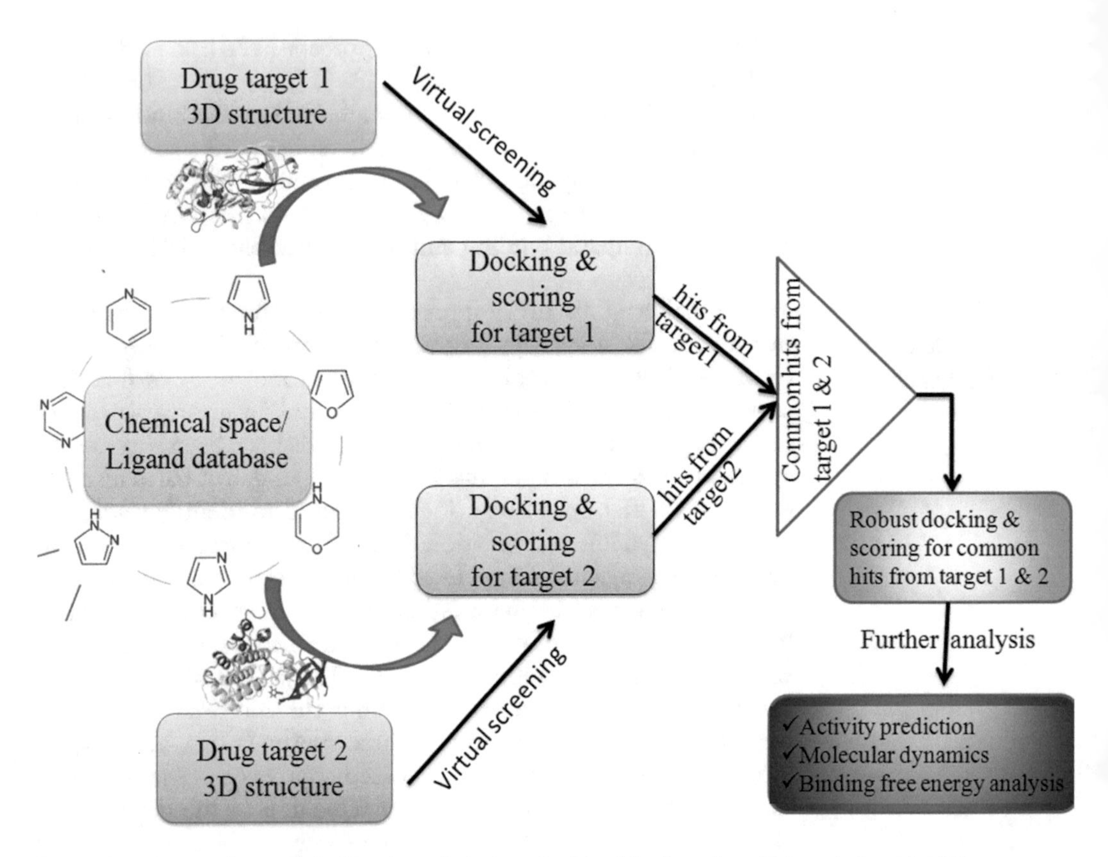

Fig. 2 Virtual screening and docking-based strategy for identification of multi-target direct inhibitors

discussed: (1) docking- and scoring-based virtual screening (Fig. 2), (2) pharmacophore-based screening (Fig. 3), and (3) machine learning-based methods (Fig. 4).

3.8.1 Virtual Screening and Docking-Based Methods to Design Dual Inhibitors

In this method if the retrieval rate against individual target is 50 ~ 70%, the collective retrieval rate for multi-target agents against two targets may be statistically reduced to 25 ~ 49% [196]. Thus, retrieval rate against individual target must be high enough. Virtual screening and docking-based methods are useful when the 3D crystallography structures are available. Docking is a widely used method and does not require prior knowledge of active ligands for particular targets. However, prior knowledge of 3D structures for both the targets is a must in this case of dual inhibitor designing. After the retrieval of the 3D structure, one needs a well-defined active site or site where ligand may bind to achieve the desired biological response. It is good to have a target 3D X-ray structure with an inhibitor or substrate bound into the active site because docking process is more reliable when the drug target has a well-defined active site.

Virtual screening and further docking may be performed on the same binding site after removing the bound ligand or substrate.

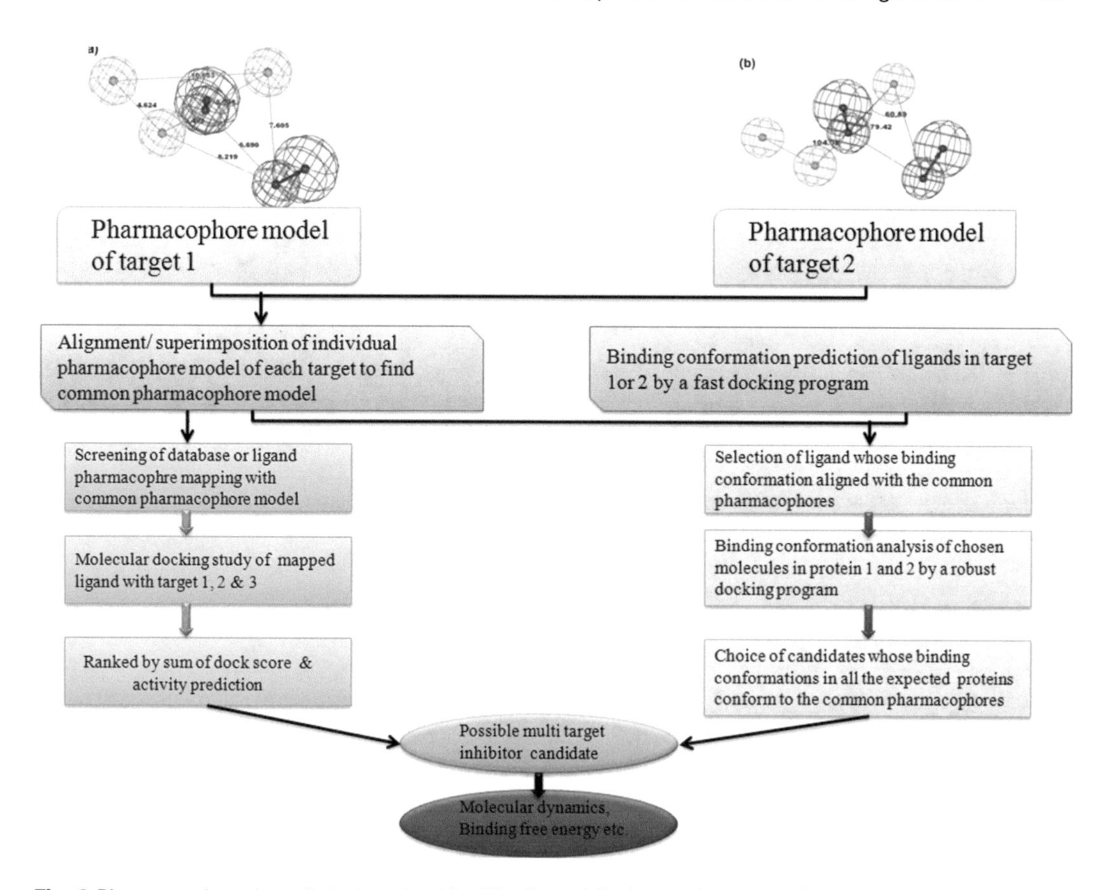

Fig. 3 Pharmacophore-based strategy for identification of dual or multi-target direct inhibitors

If target is recently discovered and the active site is not known yet, blind docking may be performed in which grid is placed to cover the entire protein, and then docking is performed to identify sites where ligand may possibly bind [197]. If the protein is large enough or exact binding site is not known, we need to cover the whole protein with the grid. In this case, we may adopt two strategies: first, a large grid spacing of 1 Å may be used (in the case of AutoDock), allowing the entire target to fit into the map space. However, this may cause problems with accuracy. Second, to cover the entire protein surface, we may place regular small size grid in an overlapping manner and then perform docking to know the site where ligand binds tightly as compared to the other sites. We may take the help of different active site or binding or cavity site prediction software and/or web server. Further, we can go for sequence alignment to identify conserved residues within the cavity or probable binding site since mostly protein active site residues are conserved within the same class of protein or functionally related proteins.

After retrieval of protein 3D structure and active site validation, virtual screening is performed to separate the binders from

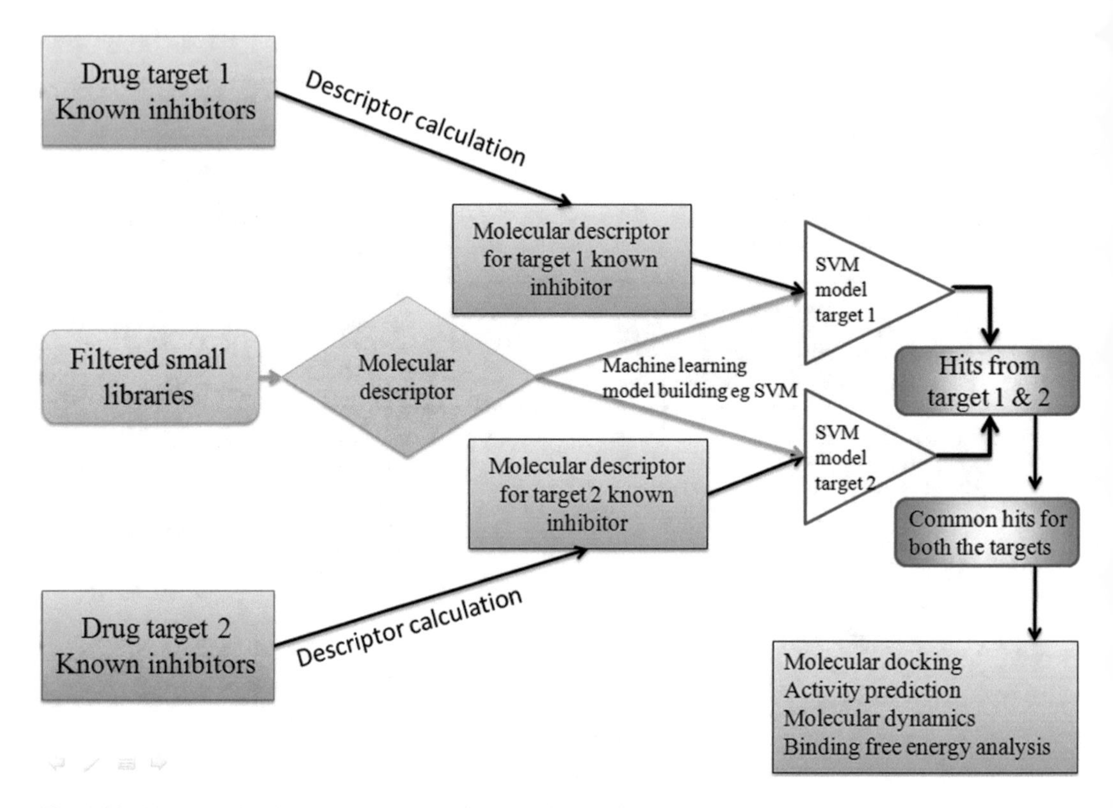

Fig. 4 Machine learning-based scheme for dual inhibitor design

nonbinders. Further to filter virtual screening results, top hits are docked with a robust method into the active site of both the receptors and in order to estimate the binding energy or binding affinity in term of docking scores given by docking software. Each docking software uses different algorithms and scoring functions. Thus, it is a good practice to validate the protocol (algorithm and scoring function) for the given set of proteins/drug targets. Redocking can be used for the primary validation of the docking protocol. Redocking is the process of removing the ligand molecule from 3D X-ray or NMR structure of the drug target and then dock it into the same active or binding site to generate docked pose [197]. Further, the ligand predicted dock pose and X-ray or NMR experimental pose are compared in terms of RMSD. Low RMSD indicates that the software or docking algorithm is able to predict the correct pose for the given protein-ligand complex. It was suggested that <2 Å RMSD is considerably good. The RMSD value also depends upon the number of rotatable bonds. The higher is the number of the rotatable bonds, the greater will be the RMSD. Other than redocking, we may use the cross docking method. In cross docking, the target 3D structure is obtained from a PDB file where another ligand is removed, and it is used for docking of the small molecule. In redocking and docking experiments, one should take into consideration that most computational docking techniques have standard deviation values. AutoDock has a standard deviation of free-energy prediction of about 2–3 kcal/mol [198]. The docking score for both the targets will be combined, and top hits may be selected. On the other hand, one may screen databases for the first target and then select the "filter" compound. These filter compounds are screened for other targets. In one of the studies, five hits were screened as multi-targeted agents for the treatment of AD using virtual screening method to identify dual inhibitors against GSK-3β and CDK5 using AutoDock and AutoDock Vina tools and then further validated [199]. After molecular docking experiment initially, we get an estimate binding free energy of the top hit ligands for both the targets. We can take top hits from both the targets and dock with another target and sum the docking scores and select the top scoring ligands as potential dual inhibitors. This can be further analyzed for the activity prediction with the help of QSAR and/or 3D-QSAR pharmacophore. Strength of protein-ligand complex and various interaction energies can be derived with molecular dynamics simulation. Protein conformational changes due to ligand-induced binding can be studied by normal mode analysis and MD simulations.

3.8.2 Pharmacophore-Based Methods to Design Dual Inhibitors

Pharmacophore-based methods complement molecular docking by enabling the selection of drug leads with higher structural flexibility, while molecular docking provides more accurate evaluation of specific binding interactions [200]. One of the studies of dual inhibitor design applied docking and pharmacophore to screen quinoxaline-based hybrid compounds to identify hit against AChE, H3R, and BACE-1 [201]. In this study, a virtual database of quinoxaline derivatives was filtered on the basis of pharmacophore model of BACE-1 inhibitors; further docking with AChE was performed to identify top hits on the basis of docking scores, and then the hits were evaluated for their biological activities. Several reports are available on pharmacophore application in drug discovery [202–204]. Pharmacophore study was also applied to identify dual inhibitors against BACE-1 with metal chelator properties against Alzheimer's disease [205].

Once we have pharmacophore model, its predictive ability and validation can be done on test and external test set for both the targets. Pharmacophore models for target 1 and target 2 are superimposed on each other to generate a common pharmacophore model. This model contains feature for both the targets. One should choose the features carefully according to earlier reports. If dual inhibitor is already reported for both the targets, then one should create a pharmacophore model from the reported dual inhibitor by simply aligning the most active molecules. Already reported dual inhibitor data set can also be used as an external test set for common pharmacophore model generated from individual methods. Both docking and pharmacophore models may be used to identify the dual inhibitors. In this method, conformations of the ligands docked into the active sites of all the targets are aligned to each pharmacophore model for the targets. If each ligand conformation generated after docking fits into the pharmacophore model of both the targets, the ligand may be considered as MTDL. Pharmacophore models can also be used to screen ligand libraries. Top hits can be further screened by robust docking approach to select the potential lead molecules. In one of the earlier reported studies, molecular docking and pharmacophore model were used to design dual inhibitor for BACE-1 and AChE enzymes [206]. Further structural characteristics of the ligands were studied using a novel group-based QSAR analysis method. In one of the studies, silibinin was reported as inhibitor of acetylcholinesterase and amyloid β peptide with the help of docking and MD simulation [207].

3.8.3 Machine Learning-Based Approach to Design Dual Inhibitors

Machine learning is a very useful tool in drug discovery. Structure-activity relationship can be effectively predicted by applying various algorithms. QSAR and machine learning techniques are very successful in single-target activity prediction. Moreover, multi-target QSAR models and chemoinformatics approach have also been successfully applied to identify ligands which are able to hit multiple targets in neurological disorder [35, 170–172]. These QSAR models based on multi-target-dependent molecular descriptors or species-dependent molecular descriptor combined with various algorithms like stochastic Markov drug-binding process models may achieve high retrieval rates and low false hit rates [196]. For the good model generation, data set of known inhibitor must be wide enough and cover larger range of inhibitor activities for selected targets. Moreover, the molecular size of the testing drugs needs to be in a certain range for accurate computation of multi-target-dependent or species-dependent molecular descriptors, which in some cases may also affect its capability for developing multi-target QSAR models [196, 200, 211]. A good QSAR model can be generated if factors such as compatibility of concepts, representativeness of assay data, influence of data outliers, selection of molecular descriptors, fitness of developed quantitative relationships, occurrence of chance correlations, and starting geometry are considered. Advancement of QSAR in identification of anti-AD lead has been discussed [129].

3.9 Various MTDL Moieties in Alzheimer's Disease Treatment

Due to multifactorial nature of AD and discovery of various disease-modifying drug targets, MTDL approach has been applied in search for better therapeutic ligands. Discovery of MTDLs for AD gained pace in recent years [208–210, 212], and few molecules reached up to clinical phase II. One of such examples is ladostigil, which is a AChE/BuChE and brain-selective monoamine oxidase A and B inhibitor with in vivo neuroprotective properties [213–215]. Tarcine combined with multi-alkoxybenzene moiety using a long-chain linker acts as an inhibitor of cholinesterases and β amyloid inhibitor [216], and the same chemical moiety was reported as multi-target inhibitor [217]. Donepezil derivatives combined with huprine and other chemical moieties were reported as inhibitors of AChE and AChE-induced Aβ aggregation, and some compounds showed BACE-1 inhibitory activity [218, 219]. Various xantostigmine derivatives were also reported to act as an AChE and Aβ aggregation inhibitors [220]. Other chemical moieties such as benzofuran-based hybrids, polycyclic benzophenone, 1-substituted (benzylidene-hydrazono)-1, 4-dihydropyridines, benzothiadiazine, and dihydropyridine derivatives are also reported as MTDLs against AD [222]. Other than synthetic compounds, a large number of natural molecules such as infractopicrin, resveratrol, vistin A, and heyneanol A have been reported as multifunctional agents [221, 222].

3.10 Challenges in the Design of Multi-target Drugs Against Multifactorial Diseases

For the treatment of multifactorial diseases like cancer, metabolic disorders, and CNS diseases, the MTDL method gained popularity as therapeutic agents. Some of the compounds showed success and are available for clinical trials. Perhaps the most successful examples of recently discovered multi-target inhibitors are kinase inhibitors as anti-angiogenesis anticancer agents [223, 224, 227, 228]. MTDLs are easy to design for kinase because mostly all the kinases have similar ATP binding site. One of the examples is sunitinib [225] which was first designed to selectively inhibit VEGFR2. Later, experiments showed that a great number of kinases were inhibited by this drug [186–188]. But the issue is how one can design selective inhibitor for only two to three kinases. Another problem is if ligand binds to two to three targets, it might activate the other pathway which may lead to an increase in the disease condition. Thus, only a full understanding of the networks present in an AD brain and neuronal cell can assist the proper selection of drug target and minimize the off-target side effect. To overcome off-target side effects, fragment-based methods were used. Computational methods are also available for fragment-based drug discovery [223]. This issue probably will be successfully reduced in the future through in silico procedures, since computational chemistry is developing strategies to predict the promiscuous binding propensities of drug molecules [226]. Network polypharmacology may be the better option for identifying new targets in multifactorial diseases which have greater impact in improving disease conditions [190]. Another problem of MTDL development is to design inhibitors for different or unrelated classes of targets/proteins because dissimilarity of active site residues and pharmacophore requirement for both the proteins are different. Thus, achieving selectivity for multiple targets is difficult.

4 Notes

4.1 Considerations During Virtual Screening Protocol Experiment

1. Random databases for VS protocol need to be avoided, search for the literature and properties needs to match the biological target, and there are various focused libraries for VS for specific targets.
2. Ligand preparation is a crucial starting step for effective VS protocol. Correct 3D structure of ligand, geometry optimization, bond length, bond order, etc., should be chosen.
3. Duplicate structure, toxicophores, or metabolically likely moieties should be removed from the libraries.
4. Filtering of compound data sets for chemically reactive and assay interfering as they appear in hit, such as the pan assay interference and the ALARM-NMR compounds. The new Zinc 15 version included these features.
5. Conversion of one format to another format should be done carefully and after that check for missing or broken link and correct them.
6. Filtering large libraries for lead-like properties, drug likeness, or optimal ADME/T properties before virtual screening should be done.
7. Filtering for desired pharmacological properties, for example, in AD drug discovery, molecule must have BBB penetration, may be performed.
8. It is advisable to use decoy set and benchmark the VS protocol because virtual screening software/tools are applicable and successful in specific cases.
9. One should consider the receptor structural flexibility and water molecules in docking computations. AD drug target mostly presents in the brain, where pH conditions are different, so receptor protein may adopt different conformation.
10. Potent leads are rarely identified through VS. VS is aimed at screening binders from nonbinders and not for the discovery of most potent molecule.
11. Scoring is still challenging in predicting accurately the correct binding pose; thus, to overcome this, the use of multiple docking program and manual inspection is necessary. Predicted protein structures from homology modeling and predicted protein-ligand complexes may result in increased rates of false-positive/false-negative results.

4.2 Considerations During Docking Experiment

1. Redocking and cross docking experiment may be performed to validate the docking protocol.

2. Two of the major considerations are sampling and scoring in protein-ligand docking. One should generate large number of conformations and cluster them on the basis of RMSD.
3. Docking studies are particularly challenging in case of structural heterogeneity in solution, making it difficult to choose a suitable structure or set of structures that could serve as a receptor for docking such as Aβ peptide.
4. Need to account structural water in approximate models of the solvent in docking experiment or consider fewer water molecules which are present only into the active site in docking experiment.
5. Off-target effect binding may be considered. One should dock the ligand with similar protein or receptor to check the off-target binding. This is especially important for minimizing side effects and toxicity.
6. In protein-ligand binding, the effect of entropy may be taken into account.
7. Docking into multiple 3D structures of the same receptor should be performed to ensure the good binding.

4.3 Considerations During Molecular Dynamics Experiment

1. The limitations of docking (rigidity and sampling) are complemented by MD.
2. The major challenge in MD simulation of protein-ligand complex starting structures is very important. The use of X-ray crystallographic structure is advisable if available.
3. Choice of force field. Various atomistic force fields are available and provide a better detail, but they are computationally time-consuming.
4. Topology parameters and files must be compatible with force field.
5. The simulations may be conducted for longer time scale and with replicates.
6. RMSD graph and potential energy graph should be checked for the stability of the system.
7. Binding free-energy calculation for protein-ligand system may be done to know the individual component of energy and per residue energy contribution.

5 Conclusions

The number of people with AD is increasing worldwide. In 2050, AD population is projected to be three times, and the current treatments of AD are limited. Thus, there is a need to find solutions which slow down or completely halt the AD. However, the process of CNS drug development is slow, complex, and costly. In recent years, few of the compounds that entered into clinical trials were not able to make it to the market. In this alarming situation and high rate of clinical failure and due to multifactorial nature of AD, computational methods such as virtual screening ADME/T prediction and MTDL approach seem promising to design new potential therapeutics like dual inhibitor with antioxidant or anti-inflammatory properties. The MTDL field has already passed the transition from serendipity to a new era where more rational target selections for MTDLs are feasible. Computational methods are very useful in designing more rational molecules against various targets which belong to different classes or their active sites are completely different. Virtual screening, docking, pharmacophore, and QSAR model are very useful for initial hit identification and activity prediction for top hits. MD simulations are the important methods which provide the molecular insight into protein conformational changes and atomic details and protein-ligand dynamics interactions as well. One of the major challenges of lead optimization can also be addressed by various computational methods. One area that needs more attention is the optimal activity ratio which has been ignored in most multi-target drug discovery. Recently, new computational approaches have emerged to address such issues. Researchers in this area will continue to build momentum and over the time will provide a solid foundation for addressing unmet clinical needs with the help of computational methods in AD drug discovery.

References

1. Hippius H, Neundörfer G (2003) The discovery of Alzheimer's disease. Dialogues Clin Neurosci 5:101–108
2. Lautenschlager NT, Martins RN (2005) Common versus uncommon causes of dementia. Int Psychogeriatr/IPA 17(Suppl 1):S27–S34
3. 2016 Alzheimer's disease facts and figures. http://www.sciencedirect.com/science/article/pii/S1552526016000856
4. Goedert M, Spillantini MG (2006) A century of Alzheimer's disease. Science (New York, NY) 314:777–781
5. Jellinger KA (2003) General aspects of neurodegeneration. J Neural Transm Suppl 65:101–144
6. Karran E, Mercken M, De Strooper B (2011) The amyloid cascade hypothesis for Alzheimer's disease: an appraisal for the development of therapeutics. Nat Rev Drug Discov 10:698–712
7. BABIC T, FRANCIS P, PALMER A et al (1999) The cholinergic hypothesis of Alzheimer's disease: a review of progress. J Neurol Neurosurg Psychiatry 67:558
8. Craig LA, Hong NS, McDonald RJ (2011) Revisiting the cholinergic hypothesis in the development of Alzheimer's disease. Neurosci Biobehav Rev 35:1397–1409
9. Bezprozvanny I, Mattson MP (2008) Neuronal calcium mishandling and the pathogenesis of Alzheimer's disease. Trends Neurosci 31:454–463
10. Praticò D (2008) Oxidative stress hypothesis in Alzheimer's disease: a reappraisal. Trends Pharmacol Sci 29:609–615
11. Bonda DJ, Lee H, Blair JA et al (2011) Role of metal dyshomeostasis in Alzheimer's disease. Metallomics Integr Biometal Sci 3:267–270
12. Trepanier CH, Milgram NW (2010) Neuroinflammation in Alzheimer's disease: are NSAIDs and selective COX-2 inhibitors the

next line of therapy? J Alzheimers Dis JAD 21:1089–1099
13. Hardy J, Selkoe DJ (2002) The amyloid hypothesis of Alzheimer's disease: progress and problems on the road to therapeutics. Science (New York, NY) 297:353–356
14. Hamaguchi T, Ono K, Yamada M (2006) Anti-amyloidogenic therapies: strategies for prevention and treatment of Alzheimer's disease. Cell Mol Life Sci 63:1538–1552
15. Chang Y-J, Chen Y-R (2014) The coexistence of an equal amount of Alzheimer's amyloid-β 40 and 42 forms structurally stable and toxic oligomers through a distinct pathway. FEBS J 281:2674–2687
16. Jan A, Gokce O, Luthi-Carter R et al (2008) The ratio of monomeric to aggregated forms of Abeta40 and Abeta42 is an important determinant of amyloid-beta aggregation, fibrillogenesis, and toxicity. J Biol Chem 283:28176–28189
17. Jellinger KA, Bancher C (1998) Neuropathology of Alzheimer's disease: a critical update. J Neural Transm Suppl 54:77–95
18. Vassallo N (ed) (2015) Natural compounds as therapeutic agents for Amyloidogenic diseases. Springer International Publishing, Cham
19. Kumar P, Pillay V, Choonara YE et al (2011) In silico theoretical molecular modeling for Alzheimer's disease: the nicotine-curcumin paradigm in neuroprotection and neurotherapy. Int J Mol Sci 12:694–724
20. Nie Q, Du X, Geng M (2011) Small molecule inhibitors of amyloid β peptide aggregation as a potential therapeutic strategy for Alzheimer's disease. Acta Pharmacol Sin 32:545–551
21. Kumar A, Srivastava S, Tripathi S et al (2015) Molecular insight into amyloid oligomer destabilizing mechanism of flavonoid derivative 2-(4′ benzyloxyphenyl)-3-hydroxy-chromen-4-one through docking and molecular dynamics simulations. J Biomol Struct Dyn 0:1–12
22. Verma A, Kumar A, Debnath M (2016) Molecular docking and simulation studies to give insight of surfactin amyloid interaction for destabilizing Alzheimer's Aβ42 protofibrils. Med Chem Res 25:1–7
23. Singh SK, Sinha P, Mishra L et al (2013) Neuroprotective role of a novel copper Chelator against Aβ42 induced neurotoxicity. Int J Alzheimers Dis 2013:567128
24. Singh SK, Gaur R, Kumar A et al (2014) The flavonoid derivative 2-(4′ Benzyloxyphenyl)-3-hydroxy-chromen-4-one protects against Aβ42-induced neurodegeneration in transgenic drosophila: insights from in silico and in vivo studies. Neurotox Res 26:331–350
25. Goedert M, Spillantini MG, Crowther RA (1991) Tau proteins and neurofibrillary degeneration. Brain Pathol 1:279–286
26. Iqbal K, del C. Alonso A, Chen S et al (2005) Tau pathology in Alzheimer disease and other tauopathies. Biochim Biophys Acta (BBA) - Mol Basis Dis 1739:198–210
27. Chun W, Johnson GVW (2007) The role of tau phosphorylation and cleavage in neuronal cell death. Front Biosci 12:733–756
28. Deane R, Zlokovic BV (2007) Role of the blood-brain barrier in the pathogenesis of Alzheimer's disease. Curr Alzheimer Res 4:191–197
29. França MB, Lima KC, Eleutherio ECA (2016) Oxidative stress and amyloid toxicity: insights from yeast. J Cell Biochem 118:1142
30. Islam MT (2017) Oxidative stress and mitochondrial dysfunction-linked neurodegenerative disorders. Neurol Res 39:73–82
31. Cristóvão JS, Santos R, Gomes CM (2016) Metals and neuronal metal binding proteins implicated in Alzheimer's disease. Oxidative Med Cell Longev 2016:9812178
32. Kenche VB, Barnham KJ (2011) Alzheimer's disease & metals: therapeutic opportunities. Br J Pharmacol 163:211–219
33. Killin LOJ, Starr JM, Shiue IJ et al (2016) Environmental risk factors for dementia: a systematic review. BMC Geriatr 16:175
34. Anand R, Gill KD, Mahdi AA (2014) Therapeutics of Alzheimer's disease: past, present and future, neuropharmacology. Neuropharmacology 76 Pt A:27–50
35. Youdim MBH, Buccafusco JJ (2005) Multifunctional drugs for various CNS targets in the treatment of neurodegenerative disorders. Trends Pharmacol Sci 26:27–35
36. Rosini M (2014) Polypharmacology: the rise of multitarget drugs over combination therapies. Future Med Chem 6:485–487
37. Boran ADW, Iyengar R (2010) Systems approaches to polypharmacology and drug discovery. Curr Opin Drug Discov Devel 13:297–309
38. Bolognesi ML, Melchiorre C, der Schyf CJV, et al (2012) Chapter 18: discovery of multitarget agents for neurological diseases via ligand design, Presented at the March 28
39. Klotz U (2007) The elderly—a challenge for appropriate drug treatment. Eur J Clin Pharmacol 64:225–226
40. Evans BE, Rittle KE, Bock MG et al (1988) Methods for drug discovery: development of

potent, selective, orally effective cholecystokinin antagonists. J Med Chem 31:2235–2246
41. Hu Y, Bajorath J (2013) How promiscuous are pharmaceutically relevant compounds? A data-driven assessment. AAPS J 15:104–111
42. Keiser MJ, Setola V, Irwin JJ et al (2009) Predicting new molecular targets for known drugs. Nature 462:175–181
43. Morphy R (2010) Selectively nonselective kinase inhibition: striking the right balance. J Med Chem 53:1413–1437
44. Millan MJ (2009) Dual- and triple-acting agents for treating core and co-morbid symptoms of major depression: novel concepts, new drugs. Neurotherapeutics J Am Soc Exp NeuroTherapeutics 6:53–77
45. Zeng H, Wu X (2016) Alzheimer's disease drug development based on computer-aided drug design. Eur J Med Chem 121:851
46. Cavalli A, Bolognesi ML, Minarini A et al (2008) Multi-target-directed ligands to combat neurodegenerative diseases. J Med Chem 51:347–372
47. Schmitt B, Bernhardt T, Moeller H-J et al (2004) Combination therapy in Alzheimer's disease: a review of current evidence. CNS Drugs 18:827–844
48. Hopkins AL (2008) Network pharmacology: the next paradigm in drug discovery. Nat Chem Biol 4:682–690
49. Yildirim MA, Goh K-I, Cusick ME et al (2007) Drug-target network. Nat Biotechnol 25:1119–1126
50. Anighoro A, Bajorath J, Rastelli G (2014) Polypharmacology: challenges and opportunities in drug discovery. J Med Chem 57:7874–7887
51. Peters J-U (2013) Polypharmacology - foe or friend? J Med Chem 56:8955–8971
52. Oprea TI, Mestres J (2012) Drug repurposing: far beyond new targets for old drugs. AAPS J 14:759–763
53. Reddy AS, Zhang S (2013) Polypharmacology: drug discovery for the future. Expert Rev Clin Pharmacol 6:41–47
54. Oprea TI, Nielsen SK, Ursu O et al (2011) Associating drugs, targets and clinical outcomes into an integrated network affords a new platform for computer-aided drug repurposing. Mol Inf 30:100–111
55. Achenbach J, Tiikkainen P, Franke L et al (2011) Computational tools for polypharmacology and repurposing. Future Med Chem 3:961–968
56. Durrant JD, Amaro RE, Xie L et al (2010) A multidimensional strategy to detect Polypharmacological targets in the absence of structural and sequence homology. PLoS Comput Biol 6:e1000648
57. Miller G (2010) Is pharma running out of brainy ideas? Science (New York, NY) 329:502–504
58. Berger JR, Choi D, Kaminski HJ et al (2013) Importance and hurdles to drug discovery for neurological disease. Ann Neurol 74:441–446
59. Pammolli F, Magazzini L, Riccaboni M (2011) The productivity crisis in pharmaceutical R&D. Nat Rev Drug Discov 10:428–438
60. Jorgensen WL (2004) The many roles of computation in drug discovery. Science (New York, NY) 303:1813–1818
61. Owens J (2007) Determining druggability. Nat Rev Drug Discov 6:187–187
62. Cheng AC, Coleman RG, Smyth KT et al (2007) Structure-based maximal affinity model predicts small-molecule druggability. Nat Biotechnol 25:71–75
63. Hopkins AL, Groom CR (2002) The druggable genome. Nat Rev Drug Discov 1:727–730
64. Dixon SJ, Stockwell BR (2009) Identifying druggable disease-modifying gene products. Curr Opin Chem Biol 13:549–555
65. Silva T, Reis J, Teixeira J et al (2014) Alzheimer's disease, enzyme targets and drug discovery struggles: from natural products to drug prototypes. Ageing Res Rev 15:116–145
66. Bohacek RS, McMartin C, Guida WC (1996) The art and practice of structure-based drug design: a molecular modeling perspective. Med Res Rev 16:3–50
67. Reymond J-L, Awale M (2012) Exploring chemical space for drug discovery using the chemical universe database. ACS Chem Neurosci 3:649–657
68. Harvey AL, Edrada-Ebel R, Quinn RJ (2015) The re-emergence of natural products for drug discovery in the genomics era. Nat Rev Drug Discov 14:111–129
69. Huang L, Su T, Li X (2013) Natural products as sources of new lead compounds for the treatment of Alzheimer's disease. Curr Top Med Chem 13:1864–1878
70. Williams P, Sorribas A, Howes M-JR (2011) Natural products as a source of Alzheimer's drug leads. Nat Prod Rep 28:48–77
71. McLaurin J, Kierstead ME, Brown ME et al (2006) Cyclohexanehexol inhibitors of Abeta aggregation prevent and reverse Alzheimer

phenotype in a mouse model. Nat Med 12:801–808
72. Popik P, Bobula B, Janusz M et al (1999) Colostrinin, a polypeptide isolated from early milk, facilitates learning and memory in rats. Pharmacol Biochem Behav 64:183–189
73. Forloni G, Colombo L, Girola L et al (2001) Anti-amyloidogenic activity of tetracyclines: studies in vitro. FEBS Lett 487:404–407
74. Kim S, Thiessen PA, Bolton EE et al (2016) PubChem substance and compound databases. Nucleic Acids Res 44:D1202–D1213
75. Pence HE, Williams A (2010) ChemSpider: an online chemical information resource. J Chem Educ 87:1123–1124
76. Irwin JJ, Sterling T, Mysinger MM et al (2012) ZINC: a free tool to discover chemistry for biology. J Chem Inf Model 52:1757–1768
77. Sterling T, Irwin JJ (2015) ZINC 15 – ligand discovery for everyone. J Chem Inf Model 55:2324–2337
78. Chen CY-C (2011) TCM database@Taiwan: the World's largest traditional Chinese medicine database for drug screening in Silico. PLoS One 6:e15939
79. Banerjee P, Erehman J, Gohlke B-O et al (2015) Super natural II—a database of natural products. Nucleic Acids Res 43:D935–D939
80. Ntie-Kang F, Zofou D, Babiaka SB et al (2013) AfroDb: a select highly potent and diverse natural product library from African medicinal plants. PLoS One 8:e78085
81. Lin Y-C, Wang C-C, Chen I-S et al (2013) TIPdb: a database of anticancer, antiplatelet, and Antituberculosis phytochemicals from indigenous plants in Taiwan. Sci World J 2013:e736386
82. Gu J, Gui Y, Chen L et al (2013) Use of natural products as chemical library for drug discovery and network pharmacology. PLoS One 8:e62839
83. Valli M, dos Santos RN, Figueira LD et al (2013) Development of a natural products database from the biodiversity of Brazil. J Nat Prod 76:439–444
84. Nakamura K, Shimura N, Otabe Y et al (2013) KNApSAcK-3D: a three-dimensional structure database of plant metabolites. Plant Cell Physiol 54:e4
85. Maeda MH, Kondo K (2013) Three-dimensional structure database of natural metabolites (3DMET): a novel database of curated 3D structures. J Chem Inf Model 53:527–533
86. Shoichet BK (2004) Virtual screening of chemical libraries. Nature 432:862–865
87. Christofferson AJ, Huang N (2012) How to benchmark methods for structure-based virtual screening of large compound libraries. Methods Mol Biol (Clifton, NJ) 819:187–195
88. Xia J, Jin H, Liu Z et al (2014) An unbiased method to build benchmarking sets for ligand-based virtual screening and its application to GPCRs. J Chem Inf Model 54:1433–1450
89. Lagarde N, Zagury J-F, Montes M (2015) Benchmarking data sets for the evaluation of virtual ligand screening methods: review and perspectives. J Chem Inf Model 55:1297–1307
90. Evers A, Hessler G, Matter H et al (2005) Virtual screening of biogenic amine-binding G-protein coupled receptors: comparative evaluation of protein- and ligand-based virtual screening protocols. J Med Chem 48:5448–5465
91. Ferrari AM, Wei BQ, Costantino L et al (2004) Soft docking and multiple receptor conformations in virtual screening. J Med Chem 47:5076–5084
92. Jiang F, Kim SH (1991) "soft docking": matching of molecular surface cubes. J Mol Biol 219:79–102
93. Lavecchia A, Di Giovanni C (2013) Virtual screening strategies in drug discovery: a critical review. Curr Med Chem 20:2839–2860
94. Amaro RE, Baron R, McCammon JA (2008) An improved relaxed complex scheme for receptor flexibility in computer-aided drug design. J Comput Aided Mol Des 22:693–705
95. López-Vallejo F, Caulfield T, Martínez-Mayorga K et al (2011) Integrating virtual screening and combinatorial chemistry for accelerated drug discovery. Comb Chem High Throughput Screen 14:475–487
96. Huang S-Y, Zou X (2010) Advances and challenges in protein-ligand docking. Int J Mol Sci 11:3016–3034
97. Teague SJ (2003) Implications of protein flexibility for drug discovery. Nat Rev Drug Discov 2:527–541
98. Cozzini P, Kellogg GE, Spyrakis F et al (2008) Target flexibility. J Med Chem 51:6237–6255
99. Totrov M, Abagyan R (2008) Flexible ligand docking to multiple receptor conformations: a practical alternative. Curr Opin Struct Biol 18:178–184

100. Apostolakis J, Plückthun A, Caflisch A (1998) Docking small ligands in flexible binding sites. J Comput Chem 19:21–37
101. Lionta E, Spyrou G, Vassilatis DK et al (2014) Structure-based virtual screening for drug discovery: principles, applications and recent advances. Curr Top Med Chem 14:1923–1938
102. Halperin I, Ma B, Wolfson H et al (2002) Principles of docking: an overview of search algorithms and a guide to scoring functions. Proteins Struct Funct Bioinf 47:409–443
103. Dias R, de Azevedo WF (2008) Molecular docking algorithms. Curr Drug Targets 9:1040–1047
104. Morris GM, Huey R, Lindstrom W et al (2009) AutoDock4 and AutoDockTools4: automated docking with selective receptor flexibility. J Comput Chem 30:2785–2791
105. Allen WJ, Balius TE, Mukherjee S et al (2015) DOCK 6: impact of new features and current docking performance. J Comput Chem 36:1132–1156
106. Pierce BG, Wiehe K, Hwang H et al (2014) ZDOCK server: interactive docking prediction of protein-protein complexes and symmetric multimers. Bioinformatics (Oxford, England) 30:1771–1773
107. Sauton N, Lagorce D, Villoutreix BO et al (2008) MS-DOCK: accurate multiple conformation generator and rigid docking protocol for multi-step virtual ligand screening. BMC Bioinformatics 9:184
108. Liu M, Wang S (1999) MCDOCK: a Monte Carlo simulation approach to the molecular docking problem. J Comput Aided Mol Des 13:435–451
109. Neves MAC, Totrov M, Abagyan R (2012) Docking and scoring with ICM: the benchmarking results and strategies for improvement. J Comput Aided Mol Des 26:675–686
110. Hartshorn MJ, Verdonk ML, Chessari G et al (2007) Diverse, high-quality test set for the validation of protein–ligand docking performance. J Med Chem 50:726–741
111. Spitzer R, Jain AN (2012) Surflex-dock: docking benchmarks and real-world application. J Comput Aided Mol Des 26:687–699
112. Hindle SA, Rarey M, Buning C et al (2002) Flexible docking under pharmacophore type constraints. J Comput Aided Mol Des 16:129–149
113. Pang Y-P, Perola E, Xu K et al (2001) EUDOC: a computer program for identification of drug interaction sites in macromolecules and drug leads from chemical databases. J Comput Chem 22:1750–1771
114. Miller MD, Kearsley SK, Underwood DJ et al (1994) FLOG: a system to select "quasi-flexible" ligands complementary to a receptor of known three-dimensional structure. J Comput Aided Mol Des 8:153–174
115. Verdonk ML, Chessari G, Cole JC et al (2005) Modeling water molecules in protein–ligand docking using GOLD. J Med Chem 48:6504–6515
116. Friesner RA, Banks JL, Murphy RB et al (2004) Glide: a new approach for rapid, accurate docking and scoring. 1. Method and assessment of docking accuracy. J Med Chem 47:1739–1749
117. Schnecke V, Kuhn LA (2000) Virtual screening with solvation and ligand-induced complementarity. Perspect Drug Discovery Des 20:171–190
118. Cavasotto CN, Abagyan RA (2004) Protein flexibility in ligand docking and virtual screening to protein kinases. J Mol Biol 337:209–225
119. Langer T, Hoffman RD (2006). Pharmacophores and PharmacophoreSearches. Wiley-VCH: Weinheim
120. Wolber G, Seidel T, Bendix F et al (2008) Molecule-pharmacophore superpositioning and pattern matching in computational drug design. Drug Discov Today 13:23–29
121. Güner OF (2002) History and evolution of the pharmacophore concept in computer-aided drug design. Curr Top Med Chem 2:1321–1332
122. Khedkar SA, Malde AK, Coutinho EC et al (2007) Pharmacophore modeling in drug discovery and development: an overview. Med Chem (Shāriqah (United Arab Emirates)) 3:187–197
123. Braga RC, Andrade CH (2013) Assessing the performance of 3D pharmacophore models in virtual screening: how good are they? Curr Top Med Chem 13:1127–1138
124. Pirhadi S, Shiri F, Ghasemi JB (2013) Methods and applications of structure based pharmacophores in drug discovery. Curr Top Med Chem 13:1036–1047
125. Dixon SL, Smondyrev AM, Knoll EH et al (2006) PHASE: a new engine for pharmacophore perception, 3D QSAR model development, and 3D database screening: 1. Methodology and preliminary results. J Comput Aided Mol Des 20:647–671
126. Güner O, Clement O, Kurogi Y (2004) Pharmacophore modeling and three dimensional database searching for drug design using catalyst: recent advances. Curr Med Chem 11:2991–3005

127. Wolber G, Langer T (2005) LigandScout: 3-D pharmacophores derived from protein-bound ligands and their use as virtual screening filters. J Chem Inf Model 45:160–169
128. Martin YC, Bures MG, Danaher EA et al (1993) A fast new approach to pharmacophore mapping and its application to dopaminergic and benzodiazepine agonists. J Comput Aided Mol Des 7:83–102
129. Ambure P, Roy K (2014) Advances in quantitative structure-activity relationship models of anti-Alzheimer's agents. Expert Opin Drug Discovery 9:697–723
130. Ambure P, Roy K (2015) Exploring structural requirements of imaging agents against Aβ plaques in Alzheimer's disease: a QSAR approach. Comb Chem High Throughput Screen 18:411–419
131. Brahmachari G, Choo C, Ambure P et al (2015) In vitro evaluation and in silico screening of synthetic acetylcholinesterase inhibitors bearing functionalized piperidine pharmacophores. Bioorg Med Chem 23:4567–4575
132. Ambure P, Roy K (2017) CADD modeling of multi-target drugs against Alzheimer's disease. Curr Drug Targets 18:522–533
133. Sinko W, Lindert S, McCammon JA (2013) Accounting for receptor flexibility and enhanced sampling methods in computer-aided drug design. Chem Biol Drug Des 81:41–49
134. Tripathi S, Kumar A, Kahlon AK, et al (2016) Current trends in docking methodologies. http://services.igi-global.com/resolvedoi/resolve.aspx?doi=10.4018/978-1-5225-0115-2.ch013. pp 320–338
135. Doman TN, McGovern SL, Witherbee BJ et al (2002) Molecular docking and high-throughput screening for novel inhibitors of protein tyrosine phosphatase-1B. J Med Chem 45:2213–2221
136. Sherman W, Beard HS, Farid R (2006) Use of an induced fit receptor structure in virtual screening. Chem Biol Drug Des 67:83–84
137. Osguthorpe DJ, Sherman W, Hagler AT (2012) Exploring protein flexibility: incorporating structural ensembles from crystal structures and simulation into virtual screening protocols. J Phys Chem B 116:6952–6959
138. Osguthorpe DJ, Sherman W, Hagler AT (2012) Generation of receptor structural ensembles for virtual screening using binding site shape analysis and clustering. Chem Biol Drug Des 80:182–193
139. Polgár T, Magyar C, Simon I et al (2007) Impact of ligand protonation on virtual screening against β-Secretase (BACE1). J Chem Inf Model 47:2366–2373
140. Sastry GM, Adzhigirey M, Day T et al (2013) Protein and ligand preparation: parameters, protocols, and influence on virtual screening enrichments. J Comput Aided Mol Des 27:221–234
141. Klebe G (2006) Virtual ligand screening: strategies, perspectives and limitations. Drug Discov Today 11:580–594
142. McInnes C (2007) Virtual screening strategies in drug discovery. Curr Opin Chem Biol 11:494–502
143. Polgár T, Keserü GM (2005) Virtual screening for β-Secretase (BACE1) inhibitors reveals the importance of protonation states at Asp32 and Asp228. J Med Chem 48:3749–3755
144. Ngan CH, Bohnuud T, Mottarella SE et al (2012) FTMAP: extended protein mapping with user-selected probe molecules. Nucleic Acids Res 40:W271–W275
145. Le Guilloux V, Schmidtke P, Tuffery P (2009) Fpocket: an open source platform for ligand pocket detection. BMC Bioinformatics 10:168
146. Schmidtke P, Bidon-Chanal A, Luque FJ et al (2011) MDpocket: open-source cavity detection and characterization on molecular dynamics trajectories. Bioinformatics (Oxford, England) 27:3276–3285
147. Grove LE, Hall DR, Beglov D et al (2013) FTFlex: accounting for binding site flexibility to improve fragment-based identification of druggable hot spots. Bioinformatics (Oxford, England) 29:1218–1219
148. Athanasiadis E, Cournia Z, Spyrou G (2012) ChemBioServer: a web-based pipeline for filtering, clustering and visualization of chemical compounds used in drug discovery. Bioinformatics (Oxford, England) 28:3002–3003
149. Waszkowycz B (2008) Towards improving compound selection in structure-based virtual screening. Drug Discov Today 13:219–226
150. Malmstrom RD, Watowich SJ (2011) Using free energy of binding calculations to improve the accuracy of virtual screening predictions. J Chem Inf Model 51:1648–1655
151. Ding B, Wang J, Li N et al (2013) Characterization of small molecule binding. I. Accurate identification of strong inhibitors in virtual screening. J Chem Inf Model 53:114–122
152. Lemkul JA, Bevan DR (2010) Destabilizing Alzheimer's Aβ42 Protofibrils with Morin: mechanistic insights from molecular dynamics simulations. Biochemistry 49:3935–3946

153. Zhao J-H, Liu H-L, Elumalai P et al (2013) Molecular modeling to investigate the binding of Congo red toward GNNQQNY protofibril and in silico virtual screening for the identification of new aggregation inhibitors. J Mol Model 19:151–162
154. Jalkute CB, Barage SH, Dhanavade MJ et al (2013) Molecular dynamics simulation and molecular docking studies of angiotensin converting enzyme with inhibitor lisinopril and amyloid Beta peptide. Protein J 32:356–364
155. Kocak A, Erol I, Yildiz M et al (2016) Computational insights into the protonation states of catalytic dyad in BACE1–acyl guanidine based inhibitor complex. J Mol Graph Model 70:226–235
156. Hernández-Rodríguez M, Correa-Basurto J, Gutiérrez A et al (2016) Asp32 and Asp228 determine the selective inhibition of BACE1 as shown by docking and molecular dynamics simulations. Eur J Med Chem 124:1142–1154
157. Eskici G, Gur M (2013) Computational Design of new Peptide Inhibitors for amyloid Beta (Aβ) aggregation in Alzheimer's disease: application of a novel methodology. PLoS One 8:e66178
158. Brenk R, Klebe G (2006) "Hot spot" analysis ofprotein-bindingsites as a prerequisite forstructure-basedvirtual screening and lead optimization. In: Langer T, Hoffmann RD (eds) Pharmacophores and Pharmacophore searches.Wiley-VCHVerlag GmbH & Co. KGaA, pp 171–192, Weinheim.
159. Wermuth CG (2006) Pharmacophores: historical perspective and viewpoint from a medicinal chemist. In: Langer T, Hoffmann RD (eds) Pharmacophores and Pharmacophore searches.Wiley-VCHVerlag GmbH & Co. KGaA, pp 1–13,Weinheim.
160. Poptodorov K, Luu T, Hoffmann RD (2006) Pharmacophore model generation software tools. In: Langer T, Hoffmann RD (eds) Pharmacophores and Pharmacophore searches.Wiley-VCHVerlag GmbH & Co. KGaA, pp 15–47,Weinheim.
161. Langer T, Hoffmann RD (2006) Pharmacophores and Pharmacophore searches. John Wiley & Sons, Weinheim
162. Triballeau N, BertrandH-O,Acher F (2006) Are you sure you have a good model? In: Langer T, Hoffmann RD (eds) Pharmacophores and Pharmacophore searches.Wiley-VCHVerlag GmbH & Co. KGaA, pp 325–364,Weinheim.
163. Ju Y, Li Z, Deng Y et al (2016) Identification of novel BACE1 inhibitors by combination of Pharmacophore modeling, structure-based design and in vitro assay. Curr Comput Aided Drug Des 12:73–82
164. Kumar A, Roy S, Tripathi S et al (2016) Molecular docking based virtual screening of natural compounds as potential BACE1 inhibitors: 3D QSAR pharmacophore mapping and molecular dynamics analysis. J Biomol Struct Dyn 34:239–249
165. McCammon JA, Gelin BR, Karplus M (1977) Dynamics of folded proteins. Nature 267:585–590
166. Karplus M, McCammon JA (2002) Molecular dynamics simulations of biomolecules. Nat Struct Mol Biol 9:646–652
167. Lobanov MY, Bogatyreva NS, Galzitskaya OV (2008) Radius of gyration as an indicator of protein structure compactness. Mol Biol 42:623–628
168. Barman A, Schürer S, Prabhakar R (2011) Computational modeling of substrate specificity and catalysis of the β-Secretase (BACE1) enzyme. Biochemistry 50:4337–4349
169. Hornak V, Okur A, Rizzo RC et al (2006) HIV-1 protease flaps spontaneously open and reclose in molecular dynamics simulations. Proc Natl Acad Sci U S A 103:915–920
170. McGaughey GB, Colussi D, Graham SL et al (2007) Beta-secretase (BACE-1) inhibitors: accounting for 10s loop flexibility using rigid active sites. Bioorg Med Chem Lett 17:1117–1121
171. Kumalo HM, Bhakat S, Soliman ME (2016) Investigation of flap flexibility of β-secretase using molecular dynamic simulations. J Biomol Struct Dyn 34:1008–1019
172. Berhanu WM, Masunov AE (2015) Atomistic mechanism of polyphenol amyloid aggregation inhibitors: molecular dynamics study of Curcumin, Exifone, and Myricetin interaction with the segment of tau peptide oligomer. J Biomol Struct Dyn 33:1399–1411
173. Nguyen P, Derreumaux P (2014) Understanding amyloid fibril nucleation and aβ oligomer/drug interactions from computer simulations. Acc Chem Res 47:603–611
174. Okumura H, Itoh SG (2016) Structural and fluctuational difference between two ends of Aβ amyloid fibril: MD simulations predict only one end has open conformations. Sci Rep 6:38422
175. Melquiond A, Dong X, Mousseau N et al (2008) Role of the region 23-28 in Abeta fibril formation: insights from simulations of the monomers and dimers of Alzheimer's peptides Abeta40 and Abeta42. Curr Alzheimer Res 5:244–250

176. Lemkul JA, Bevan DR (2010) Assessing the stability of Alzheimer's amyloid protofibrils using molecular dynamics. J Phys Chem B 114:1652–1660
177. Brown AM, Bevan DR (2016) Molecular dynamics simulations of amyloid β-peptide (1-42): tetramer formation and membrane interactions. Biophys J 111:937–949
178. Kitamura K, Tamura Y, Ueki T et al (2014) Binding free-energy calculation is a powerful tool for drug optimization: calculation and measurement of binding free energy for 7-Azaindole derivatives to glycogen synthase kinase-3β. J Chem Inf Model 54:1653–1660
179. Kumari R, Kumar R, Lynn A (2014) g_mmpbsa—a GROMACS tool for high-throughput MM-PBSA calculations. J Chem Inf Model 54:1951–1962
180. Gapsys V, Michielssens S, Peters JH et al (2015) Calculation of binding free energies. Methods Mol Biol (Clifton, N.J) 1215:173–209
181. Meirovitch H, Cheluvaraja S, White RP (2009) Methods for calculating the entropy and free energy and their application to problems involving protein flexibility and ligand binding. Curr Protein Pept Sci 10:229–243
182. Roy S, Kumar A, Baig MH et al (2015) Virtual screening, ADMET profiling, molecular docking and dynamics approaches to search for potent selective natural molecules based inhibitors against metallothionein-III to study Alzheimer's disease. Methods (San Diego, Calif) 83:105–110
183. Smalley KSM, Haass NK, Brafford PA et al (2006) Multiple signaling pathways must be targeted to overcome drug resistance in cell lines derived from melanoma metastases. Mol Cancer Ther 5:1136–1144
184. Pilpel Y, Sudarsanam P, Church GM (2001) Identifying regulatory networks by combinatorial analysis of promoter elements. Nat Genet 29:153–159
185. Müller R (2004) Crosstalk of oncogenic and prostanoid signaling pathways. J Cancer Res Clin Oncol 130:429–444
186. Sergina NV, Rausch M, Wang D et al (2007) Escape from HER-family tyrosine kinase inhibitor therapy by the kinase-inactive HER3. Nature 445:437–441
187. Overall CM, Kleifeld O (2006) Validating matrix metalloproteinases as drug targets and anti-targets for cancer therapy. Nat Rev Cancer 6:227–239
188. Force T, Krause DS, Van Etten RA (2007) Molecular mechanisms of cardiotoxicity of tyrosine kinase inhibition. Nat Rev Cancer 7:332–344
189. Keith CT, Borisy AA, Stockwell BR (2005) Multicomponent therapeutics for networked systems. Nat Rev Drug Discov 4:71–78
190. Iqbal K, Grundke-Iqbal I (2010) Alzheimer disease, a multifactorial disorder seeking multi-therapies. Alzheimers Dement J Alzheimers Assoc 6:420–424
191. Bolognesi ML, Rosini M, Andrisano V et al (2009) MTDL design strategy in the context of Alzheimer's disease: from lipocrine to memoquin and beyond. Curr Pharm Des 15:601–613
192. León R, Garcia AG, Marco-Contelles J (2013) Recent advances in the multitarget-directed ligands approach for the treatment of Alzheimer's disease. Med Res Rev 33:139–189
193. Ma XH, Wang R, Yang SY et al (2008) Evaluation of virtual screening performance of support vector machines trained by sparsely distributed active compounds. J Chem Inf Model 48:1227–1237
194. Arooj M, Sakkiah S, Cao G et al (2013) An innovative strategy for dual inhibitor design and its application in dual inhibition of human Thymidylate synthase and Dihydrofolate reductase enzymes. PLoS One 8:e60470
195. Chang S-S, Huang H-J, Chen CY-C (2011) Two birds with one stone? Possible dual-targeting H1N1 inhibitors from traditional Chinese medicine. PLoS Comput Biol 7: e1002315
196. Ma XH, Shi Z, Tan C et al (2010) In-Silico approaches to multi-target drug discovery. Pharm Res 27:739–749
197. Cosconati S, Forli S, Perryman AL et al (2010) Virtual screening with AutoDock: theory and practice. Expert Opin Drug Discovery 5:597–607
198. Huey R, Morris GM, Olson AJ et al (2007) A semiempirical free energy force field with charge-based desolvation. J Comput Chem 28:1145–1152
199. Xie H, Wen H, Zhang D et al (2017) Designing of dual inhibitors for GSK-3β and CDK5: virtual screening and in vitro biological activities study. Oncotarget 8:18118
200. Ma X, Chen Y (2012) Chapter 9: In Silico Lead Generation Approaches in Multi-Target Drug Discovery, Presented at the March 28
201. Huang W, Tang L, Shi Y et al (2011) Searching for the multi-target-directed ligands against Alzheimer's disease: discovery of quinoxaline-based hybrid compounds with

AChE, H3R and BACE 1 inhibitory activities. Bioorg Med Chem 19:7158–7167
202. Yang S-Y (2010) Pharmacophore modeling and applications in drug discovery: challenges and recent advances. Drug Discov Today 15:444–450
203. Sliwoski G, Kothiwale S, Meiler J et al (2014) Computational methods in drug discovery. Pharmacol Rev 66:334–395
204. Caporuscio F, Tafi A (2011) Pharmacophore modelling: a forty year old approach and its modern synergies. Curr Med Chem 18:2543–2553
205. Huang W, Lv D, Yu H et al (2010) Dual-target-directed 1,3-diphenylurea derivatives: BACE 1 inhibitor and metal chelator against Alzheimer's disease. Bioorg Med Chem 18:5610–5615
206. Goyal M, Dhanjal JK, Goyal S et al (2014) Development of dual inhibitors against Alzheimer's disease using fragment-based QSAR and molecular docking. Biomed Res Int 2014:e979606
207. Duan S, Guan X, Lin R et al (2015) Silibinin inhibits acetylcholinesterase activity and amyloid β peptide aggregation: a dual-target drug for the treatment of Alzheimer's disease. Neurobiol Aging 36:1792–1807
208. Nikolic K, Mavridis L, Bautista-Aguilera OM et al (2014) Predicting targets of compounds against neurological diseases using cheminformatic methodology. J Comput Aided Mol Des 29:183–198
209. Fang L, Kraus B, Lehmann J et al (2008) Design and synthesis of tacrine–ferulic acid hybrids as multi-potent anti-Alzheimer drug candidates. Bioorg Med Chem Lett 18:2905–2909
210. Bautista-Aguilera OM, Esteban G, Chioua M et al (2014) Multipotent cholinesterase/monoamine oxidase inhibitors for the treatment of Alzheimer's disease: design, synthesis, biochemical evaluation, ADMET, molecular modeling, and QSAR analysis of novel donepezil-pyridyl hybrids. Drug Des Devel Ther 8:1893–1910
211. González-Díaz H, Prado-Prado FJ, Santana L et al (2006) Unify QSAR approach to antimicrobials. Part 1: predicting antifungal activity against different species. Bioorg Med Chem 14:5973–5980
212. Youdim MBH (2010) Why do we need multifunctional Neuroprotective and Neurorestorative drugs for Parkinson's and Alzheimer's diseases as disease modifying agents. Exp Neurobiol 19:1–14
213. Youdim MBH (2013) Multi target Neuroprotective and Neurorestorative anti-Parkinson and anti-Alzheimer drugs Ladostigil and M30 derived from Rasagiline. Exp Neurobiol 22:1–10
214. Sagi Y, Drigues N, Youdim MBH (2005) The neurochemical and behavioral effects of the novel cholinesterase–monoamine oxidase inhibitor, ladostigil, in response to L-dopa and L-tryptophan, in rats. Br J Pharmacol 146:553–560
215. Weinreb O, Mandel S, Bar-Am O et al (2009) Multifunctional neuroprotective derivatives of rasagiline as anti-alzheimer's disease drugs. Neurotherapeutics 6:163–174
216. Luo W, Li Y-P, He Y et al (2011) Design, synthesis and evaluation of novel tacrine-multialkoxybenzene hybrids as dual inhibitors for cholinesterases and amyloid beta aggregation. Bioorg Med Chem 19:763–770
217. Zhang C, Du Q-Y, Chen L-D et al (2016) Design, synthesis and evaluation of novel tacrine-multialkoxybenzene hybrids as multi-targeted compounds against Alzheimer's disease. Eur J Med Chem 116:200–209
218. Viayna E, Gómez T, Galdeano C et al (2010) Novel huprine derivatives with inhibitory activity toward β-amyloid aggregation and formation as disease-modifying anti-Alzheimer drug candidates. ChemMedChem 5:1855–1870
219. Piazzi L, Rampa A, Bisi A et al (2003) 3-(4-[[benzyl(methyl)amino]methyl]phenyl)-6,7-dimethoxy-2H-2-chromenone (AP2238) inhibits both acetylcholinesterase and acetylcholinesterase-induced beta-amyloid aggregation: a dual function lead for Alzheimer's disease therapy. J Med Chem 46:2279–2282
220. Rampa A, Bisi A, Valenti P et al (1998) Acetylcholinesterase inhibitors: synthesis and structure-activity relationships of omega-[N-methyl-N-(3-alkylcarbamoyloxyphenyl)-methyl]aminoalkoxyheteroaryl derivatives. J Med Chem 41:3976–3986
221. Bajda M, Guzior N, Ignasik M et al (2011) Multi-target-directed ligands in Alzheimer's disease treatment. Curr Med Chem 18:4949–4975
222. Rampa A, Tarozzi A, Mancini F et al (2016) Naturally inspired molecules as multifunctional agents for Alzheimer's disease treatment. Molecules 21:643
223. Costantino L, Barlocco D (2012) Challenges in the design of multitarget drugs against multifactorial pathologies: a new life for medicinal chemistry? Future Med Chem 5:5–7

224. Swarbrick ME (2011)Chapter 3: the learning and evolution of medicinal chemistry against kinase targets. In:Chapter 3:the learning and evolution of medicinal chemistry against kinase targets, pp 79–95, Cambridge, UK.
225. Sun CL, Christensen JG, McMahon G (2009) Discovery and development of Sunitinib (SU11248): a multitarget tyrosine kinase inhibitor of tumor growth, survival, and angiogenesis. In: Li R, Stafford JA (eds) Kinase inhibitor drugs. John Wiley & Sons, Inc, pp 1–39,Cambridge, UK.
226. Lin X, Huang X-P, Chen G et al (2012) Life beyond kinases: structure-based discovery of sorafenib as nanomolar antagonist of 5-HT receptors. J Med Chem 55:5749–5759
227. Csermely P, Agoston V, Pongor S (2005) The efficiency of multi-target drugs: the network approach might help drug design. Trends Pharmacol Sci 26:178–182
228. Hurko O, Ryan JL (2005) Translational research in central nervous system drug discovery. NeuroRx 2:671–682

Chapter 20

Neuropharmacology in Flux: Molecular Modeling Tools for Understanding Protein Conformational Shifts in Alzheimer's Disease and Related Disorders

Gerald H. Lushington, Frances E.S. Parker, Thomas H.W. Lushington, and Nora M. Wallace

Abstract

Several years of exciting discoveries finally promise to break a decades-old impasse in the treatment of many of society's most debilitating neurological disorders, including Alzheimer's disease, Parkinson's disease, amyotrophic lateral sclerosis, and Huntington's disease. These breakthroughs are beginning to paint a detailed picture of the causes and effects associated with polynucleotide-associating domains of key central nervous system proteins that misfold into energetically favored physiologically dysfunctional forms. Unfortunately, the paradigm differs so dramatically from conventional pharmacological scenarios that the translation of fundamental molecular concepts to practical therapeutic design is far more challenging than merely identifying a novel enzymatic or cellular target. This chapter seeks to grasp the fundamental physiological issues that cause neuropathies and maps out the ways in which molecular docking and molecular dynamics simulations can be brought to bear in formulating testable hypotheses that can form a basis for the systematic formulation of a new generation of medicines.

Key words Alzheimer's disease, Amyloid beta, Amyotrophic lateral sclerosis, Molecular simulations, Prion disorders, Stress granules, TDP-43, TIA-1

1 Introduction

In the course of biomedical research, there arise the occasional junctures where existing conventions and protocols have pointed in one direction for a long time, but observations begin to orient persistently elsewhere. Such junctions often pose an anxious blend of optimism and uncertainty, as some look forward with hopes for breaking long-standing impasses, while many gaze back with concern not only at old hypotheses that appear increasingly destined for abandonment but also expertise, skills, and even data that risk deprecation.

As an initial "flurry" of papers relating a broad range of neurological disorders to subtle advances in structural biology

Kunal Roy (ed.), *Computational Modeling of Drugs Against Alzheimer's Disease*, Neuromethods, vol. 132, DOI 10.1007/978-1-4939-7404-7_20, © Springer Science+Business Media LLC 2018

understanding of polynucleotide-associating proteins has gradually built toward an information "storm," it appears that neuroscience is proceeding toward an exciting new cusp in fundamental understanding that may begin to finally crack some of the most intransigent pathologies.

The immediate penalty for such a prospective advance, however, is that a true recognition of many critical underlying details is likely to lag well behind the entrenchment of new "big-picture" understanding, and a tremendous amount of basic biochemical and physiological research will be required before identified new therapeutics opportunities can be practically and effectively pursued for development. As opposed to well-established paradigms, where much more underlying detail has already been well quantified and characterized, we must be prepared for these new opportunities to be expensive in terms of both effort and capital.

As a means for enhancing the cost-effectiveness of breaking in a new subdiscipline, however, this chapter looks not only at the emerging new opportunities in neuromedicine but also at how computational methods can be productively applied to help efficiently lay the foundation for advances. In particular, part of our focus will be on the application of molecular docking and molecular dynamics simulations for identifying and computationally prioritizing molecules that may stabilize neuropathically vulnerable protein fold structures, while a second component of the chapter examines the use of molecular dynamics simulations to perceive, test, and refine peptide mimics intended to blunt the neurotoxicity of misfolded protein oligomers.

1.1 The Structural Basis for Common Neurological Disorders

Alzheimer's disease (AD), other dementias, and a number of additional neuropathic disorders such as Parkinson's disease (PD) and amyotrophic lateral sclerosis (ALS) have a dubious distinction for costly failure [1]. Namely, such neuropathies impose an exceptional economic penalty. Twenty percent of the current annual expenditures for the American Medicare system are spent on AD alone, while not a single new treatment has been approved for this uncurable, and barely treatable, disease since 2003 [2]. Such futility is a stain on the global biomedical community and is a clear indication that our core research strategies are consistently failing us for this disease class.

After well over a decade of futility, it is time to admit that the failure of currently popular paradigms for discovering new neuropathy treatments is likely to continue indefinitely, because the underlying concept does not recognize critical principles relating to the underlying physiological dysfunction that must be corrected. Presumably, then, a fundamental conceptual shift is required in order to turn the tide. Specifically, we may need to admit that the long-standing conventional medicinal principles which suffice fairly well for addressing cancer and infectious diseases are ill-suited for battling many neuropathies. In particular, it seems appropriate at this stage to reconsider whether neuropathic drug discovery research should remain grounded in standard proteocentric principles.

Over the past 50 years since the first crystallographically resolved enzyme structures began to appear, rational drug design has focused heavily on finding molecules (everything from small organics to large biologics) that modulate protein function. Such modulation treats illnesses either by disrupting the physiology of pathogens or enabling an artificial rebalancing of biochemical pathway irregularities.

Within a research culture dominated by the paradigm of protein functional modulation, other biomolecular strategies have nonetheless begun emerging to provide novel therapeutic targets. Examples include gene therapy [3] and the regulation of polynucleotide transcription factors [4]. Also outside of the primary targeting paradigms and of particular relevance to the neuroscience community is the search for schemes to mitigate harm that arises from uncontrolled protein misfolding.

A common example of uncontrolled conformational shifts involves structurally labile alpha-helical domains in peptides and proteins of importance to neurological function [5]. The most famous example of alpha-helical misfolding is the so-called prion – a proteinaceous infectious particle. The pathological implications of prion-related dysfunction initially rose to the public's attention during the late 1980s to early 1990s during Britain's mad cow disease epidemic, whereby an outbreak of the previously rare and ill-characterized neuropathy known as bovine spongiform encephalopathy (BSE) began to sweep through the United Kingdom, accompanied by revelations that consuming the meat of BSE-infected cattle was capable of inducing a variant of the generally fatal Creutzfeldt-Jakob disease (vCJD) [6] in humans.

Subsequent determination by Prusiner [5] that the primary infectious agent in BSE and vCJD was a single form of malformed protein, called a prion, roiled the biomedical community by highlighting an infectious pathogen that defied the central dogma of biology. Specifically, for such disorders, the structure, function, and reproductive capacity of the pathogen is not transmitted to proteins from a code within an organism's DNA or RNA genome but rather is manifest strictly within a single misfolded protein itself. In particular, the prion-related infection initiates from a conformational shift that transforms the predominantly α-helical PrP^{C} (cellular prion) protein into the PrP^{SC} (scrapie prion) protein, where the latter embodies a greater portion of lower energy β-sheet structure. This conformational transition occurs semi-spontaneously upon contact between healthy PrP^{C} and previously misfolded PrP^{SC} protein structures. The resulting new self-propagating PrP^{SC} protein is then no longer available to perform the fundamental physiological role of PrP^{C}. Instead, the newly formed PrP^{SC} units can join an arsenal of infective protein monomeric proteins, form extended molecular plaques [7], act via small toxic oligomers [8], or engage in some evolving combination thereof.

If the prion etiological phenomenon were only manifest within the realm of rare zoonotic diseases, it is unlikely that the incumbent molecular concepts would have ever begun to lead the field of pharmacology toward the cusp of transformative new pathological understanding and novel therapeutic strategies. In the past 15 years, however, scientists have begun to grasp that BSE and vCJD may represent the mere tip of a neuropathic iceberg. In 2002, a review by Eikelenboom et al. noted how key symptomatic similarities existed between prion diseases and Alzheimer's disease (AD), in terms of elevated expression of key inflammation markers, as well as the physical presence of protein plaques with a comparable amyloid appearance [9]. In fact, an earlier paper had already proposed a key structural analogy between Alzheimer's and BSE/vCJD, in suggesting that the amyloid plaques in AD might be a manifestation of prion-like helix-to-sheet conformational shifts within a 25 amino acid portion of a helix-coil-helix component of the $\alpha\beta^{40}$ and $\alpha\beta^{42}$ peptides cleaved from the amyloid precursor protein (APP) [10]. These convergent observations soon led to a natural extrapolation to analogous causative mechanism between prion diseases and the advent and propagation of AD symptoms [11–14].

Over the subsequent years, researchers have begun establishing possible links between the prion-like conformational transformation and the etiology of numerous other neurological (or related) disorders, including diabetes mellitus type 2 [15], Parkinson's disease (PD) [16], Huntington's disease [17], rheumatoid arthritis [18], familial amyloid polyneuropathy [19], cerebral amyloid angiopathy [20], sporadic inclusion-body myositis [21], frontotemporal dementia [22], amyotrophic lateral sclerosis (ALS) [23], chronic traumatic encephalopathy [24], and multiple sclerosis [25].

In all of these diseases, one or more misfolded forms of key protein(s) arise within our bodies. In most cases, the resulting misfolded structures are fairly biochemically inert. As opposed to the case of some compositional mutations, for example, the prion-like misfolded protein does not engage in misregulated metabolic or signaling processes. However, the misfolded structure also fails to carry out the protein's intended function, and its infectious nature means that it induces a gradual but progressively deleterious effect, generally by plundering healthy protein from within vital physiological processes, incorporating them into nonfunctioning protein oligomers with subsequent degradation toward fibrils or plaques. A worsening state of infection may produce secondary harm due to a gradual profusion of sizable inclusion bodies (larger fibrils and/or plaques) that may interfere with cell function or damage membranes. The practical implications of these findings are sobering, in that the fundamental dysfunction is not readily addressed by standard pharmaceutical research practice.

Conventional drug discovery has typically focused heavily on two distinct objectives: finding ways to kill pathogens that may have entered our bodies and finding ways to modulate biochemical imbalances that emerge intrinsically within otherwise healthy physiological processes. In this case, there is no adversarial "cell" to kill. The standard antipathogenic mindset thus encounters a dilemma of how one should go about trying to "kill" a protein especially one that is compositionally (though not conformationally) identical to proteins that perform important neurological functions for us.

Thus, historically, we have forsaken the pathogen eradication paradigm, and instead the primary pharmaceutical recourse has aimed to treat prion-like diseases by seeking to compensate for losses in key physiological functions. Unfortunately, for many neuropathies, it is often uncertain precisely what function (or set of functions) is contingent upon healthy action associated with the conformationally susceptible protein. Thus, in most cases, it has proven difficult to target one particular aberrant biochemistry that can be readily modulated. Rather, the etiology is often characterized in vague terms as an escalating loss of multiple interrelated functions.

Without some systematic way to arrest and reverse this loss of function (something which cannot be claimed by any treatments for most prion-like neuropathies [26]), the best that can be achieved therapeutically is to stumble haphazardly onto strategies for compensating for the right balance of functional loss, e.g., by amplifying the efficiency of an otherwise depleted pathway or by training a different pathway to effect some functional duplication. There are numerous such therapeutic schemes in place for incurable neuropathies (developed at great cost and effort), but none currently provide anything more than temporary palliative benefits.

Consequently, all extant treatment schemes for AD, ALS, PD, and many other comparable diseases have produced, at best, a measure of success which can be quantified in terms of lessening the symptoms or slowing disease progression [27]. The only reported exception to this arises from a treatment protocol outside of conventional pharmaceutical practice, namely, experimental stem cell treatments. On a very preliminary basis, such schemes have demonstrated promise in partial reversal of prior neurodegradation and thus a partial restoration of lost function. The effect, unfortunately, is temporary and may only be sustained through continuous cell replenishment [28].

Although the impermanent nature of palliative treatments limits their ultimate value to human health, a far greater concern arises from some therapeutic strategies under consideration that may do more harm than good. Using diabetes as analogy (a choice we believe to be appropriate since type II diabetes resembles many neurodegenerative disorders in loss of physiological function via depletion of critical molecules by misfolding), there is evidence that medical interventions aimed at bolstering availability of dysregulated functional biochemicals are ultimately self-defeating and, indeed, toxic. In the case of diabetes, the regular insulin replenishment regimen, while the only approved current way to amplify flagging carbohydrate metabolism, appears to actually accelerate disease progression, as is evinced by faulty glycemic regulation near injection sites, due to injection-related insulin-amyloidosis [29]. This may indicate that, without properly addressing the key causes of disease, any artificial amplification of healthy protein may shift the kinetic gradient further in favor of dysfunctional oligomer, fibril, tangle, or plaque growth, thus advancing and entrenching the underlying disorder.

A second strategy that can produce unexpectedly self-defeating outcomes involves anti-pathogen-like attempts to physically or chemically degrade the fibrils or plaques. In this case, breakage of the larger misfolded bodies can produce a much greater number of small but equally (or more) “pathogenic” fragments that cumulatively accelerate the net misfolding kinetics relative to the original fibril or plaque [30, 31]. Once again, an earnest attempt to treat such disorders can, without adequate foundation in physiological understanding, actually worsen a patient’s health.

While these unintended negative consequences are alarming and demoralizing, it is difficult to feel any more encouraged by the many other current medicinal strategies that, although not obviously harmful, have dubious efficacy that may not significantly exceed the value of lifestyle changes or even placebos which, at best, merely slow the disease progression.

In current practice, regardless of treatment regimen, the symptom severities of AD, ALS, PD, and HD all seem fated to increase over time. In all of these disorders, a spontaneously occurring, self-catalyzing, kinetically favored molecular transformation process is propagating through a body, unchecked by medical intervention. What can be productively attempted to arrest it?

The impasse clearly begs for new strategies. Yet; however challenging it has proven to make practical headway, the answer may be conceptually straightforward – one must find some way to identify key circumstances that enable the debilitating catalytic depletion of healthy protein, and then one must effectively shift the kinetic landscape away from those circumstances. In other words, we need some way to biochemically reengineer the interaction network to push the balance back in favor of the healthy protein conformations.

Using strategies at least somewhat akin to structure-based drug design, one can envision two distinct approaches to address this challenge. The first scenario, somewhat analogous to enzyme inhibition strategies in rational drug design, would entail blocking the interaction between the healthy and misfolded proteins, perhaps by finding binders that couple to the specific prion-like residues in diseased molecules, thus preventing them from initiating misfolding in healthy proteins. A second prospect, comparable to allosteric drug design strategies, would merely be to find molecules that stabilize the healthy conformations of proteins to the point where they are significantly less susceptible to suffering prion-like degradation.

Perhaps the most effective strategy might be embodied by a hybrid scheme that seeks to simultaneously defang the infective misfolded proteins via one therapeutic while stabilizing healthy proteins with a second agent. Given suitable efficacy by either one, or both, of these strategies, one might envision effectively halting disease progression in a way that permits gradual clearance of the misfolded proteins and a concurrent physiological restoration of a healthy concentration of the depleted protein. Such strategies may also prove effective in mitigating the worst risks associated with compensating for diminished functionality or actively promoting clearance of existing oligomers, fibrils, and plaques, thus rendering some of the existing drugs (or drug development scenarios) more effective and safely applicable.

This chapter aims to illustrate specific strategic opportunities that can be applied toward (a) stabilizing healthy proteins that, upon exposure to prion-like entities, are susceptible to induced misfolding and (b) developing special peptides that will install non-catalytic termini onto the infectious ends of misfolded beta sheets, thus hopefully diminishing (or potentially even eliminating) the core threat of the disorder. Given how novel these tasks are relative to conventional pharmaceutical practice, the chapter will also outline research methodologies that can be applied toward these goals, with a focus on molecular modeling techniques that can elucidate candidate therapeutic formulations for preliminarily testing.

1.2 Physiological Implications of Misfolding

There is substantial variation in the biochemical functions of proteins that are susceptible to prion-like misfolding and, although the protein restructuring is most commonly thought of as a transformation toward a beta-sheet structure (much more amenable to strong intermolecular coupling) from a predominantly alpha-helical conformation (dominated by intramolecular stabilization), there is a fair variety among the native conformations that are susceptible to conversion, including disordered chains, as well as mixes of short helix, coil, beta turn, and/or beta-sheet segments.

In the case of cellular prion protein, PrP^C, the core portion of the protein that is susceptible to misfolding entails the entirety of two helices, plus portions of two other helices, the entirety of a small two-strand sheet, and intervening turns and coils. The

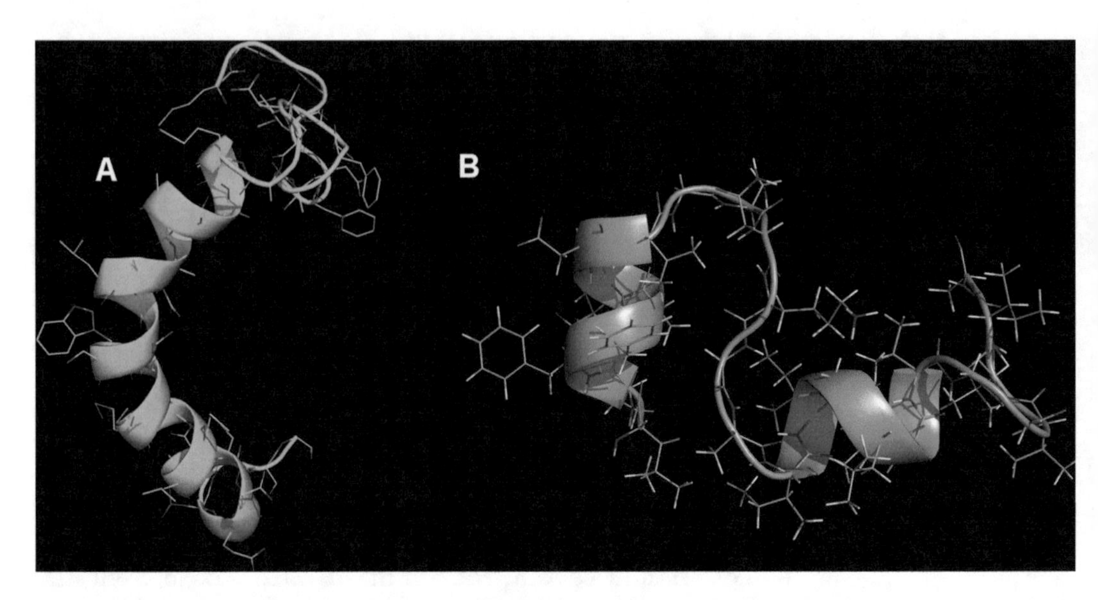

Fig. 1 Polynucleotide-associating domains of TDP-43 (**a**) and amyloid beta (**b**). The former is in a bent helix conformation already well suited to DNA or RNA major groove binding, whereas the latter is believed to conformationally assemble in the presence of a polynucleotide

physiologically critical portion of this prion domain is a helix that interacts with apurinic/apyrimidinic endonuclease in order to initiate DNA repair upon detection of genotoxic damage [32] – a function that is wholly lost upon misfolding to the PrP^{SC} conformation. Beyond the loss of PrP^{C} function, the key causes of prionic neurotoxicity remain unknown, although some have proposed a disproportionate, harmful inflammatory response [33]. Observations that administration of PrP^{SC} to PrP^{C} knockout models does not produce observable illness imply that PrP^{SC} itself may not produce a direct toxic effect and that mature fibrils are not particularly harmful [34]. The assumption, rather, is that critical toxicity manifests in some early misfolding stage, such that endogenous membrane-associated PrP^{C} proceeds first to a harmful intermediate, perhaps one that effects disruptions within critical protein-protein interaction networks on neuron surfaces [35].

Although the primary known function of healthy PrP^{C} entails fostering specific protein-protein interactions, another common role observed for other prion-like domains is in gene regulation. Specifically, the glutamine/asparagine-rich prion domain (QNPD) of TDP-43 (relevant to ALS and CTE; see Fig. 1a) is a known DNA major groove binder [36], while a number of other putative prion-like domains, such as those in the $\alpha\beta^{40}/\alpha\beta^{42}$ peptides relevant to Alzheimer's disease (Fig. 20.1b), have also been characterized as possible or probable DNA binders [37]. In the case of ALS and CTE, it appears that loss of TDP-43 function is adequate to explain the toxicological progression of the illness [38]. For Alzheimer's disease, the precise relationship between $\alpha\beta^{40}/\alpha\beta^{42}$ misfolding and the onset of neuronal death is more convoluted; however, clear evidence has emerged that the critically toxic substance entails modest sized, plasma soluble $\alpha\beta^{40}/\alpha\beta^{42}$ oligomers [39]. This oligomer-dependence has helped to foster a belief that neuropathy arises from subsequent misfolding of the tau protein, where such tau misfolding may be directly catalyzed by exposure to $\alpha\beta^{40}/\alpha\beta^{42}$ oligomers [40].

1.3 Prion Misfold Propagation

By far, the most common mechanism by which one protein molecule attains a prion-like conformation is via catalytically induced misfolding prompted by interaction with another previously misfolded protein. An excellent example of the disruptive nature of the interaction interface presented by a prion-like fold is rendered in Fig. 2 for the prion-forming portion of $\alpha\beta^{40}/\alpha\beta^{42}$ (residues 17–41).

Within Fig. 2, the outermost beta strands of a prion-folded fibril present a series of alternating backbone hydrogen acceptor (red circles) and donor (blue circles) sites for possible complexation by any amino acid sequence, except for those containing sterically blocking proline residues. In practice, though, backbone interactions alone are rarely sufficient to overcome entropic penalties associated with beta-sheet formation. Rather, the binding thermodynamics are usually only preferred over other secondary structures if augmented by reasonable side chain complementarity between any adjacent pair of strands. In Fig. 2, a number of key side chain stabilizations can be observed, including alternating aspartate-lysine salt bridges (purple arrows), and a variety of same-same interactions (brown arrows) such as stacked phenylalanine, leucine, isoleucine, and methionine hydrophobic interactions. Although

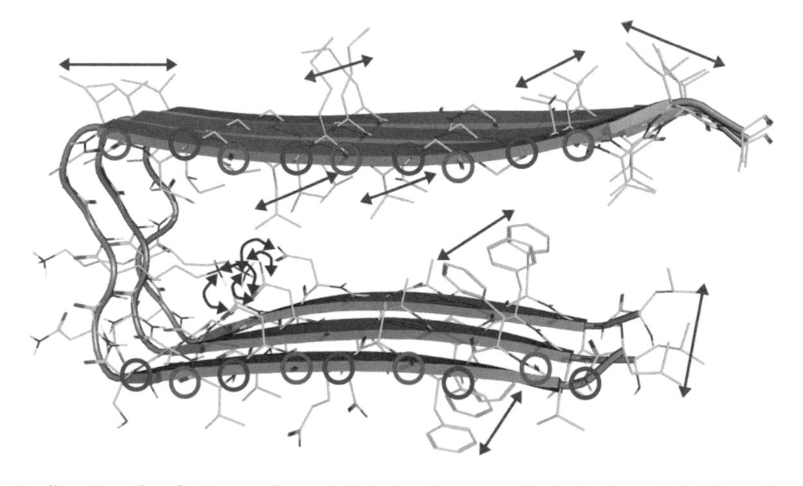

Fig. 2 Energetically attractive features of amyloid beta oligomers for inducing misfolding. *Circles* represent favorable H-bond donor (*blue*) and acceptor (*red*) sites. *Brown arrows* show attractive same-same intermolecular lipophilic coupling sites. *Magenta arrows* depict an opportunity for cross-peptide salt bridge formation

not shown in Fig. 2, it is worth mentioning that the stabilization of the TDP-43 protein's misfolded QNPD (responsible for ALS) is mediated by interstrand H-bond networks involving the amide termini of asparagine and glutamine side chains.

In general, prion-misfolded strands are characterized by a substantial number of salt bridges and complementary same-same (or same-similar) side chain stabilizations. In cases with exceptionally favorable prion associative motifs (e.g., $\alpha\beta^{40}/\alpha\beta^{42}$ [41] and TDP-43 [42]), one observes spontaneous misfolding-induced aggregation of proteins in vitro via prion seeding experiments [43], wherein wild-type protein solutions convert rapidly to insoluble fibrils upon exposure to a small amount of side-chain-compatible prion-folded protein. There is every reason to believe that similar conversion may occur efficiently in vivo.

As suggested above, perfect sequence identity is not required for a prion seed to misfold another protein, but it seems that a certain minimum (albeit not yet quantified) degree of complementarity is required. As case in point, there are several known instances of extended beta-sheet prion-folded structures (comparable to what is depicted in Fig. 2) existing naturally in healthy living creatures.

One interesting example of healthy prion-like beta assembly involves HET-s, which serves as natural and important component of cell necroptotic pathways in both mammals and prokaryotes [44]. Although HET-s proteins self-aggregate into fibrils, there is no evidence that HET-s fibrils are a key cause of any prion-like disorder as might be initiated by HET-s induced misfolding of heterogeneous proteins. Interestingly, HET-s fibrils are known to coexist with HET-s monomers, and there is little indication that the fibrils grow to an extent that yields physiological harm.

A second example that deserves attention here, and later in the chapter, is TIA-1 – a key adhesive component within neuronal stress granules. The function of TIA-1 is to respond to cellular threats (e.g., membrane disruption, oxidative stress, and invasive threats) by sequestering critical cellular components (mostly soluble RNA and RNA-binding proteins) into a discrete bundle that protects those components from harm and shelters the cell from spurious signaling. In a manner somewhat analogous to the *K'nex* children's toy, TIA-1 reversibly self-assembles into extended beta-folded prion-like fibrils while simultaneously achieving heterogeneous connectivity in a form that avails multiple intact RNA-binding sites. The final granule extension is then achieved by virtue of bound RNA coupling. A conceptual schematic depicting such a multicomponent network is shown in Fig. 3.

As with HET-s, the prion-extension evident in TIA-1-bound granules is reversible, where the latter generally dissipates once cellular homeostasis is restored. This regulated behavior of HET-s and TIA-1 prions poses a marked contrast with the conduct of

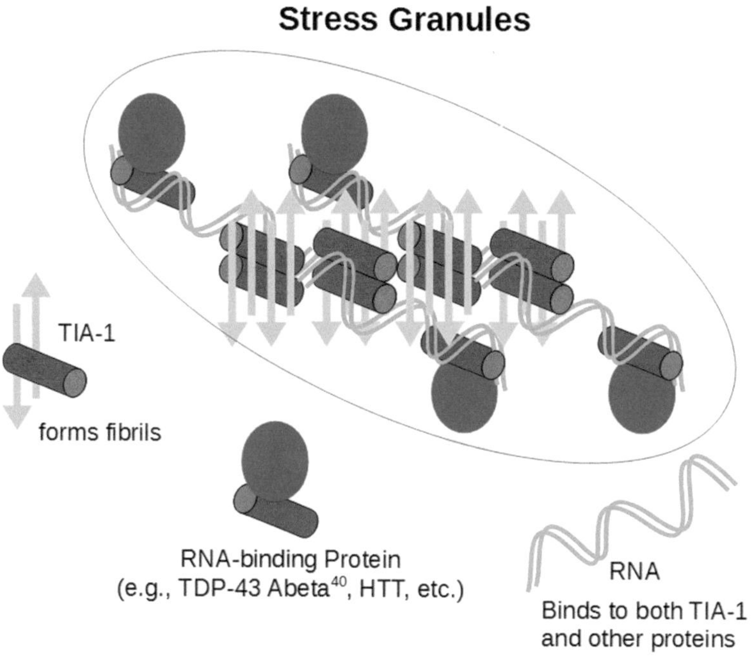

Fig. 3 Simple schematic showing known intermolecular associations fostered in stress granules. The TIA-matrix is shown as a prion backbone (*yellow arrows*) with associated non-prion helices (*red cylinders*) which bind to intracytosolic RNA (*green curves*). RNA tails also bind numerous different forms of neurologically relevant proteins (*blue ovoids*, with magenta RNA-binding helices)

more harmful analogs such as PrP^{SC}, misfolded $\alpha\beta^{40}/\alpha\beta^{42}$, and the TDP-43 QNPD. In these latter cases, there is no confirmed physiological benefit to the beta-misfolded conformers, while the thermokinetically favored infectious nature of the prion fold seems to preclude the likelihood of a dynamic equilibrium between the physiological and diseased states.

A critical difference between reversible, healthy prion-like constructs and unregulated pathogenic analogs is the presence of regulatory checks in the former. With TIA-1, it is believed that heat shock proteins are capable of acting upon suitable signals to refold the protein into a non-prion-like conformation, under which the stress granules dissociate [45]. In the case of HET-s, autoregulation is achieved via a capping domain located close to the C-terminus of monomeric HET-s. Cleavage of this cap leads to fibril growth, but physiological control can be exerted to restore conditions favoring retention of the C-terminal residues, thus reversing fibril-formation [44].

As will be discussed later, artificial analogies to a HET-s capping concept may afford us with insight into strategic opportunities for halting the disease progression of Alzheimer's disease, ALS, and numerous other prion-related conditions.

1.4 Prion Misfold Initiation

While the processes of oligomer and fibril extension have been characterized experimentally [45] and have been rationalized in molecular dynamic simulations [46], one biomedically critical part of prion disease etiology remains contentious – the precise origin of initially misfolded seeds from which misfolded structures may arise in the first place.

While vCJD cases are generally attributable to human exposure to misfolded exogenous proteins arising from prion-infected animals [47], there is no evidence for the inter-organism transmission of many other misfolded proteins such as Alzheimer's $\alpha\beta^{40}/\alpha\beta^{42}$ [48] and the TDP-43/QNPD in ALS. What, then, is the molecular basis by which AD, ALS, and many other prionic neuropathies may take hold in a human body in the first place?

The simplest hypothesis to entertain for the creation of misfolded prion domains is simply that, by chance, a properly folded protein might stochastically vibrate its way into a prion-like beta-strand conformation. To support this prospect, it can be argued that there may exist environmental conditions under which an innate entropic preference for weak helices within prion-like proteins may tip toward the enthalpically favored strands that can be associated into beta sheets.

Empirical observations suggest that conditions favoring beta misfolding may include the influences of oxidative stress [49] or abnormal posttranslational modifications such as hyperphosphorylation [50]. A prospective role has also been ascribed to critical genetic mutations that might enable production of proteins that are particularly susceptible to conformational shifts away from physiologically natural structures [49].

However, such conditions, by themselves, are probably insufficient to spur de novo misfolding, since isolated beta strands generally do not naturally exist in the body in much more than a transient state. Importantly, a stable beta misfold in one prion-like domain requires at least a second complementary strand to sustain the misfold via mutual stabilization. This chance concurrence of two closely juxtaposed rare events affords almost no opportunity for the original misfolded prion-like seed to arise in an infectious state through purely spontaneous means.

Resolution of this dilemma may remain elusive for some time, but the informational basis for a comprehensive theory is almost certainly already beginning to accrue. In particular, it is reasonable to assume that a fundamental understanding of the mechanistic origin of prion dysfunction may be illuminated by a very clear account of how prions "function," and an appreciation for how such function may be turned against the protein.

In this case, it may prove useful to further examine the fact that a fair number of prion-like domains, as previously mentioned, perform key DNA-binding roles within our nervous system. It may thus be useful to definitively characterize this aspect of the behavior of $\alpha\beta^{40}/\alpha\beta^{42}$ [37] and TDP-43/QNPD [51], as well as to seek for analogous physiological function evident among other proteins containing potentially neuropathic prion-like domains.

If we extend our survey to the most famous prion, namely, PrP^{C}, we encounter an interesting ambiguity. Although there have been no clear in vivo demonstrations of the endogenous cellular prion protein having a key physiological role as a DNA binder, the prospect of meaningful PrP^{C}-polynucleotide interactions has been highlighted by studies finding that the KKRPK motif within recombinant PrP^{C} serves as a nuclear localization signal and is observed to complex with both DNA and RNA [30].

The ability of recombinant PrP^{C} to bind to short chain DNA and RNA may be an important clue regarding the commonalities among prion disease etiologies and may shed light on the underlying molecular causative basis shared by these neuropathies. From this perspective, we can draw functional analogies between $\alpha\beta^{40}/\alpha\beta^{42}$, TDP-43/QNPD, and PrP^{C}. We can also seek the implications of prion proteins that may interact with both DNA and RNA (a class that, in addition to PrP^{C}, includes the TDP-43/QNPD [52], the Tau protein of relevance to AD [53, 54], and the alpha-synuclein protein in PD [55]). We may then also consider the fascinating observation that, among 210 proteins with known (ca. 2012) canonical RNA recognition motifs, 29 have prospective prion-like domains [52]. This observation seems particularly worrisome, since it raises a prospect that many types of proteins may have misfolding-based neuropathic consequences beyond those that have already been biomedically identified.

Is there some fundamental reason why such a large percentage of known polynucleotide-binding domains of proteins should be susceptible to prion-like misfolding? If so, does this conceptual linkage have implications toward the cause of prion-related diseases?

As with many pairs of correlated observations, it is not clear precisely what the causative relationship between the presence of a polynucleotide-binding domain and prion-like misfolding might be. Are there some structural or dynamic properties of prion domains that make them independently both amenable to polynucleotide-binding and susceptible to prion-like misfolding? Or, do polynucleotide-binding events somehow predispose the normal domain structure to misfold?

In arguing for the first scenario, it is fairly well established that, within (and proximal to) polynucleotide-binding protein helices, conformational flexibility helps to enable the protein to couple strongly with the curving structural cavity associated with double-stranded polynucleotide major or minor grooves [56]. Such flexibility typically yields secondary structure features whose conformational dynamics reside near a stability cusp that differentiates transient structures (those that spontaneously arise and vanish into plasma-soluble "unstructured" coils) from crystallographically distinguishable folds such as transmembrane-spanning domains and core helices.

Unfortunately, as mentioned earlier, even if a helix is weak, the prospect for transitioning to a stable beta strand is minimal without a preexisting sheet template to associate with. Thus, while conformational flexibility may facilitate the capacity of a prion domain to participate in fibril extension, the flexibility alone is not sufficient to initiate a misfold.

The second hypothesis, whereby some aspect of RNA- or DNA-binding helps to trigger a misfold, is more likely. Plausible corroboration for this concept has been building in recent years. In particular, a hypothetical mechanistic basis for polynucleotide-facilitated prion-domain misfolding is beginning to take shape. As opposed to assertions that RNA or DNA is a direct cause of the conformational transition (e.g., see ref. [57]), evidence is building for a plausible role involving the aforementioned natural cellular process known as stress-granule formation.

Stress granules are small, non-lipid subcellular compartments formed under conditions of perturbed cellular homeostasis. The primary function of stress granules is believed to entail temporary sequestration of much of a cell's mRNA, as a means to prevent the mRNA from either abetting the invasion of an external pathogen or contributing to unintentional further cellular degradation by signaling for structural or chemical deconstruction [58]. In this sense, stress granules are assumed to be a temporary cellular defense mechanism aimed at weathering adverse conditions such as oxidative stress, inflammation, and microbial challenge.

Because stress granules are intended, physiologically, as short-duration constructs, our body does not accord them the stability that comes with having a lipid membrane. Rather, the granules aggregate within a matrix whose primary adhesive is the RNA-binding protein TIA-1. From the TIA-1 structure shown in Fig. 4, one sees precisely the same sort of two-helix/four strand X-bundle that has been predicted as the most plausible structure of

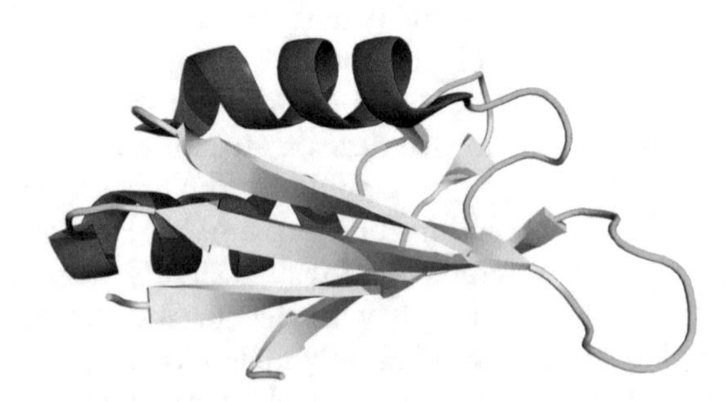

Fig. 4 Structure of TIA-1 showing a bundle of alpha helices (*red*) associating with a small beta sheet (*yellow*)

the known misfolded prion, PrP^{SC} [5]. Indeed, this is likely not coincidental, because TIA-1 is known to have an innate natural capacity for weakly prion-like aggregation, courtesy of an oligomerically extensible beta sheet depicted (yellow strands) in the foreground of Fig. 4, immediately below the primary RNA-associative domain (red helix at the top of the graphic).

Conditions of physiological stress are assumed to encourage TIA-1 proteins to assemble into a network that presents numerous sites for mRNA association. Numerous RNA-binding peptides and proteins, such as $\alpha\beta^{40}/\alpha\beta^{42}$, TDP-43, Huntingtin [59], alpha-synuclein [60], and so forth, may be secondarily recruited (as rendered schematically in Fig. 3) into the granule due to their tendency to also associate with the mRNA molecules that are concentrated within the network.

As articulated in the seminal review by Waris et al. [61], this close juxtaposition of a weak natural prion (TIA-1) and potentially neuropathic prion-susceptible molecules may constitute a risk factor for prion-related neurodegenerative disorders. In particular, the compact, concentrated granule environment (so many prion-susceptible proteins in close conjunction to so many transient prion-like oligomers) may almost functionally approximate the conformationally persuasive pressures of molecular chaperones, with the exception that stress granules steer toward neuropathically risky structures.

At this point, the quantification of such risks is mere speculation. Based on empirical logic, however, one might surmise that the prion-initiation risk associated with stress-granule formation is generally acceptable (i.e., many people suffer countless episodes of cellular stress without acquiring severe misfolding-related neuropathies), but the opportunities for problems may be exacerbated by extenuating circumstances such as oft-repeated stress-granule formation, as a function of:

1. Physicochemical stress (e.g., recurring physical [62] or toxic trauma [63])
2. Biochemical imbalance (e.g., metabolic disorders) [64]
3. Microbial challenges [65–67]
4. Sheer chronological age [68]

Although there is no direct evidence yet to definitively conclude that TIA-1 is an initiator of toxic fibrillogenic oligomers of prion-susceptible proteins, the conceptual logic underlying such a cause is compelling. Firstly, common neuropathic circumstantial biomarkers (e.g., cerebral trauma and inflammation, viral infections, and age; possibly also the diabetes) correlate well with scenarios likely to produce elevated stress-granule exposure. Secondly, stress granules colocate those proteins most commonly associated with prion-like neuropathies, including $\alpha\beta^{40}/\alpha\beta^{42}$, TDP-43,

Huntingtin, and alpha-synuclein as mentioned above. Tau also factors critically in stress granules, since granules generally associate with Tau proteins in microtubules. Finally, we have the conceptual argument that stress granules may provide a rare opportunity for prion-susceptible domains to have prolonged contact with existing prion-like folds of TIA-1.

This begs the question of whether an exposed beta strand within a TIA-1 fibril can induce misfolding with prion-susceptible domains of compositionally distinct proteins. TIA-1 does not have close sequence homology or beta-complementarity with other neuropathic prion-like domains, although there is an octapeptide unit within TIA-1 (sequence: VQQQNQIG) that does have exceptional local homology with an octapeptide component of the TDP-43 (ASQQNQSG) prion domain believed to play a critical role in the onset of ALS. At the time of writing, the authors have not yet confirmed that a high-identity octapeptide is adequate to induce a conformational transition in the TDP-43 helix; however, molecular dynamics simulations are planned to address this question. Comparable simulations can also be formulated to probe the prospective misfolding kinetics of other prion-susceptible proteins with lesser side chain complementarity.

This hypothesis purports to crack the lock on an urgent and hitherto unsolved medical issue and may furthermore lead to many novel opportunities for neuropathic drug discovery. Much study remains before the notion can be accepted as probable fact; however, the many plausible aspects of the concept do suggest a legitimate concern that extensive colocation of prion-susceptible proteins and beta-oligomeric TIA-1 within stress granules will amplify misfolding chances and that repeated stresses may eventually reach a statistical tipping point beyond which misfolding becomes likely.

2 Methods

2.1 Possible Misfold Mitigation Strategies

Given strong evidence that the etiology of prion-related neuropathies is at least somewhat related to a stable conformational transformation of prion domains (of various different native structures) into a pathological extended beta-sheet network, it is reasonable to assume that numerous possible corrective opportunities exist. In particular, for any conformational degeneration step that we may perceive as leading to an initial misfold (or to an oligimeric extension thereof), it should be possible to identify plausible molecular intervention strategies to hinder the degradation or potentially halt it outright. Consequently, we may imagine formulating possible therapeutic targets based on any of the following possible intervention scenarios:

1. Prevention of potentially deleterious beta-induction within stress granules upon prion-susceptible peptides and proteins, as effected by weakly prion-like proteins such as TIA-1
2. Conformational stabilization of native/healthy DNA- or RNA-associating helices in peptides, and proteins such as $\alpha\beta^{40}/\alpha\beta^{42}$, TDP-43, Huntingtin, and alpha-synuclein
3. Structurally capping misfolded prion-protein oligomers, fibrils, and plaques to halt disease progression

Among these three mitigation scenarios, the first is the least likely to bear therapeutic fruit, in that although it seems appealing to try to correct a prospective misfolding indication very early in the etiology, the circumstances defy conventional prescription protocols. In particular, we do not yet have the means to predict whether any given granulation event (very common happenstance in all eukaryotes) is likely to initiate a new pathologic indication (very rare, even for subjects classified by multiple neuropathic risk factors). In other words, it is very different to envision what logical criteria one might set for useful administration of stress-granule modulation therapeutics, and we are unaware of any current compelling biomarkers that might reliably dictate blanket anti-neuropathic prophylaxis campaigns.

Helical stabilization seems to afford a more feasible goal, especially considering there are known alpha-helical stabilizing agents within the broader pharmacopeia [69, 70]. The original notion of stabilizing $\alpha\beta^{40}$ or $\alpha\beta^{42}$ helices for prospective AD treatment was raised from observations by Fezoui and Teplow that coadministration of trifluoroethanol (TFE) to in vitro quantities of helical $\alpha\beta^{42}$ tended to significantly stabilize the helices upon challenge by the standard fibrillization assay [71]. Attempts to leverage these findings toward practical in vivo applications have been presented for AD in the patent WO2002041002 A3 [72] and the modeling paper of Li et al. [73].

Although the concept of helical stabilization seems to conceptually afford a possible benefit in treating prion-like neurological disorders, the authors note that the recent clinical trial efficacy failure of Alzheimer's drug candidate bapineuzumab (a known antibody-based stabilizer for $\alpha\beta^{40}/\alpha\beta^{42}$ helices) presents a sobering disappointment. Thus, while our group had engaged in a prior research effort to apply molecular modeling techniques to conduct a fairly involved search for conceptually analogous small organic molecules capable of stabilizing the QNPD within the ALS target TDP-43, we have opted to not present this (still unpublished) work in the current chapter until a greater understanding is achieved regarding the fundamental causes of bapineuzumab failure.

One speculative rationalization for bapineuzumab's failure could revolve around the fact that the therapeutic, although demonstrably effective at clearing $\alpha\beta^{40}/\alpha\beta^{42}$ amyloid deposits, did not confront the concurrent pathogenic progress of a secondary prion-susceptible protein, Tau, which is gradually gaining attention as a potentially common thread-linking AD etiology to that of other non-αβ neurodegenerative disorders such as Pick's disease, corticobasal degeneration, and progressive supranuclear palsy [74]. In other words, the drug candidate may simply have corrected the dysfunction of a protein that was not at the causative heart of AD symptoms.

By analogy, in our own ALS studies, we have now thus revised our scope to simultaneously explore the targeting of the secondary prion-susceptible protein FUS, which has long been considered a possible ALS drug target [75] and whose mutated form is a known biomarker for familial ALS [76].

Nonetheless, since we believe that the fundamental concept of helical stabilization is of value to the discovery and design of prospective new therapeutics for prion-like neurodegeneration, we do provide methodological recommendations on computational protocols that can be put to this task in the upcoming molecular modeling methods Sect. 2.2.

The final prospective avenue for treatment that we recognize from the prion-like misfolding concept is finding ways to mute the beta-inductive nature of proteins that have already become misfolded into prion oligomers. In particular, sheet-capping peptidomimetics provide a very straightforward, intuitive path to designing specific compounds that will bind to existing beta oligomers, or fibrils, but will not encourage further neurodegradative extension.

As an illustration for how one might approach this, we propose herein a novel peptidomimetic intended to bind to an oligomer of the misfolded QNPD of the ALS target TDP-43. Note that while the primary focus of this edition is Alzheimer's research, we chose this ALS system as our primary application example because (*a*) specific doubts have been cast regarding the direct value of correcting $\alpha\beta^{40}/\alpha\beta^{42}$ misfolding, and (*b*) we are not aware of a suitable crystallographic or NMR representation of the misfolded portion of Tau.

Based on observations that the QNPD of TDP-43 misfolds into a beta turn, such that residues 347–351 (SGPSG) comprises the turn, flanked by short (six amino acid) asparagine- and glutamine-rich beta strands, we have rationalized a hyper-stabilized putative amyloid core portion of extended TDP-43 oligomers or fibrils according to the structure represented in Fig. 5a. A key oligomer stabilizing feature is the extensive network of intra- and intermolecular H-bonds arising from placement of two asparagine and two glutamine residues within the compact interior space of the fibril. Additional oligomer stabilization should derive from complex H-bonding networks arising with an additional six amphoteric side chains – three glutamines, two asparagines, and one serine.

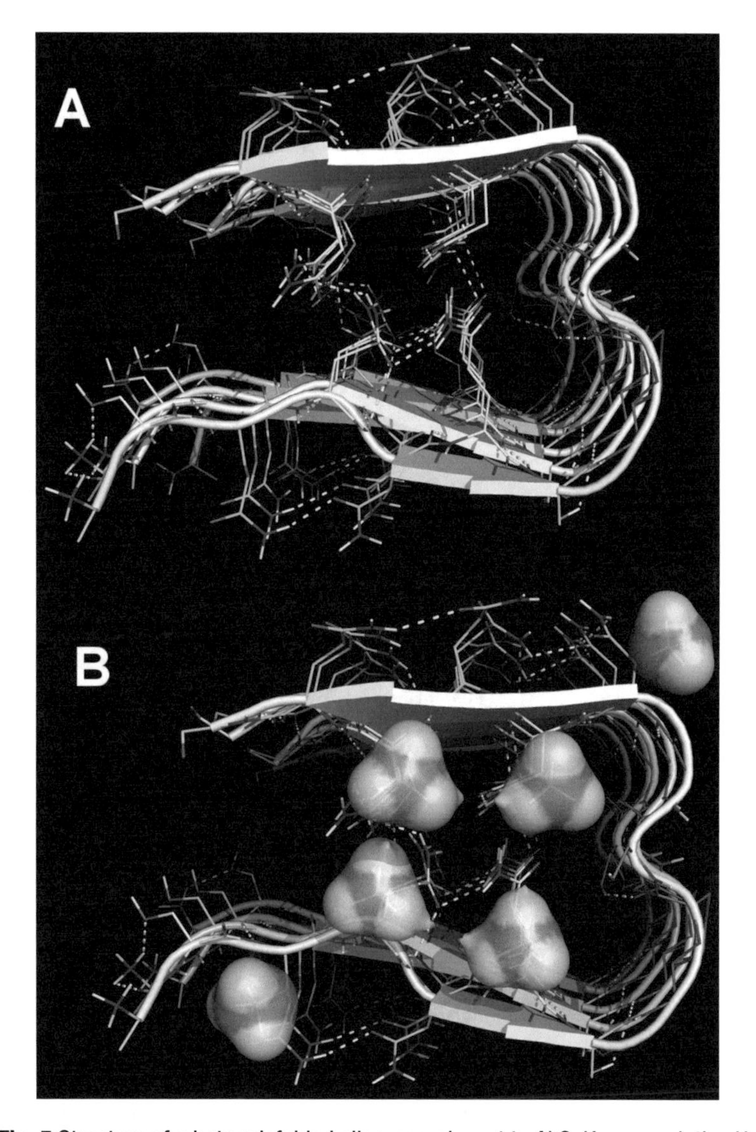

Fig. 5 Structure of a beta misfolded oligomer relevant to ALS. Key associative H-bonds (*yellow pulses*) are shown in **a**, whereas selectively disruptive beta methyls (*white/orange* space-filled groups) are superimposed in **b**.

The favorable interactions presented at the face of a TDP-43 prion-like oligomer prove immensely attractive to the properly folded QNPD structures in healthy TDP-43 [77] and even to structurally fairly dissimilar molecules such as $\alpha\beta^{40}/\alpha\beta^{42}$, whose misfolding may theoretically be initiated by TDP-43 oligomers [78]. If so, one might assume that special considerations such as localized pH might play a key role in initiation since the right conditions of ionicity could render the highly amphoteric QNPD oligomer a good coupling match for the anionic $\alpha\beta^{40}/\alpha\beta^{42}$ structures.

The key to devising a peptide mimic to disrupt the attractive interface of the misfolded QNPD is to find a structure that is capable of facing inwardly toward the prion oligomer with structural complementarity competitive to healthy TDP-43 amino acids but, once bound, presenting an outward surface that lacks the highly electrostatically attractive profile of a QNPD misfold itself. Such a prospective oligomer capping peptide mimic is depicted in Fig. 5b and can be represented according to the following sequence:

$$\mathrm{AS}(\mathrm{Q}^*)(\mathrm{Q}^*)\mathrm{N}(\mathrm{Q}^*)\mathrm{SGPSG}(\mathrm{N}^*)(\mathrm{N}^*)\mathrm{Q}(\mathrm{N}^*)\mathrm{QG}$$

In the above, each asterisk indicates amino acid side chain whose beta carbon is methylated at a stereochemical position chosen to have minimal effect on interactions with the existing oligomer but rather to project outward in a manner that effects a steric dissuasion against binding any subsequent TDP-43 molecules. In the example shown in Fig. 5b, the beta-carbon stereocenter is in *S* conformation.

Note that the upstream and downstream faces of the oligomer are structurally distinct; thus, the precise nature of the capping molecule for the opposite end of the oligomer will differ from that depicted in Fig. 5b, with a sequence instead represented as:

$$\mathrm{ASQ}(\mathrm{Q}^{**})(\mathrm{N}^*)(\mathrm{Q}^{**})\mathrm{SGPSGN}(\mathrm{N}^{**})(\mathrm{Q}^*)(\mathrm{N}^{**})(\mathrm{Q}^*)\mathrm{G}$$

Note that in the above, the double asterisk indicates that the methyl should be in a stereochemical conformation (*R*) opposite of what was rendered in Fig. 5b.

This general logic of preserving inward-looking interactions while diminishing outward ones can be readily applied to developing caps to misfolded oligomers for proteins and peptides other than TDP-43. The precise choice of tools can be productively altered, of course. For any beta-strand residue whose side chain plays little role in binding inwardly to the oligomer, one can consider employing a proline, as these are natural beta-sheet disruptors. Regular salt-bridge networks may be truncated by using asparagine instead of a bridging aspartate, or glutamine instead of a bridging glutamate. Pi-stacking instances may be disrupted through the use of alpha-disubstituted amino acid variants (i.e., analogous to the alpha-aminoisobutyric (Aib) residue) of the stacked residue. Aliphatic hydrophobic coupling may be capped by substituting the aliphatic residue (valine, leucine, or isoleucine) with the amphipathic threonine. Other noncanonical amino acids may be considered as cases warrant.

The conceptual design of such peptidomimetic caps may be initiated using the visual logic provided by a great many different molecular visualization tools but, as shall be discussed in the next section, the likelihood of successful translation from visual concept to promising lead can be greatly enhanced through the application of more judicious and rigorous molecular modeling tools.

2.2 Molecular Modeling Methods for Discovering Helix-Stabilizing Leads

For most of the prion-like proteins believed to play a key role in some neuropathology, the prion-like conformational shift involves a transition from an intramolecularly stabilized alpha helix to an intermolecularly extended beta sheet. There are some exceptions (most notably alpha-synuclein, of relevance to Parkinson's disease, where the misfold affects a twisted beta hairpin becoming pulled out to form a single beta strand), but most prion-like neurological disorders might benefit tangibly from helical stabilizer therapeutics. Due to the transient nature of the protection provided, it is unlikely that helical stabilization will form the basis for a standalone cure for diseases such as Alzheimer's or ALS; however, it may be a crucial adjunct for sustaining patient health through currently risky treatments such as those that, toward an eventual goal of clearing fibrils and plaques, temporarily flood the body with toxic misfolded oligomers.

How, then, does one plan to stabilize a helix? Fundamentally, this requires recognizing that, in the presence of a potentially destabilizing beta sheet, one or more of the intramolecularly stabilizing H-bonds of the healthy helix will break in order to instead begin forming an intermolecular coupling with the sheet. If one identifies which helix-sustaining H-bond runs the risk of initiating a conversion, one may theoretically circumvent the process by somehow shifting the kinetic or thermodynamic landscape to prevent the breach. Ideally it should be possible to achieve this in a reversible manner with enough stability to sustain most of the body's complement of healthy protein for a therapeutically relevant duration (potentially measured in days).

Although some prior effort has been invested in biomedical studies aimed at discovering molecules that can bolster the conformational stability of protein secondary structure features that serve key physiological functions (e.g., in stabilizing the GACKIX domain of the master coactivator CBP/p300 [79]), the concept of attempting to therapeutically modulate only a localized structural feature within a protein remains rather peripheral in the modern pharmaceutical research practice. Consequently, unlike many conventional drug targets, there are few (if any) screening sets of known helical stabilizers to provide the interesting binding hits that may then conceptually inspire leads for a given phenotype. For many key measurements, we are also lacking in well-established, textbook-level protocols upon which we can sculpt a research program to progress smoothly from hypothesis toward real-world milestones. Fortunately, we can compensate for some of this paucity of conceptual underpinning by using molecular

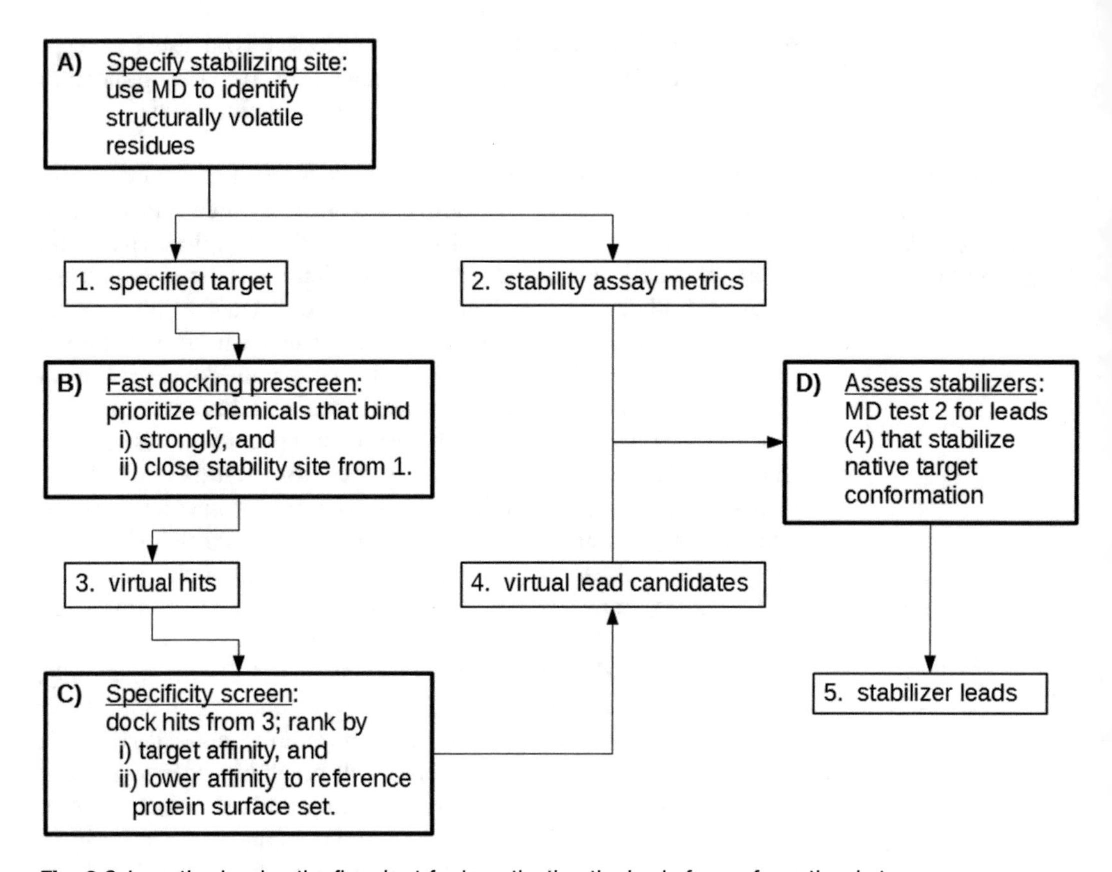

Fig. 6 Schematic showing the flowchart for investigating the basis for conformational st.

modeling experiments to attain suitable mechanistic insight and guidance on specific chemotypes that might be effective structural stabilizers.

Our strategy for applying computational tools to uncover plausible helical stabilizers is outlined schematically in Fig. 6. The process entails four stages: (a) the application of molecular dynamics simulations to identify target sites of conformationally susceptible proteins and simultaneously establish reference metrics by which to gauge successful conformational stabilization, (b) the use of a fast molecular docking utility such as DockBlaster [80] to identify small molecules that are predicted to bind to the right area of the target protein with promising affinity, (c) more sophisticated docking simulations to verify that top virtual hits bind to susceptible regions of the target with greater affinity than they bind to a range of other reference protein surfaces, followed finally by (d) molecular dynamics simulations that can confirm (to a reasonable level of confidence whether the candidate stabilizers actually do serve to functionally stabilize the target, in the presence of a destabilizing influence such as a neuropathic fibril.

The critical first step in this process (step A in Fig. 6) is to identify a stabilizer-binding target. In many cases, the choice of target may be fairly readily identified through prior structural biology knowledge (see Sect. 2.2.2 on TDP-43 helix stabilization); however, in scenarios where prior knowledge offers no clear guidance, Sect. 2.2.1 outlines some basic analyses that can provide the basis for a reasonable initial guess.

2.2.1 Helix Hot Point Perception

If one is unaware of the mechanistic details for how a given helix might unravel in order to bind to a beta strand or beta turn, one may turn to molecular dynamics (MD) simulations to glimpse which backbone H-bonds in a structurally labile helix are most conformationally susceptible.

MD constitutes an exceptionally general and generalizable simulation tool that can be adapted to a tremendous diversity of different applications at broadly varying levels of rigor, scope, and complexity. The precise choices that one may make in specifying levels of rigor and exactitude are dictated by questions such as whether one is intending to make a complex, high-level prediction, or whether one is instead seeking to generate rough, testable hypotheses.

In our case, a basic sampling of conformational purity for a single amino acid helix of moderate length implies no great need for advanced rigor and exactitude, since there will be ample opportunities to verify and refine any original guesses regarding the practical selection of helix-stabilizing binding sites.

In that sense, one may make numerous approximations in order to quickly achieve rough insight. One may thus specify a very basic modeling protocol, something along the lines of:

1. Generate a preliminary structural model for your helix, either by extracting suitable atomic coordinates from a crystallographic or NMR structure file from the PDB [81] or else by building a peptide directly into alpha-helical form using a molecular builder such as PyMol [82].
2. Preparing a MD simulation requires specifying a force-field parametrization that recognizes empirically how the constituent atoms within peptides will move as a function of temperature and time. Assessing time-dependent motions is effected by Newtonian motion solvers coupled with stochastic (e.g., Poisson-Boltzmann (PB)) motion sampling. Effective visualization and analysis requires a graphical and plotting interfaces. There are numerous possible computational resources available to address each of these needs; however, for brevity herein, we will refer to our own experiences using of the Charmm force field [83], the NAMD molecular dynamics simulation program [84], and the VMD molecular graphics program [85].

3. Using VMD to set up our MD simulation, we load the helix structure created in step 1) and use the "automated PSF builder" modeling extension to convert the structure into a form that NAMD can interpret. For a simple simulation on a standard protein or peptide, one may frequently employ only default (automatically prespecified) settings at this step, although customization may be required for the use of nonstandard amino acids or posttranslational modifications.
4. In VMD, we use the "NAMD graphical interface" simulation extension to launch and monitor the simulation. For very simple simulations such as described in this subsection, we opt to change only a few settings from default values, including (a) adding 10,000,000 molecular dynamics time steps to the existing 1000 molecular mechanics steps (this means we plan to model 10 nanoseconds of peptide conformational sampling), (b) we change the background dielectric constant to 80.4, to reflect a polar solvent environment somewhat analogous to plasma, and (c) in order to not get overwhelmed by excessive detail, we scale back the "output frequency" settings to once every 200 time steps.
5. Once the MD simulation is submitted, a message near the bottom of the interface window in step 4) will indicate that the calculation is running (we would estimate perhaps 1–2 hours for a 40 amino acid peptide such as $\alpha\beta^{40}/\alpha\beta^{42}$ and a decent speed processor, ca. 2016) and will revert to the message "ready" when the simulation is complete.
6. If we load the conformational trajectory (written to a file with the extension *.dcd) from step 5) into VMD (these will be added as new "time steps" onto the previously loaded helix structure file), we can then analyze the backbone conformational distribution via the "Ramachandran plot" analysis extension. As per Fig. 7, at the outset of the simulation, a structure in a roughly helical conformation will have most of its amino acid residing in the lower left quadrant of the Ramachandran plot. Over the course of the simulation, residues may lose their conformational purity and distribute through conformation space. Those that drift most prominently from the lower left toward the upper left are most relevant to the current study, since this represents a shift from alpha helix toward beta-strand conformations. Note that clicking on a point on the Ramachandran Plot reveals the amino acid identity (and its torsional distribution) of the point.
7. Once some possible alpha-to-beta transforming (hot point) residues have been identified from step 6), it is important to derive some measure of the reproducibility of these observations by running replicate simulations according to step 5). When

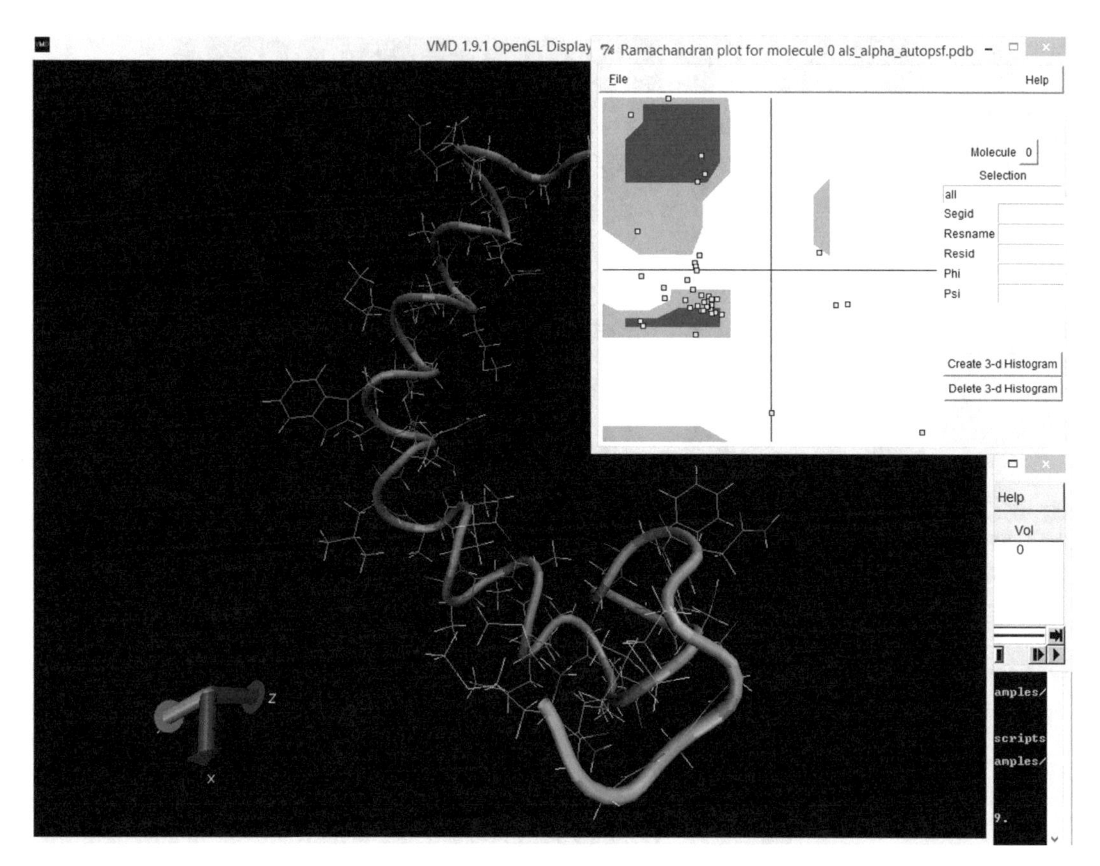

Fig. 7 Prion domain of TDP-43 rendered in VMD, showing inset Ramachandran plot (*upper right*). Within the Ramachandran plot, residues (*small yellow squares*) are classified as having helical conformation (*lower left quadrant*) or beta-strand structure (*upper left quadrant*)

using NAMD, initiating a simulation on the exact same set of files a second time (or any number of times) will produce somewhat different results, because (unless otherwise instructed) the initial atomic motions are assigned randomly via PB-sampling queued to the computer's clock. Thus, one may find (in running replicate simulations) that a residue whose helical conformation appeared to weaken early in one simulation might remain stronger on others.

8. It is difficult to intuit the precise alpha-instability behavior that is most likely to produce a permanent shift to a beta misfold, but there are two behaviors (evident from step 7) that, through empirical observations, we recommend paying the most attention to: (a) residues that almost always tend to show alpha lability regardless of which replicate one is examining or (b) those whose shift to beta may be more rare, but which appear to be fairly irreversible (i.e., the residue tends to get stuck in a plausible pro-strand conformation).

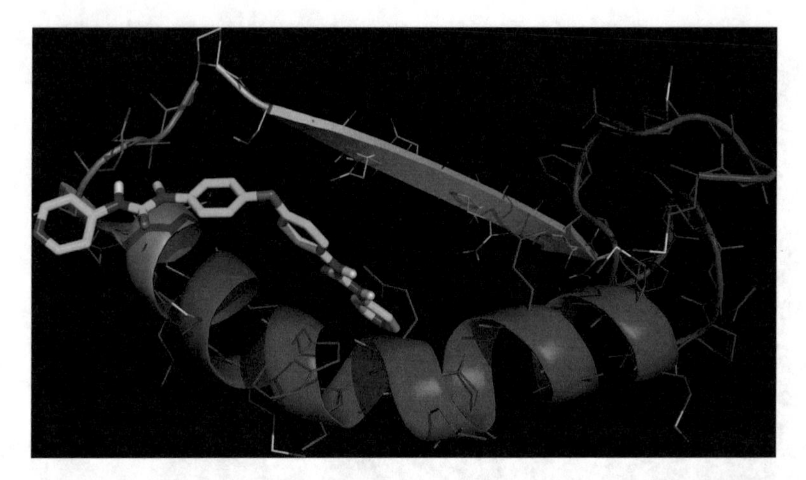

Fig. 8 Structure of the prion domain of TDP-43, showing the conformationally critical (*orange*) trigger zone at the boundary of the native RNA-binding helix (*red*) and native beta strand (*yellow*). A prospective helix-stabilizing ligand (*green sticks*) is shown bound to the trigger region

2.2.2 Structural Inference

Part of the determination of plausible alpha-to-beta susceptible residues may already be evident from prior structural biology characterizations. For example, in the case of the susceptible QNPD of TDP-43 (see Fig. 8), we know that among those residues (341–357) observed to misfold as a beta turn in ALS etiology, a critical hexapeptide unit (residues Ala350 – Gln-355) is present as part of the polynucleotide-associating helix in the healthy conformation. Identification of such residues, in addition to establishing a macromolecular target to stabilize with small molecules, also provides the basis for establishing performance metrics by which to gauge effective stabilization. In particular, establishing conformational ranges by which structural stability is manifest (and beyond which, structural degradation occurs) provides testable criteria by which (through subsequent dynamics simulations, circular dichroism, or NMR) might verify effective stabilization.

2.2.3 Stabilizer Searches

While it is challenging to imagine a molecular intervention that is capable of stabilizing an unstructured coil or a preexisting beta strand in a way as to prevent either from associating with a prion-like beta oligomer, the retention of clearly identified amino acids in a healthy helical form provides a very tangible basis for the design of a suitable structural modulation therapeutic – a molecule that binds to this site in a manner that anchors and sustains the existing structure, even when external forces may exert denaturing forces. In other words, we would theoretically seek a small molecule with a shape, size, and electrostatic/lipophilic profile that is well suited to binding in the region shown in Fig. 8 in such a way as to complement and bolster the current shape. A plausible approach to address this search is identified in step B of Fig. 6.

The single most efficient research tool available to biomedical science for initiating extensive chemically agnostic searches for a molecule of specific shape, size, and electrostatic/lipophilic profile dictated by a given binding target is molecular docking. As was stated previously for MD simulations, there are a wide range of different computational methods and strategies available for docking. Like MD, the rigor levels vary from method to method, but unlike MD, the relative aptitude of different docking packages varies substantially over specific classes of molecular targets. For example, some docking programs and parameter sets produce more accurate results for ligands binding to kinase targets than they do for hydrolases, because most docking methods tend to incorporate trained free energy estimation parameters whose training tends to rely on preexisting binding data, which can thus produce results biased to best reproduce systems most similar to those best represented in their core training sets.

There is little current basis for a clear recommendation on which docking method will best predict the binding of organic molecules to the surfaces of bare helices. The obvious reason for this is that the convex topology associated with such an exposed binding side is structurally very dissimilar to the enzyme cavities that such methods are most often optimized to consider. As a result, docking trends predicted for our proposed alpha stabilizers are unlikely to correlate as well to eventual experimentally assessed trends, when contrasted with predictive performance for more conventional targets. Nonetheless, docking programs are of great value in sampling over not only large numbers of molecules but over diverse conformational degrees of freedom for each molecule. This fosters a fresh, unbiased search for chemical entities whose conformationally-dependent shape and charge profile is compatible with a specific surface location on a peptide helix.

Given the broad range of different docking resources available to a researcher intending to search for alpha-stabilizing compounds from within compendia of organic compounds, it is our recommendation (in the absence of rigorous experiential data) to simply employ whichever docking suite one is already most familiar with, in the sense that this familiarity should carry with it a somewhat greater chance of accuracy that comes with already knowing the strengths, limitations, and realistic expectations for that software. The broad scale search, such as specified as Step B in Fig. 6, is best effected by a program with high-computational efficiency and an exceptional amenability to automation. In particular, a large number of ligands must be fed to a batch submission, and one single specification of execution parameters should suffice for all ligand docking simulations, so that users are not required to continually adjust the program during its lengthy screening process.

For those researchers who do not yet have a favorite docking program, it is our recommendation to try the exceptional simplicity, efficiency, and low bias associated with the DockBlaster service [80]. The DockBlaster website offers various mechanisms to spatially steer a screening simulation (i.e., tutorials on the site delineate methods, such as uploading lists of critical substructures, for users to explicitly guide the docking simulations toward a specific target binding subsite); however, the simplest protocol required in order to run the DockBlaster service is elementary: one merely chooses a given chemical collection to analyze and uploads the target of interest. In this case, one makes no attempt to steer the focus onto ligands binding closely to the conformationally sensitive residues identified in Sects. 2.2.1 or 2.2.2. In our particular case of interest, wherein we are screening for ligands that have a strong preference for specific hot point residues, the deliberate avoidance of site bias can actually be considered a useful feature for enhancing predictive rigor. Instead of the up-front site bias which may steer ligands toward a site that they don't truly prefer to bind to, one may instead permit completely unbiased ligand positional sampling, after which one may graphically parse top scoring DockBlaster hits to determine, post-simulation, which ones appear to bind to the right area. This spatial analysis, readily accomplished using visualization software such as PyMol [82], enables the informed elimination of compounds that, although potentially somewhat suited to binding to the right region of the helix, likely have stronger preferences for binding elsewhere. Such compounds may be inherently promiscuous and present challenges for high-specificity targeting to hot point residues.

One other feature of DockBlaster that may help to steer an effective high-throughput virtual screen of possible conformational stabilizers is that it is capable of a degree of intelligent selection. Specifically, as is described in tutorials on the DockBlaster site [80], one may provide the service with a list of ligands that are known to be active for the application of interest, plus a list of ligands that are known to be inactive. Based on this information, DockBlaster may glean some insight into relevant (or deleterious) ligand pose locations, and orientations that can complement the original residue hot point information derived from Step A in Fig. 6, thus producing a degree of screening discrimination that might exceed the value of what one might achieve from pose spatial selection alone.

Although we have little by the way of procedural detail to recommend in initiating docking searches for plausible helix-stabilizer virtual hits, we would offer one cautionary note on the use of DockBlaster. Namely, the service is very powerful and potentially very useful, but the calculations are run on a centralized server which can (under conditions of heavy load) dictate that execution will be slow. As a result, researchers are strongly advised to submit small scale test jobs (perhaps sampling several thousand structurally diverse ligands) to begin with. These will run relatively quickly and give some interesting preliminary results to examine while one waits (potentially days) for comprehensive production runs over millions of potential binders. Such preliminary results may enable the fine-tuning of spatial selection criteria to specify when analyzing the eventual larger data set.

2.2.4 Stabilizer Validation

In all of the above steps, there are inherent approximations made whose true numerical uncertainty is ultimately difficult to quantify. Fortunately, even before one invests in the expense of procuring compounds and developing a simple assay with which to test them, there are additional computational tests that may be performed to gauge, with reasonable accuracy, whether the prior predictive uncertainties have produced meaningless value or whether some or all of the preliminary virtual hits from Sect. 2.2.3 are worth further consideration.

Since the DockBlaster utility has been developed more for speed than accuracy, it may prove helpful to pursue a modestly more exacting test to gauge the plausibility of virtual hits from Step B of Fig. 6, which could involve application in Step C of Fig. 6 of a refined molecular docking method such as the Vina program [86] to verify that the molecules selected in the prior step generally do prefer to bind to the intended target rather than simply being chemotypes with general affinity toward a broad range of targets. Although there are numerous different molecular docking suites of reasonable repute, the value of a program such as Vina toward an application such as alpha-stabilizer searches is that it generally reproduces binding-affinity trends and ligand-bound structural characterizations across a rather very broad range of different macromolecular targets – a key attribute to consider given the fact that our goal herein (finding molecules that bind not to standard enzymatic active site but rather to the surface of an exposed protein helix) differs from most common in silico drug design activities. In other words, whereas other docking software may give more accurate assessments of affinity and docking poses for more conventional drug targets, the generality of a program like Vina should afford a reasonable evaluation of whether molecules of interests truly do prefer to bind to exposed neurologically relevant helices.

A suitable test of plausible ligand specificity would be to dock the hits of interest (derived from Fig. 6, Step B) to the helix-stabilization target and compare the resulting Vina-docking scores with Vina scores for a range of common drug targets, such as might include a representative GCPR, kinase, protease, esterase, ion channel, hormone receptor, membrane transport protein, etc. Higher affinity for the intended helix target relative to the common druggable control targets would provide an assurance in Step C that the compound identified in Step B is reasonably selective for helix stabilization.

The precise protocol for assembling and running Vina-docking simulations is trivial. Any receptor may be converted from standard PDB format into a Vina-compliant PDBQT format using the ADT program [87] at default receptor preparation settings. Ligands may be similarly generated by ADT for Vina screens in an automated fashion. From this point, it is merely a matter of specifying a pose search zone – a coordinate center relative to the intended receptor, surrounded by a cube of specified dimension.

As shown in Step D of Fig. 6, the prospective alpha stabilization value of those ligand candidates identified through docking simulations can then be accomplished via MD simulations. A preliminary round of such simulations can be specified in a manner very similar to those discussed in Sect. 2.2.1, while subsequent rounds of more rigorous testing may be performed.

In the first MD simulation round, one can anticipate using the exact same protocol as described earlier for conformationally labile residue discovery, with the exception that the pure helix is replaced by the helix complexed by one of the ligands discovered in Sect. 2.2.3. The only difference in the setup is a requirement that the ligand, being something other than a simple canonical biomolecule, must be accorded a set of parameters so that the MD solver is able to recognize how the ligand structure interacts with the helix and conformationally evolves as a function of temperature, time, and any intermolecular coupling with the peptide.

The final analysis conceptually also identical to what was described in Sect. 2.2.1, with a critical emphasis on determining whether any amino acids susceptible to alpha to beta conformational shift appear to be tangibly stabilized by the presence of the ligand. The difference in this more rigorous analysis is that a more detailed and realistic model is constructed, with a capacity for accommodating structural perturbations arising from a realistic physiological environment.

One physiological feature that may have a legitimate effect on the propensity of a helix to sustain its shape is an explicit solvent, which can be included within MD simulations by extending the simple unimolecular or bimolecular simulations discussed thus far (in which solvent effect is modeled only with an implicit dielectric perturbation).

In order for solvent to be incorporated realistically into a MD model, one typically must embed the system of interest within an infinitely repeating 3D unit cell, across whose boundaries atoms and molecules may pass, with the proviso that (through periodicity) identical matter enters the unit cell at the diametrically opposed cell location. Cross-cell forces can be efficiently and seamlessly approximated via mathematical constructs such as Taylor series expansions.

From an implementation perspective, common practice typically dictates embedding an important non-solvent molecule (in this case, our conformationally labile helix in complex with one or more small helix stabilizers) within a solvent-buffered cuboid, where the term "cuboid" indicates any three-dimensional shape for which all defining surfaces are rectangles or squares connected by right angles and where "buffered" means that each side is extended by a certain distance (often 10–20 Å) beyond the minimum distance needed to fully enclose the central molecule (or complex) itself.

Once, a periodically replicated cuboid has been established around the central molecule(s), all intervening space is filled (in an automated fashion via software like VMD) with periodically replicated solvent molecules. Frequently, software is then employed to replace a small number of these solvent molecules with charged entities (e.g., Na^+ and Cl^- ions) that can achieve an electrostatically balanced environment which may reasonably simulate a system such as physiological plasma.

The general protocol for running such solvated MD simulations is functionally similar to the simple schemes described earlier for non-explicitly solvated models, with the key proviso that the automatically constructed solvent environment is not considered to be physically realistic until it has been subjected to rigorous "conditioning." In other words, before realistic assessment of the structural behavior of the central molecule(s) is possible, the entire system must be thermodynamically relaxed using a preliminary MD simulation that permits gradual thermal warming up to the desired simulation conditions (e.g., room temperate = 298 K or physiological temperature = 310 K). While this equilibration is carried out, the accurate representation conditions are further improved by permitting the unit cell to adjust its dimensions to match a reasonable physiological pressure (typically 1.0 atmosphere). Frequently, the modeler will employ between 20,000–100,000 MD steps in order to achieve this equilibration in a gentle fashion that will not artificially perturb the central molecule(s).

Finally, as one progresses through toward more elaborate validation tests such as those provided by explicitly solvated MD simulations, it is often useful to consider longer simulation times, potentially spanning 10–100 ns (simulated time) in order to attain a greater degree of confidence in any assessments being made from the model. This will increase the simulation length in a linear manner (a 100 ns simulation will take 100 times as long to execute as a 1 ns run); however, modelers may save some effort by specifying a longer (2 femtosecond = 2 fs) time step than the default (1 fs) without any expected loss of accuracy. One may also conserve some resources by requesting longer intervals (perhaps up to 1000 time steps) between reporting simulation energy or current atomic coordinate updates.

2.3 Molecular Modeling Methods for Capping Beta Oligomers

Whereas the alpha-stabilizer search is dominated by the de novo discovery of new chemotypes (which the investigator might not have readily guessed in advance) exhibiting novel helix-binding modes, the development of beta-capping peptides is based much more on practical logic and existing knowledge.

The complement-based capping strategy outlined in Sect. 2.1 produces one form of rational binder that should (given careful design) be expected to bind strongly to ends of a misfolded oligomer but not permit further oligomer propagation. A second possibility, mentioned earlier, is the natural capping scheme inspired from the reversibly prionic HET-s protein. As rendered in Fig. 9, one sees that HET-s caps a prion by virtue of ten amino acids that readily form an oligomer-binding beta strand (yellow), connected by a short turn to 22 residues that rigorously resist misfolding from helical state. The resulting short-tethered helix provides an excellent steric barrier to beta complexation by any other conformationally labile proteins or peptides.

While a logical structural basis exists for intuiting either of the above capping schemes, some degree of experimentation will be required in order to posit strong candidate caps for any of the neuropathic targets of therapeutic interest. Molecular modeling can greatly augment and expedite this refinement process, by providing a low-cost, high-insight generator within which to

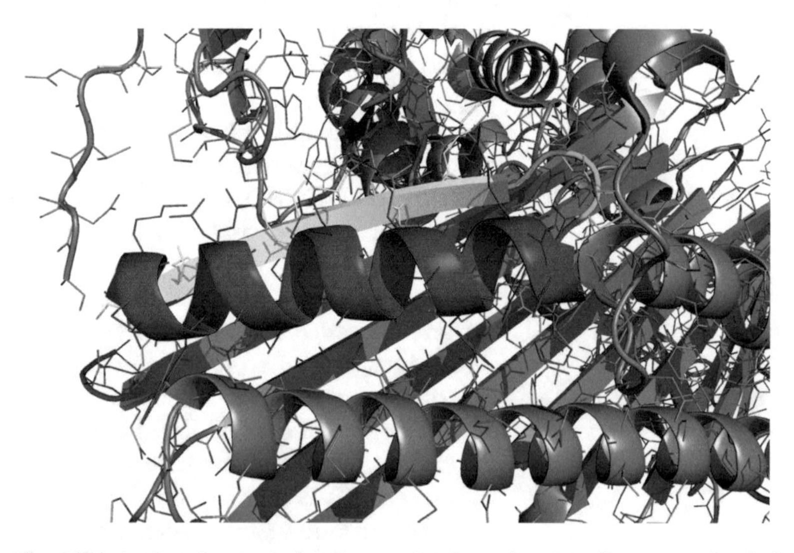

Fig. 9 TIA-1 prion structure, showing a natural capping domain compromised of an associative beta strand (*yellow*) and a disruptive adjacent helix (*red*)

attempt novel structural modifications and observe meaningful consequences of any structural modifications.

Once again, our main tool for such analysis will be MD simulations. We would further recommend the two-tiered evaluation strategy outlined in Sect. 2.2.4 as a mechanism for evaluating the stability of the capped beta oligomer complex first in an approximate implicit solvation environment and then, more rigorously, with explicit solvation and charge balancing. This covers effectively the same analytical protocols as have been associated with Steps A and D of Fig. 6, where Steps B and C may be omitted, since beta-capping molecules are likely to be devised rationally, as opposed to alpha-stabilizer molecules which may be best discovered through agnostic screening followed by multiple steps of validation.

Unlike the analysis discussed for alpha-stabilizer validation, our first objective in assessing beta-capping peptides is to verify that the designed molecules do bind to the sheet edge. A useful test is to compare the strength of relevant beta sheet H-bonds sustained by the cap, versus those that would be formed by joining an additional prion-like monomer to the end, such that the capping peptide should bind with a strength comparable to that of the prion.

The final form of simulation used to gauge the aptitude of a peptide cap design would be to model the ability of the capped peptide to effectively dissuade complexation by a known conformationally susceptible prion domain, where the latter may be positioned close to the expected sheet extension site, in a conformation that might be expected to bias the susceptible domain toward stable oligomer extension. In general, we would primarily advocate using the more rigorous explicit solvation representation for this final step of computational validation for the simple reason that the implicitly solvated system may be prone to false negatives. That is, the simpler simulation may fail to recognize a viable complexation by the prion-susceptible domain, since implicit solvation provides an unrealistically low barrier to dissociation by partially complexed binders, even if the simulation was constructed to bias them toward binding.

It should be noted, finally, that the value of MD simulations toward assessing competitive binding scenarios (i.e., whether a prion-susceptible domain can attach to a peptide cap or potentially even supplant it) is often enhanced proportionally to simulation duration. This is to say that the longer one is able to execute a simulation, the less chance there is to miss evidence of a relevant conformational shift that may, in real time, potentially occur on a time scale that challenges the applicability domain of MD modeling. In practice, we are limited by the amount of time we can devote to running the calculations; thus, we can never be certain that we have detected any plausible structural adaptations that could potentially be resolved through longer run-run-time, but a pragmatic compromise is to set a real-world time limit (e.g., a week or a month) in which to exhaustively test given candidate formulations in silico, then choose to perform faster in vitro validations on any that appear (within the limited time frame) to optimally address our requirements for binding strongly to beta oligomers and/or preventing additional complexation by prion-susceptible protein domains or peptides.

3 Notes

Molecular-docking studies are an exceptionally expeditious tool to roughly sample small molecules for potentially useful intermolecular coupling. As discussed in the previous section, docking assessments are inherently rather approximate, but the value of any promising results can be enhanced by validating the predicted binding with higher confidence MD simulations.

As also mentioned in the previous section, a key limitation of MD simulations is that the combination of practical time investment limits and the computational complexity of the technique may conspire to preclude full perception of all relevant structural adaptations that might be evinced in a longer duration calculation. As a result, a simulation is best regarded in terms of qualitative guidance. One may not be able to state definitely that peptide A will absolutely never dissociate from its therapeutic complex to a toxic oligomer; however, one would hopefully have (based on defensible statistics) the basis to make defensible assertions along the lines of "we believe, with at least 95% confidence, that peptide A binds with greater stability than control compound B." In other words, the predictive strength of simulations is often best when one can frame the assessments in qualitative terms relative to verifiable reference points.

It should further be noted that, even when simulations uncover candidate ligands that demonstrate very promising behavior which can be verified in wet lab (in vitro) experiments, one may face major intervening hurdles between the preliminary lead formulation and any eventual success in intracellular or whole organism tests. Some key obstacles are blood-brain barrier (BBB) permeability of the therapeutic candidate as well as determining whether a given chemical meets minimal requirements in metabolic stability in order to have suitable therapeutic effect before degrading.

BBB permeability is a key precondition for efficacy of most neuroactive therapeutics, as molecules must successfully transit the body's natural protective barriers that isolate our central nervous system from common toxins and microbes. BBB permeation of small organic molecules such as those explored in Sect. 2.2 can be fairly well estimated with simple, online calculators such as CBLig [88]. Most peptides are at least somewhat BBB permeable; however, comparable computational methods do not yet appear to exist for quantifying or refining peptide BBB permeability; thus, serious development efforts should seek guidance in preparative schemes for BBB delivery, such as those described in reviews such as that of Egleton and Davis [89].

Metabolic stability of small, drug-like organic molecules can be effective predicted using any of the numerous predictive software tools that exist, of which a reasonable compilation has been provided in Table 2 of the review by Kirschmair et al. [90]. Once again, there does not appear to be a single simple scheme for predicting the metabolic stability of peptide formulations, although a comprehensive review is available to address general empirical knowledge regarding in vivo stability and other therapeutically relevant properties [91].

3.1 Do's and Don'ts

In applying molecular docking and molecular dynamics simulations in the search for therapeutic interventions that can prevent neuropathically harmful misfolding, some aspects of basic wisdom have come to mind, which we share herein:

Do not use docking score alone as the key measure for prioritizing a helix stabilizer or a fibril capper candidate. A high docking score may reflect binding promiscuity – a tendency to stick to a wide range of different proteins.

Do establish a baseline reference of different protein surfaces to compare your helix stabilizer or fibril capper-docking scores to. In order to attain confidence that a given molecule is interesting, it should have a docking score to the intended surface that exceeds the scores for the same ligand on the reference surfaces by a statistically significant margin.

Do not blindly assume that molecular dynamics simulation should be run in plasma (high-dielectric) conditions.

Do experiment, upfront, with different solvent dielectrics to determine which conditions lead to the formation of stable fibrils that are capable of devolving healthy protein structures. Such conditions are likely to approximately mimic those biological circumstances in which neuropathies propagate, thus providing a strong test for therapeutic discovery. Molecules that successfully counter propagation in such an in silico environment will stand an elevated chance of displaying real in vivo efficacy.

Acknowledgments

The authors gratefully acknowledge Dr. Anthony Barnes for many technical suggestions, especially including the notion of beta-capping peptides. We also greatly appreciate the ALS advocacy work by Dr. Ashley Beason-Manes as well as her tip regarding recent neuromedical stem cell treatment advances. Finally, we acknowledge Kansas City Science Pioneers for providing the initial impetus and encouragement for our activities in this field.

References

1. Geldmacher DS (2008) Cost-effectiveness of drug therapies for Alzheimer's disease: a brief review. Neuropsychiatr Dis Treat 4:549–555
2. Campbell T (2015) The truly staggering cost of Alzheimer's disease. http://www.fool.com/investing/general/2015/04/04/the-truly-staggering-cost-of-alzheimers-disease.aspx. Accessed 21 Nov 2016
3. Verma IM, Weitzman MD (2005) Gene therapy: twenty-first century medicine. Annu Rev Biochem 74:711–738
4. Darnell JE Jr Transcription factors as targets for cancer therapy. Nature Reviews Cancer 2:740–749
5. Huang H, Prusiner SB, Cohen FE (1996) Scrapie prions: a three-dimensional model of an infectious fragment. Fold Des 1:13–19
6. Dillner L (1996) BSE linked to new variant of CJD in humans. The BMJ 312:795
7. Scott M et al (1989) Transgenic mice expressing hamster prion protein produce species-specific scrapie infectivity and amyloid plaques. Cell 59:847–857
8. Barz B, Wales DJ, Strodel B (2014) A kinetic approach to the sequence–aggregation relationship in disease-related protein assembly. J Phys Chem B 118:1003–1011
9. Eikelenboom P et al (2002) Neuroinflammation in Alzheimer's disease and prion disease. Glia 40:232–239
10. Serpell LC (2007) Alzheimer's amyloid fibrils: structure and assembly. Biochim Biophys Acta 1:16–30
11. Petkova AT et al (2005) Self-propagating, molecular-level polymorphism in Alzheimer's ß-amyloid fibrils. Science 307:262–265
12. Gunnar K, Gouras GK, Almeida CG, Takahashi RH (2005) Intraneuronal Aß accumulation and origin of plaques in Alzheimer's disease. Neurobiol Aging 26:1235–1244
13. Meyer-Luehmann M et al (2006) Exogenous induction of cerebral ß-Amyloidogenesis is governed by agent and host. Science 313:1781–1784
14. Chesebro B et al (2005) Anchorless prion protein results in infectious amyloid disease without clinical Scrapie. Science 308:1435–1439
15. Haataja L et al (2008) Islet amyloid in type 2 diabetes, and the toxic oligomer hypothesis. Endocr Rev 29:303–316
16. Irvine GB et al (2008) Protein aggregation in the brain: the molecular basis for Alzheimer's and Parkinson's diseases. Mol Med 14:451–464
17. Truant R et al (2008) Huntington's disease: revisiting the aggregation hypothesis in polyglutamine neurodegenerative diseases. FEBS J 275:4252–4262
18. Targońska-Stępniak B (2014) Serum Amyloid A as a marker of persistent inflammation and an indicator of cardiovascular and renal involvement in patients with Rheumatoid Arthritis. Mediators of inflammation 2014, Article ID 793628, 7 pages
19. Dotti CG, De Strooper B (2009) Alzheimer's dementia by circulation disorders: when trees hide the forest. Nat Cell Biol 11:114–116
20. Bannykh SI et al (2013) Formation of gelsolin amyloid fibrils in the rough endoplasmic reticulum of skeletal muscle in the gelsolin mouse model of inclusion body myositis. Comparative analysis to human sporadic inclusion body myositis. Ultrastruct Pathol 37:304–311
21. Vattemi G et al (2009) Amyloid-beta42 is preferentially accumulated in muscle fibers of patients with sporadic inclusion-body myositis. Acta Neuropathol 117:569–574
22. Mackenzie IR et al (2011) A harmonized classification system for FTLD-TDP pathology. Acta Neuropathol 122:111–113
23. Neumann M et al (2006) Ubiquitinated TDP-43 in frontotemporal lobar degeneration and amyotrophic lateral sclerosis. Science 314:130–133
24. Schwarz A (2010) Study says brain trauma can mimic A.L.S. The New York Times. Accessed 2 Dec 2016
25. Dal BA et al (2008) Multiple sclerosis and Alzheimer's disease. Ann Neurol 63:174–183
26. Morris DZ (2016) Biogen drug shows early promise against Alzheimer's disease. Fortune Magazine. http://fortunecom/2016/09/03/biogen-drug-alzheimers-disease/. Accessed 18 Nov 2016
27. Mayo Clinic Staff; authors not identified (2014) Alzheimer's still has no cure, but two different types of drugs can help manage symptoms of the disease. http://wwwmayoclinicorg/diseases-conditions/alzheimers-disease/in-depth/alzheimers/art-20048103. Accessed 14 Dec 2016
28. Ashfaq Ul Hassan AUI, Hassan G, Rasool Z (2009) Role of stem cells in treatment of neurological disorder. Int J Health Sci (Qassim) 3:227–233
29. Yashdeep GY, Singla G, Singla R (2015) Insulin-derived amyloidosis. Indian J Endocrinol Metab 19:174–177

30. Schreck JS, Yuan J-M (2013) A kinetic study of amyloid formation: fibril growth and length distributions. J Phys Chem B 117:6574–6583
31. Kayed R, Lasagna-Reeves CA (2013) Molecular mechanisms of amyloid oligomers toxicity. J Alzheimers Dis 33:S67–S78
32. Anne BA et al (2014) The prion protein is critical for DNA repair and cell survival after genotoxic stress. Nucleic Acids Res 43:904–916
33. Claudio SC, Satani N (2011) The intricate mechanisms of neurodegeneration in prion diseases. Trends Mol Med 17:14–24
34. Hu PP, Huang CZ (2013) Prion protein: structural features and related toxicity. Acta Biochim Biophys Sin 45:435–441
35. Chiesa R (2015) The elusive role of the prion protein and the mechanism of toxicity in prion disease. PLoS Pathog 11:e1004745
36. Tripathi VB et al (2014) Tar DNA-binding protein-43 (TDP-43) regulates axon growth in vitro and in vivo? Neurobiol Dis 100:25–34
37. Camero S et al (2013) Specific binding of DNA to aggregated forms of Alzheimer's disease amyloid peptides. Int J Biol Macromol 55:201–206
38. Chunxing YC et al (2016) Partial loss of TDP-43 function causes phenotypes of amyotrophic lateral sclerosis. PNAS 111:E1121–E1129
39. Sengupta U, Nilson AN, Kayed R (2016) The role of amyloid-ß oligomers in toxicity, propagation, and immunotherapy. EBioMedicine 6:42–49
40. Goure WF et al (2014) Targeting the proper amyloid-beta neuronal toxins: a path forward for Alzheimer's disease immunotherapeutics. Alzheimers Res Ther 6:42
41. Morales R, Moreno-Gonzalez I, Soto C (2013) Cross-seeding of misfolded proteins: implications for etiology and pathogenesis of protein misfolding diseases. PLoS Pathog 9: e1003537
42. Maniecka Z, Polymenidou M (2001) From nucleation to widespread propagation: a prion-like concept for ALS. Virus Res 207:94–105
43. Westlind-Danielsson A, Arnerup G (2001) Spontaneous in vitro formation of supramolecular beta-amyloid structures, "betaamy balls", by beta-amyloid 1-40 peptide. Biochemistry 40:14736–14743
44. Daskalov A, Dyrka W, Saupe SJ (2015) Theme and variations: evolutionary diversification of the HET-s functional amyloid motif. Sci Rep 5:12494
45. Wolozin B (2012) Regulated protein aggregation: stress granules and neurodegeneration. Mol Neurodegener 7:56
46. Xu W et al (2012) Molecular dynamics simulation study on the molecular structures of the amylin fibril models. J Phys Chem B 116:13991–13999
47. Brown P (2001) Bovine spongiform encephalopathy and variant Creutzfeldt-Jakob disease: background, evolution, and current concerns. Emerg Infect Dis 7:6–16
48. Underwood E (2015) Is the Alzheimer's protein contagious? Science Magazine http://wwwsciencemagorg/news/2015/09/alzheimers-protein-contagious Accessed 7 Dec 2016
49. Zhao Y, Zhao B (2013) Oxidative stress and the pathogenesis of Alzheimer's disease. Oxidative medicine and cellular longevity. 2013, 316523, and references therein
50. Ciechanover A, Kwon YT (2015) Degradation of misfolded proteins in neurodegenerative diseases: therapeutic targets and strategies. Exp Mol Med 47:e147
51. Ou S-HI et al (1995) Cloning and characterization of a novel cellular protein, TDP-43, that binds to human immunodeficiency virus type 1 TAR DNA sequence motifs. J Virol 69:3584–3596
52. King OD, Gitler AD, Shorter J (2012) The tip of the iceberg: RNA-binding proteins with prion-like domains in neurodegenerative disease. Brain Res 1462:61–80
53. Zeyni MZ et al (2016) Loss of Tau protein affects the structure, transcription and repair of neuronal pericentromeric heterochromatin. Sci Rep 6:1–16
54. Wang X et al (2006) The proline-rich domain and the microtubule binding domain of protein tau acting as RNA binding domains. Protein Pept Lett 13:679–685
55. Snead D, Eliezer D (2014) Alpha-Synuclein function and dysfunction on cellular membranes. Exp Neurobiol 23:292–313
56. Remo RR et al (2010) Origins of specificity in protein-DNA recognition. Annu Rev Biochem 79:233–269
57. Hegde ML, Jagannatha RK (2007) DNA induces folding in a-synuclein: understanding the mechanism using chaperone property of osmolytes. Arch Biochem Biophys 464:57–69
58. Protter DSW, Parker R (2016) Principles and properties of stress granules. Trends Cell Biol 26:668–679

59. Ratovitski T et al (2006) Huntingtin protein interactions altered by polyglutamine expansion as determined by quantitative proteomic analysis. Cell Cycle 11:2006–2021
60. Furukawaa Y, Nukina N (2013) Functional diversity of protein fibrillar aggregates from physiology to RNA granules to neurodegenerative diseases. Biochim Biophys Acta (BBA) – Mol Basis Dis 8:1271–1278
61. Saboora WS, Wilce MCJ, Wilce JA (2014) RNA recognition and stress granule formation by TIA proteins. Int J Mol Sci 15:23377–23388
62. Cruz DC et al (1999) Physical trauma and family history of neurodegenerative diseases in amyotrophic lateral sclerosis: a population-based case-control study. Neuroepidemiology 18:101–110
63. Cannon JR, Greenamyre JT (2012) The role of environmental exposures in neurodegeneration and neurodegenerative diseases. Toxicological Sciences 124(2):225–250
64. Pierre G (2013) Neurodegenerative disorders and metabolic disease. Arch Dis Child 98:618–624
65. Maheshwari P, Eslick GD (2015) Acterial infection and Alzheimer's disease: a meta-analysis. J Alzheimers Dis 43:957–966
66. Alonso R et al (2015) Evidence for fungal infection in cerebrospinal fluid and brain tissue from patients with amyotrophic lateral sclerosis. Int J Biol Sci 11:546–558
67. Nicolson GL (2008) Chronic bacterial and viral infections in neurodegenerative and neurobehavioral diseases. Lab Med 39:291–299
68. Cao K et al (2010) Age-correlated gene expression in normal and neurodegenerative human brain tissues. PLoS One 5:e13098
69. Rovó P et al (2014) Rational design of a-helix-stabilized exendin-4 analogues. Biochemistry 10:3540–3552
70. Miles LA et al (2013) Bapineuzumab captures the N-terminus of the Alzheimer's disease amyloid-beta peptide in a helical conformation. Sci Rep 3:1–5
71. Fezoui Y, Teplow DB (2002) Kinetic studies of amyloid beta-protein fibril assembly. Differential effects of alpha-helix stabilization. J Biol Chem 277:36948–36954
72. Jan Johansson J, White MP (2003) Discordant helix stabilization for prevention of amyloid formation. Patent publication WO2002041002 A3
73. Li J et al (2011) Alzheimer's disease drug candidates stabilize A-b protein native structure by interacting with the hydrophobic Core. Biophys J 100:1076–1082
74. Himmelstein D et al (2012) Tau as a therapeutic target in neurodegenerative disease. Pharmacol Ther 136:8–22
75. Kryndushkin D, Shewmaker F (2011) An efficient approach for studying protein aggregation and toxicity. Prion 5:250–257
76. Kwiatkowski TJ Jr et al (2009) Mutations in the FUS/TLS gene on chromosome 16 cause familial amyotrophic lateral sclerosis. Science 323:1205–1208
77. Furukawa Y et al (2011) A seeding reaction recapitulates intracellular formation of Sarkosyl-insoluble transactivation response element (TAR) DNA-binding protein-43 inclusions. J Biol Chem 286:18664–18672
78. Josephs KA et al (2014) Staging TDP-43 pathology in Alzheimer's disease. Acta Neuropathol 127:441–445
79. Wang N et al (2013) Ordering a dynamic protein via a small-molecule stabilizer. J Am Chem Soc 135:3363–3366
80. Irwin JJ et al (2009) Automated docking screens: a feasibility study. J Med Chem 52:5712–5720
81. Berman HM et al (2000) The Protein Data Bank. Nucleic Acids Res 28:235–242 and www.rcsb.org
82. The PyMOL Molecular Graphics System (2015) Version 1.8, Schrödinger, LLC
83. Mackerell AD Jr et al (2004) Extending the treatment of backbone energetics in protein force fields: limitations of gas-phase quantum mechanics in reproducing protein conformational distributions in molecular dynamics simulations. J Comput Chem 25:1400–1415
84. Phillips JC et al (2005) Scalable molecular dynamics with NAMD. J Comput Chem 26:1781–1802
85. Humphrey W, Dalke A, Schulten K (1996) VMD – Visual molecular dynamics. J Molec Graphics 14:33–38
86. Trott O, Olson AJ (2010) AutoDock Vina: improving the speed and accuracy of docking with a new scoring function, efficient optimization and multithreading. J Comput Chem 31:455–461

87. Morris GM et al (2009) Autodock4 and AutoDockTools4: automated docking with selective receptor flexiblity. J Computational Chemistry 16:2785–2791
88. Wang L et al (2012) Linear and non-linear support vector machine for the classification of human 5-HT1A ligand functionality. Molecular Informatics 31:85–95
89. Egleton RD, Davis TP (2005) Development of neuropeptide drugs that cross the blood-brain barrier. NeuroRx 2:44–53
90. Kirchmair J et al (2015) Predicting drug metabolism: experiment and/or computation? Nat Rev Drug Discov 14:387–404
91. Di L (2014) Strategic approaches to optimizing peptide ADME properties. AAPS J 17:134–143

Chapter 21

Computational Nanotechnology: A Tool for Screening Therapeutic Nanomaterials Against Alzheimer's Disease

R. Navanietha Krishnaraj, Dipayan Samanta, and Rajesh K. Sani

Abstract

Alzheimer's disease (AD) is a progressive neurodegenerative disorder leading to several structural, biochemical, or electrical abnormalities in the brain. Though it is mainly caused by the mutation in the genes, the other factors such as environment, health, and lifestyle also contribute to the disease. Recent interest has been turned toward harnessing the potential of nanomaterials for the treatment of AD. Assessing the therapeutic potential of the nanomaterials toward AD using in vivo and in vitro methods suffers from several limitations. The conventional in vivo and in vitro methods are laborious and time consuming and it requires sophisticated facilities. Herein we report the computational nanotechnology approaches for modeling the different nanomaterials, assessing the toxicity of the nanomaterials, and strategies to investigate the therapeutic efficacy of these materials for treating AD. This chapter discusses on using Lipinski's rule of five and rule of three for assessing the drug-like and lead-like characteristics of the nanomaterials. The chapter also addresses the advantages of computational analysis of ADME (absorption, distribution, metabolism, and excretion) characteristics, drug likeliness of nanomaterials, and the role of molecular docking techniques for assessing the therapeutic efficacy of the nanomaterials.

Key words Nanomaterials, ADME analysis, Molecular docking, Lipinski rule, Lead-like characteristics

1 Introduction

Alzheimer disease is an irreversible, progressive, and devastating neurodegenerative disorder which is one of the top five leading causes of death worldwide. It affects almost 4% of the world's population that includes two-third of population above the age of 65 and one-third of population above the age of 85 [1]. However, much progress in research into its symptoms, causes, risk factors, and treatment has accelerated only in the recent decades [2]. The major functional characteristics of this disease are the loss of connections between neurons in the brain at the synapse. The neurons lose the ability to transmit messages from one neuron to the other. The transmission of information is arrested between different parts of the brain as well as from the brain to muscles and other organs

Kunal Roy (ed.), *Computational Modeling of Drugs Against Alzheimer's Disease*, Neuromethods, vol. 132, DOI 10.1007/978-1-4939-7404-7_21, © Springer Science+Business Media LLC 2018

[3]. The disease leads to accumulation of fibrillary protein such as beta-amyloid and hyperphosphorylated Tau. The overaccumulation of the fibrillary protein aggregates results in synaptic dysfunction and neurodegeneration [4]. There are three general stages of disease, namely, the early, middle, and late onset of AD. Familial history is also a risk factor for the disease. Both early onset and late onset AD disease are reported in patients with a positive family history of AD [5]. There are three different forms of autosomal-dominant early-onset familial AD that have been reported in the literature [6]. The current state of knowledge about the pathways and molecular mechanisms of pathogenesis of AD is still limited.

As per the statistical reports of Alzheimer's association, more than 5.4 million cases of AD are reported in the USA. It is expected that the number of AD cases will increase to 74.7 million in 2030 and 131.5 million in 2050. Cholinesterase inhibitors and N-methyl-D-aspartate (NMDA) receptor antagonist are promising candidates for the treatment of neurodegenerative disorders. Several synthetic drugs and phytochemicals have been reported for treatment of AD. Nanomaterials have several advantages such as high surface area and ease of functionalization that confers therapeutic efficacy and drug delivery for treatment of AD. Several nanomaterials with different dimensions and functionalization strategies have been reported in the literature. There is a growing demand for identifying a new drug that confers therapeutic potential to AD by fighting against AD etiology unlike the conventional drugs that work based on the symptoms. The drugs for AD are currently categorized into three different groups based on the mechanisms to treat cognitive symptoms. The first category includes cholinesterase inhibitors that prevent the breakdown of acetylcholine, a chemical messenger for memory and learning. Donepezil, rivastigmine, and galantamine are some of the examples of cholinesterase inhibitors. The second group of drugs, NMDA (N-methyl-D-aspartate) receptor antagonist that acts by regulating the activity of glutamate, a different messenger chemical involved in information processing. Memantine is an example in this category. The third strategy works by a combination of cholinesterase inhibitor and a glutamate regulator. The drugs that are currently available for the treatment of AD will help to improve symptoms, but these drugs do not have profound neuroprotective effect and disease-modifying effects [7]. Reports are available on the side effects of some of these AD drugs. For instance, cholinesterase inhibitors such as physostigmine and tacrine which are used for the treatment of AD have been banned recently in many countries including the USA because of their liver toxicity [8]. Other side effects of cholinesterase inhibitors include increase in acetylcholine activity in the peripheral nervous system, gastrointestinal (GI) upset, diarrhea, nausea, muscular weakness, and syncope [9]. Memantine causes side effects such as delusions and hallucinations in dementia [10].

Different phytochemicals such as phenols, alkaloids, flavonoids, saponins, and terpenes possess neuroprotective activity and have been reported for the treatment of AD [11]. Attempts have also been made to develop synthetic drugs such as rivastigmine with structure of physostigmine (a natural drug) from *Physostigma venenosum* as the template [12]. There is a constant search out for new approaches for developing drugs for AD and disease-modifying agents such as amyloid β (Aβ) deposition inhibitors, γ-secretase inhibitors, and Tau protein aggregation inhibitors [13]. Attempts are underway for the development of Aβ vaccines for AD, but it suffers from several shortcomings [14].

In the recent decades, research interest has spurred toward harnessing the therapeutic potential of nanomaterials for the treatment of AD (Fig. 1). Nanomaterials have interesting physical and chemical characteristics which makes them promising candidates for the treatment of AD. Reports are available on the use of nanomaterials for diagnosis, prevention, and treatment of AD [15]. Several nanomaterials such as carbon nanotubes, graphene, metal, and metal oxide nanomaterials have reported in the literature for the treatment of AD [16–18]. However, assessing the efficacy of nanomaterials for the treatment, drug delivery, and diagnosis for the treatment of AD remains a challenge. The currently available in vitro and in vivo methods to identify efficacy of the therapeutic molecules (including nanomaterial drugs) for AD are time consuming, and they demand sophisticated facilities.

2 Computational Nanotechnology

Computational nanotechnology is the field of nanotechnology that develops and makes use of computer-based models for understanding, analyzing, and predicting the characteristics of the nanomaterials/nanosystems [19]. Computational nanotechnology helps in the structural and functional analysis of the nanomaterials. Nanoinformatics is an emerging area of computational nanotechnology and has great potential for assessing the therapeutic efficacy of nanomaterials. It lies at the interface of bioinformatics, computational chemistry, and nanobiotechnology. Nanoinformatics is at the intersection of classical bioinformatics and computational chemistry approaches to store, standardize, investigate, and visualize nanobiotechnological information. It includes development of databases for organizing the data related to nanobiotechnology [20]. These databases contain modeled structures of nanomaterials and molecular data such as the structure of the biomolecules related to the effect of nanomaterials on biological systems. However, the databases for storing or retrieving the structure of nanomaterials are limited. The structures of the nanomaterials will be very helpful for studying the interaction of nanomaterials with biomolecules/

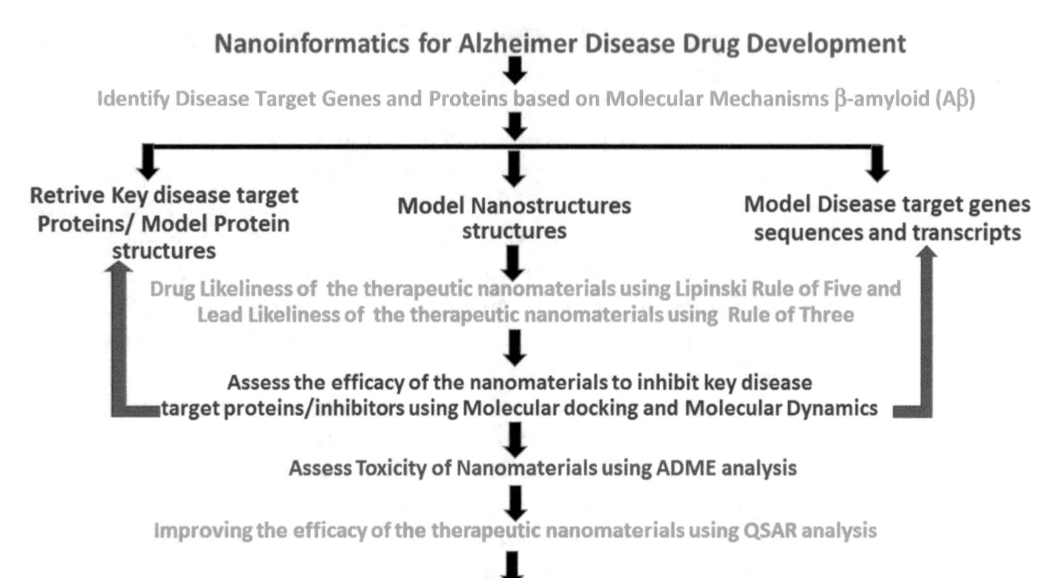

Fig. 1 Schematic diagram of for computational nanotechnology approaches for screening therapeutic nanomaterials against Alzheimer's disease. The reliability of in vitro techniques is limited, whereas the in vivo techniques raise ethical concerns. The use of computational nanotechnology approaches can efficiently help in screening the nanomaterials for treatment of AD in a shorter time. In this book chapter, the demand for computational nanotechnology approaches for assessing the therapeutic efficacy of the nanomaterials for AD and various techniques such as modeling the nanomaterials, molecular docking, molecular dynamics, Absorption Distribution Metabolism and Excretion (ADME) analysis, Quantitative Structure-Activity Relationship (QSAR), and computer-aided drug designing are discussed

biological systems. Computational nanotechnology also includes development of tools, programs, and algorithms for the prediction of the characteristics of nanomaterials for biological applications.

Nanoinformatics also makes use of the developed computational tools/programs for assessing the efficacy of the nanomaterials for biomedical applications. Nanobiotechnology makes use of the nanoscale materials or devices that interact with biological systems at an omics level and perform the agonistic/antagonistic activities, thereby causing the physiological changes leading to therapeutic applications for a wide range of diseases with minimal/no side effects [21]. There are a wide range of nanomaterials that are of great interest to biomedical researchers because of their extraordinary physical, chemical, and mechanical properties [22]. Properties of the nanomaterials such as nano-size (range of 2 to 100 nm), high surface to volume ratio, and quantum confinement are the key characteristics that confer unique properties to the nanomaterials. It includes zero-dimensional materials, one-dimensional materials (nanorod and nanowire), and two-dimensional materials (nanodisks, films, nanobelts, and nanosheets) [23]. Nanomaterials have been shown to be useful for the wide range of biomedical applications because of their antibacterial, antifungal, antiviral, antiplatelet, and anticancer activities [24–26].

There are also several reports documented in the literature on the neuroprotective activity as well as their therapeutic potential to treat neurodegenerative disorders. Schubert et al. (2006) reported the potential of cerium oxide (CeO_2) or yttrium oxide (Y_2O_3) nanoparticles to protect nerve cells from oxidative stress and confer neuroprotective activity. It was also shown that the neuroprotective activity is independent of the size of the (CeO_2) or (Y_2O_3) nanoparticles [27]. The neuroprotective activity of different combinations of cerium oxide and yttrium oxide nanoparticles were studied against acute lead-induced neurotoxicity in rat hippocampus. The results showed that the cerium oxide and yttrium oxide nanoparticles helped to reduce the activities of lead-induced superoxide dismutase and catalase and restore the ADP/ATP ratio in rat hippocampus. The cerium oxide and yttrium oxide nanoparticles also helped to downregulate the oxidative stress markers such as reactive oxygen species (ROS), lipid peroxidation, and apoptosis indexes such as caspase-3 protein expression [28]. The nanomaterials confer neuroprotective activity by scavenging the ROS that is induced by the oxidative stress. Similar reports are also shown on the neuroprotective effects of cerium and yttrium oxide nanoparticles against the high glucose-induced oxidative stress and apoptosis in undifferentiated rat pheochromocytoma (PC12) cells. These nanoparticles upregulated the expression of ROS, lipid peroxidation (LPO), apoptotic Bax, and caspase-3 proteins, whereas apoptotic Bcl-2 and total thiol molecules (TTM) downregulated the high glucose-induced oxidative stress [29].

Xue et al. (2014) reported that the single-walled carbon nanotubes restore the autophagic and lysosomal defects in primary glia in a mouse model of AD. The neuroprotective activity of the carbon nanotube is conferred by reversing abnormal activation and aid in elimination of autophagic substrates [16]. Investigations have been made on the use of platinum-based complexes as inhibitors of Aβ peptide, which is key causative agent for AD. The results of these studies with mouse hippocampal slices have shown that platinum-based complexes inhibit neurotoxicity and recovered Aβ peptide-induced synaptotoxicity [30].

Graphene and graphene oxide nanomaterials are shown to effectively inhibit Aβ peptide monomer fibrillation and amyloid fibrils, which confirms the therapeutic efficacy of the graphene and graphene oxide nanosheets for the treatment of AD [17]. Similar studies were also made on the inhibition of Aβ peptide aggregation using graphene dots for development of drugs for AD. In addition, the quantum dots, because of its smaller size, also help in crossing the blood-brain barrier [31]. The nanomaterials were also shown to have applications for drug delivery applications. Nanomaterials are promising candidates as carriers for targeted drug delivery. The major limitation with the conventional drug delivery systems is the time span required and the reactivity of the drug. The other limitation for the neurotherapeutic drugs is their inability to cross blood-brain barrier.

The nanomaterials are promising candidates that help in the delivery of the drugs as they can cross the blood-brain barrier. The entry of nanomaterial across blood-brain barrier is through receptor-mediated endocytosis mechanism in brain capillary endothelial cells [32]. The uptake of the drug-loaded nanoparticles across the blood-brain barrier is mediated by the surface functionalization of nanoparticle surface by covalently linking the targeting ligands or surfactants. In addition to treatment against AD and drug delivery applications, the nanomaterials were also shown to be useful for the diagnostic applications. Kang et al. 2015 developed a label-free detection of β-amyloid (Aβ) aggregation using gold nanoparticles based on the principles of localized surface plasmon resonance [33]. Nanomaterials such as graphene have been used in the development of immunobiosensor for noninvasive diagnosis of AD. The magnetic nitrogen-doped graphene (MNG) was used for the modification of Au electrode for the detection of Aβ peptide up to a detection limit of 5 pg mL^{-1} [34]. The scientific and industrial community has realized the tremendous potential of the nanomaterials, and the rapid growth in nanotechnology is evident from the recent growth in research activities. Reports have documented that the total number of publications in nanotechnology from the USA was 160,870 with total citations of 4,056,278 from 2003 to 2013 [35]. However, the experimental techniques have several limitations to identify the type of the nanomaterial, its size, shape, and other properties which confer therapeutic potential effectively.

3 Advantages of Nanoinformatics for Drug Discovery

Nanoinformatics approaches have several advantages over the experimental methods for developing drugs/therapeutic strategies for AD. In vivo and in vitro techniques have been established for research activities in AD. However, animal models and cellular models of AD suffer from several limitations [36].

There are several ethical issues in using these animal models because of the immense pain, distress, and death caused by the experiments/research activities to these animals. Animals are subject to investigation or modified using techniques such as transgenic approaches to induce diseased state for understanding the disease mechanisms or to evaluate the efficacy of the therapeutic molecules against the disease.

Certain animal models have the same physiological and genetic characteristics as humans; however, they fail to mimic all the molecular characteristics of AD. The ethicists claim that animals should not be used as tools for broadening knowledge about the disease. They insist on ethical procedures in handling animals, including good laboratory conditions for animal models and the use of anesthesia, and use trained personnel for handling the animal models. Hence, there is an urging need to avoid animal models as much as possible. Models which are less complex such as bacteria, fruit flies, and yeast are much preferred over animal models for investigations in AD. However, these models fail to mimic different proteomic and genomic markers of AD similar to those of humans. Besides these, chick embryo, *Caenorhabditis elegans*, zebra fish, and yeast are also used as models for AD. There are several discrepancies that arise with the currently available animal models and disease in humans. This is one of the major reasons for the failure during the translation of drugs from the laboratory studies to clinical studies. Many drugs that are shown to be very effective fail to be successful during translation at preclinical and clinical stages. Besides the decrease/loss of efficacy between the laboratory and clinical studies, there are also drastic differences in side effects, doses, and pharmacokinetics. It is practically difficult and laborious to screen the optimized dimension of the nanomaterials that has the maximum therapeutic potential. Computational methods offer a chance to precisely tailor the nanomaterials with the desired characteristics. The nanoinformatics approach also provides an advantage to study the specific function based on the interactions of the nanomaterials on individual protein and receptor.

4 Modeling the Structure of Nanomaterials

Different nanomaterials such as carbon nanomaterials, metal nanoparticles, and metal oxide nanoparticles are shown to be promising candidates as drugs as well as drug carriers (for drug delivery) for the treatment of neurodegenerative diseases including AD. Besides these, the properties of the nanomaterials such as size, shape, etc. have impacts on the therapeutic efficacy. There are several computational tools for modeling the structure of nanomaterials. Nanoinformatics approaches make use of different modeling techniques and computational approaches such as ab initio methods for analysis of not only their electronic structure and properties but also their interactions with biomolecules and the impacts on biological systems including therapeutic applications. The techniques of statistical thermodynamics can be used for simulations of large molecular ensembles of nanomaterials, proteins, and protein-nanomaterial complexes [37].

"Avogadro" is useful for modeling the structure of nanomaterials such as carbon nanotubes [38]. Navanietha Krishnaraj et al. (2014) modeled the chiral, armchair, and zigzag carbon nanotubes with different number of layers using Avogadro and showed their

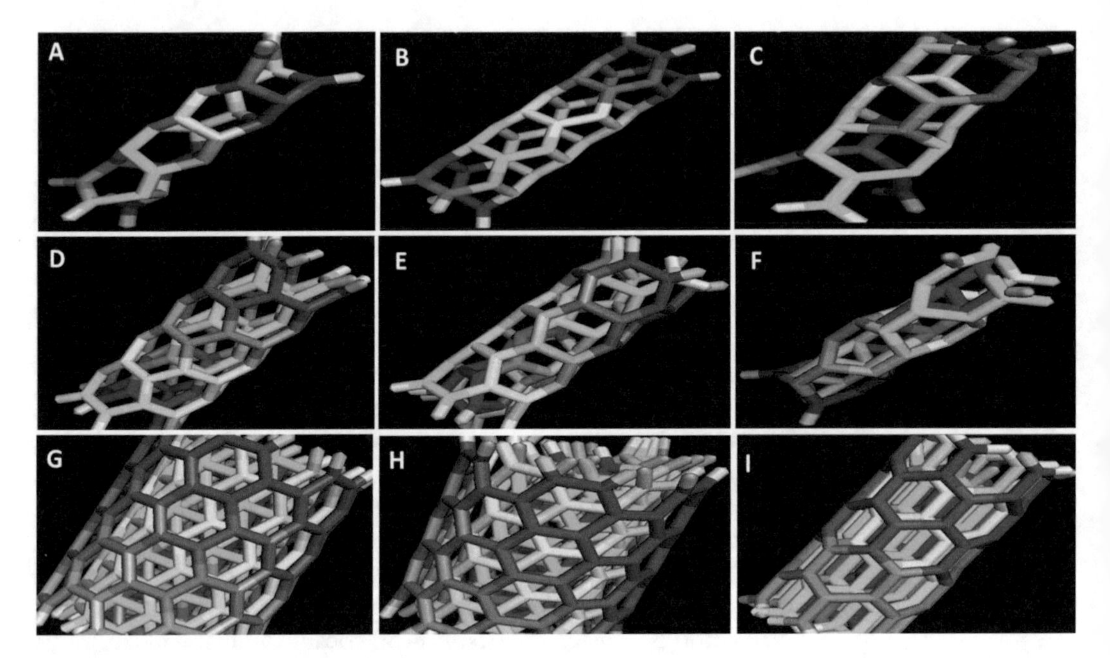

Fig. 2 Modeled structure of **(a)** armchair, **(b)** chiral, and **(c)** zigzag single-walled carbon nanotubes (length 1 nm); **(d)** armchair, **(e)** chiral, and **(f)** zigzag double-walled carbon nanotube (length 1 nm); and **(g)** armchair, **(h)** chiral, and **(i)** zigzag multi (five)-walled carbon nanotube (length 1 nm)

molecular investigations with Human immunodeficiency virus-Viral Protein R (HIV-Vpr), Human immunodeficiency virus-Negative Regulatory Factor (HIV-Nef), and Human Immunodeficiency Virus-Group-specific antigen (HIV-gag) proteins (Fig. 2) [39]. Three different types of single-walled carbon nanotubes, namely, armchair, chiral, and zigzag of different dimensions, can be modeled using Avogadro. Their molecular weight and dipole moments can also be calculated using this tool.

The modeled carbon nanotubes of 1 nm dimension and their effect on increasing the number of walls in carbon nanotubes on the number of atoms, bonds, molecular weight, and dipole moment were analyzed and shown in Table 1. The armchair carbon nanotubes had higher molecular weight and dipole moment than chiral and zigzag carbon nanotubes. The increase in the dipole moment of the molecules/materials will in turn increase the hydrophobicity. The higher the lipophilicity of the therapeutic molecule/nanomaterial, the better the efficacy of the drug to bind the key disease target protein.

These modeled structures will be useful for understanding the molecular interaction of these nanoparticles with biomolecules such as proteins and nucleic acids. Several reports are documented in the literature on the different applications of carbon nanotubes, which are shown to have antimicrobial [40], anticancer [41], antioxidant [42], antiviral [43], and platelet aggregation properties

Table 1
Characteristics of modeled carbon nanotube

Single-walled carbon nanotube (*length = 1 nm*)			
	ARMCHAIR	*CHIRAL*	*ZIGZAG*
N	2	1	3
M	2	2	0
Atoms	44	32	36
Bonds	58	41	45
Molecular wt.(g/mol)	440.449	307.323	333.360
Dipole moment(D)	162.835	139.111	143.398
Double-walled carbon nanotube (*length = 1 nm*)			
Atoms	110	84	56
Bonds	145	109	80
Molecular wt.(g/mol)	1101.122	821.852	551.569
Dipole moment(D)	138.405	121.039	136.944
Multi (five)-walled carbon nanotube (*length = 1 nm*)			
Atoms	440	360	258
Bonds	580	469	327
Molecular wt.(g/mol)	4404.487	3542.656	2361.576
Dipole moment(D)	138.202	120.395	849.015

[44]. The use of these computational models will further enrich our understanding toward specific mechanisms by which the carbon nanotubes confer different activities in biological systems.

Similarly, several reports have been documented on the wide applications of graphene because of its unique physical, chemical, biological properties [45]. They are shown to be promising for biomedical applications such as diagnostics, drug immobilization, and drug delivery applications [46, 47]. The ease of functionalization, high surface area, and good electrical properties altogether make graphene a promising candidate for theranostic applications [48]. However, the detailed investigations on the mechanisms by which these materials gain therapeutic potential are still at infancy. Hence, these modeled structures of graphene will help to make a detailed computational investigation whenever the experimental investigations have limitations in providing the desired information. The structures of the graphene modeled using Avogadro with dimensions 110 nm × 110 nm, 140 nm × 140 nm, and 190 nm × 190 nm are shown in Fig. 3.

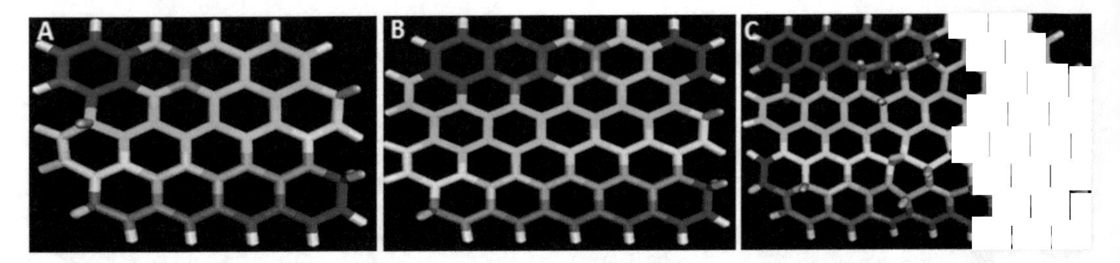

Fig. 3 Modeled structure of graphene with dimensions **(a)** 110 nm × 110 nm, **(b)** 140 nm x 140 nm, and **(c)** 190 nm × 190 nm

Table 2
Properties of the modeled graphene structures

Dimension of graphene	110 × 110 nm	140 × 140 nm	190 × 190 nm
Number of atoms	70	96	156
Number of bonds	118	166	286
Molecular wt.(g/mol)	598.6883	866.9554	1367.5422

The properties of the modeled graphene structures such as number of atoms, bonds, and molecular weight are discussed in Table 2.

Further, investigations have shown that the increase in number of layers and dimension of graphene resulted in an increase in binding affinity on the key disease targets. The poor molecular flexibility with rigid rotatable bonds is one of an interesting feature of graphene that confers ideal drug oral bioavailability and makes it promise for drug delivery applications. Aghili, Z et al. (2016) modeled Fe nanoparticle using Materials Studio 5.0 (developed by Accelrys Software Inc.) and studied their interactions with lysozyme. The geometry of the structures that was made with 13 atoms in a cluster was optimized using the CASTEP module. The Fe NP was optimized and the cluster had 12 surface atoms. Every surface atom was considered to have one adsorbing water molecule. Further, addition of water molecules in the model of the nanomaterial will cause lowering of binding energy for complex because of the steric effect between water molecules [49]. Tokarskýa et al. (2010) made a detailed force field calculation on the adhesion of silver nanoparticles with montmorillonite and kaolinite using molecular modeling technique which helped in the prediction of the orientation of the nanoparticles, structure, and stability of the nanoparticle [50]. Cagin et al. (2000) modeled the molecular structure of PAMAM dendritic supermolecular assemblies with continuous configurational Boltzmann biased (CCBB) direct Monte Carlo method and studied their energetic and structural properties using molecular dynamics after annealing these molecular representations. The CCBB direct MC method for modeling nanomaterial is based on the independent rotational sampling technique. In this technique, the torsional degrees of the polymer chains are sampled by weighting function on the basis of the Boltzmann factor of the torsion energy [51].

5 Molecular Docking

Nanomaterials have been shown to be having therapeutic potential against AD besides their applications in site-directed drug delivery and controlled drug release. Nanobiotechnology explores the use of the nanoscale materials that interact with biological systems at an omics level, thereby conferring therapeutic potential. Nanomaterials perform agonistic/antagonistic activities thereby causing the physiological changes in biological systems. The molecular interactions between the nanomaterial and the proteins will provide a valuable information about the molecular mechanisms and the toxicity/therapeutic potential of the nanomaterials. Molecular docking is a computational procedure which is helpful for predicting the non-covalent binding of macromolecules or interactions of a receptor to a nanomaterial [52]. It is useful to predict the bound conformations and the binding affinity of the nanomaterials with target genes/proteins. Molecular docking is used for the prediction of interaction of the enzyme to the substrate for different industrial and biomedical applications including biological fuel cells [53–56]. Navanietha Krishnaraj et al. (2015) reported the use of the computational approaches for screening electroactive microorganisms and have shown to be advantageous over the conventional electrochemical methods [57–60]. Computational approaches helps in the system biology investigations which are advantageous over the reductionist approaches [61, 62]. The use of molecular docking techniques to screen the inhibitors of key disease target proteins is well documented in the literature. The acetylcholinesterase (AChE) inhibitors are widely used for the treatment of AD. AChE inhibitors help to prevent/decrease the breakdown of acetylcholine (ACh) and prevent the death of cholinergic neurons. Navanietha Krishnaraj et al. (2014) reported the antagonistic molecular interactions of photosynthetic pigments with molecular disease targets of AD, namely, P53 kinase receptor, EphA4, and histone deacetylase [55].

The interactions of natural pigments such as β-carotene, chlorophyll a, chlorophyll b, phycoerythrin, and phycocyanin on the key disease targets such as P53 kinase receptor (4AT3), EphA4 (3CKH), and histone deacetylase (3SFF) have been analyzed using molecular docking technique. Among the five different pigments, β-carotene had the least binding affinity of −10.1 kcal/mol and −8.3 kcal/mol with the p75 neurotrophin receptor and EphA4 receptor, respectively. The phycocyanin and the phycoerythrin ligands have the least binding affinities of −7.9 kcal/mol and −8.2 kcal/mol with HDAC receptor.

These three proteins play a crucial role in disease pathogenesis, and inhibition of these targets confers therapy toward AD. The surfaces of the neurons have two types of receptors which decide the fate of the cell based on the interaction of neurotrophins. Upon binding of neurotrophins, the TRK receptors send the signal to the neuron to promote its survival and growth. The p75 neurotrophin receptor is the second receptor that has the opposite activity of the TRK receptor. When the neurotrophins bind onto the p75 kinase receptor, they cause apoptosis of the neuron [63]. Binding of neurotrophin to the p75 kinase receptor is one of the major reasons for the neurodegeneration and neuronal cell death in AD [64]. There are several reports on the programmed mechanism of cell death resembling apoptosis in the AD [65].

Erythropoietin-producing hepatocellular A4 (EphA4) is another key target for the treatment of AD, and inhibitors of EphA4 are promising for treatment of AD. The upregulations of EphA4 have shown to inhibit the axonal growth of motor neurons. Recent reports showed that EphA4 mediates hippocampal synaptic dysfunctions in AD and the inhibition of EphA4 was shown to revert synaptic impairment in mouse models of AD [66]. Rosenberger et al., 2014, reported the reduced availability of EphA4 which leads to synaptic dysfunction in AD. The immunohistochemical studies on the localization of EphA4 in hippocampus showed that there is an altered distribution in AD when compared with the control [67].

Histone deacetylase (HDAC) inhibitors are another group of inhibitors that are shown to be promising for the treatment of AD and other neurodegenerative diseases [68].

Molecular docking is an important tool to identify suitable nanomaterials for the treatment of AD based on the interactions of the nanomaterials with the key disease target proteins/nucleic acids. The screening of nanomaterials as inhibitors of the key diseases of AD can be made with the help of molecular docking and dynamics. AutoDock Vina is a molecular modeling simulation software which is especially effective for protein-ligand docking. AutoDock consists of two main programs which include docking of the ligand to a set of grids describing the target protein and precalculating these grids [69]. However, the scoring functions in molecular docking have limitations that they do not correlate well with the experimental data. It may be due to several reasons such as solubility of hits, accessibility of the ligand to the target, molecular interaction with other proteins, modification of hits by enzymes, or any other reactions in the in vivo system.

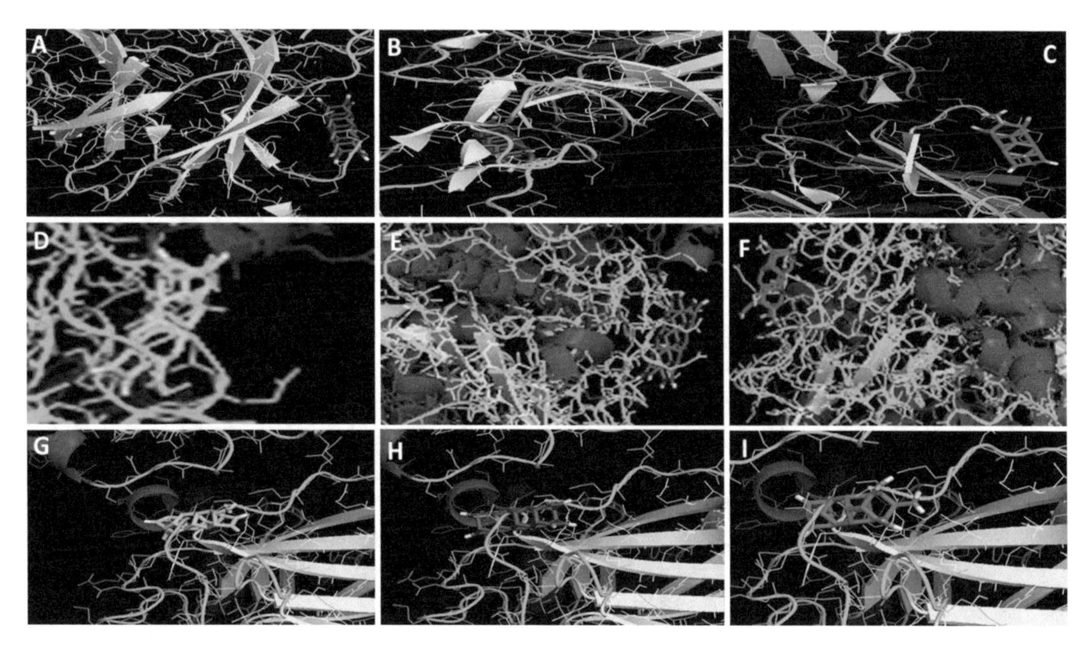

Fig. 4 Molecular interactions of single-walled carbon nanotubes with (**a**) P53 kinase receptor, (**b**) EphA4, (**c**) histone deacetylase; molecular interactions of double-walled carbon nanotubes with (**d**) P53 kinase receptor, (**e**) EphA4, (**f**) histone deacetylase; molecular interactions of multi-walled carbon nanotubes with (**g**) P53 kinase receptor, (**h**) EphA4, (**i**) histone deacetylase

Molecular docking investigations were carried out using AutoDock Vina algorithm to study the interactions of armchair, chiral, and zigzag carbon nanotubes on the key disease target proteins of AD, namely, P53 kinase receptor (4AT3), EphA4 (3CKH), and histone deacetylase (3 SFF). The single-walled carbon nanotubes modeled using Avogadro were used as the ligand for these experiments. The modeled structures of carbon nanotubes were processed by adding hydrogen atom using MGL tools. The energy of the processed carbon nanotubes were minimized by calculating Gasteiger charges and storing in the PDBQT format. The structures of the p75 neurotrophin receptor, EphA4, and histone deacetylase collected from PDB ids 4AT3, 3CKH, and 3SFF, respectively (shown in Fig. 4), are used as disease targets for the molecular docking experiments. The retrieved structures of the proteins were processed by removing the native ligands from their protein structure. Further, the conformation of the modeled structures was evaluated after removing the ligand by superimposing the structure of the key disease target protein before and after removing the ligand, and the RMSD (root-mean-square deviation) value was calculated between them using the PYMOL viewer. Once the structures of the protein have been confirmed, the heteroatoms including water molecules are removed from the protein structures. The structures of the key disease target proteins were further processed by adding polar hydrogens and applying Kollman's partial atomic charges to minimize the energy. The processed structures of key disease target protein were saved in PDBQT file format which has hydrogen's in all polar residues.

The active sites of the key disease target proteins was identified based on the previous reports or predicted using computational tools. The active binding region of disease target protein for docking was chosen in such a way that the entire carbon nanotube binding region of the target protein is completely covered by the GRID. The AutoDock tool was used to select the active region where the nanomaterial interacts. The dimensions of the grid depend on the size of the ligand, protein, and their active region, and it need not be uniform along all the three directions. AutoDock Vina helps to perform molecular docking with improved speed and accuracy with a new scoring function, efficient optimization, and multithreading. Docking investigations between modeled carbon nanotubes were carried out using AutoDock Vina with AMBER force field and Monte Carlo simulated annealing. The structures of the processed disease target proteins were kept rigid throughout the docking process, and the binding energy of the docked carbon nanotube disease target proteins of AD was identified. The armchair, chiral, and zigzag carbon nanotubes had the binding affinities of −10.2, −8.1, and −7.3 kcal/mol with P53 kinase receptor. Similar investigations between armchair, chiral, and zigzag carbon nanotubes and EphA4 also showed that the nanotubes have very high binding affinities of −9.5, −7.5, and −7.0 kcal/mol, respectively. The results of the docking experiments of armchair, chiral, and zigzag carbon nanotubes with histone deacetylase have a very high binding affinities of −11.5, −10.5, and −7.8 kcal/mol, respectively [70]. The interactions of armchair, chiral, and zigzag carbon nanotubes with the key disease target proteins of AD, namely, P53 kinase receptor (4AT3), EphA4 (3CKH), and histone deacetylase (3SFF), are shown in Fig. 4.

Similar procedures were followed for accessing the molecular interaction of the modeled graphene ligands with disease target proteins of AD. The modeled structure of the graphene with different dimensions was processed as in the carbon nanotubes discussed in this section. The docking investigations have showed that the modeled structure of graphene with 110 × 110 nm, 140 × 140 nm, and 190 × 190 nm had the binding affinities of −11.2, −12.5, and −16.1 kcal/mol with P53 kinase receptor. The results of the molecular interactions between EPHA4 and graphene with dimensions 110 × 110 nm, 140 × 140 nm, and 190 × 190 nm had the binding affinities of −10.0, −13.7, and −15.6 kcal/mol. Similarly, the binding affinities between graphene with 110 × 110 nm, 140 × 140 nm, and 190 × 190 nm dimensions and histone deacetylase were found to be −12.6, −13.7, and −17.3 kcal/mol, respectively [70]. Molecular interactions of different dimensions of graphene with key disease target proteins of AD are shown in Fig. 5.

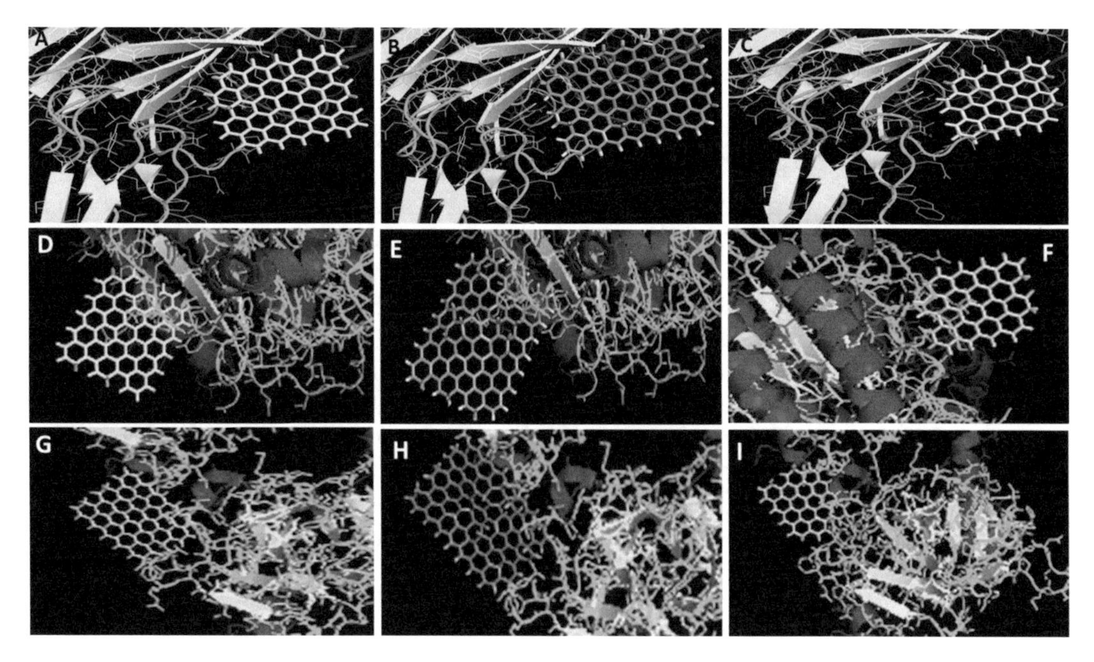

Fig. 5 Molecular interactions of modeled graphene (of dimension 110 × 110 nm) with (**a**) P53 kinase receptor, (**b**) EphA4, (**c**) histone deacetylase (helix in *red*, sheet in *yellow*, and loop in *green*); molecular interactions of modeled graphene (of dimension 140 × 140 nm) with (**d**) P53 kinase receptor, (**e**) EphA4, (**f**) histone deacetylase; molecular interactions of modeled graphene (of dimension 190 × 190 nm) with (**g**) P53 kinase receptor, (**h**) EphA4, (**i**) histone deacetylase

The binding interactions between the nanomaterials and disease targets of AD provide insights about the mechanisms by which the nanomaterials lessen the symptoms of the disease. These results support the previous experimental investigations on the use of carbon nanotubes and graphene for AD therapy. Yang et al. 2015 [17] showed the inhibition of Aβ peptide monomer fibrillation and clear mature amyloid fibrils using graphene for reverting AD pathogenesis using computational and experimental evidences. The binding affinities of the nanomaterials to the three different target proteins P53 kinase receptor, EphA4, and histone deacetylase give a clue regarding the therapeutic potential of the modeled graphene.

6 ADME Analysis and Drug Likeliness of Nanomaterials

Although several reports focus on the antagonistic/agonistic interaction of nanomaterials with specific proteins/nucleic acids against AD, it is equally important to study the ADME analysis of nanomaterials which is very important toward drug development. The neurotoxicity assays can be used to assess the nanomaterials, and also a variety of experimental techniques such as transcriptomics and proteomics can be used to study the effect of nanomaterials on gene expression and regulation, epigenetic modification, genotyping, and signal transduction pathway activation [71, 72]. The nanomaterials may have very high therapeutic effects on AD or may be promising for drug delivery applications. However, if the rates of absorption, distribution, metabolism, and excretion of nanomaterials are not regular, the nanomaterials will purpose. The ADME analysis of nanomaterials is an important phase in the drug development process. It provides the key understanding on acceptance of these nanomaterials for the biological system. The main reason for the failure of major drugs in clinical trials is due to their unfavorable toxicology and poor ADME characteristics. The use of computational methods is promising for the ADME analysis of the drugs and nanomaterials (especially to study the distribution to the brain).

Absorption studies of the nanomaterials by the biological systems will help in identifying the suitable route of delivery into the body. The developed therapeutic compounds can be delivered by different routes such as oral, intravenous, intramuscular, and nasal. The absorption of the nanomaterial in biological system and its efficacy greatly depend on its routes of delivery. In the case of AD drug development, suitable drug delivery systems that target the drug to the brain should be chosen. The delivery of drug to the brain depends on two major factors, namely, the route of delivery and the chemical composition of the drug. The drugs delivered through oral routes are greatly affected by release of the acids by the oxyntic cells, and they are stabilizing agents prone to be affected by the enzymes. The effect of these acids and enzymes affects the surface characteristics of the nanomaterial and thereby alters the efficacy of the nanomaterial in the biological system. Different parameters of the therapeutic nanomaterials such as solubility, intestinal transit time, gastric emptying time, stability of the nanomaterial in the stomach (with reference to structure and surface characteristics), and the rates of permeation of the nanomaterials into intestinal wall influence the rates at which these nanomaterials are being absorbed after oral uptake. The nanomaterials which are poorly absorbed by the body should be delivered by different routes to improve the rates of the efficacy. The distribution of the nanomaterials from one compartment to another compartment is a major challenge, and it depends on several parameters including rates of regional blood flow, molecular size of the nanomaterial, polarity of the nanomaterial, and its binding to different proteins in the body. The distribution of nanomaterials depends on different mass transport characteristics such as migration, convection, and diffusion. The alterations of the nanomaterials caused by the acids/ enzymes produced by the body also affect the rates of distribution of the nanomaterial in the biological system.

The other important factor is the efficacy of the therapeutic nanomaterial to metabolize the nanomaterial. Once the nanomaterial is delivered into the biological system, the metabolic process starts, even before the absorption and distribution process. Generally, the drugs are metabolized by the drug-metabolizing enzymes such as cytochrome P450 enzymes. These drug-metabolizing enzymes either act as inhibitors, or after getting metabolized, the metabolites mediate therapeutic effects. However, the mechanisms by which the nanomaterial confers therapeutic efficacy are different from that of the conventional drugs. The nanomaterials being particles of small size make it difficult for the biological system to metabolize them; instead the nanomaterial affects the metabolic process and physiology. However, the surface characteristics of the nanomaterials will be affected by the enzymatic machinery of the biological system. Excretion of the nano-drugs or nano-drug carriers is another important criterion. It depends on the mass transport across the various membranes through different routes. Immediate excretion of the nano-drugs will lower the activity of the drug on the biological system. However, it should also be noted that these drugs are not getting accumulated at specific sites because of any biochemical parameters such as pH. Accumulation of nanomaterials in biological systems can have negative impacts including minor side effects to even fatal conditions. The major aspect in the ADME analysis in developing drugs for neurodegenerative diseases including AD is the limitation of the nanomaterials to cross the blood-brain barrier. Distribution can be a serious problem at some natural barriers like the blood-brain barrier. This can be circumvented by the use of suitable nanomaterials or suitable size and properties which can cross the blood-brain barrier. Further, the use of surface functionalization strategies can also help to improve the flux across the blood-brain barriers. Computational investigations between the interaction of nanomaterials and the functionalizing agents can be used for choosing suitable nanomaterials and developing suitable functionalization strategies.

The suitability of the nanomaterials to act as drugs can be assessed by drug likeliness. Computation tools are also widely used to design a specific drug or nanomaterials and to assess the therapeutic properties of nanomaterials. The computational approaches also help in accessing bioactivity scores and their potential to combat different diseases.

One very common technique for assessing the drug likeliness is using the Lipinski's rule of five. This rule stressed the therapeutic potential of the drug based on the size of the molecule and the lipophilicity of the drug. This rule also provides insights about the molecular properties of the drugs including nanomaterials which influence the pharmacokinetics. The rule also provides details about absorption, distribution, metabolism, and excretion. The rule will be very helpful for the assessing the therapeutic potential of the nanomaterials. It will be helpful to develop drug for the AD that can cross blood-brain barrier. It will also be useful for designing the nanomaterials and their functionalization to increase the activity. This rule can also be useful for studying the selectivity of the nanodrug molecules and to evaluate drug-like physicochemical properties. Another similar rule is the rule of three which gives the insights about the lead-like characteristics of the therapeutic molecule. The rule of five can be extended for predicting the drug likeliness of the therapeutic nanomaterials, and the rule of three can be extended for predicting the lead likeliness of the therapeutic nanomaterials [73].

The armchair, chiral, and zigzag carbon nanotubes have the Log P values of 4.80, 5.67, and 5.29, respectively. The mass of the armchair, chiral, and zigzag carbon nanotubes was found to be 468, 332, and 360, respectively. The molar refractivity values of armchair, chiral, and zigzag carbon nanotubes were found to be 134.28, 95.07, and 104.162, respectively. Among the three different carbon nanotubes studied, the chiral carbon nanotubes have the minimum mass, minimum Log P values, and minimum molar refractivity. The modeled graphene structures with dimensions 110 $\times$ 110 nm, 140 $\times$ 140 nm, and 190 $\times$ 190 nm had the Log P values of 8.73, 11.68, and 16.56, respectively. The mass of the graphene structures of dimension 110 $\times$ 110 nm, 140 $\times$ 140 nm, and 190 $\times$ 190 nm was found to be 600, 868, and 1378, respectively. The graphene structures of dimension 110 $\times$ 110 nm, 140 $\times$ 140 nm, and 190 $\times$ 190 nm had the molar refractivity values of 184.31, 262.46, and 401.21, respectively. The values of mass, Log P, and molar refractivity linearly increase with increase in the dimension.

7 Nanoinformatics for Nanomaterial Functionalization

The nanomaterials are widely used for the delivery of the drugs because of their large surface area which confers high loading capacity of the drug. To improve the therapeutic efficacy and ADME characteristics, the surface of the nanomaterial can be functionalized using suitable drugs or stabilizing compounds. The functionalization strategies also aid in increased therapeutic activity because of the synergistic activity between the nanomaterial and the functionalized compound [74]. There are several functionalization strategies that are reported in the literature to improve the efficacy of AD drugs. The nanomaterials can be modified with the functionalized molecules either based on the adsorption or using covalent bonding. Poor functionalization strategies fail to confer the stability, and also the drug gets desorbed from the nanomaterial before reaching the target site. Suitable functionalization strategies will help in targeted drug delivery. Several reports are available in the literature on the use of monoclonal antibodies for the treatment of AD. Modifying the nanomaterial with monoclonal antibody will

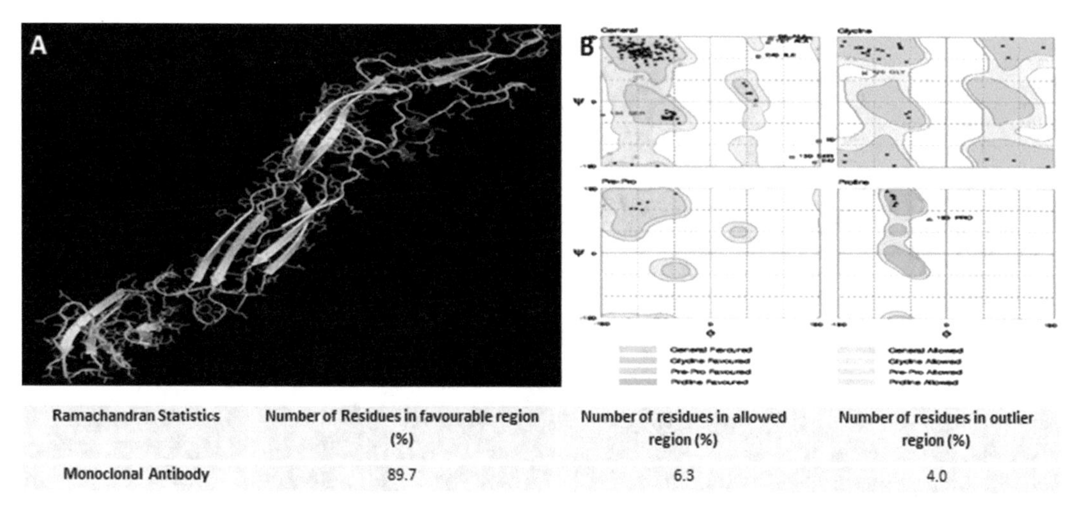

Ramachandran Statistics	Number of Residues in favourable region (%)	Number of residues in allowed region (%)	Number of residues in outlier region (%)
Monoclonal Antibody	89.7	6.3	4.0

Fig. 6 (a) Modeled structure of anti-beta amyloid 40 monoclonal antibody and its **(b)** Ramachandran plot statistics

help in targeted drug delivery to the site of the pathogenesis [75, 76].

The sequence of the anti-beta amyloid 40 monoclonal antibody (accession no.: AF502169) has been derived from IMGT website against the beta protein. The retrieved sequence had the length of 768 base pairs comprised of 255 amino acids. The structure of the anti-beta amyloid 40 monoclonal antibody was modeled using homology modeling technique and validated using Ramachandran plot analysis. The modeled structure and the Ramachandran plot statistics of the anti-beta amyloid 40 monoclonal antibody are shown in Fig. 6. The monoclonal antibody had 89.6% of the residues in favorable region and 6.3% of the residues in the allowed region. This indicates the good stereochemistry of the structure of the monoclonal antibody.

Investigations were carried out to study the interactions of the modeled anti-beta amyloid 40 monoclonal antibody with the modeled structures of the carbon nanotubes and different dimensions of modeled graphene (shown in Fig. 7). The armchair, chiral, and zigzag carbon nanotubes had the binding affinities of −8.5, −7.3, and −7.2 kcal/mol, respectively. The modeled graphene structures with dimensions 110 × 110 nm, 140 × 140 nm, and 190 × 190 nm had the binding affinities of −8.9, −9.0, and −10.4 kcal/mol, respectively (our computational data). These investigations aid in choosing suitable nanomaterials for immobilizing monoclonal antibodies for targeted drug delivery applications.

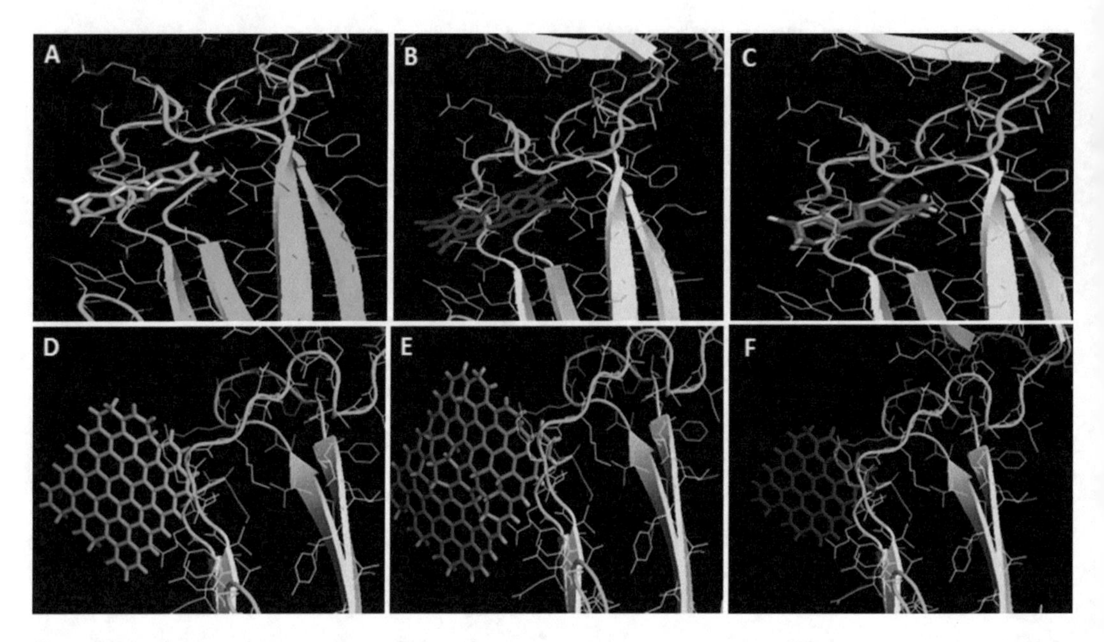

Fig. 7 Molecular interactions of modeled anti-beta amyloid 40 monoclonal antibody (helix in *red*, sheet in *yellow*, and loop in *green*) with **(a)** armchair, **(b)** chiral, and **(c)** zigzag carbon nanotubes; graphene structures with dimensions **(d)** 110 × 110 nm, **(e)** 140 × 14 Å, and **(f)** 19 × 19 Å

8 Conclusion

Rapid advancements in nanobiotechnology research and the growing need for new promising therapeutic drugs for AD demand techniques for screening the drugs with good efficacy precisely in a shorter time. Computational nanoinformatics is an ideal option for analyzing the therapeutic potential of the drugs as well as to design the drugs based on the structure-analysis relationship or structure-property relationship or structure-reactivity relationship. Further, nanoinformatics approaches also help in avoiding the ethical issues that arise from the use of animal models and the limitations of using cellular models of AD.

References

1. Alzheimer's Association (2016) 2016 Alzheimer's disease facts and figures. Alzheimers Dement 12(4):459–509
2. Berrios GE (1991) Alzheimer's disease: a conceptual history. Int J Geriatr Psychiatry 5:355–365
3. Brundin P, Melki R, Kopito R (2010) Prion-like transmission of protein aggregates in neurodegenerative diseases. Nat Rev Mol Cell Biol 11(4):301–307
4. Bachurin SO, Bovina EV, Ustyugov AA (2017) Drugs in clinical trials for Alzheimer's disease: the major trends. Med Res Rev. doi:10.1002/med.21434
5. Bekris LM, Yu CE, Bird TD, Tsuang DW (2010) Genetics of Alzheimer disease. J Geriatr Psychiatry Neurol 23(4):213–227
6. Bird TD (2008) Genetic aspects of Alzheimer disease. Genet Med 10:231–239

7. Skaper SD (2012) Alzheimer's disease and amyloid: culprit or coincidence? Int Rev Neurobiol doi 102:277–316
8. Yoshida S, Suzuki N (1993) Antiamnesic and cholinomimetic side-effects of the cholinesterase inhibitors, physostigmine, tacrine and NIK-247 in rats. Eur J Pharmacol 250(1):117–124
9. Casey DA, Antimisiaris D, O'Brien J (2010) Drugs for Alzheimer's disease: are they effective? P T 35(4):208–211
10. Ridha BH, Josephs KA, Rossor MN (2005) Delusions and hallucinations in dementia with Lewy bodies: worsening with memantine. Neurology 65:481–482
11. Kumar GP, Khanum F (2010) Neuroprotective potential of phytochemicals. Pharmacogn Rev 6(12):81–90
12. Howes MJ, Houghton PJ (2012) Ethnobotanical treatment strategies against Alzheimer's disease. Curr Alzheimer Res 9(1):67–85
13. Galimberti D, Scarpini E (2011) Disease-modifying treatments for Alzheimer's disease. Ther Adv Neurol Disord 4(4):203–216
14. Marciani DJ (2016) Rejecting the Alzheimer's disease vaccine development for the wrong reasons. Drug Discov Today 2017 22 (4):609–614
15. Doggui S, Dao L, Ramassamy C (2012) Potential of drug-loaded nanoparticles for Alzheimer's disease: diagnosis, prevention and treatment. Ther Deliv 3(9):1025–1027
16. Xue X, Wang LR, Sato Y, Jiang Y, Berg M, Yang DS, Nixon RA, Liang XJ (2014) Single-walled carbon nanotubes alleviate autophagic/lysosomal defects in primary glia from a mouse model of Alzheimer's disease. Nano Lett 14 (9):5110–5117
17. Yang Z, Ge C, Liu J, Chong Y, Gu Z, Jimenez-Cruz CA, Chai Z, Zhou R (2015) Destruction of amyloid fibrils by graphene through penetration and extraction of peptides. Nanoscale 7 (44):18725–18737
18. Soltani N, Gholami MR (2016) Increase in the Beta-sheet character of an Amyloidogenic peptide upon adsorption onto gold and silver surfaces. Chemphyschem 29. doi:10.1002/cphc.201601000
19. Merkle RC (1991) Computational nanotechnology. Nanotechnology 2(3):134–141
20. González-Nilo F, Pérez-Acle T, Guínez-Molinos S, Geraldo DA, Sandoval C, Yévenes A, Santos LS, Laurie VF, Mendoza H, Cachau RE (2011) Nanoinformatics: an emerging area of information technology at the intersection of bioinformatics, computational chemistry and nanobiotechnology. Biol Res 44 (1):43–51
21. Zdrojewicz Z, Waracki M, Bugaj B, Pypno D, Cabała K (2015) Medical applications of nanotechnology. Postepy Hig Med Dosw 69:1196–1204
22. Balasundaram G, Webster TJ (2006) Nanotechnology and biomaterials for orthopedic medical applications. Nanomedicine (Lond) 1 (2):169–176
23. Tiwari JN, Tiwari RN, Kim KS (2012) Zero-dimensional, one-dimensional, two-dimensional and three-dimensional nanostructured materials for advanced electrochemical energy devices. Prog Mater Sci 57(4):724–803
24. Kevin John PA, Muralidharan M, Muniyandi J, Navanietha Krishnaraj R, Sangiliyandi G (2014) Synthesis of silver nanoparticles using pine mushroom extract: a potential antimicrobial agent against E.Coli and B. Subtilis. J Ind Eng Chem 20(4):2325–2321
25. Navanietha Krishnaraj R, Berchmans S (2013) In vitro antiplatelet activity of silver nanoparticles synthesized using the microorganism Gluconobacter roseus: an AFM-based study. RSC Adv 3:8953–8959
26. Pramanik A, Kole AK, Navanietha Krishnaraj R, Biswas S, Tiwary CS, Varalakshmi P, Rai SK, Kumar A, Kumbhakar P (2016) A novel technique of synthesis of highly fluorescent carbon nanoparticles from broth constituent and in-vivo bioimaging of *C. elegans*. J Fluoresc 26 (5):1541–1548
27. Hosseini A, Sharifi AM, Abdollahi M, Najafi R, Baeeri M, Rayegan S, Cheshmehnour J, Hassani S, Bayrami Z, Safa M (2015) Cerium and yttrium oxide nanoparticles against lead-induced oxidative stress and apoptosis in rat hippocampus. Biol Trace Elem Res 164 (1):80–89
28. Ghaznavi H, Najafi R, Mehrzadi S, Hosseini A, Tekyemaroof N, Shakeri-Zadeh A, Rezayat M, Sharifi AM (2015) Neuro-protective effects of cerium and yttrium oxide nanoparticles on high glucose-induced oxidative stress and apoptosis in undifferentiated PC12 cells. Neurol Res 37(7):624–632
29. Schubert D, Dargusch R, Raitano J, Chan SW (2006) Cerium and yttrium oxide nanoparticles are neuroprotective. Biochem Biophys Res Commun 342(1):86–91
30. Barnham KJ, Kenche VB, Ciccotosto GD, Smith DP, Tew DJ, Liu X, Perez K, Cranston GA, Johanssen TJ, Volitakis I, Bush AI, Masters CL, White AR, Smith JP, Cherny RA, Cappai R (2008) Platinum-based inhibitors of amyloid-beta as therapeutic agents for Alzheimer's disease. Proc Natl Acad Sci U S A 105 (19):6813–6818

31. Liu Y, Xu LP, Dai W, Dong H, Wen Y, Zhang X (2015) Graphene quantum dots for the inhibition of β amyloid aggregation. Nanoscale 7 (45):19060–19065
32. Wohlfart S, Gelperina S, Kreuter J (2012) Transport of drugs across the blood-brain barrier by nanoparticles. J Control Release 161 (2):264–273
33. Kang MK, Lee J, Nguyen AH, Sim SJ (2015) Label-free detection of ApoE4-mediated β-amyloid aggregation on single nanoparticle uncovering Alzheimer's disease. Biosens Bioelectron 72:197–204
34. Li SS, Lin CW, Wei KC, Huang CY, Hsu PH, Liu HL, Lu YJ, Lin SC, Yang HW, Ma CC (2016) Non-invasive screening for early Alzheimer's disease diagnosis by a sensitively immunomagnetic biosensor. Sci Rep 6:25155
35. Haiyan D, Yu G, Patrick JS, Zaisheng W, Jianguo X, Lee J (2016) The nanotechnology race between China and USA. Nano Today 11 (1):7–12
36. Doke SK, Dhawale SC (2015) Alternatives to animal testing: a review. Saudi Pharm J 23 (3):223–229
37. Maojo V, Fritts M, de la Iglesia D, Cachau RE, Garcia-Remesal M, Mitchell JA, Kulikowski C (2012) Nanoinformatics: a new area of research in nanomedicine. Int J Nanomedicine 7:3867–3890
38. Hanwell MD, Curtis DE, Lonie DC, Vandermeersch T, Zurek E, Hutchison GR (2012) Avogadro: an advanced semantic chemical editor, visualization, and analysis platform. J Chem 4:17
39. Navanietha Krishnaraj R, Chandran S, Pal P, Berchmans S (2014) Investigations on the antiretroviral activity of carbon nanotubes using computational molecular approach, combinatorial chemistry and high throughput screening 17(6):531–5
40. Vecitis CD, Zodrow KR, Kang S, Elimelech M (2010) Electronic-structure-dependent bacterial cytotoxicity of single-walled carbon nanotubes. ACS Nano 4:5471–5479
41. Samorì C, Ali-Boucetta H, Sainz R, Guo C, Toma FM, Fabbro C, da Ros T, Prato M, Kostarelos K, Bianco A (2010) Enhanced anticancer activityof multi-walled carbon nanotube-methotrexate conjugates using cleavable linkers. Chem Commun (Camb) 46:1494–1496
42. Lucente-Schultz RM, Moore VC, Leonard AD, Price BK, Kosynkin DV, Lu M, Partha R, Conyers JL, Tour JM (2009) Antioxidant single-walled carbon nanotubes. J Am Chem Soc 131(11):3934–3941
43. Ji H, Yang Z, Jiang W, Geng C, Gong M, Xiao H, Wang Z, Cheng L (2008) Antiviral activity of nano carbon fullerene lipidosome against influenza virus in vitro. J Huazhong Univ Sci Technolog Med Sci 28:243–246
44. Radomski A, Jurasz P, Alonso-Escolano D, Drews M, Morandi M, Malinski T, Radomski MW (2005) Nanoparticle-induced platelet aggregation and vascular thrombosis. Br J Pharmacol 146(6):882–893
45. Novoselov KS, Fal'ko VI, Colombo L, Gellert PR, Schwab MG, Kim K A roadmap for graphene. Nature 490:192–200
46. Navanietha Krishnaraj R, Chandran S, Pal P, Varalakshmi P, Malliga P (2014) Molecular interactions of Graphene with HIV-Vpr, Nef and gag proteins. Korean J Chem Eng Springer 31(5):744–747
47. Liu Y, Dong X, Chen P (2012) Biological and chemical sensors based on graphene materials. Chem Soc Rev 41(6):2283–2307
48. Yang K, Feng L, Shi X, Liu Z (2013) Nanographene in biomedicine: theranostic applications. Chem Soc Rev 42(2):530–547
49. Aghili Z, Taheri S, Zeinabad HA, Pishkar L, Saboury AA, Rahimi A, Falahati M (2016) Investigating the interaction of Fe nanoparticles with lysozyme by biophysical and molecular docking studies. PLoS One 11(10): e0164878
50. Tokarský J, Čapková P, Rafaja D, Klemm V, Valášková M, Kukutschová J, Tomášek V (2010) Adhesion of silver nanoparticles on the clay substrates; modeling and experiment. Appl Surf Sci 256(9):2841–2848
51. Cagın T, Wang G, Martin R, Breen N, Goddard WA III (2000) Molecular modelling of dendrimers for nanoscale applications. Nanotechnology 11(2000):77–84
52. Navanietha Krishnaraj R., Chandran S., Pal P., Berchmans S. (2013) Screening of photosynthetic pigments for herbicidal activity with a new computational molecular approach, Combinatorial chemistry and high throughput screening, 16, 777–781
53. Navanietha Krishnaraj R, Chandran S, Pal P, Berchmans S (2014) Molecular Modeling and assessing the catalytic activity of glucose dehydrogenase of *Gluconobacter suboxydans* with a new approach for power generation in a microbial fuel cell. Curr Bioinforma 9(3):327–330
54. Dodda SR, Sarkar N, Aikat K, Navanietha Krishnaraj R, Bhattacharjee S, Bagchi A, Mukhopadhyay SS (2016) Insights from the moleculardynamics simulation of Cellobiohydrolase Cel6A molecular structural model

from Aspergillus fumigates NITDGPKA3. Comb Chem High Throughput Screen 19:000
55. Navanietha Krishnaraj R, Kumari SS, Mukhopadhyay SS (2016) Antagonistic molecular interactions of photosynthetic pigments with molecular disease targets-a new approach to treat AD and ALS. J Recept Signal Transduction 36:67–71
56. Navanietha Krishnaraj R, Samanta S, Kumar A, Sani R (2017) Bioprospecting of the thermostable cellulolytic enzyme through modeling and virtual screening method. Can J Biotechnol 1(1):19–25
57. Navanietha Krishnaraj R, Pal P (2017) Enzyme-substrate interaction based approach for screening Electroactive microorganisms for microbial fuel cell applications. Indian J Chem Technol 24(1):93–96
58. Navanietha Krishnaraj R, Berchmans S, Pal P (2015) The three-compartment microbial fuel cell: a new sustainable approach to bioelectricity generation from lignocellulosic biomass. Cellulose 22:655–662
59. Bhuvaneswari A, Navanietha Krishnaraj R, Berchmans S (2013) Metamorphosis of pathogen to electrigen at the electrode/electrolyte interface: direct electron transfer of *Staphylococcus aureus* leading to superior electrocatalytic activity. Electrochem Commun 34:25–28
60. Navanietha Krishnaraj R, Karthikeyan R, Berchmans S, Chandran S, Pal P (2013) Functionalization of electrochemically deposited chitosan films with alginate and Prussian blue for enhanced performance of microbial fuel cells. Electrochim Acta 112:465–472
61. Navanietha Krishnaraj N, Yu JS (2014) Systems biology approaches for microbial fuel cell applications. Bioenergy: opportunities and challenges. Apple Academic Press, New Jersey, pp 125–139. doi:10.1201/b18718-8. ISBN-10: 1771881097.
62. Mahato D, Samanta D, Mukhopadhyay SS, Navanietha Krishnaraj R (2016) A systems biology approach for elucidating the interaction of curcumin with Fanconi anemia FANC G protein and the key disease targets of leukemia. J Recept Signal Transduct 8:1–7
63. Coulson EJ, May LM, Osborne SL, Reid K, Underwood CK, Meunier FA, Bartlett PF, Sah P (2008) p75 neurotrophin receptor mediates neuronal cell death by activating GIRK channels through phosphatidylinositol 4,5-bisphosphate. J Neurosci 28:315–324
64. Kalb R (2005) The protean actions of neurotrophins and their receptors on the life and death of neurons. Trends Neurosci 28:5–11
65. Ma L, Ohyagi Y, Miyoshi K, Sakae N, Motomura K, Taniwaki T, Furuya H, Takeda K, Tabira T, Kira J (2009) Increase in p53 protein levels by presenilin 1 gene mutations and its inhibition by secretase inhibitors. J Alzheimers Dis 16(3):565–575
66. Fu AK, Hung KW, Huang H, Gu S, Shen Y, Cheng EY, Ip FC, Huang X, Fu WY, Ip NY (2014) Blockade of EphA4 signaling ameliorates hippocampal synaptic dysfunctions in mouse models of Alzheimer's disease. Proc Natl Acad Sci U S A 111(27):9959–9964
67. Rosenberger AF, Rozemuller AJ, van der Flier WM, Scheltens P, van der Vies SM, Hoozemans JJ (2014) Altered distribution of the EphA4 kinase in hippocampal brain tissue of patients with Alzheimer's disease correlates with pathology. Acta Neuropathol Commun 2:79
68. Schmalbach S, Petri S (2010) Histone deacetylation and motor neuron degeneration. CNS Neurol Disord Drug Targets 9:279–284
69. Trott O, Olson AJ (2010) AutoDock Vina: improving the speed and accuracy of docking with a new scoring function, efficient optimization, and multithreading. J Comput Chem 31:455–461
70. Navanietha Krishnaraj R., Sani R., (2017) Molecular interactions of carbon nanomaterials with disease target proteins of Alzheimer's disease. (our unpublished data)
71. Alavijeh MS, Chishty M, Qaiser MZ, Palmer AM (2005) Drug metabolism and pharmacokinetics, the blood-brain barrier, and central nervous system drug discovery. NeuroRx 2 (4):554–571
72. Pardridge WM (2012) Drug transport across the blood–brain barrier. J Cereb Blood Flow Metab 32(11):1959–1972
73. Congreve M, Carr R, Murray C, Jhoti H (2003) A 'rule of three' for fragment-based lead discovery? Drug Discov Today 8 (19):876–877
74. Agyare EK, Curran GL, Ramakrishnan M, Yu CC, Poduslo JF, Kandimalla KK (2008) Development of a smart nano-vehicle to target cerebrovascular amyloid deposits and brain parenchymal plaques observed in Alzheimer's disease and cerebral amyloid angiopathy. Pharm Res 25(11):2674–2684
75. Panza F, Frisardi V, Solfrizzi V, Imbimbo BP, Logroscino G, Santamato A, Greco A, Seripa D, Pilotto A (2012) Immunotherapy for Alzheimer's disease: from anti-β-amyloid to tau-based immunization strategies. Immunotherapy 4(2):213–238
76. Rygiel K (2016) Novel strategies for Alzheimer's disease treatment: an overview of anti-amyloid beta monoclonal antibodies. Indian J Pharm 48(6):629–636

Index

A

Kunal Roy (ed.), *Computational Modeling of Drugs Against Alzheimer's Disease*, Neuromethods, vol. 132, DOI 10.1007/978-1-4939-7404-7, © Springer Science+Business Media LLC 2018

Index

B

C

D

E

F

G

H

I

J

K

L

O

P

T

V

W

X

Printed in the United States
By Bookmasters